Advances in Intelligent Systems and Computing

Volume 1323

AF806303

Series Editor

Janusz Kacprzyk, Systems Research Institute, Polish Academy of Sciences, Warsaw, Poland

Advisory Editors

Nikhil R. Pal, Indian Statistical Institute, Kolkata, India

Rafael Bello Perez, Faculty of Mathematics, Physics and Computing, Universidad Central de Las Villas, Santa Clara, Cuba

Emilio S. Corchado, University of Salamanca, Salamanca, Spain

Hani Hagras, School of Computer Science and Electronic Engineering, University of Essex, Colchester, UK

László T. Kóczy, Department of Automation, Széchenyi István University, Gyor, Hungary

Vladik Kreinovich, Department of Computer Science, University of Texas at El Paso, El Paso, TX, USA

Chin-Teng Lin, Department of Electrical Engineering, National Chiao Tung University, Hsinchu, Taiwan

Jie Lu, Faculty of Engineering and Information Technology, University of Technology Sydney, Sydney, NSW, Australia

Patricia Melin, Graduate Program of Computer Science, Tijuana Institute of Technology, Tijuana, Mexico

Nadia Nedjah, Department of Electronics Engineering, University of Rio de Janeiro, Rio de Janeiro, Brazil

Ngoc Thanh Nguyen, Faculty of Computer Science and Management, Wrocław University of Technology, Wrocław, Poland

Jun Wang, Department of Mechanical and Automation Engineering, The Chinese University of Hong Kong, Shatin, Hong Kong

The series "Advances in Intelligent Systems and Computing" contains publications on theory, applications, and design methods of Intelligent Systems and Intelligent Computing. Virtually all disciplines such as engineering, natural sciences, computer and information science, ICT, economics, business, e-commerce, environment, healthcare, life science are covered. The list of topics spans all the areas of modern intelligent systems and computing such as: computational intelligence, soft computing including neural networks, fuzzy systems, evolutionary computing and the fusion of these paradigms, social intelligence, ambient intelligence, computational neuroscience, artificial life, virtual worlds and society, cognitive science and systems, Perception and Vision, DNA and immune based systems, self-organizing and adaptive systems, e-Learning and teaching, human-centered and human-centric computing, recommender systems, intelligent control, robotics and mechatronics including human-machine teaming, knowledge-based paradigms, learning paradigms, machine ethics, intelligent data analysis, knowledge management, intelligent agents, intelligent decision making and support, intelligent network security, trust management, interactive entertainment, Web intelligence and multimedia.

The publications within "Advances in Intelligent Systems and Computing" are primarily proceedings of important conferences, symposia and congresses. They cover significant recent developments in the field, both of a foundational and applicable character. An important characteristic feature of the series is the short publication time and world-wide distribution. This permits a rapid and broad dissemination of research results.

Indexed by DBLP, EI Compendex, INSPEC, WTI Frankfurt eG, zbMATH, Japanese Science and Technology Agency (JST), SCImago.

All books published in the series are submitted for consideration in Web of Science.

More information about this series at http://www.springer.com/series/11156

Rafik Aziz Aliev ·
Nodirbek Rustambekovich Yusupbekov ·
Janusz Kacprzyk · Witold Pedrycz ·
Fahreddin M. Sadikoglu
Editors

11th World Conference "Intelligent System for Industrial Automation" (WCIS-2020)

Springer

Editors
Rafik Aziz Aliev
Azerbaijan State Oil and Industry University
Baku, Azerbaijan

Janusz Kacprzyk
Polish Academy of Sciences
System Research Institute
Warsaw, Poland

Fahreddin M. Sadikoglu
Department of Computer Engineering
Near East University
Nicosia, Turkey

Nodirbek Rustambekovich Yusupbekov
Department Automation
of Production Processes
Tashkent State Technical University
Tashkent, Uzbekistan

Witold Pedrycz
Department of Electrical
and Computer Engineering
University of Alberta
Edmonton, AB, Canada

ISSN 2194-5357 ISSN 2194-5365 (electronic)
Advances in Intelligent Systems and Computing
ISBN 978-3-030-68003-9 ISBN 978-3-030-68004-6 (eBook)
https://doi.org/10.1007/978-3-030-68004-6

© The Editor(s) (if applicable) and The Author(s), under exclusive license
to Springer Nature Switzerland AG 2021
This work is subject to copyright. All rights are solely and exclusively licensed by the Publisher, whether the whole or part of the material is concerned, specifically the rights of translation, reprinting, reuse of illustrations, recitation, broadcasting, reproduction on microfilms or in any other physical way, and transmission or information storage and retrieval, electronic adaptation, computer software, or by similar or dissimilar methodology now known or hereafter developed.
The use of general descriptive names, registered names, trademarks, service marks, etc. in this publication does not imply, even in the absence of a specific statement, that such names are exempt from the relevant protective laws and regulations and therefore free for general use.
The publisher, the authors and the editors are safe to assume that the advice and information in this book are believed to be true and accurate at the date of publication. Neither the publisher nor the authors or the editors give a warranty, expressed or implied, with respect to the material contained herein or for any errors or omissions that may have been made. The publisher remains neutral with regard to jurisdictional claims in published maps and institutional affiliations.

This Springer imprint is published by the registered company Springer Nature Switzerland AG
The registered company address is: Gewerbestrasse 11, 6330 Cham, Switzerland

11th International Conference on Intelligent Systems for Industrial Automation, WCIS-2020, Tashkent, Uzbekistan, November 26–28

Preface

The 11th International Conference on Intelligent Systems for Industrial Automation, WCIS-2020, is the premier international conference organized by Ministry of Innovational Development of the Republic of Uzbekistan, the Ministry of Higher and Secondary-Specialized Education of the Republic of Uzbekistan, the Academy of Sciences of the Republic of Uzbekistan, Tashkent State Technical University (Uzbekistan), Azerbaijan Association of "Zadeh's Legacy and Artificial Intelligence" (Azerbaijan), Azerbaijan State Oil and Industry University (Azerbaijan), University of Texas, San Antonio (USA), Georgia State University (Atlanta, USA), Near East University (North Cyprus), Korea Multimedia Society (South Korea), Honeywell Company, Uzkimyosanoat Joint Stock Company (Uzbekistan), Uzbekneftegaz National Holding Company (Uzbekistan), ZIFIT, Zadeh Institute for Information Technologies (Berkeley, USA).

This volume presents an edited selection of the presentations from the 11th International Conference on Intelligent Systems for Industrial Automation, WCIS-2020, which was held in Tashkent, Uzbekistan, November 26–28, 2020. WCIS-2020 is held as a meeting for the communication of research on integrated computer-aided industrial systems, Industry 4.0, Internet of things, cloud technologies, neural computing, fuzzy and evolutionary computation, information security and other fields. The conference provided an opportunity to present and discuss state-of-the-art research in this expanding domain.

This volume will be a useful guide for academics, practitioners, and graduates in the scope of intelligent industrial system design. It will allow for increasing of interest in development and applying methods of development of intelligent control and information systems in various real-life fields.

November 2020

Nodirbek Yusupbekov
Rafik Aliev
Chairmen of WCIS-2020

Organization

Co-chairmen and Editors

R. A. Aliev	Azerbaijan State Oil and Industry University, Azerbaijan
N. Yusupbekov	Tashkent State Technical University, Uzbekistan
M. B. Babanli	Azerbaijan State Oil and Industry University, Azerbaijan
S. M. Turabdjanov	Tashkent State Technical University, Uzbekistan
J. Kacprzyk	Systems Research Institute Polish Academy of Sciences, Poland
W. Pedrycz	University of Alberta, Canada
F. Sadikoglu	Near East University, Turkey

International Program Committee

A. Averkin	Plekhanov Russian University of Economics, Russia
A. Khashman	Final International University, Turkey
A. Marahimov	National University of Uzbekistan, Uzbekistan
A. Verlan	Norwegian University of Science and Technology (NTNU), Norway
A. Yusupbekov	Tashkent State Technical University, Uzbekistan
B. Dozortsev	Institute of Control Sciences RAS, Russia
B. Fazlollahi	Georgia State University, USA
V. Kreinovich	University of Texas, USA
D. Dubois	Université Paul Sabatier, IRIT, France
D. Mukhitdinov	Tashkent State Technical University, Uzbekistan
E. Babaei	University of Tabriz, Iran
F. Adilov	LLC "Himavtomatika", Uzbekistan
F. Aminzadeh	University of Houston, USA

F. Sadikoglu	Near East University, Turkey
T. Takagi	Meiji University, Japan
Jaegeol Yim	Dongguk University, South Korea
Jaehun Joo	Dongguk University, South Korea
Jeong Yong Byun	Dongguk University, South Korea
K. Wang	Shanghai University, China
Kang-Hyun Jo	University of Ulsan, South Korea
Kh. Igamberdiev	Tashkent State Technical University, Uzbekistan
Kh. Vafai	Sharif University of Technology, Tehran, Iran
L. Sorkin	Moscow Institute of Physics and Technology, Moscow, Russia
L. Zeynalova	Azerbaijan State Oil and Industry University, Azerbaijan
M. Chen	Institute of Automation Shandong Academy of Science, China
M. Kamilov	Scientific and Innovation Center of Information and Communication Technologies (SIC ICT) at the Tashkent University of Information Technologies, Uzbekistan
M. Karimov	State Test Center under the Cabinet of Ministers of the Republic of Uzbekistan, Uzbekistan
M. Nikravesh	University of California, Berkeley, USA
M. Sugeno	Tokyo Institute of Technology, Japan
O. Zaripov	Tashkent State Technical University, Uzbekistan
R. Aliyev	Eastern Mediterranean University, North Cyprus
R. Yager	Iona College, USA
S. Ulyanov	Institute of Gene Biology Russian Academy of Sciences, Russia
E. Nematov	Tashkent Chemical Technological Institute, Uzbekistan
Sh. Gulyamov	Tashkent State Technical University, Uzbekistan
T. Aripov	Tashkent Institute of Postgraduate Medical Education, Uzbekistan
T. Bekmuratov	Tashkent University of Information Technologies named after Muhammad al-Khwarizmi, Uzbekistan
U. Begimkulov	Head scientific and methodological center, Uzbekistan
U. Mannanov	Tashkent State Technical University, Uzbekistan
U. Mukhamedkhanov	Tashkent State Technical University, Uzbekistan
V. Nourani	Near East University, Turkey
V. Tarassov	Bauman Moscow State Technical University, Russia

Organizing Committee

Chairmen

N. Yusupbekov	Tashkent State Technical University, Uzbekistan
U. Eberhardt	University of Siegen, Germany

Co-chairmen

L. Gardashova	Azerbaijan State Oil and Industry University, Azerbaijan
T. Abdullayev	Odlar Yurdu University, Azerbaijan
Sh. Gulyamov	Tashkent State Technical University, Uzbekistan
Kh. Igamberdiev	Tashkent State Technical University, Uzbekistan

Members

A. Alizadeh	Azerbaijan State Oil and Industry University, Azerbaijan
B. Guirimov	State Oil Company of Azerbaijan Republic, SOCAR, Azerbaijan
N. Adilova	Azerbaijan State Oil and Industry University, Azerbaijan
A. Guliyev	Azerbaijan State Oil and Industry University, Azerbaijan
R. Rzayev	Institute of Control Systems of ANAS, Azerbaijan
M. Elamin	Near East University, North Cyprus
M. Salahli	Çanakkale Onsekiz Mart University, Turkey
V. Shamsutdinova	Tashkent State Technical University, Uzbekistan
U. Kholmanov	Tashkent State Technical University, Uzbekistan
N. Matyokubov	Tashkent State Technical University, Uzbekistan

Conference Organizing Secretariats

1. Azadlig Ave. 20, AZ 1010 Baku, Azerbaijan
Phone: +99 412 493 45 38, Fax: +99 412 598 45 09
E-mail: latsham@yandex.ru, adilovanigarr@gmail.com
2. Matyokubov Nurbek, 2, University Str., 100095, Tashkent, Uzbekistan
Phone: +998 97 771 15 51
E-mail: nm85@mail.ru
Shamsutdinova Vinera
Phone: +998 90 174 77 63
Kholmanov Utkir

Contents

Prediction Intervals for the Artificial Neural Network (ANN) and Adaptive Neuro-Fuzzy Inference System (ANFIS) via the LUBE Method 1
Vahid Nourani, Fahreddin Sadikoglu, Nardin Jabbarian Paknezhad, and Elnaz Sharghi

Analysis and Modeling of Information Security Systems in Industry 4.0 10
Mastura Zaynutdinova and Sherzod Sayfullaev

Using Non-linear Integral Models in Automatic Control and Measurement Systems for Sensors' Input Signals' Recovery 18
Andriy Verlan, Volodymyr Fedorchuk, Vitalii Ivaniuk, and Jo Sterten

Neural Network Method and Algorithm for Document Detection Based on Signaling Analysis 26
Bekmuratov Tulkun Fayziyevich and Botirov Fayzullajon Baxtiyorovich

Using Fuzzy Probabilistic Implication in Z-set Based Inference 33
Latafat A. Gardashova

Accounting Experience Between Fuzzy Integral and Z-numbers 40
H. A. Primova, F. N. Iskandarova, and Q. M. Gaybulov

A Recurrent Method for Structural-Parametric Identification of Fuzzy Neural Networks 47
Marakhimov Avazjon Rakhimovich and Khudaybergenov Kabul Kadirbergenovich

Voltage Control System for Electrical Networks Based on Fuzzy Sets 55
Nurali Yusifbayli, Huseyngulu Guliyev, and Aydin Aliyev

Algorithms for the Synthesis of Optimal Linear-Quadratic Stationary Controllers 64
J. U. Sevinov, A. R. Mallaev, and S. N. Xusanov

Toward Ordering of n-Tuples of Z-Numbers 72
Akif V. Alizadeh

Determining the Activity of Fullerene Nanoparticles Using QSAR Models 81
Alisher Ikramov, Kamoliddin Nabijonov, and Bakhtiyor Rasulev

Selection of the Optimal Class of Features for Recognition of the Azerbaijani Handprinted Characters 96
Araz R. Aliev and Elviz A. Ismayilov

About One Feature Selection Algorithm in Pattern Recognition 103
A. H. Nishanov, B. B. Akbaraliev, and Sh. Kh. Tajibaev

The Neural Network Based Decision Support System Using Aggregated Opinions of Public Observers 113
Elchin Aliyev, Ramin Rzayev, and Khanmurad Abdullayev

Multi-model Virtual Analysers of Parameters of Technological Processes 121
A. N. Yusupbekov, U. A. Ruziev, and M. K. Shodiev

Fuzzy Assessment of Technological Risks in the Main Oil Pipeline 127
H. G. Ismayilova, Z. I. Farzalizada, J. R. Damirova, Y. Z. Alakbarov, and M. E. Shahlarli

The Efficiency of the FSV Model in an Intelligent Search Engine 132
Bahodir Boltayevich Muminov and Ulugbek Bakhrom Bekmurodov

Neural-Network Based Approach to Mechanism of Liquid Extraction 141
A. A. Hasanov

Predicting the Performance of Marital Gas Pipelines with Different Coatings 147
Iravanli Konul Babek and Habibov Ibrahim Abulfas

Enhancing Teaching Approach with 3D Primitives in Virtual and Augmented Reality 155
F. M. Nuraliev, U. E. Giyosov, and Yoshihiro Okada

Modelling of Gas Lift Dynamic Processes Using ANFIS and ANN 164
I. Salahaddin Yusifov, B. Allahverdi Hasanov, S. Rza Safarov, and O. Azim Logmanli

Recognition and Classification of Pollen Grains Based on the Use of Statistical, Dynamic Image Characteristics, and Unique Properties of Neural Networks 170
Jumanov Isroil Ibragimovich, Djumanov Olim Isroilovich, and Safarov Rustam Abdullayevich

Fuzzy Type-2 Decision Making Method on Project Selection 180
Kamala Aliyeva

IoT and Intelligent Wireless Sensor Network for Remote Monitoring Systems of Solar Power Stations 186
Ilkhom Siddikov, Khalim Khujamatov, Doston Khasanov, and Ernazar Reypnazarov

Path Planning Collision Free for Warehouse Mobile Robot 196
A. B. Sultanova and M. Y. Abdullayeva

Simulation Model of the Multi-position Mechatronic Module Functioning of an Intelligent Robot 201
Kh. N. Nazarov, N. R. Matyokubov, and T. O. Rakhimov

Decision-Making in the Production of Hard-To-Recover Oil Reserves Under Uncertainty 207
G. M. Efendiyev, R. H. Kuliyev, M. K. Karazhanova, I. A. Piriverdiyev, D. A. Akhmetov, and L. B. Zhetekova

Automatic Recognition of Uzbek Speech Based on Integrated Neural Networks 215
Muhammadjon Musaev, Ilyos Khujayorov, and Mannon Ochilov

Software Selection Problem by Using Single Valued Neutrosophic (SVN)-Based TOPSIS Method 224
Nihad Mehdiyev

Neural Network Model for Adaptive Control of Nonlinear Dynamic Object 229
N. R. Yusupbekov, D. P. Mukhitdinov, and O. U. Sattarov

Identification and Ranking of Key Factors for Pattern of Consumer Buying Decisions in Digital Marketing 237
Gunay Imanova

The Possibilities of Using Cloud Computing for Applications of Industrial Automation 246
N. R. Yusupbekov, Sh. M. Gulyamov, and N. B. Usmanova

Application of Fuzzy AHP Method to Material Selection Problem 254
Mustafa Babanli and Tural Gojayev

Problems of Visualization and 3d Modeling in Chemistry: Analysis of Electronodonor Properties and Spatial Complex-Forming Structure of the Obtained Sulfocationite 262
S. M. Turabdzhanov, A. G. Eshimbetov, A. A. Nabiev, and L. S. Rakhimova

Application of Modern Technologies for Planning Improvement and Saving on Costs in the Enterprise of the Industry 4.0 272
Nazim Tofiq oglu Huseynov

The Problem of Classifying and Managing Risk Situations in Poorly Formed Processes 280
Kamilov Mirzayan, Muxamedieva Dilnoz, and Solieva Barno

Z-information Based PROMETHEE Method 287
K. I. Jabbarova and Aynur I. Jabbarova

Integrated Risk Management System for Sanitary and Epidemiological Control at the Customs Border of the Republic of Uzbekistan 294
A. A. Saidov, L. T. Pulatova, and F. A. Xakimova

Estimation of Linguistically Described Weights of Criteria 303
Shamil A. Ahmadov

Modeling of the Technological Process of Ammonization of Soda Production Under Conditions of Changing Initial Data 309
B. T. Kaipbergenov, B. A. Fayzullaev, and M. K. Elmuratov

Analyisis of the Influence of the Shape of the Matrix on the Compaction of Porous Powder Bodies of "Iron-Cast Iron-Glass" Under Imprecise Information 316
T. G. Jabbarov

Regular Algorithms for the Parametric Estimation of the Uncertain Object Control 322
H. Z. Igamberdiev and U. F. Mamirov

Decision Making Based on Case Company Target by Using GP Model 329
Rafig R. Aliyev

Automatic Control of Fuel Efficiency in Gas Burning Furnaces 334
U. U. Kholmanov and A. T. Rajabov

Synthesis of Compound Esters on the Basis of Acid Chloride of Low-Molecular Petroleum Acids of Baku Marine Oil Fields and Chiral Unsaturated Alcohols 342
A. N. Baghirli, S. A. Mamedkhanova, and G. M. Talybov

Radiation Resistance of Bituminous Hydroisolation Materials 348
Islam Mustafayev, Gulara Imanova, Gulnara Hatamkhanova, and Nigar Guliyeva

Analysis of Influence of Imagination, Fantasy, Exaggeration, and Hoax on a Level of Lie Under Linguistic Information 358
Farida Huseynova

Governance Model the Stochastic Process of Rectification of Multicomponent Mixtures Based on Fuzzy Logic 364
Avazov Yusuf Shodiyevich

Mental-Computing Based Investigation of Kinetic and Mechanisms of Homogeneous Gas-Phase Oxidation of Methane by Hydrogen Peroxide 377
L. M. Gasanova

Fuzzy Assessment of Corrosion Risks 385
H. G. Ismayilova, F. B. Ismayilova, E. Kh. Iskandarov, and S. G. Hajizade

Intellectual Decision-Making Support System 389
Yusupbekov Azizbek Nodirbekovich and Turaev Khurshid Salim Ugli

A Strategy of Prosthesis Control Using Artificial Intelligence and Execution on Robotic Fingers 394
Elbrus Imanov and Ehsan Faramarzi Asli

Analysis of the Main Trends and Directions of Automation in the Formation of the "Smart Archives" System 402
Ahmedov Barat and A. Abdullayev

Modelling of Environmental and Economic Factors in the Azerbaijani Oil Industry in the Context of Sustainable Development 409
Ata Babayev

Method of Managing Configurations of Engineering Systems 418
Mikhail Ovsiannikov, Sergey A. Bukhanov, and Sergey A. Podkopaev

Iodinated Multi-walled Carbon Nanotubes and Their Antioxidant Profile by Using ANFIS 427
E. B. Zeynalov, M. M. Agahuseynova, A. B. Huseynov, and N. I. Salmanova

Algorithm for the Synthesis of a Self-tuning Neural Network Control System for Multicomponent Dynamic Objects 435
Siddikov Isamiddin Hakimovah and Iskandarov Zohid Ergashbaevich

Modeling the Complex Formation of Rhenium with Organic Nitrogen-Containing Ligands by Using Fuzzy Inference 442
F. D. Gudratova and M. M. Agaguseynova

Traffic Sign Recognition Using Histogram of Oriented Gradients and Convolutional Neural Networks 452
Sanan Abizada

Algorithms for the Synthesis of a Neural Network Regulator for Control of Dynamic Objects 460
H. Z. Igamberdiev and T. V. Botirov

A Generic Model-Based on Probabilistic Reasoning for Path Loss of Wireless Indoor Channel 466
Bülent Bilgehan and Fahreddin Sadikoglu

Ecological Cybernetics of Oil Processing and Petrochemical Processes 472
A. I. Babayev

Industry 4.0 Technologies: Impact on Increasing the Productivity of Industrial Enterprises 479
O. O. Zaripov and U. Sh. Khamrakulov

Fuzzy C-Means Based Synthesis of Ion-Exchange Materials 487
F. A. Amirov, G. A. Aliyeva, F. R. Rahimova, and G. F. Rustamova

The Impacts of Genetic Studying on the Cost Analysis Based on Labour Market 496
Hüseyin Özdeşer and Günay Kuşaf

Drive Control Algorithm in Conditions of Insufficient Information 504
F. O. Qosimov, E. H. Nematov, A. H. Ahmedov, and M. A. Ismailov

Fuzzy Clustering of Influence of Oxygen on Oxidation of Isobutylene in Pulse Mode on Tin-Antimony Catalyst 512
A. N. Guliyeva and K. A. Gazvini

Toward Analysis of a Process of Obtaining High-Dispersive Stable Oil/water Emulsions 520
G. S. Hasanov, R. G. Hasanov, N. F. Aliyeva, and V. K. Valiyev

Multidimensional Classifier of Risk Analysis Methods 529
Chereshkin Dmitriy, Royzenson Gregory, and Britkov Vladimir

Modelling of the Droplet Velocity of the Dispersed Phase Components in Liquid Extraction 537
A. A. Hasanov

Implementation in Analytical Cloud the Algorithm of Optimization of MDEA Consumption 545
Farukh Adilov, Fakhritdin Abdurasulov, Andrey Korotaev, Fedor Davidov, and Sitora Isakova

The Concept of Connectivity Model Measurements of Geographic and Information Space of the Region . . . 553
Vladimir V. Khramov

Development of Criteria for Synthesis of the Optimal Structure of Monitoring and Control Systems . . . 559
E. Uljayev, U. M. Ubaydullayev, G. T. Tadjitdinov, and Sh.N. Narzullayev

On the Use of Fuzzy Logic in Criminal Justice . . . 564
Ed. A. Vasil'ev and A. I. Fomenko

Development of an Integrated Security Monitoring System Based on the Analysis of the Goals of Actions of Subjects of Computer Systems . . . 571
M. M. Karimov and S. M. Khamdamova

Application of Fuzzy Models in Building a Digital State in the Russian Federation . . . 580
N. V. Mamitova and M. A. Makhotenko

Control of Deployment and Development of Cluster and Virtual Structures in Industry . . . 588
A. S. Khokhlov, D. Y. Mishutin, and E. S. Baulin

Modeling of Dehydrogenation and Oxidative Dehydrogenation of Isopropyl Alcohol Over Ti-V-O Catalyst . . . 596
N. I. Dushdurova and S. A. Mammadkhanova

Regression Model of Composites Based on Epoxy-Diane Functionalized Oligomer . . . 604
T. M. Naibova, Ahmet Tutar, F. A. Amirov, G. S. Hasanov, and A. A. Mammadova

Author Index . . . 613

Prediction Intervals for the Artificial Neural Network (ANN) and Adaptive Neuro-Fuzzy Inference System (ANFIS) via the LUBE Method

Vahid Nourani[1], Fahreddin Sadikoglu[2], Nardin Jabbarian Paknezhad[1](✉), and Elnaz Sharghi[1]

[1] Department of Water Resources Engineering, Faculty of Civil Engineering, University of Tabriz, Tabriz, Iran
{nourani,n.jabbarian,sharghi}@tabrizu.ac.ir, vnourani@yahoo.com, jabbarian.nardin@gmail.com, elnaz_sharghi@yahoo.com

[2] Department of Electrical and Electronic Engineering, Faculty of Engineering, Near East University, PO BOX: 99138, Nicosia, TRNC, Mersin 10, Turkey
fahreddin.sadikoglu@neu.edu.tr

Abstract. In this study, the uncertainty associated with Artificial Neural Network (ANN) and Adaptive Neuro-Fuzzy Inference System (ANFIS) was investigated. The uncertainty of rainfall-runoff modeling on daily and monthly scales was evaluated by Prediction Intervals (PIs) for two watersheds, West Nishnabotna River basin in the United States and Lighvanchai River basin in Iran. Upper Lower Bound Estimation (LUBE) method was applied to construct the PIs. In the LUBE method, the ANN and ANFIS were trained by minimizing the objective function via the genetic algorithm optimization method, and the objective function contains the width and coverage criteria of PIs evaluation. PIs coverage probability and PIs width values, respectively were up to 10% higher and 39% lower for the PIs of the ANFIS compared to the those of ANN. Moreover, the CWC measure was lower for ANFIS PIs than that for the ANN. Also, the Lighvanchai basin modeling showed more accurate results than the West Nishnabotna River basin, which is due to four well defined regular seasons of the Lighvanchai basin.

Keywords: Rainfall-runoff · Prediction interval · Artificial Neural Network · Adaptive Neuro-Fuzzy Inference System · Upper Lower Bound Estimation

1 Introduction

Accurate rainfall-runoff modeling is one of the main challenges in water and watershed management. True estimation of runoff is efficient in many cases like urban planning, land use, reduction in the impact of flood, and damage to the environment. The relationship between the rainfall and runoff values has been a complicated issue and hot topic in hydrology, due to spatial and temporal changes of factors that influence runoff. Therefore, many complicated hydrological models were applied for simulation of this stochastic process. As regards to the uncertainty and complication of considering the

© The Author(s), under exclusive license to Springer Nature Switzerland AG 2021
R. A. Aliev et al. (Eds.): WCIS 2020, AISC 1323, pp. 1–9, 2021.
https://doi.org/10.1007/978-3-030-68004-6_1

physical parameters, it is obvious that the application of black-box modeling can be an effective alternative because, in black-box modeling, the relation between inputs and output was obtained, independent of any information of the procedure. Recently, the Artificial Neural Networks (ANNs) and Adaptive Neuro-Fuzzy Inference System (ANFIS) are applied in different hydrological modeling to identify the complicated non-linear relation between input and output. The ANN and ANFIS are capable of modeling rainfall-runoff and it has been expressed in some researches (e.g., see, [1, 2]).

Despite the appropriate ability of ANN and ANFIS in hydrological modeling, the point predictions of the ANN and ANFIS models do not express the error and uncertainty of modeling, which is involved in the system being modeled [2]. With this regard, quantifying these models reliability will be helpful in hydrological studies. Prediction Intervals (PIs) are appropriate tools to quantify the uncertainty of modeling. Decision makers should decide on the management of the critical issues associated to the rainfall. Therefore, by applying the PIs, uncertainties are considered and the achieved results present the important information to properly manage, decide and design the equipment's associated to the model. PIs express that if the issue is critical so pessimistic decision making is used and decision is made according to upper bound but it may not be economical; however, optimistically decision and consideration of the lower bound leads to high risk. There are several methods that calculate the PIs like delta [3], Bayesian [4], and Bootstrap [5]. In some previous studies Lower Upper Bound Estimation (LUBE) method was applied to evaluate the uncertainty of ANN modeling (e.g. see [6]).

In this paper, the LUBE technique was used to construct the PIs. This technique is independent of any knowledge about the bounds of PIs or data distribution and it has a good coverage of observed values [7]. Quantitatively comparing the constructed PIs states that the LUBE method has higher speed and lower uncertainty. The ANN and ANFIS based LUBE method includes outputs correspond to PIs in contrast to the classic ANNs and ANFIS, which consider one output as point prediction.

2 Methods

2.1 Study Area and Data Sets

The data from the West Nishnabotna River (a sub-basin of Missouri River in United States) and the Lighvanchai River (a sub-basin in Iran) which have various geomorphology were considered in this study. The first set of data used in this paper is for Lighvanchai watershed, placed in northwest Iran at Azerbaijan province. The time series data obtained from Iran Water & Power Resources Development Co. (IWPC). This basin is placed at the latitude of 37°43′ and 37°50′ and longitude of 46°22′ and 46°28′, in the northern slope of Sahand Mountain (northwestern Iran). The area of the basin is 142 km^2 (Fig. 1). Its height is in the range of 1263 m and 3679 m. The length of the Lighvanchai River is 28.5 km. Lighvanchai River is a perennial flow as the regime of the river is snowy and the watershed has a permanent snow cover in different altitudes. Thus the snowmelt has a significant influence on the river flow and consequently, its drainage basin density is low. This basin weather is mostly wet and humid and has four well defined seasons. The topography is steep with an average slope of 11%. Consequently, the soils are disposed to erosion to some extent. The second basin, the West Nishnabotna River watershed is the main sub-basin of the Missouri River watershed that is placed in southwestern Iowa (Fig. 1). The river is located in the longitude 95°67′ W and latitude 40°51′ N. Also, the area of the watershed is 7600 km^2. This watershed consists of the wide highlands and valleys. The data for this basin were extracted from waterwatch.usgs.gov web site.

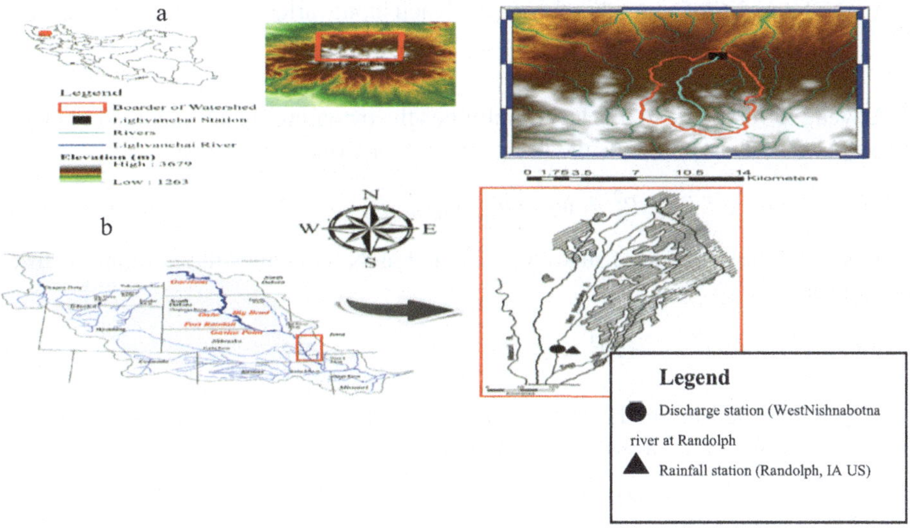

Fig. 1. Case study maps a) Lighvanchai b) West Nishnabotna River.

2.2 Artificial Neural Network (ANN)

ANN contains an input layer, a hidden layer, and an output layer. The number of the neurons in the hidden layer is obtained via the trial and error process. Weights connect the neurons of the layers and optimum values of the weights are achieved by training the ANN. Then by comparing the outputs of the ANN and the target values the error is calculated and back-propagated to update the weights. The output of the ANN is calculated via the Eq. (1) as [8]:

$$\hat{y}_j = f_j\left[\sum\nolimits_{h=1}^{m} W_{jh} \times f_h\left(\sum\nolimits_{i=1}^{n} W_{hi}X_i + W_{hb}\right) + W_{jb}\right] \quad (1)$$

where i, h, j, b and W express the input, hidden, and output layers neurons, bias and the applied weight (or bias) via the neuron, respectively; f_h, f_j indicated activation function for the hidden and output layers, respectively; X_i, n, m demonstrates, respectively input layer variable, number of input and number of hidden neurons; and y, $\hat{y}_j$ represented the observed and computed values of the output neuron, respectively. The values of the hidden and output layers and their weights were obtained via the training and verifying process.

2.3 Adaptive Neuro-Fuzzy Inference System (ANFIS)

ANFIS is a hybrid artificial intelligence technique which is the combination of fuzzy logic and ANNs. In fuzzy logic, the control signal is generated from firing the rule base. This rule base is drawn on historical data and is random in nature. ANFIS consists of 5 layers.

Layer 1 consists of a member function of each input variable, which can be adjusted.

$$O_{1,i} = \mu_{Ai}(x) \tag{2}$$

Layer 2: Every node in this layer multiplies the incoming signals, denoted as Q, and the output OP_1^2 that represents the firing strength of a rule is computed as,

$$OP_1^2 = w_i = \mu_{Ai}(x)\mu_{Bi}(y), \quad i = 1,2 \tag{3}$$

Layer 3: The ith node of this layer, labeled as N; computes the normalized firing strengths as,

$$OP_1^3 = \overline{w_i} = \frac{w_i}{w_1 + w_2}, \quad i = 1,2 \tag{4}$$

Layer 4: Node i in this layer computes the contribution of the ith rule towards the model output, with the following node function

$$OP_1^4 = \overline{w_i}f_i = \overline{w_i}(p_i x + q_i y + r_i) \tag{5}$$

Layer 5: The output layer of the ANFIS is calculated as:

$$OP_1^5 = Overall\ output = \sum\nolimits_i \overline{w_i}f_i = \frac{\sum_i w_i f_i}{\sum_i w_i} \tag{6}$$

Evaluation Criteria

Point Prediction Evaluation. One of the frequent measures to assess the performance of the modeling is the Correlation Coefficient (CC), which reveals the linear relativity between two variables and can range from −1 to 1. The values more than 0 present a positive dependence and values lower than 0 express a negative relevant, while the value of 0 means there isn't any correlation between the two variables. CC is calculated as Eq. 7.

$$CC = \frac{\sum(R_i - \overline{R}) - \sum(Z_i - \overline{Z})}{\sqrt{\sum(R_i - \overline{R})^2 \sum(Z_i - \overline{Z})^2}} \tag{7}$$

As well, to assess the exactness of modeling, the Determination Coefficient (DC) is applied. DC explains the agreement of the observed value and the output of the modeling and it is between 0 to 1. The higher DC, presents the adaptation between the output and observed values. DC is measured as Eq. 8.

$$DC = 1 - \frac{\sum_{i=1}^{N}(R_i - Z_i)^2}{\sum_{i=1}^{N}(R_i - \overline{R})^2} \tag{8}$$

PIs Assessment Measures. In order to quantify the PIs, the criteria of PI coverage probability (PICP) and mean PI width (MPIW) were used. The first criterion corresponds to the encompassments of the intervals. The wider PIs led to higher values of PICP [7]. PICP is calculated as Eq. 9.

$$PICP = \frac{1}{N}\sum_{i=1}^{N} c_i \tag{9}$$

$$if \quad L(X_i) < x_i < U(X_i) \rightarrow c_i = 1; \quad else \rightarrow c_i = 0$$

where $U(X_i)$ and $L(X_i)$ present respectively the upper and lower bounds of PIs corresponding to the ith sample.

Another criterion for quantification of PIs is their width. Equation (10) shows the normalized width [7].

$$NMPIW = \frac{1}{nR}\sum\nolimits_{i=1}^{n} U(X_i) - L(X_i) \tag{10}$$

where R is span of maximum and minimum of observed values. NMPIW is a dimensionless measure presenting the average width of PIs and the targets range.

Each of these criteria separately can't lead to a clear conclusion due to their inverse relationship. Therefore, the combinational coverage width-based criterion (CWC) is introduced, which contains both above criteria to evaluate the PIs [7] as Eq. 11:

$$CWC = NMPIW(1 + \gamma(PICP)e^{-\eta(PICP-\mu)} \tag{11}$$

$$\gamma = \begin{cases} 0, & PICP \geq \mu \\ 1, & PICP < \mu \end{cases}$$

The constants η and μ are two hyperparameters determine the penalty associated with lower PICP. μ shows the confidence level of the PIs. η magnifies the low variation between PICP and μ. The value of η is considered 10 in this study.

2.4 PIs Construction by LUBE Method

The procedure of proposed method is to obtain two outputs, which indicate the upper and lower bounds. The first step for construction of both ANN and ANFIS PIs is driving the data into subsets of train and verify. Then for ANN assigned weights are considered as initial weights. Then Genetic Algorithm (GA) starts generating new sets of weights by using the elitism, cross over, and mutation operators. Elitism determines number of the individuals with the lowest CWC. A selection operator is used for selecting the parents for the next generation. The mutation mechanism randomly generates new populations. For the construction of the ANFIS same process are done and GA produces optimum parameters to achieve the most reliable PIs according to lowest cost function (CWC). The GA optimization ended when the number of iterations achieved the intended number or when the successive iterations didn't improve.

3 Results

Since the performance of any data-driven method depends on the inputs of the model, selecting dominant inputs is critical in any ANN and ANFIS based modeling. For selecting effective inputs of ANN, Mutual Information (MI) between target and time series of rainfall and runoff up to 20 and 13 lags, respectively for daily and monthly modeling were calculated and time series, which led to higher MI were chosen as inputs. For the West Nishnabotna River, on daily scale, the modeling inputs were only runoff from 1 to 5 time steps lag times. Also on monthly scale, the inputs of the modeling were combination of rainfall with one step lag time and runoff time series up to two steps lag times. For the Lighvanchai on daily scale runoff time series up to three lag times were used as inputs of the modeling. On monthly scale combination of rainfall with 1 step lag time with time series of runoff up to three time steps and 12th and 13th lag times for monthly scales were used as inputs. Therefore, four ANNs and ANFIS models with different inputs were trained in this study on monthly and daily scales; the used time series as inputs of modeling were tabulated in Table 1. The 75% and 25% of time series were splitted to train and verification subsets, respectively. Therefore, the data in periods 1989–2007 and 1978–2005 were used for the training, respectively for the Lighvanchai and the West Nishnabotna River and in periods 2008–2015 and 2006–2015 for the validation purpose. Since the point prediction of ANN didn't give any description about the accuracy of modeling, so, PI construction method (the LUBE method) was used. The values of the PIs accuracy measurements were tabulated in Table 2.

Table 1. Point prediction results for two watersheds in both daily and monthly scales.

Scale	Model	Basin	Input	ANN				ANFIS			
				DC train	DC verification	CC train	CC verification	DC train	DC verification	CC train	CC verification
Daily	1	Lighvanchai	$Q_{t-1}, Q_{t-2}, Q_{t-3}$	0.93	0.9	0.96	0.95	0.94	0.91	0.97	0.95
	2	West Nishnabotna River	$Q_{t-1}, Q_{t-2}, Q_{t-3}, Q_{t-4}, Q_{t-5}$	0.7	0.7	0.84	0.83	0.71	0.7	0.84	0.81
Monthly	3	Lighvanchai	$I_{t-1}, Q_{t1}, Q_{t2}, Q_{t-3}, Q_{t-12}, Q_{t-13}$	0.71	0.64	0.84	0.8	0.95	0.57	0.97	0.75
	4	West Nishnabotna River	$I_{t-1}, Q_{t-1}, Q_{t-2}$	0.55	0.5	0.67	0.69	0.89	0.52	0.94	0.79

Table 2. Criteria of PIs quantification via LUBE method.

Scale	Model	ANN			ANFIS		
		PICP	NMPIW	CWC	PICP	NMPIW	CWC
Daily	1	0.93	0.25	0.25	0.93	0.22	0.22
	2	0.88	0.05	0.1	0.9	0.03	0.03
Monthly	3	0.69	0.17	1.12	0.79	0.28	0.79
	4	0.7	0.17	1.04	0.8	0.25	0.8

4 Discussion

A three-layer feed-forward ANN with back propagation training algorithm of levenberg-Marquardt and the tangent sigmoid as the activation function were used in the models. To distinguish the appropriate ANN model, the epoch and hidden neuron numbers were determined by trial and error procedure. A Sugeno fuzzy inference system and Gaussian membership functions were applied in this study for the ANFIS. From 10 to 200 and 2 to 4 epochs and membership functions respectively were tested in modeling via the ANFIS. DC, and CC measures were applied for evaluation of the models implementation. In the Table 2, the results of modeling were shown. As tabulated in Table 2, ANN modeling on daily scale led to better performance and DC of validation sets were 10% and 15% higher than monthly scale respectively for the Lighvanchai and the West Nishnabotna River watersheds. Moreover, ANFIS modeling on daily scale led to better performance and DC of validation sets were 10% and 15% higher than monthly scale respectively for the Lighvanchai and the West Nishnabotna River watersheds. This may be due to the consideration of the basin soil moisture, which causes the lower water infiltration, so the estimation of runoff values may be more accurate. In addition, stronger autoregressive time series of daily scale modeling compared to monthly scale modeling, which causes a better performance of modeling on daily scale. In addition, more samples on the daily data with regard to the monthly date cause better model calibration on daily scale. For the Lighvanchai and the West Nishnabotna River watersheds modeling on the daily scale leading to appropriate results. The Lighvanchai on both monthly and daily scales showed higher performance with validation DC of about 20% and 26% higher than the West Nishnabotna River, respectively for monthly and daily scales modeling. Low efficiency of the modeling in West Nishnabotna River may be due to highly variable weather and rainfall patterns for this watershed and also vast area of this basin compared to the Lighvanchai basin. Moreover, the construction of dams in this watershed might impact on the river regime.

Comparison between the results of two applied methods indicated that modelling via ANFIS may lead to higher DC especially in training data sets and DC of training data set of ANFIS was up to 39% higher than that of the ANN. But in verification data sets DC was only up to 1% higher than that of the ANN. Similar results associated with the superiority of ANFIS compared to ANN were obtained in some previous studies (e.g. see [9]).

As shown in Table 3, for PIs of the ANN, the values of PICP for the daily estimation were 29% and 18% higher than monthly estimation for the Lighvanchai, and the West Nishnabotna River basins, respectively and the values for NMPIW were 20% and 17% narrower for daily modeling in the Lighvanchai and the West Nishnabotna River basins, respectively. In addition, for PIs of the ANFIS, the values of PICP for the daily estimation were 14% and 10% higher than monthly estimation for the Lighvanchai, and the West Nishnabotna River basins, respectively and the values for NMPIW were 21% and 88% narrower for daily modeling in the Lighvanchai and the West Nishnabotna River basins, respectively. Considering the values of CWC indicated that on daily scale it was considerably lower than monthly scale, which showed the higher reliability on daily scale modeling. The overall outcome for PIs construction corresponded to the results of the point prediction. The length of the dataset in training may have important role in the performance of the PIs. On daily scale, the value of PICP for the Lighvanchai basin was up to 9% higher than the West Nishnabotna River; but on monthly scale, both watersheds approximately showed similar modeling performance. Consideration of the PIs evaluation criteria showed at the CWC values were up to 29% lower for the constructed PIs of the ANFIS compared to that for the ANN. This results confirmed more reliable performance of the ANFIS, which indicated the lower uncertainty of the hybrid models. Analyzing the values of the PICP and NMPIW denoted that the constructed PIs of the ANFIS method led to higher values of PICP than those for ANN PIs and PICP for the ANFIS was up to 10% higher than that for the ANN, but NMPIW of both models were approximately the same.

5 Conclusions

In this paper, ANN and ANFIS modeling, for rainfall-runoff modeling for two watersheds of Lighvanchai and West Nishnabotna River, were carried out on monthly and daily time scales. The obtained PIs via LUBE method were evaluated via PIs criteria. Analysis indicated that daily scale modeling led to more reliable results. Values of PICP, respectively were up to 29% and 21%, higher on daily scale compared to monthly scale for constructed PIs of Lighvanchai and West Nishnabotna River basin modeling respectively. Moreover, the value of CWC was significantly lower on daily scale modeling. Results indicated that constructed PIs of ANFIS led to higher quality, values of PICP for constructed PIs of ANFIS model were up to 10% higher than those constructed for ANN PIs, which denoting the superiority of hybrid methods with regard to the ANN model. Comparison between the performance of the two basins, indicated that the Lighvanchai showed better performance compared to the West Nishnabotna River basin, PICP values for constructed PIs of the Lighvanchai basin modeling were up to 3% higher than those for the West Nishnabotna River basin. The regular four seasons for the Lighvanchai basin cause better performance of its modeling.

References

1. Wang, W.C., Chau, K.W., Cheng, C.T., Qiu, L.: A comparison of performance of several artificial intelligence methods for forecasting monthly discharge time series. J. Hydrol. **374**(3–4), 294–306 (2009). https://doi.org/10.1016/j.jhydrol.2009.06.019

2. Nayak, P.C., Sudheer, K.P., Rangan, D.M., Ramasastri, K.S.: A neuro-fuzzy computing technique for modeling hydrological time series. J. Hydrol. **291**(1–2), 52–66 (2004). https://doi.org/10.1016/j.jhydrol.2003.12.010
3. Chryssolouris, G., Lee, M., Ramsey, A.: Confidence interval prediction for neural network models. IEEE Trans. Neural Netw. **7**(1), 229–232 (1996). https://doi.org/10.1109/72.478409
4. MacKay, D.J.C.: A practical Bayesian framework for backpropagation networks. Neural Comput. **4**(3), 448–472 (1992)
5. Efron, B., Tibshirani, R.J.: An Introduction to the Bootstrap. CRC Press, Boca Raton (1994)
6. Nourani, V., Paknezhad, N.J., Sharghi, E., Khosravi, A.: Estimation of prediction interval in ANN-based multi-GCMs downscaling of hydro-climatologic parameters. J. Hydrol. **579**, 124226 (2019). https://doi.org/10.1016/j.jhydrol.2019.124226
7. Khosravi, A., Nahavandi, S., Creighton, D., Atiya, A.F.: Lower upper bound estimation method for construction of neural network-based prediction intervals. IEEE Trans. Neural Netw. **22**(3), 337–346 (2011). https://doi.org/10.1109/TNN.2010.2096824
8. Nourani, V., Alami, M.T., Vousoughi, F.D.: Wavelet-entropy data pre-processing approach for ANN-based groundwater level modeling. J. Hydrol. **524**, 255–269 (2015). https://doi.org/10.1016/j.jhydrol.2015.02.048
9. Okwu, M.O., Adetunji, O.: A comparative study of artificial neural network (ANN) and adaptive neuro-fuzzy inference system (ANFIS) models in distribution system with nondeterministic inputs. Int. J. Eng. Bus. Manag. **10**, 1847979018768421 (2018). https://doi.org/10.1177/1847979018768421

Analysis and Modeling of Information Security Systems in Industry 4.0

Mastura Zaynutdinova and Sherzod Sayfullaev(✉)

Tashkent University of Information Technologies name after Muhammad al-Khwarizmi, Tashkent, Uzbekistan
dilfuza65@gmail.com, sherzodsay@gmail.com

Abstract. This paper analyzes the vulnerability and identification of attacks on corporate networks, the properties of adaptability and development, security model, studied the methods that will reduce the costs of implementing the information security system, consider the analogy Bio-systems security. The influence of Industry 4.0 on information security and the processes of modeling information security systems has been studied. The article examines the impact of the Industry 4.0 revolution on information security and its consequences. In particular, changes and transformations of information security in corporate systems are studied. The systematization of threats to corporate systems was carried out and the modeling of information security systems was studied. Modeling of information security systems and development of security characteristics of IT systems is an important period for automating vulnerability analysis procedures and identifying attacks on a corporate system in order to give IT evolutionary properties of adaptability and development.

Keywords: Information security system · IT · Bio-system · Immune defense · Simulation models · Industry 4.0 · Cybernetic systems

1 Introduction

In 2020, attention to the development and implementation of information technology, as well as the development of the digital economy in Uzbekistan increased sharply. This is evidenced by the fact that the President of Uzbekistan declared 2020 "The Year of Science, Education and the Development of the Digital Economy."

In this regard, digitalization in the corporate sector of Uzbekistan has accelerated with the rapid introduction of information technology, in particular the revolutionary technology Industry 4.0. It also increases the need for information security systems.

2 Analysis and Modeling

Information attacks are a very dangerous process that can occur in corporate systems around the world. Preventing this is an emergency measure. According to statistics from

© The Author(s), under exclusive license to Springer Nature Switzerland AG 2021
R. A. Aliev et al. (Eds.): WCIS 2020, AISC 1323, pp. 10–17, 2021.
https://doi.org/10.1007/978-3-030-68004-6_2

"Kaspersky Lab", in the first quarter of 2020, Uzbekistan ranked first in terms of the number of attacks on corporate systems (Table 1) (Fig. 1) [13].

Modeling an information security system (ISS) and assessing the value of security of information technology (IT) systems is an important period for automating vulnerability analysis procedures and identifying attacks on a corporate system in order to give IT evolutionary qualities of adaptability and formation [1–3].

Table 1. Statistics of Kaspersky Lab on going information attacks by country (Top 10).

№	Country	%
1	Uzbekistan	10,5
2	Tajikistan	6,9
3	Turkmenistan	5,5
4	Afghanistan	5,1
5	Yemen	3,1
6	Kazakhstan	3,0
7	Guatemala	2,8
8	Syria	2,4
9	Sudan	2,1
10	Kyrgyzstan	2,1

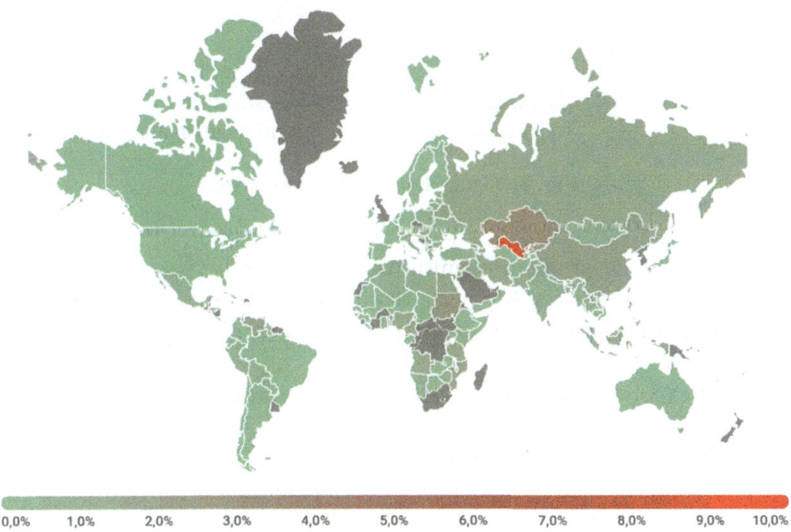

kaspersky

Fig. 1. Analysis of the geography of malicious attacks by Kaspersky Lab in the 1st quarter of 2020

Because of the development of information technology, the flow of information also strengthens on its own. And information has its value. This value decreases with time. Failure to deliver information on time will make it unnecessary later, that is, information that is not delivered on time will lose its value. It is very important that the information is delivered within a certain time, regardless of the distance from the sender to the recipient. This indicates that any process at an intermediate distance should be completed quickly. Information undergoes many processing processes throughout the life cycle. Corporate information is processed under the influence of technologies introduced by the enterprise. These technologies include Big Data, cloud computing, information security, and more. These are technology trends based on cyber-physical systems and part of Industry 4.0. All of the above technologies have specific information security systems.

Corporate systems have undergone significant changes as a result of computerization. Industry 4.0. Can be divided into two stages that generalize the industry for both modern corporate systems and conventional systems (Fig. 2).The first step is to create a digital copy of the product. In this case, general information security measures are applied: creating a reliable environment, controlling external access, and so on.

The second stage is modern additive manufacturing. It is at this stage that much attention is paid to information security [14].

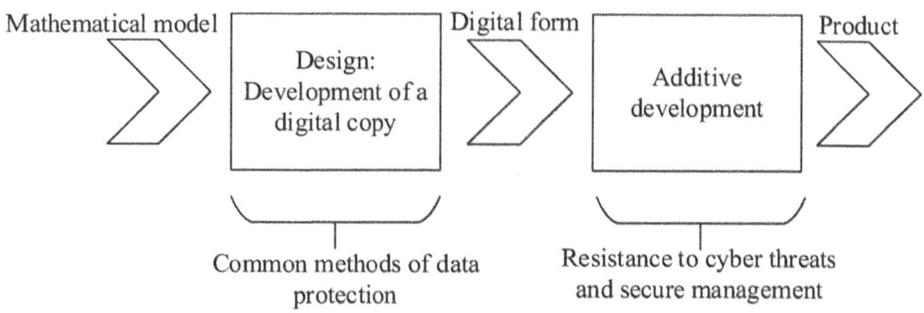

Fig. 2. Generalized production steps

As noted above, the problem of information security is critical, especially in the processes that occur during the life cycle of corporate information. Therefore, various information security systems, models and methods of its development are analyzed at different levels.

Studies and experiments on the construction of information security systems show that it is advisable to build simulation models and bio-systems as an analogue for their construction. They are highlighted below.

There are reports in the press about the development of effective methods that can reduce the costs of introducing information security systems (ISS), for example, using simulation models [4]. The method is aimed at concluding the task of creating an economically good information security system (ISS) in the context of investments that minimize aortal harm in violation of information security (IS). The use of relatively cheap methods and means of providing information security (anti-virus programs, organizational limits, etc.) significantly reduces the artillery damage. Because of this, the investments of information security systems (ISS) in relatively small volumes are effective in small organizations that are not subjected to special computer attacks. For vibrant firms operating in a competitively changing environment, rising costs of information security systems (ISS) does not always result in less damage from attacks on the corporate system.

Often, defense models are considered part of the risk management system and provide for these characteristics, such as burning dangers, misses in the software, significance, interruption and downtime of all kinds of resources, the possibility of attack, defense options and the probable amount of damage. The risk management system in the IT system allows you to calculate the dangers occurring, simulate a suitable ensemble of countermeasures, mechanically develop a defense profile and assess residual dangers [6].

It is necessary to systematize threats to corporate system security models in the enterprise. At the same time, it is necessary to obtain the technologies that are most relevant for the corporate system. Therefore, these technologies can be virtualization, cloud technologies, mobile systems and the Internet of things associated with the revolution in Industry 4.0 (Table 2) [14].

Table 2. Systematization of threats in corporate systems

Technology	Weak element	The consequence of the threat	Security specifics
Virtualization	Hypervisor	Take control	Security of the "last part" - a successful attack is impossible
Cloud technologies	Restriction mechanisms between virtual machines	Cloud destruction	The need to limit personal resources
Mobile systems	Authentication and encryption mechanisms (not applicable or partially applicable)	Gathering, disabling emergency information and taking control into their own hands	Lack of energy
Internet of things	Software Integrity, Insecure Internet Connection	Gathering, disabling emergency information and taking control into their own hands	Inability to interact with users interactively

The biosystematics analogy in the structure of defense of IT systems is based on the hierarchy of information security systems (ISS), integrated mechanisms of immune defense and skill accumulation. The familiar information security systems (ISS) as a rule, is limited to the implementation of the functions of the lower value of the defense system and the anti-virus trend of immune defense equipment. Accordingly [5] within 70% of virus attacks it is executed externally through the entry point to the protected network and only within 30% from the inside. The first ones can be attributed to external threats to the life support of the system, the second - to internal ones. In both cases, the immune defense of the bio-system is activated. The implementation of the idea of an informational immune system was made in the fact that in case of detection of symptoms of infection on the network, a fresh microbe sample is sent to the antivirus center, from where, after some time, they receive an update of the antivirus database, which they distribute over the corporate network before that time microbe spread.

The abovementioned scenario comes into objection with a bio-system analogy, in particular, with the intra-system implementation of immune defense, because in the considered anti-virus defense system (unlike the bio-system), a huge proportion of the immune defense devices are located in the anti-virus center located outside the corporate network.

Placing an anti-virus center outside the protected IT system allows villains: first, under the guise of updating the anti-virus database, create a channel for downloading microbes and Trojan stallions, secondly, in the case of self-sending files accused for the presence of a microbe to test for access to secret information. In addition, the response time of a similar immune defense is measured in the best case over several hours, which, along with the listed probabilities of the implementation of unauthorized access channels (NSA), is not tolerable enough for most critical applications. This means that the scope of use of a similar layout is limited only by the restoration of a disabled corporate network (an analogue of the resuscitation process, bedridden to a bio-system with injection support). In bio-systems, the functions of immune defense are realized through.

- internal mechanisms of an operational response to dangers and destabilizing influences, distributed among the levels of the information security systems (ISS) hierarchy,
- long-term processes of accumulation of the actual skill, bearing evolutionary character [7, 8].

The bio-system analogy of IT systems in evolutionary processes is based on the implementation of a set of inheritance, formation, habituation and selection devices characteristic of bio-systems. While the developed mental methods of detecting attacks and unauthorized information processes in the corporate network, the main concern is given only to the adaptability property when building promising SIS [9–11]. At the same time, the information security systems (ISS) values of mail gateways and firewalls are more aimed at detecting external attacks, and the information security systems (ISS) server values are aimed at neutralizing internal dangers in the corporate system.

Information security has been transformed as a result of the Industry 4.0 revolution. With the advent of cyber physical systems, the concept of information security has been added to information security, which is the flagship of the digital industry. Supply methods and evaluation of this area have already been developed. When you integrate this with business processes, the concept of user information security can be expanded. This is difficult to predict at present.

The security concept is formed in accordance with the system: information security in information systems, security of cybernetic systems in cyber-physical systems. But digital production is not yet clear. This may be related to the safety of biosystems, since biosystems are considered and proposed (Fig. 3) [14].

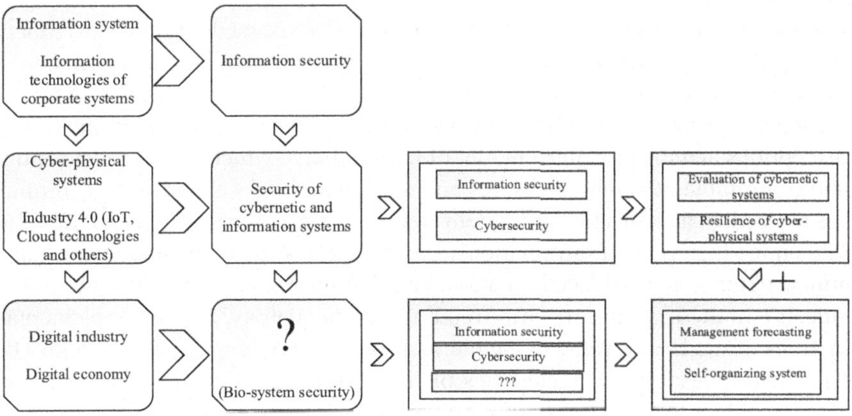

Fig. 3. Transformation of information security

The familiar information security systems (ISS) [12], as a rule, will only be sold by the mechanisms of prompt reaction and neutralization of IT system vital dangers, literally not paying attention to the coordinating role played by the nervous system - the upper level of the bio-system defense hierarchy in the implementation of the evolutionary process of accumulating the current system skill (long-term storing system information). The processes of gradual habituation of the hierarchical life support and defense system with the introduction of the entire arsenal of means of evolutionary processes have space in bio-systems.

In IT systems, in addition to the immune value of information security systems, (ISS) the hierarchy of defense values is important and, above all, the presence of upper values of information security systems (for example, the receptor value of security tools) that perform the functions of the nervous system of a bio-system for accumulating relevant skills, coordinating and establishing associative (long-term) connections between the processes occurring at the lower levels of information security systems - attacks and changing the danger field. In other texts, in IT systems, in particular, a corporate or local network, a hierarchical degree of accumulation of the actual skill to neutralize attacks, presented in the form of structured information fields that are comfortable for inheritance in further implementations of the system, is needed.

3 Conclusion

The article examines the impact of the Industry 4.0 revolution on information security and its consequences. In particular, changes and transformations of information security in corporate systems are studied. The systematization of threats to corporate systems was carried out and the modeling of information security systems was studied.

Modeling of information security systems and development of security characteristics of IT systems is an important period for automating vulnerability analysis procedures and identifying attacks on a corporate system in order to give IT evolutionary properties of adaptability and development.

The test recommended, in fact, that to develop effective methods that can reduce the cost of introducing information security systems (ISS), they use simulation models, risk management system models that provide for the relevance of dangers, misses in software, significance, interruption and downtime of all kinds of resources, the possibility of attack, defense options and the probable amount of harm, in fact, which allows you to simulate a suitable ensemble of countermeasures and mechanically develop a defense profile.

It is noticed, in fact, that the bio-system analogy in the defense structure of IT systems is based on the hierarchy of information security systems (ISS), integrated mechanisms of immune defense and skill accumulation. Familiar information security systems (ISS), as a rule, are limited by the anti-virus trend of immune defense and the implementation of functions of the lowest value in the hierarchy of information security systems (ISS).

The existing security characteristics of the IT system reflect the static position of the defense object, based on the presence of defense devices, and do not provide for the vigor of defense devices to neutralize the results of attacks, the dynamics of the configuration of the danger field, the likelihood of getting used to information security systems (ISS), to a change in the danger field, do not highlight indications of a change in the composition of defense devices and the structure of multi-level information security systems (ISS).

References

1. Nesteruk, F.G., Nesteruk, G.F., Harchenko, A.F.: Modeling of adaptive processes for protecting the information resources of economic objects. In: Sat. Doc. Int. NIJ Global Trends in Statistics and Mathematical Methods in the Economy, St. Petersburg, pp. 218–220 (2004)
2. Nesteruk, G.F., Osovetsky, L.G., Nesteruk, F.G.: Adaptive model of neural network information security systems. Perspect. Inform. Tech. Intel. Syst. **3**, 45–56 (2003)
3. Nesteruk, Ph., Kharchenko, A., Nesteruk, G.: Information safety in electronic business: adaptive model of systems safety of information technologies. In: Proceedings of the International Conference on Information Technology in Business, St. Petersburg, pp. 124–128 (2003)
4. Karpychev, V.Yu., Minaev, B.A.: Price Information Security II Security Systems, vol. 5, pp. 128–130 (2003)
5. Kaspersky, K.: Attack on Windows NT. Tab "Overview of antivirus products from AIDSTEST to the information immune system". J. Netw. Solutions 88–95 (2000)
6. Korzhov, V.: Security automation. Computerworld Russ. **53**, 17–18 (2004)
7. Gaynazyan, M.: Information Processes and Reality. Nauka, Moscow, 192 p. (1998)
8. Yakovlev, N.N.: Life and environment: molecular and functional fundamentals of the organism's adaptation to environmental conditions. - L.: Nauka (1986)
9. Grinyaev, S. N.: Intellectual Opposition to Information Weapons. SINTEG, Moscow (1999)
10. Kotenko, I.V.: Models of confrontation between teams of agents for the implementation and security against distributed attacks "Denial of service". In: Tr. International Science and Technology Conference. IEEE AIS 2003 and CAD-2003, vol. 1, pp. 422–428. Fizmatlit, Moscow (2003)

11. Korneev, V.V., Maslovich, A.I., et al.: Recognition of software modules and detection of unauthorized actions using the apparatus of neural networks. Inform. Tech. 10 (1997)
12. Tan, K.: The application of neural networks to UNIX computer security. In: Proceedings of IEEE International Conference Neural Networks, vol. 1, pp. 476–481 (1995)
13. Electronic resource: The development of information threats in the first quarter of 2020. Statistics. https://securelist.ru/it-threat-evolution-q1-2020-statistics/96202/
14. Zegzhda, D.P., Vasiliev, Yu.S., Poltavtseva, M.A., Kefeli, I.F., Borovkov, A.I.: Cybersecurity of progressive production technologies in the era of digital transformation. https://doi.org/10.21681/2311-3456-2018-2-2-15

Using Non-linear Integral Models in Automatic Control and Measurement Systems for Sensors' Input Signals' Recovery

Andriy Verlan[1(✉)], Volodymyr Fedorchuk[2], Vitalii Ivaniuk[2], and Jo Sterten[1]

[1] Norwegian University of Science and Technology (NTNU), Gjovik, NTNU Gjøvik Postboks 191NO, 2802 Gjøvik, Norway
{andriy.verlan,jo.sterten}@ntnu.no
[2] Kamianets-Podilskyi National Ivan Ohiienko University, Symona Petliury, 1, Kamianets-Podilskyi 32302, Ukraine
{fedvolod,wivanyuk}@kpnu.edu.ua

Abstract. The article is devoted to developing methods of dynamic correction of signals registered at the output of a non-linear measuring transducer for the purpose to recover its input signal. Dynamic correction leads to decreased inertia/time response of the measuring transducer and smoothes its non-linearity. Signal recovery problem is resolved based on the non-linear model of measuring transducer as a first kind, second degree polynomial Volterra integral equation. As the problem is ill-posed, the model regularization method is used. The article offers a differential regularization operator transforming the input polynomial integral equation into a polynomial integro-differential equation. For numerical realization of the given models' type an algorithm based on the difference and quadrature methods is offered. The key problem of limited modelling time when using the kernel, given by the tabular method, is resolved by applying a procedure of restarting of the calculation process with offset of the time interval the integral equation kernel defined on. To resolve the issue of significant short-period errors in calculations after the restart of the calculation process, we offer a method for obtaining solution from two parallel computing processes with tabular kernels offset in time at a half of the time interval they are defined on. A feature of integral models is their resistance to high-frequency interference, which are present in real engineering systems. The obtained results can be used in dynamic correction devices of measuring transducers of automatic control and measurement systems.

Keywords: Dynamic correction of non-linear measuring transducer · Polynomial integral models · Tabular kernel · Signal recovery

1 Introduction

Intensive development of modern controlled engineering systems demands constant improvement of measuring devices as the basic components of such systems. Computerized measuring transducers play a special role in the development of controlled engineering systems. They are usually composed of a sensor, AD transducer and dynamic

© The Author(s), under exclusive license to Springer Nature Switzerland AG 2021
R. A. Aliev et al. (Eds.): WCIS 2020, AISC 1323, pp. 18–25, 2021.
https://doi.org/10.1007/978-3-030-68004-6_3

correction device [4, 8]. A dynamic correction device is designed based on mathematic models of the researched processes. To a great extent it depends on the difficulty of algorithmic signal processing relying upon these models. The difficulty also arises due to the fact that digital signals shall be processed in real time with limited hardware as such measuring systems are usually represented as mounted units [5].

The key issue arising in measurement process is recovering the input signals, which is classified as an inverse ill-posed problem [9, 10]. Models of different types can be used to describe the dynamic features of measuring transducers [8, 10]. To provide for the necessary level of adequacy of multiple practical problems, measuring transducer mathematic models shall account for different features, specifically, non-linearity of processes and space distribution of parameters. As such features are considered, models grow increasingly more complicated together with the dynamic correction device software. Using operator models in integral form with a range of positive features (multipurposeness, smoothing measuring data, efficient macromodels, etc. [2, 10]) appears promising. What concerns recovery, these models acquire the form of the first kind Volterra integral equation. Hence, the basis integral form of signal recovery model is.

$$\int_{0}^{t} K(\xi, \zeta, t, s)x(\zeta, s)ds - y(\xi, t), \tag{1}$$

where $K(\xi, \zeta, t, s)$ is the kernel of integral operator, $y(\xi, t)$ is the set input signal in point ξ, and $x(\zeta, t)$ is the sought input signal in point ζ, $t \in [0, T]$. Such models can be used only to describe linear processes; in its turn, to consider the non-linear features, the basic model may acquire a form of first kind polynomial Volterra integral equation [1, 6, 9].

$$\sum_{m=1}^{n} \underbrace{\int_{0}^{t} \ldots \int_{0}^{t}}_{m} K_m(\xi, \zeta, t, s_1, \ldots, s_m) \prod_{i=1}^{m} x(\zeta, s_i)ds_i = y(\xi, t), \tag{2}$$

where $K_m(\xi, \zeta, t, s_1, \ldots, s_m)$ are multidimensional kernels, $x(\zeta, t)$, $y(\xi, t)$ are, correspondingly, the input and output signals applied at points ζ and ξ of the distributed object, n is a certain integer, and T is the transition process time.

Many publications [2, 10] dwell upon solving inverse problems based on the (1), with regularization methods being the most efficient. Researches [1, 9] are devoted to developing methods for solving the Eqs. (2), which usually based upon direct or iterative methods. A range of factors complicate the use of the current methods to develop mathematical means and software for dynamic correction devices of measuring transducers, particularly, accumulation of calculations during numerical implementation of multidimensional integral models and ill-posedness of the problem, which does not allow to sufficiently recover the signals in the case of input data noise interference.

Thus, the task of developing methods for solving polynomial integral equations, which will serve as a foundation for software of computerized dynamic correction devices for measuring transducers, becomes topical.

2 Methods for Solving Polynomial Integral Equations

Let us consider a partial case (2) as a model of non-linear sensor – first kind, second degree polynomial Volterra integral equation

$$\int_0^t K_1(\xi,\zeta,t,s)x(\zeta,s)ds + \int_0^t\int_0^t K_2(\xi,\zeta,t,s_1,s_2)x(\zeta,s_1)x(\zeta,s_2)ds_1ds_2 = y(\xi,t), \tag{3}$$

where $y(\xi, t)$ is the value obtained from the sensor, $x(\zeta, t)$ is the measuring value sought, $K_1(\xi, \zeta, t, s)$ is the first order kernel, and $K_2(\xi, \zeta, t, s_1, s_2)$ is the second order kernel. Integral equation kernels (3) depending on the used method may have different representation: analytical representation when using analytical methods for transformation of other models' types; tabular representation when using the identification methods [3, 7]. In our case, kernel $K_1(\xi, \zeta, t, s)$ is presented as a table (vector) and kernel $K_2(\xi, \zeta, t, s_1, s_2)$ is presented as a matrix.

Let us consider a method for solving the (3) based on model regularization.

2.1 Model Regularization

As the signal recovery problem is ill-posed, and the use of known methods and algorithms does not provide a solution of necessary accuracy in case of interference of input data in limited time, it is proposed to create regularization algorithms using the differential regularization operator.

$$Dx(\zeta,t) = \rho_2\alpha^2\frac{d^2x(\zeta,t)}{dt^2} + \rho_1\alpha\frac{dx(\zeta,t)}{dt} + \rho_0\alpha x(\zeta,t), \tag{4}$$

where α is regularization parameter and ρ_i are regularization coefficients defining the operator order; they can equal 0 or 1.

The model experiments' method revealed that the first order differential regularization operator yields the best results. Thus, solving first kind Volterra integral equation is reduced to solving a polynomial integro-differential equation:

$$\begin{gathered}\alpha\frac{dx(\zeta,t)}{dt} + \int_0^t K_1(\xi,\zeta,t,s)x(\zeta,s)ds \\ + \int_0^t\int_0^t K_2(\xi,\zeta,t,s_1,s_2)x(\zeta,s_1)x(\zeta,s_2)ds_1ds_2 = y(\xi,t),\end{gathered} \tag{5}$$

where α is a regularization parameter proposed to be found based on the model experiments' method [10].

2.2 Algorithms for Solving Polynomial Integral Equations

It is proposed to solve the given task by replacing the integrals with quadrature sums, which provides several advantages, in particular simple implementation and high stability of computational algorithms due to regularization features of sampling interval selection [10].

By introducing fixed-rate sampling ($i = \overline{1...n}, h = t_i - t_{i-1}$), applying the method of trapezoids and first order difference formula to the (5) and grouping the sought $x(\zeta, t_i)$ by degrees, we obtain the following equation:

$$
\begin{aligned}
&\frac{1}{4}h^2 K_2(\xi, \zeta, t_i, t_{i,}, t_i)x(\zeta, t_i)x(\zeta, t_i) \\
&+\Bigg(\frac{1}{4}h^2(K_2(\xi, \zeta, t_i, t_i, t_0) + K_2(\xi, \zeta, t_i, t_0, t_i))x(\zeta, t_0) \\
&+\frac{1}{2}h^2 \sum_{j=1}^{i-1} (K_2(\xi, \zeta, t_i, t_{i,}, t_j) + K_2(\xi, \zeta, t_i, t_{j,}, t_i))x(\zeta, t_j) \\
&+\frac{1}{2}hK_1(\xi, \zeta, t_i, t_i) + \frac{\alpha}{h}\Bigg)x(\zeta, t_i) + \sum_{j=1}^{i-1} hK_1(\xi, \zeta, t_i, t_j)x(\zeta, t_j) \\
&+\frac{1}{2}hK_1(\xi, \zeta, t_i, t_0)x(\zeta, t_0) + \frac{1}{4}h^2 K_2(\xi, \zeta, t_i, t_0, t_0)x(\zeta, t_0)x(\zeta, t_0) \\
&+\frac{1}{2}h^2 \sum_{j=1}^{i-1} (K_2(\xi, \zeta, t_i, t_{0,}, t_j) + K_2(\xi, \zeta, t_i, t_{j,}, t_0))x(\zeta, t_0)x(\zeta, t_j) \\
&+h^2 \sum_{j=1}^{i-1}\sum_{g=1}^{i-1} K_2(\xi, \zeta, t_i, t_{j,}, t_g)x(\zeta, t_j)x(\zeta, t_g) - \frac{\alpha}{h}x(\zeta, t_{i-1}) - y(\xi, t_i) = 0.
\end{aligned}
\tag{6}
$$

Let us introduce the following notation:

$$
A_i = \frac{1}{4}h^2 K_2(\xi, \zeta, t_i, t_{i,}, t_i), \tag{7}
$$

$$
\begin{aligned}
B_i = &\frac{1}{4}h^2(K_2(\xi, \zeta, t_i, t_i, t_0) + K_2(\xi, \zeta, t_i, t_0, t_i))x(\zeta, t_0) \\
&+\frac{1}{2}h^2 \sum_{j=1}^{i-1} (K_2(\xi, \zeta, t_i, t_{i,}, t_j) + K_2(\xi, \zeta, t_i, t_{j,}, t_i))x(\zeta, t_j) + \frac{1}{2}hK_1(\xi, \zeta, t_i, t_i) + \frac{\alpha}{h},
\end{aligned}
\tag{8}
$$

$$C_i = \sum_{j=1}^{i-1} hK_1(\xi, \zeta, t_i, t_j)x(\zeta, t_j)$$
$$+\frac{1}{2}hK_1(\xi, \zeta, t_i, t_0)x(\zeta, t_0) + \frac{1}{4}h^2K_2(\xi, \zeta, t_i, t_0, t_0)x(\zeta, t_0)x(\zeta, t_0)$$
$$+\frac{1}{2}h^2\sum_{j=1}^{i-1}(K_2(\xi, \zeta, t_i, t_{0,}, t_j) + K_2(\xi, \zeta, t_i, t_{j,}, t_0))x(\zeta, t_0)x(\zeta, t_j) \quad (9)$$
$$+h^2\sum_{j=1}^{i-1}\sum_{g=1}^{i-1}K_2(\xi, \zeta, t_i, t_{j,}, t_g)x(\zeta, t_j)x(\zeta, t_g) - \frac{\alpha}{h}x(\zeta, t_i) - y(\xi, t_i).$$

Hence, (6) considering the notation of (7), (8), (9) has the following form.

$$A_i x(\zeta, t_i)^2 + B_i x(\zeta, t_i) + C_i = 0. \quad (10)$$

Quadratic equations of the system (10) are solved in sequence, and for selection of one of the routes, the following algorithms is used: if $|x(\zeta, t_{i-1}) - x(\zeta, t_{i_1})| \geq |x(\zeta, t_{i-1}) - x(\zeta, t_{i_2})|$, then $x_i = x_{i_2}$, else $x_i = x_{i_1}$.

2.3 Solving Polynomial Integral Equations with Tabular-Type Kernels in an Infinite Period

Model (5) is distinguished by integral kernels that set tabular (vector and matrix), which imposes time limits for the model. This issue is resolved by restart of the computational process with offset of the time interval the integral model's (5) kernels defined on. Kernel displacement operation is mathematically ill-posed as the restarted computational process does not account for the initial conditions while calculating the first discretion value of solution on the offset time interval. In this connection, at the beginning of the restart, there are significant tolerance in solution calculation. However, due to regularization feature the tolerance is rapidly reduced to the level of systematic error. To obtain a solution for infinite time interval with the acceptable tolerance not exceeding systematic error, it is proposed to use an additional computational process for the Eq. (5). The peculiarity of the additional computational process is time-offset of the kernels, that set tabular, by half time interval at which they are identified. Thus, one can obtain solutions with the acceptable accuracy on the intervals with significant tolerances at the restart of the main calculation process. The resulting solution is a combination of solutions obtained from two calculation processes in two threads; it also accounts for the fragments of the obtained results with stable convergence of the solution. Fig. 1 provides for the diagrams of the obtained solutions for the type (5) equations with restarts of the time-offset computational processes. Figure 2 shows a solution diagram obtained from combining the solutions fragments of different threads.

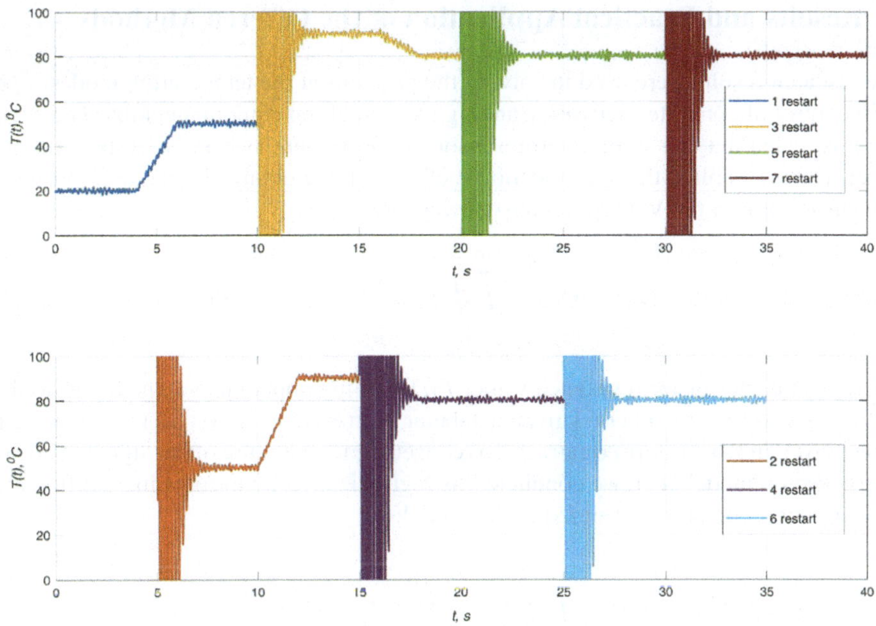

Fig. 1. Diagrams of the solutions of type (5) equations with restarts of calculation processes.

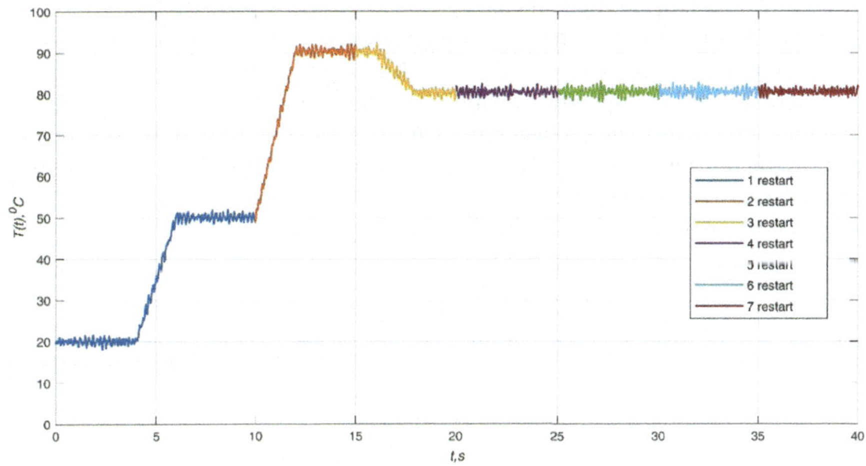

Fig. 2. Combined solutions' fragments.

3 Results and Practical Application of the Offered Methods

The research results were used in solving the problem of the temperature modes' operative control of computer network trunking and switching equipment chips. The experiment revealed that the "chip–thermosensor" measurement system is inertial and nonlinear. Having applied the identification method [7], we obtained a model represented as a partial sum of the Volterra integro-power series:

$$R(t) = k_0 + \int_0^t K_1(s)T(t-s)ds + \int_0^t \int_0^t K_2(s_1, s_2)T(t-s_1)T(t-s_2)ds_1 ds_2, \quad (11)$$

where $R(t)$ is thermistor resistance value, $T(t)$ is chip temperature, k_0 is dimensionless factor, $K_1(s)$ is first order kernel given in tabular form (vector), $K_2(s_1, s_2)$ is second order kernel given in tabular form (matrix). To recover the temperature offset signal inside the chip based on model (11), we conducted its regularization by introducing a differential regularization operator. It resulted in the model:

$$\alpha \frac{dT}{dt} + \int_0^t K_1(s)T(t-s)ds + \int_0^t \int_0^t K_2(s_1, s_2)T(t-s_1)T(t-s_2)ds_1 ds_2 = R(t) - k_0, \quad (12)$$

where α is the regularization parameter.

Model (12) is the basis of the software component of the computer subsystem for controlling the chips' temperature modes, which allows to minimize the delay of thermal sensor reaction to chip temperature change. Temperature value is defined based on the thermal sensor resistance digital data sent by the computer network to the computerized dynamic correction and control system. Fig. 3 provides for diagrams of: a) the registered signal used as a basis for recovery of the signal of temperature change inside the chip; b) the result of signal recovery (recovered and actual).

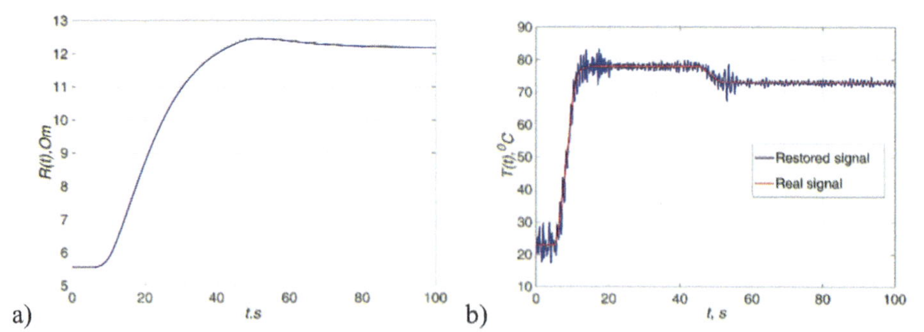

Fig. 3. Diagrams of the signals: a) registered at the thermal sensor output; b) recovered by the dynamic correction system

4 Conclusions

Thus, the developed method for solving first kind, second degree polynomial integral equations based on differential regularization operator and mechanism of restarting the computational process with offsetting the time interval the equation kernels are defined on, made it possible to solve the problem of recovery of the input signal of a non-linear measuring transducer in real time at high-frequency noise interference of input signal. A perspective direction for further research is to determine the optimal time interval the integral model kernels and regularization parameters are defined on, which provides the maximum rate of computational process convergence at restart.

References

1. Apartsyn, A.S., Solodusha, S.V., Spiryaev, V.A.: Modeling of nonlinear dynamic systems with Volterra polynomials. Int. J. Energy Optim. Eng. **2**(4), 16–43 (2013). https://doi.org/10.4018/ijeoe.2013100102
2. Hermann, B.: Volterra integral equations: an introduction to theory and applications, Cambridge University Press, **30** (2017). doi:https://doi.org/10.1017/9781316162491.
3. Doyle, F.J., Pearson, R.K., Ogunnaike, B.A.: Identification and control using Volterra models. Commun. Control Eng. (2002). https://doi.org/10.1007/978-1-4471-0107-9
4. Fraden, J.: Handbook of Modern Sensors (2016). https://doi.org/10.1007/978-3-319-19303-8
5. Giurgiutiu, V.: Mechatronics and smart structures design techniques for intelligent products, processes, and systems. Intell. Knowl. Syst. 1394–1472 (2005). https://doi.org/10.1007/978-1-4020-7829-3_39
6. Ivaniuk, V.A., Ponedilok, V.V.: Method of restoration of input signals of nonlinear dynamic object with distributed parameters. Math. Comput. Modell. Ser. Tech. Sci. **18**, 65–73 (2018). https://doi.org/10.32626/2308-5916.2018-18.65-73
7. Ivaniuk, V.A., Fedorchuk, V.A.: Adaptive method of identification of models of nonlinear dynamic systems with using integral Volterra series. Èlektronnoe Modelirovanie **41**(3), 33–42 (2019). https://doi.org/10.15407/emodel.41.03.033
8. Mukhopadhyay, S.: Next generation sensors and systems, smart sensors. Meas. Instrum. (2016). https://doi.org/10.1007/978-3-319-21671-3
9. Solodusha, S.V., Yaparova, N.M.: Numerical solving an inverse boundary value problem of heat conduction using Volterra equations of the first kind. Numer. Anal. Appl. **8**(3), 267–274 (2015). https://doi.org/10.1134/s1995423915030076
10. Verlan, A.F., Sagatov, M.V., Sytnik, A.A.: Methods of mathematical and computer modeling of measuring transducers and systems based on integral equations. Fan Publishing House of the Academy of Sciences of the Republic of Uzbekistan, Tashkent (2011)

Neural Network Method and Algorithm for Document Detection Based on Signaling Analysis

Bekmuratov Tulkun Fayziyevich and Botirov Fayzullajon Baxtiyorovich(✉)

Tashkent University of Information Technologies named after Muhammad al-Khwarizmi, Tashkent 100200, Uzbekistan
bek.tulkun@yandex.com, botirov_fz@mail.ru

Abstract. This presents a sequence article of key steps in the information processing process in an attack detection system. In order to detect attacks on information in the enterprise, the neuronal non-obscure classifier structure was selected, and a neural network method and algorithm for document detection was developed on the basis of signature analysis.

Keywords: Attack detection system · False negative · False positive · SYN flood · Neural network · IP introduction

1 Introduction

Underlying the method of signaling analysis is the idea that each attack can be described using a specific set of rules, or some formal model in the form of a semantic expression, a series of symbols, in one language or another. Attack efforts to authenticate the user to the current or incoming/outgoing traffic information stored in the database templates (signature attacks brought comparisons).

Advantages of methods based on signal analysis:

- high efficiency in detecting certain attacks;
- False negative number of false calls.

Disadvantages of these methods:

- detects only certain attacks;
- the database must be constantly replenished with new attack signatures, otherwise there is a potential risk of "missed the attack" (false positive);
- due to the obvious redundancy (attempts to take into account and remember all the possible attacks that have been observed before), the application of this technology requires a large cost [1].

© The Author(s), under exclusive license to Springer Nature Switzerland AG 2021
R. A. Aliev et al. (Eds.): WCIS 2020, AISC 1323, pp. 26–32, 2021.
https://doi.org/10.1007/978-3-030-68004-6_4

2 The Main Part

At present, modern systems for detecting attacks at the level of network use a variety of methods to search for attack signals, that is, the transmission of network traffic to a particular information system in the context of an attack. In this case, the process of information processing in the attack detection system (ADS) includes the following steps (Fig. 1):

1. Defining a set of characters (parameters);
2. Initial processing of parameters;
3. Attack recognition (classification).

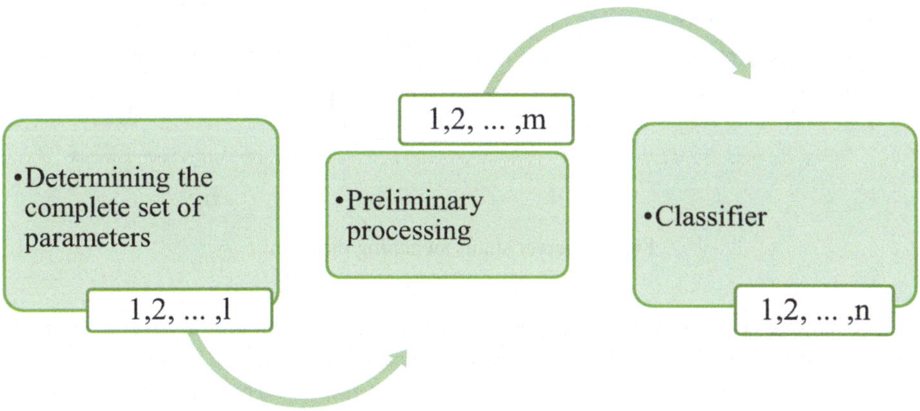

Fig. 1. Schematic of the information processing process in ADS

The first step is to capture network traffic. The collection of the required data is carried out using a special software module – a packet sniffer, which captures all packets that pass through the TCP protocol and filters them. In the current case, the value of the network packet header field serves as the primary character (parameter) that characterizes the analyzed traffic.

The second step involves sorting out the more important parameters that characterize network activity (in an incoming database) and presenting them in a form that can be used more efficiently for further processing using a classifier.

The third step is to identify and recognize the attacks. The use of neural networks at this stage is more efficient than classical attack detection systems in which simple comparison of packet header data with known attack signatures is performed because the neural network always attempts to determine how similar current network activity signs in attack images are from the read selection. In return for the ability of a neural network to be large enough, a neural network can extrapolate its knowledge of known manifestations of network attacks to unknown manifestations. Take the SYN Flood attack for example. This attack belongs to the class of "service denial" attack, that is, its main purpose is not to steal or change information, but to block the work of the server.

This can also cause significant damage, as failure to provide information in a timely manner can lead to incorrect or simple decisions in the business.

In Fig. 2 basis strong temporary change in the status of the server shown in the diagram, in accordance with the normal (a) and (b) at the time of the attack. The numbers on the C i axis indicate the server status: 1 - current status, waiting for a client request by the server; 2 - client C-SYN flags TCP-package; 3 - the answer to the client server (TCP-package), installed on the server request (C-SYN flag TCP-package shipment), customer response (C-ACK flag TCP-package); 4 - TCP-packet with S-ACK flag from the client; 5 - session set [2–4].

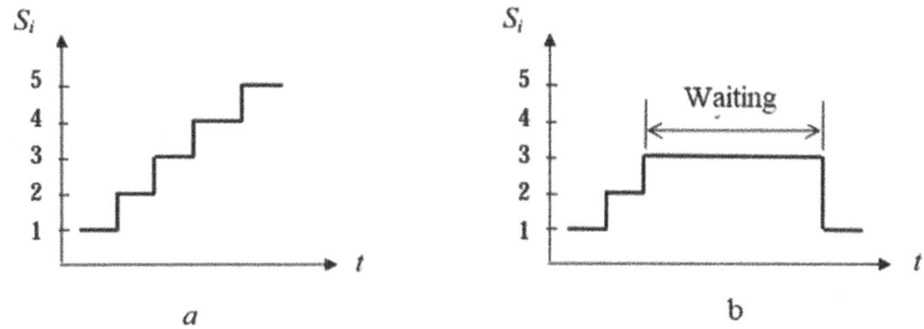

Fig. 2. Server status switching diagram

The attack symbol is a sudden increase in the number of connections over a relatively short period of time corresponding to case 3, where shown in the diagram in Fig. 3a (where N is the number of connections per watt at time t). Therefore, the difference DN between the number of connections corresponding to cases 3 and 5 can be used as the current data for the analysis. At the initial stage of the attack, a sudden increase in the value of DN is observed, which is not typical for normal operation of clients with servers.

The available methods of detecting this attack are based on comparing the value of DN with a predetermined limit (e.g., the fulfillment of the condition DN $\geq$ 100 is an attack) or comparing the ratio of DN values at a previous time moment with a given magnitude.

These methods have drawbacks. The above threshold values must be selected for each specific situation, they may be different for each server. There are no clear boundary values that correspond to a clear attack or normal. Moreover, these limits depend on the time - for example, the server's daytime performance statistics make the evening poorer. Therefore, continuous adjustment is required in the current parameters that characterize the server operation, which is almost impossible to do in these methods.

If we use a neural network in the process of detecting attacks, then we carry out the third stage of the scheme of information processing in the detection of attacks as follows.

It is assumed that the classifier input (see Fig. 1) receives some parameters that characterize the next data packet coming from the server. Important for analysis in this case are parameters such as TCP-flags, IP-addresses of the sender and receiver, forwarded packets (incoming or outgoing), packet arrival time, receiver port. The correctness of

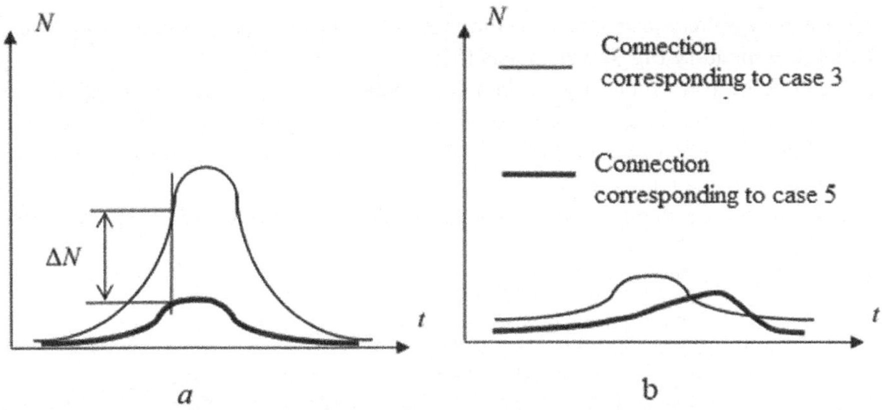

Fig. 3. Diagram of the number of simultaneous connections of the server to the client: a - attack mode; b - normal

the headers should also be checked, as the arrival of a large number of packages with the wrong header is also one of the signs of an attack. Thus, the classifier based on the results of the analysis of incoming traffic includes the following variables, including the vector of incoming parameters:

A_1 - average time of arrival of one package;
A_2- percentage of packages with different external IP- addresses;
A_3 - percentage of packages with different external ports;
A_4 - percentage of incorrectly titled packages.

The output of the classifier is a parameter B defined as the "level of reliability in attack". The above-mentioned A_1, A_2, A_3, A_4 network traffic parameters, reliability and attack B by the experts and the following:

IF Packages _ arrival _ time = " small " AND
Different _ external _ ip _ addresses _ percent " small " AND
Different _ports _ percent = ' small percentage '. THEN
Attack _ reliability _ level = **"average"**;

IF (Packages _ arrival _ time = " small " OR
Packages _ arrival _ time " average ") AND
(Different _ external _ ip _ addresses _ percent " big " OR
Different _ports _ percent = " big " OR
Incorrect _ header packages _ percent = " large "). THEN
Attack _ reliability _ level = **" high "**;

IF Packages _ Arrival _ Time " GREAT ", THEN
Attack _ reliability _ level = **" low "**.

These provisions make sense for the system to cease treatment-caused neurons to perform classificatory Fig. 4 weak basis [5].

As can be seen from the figure, the neural network shown is five-layered. The first (unwanted) layer performs the distribution function by giving the measured (calculated) values of the parameters to the neuronal inputs of the 1st latent layer..., [6, 7]. The first hidden layer neurons cease packages corresponding to the parameters, perform the function of chargeability function to calculate the values of these parameters fuzzification:

tBig - "packages_arrive_time";
tAverage - "packages_arrival_time";
tSmall - "packages_arrival_time";
tipBig - "different_external_ip_additions_center is large";
tipSmall - different_external_ip_additions_center is small;
tPortBig - "various_ports_percentage large";
tPortSmall - "various_ports_percentage is small";
tPaketBig - " incorrect _ header packages _ percent large ".

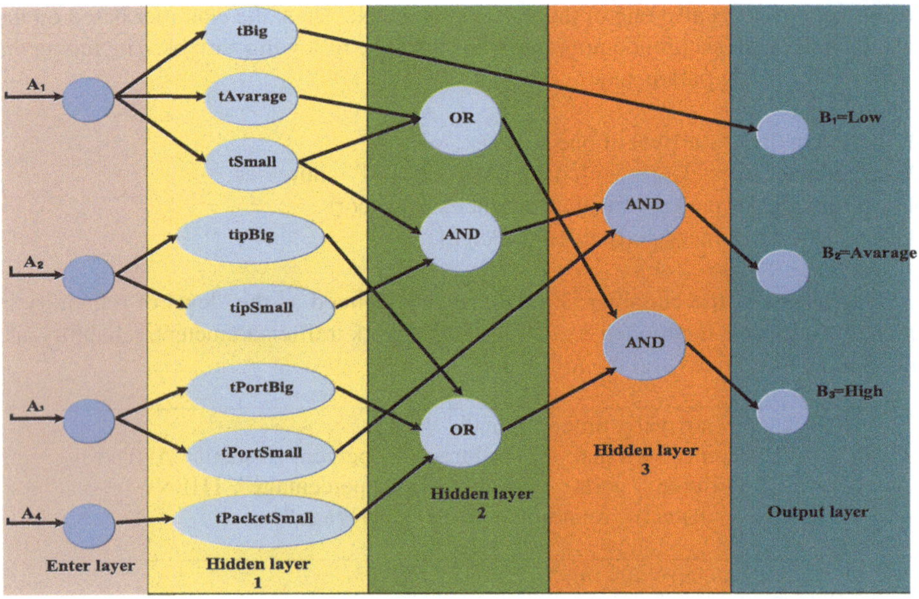

Fig. 4. Neuron obscure classifier structure

Figure 4 the second and third hidden layers cease logic based on the rules of logic (AND, OR) and increase () function neurons cease. a_i, a_j, a_k 3 palettes out with a vector or neurons, respectively, out of valuation calculated using the following formula:

$$S_l = w_{il}a_i + w_{jl}a_j + w_{kl}a_k + w_{il}a_iw_{jl}a_j + w_{il}a_iw_{kl}a - w_{il}a_iw_{jl}a_jw_{kl}a_k \quad (1)$$

Here w_{il}, w_{jl}, w_{kl}, the hidden layer 2 l-neuron hidden layer i, j, and k - connecting neurons, synaptic connections configurable weight; 1 - output of secretory layer i, j and

k - neurons. The output value for input 2 n -AND-neuron is calculated by the following formula:

$$z_n = (w_{ln} + a_l - w_{ln}a_l)(w_{mn} + a_m - w_{mn}a_m) \tag{2}$$

l and t - neurons output; - 3-hidden layer p neuron hidden layer 2 l and custom neurons between the weight.

Outgoing layer neurons are classical neurons whose synaptic connections that are adjusted during training are weighted, activating sigmoid function.

The output level of reliability in the attack = {low; average; high} determines the value of the relevance function for the exact term of the linguistic variable [8–10] [11] (Fig. 5).

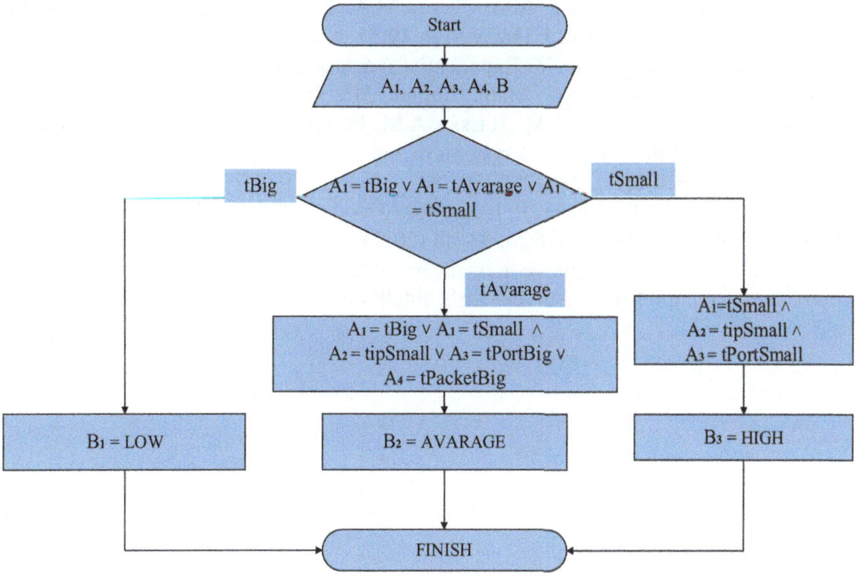

Fig. 5. Block diagram of the neural network method algorithm for attack detection

3 Conclusion

To conclude, based on the analysis of signature the process of determining the attacks will be able to determine the level of neuron network using the method of attack. The level of the neural network layer attacks, which corresponds to the position of this level of decision - making easier. For protection against the attack on the system of decision - making mechanism of the attack rate or a high level of probability that a given decision increases the reliability provided by the state. Therefore, the enterprise information attacks during the attack to determine the level of great importance. Has developed a method, based on the performance level of reliability in the attack = {low; average; high} allows you to determine the value of the relevance function for the exact term of the linguistic variable.

References

1. Medvedev, N.V., Markov A.S., Fadin A.A.: Application of staticheskogo signaturnogo analiza dlya vyyavleniya defektov bezopasnosti web- prilojeniy , N auka i obrazovanie izdatel FGBOU VPO MGTU im. N.e. baumana. № FS 77 – 48211, 299–300 (2012).
2. Mileva, A., Panajotov, B.: Covert channels in TCP/IP protocol stack - extended version. Central Eur. J. Comput. Sci. **4**(2), 45–66 (2014)
3. Mazurczyk, W., Wendzel, S., Zander, S., Houmansadr, A., Szczypiorski, K.: Information Hiding in Communication Networks: Fundamentals, Mechanisms, Applications, and Countermeasures, pp. 100–104. Wiley, Hoboken (2016)
4. Cisco Education Resource Center. https://www.cisco.com
5. Igumnov, I.V., Kutsyy, N.N.: Method Neldera - Mida in obuchenii iskusstvennoy neyronnoy seti, vxodyashchey v sostav SHIM-elementa. Nauchnыy vestnik NGTU **3**, 22–30 (2017)
6. Hassan, N., Sayed, O.R., Khalil, A.M., Ghany, M.A.: Fuzzy soft expert system in prediction of coronary artery disease. Int. J. Fuzzy Syst. **19**(5), 1546–1559 (2017)
7. Ge, C., Wang, B., Wei, X., Liu, Y.: Exponential synchronization of a class of neural networks with sampled-data control. Appl. Math. Comput. **315**, 150–161 (2017)
8. Abdel-Aleem, A., El-Sharief, A.M., Hassan, A.M., El-Sebaie, G.M.: Implementation of fuzzy and adaptive neuro-fuzzy inference systems in optimization of production inventory problem. Appl. Math. Inf. Sci. **11**(1), 289–298 (2017)
9. Ambukege, G., Justo, G., Mushi, J.: Neuro fuzzy modeling for prediction of consumer price index. Int. J. Artif. Intell. Appl. **8**(5), 33–44 (2017)
10. Karaboga, D., Kaya, E.: Adaptive network based fuzzy inference system (ANFIS) training approaches: a comprehensive survey. Artif. Intell. Rev. 1–31 (2018).

Using Fuzzy Probabilistic Implication in Z-set Based Inference

Latafat A. Gardashova(✉)

Azerbaijan State Oil and Industry University, Azadlig 35, Nasimi, Baku, Azerbaijan
l.qardashova@asoiu.edu.az, latsham@yandex.ru

Abstract. In our daily life the reasoning is characterized by imperfect information. This information is expressing by using fuzzy and probabilistic information. The main advantage of fuzzy logic is possibility of application in solving complex, nonlinear problems which described by words based on human perception. According to these in scientific literature there are fuzzy reasoning approach and probabilistic approximate reasoning approach, which can not operate imperfect information. This paper deals Z-set based inference using probabilistic implication. Numerical example with Z-set-based inference for approximate reasoning is considered.

Keywords: Fuzzy implication · Probabilistic implication · Z-set based If-Then rules · Aggregation · Z-reasoning · Fuzzy reasoning

1 Introduction

Today existing inference methods can not be made adequately by the methods which are based on two valued or on many valued logic. In the development of fuzzy inference there are critical points in scientific literature such as [1–4]. First paper on fuzzy systems is published by Zadeh. He suggested an inference rule called "compositional rule of inference". Researches proposed several methods for fuzzy reasoning, for example, linguistic approach, Takagi-Sugeno approach, Hybrid neuro-fuzzy approach and etc.

Reasoning is the one area of artificial intelligence which depends on the human judgment level and has been a dynamic research field since the 1950s.

In approximate reasoning a possible inaccurate conclusion is inferred from a collection of inexact premises. First work about approximate reasoning is given in [3] by Zadeh. Many existing methods of fuzzy reasoning are based on fuzzy implications. A comparison of fuzzy reasoning methods are discussed in [5–9]. These methods based on implication operators. Probabilistic implication-based reasoning is discussed in [10]. The presented implications are based on conditional copulas and it is proved scientifically that in some properties of copula probabilistic implication is fuzzy implication. Fuzzy probabilistic information is made relation between probability theory and fuzzy logic.

In our previous work [11] approach which able operate imperfect information consisting fuzzy and probability is discussed. We discussed a new approach to approximate

© The Author(s), under exclusive license to Springer Nature Switzerland AG 2021
R. A. Aliev et al. (Eds.): WCIS 2020, AISC 1323, pp. 33–39, 2021.
https://doi.org/10.1007/978-3-030-68004-6_5

reasoning using If-Then rules with Z-set based antecedents and consequents. An approach for reasoning with Z-rules based on copula is detailed in previous paper. Advantage of proposed method is characterized by ability to operate with combination of fuzzy and probabilistic uncertainties.

In this paper a new approach based on fuzzy probabilistic implication and Z-set is investigated.

The rest of the paper is organized as follows. Section 2 gives the preliminaries. A statement of the problem is given in Sect. 3. Section 4 considers the proposed approach to reasoning with Z-rules. In Sect. 5, a numerical example is illustrated. Finally, Sect. 6 concludes the paper.

2 Preliminaries

Definition 1 [11]. Z-Set. A Z-set is a triple $Z = (A, B, G)$, where A is a fuzzy set, B is a fuzzy restriction on probability measure of A, $P(A)$, induced by a set of distribution G:

$$G = \left\{ p_z(x) : \int_{-\infty}^{+\infty} p_Z(x)dx = 1, \int_{-\infty}^{+\infty} p_Z(x)\mu_A(x) \; is \; B, \right.$$
$$\left. \int_{-\infty}^{+\infty} xp_Z(x) = \int_{-\infty}^{+\infty} x\mu_A(x)dx / \int_{-\infty}^{+\infty} \mu_A(x)dx \right\}$$

Definition 2 [11]. Z IF-THEN Rules

If X_1 is $Z_{X_1,n} = (A_{X_1,n}, B_{X_1,n})$ and,..., and X_m is $Z_{X_m,n} = (A_{X_m,n}, B_{X_m,n})$ then Y is $Z_Y = (A_{Y,n}, B_{Y,n})$.

$Z_{X_m,n} = (A_{X_m,n}, B_{X_m,n})$ is the Z-sets based inputs and $Z_Y = (A_{Y,n}, B_{Y,n})$ is Z sets-based outputs of the rules.

Definition 3 [9]. Probabilistic Implication. A function $I_C : [0, 1[^2 \rightarrow [0, 1]$ given by

$$I_C(u, v) = \begin{cases} 1, & \text{if u} = 0 \\ \frac{\text{C(u,v)}}{\text{u}}, & \text{if u} > 0 \end{cases}$$

where C is a copula, is called a probabilistic implication based on copula.

The upper Frechet-Hoeffding bound M is given by $M(u, v) = \min\{u, v\}$. The probabilistic implication based on $M(u, v)$ is

$$I_{M(u,v)} = \begin{cases} 1, & \text{if u} = 0 \\ \frac{\min\{u,v\}}{\text{u}}, & \text{if u} > 0 \end{cases} = \begin{cases} 1, & \text{if u} \leq v \\ \frac{v}{\text{u}}, & \text{if u} > v \end{cases}$$

which is nothing else than the Goguen implication, i.e. $I_{M(u,v)} = I_{GG(u,v)}$.

3 Statement of the Problem

Given n Z-set based rules and current input, aim is to find output.

IF-Then rules is given below:

If X_1 is $Z_{X_{1,1}} = (A_{X_{1,1}}, B_{X_{1,1}})$ and, ..., and X_m is $Z_{X_{m,1}} = (A_{X_{m,1}}, B_{X_{m,1}})$ then Y is $Z_Y = (A_{Y,1}, B_{Y,1})$
If X_1 is $Z_{X_{1,2}} = (A_{X_{1,2}}, B_{X_{1,2}})$ and, ..., and X_m is $Z_{X_{m,2}} = (A_{X_{m,2}}, B_{X_{m,2}})$ then Y is $Z_Y = (A_{Y,2}, B_{Y,2})$
...
If X_1 is $Z_{X_{1,n}} = (A_{X_{1,n}}, B_{X_{1,n}})$ and, ..., and X_m is $Z_{X_{m,n}} = (A_{X_{m,n}}, B_{X_{m,n}})$ then Y is $Z_Y = (A_{Y,n}, B_{Y,n})$
and a current observation
X_1 is $Z'_{X_1} = (A'_{X_1}, B'_{X_1})$ and, ..., and X_m is $Z'_{X_m} = (A'_{X_m}, B'_{X_m})$,
find the Z-value of Y.

Here m is the number of Z-valued input variables and n is the number of rules.

4 Solution Method

Solution method consist of the following steps:

1. Using fuzzy implications such as Zadeh, Aliev, Lukasevich, Kleene-Dienes, Goguen etc. fuzzy implications to define for each rule a fuzzy relation matrix.
2. To define a composed fuzzy relation matrix R, $R = \cup R_S$, $s = 1, ..., n$ according to logical connective of fuzzy logic-conjunction.
3. To define the membership function of the output using the compositional rule of inference, $\mu_U(u) = \max_u \min[\mu_{\mathrm{E}}(e_{ic}^{c}), \mu_{\mathrm{R}}(u, e)]$
4. To calculate B.
5. To obtain set of probability distributions G over antecedents of the rules.
6. To aggregate the distributions by using probability aggregation method weighted sum: $P_G(p) = \sum_{i-1}^{n} w_i P_i(p)$.
7. To obtain set of probability distributions G over consequents of the rules.
8. To use probabilistic implication.
9. To define set of probability distributions using points of support B over new input.
10. To calculate distance between two vector by using Hellinger distance of two discrete probabilty distribution vector:

$$ZHD(P_{Z_1}, P_{Z_2}) = \min_{p_{zi} \in G}\left(\frac{1}{\sqrt{2}}\left(\sum_{j=1}^{k}(p_{z1_j}^{1/2} - p_{z2_j}^{1/2})\right)^{1/2}\right)$$

11. To calculate the similarity measure: $SM = 1/(1 + ZHD)$
12. To define support points of output B.
13. To calculate the membership value of output B.

5 Numerical Example

Let's assume that these following 7 rules are involved in Z-valued control system [12].

1. If the error *e* is [negative big (NB), small sure] THEN the control action *u* is [negative big (NB), small sure].
2. If the error *e* is [negative medium (NM), sure] THEN the control action *u* is [negative medium (NM), sure].
3. If the error *e* is [positive big (PB), medium] THEN the control action *u* is [positive big (PB), sure].

Errors for Z-valued control system are defined with *e*, control actions are described as *u*. Through rules membership function can be computed as follows:

$$\textit{small sure} = 0.3/0.7 + 1/0.73 + 0.2/0.75;\ \textit{medium} = 0.2/0.5 + 1/0.6 + 0.4/0.8;$$
$$\textit{sure} = 0.1/0.7 + 1/0.8 + 0.3/0.9;\ \textit{very sure} = 0.2/0.8 + 1/0.9 + 0.2/1$$

Rule 1

$$\mu_{E1}(e) = 1.00/-10 + 0.73/-7 + 0.34/-3 + 0.20/0 + 0.13/3 + 0.08/7 + 0.06/10$$

$$\mu_{U1}(u) = 1.00/-1 + 0.92/-0.7 + 0.67/-0.3 + 0.50/0 + 0.37/0.3 + 0.26/0.7 + 0.20/1$$

Rule 2

$$\mu_{E2}(e) = 0.73/-10 + 1/-7 + 0.61/-3 + 0.34/0 + 0.20/3 + 0.11/7 + 0.08/10$$

$$\mu_{U2}(u) = 0.91/-1 + 1.00/-0.7 + 0.86/-0.3 + 0.67/0 + 0.50/0.3 + 0.34/0.7 + 0.26/1$$

Rule 3

$$\mu_{E_3}(e) = 0.06/-10 + 0.08/-7 + 0.13/-3 + 0.20/0 + 0.34/3 + 0.73/7 + 1/10$$

$$\mu_{U_3}(u) = 0.20/-1 + 0.26/-0.7 + 0.37/-0.3 + 0.50/0 + 0.67/0.3 + 0.92/0.7 + 1.00/1$$

For given 3 rules and their two properties (errors and control actions) membership functions were evaluated. Problem is to perform Z inference.

New input:

$$(\mu_{\mathrm{E}}(e^{c}_{ic}) = 0.11/-10 + 0.17/-7 + 0.34/-3 + 0.61/0 + 0.96/3 + 0.73/7 + 0.41/10, \textit{sure})$$

Relations created by using ALI-III implication for above given 3 rules. Intersection of relations is realized by using logical connective of Ali 3 and determined the following form relation ($R = \cup R_S, s = 1, \ldots, n$), n = 3. In Table 1 is given $\mu_R(u, e)$-is the value of membership function on composed fuzzy relation.

Calculation of probability distribution for all antecedents and consequents of the rules. For example, probability distributions on antecedent and consequent of the first rules is given in Table 2.

Table 1. Composed fuzzy relation

$R = \cup R_S, s = 1, \ldots, n$						
1	0.92	0.67	0	0	0	0
1	1	0.734631	0	0	0	0
1	1	1	1	1	0	0

Table 2. Probability distributions on antecedent and consequent

Distributions (on input of third rule)			Distributions (on output of third rule)		
0.5	0.6	0.8	0.7	0.8	0.9
0	0	0,212992	0	0,2089	0,21
0	0	0	0,060591	0	0
0	−1E−06	0	0,140805	0	0
0	−1E−06	0	0,174276	0	0
0,757577	0,553792	0	0,191010	0	0
0	0,127773	0	0,203560	0,7911	0,79
0,242424	0,318437	0,787008	0,229758	0	0

Calculate similarity measurement between current vector of probability distributions with aggregated input vector of probability distributions. We have A part of Z number, aggregated probability vector(over inputs of the rule and aggregated probabilty vector over obtained distributions for all rules.

Aggregation of probability distribution of antecedent of 3 rules using probability aggregation method- a weighted sum is determined as follow (Table 3):

Table 3. Aggregation of probability distribution

p1	p2	p3
0.360862588	0.281393	0.276036123
0.312332067	0.423765	0.543411345
0.152748362	0.170579	0.028657468
0.043531749	0.006059	0
0.012607359	0	0
0.035833254	0.01124	0.004143985
0.082084621	0.106965	0.147751078

New input is given as Z number. Obtain set of probability distributions using points of support B over new input (Table 4). We are getting probability vector.

Table 4. New input distribution

Points and distributions		
0.7	0.8	0.9
0	0	0
0.042575	1E−06	1E−06
0.116257	0.038604	0
0.199474	0.274913	0.154821
0.334469	0.513237	0.819908
0.264072	0.173247	0.025272
0.043152	0	0

Calculate distance between two vector (new input based probability vector and aggregated probability vector over inputs of the rule). Obtained value of the distance:

$$Dist = 0.21938$$

Result of the calculation similarity measurement between current vector of probability distributions with aggregated input vector of probability distributions

$$\text{SM} = 0.820089$$

Calculate probability distributions in consequent part using similarity degree. We have A, similarity measure and probability distributions.

Obtained result:

$$\begin{aligned}
\text{A} &= 0.17/-1 + 0.17/-0.7 + 0.17/-0.3 + 0.3982/0 + 0.41/0.3 + 0.41/0.7 + 0.41/1 \\
\text{B} &= 0.1/0.210742181 + 0.1/0.210742377 + 0.2/0.2320789 \\
&\quad + 0.2/0.233076319 + 0.2/0.233076376 + 0.1/0.233076474 \\
&\quad + 0.2/0.242178525 + 0.2/0.242178563 + 1/0.242178721
\end{aligned}$$

6 Conclusion

In this paper, we have investigated the inference results obtained by introducing the probabilistic implication. The steps of proposed approach for reasoning with Z-rules is formulated. Advantage of this method is the possibility to operate with combination of fuzzy and probabilistic uncertainties. This reasoning method based on probabilistic implication, a distance, similarity measure, fuzzy and probability aggregations. The proposed approach can be applied on different areas of the real life. From these results we can conclude that the proposed approach which are based on the probabilistic implication is very effective tool for operating imperfect information.

References

1. Zadeh, L.A.: Outline of a new approach to the analysis complex systems and decision processes. IEEE Trans. Syst. Man Cybernet **3**, 28–44 (1973)
2. Mamdani, E.H.: Application of fuzzy logic to approximate reasoning using linguistic systems. IEEE Trans. Comput. **26**, 1182–1191 (1977)
3. Zadeh, L.A.: Fuzzy logic and approximate reasoning. Synthese **30**, 407–428 (1975)
4. Takagi, T., Sugeno, M.: Fuzzy identification of systems and its application to modelling and control. IEEE Trans. Syst. Man Cybernet. **15**, 116–132 (1985)
5. Aliev, R.A., Aliev, R.R.: Soft Computing and its Application. World Scientific, Singapore (2001)
6. Aliev, R.A., Tserkovny, A.: Systemic approach to fuzzy logic formalization for approximate reasoning. Inf. Sci. **181**(6), 1045–1059 (2011)
7. Saner, T., Gardashova, L.A., Allahverdiyev, R.A., Eyupoglu, S.Z.: Analysis of the job satisfaction index problem by using fuzzy inference. Procedia Comput. Sci. **102**, 45–50 (2017)
8. Dubois, D., Liu, W., Ma, J., Prade, H.: The basic principles of uncertain information fusion. An organized review of merging rules in different representation frameworks. Inf. Fusion **32**, 12–39 (2016)
9. Przemysław, G.: Probabilistic implications. Fuzzy Sets Syst. **226**, 53–66 (2013). https://doi.org/10.1016/j.fss.2013.01.003
10. Aliev, R.A., Pedrycz, W., Huseynov, O.H., Eyupoglu, S.Z.: Approximate reasoning on a basis of Z-number valued If-Then rules. IEEE Trans. Fuzzy Syst. **25**(6), 1589–1600 (2017)
11. Aliev, R.A., Gardashova, L.A.: Z-set based approach to control system design. Adv. Intell. Syst. Comput. 1306 (2021, in press). https://doi.org/10.1007/978-3-030-64058-3
12. Aliev, R., Mamedova, G., Aliev, R.: Fuzzy Set Theory and its Application. Tabriz, Iran (1993)

Accounting Experience Between Fuzzy Integral and Z-numbers

H. A. Primova[1(✉)], F. N. Iskandarova[1], and Q. M. Gaybulov[2]

[1] Samarkand Branch of Tashkent University of Information Technologies, Street Shoxrux Mirzo 47A, Samarqand, Uzbekistan
xolida_primova@mail.ru, feruza621988@gmail.com
[2] Samarkand State Institute of Architecture and Construction, Lolazor Street, 70, Samarkand 140140, Uzbekistan
qodirgaybulov@mail.ru

Abstract. In this paper, computational experiments using the fuzzy integral theory and the Z-number combination have been conducted using fuzzy theory. The article was considered for the selection of the best biological and technologically-based selection breeds of cotton in the case of poor decision-making, and the results of the computational work using these two methods were theoretically justified.

Keywords: Z-number · Fuzzy theory · Fuzzy integral · Fuzzy measure · Computational experiments · Decision-making

1 Introduction

Today, results are obtained using decision-making issues, but they are not always expected. For example, there are significant shortcomings in the decision-making process between the managers of the enterprise, the issues of staff training and placement, and the consideration of their strengths and weaknesses.

It is well known that the dimension has a fundamental characteristic of additiveness in classical mathematics: the number equal to the sum of several combined characteristics must be the sum of the number of corresponding characteristics. Clearly, the use of linguistic variables directly in mathematical models undermines the hypothesis that measurements are addictive, that is, there is a problem with the use and study of measurements that do not obey additivity requirements.

2 Calculation of Fuzzy Integral by the Fuzzy Measure

Let G $x_1, x_2, \ldots, x_n$ be a set of elements, $G = \{x_1, x_2, \ldots, x_n\}$. The structure of G depends on the specific problem. For example, $x_1, x_2, \ldots, x_n$ is the yield of cotton under different weather conditions G = {25, 30, 35}. In G, we construct all possible sets of components and define them by $\beta(G)$. For example, $\beta(G)$ the elements of a set are: A1 = {25}; A2

© The Author(s), under exclusive license to Springer Nature Switzerland AG 2021
R. A. Aliev et al. (Eds.): WCIS 2020, AISC 1323, pp. 40–46, 2021.
https://doi.org/10.1007/978-3-030-68004-6_6

= {30}; A3 = {35}; A4 = {25; 30}; A5 = {25; 35}; A6 = {30; 35}; A7 = G = {25; 30; 35}; [1, 2, 7].

The main feature of this is that it is closed to mergers, complements and intersections: $A_i \cup A_j \in \beta(G), \overline{A}_i \in \beta(G), A_i \cap A_j \in \beta(G)$.

For example, $A_1 \cup A_2 = A_4 = \{25;\ 30\} \in \beta(G)$ [1, 2, 8].

$\beta(G)$ Include a dimension in the set that is a function of the set.

An fuzzy measure is the function of set G, given in set $\beta(G)$ and satisfying the following conditions [1, 2, 8]:

1. Limited. $g\ (\emptyset) = 0,\ g\ (G) = 1; G = \{x_1,\ x_2,\ \ldots,\ x_n\}$
2. Monotony. If then A, B $\in \beta(\Gamma)$, $A \subseteq B$, then $g(A) \le g(B)$;
3. Continuity. Let the sequence, if $\{A_i\} \in \beta(G),\ 1 \le i \le \infty$ if $A_i \supseteq A_z \supseteq \ldots..A_n \supseteq \ldots$, do $g(\bigcap_{i=1}^{\infty} A_i) = \underset{i \to \infty}{Lim}\, g(A_i)$.

$(G, \beta(G), g)$ The triad is called a fuzzy measure space.

Sugeno $g(A \cup B)$ proposed the formation of an fuzzy measure of the association as follows [2, 9]:

$$g(A \cup B) = g(A) + g(B) + \lambda g(A) g(B).$$

Here, λ takes the values in range $-1 < \lambda < \infty$, where (3) $A \cap B = \emptyset$.

The expression given is rule λ, g_λ and the vague fuzzy is called the Sugeno measure λ.

Calculation of fuzzy integral consists of two parts: λ parameter estimation and integral self-computation on given illumination density.

1. The parameter λ applies by definition [1, 2], $(-1,\ \infty)$ to the field. Equation $\lambda = 0$, $\sum_j g(d_j) = 1$ implies that the probability criterion is complied with $\sum_j g(d_j) = 1$.
2. Calculation of integral by fuzzy density:

$$\int h(x) \circ g = \bigcup_{i=1}^{n} (h(x_i) \wedge g(E_i)),\quad E_i = \{x_i,\ x_{i+1}, \ldots, x_n\},\ E_1 \supset E_2 \supset \ldots.E_n,\ E_1 = U$$

3 Calculation on the Basis of the Combination of Z-numbers

For computational experiments on the basis of Z-numbers, we use the T-norm and the T-ratio over discrete Z numbers. $Z_1 = (A_1, B_1), Z_2 = (A_2, B_2)$ considering discrete Z numbers $Z_{T-norm}(Z_1, Z_2)$ T-norm operation is defined as [3–5]:

$$\begin{aligned} Z_{T-norm}(Z_1, Z_2) &= T((A_1, B_1), (A_2, B_2)) = (T(A_1, A_2), T(B_1, B_2)) \\ Z^{+}_{T-norm}(Z_1, Z_2) &= (T(A_1, A_2), T(R_1, R_2)). \end{aligned} \tag{1}$$

There

$$T(A_1, A_2) = \bigcup_{\alpha \in (0,1]} T(A_1^\alpha, A_2^\alpha);\ T(A_1^\alpha, A_2^\alpha) = \{T(x, y) | x \in A_1^\alpha, y \in A_2^\alpha\},$$

$$T(R_1, R_2) = R_1 \wedge R_2.$$

$Z_1 = (A_1, B_1)$, $Z_2 = (A_2, B_2)$ considering discrete Z numbers $Z_{T-conorm}(Z_1, Z_2)$ The operation of the T-beam is defined as follows [3, 4]:

$$\begin{aligned} Z_{T-conorm}(Z_1, Z_2) &= S((A_1, B_1), (A_2, B_2)) = (S(A_1, A_2), S(B_1, B_2)) \\ Z^{+}_{T-conorm}(Z_1, Z_2) &= (S(A_1, A_2), S(R_1, R_2)), \end{aligned} \tag{2}$$

there

$$S(A_1, A_2) = \bigcup_{\alpha \in (0,1]} S(A_1^\alpha, A_2^\alpha); S(A_1^\alpha, A_2^\alpha) = \{S(x, y) | x \in A_1^\alpha, y \in A_2^\alpha\},$$

$$S(R_1, R_2) = R_1 \vee R_2.$$

Step 1. Substituting each (Z_{G_i}, Z_{H_i}) pair in Z-number Z_{GH_i} based on T-norm Z_{T-norm} according to (1) [3–7]:

$Z_{H_i G_i} = Z_{T-norm}(Z_{H_i}, Z_{G_i}) = T((A_{H_i}, B_{H_i}), (A_{G_i}, B_{G_i}))$.

Step 2. In accordance with (1) T-commerce $Z_{T-conorm}$ by virtue Z_{HG_i}, Z_{agg} of the merger action [4, 7]:

$$Z_{agg} = Z_{T-conorm}(Z_{g_1 h_1}, Z_{g_2 h_2}, \ldots, Z_{g_n h_n}) = S((A_{g_1 h_1}, B_{g_1 h_1}), (A_{g_2 h_2}, B_{g_2 h_2}), \ldots, (A_{g_n h_n}, B_{g_n h_n}))$$

We will consider the proposed approach for the selection of the best biologically and technologically-selective varieties of cotton in the form of poorly made decisions, given the ill-defined initial conditions: planting and cultivation conditions (amount and composition of fertilizers, irrigation, soil type, and so on). Boundary conditions for varieties).

The problem of poor decision making is formulated as follows.

Issued by:

- Collection of alternatives: selection of cotton varieties and fertilizers depending on the soil type $\Gamma = \{x_1, x_2, \ldots, x_n\}$;
- character set: biological and technological characteristics in which the selection of the optimal $H = \{h_1, h_2, \ldots, h_k\}$ variety is made.
- $x_i \in \Gamma$ the importance of each sign in the alternative,
- Initial conditions: soil types, irrigation and fertilizer application procedures, weather conditions (sun activity: open air, cloudiness).

Required:

- Selection of the most suitable alternative: breeding for initial planting conditions (cultivation of fertilizers and agrotechnical order, irrigation, soil type and boundary conditions for the given varieties).

The experiment was conducted to determine the best $\Gamma = \{x_1, x_2, x_3, x_4\}$ four selective varieties of cotton S-4727, Tashkent 1, 159-F, 108-F with the following characteristics ($H = \{h_1, h_2, h_3, h_4\}$): yield, fiber length, strength of fiber, fat seed quality [3, 4, 7].

The importance of each sign is given and expressed by vague densities

$$g_1 = 0,66, \quad g_2 = 0,89, \quad g_3 = 0,96, \quad g_4 = 0,93$$

$$h_1 = 0,19, \quad h_2 = 0,21, \ h_3 = 0,22, \quad h_4 = 0,24$$

$0 \leq g_i \leq 1$ in the case of non-linear densities, g_λ the scale shall be constructed in accordance with rule λ:

$$g_\lambda(\{x_1, \ldots., x_k\}) = \begin{cases} \frac{1}{\lambda}(\prod\limits_{i=1}^{k} (1 + \lambda g_i) - 1), \ \lambda \neq 0) \\ \sum\limits_{i+1}^{k} g_i, \quad \lambda = 0 \end{cases}$$

$$g_\lambda(x_1, x_2, x_3, x_4) = 1.$$

Calculate the parameter $\frac{1}{\lambda}(\prod\limits_{i1}^{n} (1 + \lambda g_i) - 1) = 1$ to λ of the normalization condition [2].

$$g_1g_2g_3g_4\lambda^3 + (g_1g_2g_3 + g_1g_2g_4 + g_1g_3g_4 + g_2g_3g_4)\lambda^2$$
$$+ (g_1g_2 + g_1g_3 + g_1g_4 + g_2g_3 + g_2g_4 + g_3g_4)\lambda + g_1 + g_2 + g_3 + g_4 = 1.$$
$$0,524\lambda^3 + 2,49\lambda^2 + 4,409\lambda + 2,44 = 0.$$
$$\lambda^3 + 4,75\lambda^2 + 8,41\lambda + 4,66 = 0.$$
$$\lambda = -0,96$$

The problem of obtaining an integral estimation of the chosen strategy using an integral $\int h \circ g$ integral is raised.

$$h_1 = 0,19, \quad h_2 = 0,21, \ h_3 = 0,22, \quad h_4 = 0,24.$$

$$\int h \circ g = \bigcup_{i=1}^{4} (h(x_i) \wedge g(E_i)) = \max(0,19; \ 0,21; \ 0,22; \ 0,24) = 0,24$$

$$x_4 = 0,24$$

4 To Carry Out Computational Experiments on the Basis of Z-numbers

Computational experiments based on Z-numbers: $G = \{g_1, g_2, g_3, g_4\}$ computational experiments on selecting the best for the issue of four varieties of cotton, and $H = \{h_1, h_2, \ldots, h_4\}$ on the characteristics.

$$A_{h_1} = {}^{0.2}/_0 + {}^{0.4}/_1 + {}^{1}/_2 + {}^{0.4}/_3 + {}^{0.2}/_4 + {}^{0}/_5 + {}^{0}/_6 + {}^{0}/_7 + {}^{0}/_8 + {}^{0}/_9,$$

$$B_{h_1} = {}^0/_0 + {}^0/_{0.1} + {}^0/_{0.2} + {}^0/_{0.3} + {}^0/_{0.4} + {}^{0.1}/_{0.5} + {}^{0.3}/_{0.6} + {}^1/_{0.7} + {}^{0.7}/_{0.8} + {}^{0.6}/_{0.9};$$

$$A_{h_2} = {}^0/_1 + {}^{0.3}/_1 + {}^{0.4}/_2 + {}^{0.7}/_3 + {}^1/_4 + {}^{0.8}/_5 + {}^{0.6}/_6 + {}^0/_7 + {}^0/_8 + {}^0/_9,$$
$$B_{h_2} = {}^0/_0 + {}^0/_{0.1} + {}^0/_{0.2} + {}^0/_{0.3} + {}^0/_{0.4} + {}^{0.1}/_{0.5} + {}^{0.3}/_{0.6} + {}^1/_{0.7} + {}^{0.7}/_{0.8} + {}^{0.6}/_{0.9};$$

$$A_{h_3} = {}^0/_0 + {}^0/_1 + {}^{0.5}/_2 + {}^{0.6}/_3 + {}^{0.7}/_4 + {}^1/_5 + {}^{0.7}/_6 + {}^0/_7 + {}^0/_8 + {}^0/_9,$$
$$B_{h_3} = {}^0/_0 + {}^0/_{0.1} + {}^0/_{0.2} + {}^0/_{0.3} + {}^0/_{0.4} + {}^{0.1}/_{0.5} + {}^{0.3}/_{0.6} + {}^1/_{0.7} + {}^{0.7}/_{0.8} + {}^{0.6}/_{0.9};$$

$$A_{h_4} = {}^0/_0 + {}^0/_1 + {}^0/_2 + {}^{0.5}/_3 + {}^{0.6}/_4 + {}^{0.7}/_5 + {}^{0.8}/_6 + {}^1/_7 + {}^{0.8}/_8 + {}^0/_9,$$
$$B_{h_4} = {}^0/_0 + {}^0/_{0.1} + {}^0/_{0.2} + {}^0/_{0.3} + {}^0/_{0.4} + {}^{0.1}/_{0.5} + {}^{0.3}/_{0.6} + {}^1/_{0.7} + {}^{0.7}/_{0.8} + {}^{0.6}/_{0.9};$$

$$A_{g_1} = {}^0/_0 + {}^{0.6}/_1 + {}^{0.8}/_2 + {}^1/_3 + {}^{0.7}/_4 + {}^0/_5 + {}^0/_6 + {}^0/_7 + {}^0/_8 + {}^0/_9,$$
$$B_{g_1} = {}^0/_0 + {}^0/_{0.1} + {}^0/_{0.2} + {}^0/_{0.3} + {}^0/_{0.4} + {}^{0.1}/_{0.5} + {}^{0.3}/_{0.6} + {}^1/_{0.7} + {}^{0.7}/_{0.8} + {}^{0.6}/_{0.9};$$

$$A_{g_2} = {}^0/_0 + {}^0/_1 + {}^{0.4}/_2 + {}^{0.6}/_3 + {}^1/_4 + {}^{0.8}/_5 + {}^0/_6 + {}^0/_7 + {}^0/_8 + {}^0/_9,$$
$$B_{g_2} = {}^0/_0 + {}^0/_{0.1} + {}^0/_{0.2} + {}^0/_{0.3} + {}^0/_{0.4} + {}^{0.1}/_{0.5} + {}^{0.3}/_{0.6} + {}^1/_{0.7} + {}^{0.7}/_{0.8} + {}^{0.6}/_{0.9};$$

$$A_{g_3} = {}^0/_0 + {}^0/_1 + {}^0/_2 + {}^{0.4}/_3 + {}^{0.6}/_4 + {}^1/_5 + {}^{0.8}/_6 + {}^0/_7 + {}^0/_8 + {}^0/_9,$$
$$B_{g_3} = {}^0/_0 + {}^0/_{0.1} + {}^0/_{0.2} + {}^0/_{0.3} + {}^0/_{0.4} + {}^{0.1}/_{0.5} + {}^{0.3}/_{0.6} + {}^1/_{0.7} + {}^{0.7}/_{0.8} + {}^{0.6}/_{0.9};$$

$$A_{g_4} = {}^{0.2}/_0 + {}^{0.4}/_1 + {}^1/_2 + {}^{0.4}/_3 + {}^{0.2}/_4 + {}^0/_5 + {}^0/_6 + {}^0/_7 + {}^0/_8 + {}^0/_9,$$
$$B_{g_4} = {}^0/_0 + {}^0/_{0.1} + {}^0/_{0.2} + {}^0/_{0.3} + {}^0/_{0.4} + {}^{0.1}/_{0.5} + {}^{0.3}/_{0.6} + {}^1/_{0.7} + {}^{0.7}/_{0.8} + {}^{0.6}/_{0.9};$$

$$A_{h_1} = {}^{0.2}/_0 + {}^{0.4}/_1 + {}^1/_2 + {}^{0.4}/_3 + {}^{0.2}/_4 + {}^0/_5 + {}^0/_6 + {}^0/_7 + {}^0/_8 + {}^0/_9,$$

$$A_{g_1} = {}^0/_0 + {}^{0.6}/_1 + {}^{0.8}/_2 + {}^1/_3 + {}^{0.7}/_4 + {}^0/_5 + {}^0/_6 + {}^0/_7 + {}^0/_8 + {}^0/_9,$$

The approximate value of the cotton grade is determined by the T-norm method described in formula $p = \int_R \mu_A(u) p_x(u) du$. Consider calculating the approximate price for each grade. The minimum of the function $T(Z_H, Z_G) = \min(Z_H, Z_G)$ is used as the T-norm.

The minimum and maximum number of illuminated discrete numbers is determined by the formula $p = \int_R \mu_A(u) p_x(u) du$.

$$A_{h_1 g_1} = {}^{0.2}/_0 + {}^{0.6}/_1 + {}^1/_2 + {}^{0.4}/_3 + {}^{0.2}/_4 + {}^0/_5 + {}^0/_6 + {}^0/_7 + {}^0/_8 + {}^0/_9.$$

For each of these varieties, computations were performed, as shown below, for the second grade.

$$p_{h_2} = {}^{0}/_{0} + {}^{0.77}/_{1} + {}^{0}/_{2} + {}^{0}/_{3} + {}^{0}/_{4} + {}^{0.23}/_{5} + {}^{0}/_{6} + {}^{0}/_{7} + {}^{0}/_{8} + {}^{0}/_{9},$$

$$p_{g_2} = {}^{0}/_{0} + {}^{0.62}/_{1} + {}^{0}/_{2} + {}^{0}/_{3} + {}^{0}/_{4} + {}^{0.3}/_{5} + {}^{0}/_{6} + {}^{0}/_{7} + {}^{0}/_{8} + {}^{0.14}/_{9},$$

$$p_{hg_2} = {}^{0}/_{0} + {}^{0.897}/_{1} + {}^{0}/_{2} + {}^{0}/_{3} + {}^{0}/_{4} + {}^{0.101}/_{5} + {}^{0}/_{6} + {}^{0}/_{7} + {}^{0}/_{8} + {}^{0}/_{9}.$$

$$\sum_{k=0}^{9} \mu_{A_{h_2}}(x_{h_2,k}) \cdot p_{A_{h_2}}(x_{h_3,k}) = 0 \cdot 0 + 0.3 \cdot 0.77 + 0.4 \cdot 0 + 0.7 \cdot 0$$
$$+1 \cdot 0 + 0.8 \cdot 0.23 + 0.6 \cdot 0 + 0 \cdot 0 + 0 \cdot 0 + 0 \cdot 0 = 0.231 \cdot 0.184 = 0.415,$$

$$\sum_{k=0}^{9} \mu_{A_{g_2}}(x_{g_2,k}) \cdot p_{A_{g_2}}(x_{g_2,k}) = 0 \cdot 0 + 0 \cdot 0.56 + 0.4 \cdot 0 + 0.6 \cdot 0 + 1 \cdot 0$$
$$+0.8 \cdot 0.3 + 0 \cdot 0 + 0 \cdot 0 + 0 \cdot 0 + 0 \cdot 0.14 = 0.24.$$

Thereby,

$$b_{A_{h_2g_2}} = P(A_{h_2g_2}) = \sum_{k=0}^{9} \mu_{A_{h_2g_2}}(y_{y_i}) \cdot p_{A_{h_2g_2}}(y_i) = 0 \cdot 0 + 0.3 \cdot 0.897 + 0.4 \cdot 0 + 0.7 \cdot 0 + 1 \cdot 0$$
$$+0.8 \cdot 0.101 + 0 \cdot 0 + 0 \cdot 0 + 0 \cdot 0 + 0 \cdot 0 = 0.24 \cdot 0.080 = 0.320,$$

Calculation results for the third grade:

$$b_{A_{h_3g_3}} = P(A_{h_3g_3}) = \sum_{k=0}^{9} \mu_{A_{h_3g_3}}(y_{y_i}) \cdot p_{A_{h_3g_3}}(y_i) = 0 \cdot 0 + 0 \cdot 0.85 + 0.5 \cdot 0 + 0.6 \cdot 0 + 0.7 \cdot 0 + 0.8 \cdot 0$$
$$+0.7 \cdot 0.06 + 0 \cdot 0 + 0 \cdot 0 = 0.042,$$

Thereby,

$$b_{A_{h_3g_3}} = P(A_{h_3g_3}) = \sum_{k=0}^{9} \mu_{A_{h_3g_3}}(y_{y_i}) \cdot p_{A_{h_3g_3}}(y_i) = 0 \cdot 0 + 0 \cdot 0.85 + 0.5 \cdot 0 + 0.6 \cdot 0 + 0.7 \cdot 0 + 0.8 \cdot 0.2$$
$$+0 \cdot 0 + 0 \cdot 0 + 0 \cdot 0 = 0.16.$$

$$b_{A_{h_3g_3}} = P(A_{h_3g_3}) = \sum_{k=0}^{9} \mu_{A_{h_3g_3}}(y_{y_i}) \cdot p_{A_{h_3g_3}}(y_i) = 0 \cdot 0 + 0 \cdot 0.77 + 0.2 \cdot 0 + 0.6 \cdot 0 + 0.7 \cdot 0 + 0.8 \cdot 0$$
$$+0.7 \cdot 0.115 + 0 \cdot 0 + 0 \cdot 0 = 0.08,$$

The result is the T-dimension $Z_{T-conorm}(Z_{12}, Z_{HG_3}) = S(Z_{12}, Z_{HG_3}) = (A, B)$ to obtain the final grade:

$$A = {}^{0}/_{0} + {}^{0}/_{1} + {}^{0.5}/_{2} + {}^{0.6}/_{3} + {}^{0.7}/_{4} + {}^{1}/_{5} + {}^{0.7}/_{6} + {}^{0}/_{0} + {}^{0}/_{0} + {}^{0}/_{0},$$

$$B = {}^{0}/_{0.61} + {}^{0}/_{0.64} + {}^{0}/_{0.67} + {}^{0}/_{0.70} + {}^{0}/_{0.73} + {}^{0.01}/_{0.76} + {}^{0.14}/_{0.78}$$
$$+ {}^{0.61}/_{0.82} + {}^{1}/_{0.85} + {}^{0.6}/_{0.91}.$$

$Z_{agg} = (A, B)$ end result recommends planting 108-F cotton with a probability of 0.85.

5 Conclusion

Comparative results of computational experiments based on illuminating integral, Z-numbering and logical model based on Z-numbers are presented. The experiment was conducted to determine the best selection of cotton $X = \{x_1, x_2, x_3, x_4\}$ from four selective varieties S-4727, Tashkent 1, 159-F, 108-F with the following characteristics: yield, fiber length, fiber strength, cottonseed fat.

The results obtained from the uneven integral results show that 108-F is the best of the proposed selection varieties of cotton, with the highest value of this variety being the ill-fated set (0.24).

For the computational experiment on the basis of the Z-number association, the T-norm and T-normal methods were applied on discrete Z-numbers.

$Z_{agg} = (A, B)$ end result recommends planting 108-F cotton with a probability of 0.85.

Using the above models, it is possible to predict not only actual productivity but also potential productivity in the field.

The development of an algorithm to use discrete Z-numbers arithmetic in illuminating conclusion systems for the full implementation of Z-information can be considered as a guide for future work in the conclusion mechanism that results in the least loss of Z-information.

References

1. Sugeno, M.: Fuzzy Measures and Fuzzy Integrals: A Survey. Fuzzy Automata and Decision Processes. pp.89–102, North-Holland Publishing Company (1997)
2. Primova, H.A., Sotvoldiyev, D.M., Raximov, R.T., Bobabekova, X.: Computing fuzzy integral of the basis of fuzzy mesure. J. Phys. Conf. Ser. **1441**(1), 012161, 1–7 (2020). https://doi.org/10.1088/1742-6596/1441/1/012161
3. Zadeh, L.A.: A note on a Z-number. Inf. Sci. (USA) **181**, 2923–2932 (2011)
4. Primova, H.A., Niyozmatova, N.A.: Methodology of using Z-numbers in systems of fuzzy inference. Am. J. Math. Comput. Sci. **2**, 67–73 (2016)
5. Zadeh, L.A.: Inf. Control. 338–353 (1965)
6. Aliev, R.A., Jabbarova, K.I., Huseynov, O.H.: Eighth WORLD Conference on Intelligent Systems for Industrial Automation, Tashkent, pp.159–167 (2014)
7. Primova, H., Sotvoldiev, D., Safarova, L.: 12th International Scientific and Technical Conference "Dynamics of Systems, Mechanisms and Machines", Dynamics (2018)
8. Takagi, T., Sugeno, M.: IEEE Trans. Syst. Man Cybern. **15**(1), 116–132 (1985)
9. Mukhamedieva, D.T, Primova, Kh.A.: Approach to problem solving multicriterial optimization with fuzzy aim. Int. J. Math. Comput. Appl. Res. (IJMCAR). **4,** 55–68 (2014)

A Recurrent Method for Structural-Parametric Identification of Fuzzy Neural Networks

Marakhimov Avazjon Rakhimovich(✉) and Khudaybergenov Kabul Kadirbergenovich

National University of Uzbekistan, Tashkent, Uzbekistan
avaz.marakhimov@yandex.ru, kabul85@mail.ru

Abstract. The paper proposes a structural-parametric identification method to models in form of fuzzy rule based neural networks. The proposed method is a system identification method which includes a new method for structure and parameter initialization. Algorithm solves the problem of initial structure and parameter identification and adjustments of membership functions. Numerical experiments based on datasets are presented to illustrate the advantages of the proposed method and shows effectiveness and accuracy of the proposed approach.

Keywords: System identification · Fuzzy neural networks · Input-output clustering · Rule base · Similarity analysis

1 Introduction

Fuzzy systems are widely used in problem areas such as system identification, knowledge extraction, prediction, forecasting, modeling, system control, and decision making. Fuzzy logic and neural networks are universal approximators [1–3], since they are able to approximate any non-linear function with any given accuracy, provided that a sufficient number of hidden neurons [4] and fuzzy rules. However, these systems are used without a specific justification for the choice of the structure and parameters of the model. The indicated tasks should be solved at the stage of identifying parameters of a fuzzy model: structural-parametric algorithms and neural networks.

The results obtained in recent decades have shown that the procedure for the synthesis of these two technologies has proven to be very useful for identifying nonlinear systems [5–7]. Fuzzy models have recently become quite a powerful tool for modeling and controlling complex systems [8, 9]. In research works by Takagi and Sugeno [10] and Pedrycz [11], the identification and learning of fuzzy systems (FS) was the fundamental achievements in this field. Fuzzy modeling and identification from data sets are highly useful when input data in a fuzzy form.

Also, these techniques are very effective in modelling nonlinear dynamic systems. TS models are usually obtained by using the fuzzy modeling technique [12]. It uses the idea of linearization in a fuzzily defined region of the state space. Also, genetically-oriented identification algorithms have been successfully developed to fuzzy systems

© The Author(s), under exclusive license to Springer Nature Switzerland AG 2021
R. A. Aliev et al. (Eds.): WCIS 2020, AISC 1323, pp. 47–54, 2021.
https://doi.org/10.1007/978-3-030-68004-6_7

in [16]. Pedrycz and others presented identification techniques of fuzzy model based on information granulation and genetic algorithms [13–16]. In these research works, they study regression polynomial as consequent parts of fuzzy system and included the selection step of the order of polynomial in the structure identification process and adjusting premise parameters. Some recent applications devoted to a set of evolving fuzzy models of the midcarpal joint angles in the framework of human hand dynamics [17].

Purpose of the proposed work is to describe the possibility of combining structural and parametric selection and optimization for identifying fuzzy models.

2 Weighed Mamdani FNNs and Its Learning Problem

2.1 Weighed Mamdani Model

We are considering Mamdani type inference system with weighted rule base (1). This means that every rule in fuzzy rule base has its weighting factor (2). This type of inference system gives us an effective fuzzy neural network model. The weighted neuro-fuzzy system of the Mamdani type can be specified as follows

$$y = f(x;\theta) = \frac{\sum_{i=1}^{R} b_i w_i \prod_{j=1}^{n} \mu(x;\theta)}{\sum_{i=1}^{R} w_i \prod_{j=1}^{n} \mu(x;\theta)} \tag{1}$$

Suppose that $(x^m, y^m) \in R$ - m^{th} a pair of training data sets. θ - network parameters, $\mu(x;\theta)$ - membership function (MF).

In this case, the fuzzy rules will have the following form:

$$R_i : IF\ x_1\ is\ A_{i1}\ AND\ldots AND\ x_m\ is\ A_{im}\ with\ weight\ w_i\ THEN\ y\ is\ B_i\ \left(i = \overline{1,L}\right).$$

where w_i - the weight of i-th rule. L - the number of rules in the rule base.

In this paper, the Gaussian MF is chosen in following form:

$$\mu(x; c_j^i, \sigma_j^i) = \exp\left(-\frac{1}{2}\left(\frac{x_j - c_j^i}{\sigma_j^i}\right)^2\right) \tag{2}$$

where c_j^i and σ_j^i parameters of the selected MF. However, other types of membership functions can be chosen.

2.2 Statement of the Problem

Suppose that the input-output sample data of a given system are given as $\{x(t), y(t)\}$, $t = \overline{1..n}$, where $x(t) = (x_1(t), x_2(t), \ldots, x_m(t))$ is the input data at time t and $y(t)$ is the output data. $m-$training data set size. Structure identification of fuzzy model is to determine the number of MF's and parameter optimization. The task of structure identification and parameter optimization of weighted Mamdani fuzzy model is to partition the input space into a number of fuzzy subspaces and obtain parameter values. The approximation error is represented as mean squared error (MSE) in the following form:

$$e_m = \frac{1}{2}\left[f\left(x^m; \theta\right) - y^m\right]^2 \tag{3}$$

Using a gradient method should allow to find such a vector of parameters θ, which would provide the minimum error e_m.

3 New Approach for the Identification of Weighed Mamdani FNN

In fuzzy modeling, it is required to specify the number of the MFs which is the most important for obtaining the expected accuracy. Usually, this value is determined by experts. Let, denote the number of MFs as K. K - is the integer value which divides the input data into subintervals such that each length of the subinterval is defined by expert. The length of subinterval and the number of points in it defines MF's parameters. Usually, the number of MFs is between 2 and 7.

Let *Min* - is the lower boundary and *Max* - is upper boundary of input data respectively, that is,

$$Min = \min_{i=\overline{1..m}} (x_i(t)),\ Max = \min_{i=\overline{1..m}} (x_i(t))$$

Then,

$$X_0 = Min,\ X_K = Max$$

Other points of X_i which is from X_1 to X_{K-1} are defined by expert. By defining interval values with X_i, we do clustering of input-output data in order to initialize MFs parameters. Let C_k, $k = \overline{0..K-1}$ are the clusters of given input-output data. Elements of clusters are defined in the following form:

$$C_k \in \{x_i(t)|X_k \le x_i(t) \le X_{k+1}\},\ i = \overline{1..n},\ k = \overline{0..K-1} \tag{4}$$

Thus, in each cluster C_k there includes the elements of given data set. The number and length of each cluster can vary significantly among themselves. Further, these elements in clusters, depending on their nature, will be used to adjust the parameters of the MF. This is done in the following algorithm.

4 Structure and Parameter Initialization of the Proposed Method

The proposed algorithm consists of two stages: at the first stage, the algorithm determines the elements of the clusters using (4). The second is the calculation of the parameters of the MFs. Since a Gaussian type MF is chosen for the model (1), center and width parameters for this MF will be calculated.

Proposed algorithm's pseudo code:

```
// stage 1:
   For i=0 to K-1 do
         For j=1 to n do
                  If (x_j in C_i) Then
                  Add x_j into C_i
                  End If
         End For
   End For
   // stage 2:
   // Let be C_i = {x_{1_i}, x_{2_i}, ..., x_{n_i}},  i = 0..K-1
   For i=0 to K-1 do
         For j=1 to n_i do
```

$$c_i = \frac{1}{n_i}\sum_{j=1}^{n_i} x_{n_j}, \quad \sigma_i = \frac{1}{\sqrt{2}}\sqrt{\frac{1}{n_i}\sum_{j=1}^{n_i}\left(x_{n_j} - c_i\right)}$$

```
         End For
   End For
```

5 Parameter Learning Using a Gradient Descent Method

Let's consider the law of updating the output variables b_i. To achieve a minimum approximation error e_m we use the law of updating parameters b_i in the following form:

$$b_i(k+1) = b_i(k) - \lambda_1 \frac{\partial f(x^m; \theta)}{\partial b_i}$$

where $\lambda_1 > 0$ - parameter of learning speed for.

We denote i-th rule's firing strength as

$$\mu_i(x^m, k) = w_i \prod_{j=1}^{n} \exp\left(-\frac{1}{2}\left(\frac{x_j^m - c_j^i}{\sigma_j^i}\right)^2\right) \tag{5}$$

The error in m-th iteration is as follows

$$\varepsilon_m(k) = f\left(x^m; \theta(k)\right) - y^m$$

As a result, we obtain a recurrent algorithm for adjusting the parameters in the form

$$b_i(k+1) = b_i(k) - \lambda_1 \varepsilon_m(k) \frac{\mu_i(x^m, k)}{\sum_{i=1}^{R} \mu_i(x^m, k)} \tag{6}$$

Now, let's consider the law of updating weights:

$$w_i(k+1) = w_i(k) - \lambda_2 \frac{\partial e_m}{\partial w_i}\bigg|_k$$

where $\lambda_2 > 0$ - parameter of learning speed for weight.
Updating the centers c_i of MF's is as follows.

$$c_j^i(k+1) = c_j^i(k) - \lambda_3 \frac{\partial e_m}{\partial c_j^i}\bigg|_k$$

where $\lambda_3 > 0$ - parameter of learning speed for c_i. $i = 1..R$, $j = 1..n$ and $k \geq 0$.
Now, we consider the gradient algorithm for recurrent adjustment of δ_j^i MF.
Let, parameters σ_j^i is updated in the following form

$$\sigma_j^i(k+1) = \sigma_j^i(k) - \lambda_4 \frac{\partial e_m}{\partial \sigma_j^i}\bigg|_k$$

where $\lambda_4 > 0$ - parameter of learning speed for σ_j^i.

Thus, to update the vector of parameters θ of fuzzy system Eqs. (7), (8) and (9) can be used. It should be noted that in the Eqs. (7), (8) and (9), MFs are of the Gaussian type. By performing similar actions, we can find algorithms for setting parameter vectors and for other types of MFs.

6 Numerical Experiments

We are considering an object with two inputs $x_1, x_2 \in [-2, 2]$ and one output y dependencies, which is given below:

$$y = x_1 \times e^{-x_1^2 - x_2^2} \tag{7}$$

In Fig. 1, we can the graphic representation of the data set given in (10). It is necessary to conduct training on the above proposed model, synthesize a fuzzy model and adjust it according to a fuzzy training pattern. The adequacy of the fuzzy model must be checked by the criterion (4).

In this section comparisons of results of identification with crisp and fuzzy training patterns are given. Training patterns are given in (10). Before training of the proposed model, its weights were randomly determined. In Table 1, convergence data is shown for different values of speed of learning by λ.

Parameters of membership functions of fuzzy term of input variables before and after adjustments are shown in Tables 2 and 3, respectively. The testing of crisp and fuzzy models indicates an acceptable quality of identification of non-linear dependencies.

In Table 4 we can see the parameter values of weights w after training.

From numerical experiments it can easily be seen, that mean squared error, obtained using the proposed algorithm approaches the corresponding minimum after 200 epochs. From the experimental results, we can assume that the proposed method exceeds the crisp model on convergence.

Table 1. Performance of the models

λ	MLP [19]	Kim et al. [18]	Fuzzy-MLP [20]	Proposed method
0.30	0.0621	0.0677	0.0234	0.0220
0.40	0.0691	0.0680	0.0207	0.0200
0.45	0.0677	0.0721	0.0159	0.0155
0.50	0.0820	0.0802	0.0268	0.0132

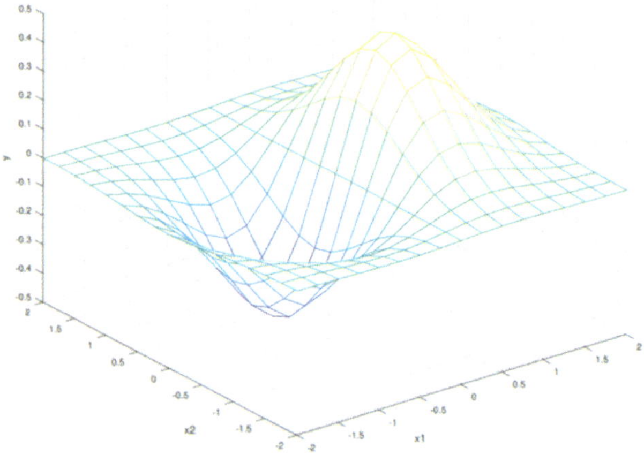

Fig. 1. Graphical representation of the given data set (7).

Table 2. Parameters of membership functions of terms for variable x^1

Terms		A1	A2	A3	A4	A5	A6	A7
Before training	c	−2.00	−1.333	−0.667	0	0.667	1.333	2.00
	σ	0.35	0.35	0.35	0.35	0.35	0.35	0.35
After training	c	−1.999	−1.311	−0.7058	0.0064	0.7176	1.3102	1.997
	σ	0.353	0.3550	0.3276	0.4588	0.3215	0.3535	0.356

Table 3. Parameters of membership functions of terms for variable x^2

Terms		B1	B2	B3	B4	B5	B6	B7
Before training	c	2.00	−1.333	−0.69	0	0.667	1.333	2.00
	σ	0.35	0.35	0.35	0.35	0.35	0.35	0.35
After training	c	−1.889	−1.4952	−0.90	0.0492	0.7038	1.5234	1.999
	σ	0.518	0.1290	0.464	0.0914	0.4664	0.2146	0.298

Table 4. Parameter values of weights w

w_{ij}	1	2	3	4	5	6	7
1	0.8822	0.8315	0.6814	0.5790	0.0099	0.2688	0.1555
2	0.8569	0.6996	0.7398	0.1040	0.0040	0.0032	0.6840
3	0.6657	0.3195	0.5458	0.3394	0.8534	0.3118	0.3660
4	0.4284	0.0320	0.5359	0.3731	0.8117	0.3317	0.1592
5	0.3909	0.4980	0.4436	0.3552	0.9384	0.4981	0.8005
6	0.8877	0.8407	0.5546	0.8556	0.9959	0.0826	0.7007
7	0.9392	0.9296	0.7598	0.8593	0.0133	0.1761	0.1709

7 Conclusion

A new structural-parametric method has been developed for identification of non-linear systems. The input data is fuzzified with the Gaussian membership function, and then passed into fuzzy model. However, other types of MF can be used. The initial parameter values of MF's are calculated according to given data set with clustering algorithm. It's very important to initialize MF parameter values according to data rather than give random values. And in learning process, for parameter and weights adjustment there used different learning speed values. After defuzzification, output of the proposed fuzzy model gives a crisp value. The obtained results show, that the proposed model converges to its minimum MSE at 200 epochs and reaches to coefficient of convergence of 93–95%. From the results of computational experiments which has been carried out, it can be assumed that the proposed fuzzy model significantly exceeds fuzzy models in terms of training speed and accuracy.

References

1. Lin, Y., Zhou, X., Gu, S., Wang, S.: The Takagi-Sugeno intuitionistic fuzzy systems are universal approximators. In: 2nd International Conference On Consumer Electronics, Communications and Networks, pp. 2214–2217 (2012)
2. Ying, H.: General interval type-2 Mamdani fuzzy systems are universal approximators. In: Proceedings of Annual Meeting of North America Fuzzy Information Processing Society, pp. 1–6 (2008). https://doi.org/10.1109/NAFIPS.2008.4531255

3. Mandal, S., Jayaram, B.: SISO fuzzy relational inference systems based on fuzzy implications are universal approximators. Fuzzy Sets Syst. **277**, 1–21 (2015)
4. Ismailov, V.E.: On the approximation by neural networks with bounded number of neurons in hidden layers. J. Math. Anal. Appl. **417**(2), 963–969 (2014)
5. Szabat, K., Orlowska-Kowalska, T.: Damping of the torsional vibration in two-mass drive system using forced dynamic control, pp. 1712–1717 (2007). https://doi.org/10.1109/EURCON.2007.4400685.
6. Khoshnejad, M., Demirli, K.: Autonomous parallel parking of a car-like mobile robot by a neuro-fuzzy behavior-based controller, pp. 814–819 (2005). https://doi.org/10.1109/NAFIPS.2005.1548645.
7. Hussain, S.A., Raju, M.: Neuro-fuzzy system for medical image processing. In: Proceedings of 2010 International Conference on Communication and Computational Intelligence, INCOCCI-2010, pp. 382–385 (2011)
8. Wang, Y., Zou, H., Tao, J., Zhang, R.: Predictive fuzzy PID control for temperature model of a heating furnace, pp. 4523–4527 (2017). https://doi.org/10.23919/ChiCC.2017.8028070
9. Omara, A., Sleptsov, M., Diab, A.: Cascaded fuzzy logic based direct torque control of interior permanent magnet synchronous motor for variable speed electric drive systems, pp. 1–6. https://doi.org/10.1109/IWED.2018.8321386
10. Takagi, T., Sugeno M.: Fuzzy identification of systems and its applications to modeling and control. IEEE Trans. Syst. Man Cybern. 116–132 (1986). https://doi.org/10.1109/TSMC.1985.6313399
11. Pedrycz, W.: An identification algorithm in fuzzy relational systems. Fuzzy Sets Syst. **13**(2), 153–167 (1984)
12. Liaoa, Q.F., et al.: Type-1 and type-2 effective Takagi-Sugeno fuzzy models for decentralized control of multi-input-multi-output processes. J. Process Control **52**, 26–44 (2017). https://doi.org/10.1016/j.jprocont.2017.01.004
13. Oh, S.K., Pedrycz, W., Park, B.J.: Relation-based neurofuzzy networks with evolutionary data granulation. Math. Comp. Model. **40**, 891–921 (2004). https://doi.org/10.1016/j.mcm.2004.10.019
14. Oh, S.K., Pedrycz, W., Park, B.J.: Self-organizing neurofuzzy networks in modeling software data. Fuzzy Sets Syst. **145**(1), 165–181 (2004). https://doi.org/10.1016/j.fss.2003.10.009
15. Pedrycz, W., Reformat, M.: Evolutionary fuzzy modeling. IEEE Trans. Fuzzy Syst. **11**(5), 652–665 (2003). https://doi.org/10.1109/TFUZZ.2003.817853
16. Liu, F., Lu, P., Pei, R.: A new fuzzy modeling and identification based on fast-cluster and genetic algorithm. Intell. Control Autom. **1**, 290–293 (2004)
17. Precup, R.E., et al.: Experiments in incremental online identification of fuzzy models of finger dynamics. Rom. J. Inf. Sci. Technol. **21**(4), 358–376 (2018)
18. Euntai, K., Park, M., Ji, S., Park, M.: A new approach to fuzzy modeling. IEEE Trans. Fuzzy Syst. **5**(3), 328–337 (1997). https://doi.org/10.1109/91.618271
19. Mu, X., Zhang, Y.: A modified gradient-based backpropagation training method for neural networks. In: IEEE International Conference on Granular Computing, 17–19 August, pp. 450–453 (2009). https://doi.org/10.1109/91.618271
20. Marakhimov, A.R., Khudaybergenov, K.K.: A fuzzy MLP approach for nonlinear system identification. Sovremennaya Matematica. Fundamentalnie Napravlenie **65**(1) (2019). https://doi.org/10.22363/2413-3639-2019-65-1-44-53

Voltage Control System for Electrical Networks Based on Fuzzy Sets

Nurali Yusifbayli[1(✉)], Huseyngulu Guliyev[2], and Aydin Aliyev[1]

[1] Azerbaijan State University of Oil and Industry, 16/21 Azadliq, Baku, Azerbaijan
yusifbayli.n@gmail.com, aaliyevadnsu@mail.ru
[2] Azerbaijan Research and Design–Prospecting Institute of Energetics, 94, Zardabi, Baku, Azerbaijan
huseyngulu@mail.ru

Abstract. Nourishing voltage intellectual management functional scheme, theoretical problems and report results in power system have been given in the presented work. Voltage management effectiveness on the basis of fuzzy technologies has also been shown.

Keywords: Power system · Distribution network · Voltage regulation · Loss of strength · Fuzzy logic

1 Introduction

To increase the efficiency of the power system (PS), including the power supply system, voltage regulation, which is one of the main regime parameters and one of the main indicators of electricity quality, is relevant to meet the growing demands of consumers [1, 2]. The use of increasingly variable renewable energy sources in the XXI century PS, along with traditional power sources, has further increased the urgency of regulating the parameters of the regime. Also, the trend of liberalization of trade relations between consumers and energy supply companies has led to the tightening of the requirements of international standards for electricity quality and the priority of providing consumers with better quality electricity [3–5]. Against the background of the above, the new conditions in the PS have made it necessary to develop and implement new principles of control of the regimes parameters.

There are many traditional approaches in the technical literature for voltage regulation in PS [6]. However, fuzzy changes in the parameters that characterize modern PSs make it difficult to build a practical method of voltage control in power grids using traditional mathematical method and reduce the effectiveness of solutions based on it, which reduces the reliability of the electricity supply of consumers [7].

In order to solve the above-mentioned difficulties of voltage quality management in PS, the authors proposed a number of methods based on modern mathematical technologies [8–15]. This paper presents the results of modeling that confirms the effectiveness of the structural scheme, algorithm and implementation of voltage regulation in electrical networks with the application of fuzzy sets and fuzzy logic theory.

© The Author(s), under exclusive license to Springer Nature Switzerland AG 2021
R. A. Aliev et al. (Eds.): WCIS 2020, AISC 1323, pp. 55–63, 2021.
https://doi.org/10.1007/978-3-030-68004-6_8

2 Problem Statement and Main Results

Automatic regulation of the feeding voltage for the distribution network using fuzzy-regulation system is depicted in Fig. 1.

The scheme includes feeding transformer that is connected to the distribution network bus, and its high voltage side is connected to the device, which changes contacts under the load condition, low voltage side of it is individually connected to the electric power consumers. It also includes analog-digital transducer (5), the most distant and the nearest of consumers buses voltages- U_2 and U_3 integrating units 6, 7 and 8 and fuzzy-regulation which separated with broken lines. The feeding line of the far end load is depicted by 3 and the nearest load feeding line is depicted by 4 in the circuit diagram.

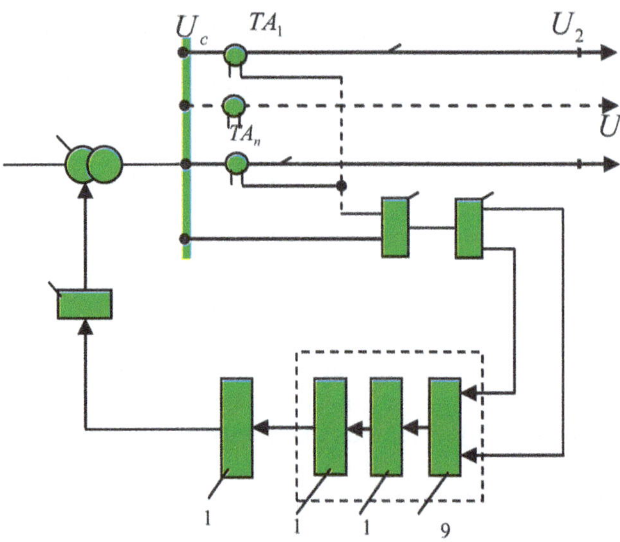

Fig. 1. Circuit diagram of automatic control of supplying voltage in the distribution network

Explicit input data enters to the fuzzifier 9, and then after logic decision unit passes to defuzzifier which transforms term on the output of unit 10 to the control signal on the output of unit 11. Determined Fuzzy relations between input and output of LGT unit consist of a quantity of fuzzy set guidance and is compiled on the base of expert evaluation. After defuzzifier the fuzzy signal changes to a distinct control signal which now passes to the transformer LTC unit for control purposes.

The voltages on the far-end (U_2) and on a nearest bus (U_3) of the radial network are related with a voltages drop on feeding lines by the following equation:

$$\underline{\dot{U}}_2 = \underline{\dot{U}}_1 - \underline{I}_3\underline{Z}_{L3}$$

$$\underline{\dot{U}}_3 = \underline{\dot{U}}_1 - \underline{I}_4\underline{Z}_{L4}$$

where U_1- voltage of the feeder busbar, I_{L3} and I_{L4} load currents from lines 3 and 4 accordingly. Z_{L3} and Z_{L4} are the full impedances.

The following voltage deviation from rating value is accepted as a quality indicator of the system.

$$|\varepsilon(U)| \leq \delta, \ldots (\delta = 4\%)$$

$\varepsilon(U)$ – voltage error. It is known that for long-term normal operation voltage may be allowed to change in the limits of -5%–$+5\%$. To provide fast and effective operation of automatic system, the maximal voltage deviation was accepted as $\pm 4\%$.

The voltage errors for the most distant and nearest consumers busbars determined from:

$$\varepsilon(U) = \frac{G - Y(U)}{G} \cdot 100\%$$

where Y(U)- current value of the changes in controlled output signal, G- task impact of the system.

E_{1i} expresses voltage error for the most distant consumers load, and linguistic fuzzy subset is defined by following algorithm:

$$\begin{aligned}
E_{11} &= NB \text{ (negative big)} && \triangleq (\varepsilon, \mu_{11}(\varepsilon(U))) \\
E_{12} &= NM \text{ (negative medium)} && \triangleq (\varepsilon, \mu_{12}(\varepsilon(U))) \\
E_{13} &= NS \text{ (negative small)} && \triangleq (\varepsilon, \mu_{13}(\varepsilon(U))) \\
E_{14} &= Z \text{ (zero)} && \triangleq (\varepsilon, \mu_{14}(\varepsilon(U))) \\
E_{15} &= PS \text{ (positive small)} && \triangleq (\varepsilon, \mu_{15}(\varepsilon(U))) \\
E_{16} &= PM \text{ (positive medium)} && \triangleq (\varepsilon, \mu_{16}(\varepsilon(U))) \\
E_{17} &= PB \text{ (positive big)} && \triangleq (\varepsilon, \mu_{17}(\varepsilon(U)))
\end{aligned}$$

E- universal set which express deviations of voltage.

Similarly, to the accepted linguistic language for the nearest consumer buses linguistic variables of the voltage errors E_{2j} fuzzy subset is set up

$$\begin{aligned}
E_{21} &= NB \text{ (negative big)} && \triangleq (\varepsilon, \mu_{21}(\varepsilon(U))) \\
E_{22} &= NM \text{ (negative medium)} && \triangleq (\varepsilon, \mu_{22}(\varepsilon(U))) \\
E_{23} &= NS \text{ (negative small)} && \triangleq (\varepsilon, \mu_{23}(\varepsilon(U))) \\
E_{24} &= Z \text{ (zero)} && \triangleq (\varepsilon, \mu_{24}(\varepsilon(U))) \\
E_{25} &= PS \text{ (positive small)} && \triangleq (\varepsilon, \mu_{25}(\varepsilon(U))) \\
E_{26} &= PM \text{ (positive medium)} && \triangleq (\varepsilon, \mu_{26}(\varepsilon(U))) \\
E_{27} &= PB \text{ (positive big)} && \triangleq (\varepsilon, \mu_{27}(\varepsilon(U)))
\end{aligned}$$

Consist of linguistic variables fuzzy subset which presented next position of LTC –tap changer U_y is expressed with the following language:

$$\begin{aligned}
U_{11} &= NI \text{ (negative direction)} && \triangleq (U, \mu_{11}(U(n))) \\
U_{12} &= D \text{ (stop)} && \triangleq (U, \mu_{12}(U(n))) \\
U_{13} &= PI \text{ (positive direction)} && \triangleq (U, \mu_{13}(U(n)))
\end{aligned}$$

The total switching bandwidth and fuzzy subset for existing errors ($-13\% \div 13\%$) and control action (± 7, 12) are determined from experience and opinion about voltage control under the load.

Fuzzy sets membership functions $\mu_{1i}(\varepsilon(U))$, $\mu_{2j}(\varepsilon(U))$, $\mu_{3\gamma}(U(n))$ consisting of voltage errors and control action for the most distant and nearest consumers busbars created using [10, 11]:

$$\left.\begin{aligned} \mu_{1i}(\varepsilon(U)) &= \exp(-g_{1i}|\varepsilon(U) - \bar{\varepsilon}_{1i}|), \quad (i = \overline{1,7}) \\ \mu_{2j}(\varepsilon(U)) &= \exp\left(-g_{2j}\left|\varepsilon(U) = \bar{\varepsilon}_{2j}\right|\right), \quad (j = \overline{1,7}) \\ \mu_{3\gamma}(U(n)) &= \exp\left(-g_{3\gamma}\left|U(n) - \bar{U}_{1\gamma}\right|\right), \quad (\gamma = \overline{1,3}) \end{aligned}\right\}$$

Membership functions are presented in Fig. 2, 3, 4. Before creation and successive usage of fuzzy control linguistic guidance table, it is necessary to define the membership degrees μ_{1i}, μ_{2j}, $\mu_{3\gamma}$.

Here $\varepsilon(U_2)$, $\varepsilon(U_3)$, $U(n)$ factors belong to E_{1i}, E_{2j}, $U_{1\gamma}$ sets.

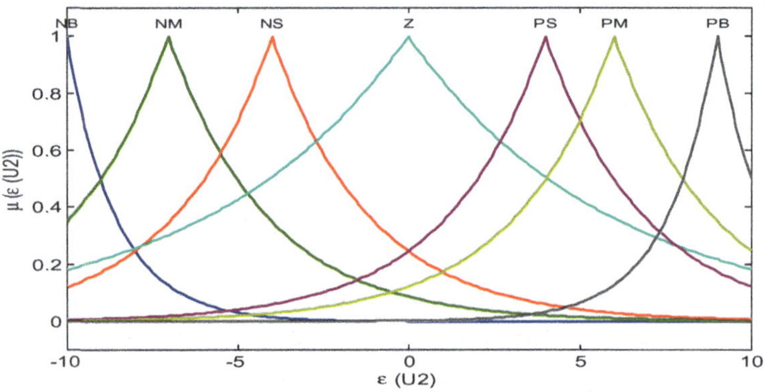

Fig. 2. Membership functions with different

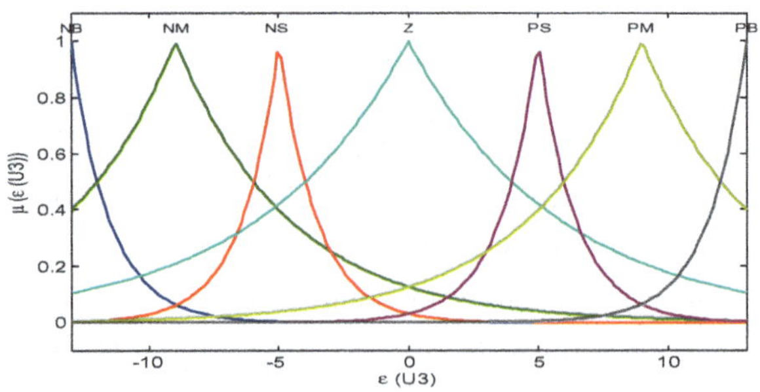

Fig. 3. Membership functions with different terms for the shortest line terms for the longest line

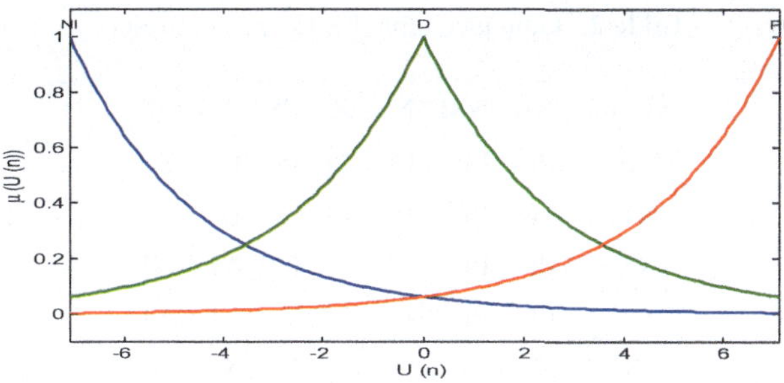

Fig. 4. Controlling impacts on terms using Membership functions

g_{1i}, g_{2j}, $g_{3\gamma}$ are constants and may be calculated from the following conditions:

$$\mu_{1i}(\varepsilon(U_2)) \geq 0,5 \Rightarrow (\varepsilon(U_2)) \in E_{1i}$$
$$\mu_{2j}(\varepsilon(U_3)) \geq 0,5 \Rightarrow (\varepsilon(U_3)) \in E_{2j}$$
$$\mu_{3\gamma}(U(n)) \geq 0,5 \Rightarrow (U(n)) \in U_{1\gamma}$$

The coefficients $\bar{E}_{1i}$ $(i = \overline{1,7})$ $\bar{E}_{2j}$ $(j = \overline{1,7})$ $\bar{U}_{1\gamma}$ $(\gamma = \overline{1,3})$ corresponds to the average of i, j, y sets of voltage errors and control actions. The membership degree of these coefficients is equal to 1. $\mu(\varepsilon(U_2)) = 1,0$; $\mu(\varepsilon(U_3)) = 1,0$; $\mu(U(n)) = 1,0$.

At Table 1 values of those coefficients g_{1i} $(i = \overline{1,7})$, g_{2j} $(j = \overline{1,7})$, $g_{3\gamma}$ $(\gamma = \overline{1,3})$, E_{1i} $(i = \overline{1,7})$, E_{2j} $(j = \overline{1,7})$, $U_{1\gamma}$ $(\gamma = \overline{1,3})$ are presented.

Table 1. Values of coefficients

Par.	*i, j, γ*						
$\overline{\varepsilon_{i1}}$	−10,0	−7,0	−4,0	0	4,0	6,0	9,0
g_{1i}	0,69	0,35	0,35	0,17	0,35	0,35	0,69
$\bar{\varepsilon}_{2j}$	−13,0	−9,0	−5,0	0	5,0	9,0	13,0
g_{2j}	0,69	0,23	0,69	0,17	0,69	0,23	0,69
$\bar{U}_{1\gamma}$			−7,12	0	7,12		
$g_{3\gamma}$			0,39	0	0,39		

In the form of (1)–(4) based on consideration table of linguistic guidance for fuzzy controller is processed. Fuzzy control algorithm with two inputs and one output is expressed in the form of orthogonal matrix (Table 2).

Table 2. Guidance table for Fuzzy controller

Terms	NB	NM	NS	Z	PS	PM	PB
NB	PI	PI	PI	PI	PI	D	D
NM	PI	PI	PI	PI	PI	D	D
NS	PI	PI	PI	D	D	NI	NI
Z	PI	PI	D	D	D	NI	NI
PS	D	D	D	D	D	NI	NI
PM	D	D	D	D	NI	NI	NI
PB	D	D	D	D	NI	NI	NI

In case of voltage control on consumer buses as a fuzzy algorithm it was selected Mamdani algorithm. While testing the fuzzy control system constructed on the base of Mamdani algorithm, it was discovered that, compared to classic methods, there is less switching actions with fuzzy logic. At the same time, fuzzy controller branch of LTC unit «take root», so the probability of switching contact to stick in extreme position is reduced. The important feature of the equipment designed on the base of fuzzy logic is that it has an ability of enhancement. It is easy to add new functions or new features to such devices. In other words, if there is a need for additional functions, there is no problem to attach them to the equipment. It is necessary to point out that voltage regulation on consumer buses with equipment using fuzzy logic approach is more appropriate than those using simple logic. Regulation is more qualitative; the number of switching operations is less, life span increases and probability of miss-operation or failure reduces.

3 Practical Results

For typical electric circuit (Fig. 1) the nearest, the most distant and for medium range power supplying lines loading currents were considered as 80, 120 and 60 A it was modeled in accordance with probability distribution, and for a certain discrete random values automatic control unit programmed with control algorithm was examined.

Voltage curves on feeder and consumer busbars are shown in the Fig. 5.

As is evident, the voltage variation during fuzzy control proceeds on more smooth curve and the values are located in the permissible area.

LTC device operation diagrams are given in the Fig. 6. As is evident, more less actions are implemented using LTC device during fuzzy control and therefore, the device reliability increases.

Voltage on 10 kV feeding busbars and voltage on consumer side are presented on Fig. 7, but distribution function. It is evident that as a result of fuzzy control voltage level is satisfying to standards demand ($\pm 5\%$) with probability of $p = 0{,}95$.

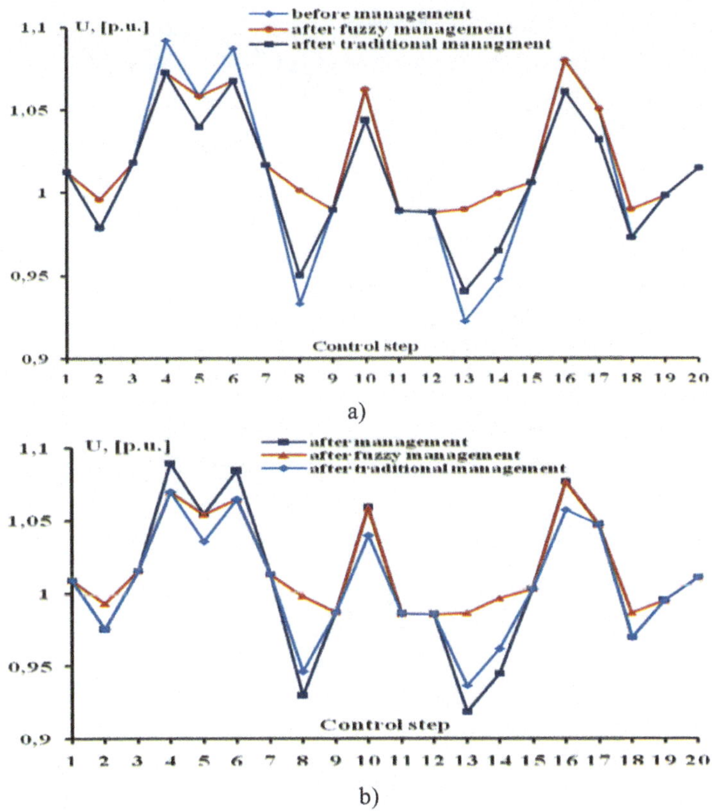

Fig. 5. Voltage curve on feeder and consumer busbars a – feeder busbar; b – consumer busbars

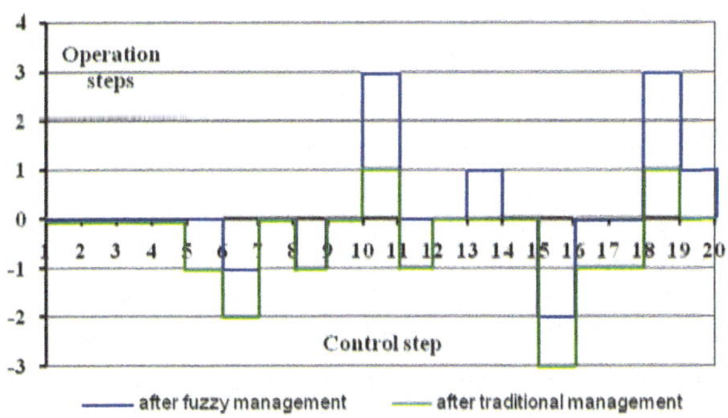

Fig. 6. LTC device operation diagram

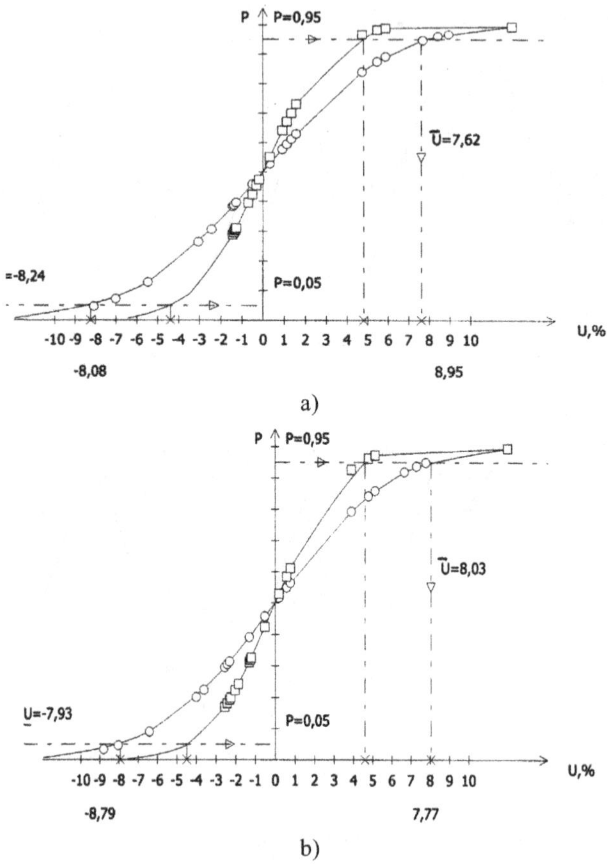

Fig. 7. Voltage distribution on supplier and consumer busbars a - Voltage on nearest line; b - Voltage on distant line

4 Conclusion

1. Efficient fuzzy logic system and automatic device for voltage control in distribution electrical network is proposed.
2. For typical electrical network modeled on PC imitation model of automatic voltage control system was examined. The results have shown that, while supply voltage is in acceptable borders, power losses reduction was established to be equal of 3.1%.
3. The used device provides more quality control, change of cycles is reduced. Operation life of LTC-unit increases and its malfunction reduces. At the same time, fuzzy controller branch of LTC unit «take root» and so the probability of switching contact to stick in one of extreme positions is reduced.

References

1. Yusifbayli, N.A., Guliyev, H.B.: Intelligent control system of voltage regimes in the electrical networks. In: IEEE PES Innovative Smart Grid Europe, Manchester, United Kingdom, 5–7 December 2011
2. Yusifbayli, N.A., Guliyev, H.B.: Intelligent voltage regulation on base of fuzzy sets theory. In: The 12th International Conference on Probabilistic Methods Applied to Power Systems. PMAPS 2012, Istanbul, Turkey, 10–14th June 2012
3. Hashimov, A.M., Rahmanov, N.R., Guliyev, H.B.: Criteria for determination of membership function type in fuzzy management of regime parameters of electric networks. Int. J. Technical Phys. Probl. Eng. (IJTPE) **8**(3) (2016). Published by International Organization of IOTPE. Issue 28. Serial No: 0028-0803-0916
4. Guliyev, H.B.: Assessment of the state of the electric power system models of probabilistically fuzzy systems. Electroenergetics, Electrotechnics Electromechanics + Control (EEEC), Sci. – Ind. J. **9**(2), 27–42 (2019)
5. Hashimov, A.M., Guliyev, H.B., Babayeva, A.R., Ibrahimov, F.Sh.: A procedure for placing shunt reactors in high-voltage networks and justification of its efficiency. Energy Syst. Res. Int. Sci. Peer-Reviewed J., Melentiev Energy Syst. Inst. Siberian Branch Russ. Acad. Sci. **3**(1), 70–78 (2020)
6. Nayeripour, M., Khorsand, H., Roosta, A.R., Niknam, T.: A new approach based on fuzzy controller for Volt/Var control in distribution system. Aust. J. Basic Appl. Sci. **4**(3), 468–480 (2010)
7. Zhao, J., Jiang, L.: Forecasting of Type-2 Fuzzy Electric Power System based on Phase Space Reconstruction model. Int. J. Grid Distrib. Comput. **8**(5), 263–272 (2015)
8. Zade, L.A.: Concept of a linguistic variable and its application to approached decisions acceptance. M, Mir, p. 178 (1976). (in Russian)
9. Diego, A.-A., Julio, P., Luis, M., Pablo, S., Francesc, G.: Fuzzy logic-based energy management system design for residential grid-connected microgrids. IEEE Trans. Smart Grid 530–543 (2018)
10. Tomsovic, K., Chow, M.Y.: Tutorial on fuzzy logic applications in power systems. IEEE - PES Winter Meeting in Singapore, January 2000
11. Ross, T.J.: Fuzzy Logic with Engineering Applications. 3 edn. John Wiley & Sons Ltd., Hoboken, p. 585 (2010). ISBN: 978-0-470-74376-8
12. Shtovba, S.D.: Designing of indistinct systems means MATLAB. M, Hot line (2007). (in Russian)
13. Momen, B., Nair, N.-K.C.: Intelligent control of on-load tap changing transformer. IEEE Trans. Smart Grid **5**(5), 2255–2263 (2014)
14. Guliyev, H.B.: Method and algorithm of fuzzy control of reactive capacity and voltage providing regime reliability of electric networks. In: Reliability: Theory & Applications, vol. 11, no. 2(41), USA, San Diego (2016)
15. Rahideh, A.K., Gitizadeh, M., Rahideh, A.B.: Fuzzy logic in real time voltage/reactive power control in FARS regional electric network. Electric Power Syst. Res. **76**(11), 996–1002 (2006)

Algorithms for the Synthesis of Optimal Linear-Quadratic Stationary Controllers

J. U. Sevinov[1(✉)], A. R. Mallaev[2], and S. N. Xusanov[2]

[1] Tashkent State Technical University, Tashkent, Uzbekistan
jasur.sevinov@tdtu.uz, sevinovjasur@gmail.com
[2] Karshi Engineering-Economic Institute, Karshi, Uzbekistan
sayyod@rambler.ru, suban1983@mail.ru

Abstract. Algorithms for the synthesis of optimal linear-quadratic stationary controllers in adaptive control systems for technological objects are presented. To construct the control system, the state parameter space method was used. To determine the optimal control of the problem under consideration in various situations, it is proposed to use the solutions of the algebraic Riccati equation. The given computational schemes made it possible to effectively solve the problems of parametric synthesis of optimal controllers in control systems for industrial technological processes.

Keywords: Optimal linear-quadratic stationary controller · Synthesis algorithms · Optimization · Parameter estimation · Parametric and nonparametric models

1 Introduction

When implementing various principles of adaptive control of an object, the question arises of how to choose the structure of controllers for coordinate and parametric control and adaptation devices that change the parameters of the controllers and the observer [1, 2]. It should be noted that at present there is still no unified theory of non-stationary control systems capable of functioning in different conditions, accumulating experience about the effectiveness of their actions, adapting to these conditions and achieving the goal of control, and the existing individual methods of their design are often heuristic, weak are related to each other and to the main method of designing systems with complete information - analytical design.

The development of the method of analytical design of optimal control systems [3] for the design of suboptimal (both deterministic and stochastic) control systems for objects with incomplete information about their state, parameters and characteristics of external influences is the method of algorithmic design [1, 2, 4–10], which allows one to optimize systems, in the sense of the assigned quality functional, to approach the synthesis of regulators and adaptation devices. The main feature of the algorithmic design method is the use of the main analytical results of the theory of optimal control for the formation of the control structure and algorithms for optimizing the system parameters.

© The Author(s), under exclusive license to Springer Nature Switzerland AG 2021
R. A. Aliev et al. (Eds.): WCIS 2020, AISC 1323, pp. 64–71, 2021.
https://doi.org/10.1007/978-3-030-68004-6_9

The method of algorithmic design of regulators is used to solve problems of suboptimal control of linear nonstationary plants with Gaussian disturbances [1, 11–15].

The real operating conditions of automatic systems are characterized by the presence of unmeasurable limited external disturbances, which lead to the appearance of control errors. Providing specified tolerances for control errors is traditionally one of the most important tasks of the theory and practice of automatic control. Under deterministic disturbances, the accuracy of the control system, as a rule, is estimated by the maximum modulus of deviation of the controlled variable from zero in the steady-state operating mode. In case of random disturbances, this characteristic is usually the root-mean-square value of the controlled variable (variance). To describe the dynamics of a controlled object under conditions of external disturbances, a linear difference or differential equation with an additive noise is usually used. This equation, which connects the output of the object with the control, is completely determined by the transfer function of the object, as well as the characteristics of the interference. Suppose that the control and the output are scalars, the transfer function is fixed, and the control is constructed in order to minimize the absolute value of the object's output.

The solution of this problem in a stochastic setting is known in the case when the interference is considered either white noise or colored noise with a known density. The solution, called stochastic optimal control [2, 9, 10, 14], is determined by a linear regulator that optimizes the root mean square output function. The convenience of the technical implementation of linear controllers has led to their widespread use. It should be noted that the stochastic formulation of the problem has a number of disadvantages. If the controlled process operates one or more times, then the assumption about the ensemble of trajectories and its averaging, which underlies the concept of a random stationary process, ceases to correspond to the problem. In addition, the assumption that the processes corresponding to white noise in the object's equation are uncorrelated is strong and difficult to verify. Based on the foregoing, we can conclude that the algorithms for the synthesis of an optimal linear-quadratic stationary controller in the presence of noise become very important.

2 Problem Definition

Consider a stationary system described by the equations [6, 7, 9, 16]:

$$\dot{x} = Ax + Bu, \quad x(0) = x_0, \tag{1}$$

$$z = Dx, \tag{2}$$

together with the functional:

$$J_{\epsilon}(x_0) = \min_{u} \int_{0}^{\infty} \left(\|z\|^2 + \epsilon^2 \|u\|^2 \right) dt, \quad \epsilon > 0, \tag{3}$$

here u, x and z are real clear dimensional control, state and output vectors.

Consider the nature of the optimal state and control trajectories [16, 17], say $x_{\in}(t)$ and $u_{\in}(t)$, at $\in\downarrow 0$.

The bounding paths, for example $x(t)$ and $u(t)$, behave as follows. There is a subspace of the state space called a degenerate hyperplane and, regardless of the fact that the initial condition $x_0,\ x(0+)$ is located on this degenerate hyperplane, $x(t)$ is displaced relative to it [6, 9, 16, 18].

Thus, about $t = 0,\ \ x(t)$ is degenerate. The optimal control $u(t)$ is correspondingly degenerate for $t = 0$.

Let us now consider for which initial conditions $x_0,\ J_{\in}(x_0)$ tends to zero at $\in\downarrow 0$. For the case when the transition matrix [16]:

$$G(s) \triangleq D(s - A)^{-1}B,$$

is square and invertible, it is proved that

$$\lim_{\in\downarrow 0}(t)J_{\in}(x_0) = 0 \text{ for all } x_0, \tag{4}$$

if and only if $G(s)$ is in the minimum phase.

Consider the linear A, B and D what are in linear feature clearly linear measures X, Y and Z A: X→X, B: U→X, D: X→Z.

The transposition will be denoted by $'$, the complex conjugate transposition by *, and the Laplace transform by ^.

3 Control Law

The optimal control law is $u = F_{\in}x$, where $F_{\in} = -\frac{1}{\in^2}B'P_{\in}$, and $F_{\in}$ is the only positive semidefinite solution to the algebraic Riccati equation [6, 16]:

$$A'P_{\in} + P_{\in}A + D'D = \frac{1}{\in^2}P_{\in}B'BP_{\in}. \tag{5}$$

It is known that $P_{\in}$ does not increase with decreasing $\in$ and that exists: $P = \lim_{\in\downarrow 0} P_{\in}$. The closed loop under optimal control is [16, 19, 20]: $x = (A + BF_{\in})x,\quad x(0) = x_0$.

The transition matrix is $T_{\in}(t) \triangleq \exp[t(A + BF_{\in})]$ and the optimal state and control paths are given as [16]:

$$x_{\in}(t)T_{\in}(t)x_0,\quad u_{\in}(t) = W_{\in}(t)x_0, \tag{6}$$

where $W_{\in}(t) \triangleq F_{\in}T_{\in}(t)$.

Consider the Laplace transforms $\hat{T}_{\in}(s)$ and $\hat{W}_{\in}(s)$ from $T_{\in}(t)$ and $W_{\in}(t)$ at $\in\downarrow 0$.

Any non-zero rational matrix $H(s)$ can be represented as $H(s) = \sum_{-\infty}^{m} s^n H_n,\quad H_m \neq 0$.

Suppose that $\#(H) = m$ is positive and represents the number of differentiating links needed to realize $H(s)$. Suppose also that $\#(0) = -\infty$.

4 Solution

Let us define the sequence $\{\Gamma_i\}$ of the subspace of the space X:

$$\Gamma_0 = \mathrm{B}, \quad \Gamma_{i+1} = \mathrm{B} + A(\ker D \cap \Gamma_i), \quad i \geq 0. \tag{7}$$

This sequence is non-decreasing, i.e. $\Gamma_i \subset \Gamma_{i+1}$ and reaches the limit of Γ in a finite number of steps. Subspace Γ is the smallest controllable subspace obtained by feeding the output to the input, i.e. it is the smallest subspace that is manageable subspace $(A + KD,\ B)$ for some K [6, 10, 16].

Let ν denote the smallest integer i such that $D\Gamma_i = D\Gamma$ [16].

1. At $\in\downarrow 0$ $\hat{T}_{\in}(s)$ and $\hat{W}_{\in}(s)$ converge to rational matrices $\hat{T}(s)$ and $\hat{W}(s)$ and

$$\#(\hat{W}) = \#(\hat{T}) + 1 = \nu. \tag{8}$$

Operators $\hat{T}_{\in}(s)$ and $\hat{W}_{\in}(s)$ converge in the open right half-plane.

Matrix $T_{\in}(t)$ converges to a distribution in the form $\delta^{(\nu-1)}Y_0 + \ldots + \delta Y_{\nu-1} + T_a(t)$ [16], where Y_i are constant matrices and $T_a(t)$ is the smoothed polynomial growth function over time t. Similar reasoning is valid for $T_{\in}(t)$.

To consider a system with many inputs and outputs, take $G(s)$ in Q^*- canonical form, which is based on the canonical Morse form for the three-dimensional matrix (D, A, B) [6, 16, 21]. With the feedback provided by applying the output to the input and changing the bases in U and Z, $G(s)$ it is equivalent to the transition matrix in the form

$$\left[\begin{array}{c|ccc} 0 & & 0 & \\ \hline & \frac{1}{s^{n_1}} & & \\ & & \cdots & \\ 0 & & & \frac{1}{s^{n_r}} \end{array}\right],$$

where $n_1 > \ldots > n_r$. It can be proved that $\nu + 1 = n_1$.

So, $\#(\hat{W}) = n_1 - 1$.

Now let's define ν. If

$$M_i \triangleq \begin{bmatrix} DA^iB & DA^{i-1}B & \ldots & DB \\ DA^{i-1}B & DA^{i-2}B & \ldots & 0 \\ \vdots & \vdots & \ldots & 0 \\ DB & 0 & \ldots & 0 \end{bmatrix},$$

then ν is the smallest integer i such that $\mathrm{rank}M_i = \mathrm{rank}G(s)$.

To solve the problem of perfect regulation, we take the sequence $\{\delta_i\}$ as the subspace X [9, 16, 21]:

$$\delta_0 = \ker D, \quad \delta_{i+1} = \ker D \cap A^{-1}(\mathrm{B} + \delta_i), \quad i \geq 0. \tag{9}$$

This sequence is non-increasing and reaches the limit δ in a finite number of steps [16]; δ is the largest unobservable subspace obtained using feedback, i.e. this is the largest subspace that is the unobservable subspace $(D,\ A+BF)$ for some F. We represent the complex plane C as

$$C = C^{-} \cup C^{0} \cup C^{+},$$

where C^{-} is the open left half-plane, C^{0}-imaginary axis and C^{+} – is the open right half-plane. Then δ gives rise to three corresponding subspaces [16, 17]: $\delta^{-},\ \delta^{0},\ \delta^{+}$.

Let's choose $F : \mathrm{X} \to \mathrm{U}$ so that δ is $(A+BF)$ - invariant and let $\tilde{\alpha}(s)$ denote the minimum polynomial from $(A+BF)|\delta$. Then

$$\alpha(s) = \alpha^{-}(s)\alpha^{0}(s)\alpha^{+}(s),$$

where zeros from α^{-}, α^{0} and α^{+} are in C^{-}, C^{0} and C^{+}, respectively.

Let us define

$$\delta^{-} = \delta \cap \ker \alpha^{-}(A+BF) + \delta \cap \Gamma;$$

$$\delta^{0} = \delta \cap \ker \alpha^{0}(A+BF) + \delta \cap \Gamma;$$

$$\delta^{+} = \delta \cap \ker \alpha^{+}(A+BF) + \delta \cap \Gamma.$$

These three subspaces are independent of F and

$$\delta = \delta^{-} + \delta^{0} + \delta^{+}.$$

2. $J_{\in}(x_0)$ tends to zero at $\in\downarrow 0$ if and only if

$$x_0 \in \delta^{-} + \delta^{0} + \Gamma. \tag{10}$$

$J_{\in}(x_0)$ tends to zero at $\in\downarrow 0$ for every $x_0 \in \mathrm{X}$ if and only if $G(s)$ is inverted to the right and is in minimum phase.

Consider a stochastic control problem where the system is described as [6, 16, 17]

$$\dot{x} = Ax + Bu + E\xi;\ \ y = Cx + \eta;\ \ z = Dx.$$

Here u, x and z are real finite dimension vectors of control, state and output [16]; y is the observability vector and ξ and η are independent means, having zero values, of Gaussian white noise with covariances of $\delta(t)I$ and $\rho^2\delta(t)I$, respectively, where ρ-small positive real number [2, 16].

The price functional has a limit at $t_0 \to -\infty$ and $t_1 \to \infty$ with a mathematical expectation of

$$\frac{1}{t_1 - t_0}\int_{t_0}^{t_1}\left(\|z\|^2 + \in^2 \|u\|^2\right)dt.$$

The resulting minimum value of the functional is

$$J_{\in p} = \operatorname{trace}\left(\frac{1}{\in^2}N_\rho P_{\in}BB'P_{\in} + EE'P_{\in}\right), \tag{11}$$

where N_ρ is a positive semidefinite solution to equation

$$AN_\rho + N_\rho A' + EE' = \frac{1}{\rho^2} N_\rho CC' N_\rho.$$

The problem of perfect stochastic control is solvable when $J_{\in\rho}$ tends to 0 at $\in\downarrow 0$ and $\rho \downarrow 0$ [16].

For this case, there are two three-dimensional matrices $(D,\ A,\ B)$ and $(C,\ A,\ E)$. For $(D,\ A,\ B)$ the subspaces were defined earlier, so we define the corresponding subspaces for $(C,\ A,\ E)$ [16, 17]. We define

$$\mathrm{V}_0 = \ker C,\quad \mathrm{V}_{i+1} = \ker C \cap A^{-1}(\mathrm{E} + \mathrm{V}_i),\quad i \geq 0,$$

$$\mathrm{W}_0 = \mathrm{E},\quad \mathrm{W}_{i+1} = \mathrm{E} + \mathrm{A}(\ker C \cap \mathrm{W}_i),\quad i \geq 0.$$

Then $\mathrm{V} = \lim \mathrm{V}_i$ is the largest unobservable subspace and $\mathrm{W} = \lim \mathrm{W}_i$ is the smallest manageable subspace for the three-dimensional matrix $(C,\ A,\ E)$.

Let's choose K so that W is $(A + KC)$ - invariant and the minimum polynomial $\beta(s)$ out of $(A + KC)|\mathrm{W}$ is represented as [16, 17]:

$$\beta(s) = \beta^-(s)\beta^0(s)\beta^+(s).$$

Let us define

$$\mathrm{W}^- = [\mathrm{W} + \mathrm{Im}\, \beta^-(A + KC)] \cap (\mathrm{W} + \mathrm{V});$$

$$\mathrm{W}^0 = [\mathrm{W} + \mathrm{Im}\, \beta^0(A + KC)] \cap (\mathrm{W} + \mathrm{V});$$

$$\mathrm{W}^+ = [\mathrm{W} + \mathrm{Im}\, \beta^+(A + KC)] \cap (\mathrm{W} + \mathrm{V}).$$

3. $J_{\in\rho}$ tends to zero at $\in\downarrow 0$, $\rho \downarrow 0$ if and only if

$$\mathrm{E} \subset \delta^- + \delta^0 + \Gamma \tag{12}$$

and

$$\mathrm{V} \cap \mathrm{W}^- \cap \mathrm{W}^0 \subset \ker L. \tag{13}$$

Here $L = \lim_{\in\downarrow 0} \in F_\in$.

5 Conclusion

Condition (12) is conditioned by the requirement of perfect control, and condition (13) - by the requirement of perfect filtration. Note also that the sufficient conditions for $J_{\in\rho}$ are that $D(s-A)^{-1}B$ is invertible to the right and is in the minimum phase, and $C(s-A)^{-1}E$ is invertible to the left and is also in the minimum phase [16]. The algorithms presented help to improve the accuracy of the parameters of the considered class of dynamic control objects and to effectively solve the problems of parametric synthesis of optimal controllers in control systems of industrial technological processes.

References

1. Aliev, R.A., Kacprzyk, J., Pedrycz, W., Jamshidi, M., Babanli, M., Sadikoglu, F.M.: 10th International Conference on Theory and Application of Soft Computing, Computing with Words and Perceptions - ICSCCW-2019, p. 970, Springer Science and Business Media LLC (2020). https://doi.org/10.1007/978-3-030-35249-3
2. Krasovskogo, A.A., Nauka, M.: Automatic Control Theory Handbook (1987). (in Russian)
3. Averyanov, V.K., Sidorenko, G.I., Tolmachev, V.N., Shapovalo, A.A.: Methodology of synthesis of algorithms for optimal control of adaptive power supply systems with wind power plants. In: IOP Conference Series: Materials Science and Engineering (2019). https://doi.org/10.1088/1757-899X/666/1/012031
4. Egupov, N.D., Pupkov, K.A.: Methods of classical and modern theory of automatic control, vol. 5, Textbook, Baumana (2004). (in Russian)
5. Antonov, V., Terexov, V., Tyukin, I.: Adaptive control in technical systems, Tutorial, Sankt-Peterburg (in Russian). (2001)
6. Fomin, V.N., Fradkov, A.L., Yakubovich, V.A.: Adaptive management of dynamic objects, The Science (1981). (in Russian)
7. Igamberdiyev, H.Z., Sevinov, J.U., Zaripov, O.O.: Regular methods and algorithms for the synthesis of adaptive control systems with tunable models, TashGTU (2014). (in Russian)
8. Igamberdiev, H.Z., Sevinov, J.U.: Algorithms for regular synthesis of adaptive systems management of technological objects based on the concepts of identification approach. J. Chem. Tech. Cont. Manag. **6**, 42–50 (2019)
9. Briand, D.O., Anderson, J.B.: Moore. In: Optimal Control Linear Quadratic Methods, p. 380, Prentice-Hall, Inc., USA, Canberra (1989). ISBN 0-13-638651-2
10. Barabanov, A.E.: Using the method of least squares in design of adaptive optimal control for a linear dynamic process. Autom. Remote Control **44**(12), 1574–1581 (1983)
11. Igamberdiev, H.Z., Yusupov, E.A., Sotvoldiev, H.I., Azamxonov, B.S.: Sustainable algorithms for the synthesis of a suboptimal dynamic object management system. In: Advance Intelligence System Computing, pp. 902–907, Springer Science and Business Media LLC (2020)
12. Loan, N.T., Shon, X.X.: On the solution of an ill-posed problem of optimal linear filtering with correlated noise. Autom. Telemechanics **4**, 58–73 (1983)
13. Yusupbekov, N.R., Igamberdiev, H.Z., Mamirov. U.F.: Algorithms of sustainable estimation of unknown input signals in control systems. J. Mult. Val. Logic Soft Comp. **33**(1-2), 1–10 (2019). https://www.oldcitypublishing.com/pdf/9291
14. Mallaev, A.R., Xusanov, S.N., Sevinov, J.U.: Algorithms of nonparametric synthesis of discrete one-dimensional controllers. Int. J. Adv. Sci. Tech. **29**(5), 1045–1050 (2020)
15. Sevinov, J.U., Zaripov, O.O., Zaripova, S.O.: The algorithm of adaptive estimation in the synthesis of the dynamic objects control systems. Int. J. Adv. Sci. Tech. **29**(5), 1096–1100 (2020). http://sersc.org/journals/index.php/IJAST/article/view/7887
16. Francis, B.: The optimal linear-quadratic time invariant regulator with cheap control. IEEE Trans. Autom. Control (1979)
17. Francis, B.: Perfect regulation and feedforward control of linear multivariable systems. In: IEEE Conference Decision Control including the 16th Symposium on Adaptive Processes and a Special Symposium on Fuzzy Set Theory and Applications (1977)
18. Igamberdiyev, H.Z., Yusupbekov, A.N., Zaripov, O.O., Sevinov, J.U.: Algorithms of adaptive identification of uncertain operated objects in dynamical models. Procedia Comput. Sci. **120**, 854–861 (2017). https://doi.org/10.1016/j.procs.2017.11.318
19. Igamberdiev, H.Z., Sevinov, J.U., Yusupbekov, A.N.: Regular algorithms for identifying the parameters of an object and a controller in a closed-loop control system. J. Chem. Tech. Cont. Manag. **6**, 50–54 (2017)

20. Zaripov, O.O., Shukurova, O.P., Sevinov, J.U.: Algorithms for identification of linear dynamic control objects based on the pseudo-concept concept. Inter. J. Psy. Rehab. **24**(3), 261–267 (2020). https://doi.org/10.37200/IJPR/V24I3/PR200778
21. Igamberdiev, H.Z., Boeva, O.H., Sevinov, J.U.: Sustainable algorithms for selecting feedback in dynamic object management systems. J. Adv. Res. Dyn. Control Syst. **12**(07), 2162–2166 (2020). https://doi.org/10.5373/JARDCS/V12SP7/20202337

Toward Ordering of n-Tuples of Z-Numbers

Akif V. Alizadeh(✉)

Department of Control and Systems Engineering, Azerbaijan State Oil and Industry University, 20 Azadlig Avenue, Baku AZ1010, Azerbaijan
a.alizadeh@asoiu.edu.az, akifoder@gmail.com

Abstract. A Z-definition of optima, being a generalization of fuzzy Pareto-optimality, is applied to the problem. The result of such an approach is that subsets of Pareto-optimal solutions can be properly selected because of inputs from the designer. The obtained optimal solutions are compared with the reference alternative and with the optima previously obtained with design of experiments techniques and different MO optimization strategies.

Keywords: Z-optimality · Pareto optimality · Degree of optimality · Z-distance

1 Introduction

In the realm of formal methods of decision-making, the general approach is the use of the utility theories. The main disadvantage of the utility theories is that they are based on an evaluation of vector-valued alternatives by means of a scalar-valued quantity. This transformation always leads to loss of information and is counterintuitive. Human being never separately transforms vector of attributes' values to a scalar value for comparison in reasoning or decision-making. Although approaches based on a vector-valued utility function exist, a fundamental axiomatic theory is absent. From the other side, preferences as human judgments are often vague and cannot be described by exact numerical values. However, the existing works on vector-valued utility approaches are devoted to the situations characterized by perfect decision-relevant information, which are rarely met in real-life decision making. At the same time, utility-based approaches rely on various restrictive assumptions like independence and its various relaxations, completeness, transitivity, orderings etc. Very restrictive assumptions make the resulting utility model simple but not adequate; less restrictive and, as a result, more adequate ones make a utility model more complicated.

The other main counterargument against utility theories is that there also exist situations for which utility function cannot be applied.

The above-mentioned circumstances mandate necessity to develop new decision approaches which, from one side may be based on direct pairwise comparison of vector-valued alternatives. From the other side, new approaches should be based on linguistic comparison of alternatives to be able to deal with vague vector-valued alternatives because real-life alternatives are almost always matter of a degree. Linguistic modeling of preferences will help to reduce Pareto optimal set of alternatives arriving at one

© The Author(s), under exclusive license to Springer Nature Switzerland AG 2021
R. A. Aliev et al. (Eds.): WCIS 2020, AISC 1323, pp. 72–80, 2021.
https://doi.org/10.1007/978-3-030-68004-6_10

optimal alternative or a narrowed subset of optimal alternatives when all relevant information is described in NL. For this purpose, a fuzzy optimality concept [18] can be used as obtained from the ideas of CW-based redefinitions of the existing scientific concepts.

The layout of the rest of the paper as follows. Section 2 includes state of the art. In Sect. 3 we give preliminaries. Section 4 includes statement of the problem and method of solution. In Sect. 5, conclusion is given.

2 State of the Art

The existing classical decision theories, especially in economics, use utility function approach which is based on transformation of a vector to an appropriate scalar value. For many real-life decision-making problems, information is judgment-based, it is vague, incomplete, imprecise, partially true etc. Preferences of a DM in this case are too inconsistent or ambiguous to permit a utility function. From the other side, in the existing theories there are so much restrictive assumptions that facilitate decision making where, really, utility function does not exist. For such cases, scoring of alternatives may be conducted by direct ranking of alternatives. Then, as a rule, instead of a utility function it is used binary relations that provide finding an optimal or near to optimal alternative(s). There exists a spectrum of works in this area.

For the performing outranking of alternatives, some methods had been developed for the traditional multi-attribute decision making (MADM) problem [1, 2]. One of the first among these methods was ELECTRE method [3–5]. The general scheme of this method can be described as follows. A DM assigns weights for each criterion and concordance and discordance indices are constructed. After this, a decision rule is constructed including construction of a binary relation on the base of information received form a DM. Once a binary relation is constructed, a DM is provided with a set of non-dominated alternatives and a DM chooses an alternative from among them as a final decision.

Similar methods like TOPSIS [6–8], VIKOR [8–10] are based on the idea that optimal vector-valued alternative(s) should have the shortest distance from the positive ideal solution and farthest distance from the negative ideal solution. Optimal alternative in the VIKOR is based on the measure of "closeness" to the positive ideal solution. TOPSIS and VIKOR methods use different aggregation functions and different normalization methods. TOPSIS and VIKOR methods were also extended for interval and fuzzy numbers-valued attributes [1, 11–15].

In [16] it is suggested fuzzy balancing and ranking method for MADM. They appraise the performance of alternatives against criteria via linguistic variables as TFNs.

Main drawbacks of the approaches mentioned above are the following: 1) detailed and complete information on alternatives provided by a DM and intensive involving the latter into choosing an optimal or suboptimal alternative(s) are required; 2) TOPSIS, ELECTRE, VIKOR and similar approaches require to pose the weights of effective decision criteria that may not be always realizable; 3) Almost all these methods are based on the use of numerical values to evaluate alternatives with respect to criteria; 4) Classical Pareto optimality principle which underlies these methods makes it necessary to deal with a large space of non-dominated alternatives that sufficiently complicates the choice; 5) All these methods are developed only for MADM problems and not for problems of decision making under risk or uncertainty. Although there exists parallelism between these problems, these methods cannot be directly applied for decision making under risk or uncertainty.

3 Preliminaries

Definition 1 A Z-number [24]. A Z-number is an ordered pair $Z = (A, B)$, where A is a fuzzy number playing a role of a fuzzy constraint on values that a random variable X may take:

$$X \ is \ A$$

and B is a fuzzy number with a membership function $\mu_B : [0, 1] \to [0, 1]$, playing a role of a fuzzy constraint on the probability measure of A:

$$P(A) \ is \ B.$$

$P(A)$, induced by a set of distributions G:

$$G = \left\{ p_Z(x) : \int_X p_Z(x)dx = 1, \int_X p_Z(x)A(x)dx \ is \ B, x \in X \right\}.$$

Definition 2 (Join of Z-numbers). Join ($\sqcup$) of Z-numbers $Z_i = (A_i, B_i)$, $i = 1, 2$ is defined as

$$Z_{12} = Z_1 \sqcup Z_2 = \max(Z_1, Z_2) = (A_{12}, B_{12}).$$

The join $A_{12} = MAX(A_1, A_2)$ of the fuzzy numbers A_1 and A_2 is defined as follows:

$$A_{12}(z) = MAX(A_1, A_2)(z) = \sup_{z=\max(x,y)} \min[A_1(x), A_2(y)].$$

The set G_{12} of resulting distributions p_{12} is defined as

$$G_{12} = G_1 \sqcup G_2,$$

where

$$G_i = \left\{ p_{Z_i}(x) : \int_X p_{Z_i}(x)dx = 1, \int_X p_{Z_i}(x)A_i(x)dx \ is \ B_i, x \in X \right\}$$

for $i = 1, 2$.

Convolution $p_{12} = p_1 \circ_{\max} p_2$ of probability distributions is defined as

$$p_{12}(x) = p_1(x)F_1(x) + p_2(x)F_2(x)$$

where F_1 and F_2 are cumulative distribution functions:

$$F_1(x) = \int_{-\infty}^{x} p_1(x)dx, \ F_2(x) = \int_{-\infty}^{x} p_2(x)dx.$$

Then

$$G_{12} = \{p_{12}(x) : p_{12}(x) = p_1(x)F_1(x) + p_2(x)F_2(x), p_i(x) \in G_i\},$$

$$B_{12} = \{(\mu_{p_{12}}(p_{12}), \mu_{A_{12}} \cdot p_{12}) : p_{12} \in G_{12}\}.$$

Definition 3 (Meet of Z-numbers). Meet ($\sqcap$) of Z-numbers $Z_i = (A_i, B_i)$, $i = 1, 2$ is defined as

$$Z_{12} = Z_1 \sqcap Z_2 = \min(Z_1, Z_2) = (A_{12}, B_{12}).$$

The meet $MIN(A_1, A_2)$ of the fuzzy numbers A_1 and A_2 is defined as follows:

$$A_{12}(z) - MIN(A_1, A_2)(z) = \sup_{z=\min(x,y)} \min[A_1(x), A_2(y)].$$

The set G_{12} of resulting distributions p_{12} is defined as

$$G_{12} = G_1 \cap G_2,$$

where G_i defined as (7).

Convolution $p_{12} = p_1 \circ_{\min} p_2$ of probability distributions is defined as

$$p_{12}(x) = p_1(x) + p_2(x) - p_1(x)F_1(x) - p_2(x)F_2(x),$$

where F_1 and F_2 are cumulative distribution functions:

$$F_1(x) = \int_{-\infty}^{x} p_1(x)dx, \ F_2(x) = \int_{-\infty}^{x} p_2(x)dx.$$

Then

$$G_{12} = \{p_{12}(x) : p_{12}(x) = p_1(x) + p_2(x) - p_1(x)F_1(x) - p_2(x)F_2(x), p_i(x) \in G_i\}.$$

Thus,

$$B_{12} = \{(\mu_{p_{12}}(p_{12}), \mu_{A_{12}} \cdot p_{12}) : p_{12} \in G_{12}\}.$$

Definition 4 (Order relation for Z-numbers). Order relation for Z-numbers $Z_i = (A_i, B_i)$, $i = 1, 2$ is defined as

$$Z_1 \sqsubset Z_2 \text{ iff } Z_1 \sqcap Z_2 = Z_1 \ or \ Z_1 \sqcup Z_2 = Z_2.$$

4 Statement of the Problem and Method of Solution

Vague preferences over a set of imprecise alternatives is modeled by linguistic preference relation over V. For this purpose, it is adequate to introduce a linguistic variable "degree of preference" [11, 21] with term-set $\mathrm{T} = (T_1, \ldots, T_n)$. Terms can be labeled as, for example "equivalence", "little preference", "high preference", and each can be described by a fuzzy number defined over some scale, for example [0,1] or [0,10] etc. The fact that $\tilde{f}$ is linguistically preferred to $\tilde{g}$ is written as $\tilde{f} \succsim_l \tilde{g}$. The latter means that there exist some $T_i \in \mathrm{T}$ as a linguistic degree $Deg(\tilde{f} \succsim_l \tilde{g})$ to which $\tilde{f}$ is preferred to $\tilde{g}$: $Deg(\tilde{f} \succsim_l \tilde{g}) \approx T_i$.

The solution of the considered problem consists in determination of a linguistic degree of preference of $\tilde{f}$ to $\tilde{g}$ for all $\tilde{f}, \tilde{g} \in A$ by direct comparison of $\tilde{f}$ and $\tilde{g}$ as vector-valued alternatives. In the suggested approach, as a fundamental basis for determination of $\succsim_l$ a fuzzy optimality concept [18] will be used. However, in that work fuzzy optimality relation is formalized for perfect information structure, i.e. provided that all the decision relevant information is represented by precise numerical evaluations. From the other side, this approach is developed for multi-attribute decision making. We will extend the fuzzy optimality concept-based approach for our framework of decision making with imperfect information. The method of solution is described below.

Fuzzy optimality concept. In [18] they suggested the notion of fuzzy optimality for MADM problems as an extension of classical notion of Pareto optimality (PO) by grading optimality of each alternative within Pareto optimal set. This helps to differentiate more optimal alternatives from less optimal ones. Below in this section we give the main notions of the formalism in [18] to be used in the sequel.

Denote $\mathcal{V} = \{Z_1, Z_2, \ldots, Z_N\}$ the set of alternatives and $\mathcal{C} = \{C_1, C_2, \ldots, C_M\}$ the set of criteria.

Definition 5 (Z Pareto Dominance). For any two points $Z_1, Z_2 \in V$, Z_1 is said to dominate Z_2 in the Pareto sense if and only if the following conditions hold:

$$Z_{1k} \leq Z_{2k} \quad \textit{for all} \;\; k \in \{1, 2, \ldots, M\},$$

$$Z_{1j} < Z_{2j} \text{ for at least one } j \in \{1, 2, \ldots, M\}.$$

Definition 6 (Z-Pareto Optimality). $Z^* \in V$ is PO if there is no $Z \in V$ such that Z P-dominates Z^*.

Definition 7 (Pareto Set and Front). We call PO set S_P and PO front F_P the set of PO solutions in design domain and objective domain respectively.

In [18], by directly comparing alternatives, they arrive at total degrees to which one alternative is better than, is equivalent to and is worse that another one. In order to do this, for each criterion they determine difference of between its values for considered alternatives. For obtaining the mentioned total degrees, these relative improvements are graded by membership degrees of predefined fuzzy sets "greater than 0", "equal to 0", "less than 0" and summed. Formally, these total degrees denoted *nbZ*, *neZ*, and *nwZ* are determined as follows:

$$nbZ(\overrightarrow{Z_i}, \overrightarrow{Z_j}) = \sum_{k=1}^{n} \mu_b^k(v_k^{ij}),$$

$$neZ(\vec{Z_i},\ \vec{Z_j}) = \sum_{k=1}^{n} \mu_e^k(v_k^{ij}),$$

$$nwZ(\vec{Z_i},\ \vec{Z_j}) = \sum_{k=1}^{n} \mu_w^k(v_k^{ij}).$$

Where $\delta_k^{ij} = Z_{ik} - Z_{jk}, v_k^{ij} = \delta_k^{ij} \cdot w_k$ and w_k is importance degree of component k, $\mu_b^k(v_k^{ij}) = 1 - d(v_k^{ij}, Z_b)$.

$\mu_b^k, \mu_e^k, \mu_w^k$ are similarity degree of v_k^{ij} and linguistic evaluations Z_b-"Z-better", Z_e-"Z-equivalent" and Z_w-"Z-worse" respectively.

The similarity degrees $\mu_b^k, \mu_e^k, \mu_w^k$ are constructed such that following condition holds:

$$nbZ + neZ + nwZ = \sum_{i=k}^{M} (\mu_b^k + \mu_e^k + \mu_w^k) = M.$$

Definition 8 ((1 − kZ)-dominance). Z_1 is said to (1 kZ)-dominates Z_2 if and only if:

$$neZ < M,\ nbZ \geq \frac{M - neZ}{kZ + 1},$$

where $0 \leq kZ \leq 1$.

A function d with the following property is defined: for any couple of alternatives $Z_1, Z_2 \in V$, $1 - d(Z_1, Z_2)$ is the greatest kZ value such that Z_1 $(1 - kZ)$-dominates Z_2:

$$d(Z_1, Z_2) = \begin{cases} 0, & if\ nbZ \leq \frac{M - neZ}{2} \\ \frac{2 \cdot nbZ + neZ - M}{nbZ}, & otherwise \end{cases}$$

$d(A_1, A_2) = 1$ shows that Z_1 Pareto dominates Z_2; $d(Z_1, Z_2) = 0$ shows that Z_1 does not Pareto dominates Z_2.

Definition 9 (kZ-optimality). Z^* is kZ-optimum if and only if there is no $Z \in V$ such that A $(1 - kZ)$-dominates Z^*.

A function $do : V \to [0, 1]$ is defined as a degree of optimality of $Z \in V$ as follows:

$$do(Z^*) = 1 - \max_{Z \in V} d(Z, Z^*).$$

In [18] it is mentioned that function do can be considered as the membership function of a fuzzy set describing the notion of kZ-optimality.

Definition 10 (kZ-Optimal Set and Front). We call kZ-Optimal Set $\mathcal{S}_{kZ}$ and kZ-Optimal Front $\mathcal{F}_{kZ}$ the set of kF-Optimal solutions in design domain and objective domain respectively.

Definition 11 (Z dominance). Let $\mu_D(Z_1, Z_2)$ be a membership function defined as follows:

$$\mu_D(Z_1, Z_2) = \varphi_{\mu_D}(nbZ(Z_1, Z_2), neZ(Z_1, Z_2), nwZ(Z_1, Z_2)).$$

Then $\mu_D(Z_1, Z_2)$ is a fuzzy dominance relation if for any $\alpha \in [0, 1]$ $\mu_D(Z_1, Z_2) > \alpha$ implies that Z_1 $(1 - kZ)$-dominates Z_2.

In [18], particularly, they suggest defining φ_{μ_D} as follows:

$$\varphi_{\mu_D} = \frac{2 \cdot nbZ(Z_1, Z_2) + neZ(Z_1, Z_2)}{2M}.$$

Definition 12 (Z optimality). A membership function $\mu_D(Z_1, Z_2)$ represents the Z optimality relation if for any $0 \leq kZ \leq 1$, Z^* belongs to the kZ-cut of μ_D, if and only if there no $Z \in V$ such that

$$\mu_D(Z, Z^*) > kZ.$$

Ranking of Discrete Z-Numbers. Ranking of discrete Z-numbers is a necessary operation in arithmetic of Z-numbers and is a challenging practical issue. Zadeh addresses the problem of ranking Z-numbers as a very important problem [24]. In contrast to real numbers, Z-numbers are ordered pairs, for ranking of which there can be no unique approach. We suggest considering comparison of Z-numbers based on fuzzy optimality principle. First, it is needed to calculate the functions *nbZ*, *neZ*, and *nwZ* which evaluate how much one of the Z-n-tuple is better, equivalent, and worse than the other one [23]:

Direct comparison of vector-valued alternatives

$$\vec{Z_i} = (Z_{i1}, Z_{i2}, \ldots, Z_{in}) = ((A_{i1}, B_{i1}), (A_{i2}, B_{i2}), \ldots, (A_{in}, B_{in}))$$

and

$$\vec{Z_j} = (Z_{j1}, Z_{j2}, \ldots, Z_{jn}) = ((A_{j1}, B_{j1}), (A_{j2}, B_{j2}), \ldots, (A_{jn}, B_{jn}))$$

may, particularly, supposes to determine the total degrees of component-wise superiority, equivalence and inferiority of $\vec{Z}_i$ with respect to $\vec{Z}_j$. We suggest to determine the total degrees of component-wise superiority, equivalence and inferiority, denoted *nbZ*, *neZ*, and *nwZ* respectively, as follows:

$$nbZ(\vec{Z_i}, \vec{Z_j}) = \sum_{k=1}^{n} \mu_b^k(v_k^{ij}),$$

$$neZ(\vec{Z_i}, \vec{Z_j}) = \sum_{k=1}^{n} \mu_e^k(v_k^{ij}),$$

$$nwZ(\vec{Z_i}, \vec{Z_j}) = \sum_{k=1}^{n} \mu_w^k(v_k^{ij}).$$

Where $\delta_k^{ij} = Z_{ik} - Z_{jk}$, $v_k^{ij} = \delta_k^{ij} \cdot w_k$ and w_k is importance degree of component k. μ_b^k, μ_e^k, μ_w^k are similarity degree of v_k^{ij} to linguistic evaluations Z_b-"Z-better", Z_e-"Z-equivalent" and Z_w-"Z-worse" respectively, and defined as

$$\mu_b^k(v_k^{ij}) = 1 - distC(v_k^{ij}, Z_b),$$

$$\mu_e^k(v_k^{ij}) = 1 - distC(v_k^{ij}, Z_e),$$

$$\mu_w^k(v_k^{ij}) = 1 - distC(v_k^{ij}, Z_w).$$

By following that approach, on the base of values of nbZ, neZ, and nwZ, the value of degree of optimality $do(\overrightarrow{Z_i})$, as a degree of membership to fuzzy Pareto optimal set, is determined for all $\overrightarrow{Z_i} \in V$.

$do()$ allows for justified determination of linguistic preference relation $\succsim_l$ over V.

On the base of $do()$ the degree of preference of $\overrightarrow{Z_i}$ to $\overrightarrow{Z_j}$ for any $\overrightarrow{Z_i}, \overrightarrow{Z_j} \in V$ can be determined from $do()$ as follows:

$$Deg(\overrightarrow{Z_i} \succsim \overrightarrow{Z_j}) = do(\overrightarrow{Z_i}) - do(\overrightarrow{Z_i}).$$

5 Conclusion

In Multi Criteria Decision Making with linguistic variables, the DMs may have vague information, limited attention and different information processing capabilities.

This paper proposes a new Z decision-making method which allows Z-preferences in linguistic terms for alternative selection.

The approach is computationally simple, and its underlying concept is logical and comprehensible, thus facilitating its implementation in a computer-based system.

References

1. Lu, J., Zhang, G., Ruan, D., Wu, F.: Multi-Objective Group Decision Making: Methods, Software and Applications with Fuzzy Set Techniques. Series in Electrical and Computer Engineering, vol. 6. Imperial College Press, London (2007)
2. Zeleny, M.: Multiple Criteria Decision Making. McGraw-Hill, New York (1982)
3. Roy, B.: Multicriteria Methodology for Decision Aiding. Kluwer, Dordrecht, The Netherlands, pp. 32–48
4. Roy, B., Berlier, B.: La Metode Electre II. In: Sixieme Conference Internationale de Rechearche Operationelle, Dublin
5. Huang, W.-C., Chen, C.-H.: Using the electre ii method to apply and analyze the differentiation theory. In: Proceedings of the Eastern Asia Society for Transportation Studies, vol. 5, pp. 2237–2249 (2005)
6. Hwang, C.L., Yoon, K.: Multiple Attribute Decision Making. Lecture Notes in Economics and Mathematical Systems, vol. 186, pp. 20–41. Springer-Verlag, Berlin

7. Yoon, K.: A reconciliation among discrete compromise solutions. J. Oper. Res. Soc. **38**(3), 272–286
8. Opricovic, S., Tzeng, G.H.: Compromise solution by MCDM methods: a comparative analysis of VIKOR and TOPSIS. Eur. J. Operat. Res. **156**(2), 445–455 (2004)
9. Opricovic, S.: Multicriteria Optimization of Civil Engineering Systems. Faculty of Civil Engineering, Belgrade (1998)
10. Opricovic, S., Tzeng, G.H.: Extended VIKOR method in comparison with outranking methods. Eur. J. Operat. Res. **178**(2), 514–529 (2007)
11. Liu, W.J., Zeng, L.: A new TOPSIS method for fuzzy multiple attribute group decision making problem. J. Guilin Univ. Electron. Technol. **28**(1), 59–62 (2008)
12. Jahanshahloo, G.R., Lotfi, F., Hosseinzadeh, I.M.: An algorithmic method to extend TOPSIS for decision-making problems with interval data. Appl. Math. Comput. **175**(2), 1375–1384 (2006)
13. Zhu, H.P., Zhang, G.J., Shao, X.Y.: Study on the application of fuzzy TOPSIS to multiple criteria group decision making problem. Ind. Eng. Manag. **1**, 99–102 (2007)
14. Wang, Y.M., Elhag, T.M.S.: Fuzzy TOPSIS method based on alpha level sets with an application to bridge risk assessment. Expert Syst. Appl. **31**(2), 309–319 (2006)
15. Liu, P., Wang, M.: An extended VIKOR method for multiple attribute group decision making based on generalized interval-valued trapezoidal fuzzy numbers. Sci. Res. Essays **6**(4), 766–776, 18 February 2011
16. Vahdani, B., Zandieh, M.: Selecting suppliers using a new fuzzy multiple criteria decision model: the fuzzy balancing and ranking method. Int. J. Prod. Res. **18**, 5307–5326, 15 September 2010
17. Lee, E.S., Li, R.-J.: Comparison of fuzzy numbers based on the probability measure of fuzzy events. Comput. Math Appl. **15**(10), 887–896 (1988)
18. Farina, M., Amato, P.: A fuzzy definition of "optimality" for many-criteria optimization problems. IEEE Trans. Syst. Man Cybern. Part A Syst. Hum. **34**(3), 315–326 (2004)
19. Zadeh, L.A., Aliev, R.A., Fazlollahi, B., Alizadeh, A.V., Guirimov, B.G., Huseynov, O.H.: Decision theory with imprecise probabilities. In: Contract on "Application of Fuzzy Logic and Soft Computing to Communications, Planning and Management of Uncertainty", p. 95. Berkeley, Baku (2009). http://www.raliev.com/report.pdf
20. Aliev, R.A., Alizadeh, A.V., Guirimov, B.G., Huseynov, O.H.: Precisiated information-based approach to decision making with imperfect information. In: Proceedings of the Ninth International Conference on Application of Fuzzy Systems and Soft Computing, (ICAFS-2010), pp. 91–103. Prague, Czech Republic (2010)
21. Borisov, A.N., Alekseyev, A.V., Merkuryeva, G.V., Slyadz, N.N., Gluschkov, V.I.: Fuzzy Information Processing in Decision Making Systems. «Radio i Svyaz», Moscow (1989). (in Russian)
22. Alizadeh, A.V., Musayev, A.F., Aliev, R.R.: Application of the fuzzy optimality concept to decision making with imprecise probabilities. In: Proceedings of the Sixth International Conference on Soft Computing and Computing with Words in System Analysis, Decision and Control, (ICSCCW-2011), pp. 373–382, Antalya (2011)
23. Aliev, R.A., Alizadeh, A.V., Huseynov, O.H.: The arithmetic of discrete Z-numbers. Inf. Sci. **290**, 134–155 (2015)
24. Zadeh, L.A.: A note on Z-numbers. Inf. Sci. **181**(14), 2923–2932 (2011)

Determining the Activity of Fullerene Nanoparticles Using QSAR Models

Alisher Ikramov[1(✉)], Kamoliddin Nabijonov[1], and Bakhtiyor Rasulev[2]

[1] V.I. Romanovskiy Institute of Mathematics, Mirzo Ulugbek, 81, Tashkent 100170, Uzbekistan
a.ikramov@mathinst.uz, nabijonovkamol@gmail.com
[2] Department of Coatings and Polymeric Materials, North Dakota State University, Fargo, ND 58103, USA
rasulev@icnanotox.org

Abstract. The paper analyzes the data on 169 fullerene molecules and their activity value. We study the relationship between the properties of fullerene nanoparticles and their activity using linear regression methods. We performed feature selection, and. Out of 217 parameters in the data set, we selected the 5 most important ones using the Wrapper methods. We trained 14 different regression models and tested them on the test set. We chose 4 models with the highest performance and optimized their hyperparameters. We obtained an R^2 score of 0.90 on the test set outperforming the baseline.

Keywords: QSAR · Nano-QSAR · Activity · Properties · Outliers · Parameters · Models · Forecasts · Fullerenes

1 Introduction

The biological properties of representatives of any class of compounds are determined by their physical and chemical properties. However, this is not entirely true when applied to fullerene nanostructures. Dumpis et al. [1] describes the transition from "ordinary" systems to nanoscale systems as the problem of working with nanostructures due to fact that in the field of the nanowire, other laws apply, according to which the properties of a substance depend not only on its composition, but on the geometry, shape and size of nanoparticles.

Obtaining data for the parameters of fullerene [2] nanoparticles becomes a difficult task. To simplify this task, it is necessary to analyze the actual data and identify the most necessary features for calculating the activity of nanoparticles. Based on the results, it is necessary to train QSAR [3] regression models, which, having received parameters, can predict the activity closest to the actual data we have (Fig. 1).

Nano-QSAR is the branch of Quantitative-Structure Activity Relationship where the specter of possible descriptors is limited, while the applications are overwhelming. The nano-particles represents combination of organic and inorganic compounds.

© The Author(s), under exclusive license to Springer Nature Switzerland AG 2021
R. A. Aliev et al. (Eds.): WCIS 2020, AISC 1323, pp. 81–95, 2021.
https://doi.org/10.1007/978-3-030-68004-6_11

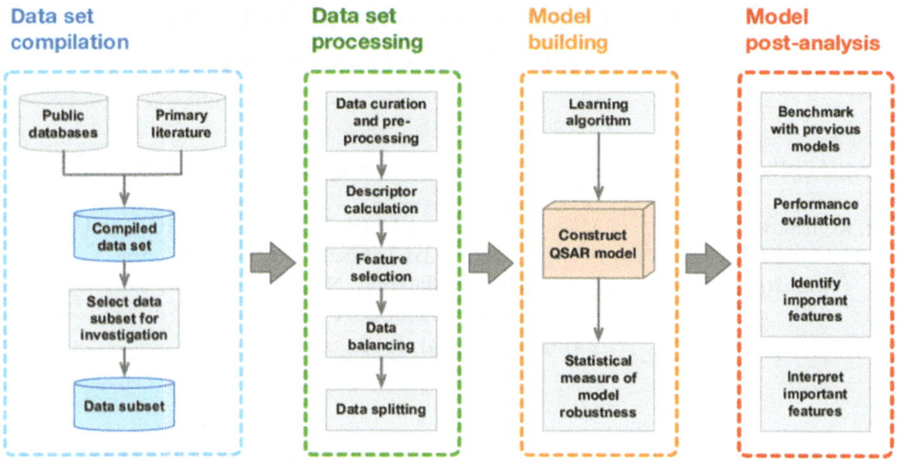

Fig. 1. Schematic representation of the QSAR modeling workflow [4].

Mikolajczyk [5] tested 34 TiO_2-based nano-photocatalysts and calculated their cytotoxicity using linear models. They checked the results through the series of experiments with Chinese hamster ovary cells. They achieved R2 score of 0.89 on external validation.

Puzyn [6] described cytotoxicity of 17 different types of metal oxide nanoparticles to bacteria Escherichia coli using nano-QSAR approach. Puzyn demonstrated that linear models can be efficiently used in tasks with small datasets.

Toropova [7] built a series of linear models to calculate mutagenicity of fullerene nano-particles using CORAL software. They used a quasi-SMILES representation to better describe nano-particles and their features. They achieved R2 of 0.68–0.72. The method behind the calculations was Monte-Carlo algorithm.

2 Methods

2.1 Data

We have extracted the data by Ahmed [8] on 169 fullerene nano-particles with their activity value against HIV-1 Reverse Transcriptase. We used descriptors generated for each molecule provided by QikProp in Schrodinger Suite, which produces 294 physicochemical features describing each nanoparticle based on its SDF.

All data is numeric, and there are no categorical data, strings, or objects among them.

2.2 Data Pre-processing

Outliers Treatment. By using Plotly histogram to plot our dataset, we decided break-point values that decide whether to keep record or delete it. For out particular case values, that are below 4000 and higher than 6000, were removed using dataframe's drop method [9].

Processing Null Values. We dropped all features that consisted mostly of zeros or were undefined for a particular compound. We used Pyplot [11] heatmap to visualize dataset with missing values.

2.3 Feature Selection

Bivariate Analysis [12]. Using the LabelEncoder [13] from the sklearn.preprocessing module and corr() method, we detected 10 positively and 10 negatively related to target variables.

Multicollinearity [14]. We used python to construct correlation matrixwe removed highly correlated features based on coefficients of the matrix. We performed visualization of the correlation matrix to determine the threshold using Seaborn [15] heatmap.

Wrapper Methods [16]. We used recursive feature elimination technique to find five most important features from our set of variables. We initialized RFE [17] and passed Ridge [18] as estimator. RFE fitted Ridge to train set using features provided, predicted values on the test set, and assigned weights to each feature.

2.4 QSAR Model Development

Model Building. We initialized 14 scikit-learn [19] regression [20] models with default parameters: Linear Regression [21], Lasso [22], Ridge [23], KernelRidge [24], Elastic Net [25], Decision Tree Regressor [26], SVR [27], K neighbors Regressor [28], Random Forest Regressor [29], Extra Trees Regressor [30], Ada Boost Regressor [31], Gradient Boosting Regressor [32], XGB Regressor [33], LGBM Regressor [34].

Evaluating and Optimizing Models. We used RMSE [35] scores to evaluate our models with default parameters on the test set. We used a Plotly scatter plot to visualize the scores. We performed a grid search to determine the best hyperparameters values [36] for each model.

Predictions. Each model was reinitialized with optimal hyperparameters. Function predict_with_optimized_models() took models and returned dataframe with predicted data using test dataset. RMSE and R^2_score [37] where used to evaluate forecasts and Pyplot barchart to plot.

3 Results

3.1 Outliers Treatment

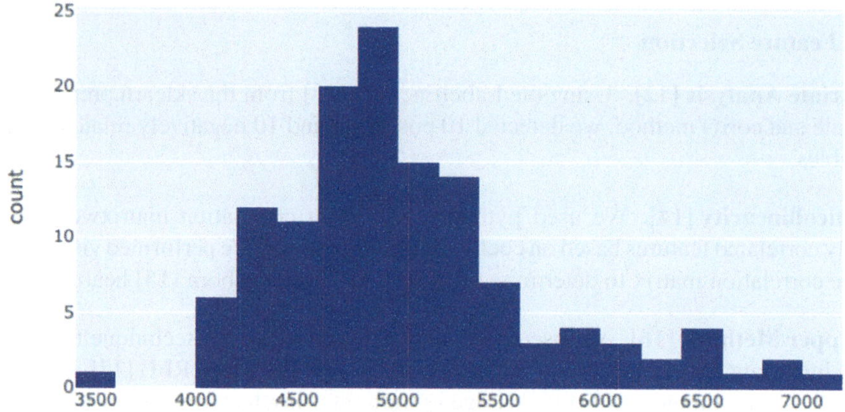

Fig. 2. Histogram representation of target values before outliers' removal.

From the histogram above, we can see that there are values that stand out and can have a bad effect on the accuracy of models. You need to delete records with a target value greater than 6000 and less than 4000. After deleting records, we reindex our table to preserve the data sequence (Figs. 2 and 3).

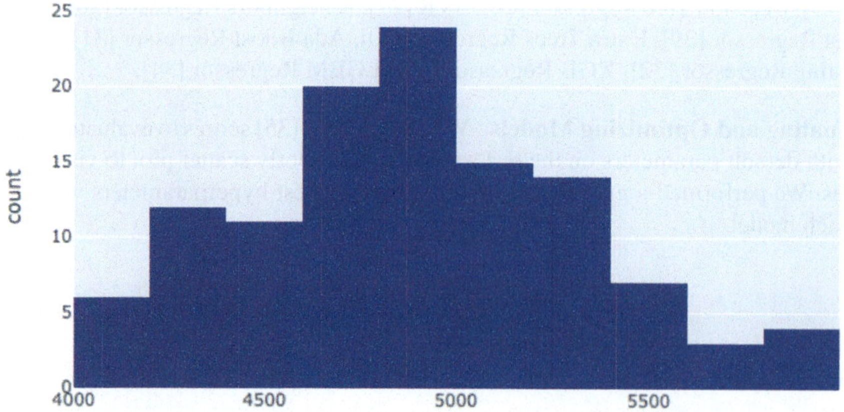

Fig. 3. Histogram representation of target values after outliers removal.

We got 116 compounds with 218 features in the training set after outliers' removal. In the process, we cleared our table from 14 outliers, which will improve the accuracy of models during training.

3.2 Processing Null Values

In our dataset, there are records that contain zeros. As a result, these values are not useful and can also impair the accuracy of our models. The easiest way to fill in these variables is to replace the missing values with the average, depending on their distribution. However, the number of features is greater than the number of records, which means we can discard some of them. This led to the decision to reject features that have at least 1 zero as an entry.

There are 218 features in the table before deleting features that have al least one zero. Before clearing the attributes, we replace all zeros with NaN so that we can use the isnull method later.

Next, we create an array of missing_columns that consists of the names of columns that have NaN using the isnull and any methods, and count the number from them (Fig. 4).

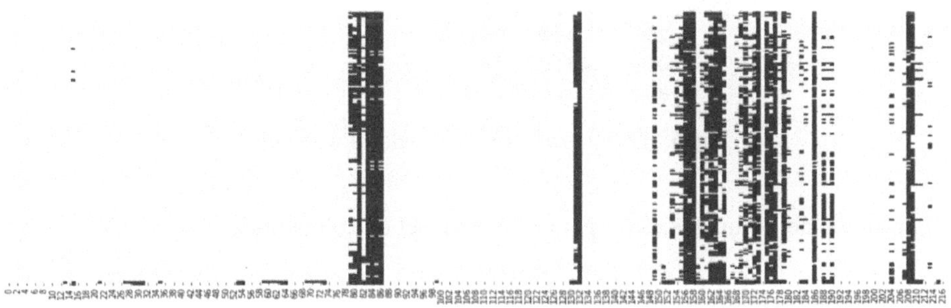

Fig. 4. Heatmap for features to display missing values.

NaN values are indicated as blue. We are deleting columns from our table using the missing_columns array that we got in the previous step.

Method shape returns tuple (155, 218) which indicates that 78 features were removed.

3.3 Bivariate Analysis

Bivariant analysis finds the relationship between two variables. We will look for a correlation or association between our variables and the target values. We will select only 20 variables that have the highest correlation scores (positive or negative).

We first need to construct an array of feature correlation. The array contains constants of the relationship of each attribute with all the others (Figs. 5 and 6).

```
target     1.000000
32         0.878608
182        0.874970
119        0.871881
192        0.861268
34         0.860962
10         0.859123
33         0.858360
187        0.831844
97         0.802718
49         0.800507
Name: target, dtype: float64
```

Fig. 5. Features positively correlated with target.

```
198   -0.419234
7     -0.440897
147   -0.530348
128   -0.532486
127   -0.641452
126   -0.688336
8     -0.741256
191   -0.752266
125   -0.779956
100   -0.854360
Name: target, dtype: float64
```

Fig. 6. Features negatively correlated with target.

From the above analysis, we get as the final set of 20 features.

3.4 Multicollinearity

The key task of regression analysis is to eliminate the relationship between each independent and dependent variable.

Multicollinearity causes the following two main types of problems:

1. Coefficient estimates may vary greatly depending on what other independent variables are in the model. Coefficients become very sensitive to small ones changes in the model.
2. Multicollinearity reduces the accuracy of the estimated coefficients, which weakens the statistical relevance of your regression model. Multicollinearity makes it difficult to interpret your coefficients and reduces your model's ability to identify independent variables that are statistically significant.

Using the corr method, and LabelEncoder we create a correlation matrix that shows the correlation coefficient for each variable in the data set.

Heatmap of the above matrix is required to better visualize and select the threshold value for the correlation constant. Variables that have a correlation value greater than this constant will be removed from the table due to multicollinearity (Fig. 7).

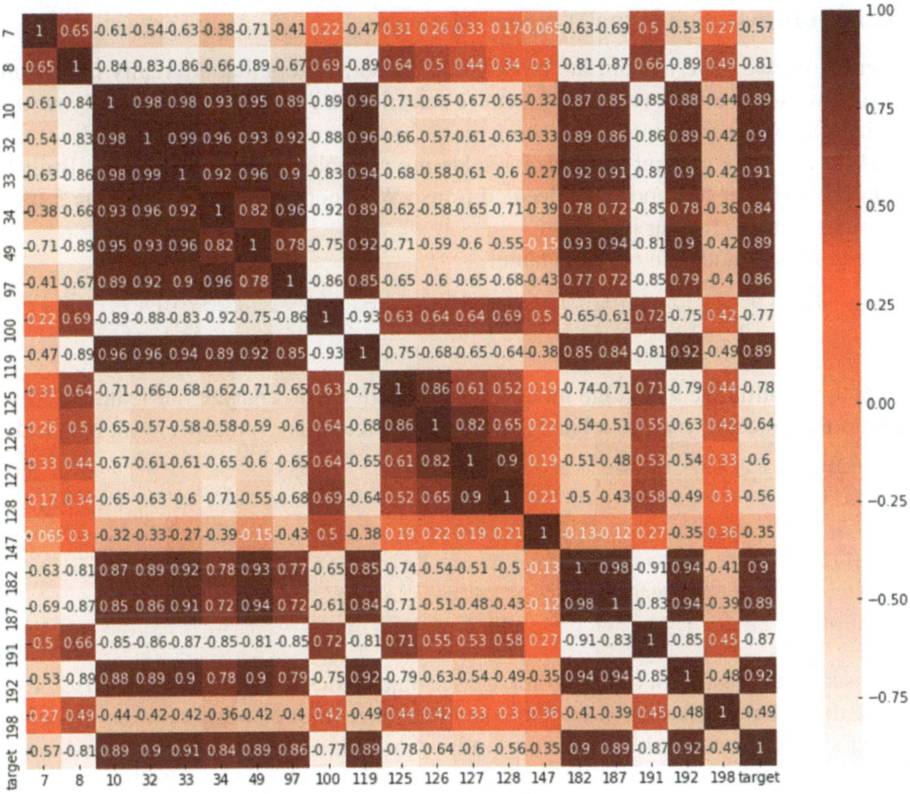

Fig. 7. Heatmap of variables and their corresponding multicollinearity constants.

From the map above, we can see that the value 0.95 is best suited for the threshold value. We create a **to_drop** array that stores the names of features that we are going to delete from the table.

List of variables to delete from the table: [10, 32, 33, 34, 182] (Table 1).

Table 1. Dataframe after removal of features with correlation higher than 0.95.

	7	8	49	97	100	119	125	126
0	2.833207	0.113151	1902.096	14	0.822973	6.054665	0.651079	0.434861
1	2.773567	0.111393	2367.615	16	0.784834	6.279611	0.643791	0.426596

	127	128	147	187	191	192	198	Target
0	0.292601	0.198028	2.589041	1798.03	0.912654	71.1467	−21.5973	5216
1	0.284526	0.19046	2.589041	2039.33	0.833685	77.8184	−19.9905	5464

3.5 Feature Selection Using Wrapper Methods

As we know, we have to train the model only on parameters that are really important. For such reason we have to perform feature selection. By doing so, we not only reduce training and evaluation time, but also increase the effectiveness of our models!

There are three general methods for selecting variables: Filtering Methods, Built-In Methods, Wrapper Methods. We decided to use Wrapper methods to select the 5 most important features for our models.

Current parameters in our data table:

```
([7, 8, 49, 97, 100, 119, 125, 126, 127, 128, 147, 187,
191, 192, 198])
```

First, we initialize the Ridge model that RFE will use to analyze our features. Next, we pass the object to our model and the number of features (5) that we want to select.

```
Number of attributes: 5
Features:[7,8,49,97,100,119,125,126,127,128,147,187,191,192,1
98]
Selected:[True False False True False False False False False
False True False True True False]
Feature rating:[1 9 11 1 5 7 2 3 4 8 1 10 1 1 6]
```

Based on the results of our program, we see that the descriptors [7, 97, 147, 191, 192] are the most important ones in our table and thus have a lot of weight for our models.

Let's create a new table consisting only of variables selected by RFE and target values (Table 2).

Table 2. Dataframe constructed from features selected by RFE.

	7	97	147	191	192	Target
0	2.833207	14	2.589041	0.912654	71.1467	5216
1	2.773567	16	2.589041	0.833685	77.8184	5464
2	2.783302	12	2.8125	0.942639	64.746	4770
3	2.817021	17	2.9	0.854002	63.2368	4952
4	2.791773	17	2.9	0.852002	63.5846	5420

Now that we have the dataset ready, we can rename our variables (numbers) to the actual object names from file called "molecu-lar_descriptors-fuller_for_FA.xlsx." (Table 3).

```
7   -> Narumi Geometric Topological (NGT)
97  -> Topological diameter (TD)
147 -> Ring complexity index (RCI)
191 -> glob
192 -> Qppolrz
```

Table 3. Dataframe after renaming features to actual descriptor names.

	NGT	TD	RCI	Glob	Qppolrz	Target
0	2.833207	14	2.589041	0.912654	71.1467	5216
1	2.773567	16	2.589041	0.833685	77.8184	5464
2	2.783302	12	2.8125	0.942639	64.746	4770
3	2.817021	17	2.9	0.854002	63.2368	4952
4	2.791773	17	2.9	0.852002	63.5846	5420

3.6 Model Building

Finally, we have processed data ready to train our regression. We initialize 14 regression models from scikit – learn with default parameters except random_state. Models are then trained on the training samples using the fit method. We got 116 training samples and 39 test samples with 5 features each.

3.7 Evaluating and Optimizing Models

At this stage, we will train our models on train and evaluate them on test groups. RMSE will be used as a metric to evaluate model's perfomance. Our script iterates through the models and calculates their corresponding RMSE on test data (Fig. 8).

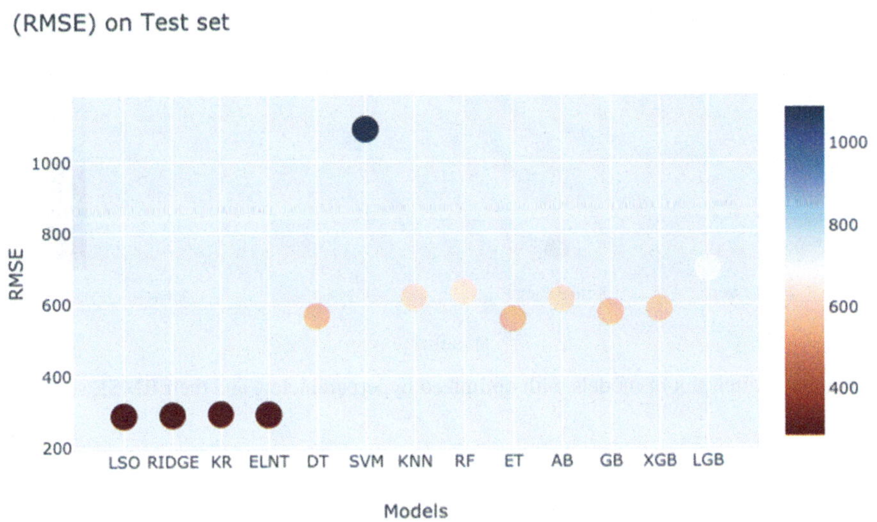

Fig. 8. Scatter plot of models and their RMSE scores on test data.

It seems that LSO, RIDGE, KR, and ELNT are best performed on our data, while SVM performs the worst.

Hyperparameter Optimizations

Our models currently have default parameters and need to be optimized to perform better on given data (Table 4 and Fig. 9).

Table 4. Models and corresponding optimal parameters.

Elastic Net best params	alpha	l1_ratio	random_state		best_score
	0.0005	0.8	43		198.66102
Kernel Ridge best params	alpha	coef0	degree	kernel	best_score
	0.3	3.5	2	polynomial	182.14643
Ridge best params	alpha	random_state			best_score
	9	43			188.90007
Lasso best params	alpha	random_state			best_score
	0.01	43			199.80077

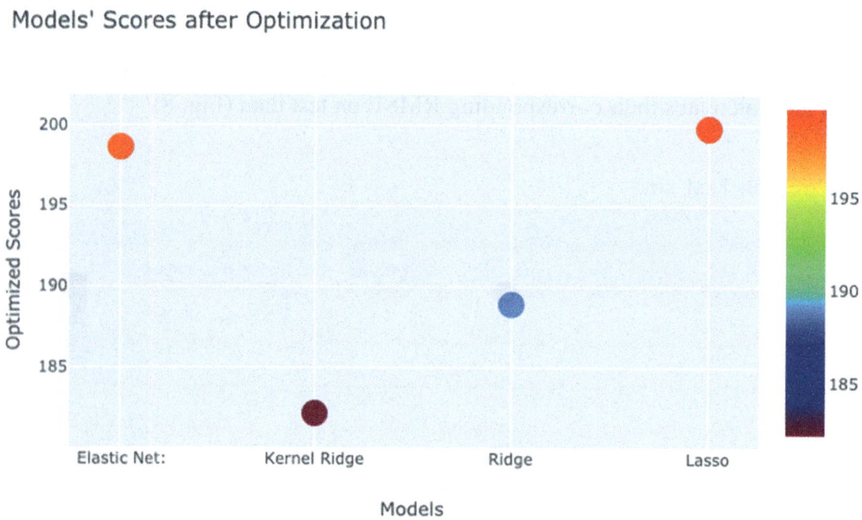

Fig. 9. Scatter plot of models with optimized hyperparameters and their RMSE scores.

3.8 Predictions

We initialize objects of the best models that we identified at the last stage [ElasticNet, KernelRidge, Ridge, Lasso], with optimal parameters (Table 5).

Table 5. Trained regression equations.

elastic_net	-151.3*(NGT)+98.5*(TD)+450*(RCI)-397*(glob)+49.7*(Qppolrz)
Ridge	-29.9*(NGT)+101*(TD)+22.7*(RCI)-4.2*(glob)+46.8*(Qppolrz)
Lasso	-146*(NGT)+97.2*(TD)+461*(RCI)-500*(glob)+49.7*(Qppolrz)

R^2 and **RMSE** are calculated using the predict_with_optimized_models and calc_r2_rmse functions (Table 6 and Fig. 10).

Table 6. RMSE and R^2 scores of models.

	RMSE	R^2
elastic_net	284.01	**0.90**
kernel_ridge	306.79	0.89
Ridge	295.43	0.89
Lasso	**283.70**	**0.90**

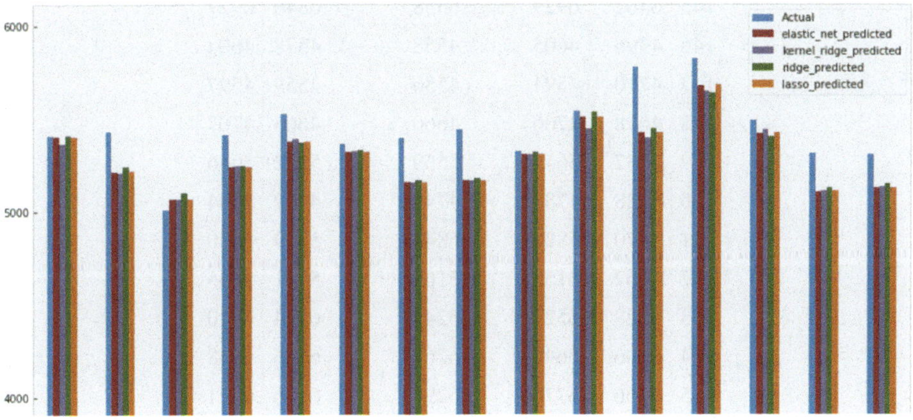

Fig. 10. Bar chart for actual and predicted values for test dataset.

We compared actual values and predicted values produced by different methods on the test set. The detailed result is presented in Table 7.

4 Discussion

We performed not only feature selection but also wrapping technique to achieve high accuracy. The resulting model outperformed model by Ahmed et al. [8] on the same

Table 7. Table of actual target values and predicted values by elastic_net, kernel_ridge, ridge and lasso for test dataset.

	Actual	elastic_net	kernel_ridge	ridge	lasso
130	5408	5383	5344	5402	5383
131	5432	5210	5192	5243	5210
132	5010	5053	5040	5101	5054
133	5418	5222	5218	5246	5222
134	5528	5362	5372	5374	5361
135	5366	5312	5308	5334	5312
136	5400	5136	5131	5169	5137
137	5444	5149	5143	5180	5150
138	5324	5298	5293	5320	5298
139	5548	5519	5448	5534	5517
140	5782	5402	5371	5441	5400
141	5828	5655	5618	5633	5657
142	5498	5407	5417	5406	5411
143	5316	5088	5080	5127	5089
144	5308	5113	5110	5153	5112
145	6402	6929	6558	6840	6927
146	4396	4605	4558	4578	4604
147	4710	4594	4536	4559	4597
148	4698	4706	4660	4663	4707
149	5232	5649	5569	5622	5646
150	4828	4789	4740	4737	4794
151	5770	5821	5843	5779	5820
152	5162	5155	5162	5141	5155
153	6528	6521	6244	6544	6520
154	6306	6647	6267	6656	6648
155	6466	6770	6298	6766	6771
156	5916	6153	6058	6200	6152
157	6062	6392	6238	6241	6397
158	4514	4603	4576	4575	4604
159	3614	3942	3824	3955	3942
160	6986	6597	6301	6451	6602
161	7304	7438	6904	7194	7453

(continued)

Table 7. (*continued*)

	Actual	elastic_net	kernel_ridge	ridge	lasso
162	8304	8510	7880	8080	8531
163	4612	4806	4882	4735	4811
164	5922	4961	4990	4899	4955
165	4852	4703	4846	4619	4710
166	4822	4780	4848	4712	4785
167	3562	3683	3527	3719	3684
168	3846	4599	4658	4593	4598

dataset. Our model uses 5 features, 2 of which are also used in [8]. Thus, there is a possibility to achieve high results on multiple targets using these 5 features instead of 2 in the baseline.

5 Conclusion

We cleaned up dataset from outliers and deleted records with zero values. We discovered 20 important features using a Bivariant analysis to determine target values. The Multicollinearity test helped us to detect and to drop dependent variables; thus, leaving only independent most important features.

Using Wrapper methods on the remaining features, we found the 5 most necessary parameters in our data table, on which we trained our regression models.

14 regression models were initialized and trained. According to their training estimates, we selected 4 and optimized their hyperparameters using 5-fold cross-validation. After optimization, the models with the best values of hyperparameters were tested on the test set.

As a result of our work, we found 5 necessary features for determining the activity of fullerene nanoparticles and have 4 regression models that, after receiving the data as input, output the target value with an R^2 of 0.90 and RMSE of 283.70.

Appendices

Program Code

Code for this paper can be acquired at: https://github.com/kamolliddin/QSAR_modeling.

References

1. Dumpis, M.A., Nikolaev, D.N., Litasova, E.V.: Biological activity of fullerenes – reality and prospects (in Russian). https://doi.org/10.17816/RCF1614-20.

2. Holkar, Ch.R., Jain, S.S., Jadhav, A.J., Pinjari, D.V.: Nanomaterials for green energy. micro and nano technologies, pp. 433–455 (2018). https://doi.org/10.1016/B978-0-12-813731-4.00014-X
3. QSAR models. ECHA. https://echa.europa.eu/support/registration/how-to-avoid-unnecessary-testing-on-animals/qsar-models
4. Schematic representation of the QSAR modeling workflow. ResearchGate. https://www.researchgate.net/figure/Schematic-representation-of-the-QSAR-modeling-workflow_fig1_338639486
5. Mikolajczyk, A., Gajewicz, A., Mulkiewicz, E., Rasulev, B., et al.: Nano-QSAR modeling for ecosafe design of heterogeneous TiO_2-based nano-photocatalysts. Environ. Sci. Nano **5**, 1150–1160 (2018)
6. Puzyn, T., Rasulev, B., Gajewicz, A., Hu, X., et al.: Using nano-QSAR to predict the cytotoxicity of metal oxide nanoparticles. Nat. Nanotechnol. **6**, 175–178 (2011)
7. Toropova, A.P., Toropov, A.A., Veselinović, A.M., Veselinović, J.B., et al.: Nano-QSAR: model of mutagenicity of fullerene as a mathematical function of different conditions. Ecotoxicol. Environ. Saf. **124**, 32–36 (2016)
8. Ahmed, L., Rasulev, B., Turabekova, M., Leszczynska, D., Leszczynski, J.: Receptor- and ligand-based study of fullerene analogs: comprehensive computational approach including quantum-chemical, QSAR and molecular docking simulations. Org. Biomol. Chem. **11**, 5798–5808 (2013)
9. How to Remove Outliers for Machine Learning. Machine Learning Mastery. https://machinelearningmastery.com/how-to-use-statistics-to-identify-outliers-in-data/
10. NumPy. https://numpy.org/
11. matplotlib.pyplot. https://matplotlib.org/3.3.1/api/_as_gen/matplotlib.pyplot.html
12. Sandilands, D.: Bivariate analysis. In: Michalos, A.C. (eds.) Encyclopedia of Quality of Life and Well-Being Research. Springer, Dordrecht (2014). https://doi.org/10.1007/978-94-007-0753-5_222
13. LabelEncoder. scikit-learn. https://scikit-learn.org/stable/modules/generated/sklearn.preprocessing.LabelEncoder.html
14. Kenton, W.: Multicollinearity. Feb 9, 2020. Investopedia. https://www.investopedia.com/terms/m/multicollinearity.asp
15. Seaborn. https://seaborn.pydata.org/
16. Luhaniwal, V.: Feature selection using Wrapper methods in Python. Towardsdatascience. Oct 4 (2019). https://towardsdatascience.com/feature-selection-using-wrapper-methods-in-python/
17. RFE. scikit-learn. https://scikit-learn.org/stable/modules/generated/sklearn.feature_selection.RFE.html
18. Kennedy, P.: A Guide to Econometrics, 5th edn. The MIT Press, Cambridge, pp. 205–206 (2003). ISBN 0-262-61183-X.
19. scikit-learn. https://scikit-learn.org/
20. Freedman, D.A.: Statistical Models: Theory and Practice. Cambridge University Press (2009). ISBN 978-1-139-47731-4
21. Rencher, A.C., Christensen, W.F.: Chapter 10, Multivariate regression – Section 10.1, Introduction. Methods of Multivariate Analysis, Wiley Series in Probability and Statistics, 3rd edn., vol. 709, Wiley, p. 19 (2012). ISBN 978-1-118-39167-9
22. Santosa, F., Symes, W.W.: Linear inversion of band-limited reflection, seismograms. SIAM J. Sci. Stat. Comput. SIAM **7**(4), 1307–1330 (1986).https://doi.org/10.1137/0907087
23. Kennedy, P.: A Guide to Econometrics, 5th edn. The MIT Press, Cambridge. pp. 205–206 (2003). ISBN 0-262-61183-X
24. Theodoridis, S.: Pattern Recognition. Elsevier B.V, p. 203 (2008). ISBN 9780080949123

25. Zou, H., Hastie, T.: Regularization and variable selection via the elastic net. J. Roy. Stat. Soc. B **67**(2), 301–320 (2005). https://doi.org/10.1111/j.1467-9868.2005.00503.x
26. Rokach, L., Maimon, O.: Data Mining with Decision Trees: Theory and Applications. World Scientific Pub Co Inc. (2008). ISBN 978-9812771711
27. Cortes, C., Vapnik, V.N.: Support-vector networks. Mach. Learn. **20**(3), 273–297 (1995). CiteSeerX 10.1.1.15.9362.https://doi.org/10.1007/BF00994018
28. Altman, N.S.: An introduction to kernel and nearest-neighbor nonparametric regression. Am. Stat. **46**(3), 175–185 (1992). https://doi.org/10.1080/00031305.1992.10475879
29. Ho, T.K.: The random subspace method for constructing decision forests. IEEE Trans. Pattern Anal. Mach. Intell. **20**(8), 832–844 (1998). https://doi.org/10.1109/34.709601
30. Geurts, P., Ernst, D., Wehenkel, L.: Extremely randomized trees. Mach. Learn. **63**, 3–42 (2006). https://doi.org/10.1007/s10994-006-6226-1
31. Boosting Algorithms: AdaBoost, Gradient Boosting and XGBoost, 5 May 2018. https://hackernoon.com/boosting-algorithms-adaboost-gradient-boosting-and-xgboost-f74991cad38c
32. Mason, L., Baxter, J., Bartlett, P.L., Frean, M.: Boosting algorithms as gradient descent. In: Solla, S.A., Leen, T.K., Müller, K. (ed.) Advances in Neural Information Processing Systems, vol. 12, pp. 512–518. MIT Press (1999)
33. Boosting algorithm: XGBoost. Towards Data Science. (2017). https://towardsdatascience.com/boosting-algorithm-xgboost-4d9ec0207d
34. An Overview of LightGBM. Avanwyk (2018). https://towardsdatascience.com/boosting-algorithm-xgboost-4d9ec0207d
35. Hyndman, R.J., Koehler, A.B.: Another look at measures of forecast accuracy. Int. J. Forecast. **22**(4), 679–688 (2006). https://doi.org/10.1016/j.ijforecast.2006.03.001
36. Bergstra, J., Bengio, Y.: Random search for hyper-parameter optimization. J. Mach. Learn. Res. **13**, 281–305 (2012)
37. Devore, J.L.: Probability and Statistics for Engineering and the Sciences, 8th edn., Cengage Learning, Boston, pp. 508–510 (2011). ISBN 978-0-538-73352-6

Selection of the Optimal Class of Features for Recognition of the Azerbaijani Handprinted Characters

Araz R. Aliev[1,2](✉) and Elviz A. Ismayilov[1]

[1] Azerbaijan State Oil and Industry University, Baku AZ1010, Azerbaijan
{alievaraz,elviz.ismailov}@asoiu.edu.az
[2] Institute of Mathematics and Mechanics of Azerbaijan National Academy of Sciences, Baku AZ1141, Azerbaijan

Abstract. The article explores the application of methods for solving binary optimization problems on the example of the problem of selecting features for recognizing handwritten characters of the Azerbaijani. The results of various algorithms for solving the binary optimization problem are analyzed, such as sequential selection forward and backward, as well as genetic algorithms for determining more informative features from a given class.

Keywords: Feature selection · PDC features · Handprinted characters · Recognition · Support vector machine · High performance computing

1 Introduction

Interpreting the visual world is one of those things that we do unconsciously. When a person sees something, he usually does not study it before he can determine whether it is a car or a tree or his friend. However, for a computer, recognizing a person (distinguishing him from a dog or a chair, not to mention another person) is a rather difficult problem. When the human eye searches for an object, it looks for information about its size, orientation, and color. The analysis of the image is carried out first superficially, then the most characteristic details are examined in more depth.

Well-known algorithms proposed for pattern recognition use similar steps:

- definition of contours;
- focusing on areas with higher or lower information content, detailed image analysis;
- search for the specified shapes, colors, appearance based on the specified features [1–3].

Nowadays, hand-printed characters recognition of different local alphabets as Cyrillic, Arabic, Hindi, Assamese, Japon, Chinese, etc. is a required research theme. Although Azerbaijani alphabet consists of widespread Latin letters, some characters ("ə", "ü", "ö", "ğ", etc.) has local nature. Therefore, Azerbaijani hand-printed characters recognition

© The Author(s), under exclusive license to Springer Nature Switzerland AG 2021
R. A. Aliev et al. (Eds.): WCIS 2020, AISC 1323, pp. 96–102, 2021.
https://doi.org/10.1007/978-3-030-68004-6_12

demands individual approach. During the last decades by different researchers, were provided recognition systems using artificial neural networks with PDC features for handwritten words [4] and features based on pixel color range for hand-printed characters [5] in Azerbaijani. Although, recognition rate of these systems did not exceed 90%, they are very important as first approaches. Recently, researchers try to improve their works and offer new approaches and methods.

2 Thinning

Skeletonization or thinning may be a exceptionally critical preprocessing step in recognition examination such as finger print acknowledgment, optical character acknowledgment and biomedical conclusion. One major advantage of diminishing is the lessening of memory space required for putting away the fundamental basic data displayed in a design [6]. Additionally it streamlines the information structure required in pattern investigation. Most of the skeletonization calculations require iterative passes through the full picture or at slightest through each pixel of the considered question. At each pass a relatively complicated examination over each pixel's neighborhood must be performed, which tends to form the method time devouring. Our calculation comes about in another realistic Figs. 1 and 2.

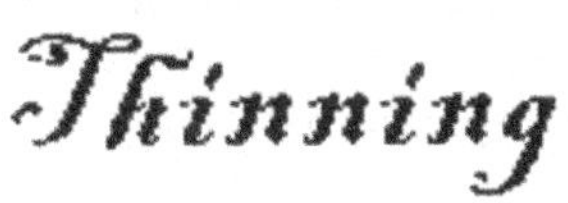

Fig. 1. Before thinning

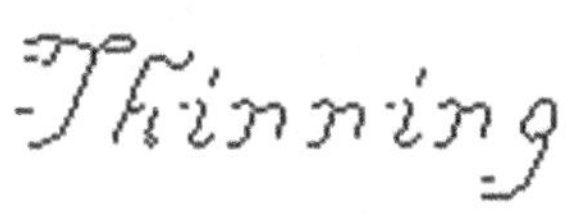

Fig. 2. After thinning

Handwritten symbols come in a variety of sizes and for the purpose of developing a recognition scheme it is imperative that all symbols be available in the same size since we need to extract an exact number of features based on their spatial distribution in a two-dimensional grid. To standardize the symbols, extra rows and columns containing only zeros are removed from all four sides of the numeral. The binary array is then fitted into a window of a size of 63 × 42. This size of the window is fixed, based on the fact that the width of the numeral is almost two-third of the height of the character (Fig. 3).

3 Extraction of Soft Features for Hand-Printed Recognition

In order to identify informative features for classification of hand-printed characters proposed different algorithms based on shape nature of figures. We used differences between structures of various characters and composed methods for calculation.

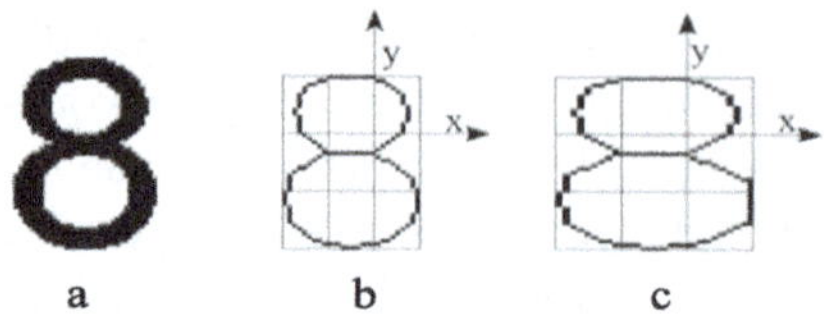

Fig. 3. a-original, b-first class features, c-second class features.

In the initial stage of process images were analyzed, characters were identified, located in the rectangles of fixed size (42 × 63) and skeletization is performed [7–9]. With the completion of these steps were calculated soft features by manipulating pixels of characters after thinning.

As soft features were used curvature degree, which is very essential in distinguishing symbols as "O" and "H", as well as identification of the closed areas, their number, location, size in the given character for classification of the symbols as "P", "B", "ə", and another features described below.

Same features use in recognition:

- Curvature degree
- Closed sector of the character
- Number of intersection points between drawn lines and character
- Number of intersection points
- Diameter of the largest closed area in character

After preprocessing of input images with characters and thinning procedure were calculated following features for hand-printed characters [10]:

Number of intersection points - different lines with given coordinates were created in rectangle where character was located. Then were defined number of intersection points between created lines and character pixels (Fig. 4).

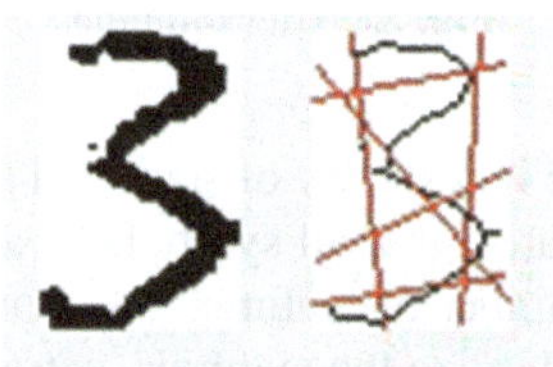

Fig. 4. Identification of intersection points of characters.

Diameter of the largest closed area in character – number of closed areas in recognized character is very important for identification, closed areas were defined by algorithm offered by author [10] and their diameters were calculated as the longest distance between the furthest points of area (Fig. 5).

Fig. 5. Location of the largest closed area in characters.

4 Results of Experiments

For the experiment, texts were collected using forms filled by 60 different authors. During experiments were compared results for three feature classes: *"soft" features offered by author, PDC features and features based on pixel colour range.*

PDC features: The Peripheral Directional Contributivity (PDC) is used as feature extraction method [11, 12]. This feature reflects well the complexity, the orientation and the relative positioning of strokes in symbols. Directional Contributivity (DC) of each point of a symbol represents 8 (or 4)-dimensional vector. Each component of a vector represents distance from this point up to a symbol border in one of possible 8 directions (or a maximum of distances on 4 directions: vertical, horizontal and two diagonal). Then values of a vector are normalized. At movement from border on one of four directions we shall meet a point in which white color passes in black. We shall name such point as the 1st order peripheral point (or the peripheral point of depth 1). If we move further, we shall meet the 2nd order peripheral point (Fig. 6).

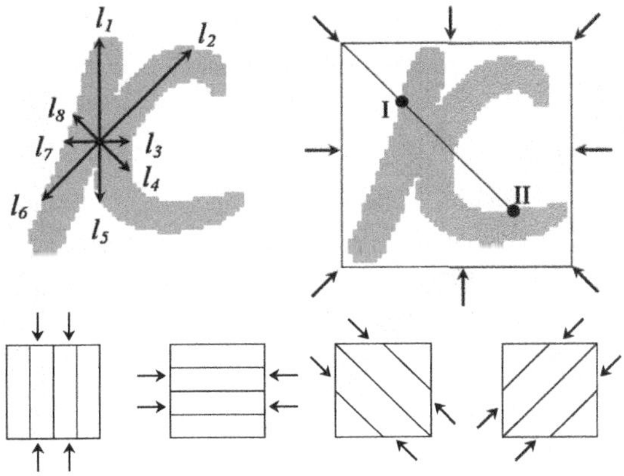

Fig. 6. Feature extraction method

As a feature, we use DC for all the 1st and 2nd order peripheral points on four directions. For reduction of dimension and sensitivity shifts, we divide the sequence of peripheral points in each direction into 8 segments and use the average DC values in each segment. So, the dimension of feature is $4 \times 4 \times 2 \times 8 = 256$ {(DC dimension) $\times$ (directions quantity) $\times$ (depth quantity) $\times$ (segments quantity)}.

Features based on pixel colour range. Were normalized to 28 × 28 and saved as grayscale in the database. One of the key tasks in recognition systems is to obtain the most informative characteristics of the studied or recognized objects. At the beginning of the study, only the pixels that make up the image were used as indicators. Thus, for each instance, you need to store 28 × 28 = 784 characters in the range from 0 to 255.

Also was investigated two classification methods: ANN (artificial neural network) with 3 layers, with 5 nodes in each layer and bootstrap SVM (support vector machine). Results of experiments were described in Table 1.

Table 1. Result table.

Type of features	Offered by author	PDC	Features identified by RGB colors of pixels
Number of features	41	256	784
SVM bootstrap	93,8%	79%	90%
ANN	79,7%	82,4%	80,3%

We can see from results that, recognition quality changes for different feature classes, so recognition quality does not depend on number of features. It is possible to observe high results with small number of features. In this situation important case is detection of more informative features for given class (Fig. 7).

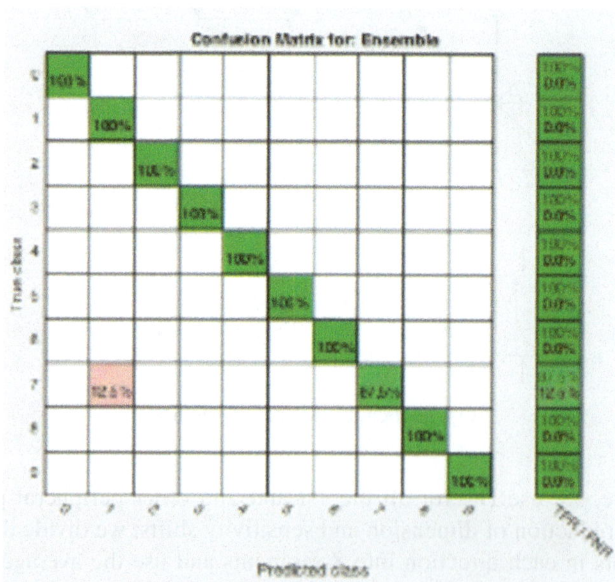

Fig. 7. Results of same handwritten digits

Also offered approach was tested for recognition of handwritten digits. Was investigated different works [13, 14]. The best result on studied papers was 98,10%, but by SVM and offered features recognition rate for handwritten digits was 98,80% (Fig. 6).

As the features grow in recognition, it becomes more difficult to choose the most informative ones. It is important to use HPC (high performance computing) when there are many features. A large number of features have also been examined as a result of experiments, and quite good results can be obtained in reducing traits using genetic algorithms for searching optimal features subset for recognition in HPC [15].

5 Conclusion

As result of numerous experiments we can say that, it is possible to apply SVM bootstrap method to recognition's problem. The main step is extraction more informative features for recognized class. We can estimate results of given paper by following items: use different feature classes, during the process of construction the system of recognition and identify more informative features by cross validation method; also we can change the class of features by adding and deleting some features from given class. Despite the fact that the number of features affects training time, it does not affect accuracy of recognition, so the most important requirement is that, feature should be informative and unique for given class.

References

1. Milewski, R., Govindaraju, V.: Binarization and cleanup of handwritten text from carbon copy medical form images. Pattern Recogn. **41**(4), 1308–1315 (2008)
2. Egmont-Petersen, M., de Ridder, D., Handels, H.: Image processing with neural networks - a review. Pattern Recogn. **35**(10), 2279–2301 (2002)
3. Jain, A.K., Duin, R.P.W., Mao, J.: Statistical pattern recognition a review. IEEE Trans. Pattern Anal. Mach. Intell. **22**(1), 4–37 (2000)
4. Aida-zade, K.R., Hasanov, J.Z.: Cursive handwritten Azerbaijani Latin text segmentation based on word vaseline. In: INISTA 2009, Trabzon, Turkey, , pp. 63–66 (2009)
5. Aida-zade, K.R., Mustafayev, E.E.: On a hierarchical handwritten forms recognition system on the basis of the neural network. In: Proceedings International Conference TAINN, Canakkale (2003)
6. Shih, F.Y. (ed.): Image Processing and Mathematical Morphology Fundamentals and Applications. CRC Press, New York, p. 416 (2009)
7. Odinets, D.N., Tatur, M.M.: Skeletization and approximation techniques applied to images of handwritten symbols based on the use of a multilevel threshold cellular automaton. Autom. Control Comput. Sci. Article **32**(3), 25–30 (1998). (in English)
8. Dhillon, A., Verma, G.K.: Convolutional neural network: a review of models, methodologies and applications to object detection. Progress Artif. Intell. Rev. Early Access 28. https://doi.org/10.1007/s13748-019-00203-0. (in English)
9. Dalton, H.: The measurement of the inequality of incomes. Econ. J. **30**(119), 348–361 (1920)
10. Mori, M., Wakahara, T., Ogura, K.: Measures for Structural and Global Shape Description in Handwritten Kanji Character Recognition (1998)
11. Hwang, Y.S., Bang, S.Y.: Recognition of unconstrained handwritten numerals by radial basis function network classifier. Pattern Recogn. Lett. **18**, 657–664 (1997)

12. Ismayilov, E., Ismayilova, N.: Fuzzy features extraction for hand-printed character/digit recognition system. In: INISTA, Italy, pp. 249–253 (2014)
13. Ouchtati, S., Redjimi, M., Bedda, M.: A set of features extraction methods for the recognition, of the isolated handwritten digits. Int. J. Comput. Commun. Eng. **3**(5), 349–355 (2014)
14. Venkateswararao, N., Srikrishna, A., Raveendrababu, B., Mohan Babu, R.: An efficient feature extraction and classification of handwritten digits using neural networks. Int. J. Comput. Sci. Eng. Appl. (IJCSEA) **1**(5), 47–56 (2011)
15. Ismayilov, E., Mammadov, R.: Parallel solution of features subset selection process for hand-printed character recognition. Azerbaijan J. High Perform. Comput. **2**(2), 170–177 (2019)

About One Feature Selection Algorithm in Pattern Recognition

A. H. Nishanov[1](✉), B. B. Akbaraliev[1], and Sh. Kh. Tajibaev[2]

[1] Tashkent University of Information Technologies named after Muhammad al-Khwarizmi, 100095 Tashkent, Republic of Uzbekistan
nishanov_akram@mail.ru, b.akbaraliev@gmail.com
[2] Karakalpak State University named after Berdakh, 100095 Nukus, Republic of Uzbekistan
tajibaevsx@gmail.com

Abstract. In this paper researched one of the problems of pattern recognition, which plays an important role in the intellectual data analysis. One of the main problem in pattern recognition is to build mathematical model classify method and algorithm. Typically, a pre-processing of the data is performed using a pre-selection option before solving this problem. One of the issues of pre-processing is normalization, and the other is the adjustment of the spatial scale of the object under study. The article deals with the second issue, the transition from a large-sized space to a more complete small-sized space. To solve the problem, the Fisher function, which is well known in the processing of research objects, is selected, and based on this function, the Fisher's type criteria are created. The Fisher's type criterion is assumed to be a goal function, on the basis of which the optimization problem is formed. In the article, methods and algorithms for solving this optimization problem are developed, compared, analyzed and tested on model examples.

Keywords: Pattern recognition · Training sample · Classification · ℓ informative systems · Fisher's type criteria · Method and algorithm

1 Introduction

As a result of the analysis of existing approaches to solving pattern recognition problems in most models of recognition algorithms: recognition of objects given in large feature space, effective solution of practical problems with small levels of complexity, building recognition models that take into account existing properties, the problem of solving taking into account the dependencies, the algorithms of selection of informative features have not been studied in depth [1–12].

Similarly, [13–21] it was pointed out that it is not always possible to form optimal ℓ informative systems for classifying objects by the proposed "Sorting" method to reduce the size of feature space relative to the criterion given in the sources.

Through this method, in the process of forming ℓ *informative systems*, the main focus is on the individual significance of each feature, but the interrelationships between the

© The Author(s), under exclusive license to Springer Nature Switzerland AG 2021
R. A. Aliev et al. (Eds.): WCIS 2020, AISC 1323, pp. 103–112, 2021.
https://doi.org/10.1007/978-3-030-68004-6_13

features are not taken into account at all. If the interrelationships between the features are significant, then the formation of *informative systems* by this method is not recommended [9, 14, 15, 20, 21].

The following is a method that takes into account both the interrelationships between features in the formation of ℓ *informative systems* and falls into the category of filter model and its implementation algorithm.

2 Methodology

2.1 Mathematical Formulation of the Problem

For this method, we also use a functionality in the form of Fisher as an informative criterion.

Hence, the problem of forming optimal ℓ *informative systems* for classifying objects will be in the form of:

$$\begin{cases} I(\lambda) = \frac{(a,\lambda)}{(b,\lambda)} \rightarrow max \\ \lambda \in \Lambda^{\ell} \end{cases} \tag{1}$$

where $a = (a_1, a_2, ..., a_N) \in R^N$ and $b = (b_1, b_2, ..., b_N) \in R^N$ are the distances between classes and within classes, respectively. The determination of a and b is given in [21].

Since the vectors a and b represent the distance between objects, the value of their components will not be negative, i.e.

$$a_i \geq 0,\ b_i \geq 0,\ \ i = \overline{1, N}.$$

If $b_i = 0$ for any i-feature, then the analysis of whether or not this feature participates in informative systems is given in detail in the source [21].

If $a_i = 0$ for any i – feature, then that feature is excluded from the feature space because it has no important information on classes.

Based on the above, we can say $a_i > 0,\ b_i > 0,\ \ i = \overline{1, N}$ without contradicting the generality.

In [11, 15, 21] a method for the formation of an optimal ℓ *informative systems* that classifies objects by solving the problem of perfection in the form (1) is proposed. With this method, only one system is formed that classifies objects for each set value of ℓ. As can be seen from the following example, the optimal ℓ information systems for each pre-given ℓ can be multiple. In particular, at $\ell = 4$, the following is appropriate:

1-case. $a = (6, 7, 5, 6, 4, 4, 2, 4, 1, 3)$ and $b = (5, 6, 6, 7.2, 5, 5.5, 5.4, 4)$ $I(\lambda_1) = I(\lambda_2) = I(\lambda_3) = \max\limits_{\lambda \in \Lambda^{\ell}} I(\lambda) = 1,$

Where $\lambda_1 = (1, 1, 1, 0, 1, 0, 0, 0)$; $\lambda_1 = (1, 1, 1, 0, 0, 0, 0, 1)$; $\lambda_3 = (1, 1, 0, 0, 1, 0, 0, 1)$.

2-case. $a' = (6, 7, 5, 7.3, 4, 4.2, 4.1, 3)$ and $b' = (5, 6, 6, 9, 5, 5.5, 5.4, 4)$ $I\left(\lambda_1'\right) = I\left(\lambda_2'\right) = I\left(\lambda_3'\right) = \max_{\lambda \in \Lambda^\ell} I(\lambda) = 1$,
where $\lambda_1' = (1, 1, 1, 0, 1, 0, 0, 0)$; $\lambda_2' = (1, 1, 1, 0, 0, 0, 0, 1)$; $\lambda_3' = (1, 1, 0, 0, 1, 0, 0, 1)$.

If the complexity of the calculations and/or the resources required to determine each feature are different, then it is important to determine all of the optimal vectors that satisfy the problem (1).

The method of forming all possible optimal ℓ *informative systems* for ℓ given on the basis of the method proposed in [15, 21] is given below.

(1) To solve the problem we use the following auxiliary vector-function:

$$\varphi(\lambda) = a(b, \lambda) - b(a, \lambda),$$
$$\varphi(\lambda) = (\varphi_1(\lambda), \varphi_2(\lambda), ..., \varphi_N(\lambda)) \in R^N.$$

This is important in determining the strategy for the formation of vector-functional information systems.

The following statement is appropriate for the vector-functional $\varphi(\lambda)$.

Statement 1 [20]. Let $\lambda, \mu \in \Lambda^\ell$ be arbitrary vectors. In this case, it is necessary and sufficient that the condition $(\varphi(\lambda), \mu) > 0$ be satisfied for $I(\mu) > I(\lambda)$.

We calculate the value of $\varphi(\lambda)$ for a $\lambda \in \Lambda^\ell$.

Since $\varphi(\lambda) \in R^N$, its components can be sorted, i.e.

$$\varphi_{j_1}(\lambda) \geq \varphi_{j2}(\lambda) \geq ... \geq \varphi_{jN} \tag{2}$$

To solve the problem (1) we can enter the gradient vector $\mu(\lambda) = (\mu_1, \mu_2, ..., \mu_N)$ determined by $\varphi(\lambda)$.

If (2) is true for $\varphi(\lambda)$, then the gradient vector $\mu(\lambda) = (\mu_1, \mu_2, ..., \mu_N)$ is defined as follows:

$$\mu_{j1}(\lambda) = \mu_{j2}(\lambda) = \ldots = \mu_{j\ell}(\lambda) = 1,$$
$$\mu_{j\ell+1}(\lambda) = \mu_{j\ell+2}(\lambda) = \ldots = \mu_{jN}(\lambda) = 0.$$

Then $\mu(\lambda) \in \Lambda^\ell$ for which the following is appropriate

$$(\varphi(\lambda), \mu(\lambda)) = \max_{\eta \in \Lambda^\ell}(\varphi(\lambda), \eta) \tag{3}$$

Property. $\forall \lambda \in \Lambda^\ell$ and $\mu(\lambda) \in \Lambda^\ell$ for $(\varphi(\lambda), \mu(\lambda)) \geq 0$.

Proof. According to (3).

$$(\varphi(\lambda), \mu(\lambda)) \geq (\varphi(\lambda), \lambda) = (a(b, \lambda) - b(a, \lambda), \lambda)$$
$$= (a, \lambda)(b, \lambda) - (b, \lambda)(a, \lambda) = 0$$

Property Has Been Proved. Based on the statement and property mentioned above, we can write the following.

Statement 2. If $I(\lambda) = I(\mu(\lambda))$, then vector $\mu(\lambda)$ is optimal solution for the problem (1).

Proof. Suppose $I(\lambda) = I(\mu(\lambda))$. Then according to Statement 1 $(\varphi(\lambda), \mu(\lambda)) = 0$. If we consider (3), then $\max_{\eta \in \Lambda^\ell}(\varphi(\lambda), \eta) = 0 \Rightarrow \forall \eta \in \Lambda^\ell$ for $(\varphi(\lambda), \eta) \leq 0 \Rightarrow$ according to Statement 1 $\Rightarrow \forall \eta \in \Lambda^\ell$ for $I(\lambda) \geq I(\eta) \Rightarrow \forall \eta \in \Lambda^\ell$ for $I(\mu(\lambda)) \geq I(\mu)$.

Statement Has Been Proved. In general, for (2), one of the following is appropriate, and the components of the vector $\mu(\lambda)$ may also vary accordingly:

$$\varphi_{j1}(\lambda) = \varphi_{j2}(\lambda) = \ldots = \varphi_{jN}(\lambda) \tag{4}$$

$$\begin{aligned} \varphi_{j1}(\lambda) \geq \varphi_{j2}(\lambda) \geq \ldots \geq \varphi_{j\ell}(\lambda) \\ > \varphi_{j\ell+1}(\lambda) \geq \ldots \geq \varphi_{jN}(\lambda) \end{aligned} \tag{5}$$

$$\begin{aligned} \varphi_{j1}(\lambda) \geq \ldots \geq \varphi_{j\ell-t}(\lambda) > \varphi_{j\ell-t+1}(\lambda) = \ldots \\ = \varphi_{j\ell+l}(\lambda) > \varphi_{j\ell+l+1}(\lambda) \geq \ldots \geq \varphi_{jN}(\lambda) \end{aligned} \tag{6}$$

where t, l – are positive integers, $\ell > t,\ \ \ell + l \leq N$.

Suppose that condition (4) is satisfied for $\varphi(\lambda)$, then the following statement is valid for (1).

Statement 3. If condition (4) is valid for the components $\varphi(\lambda)$, then $\forall \mu(\lambda) \in \Lambda^\ell$ (1) is the optimal solution of the problem.

Proof. For $\varphi(\lambda)$ condition (4) is valid. So,

$$\begin{aligned} a_1(b, \lambda) - b_1(a, \lambda) = a_2(b, \lambda) - b_2(a, \lambda) \\ = \ldots = a_N(b, \lambda) - b_N(a, \lambda) = \alpha \end{aligned}$$

Here ≥ 0. Assume the opposite, let $\alpha < 0$, then for $\forall i$

$$\begin{aligned} a_i(b, \lambda) - b_i(a, \lambda) < 0 \Leftrightarrow \\ a_i(b, \lambda) < b_i(a, \lambda) \Leftrightarrow \tfrac{a_i}{b_i} < \tfrac{(a,\lambda)}{(b,\lambda)} \end{aligned}$$

This is a contradiction, because for $\forall \lambda \in \Lambda^\ell$ the following is always true

$$\max_j \left\{ \frac{a_j}{b_j} \right\} \geq \frac{(a, \lambda)}{(b, \lambda)}.$$

Suppose that $\forall \mu(\lambda) \in \Lambda^{\ell}$. Consider the following difference with respect to this vector:

$$I(\mu(\lambda)) - I(\lambda) = \frac{(a,\mu(\lambda))}{(b,\mu(\lambda))} - \frac{(a,\lambda)}{(b,\lambda)} = \frac{(a,\lambda)+\sum\limits_{(i,j)}(a_j-a_i)}{(b,\lambda)+\sum\limits_{(i,j)}(b_j-b_i)} - \frac{(a,\lambda)}{(b,\lambda)}$$

$$= \frac{(b,\lambda)\left((a,\lambda)+\sum\limits_{(i,j)}(a_j-a_i)\right)}{(b,\lambda)(b,\mu(\lambda))} - \frac{(a,\lambda)\left((b,\lambda)+\sum\limits_{(i,j)}(b_j-b_i)\right)}{(b,\lambda)(b,\mu(\lambda))}$$

$$(b,\lambda)\sum_{(i,j)}(a_j-a_i) - (a,\lambda)\sum_{(i,j)}(b_j-b_i) = \sum_{(i,j)}\left((b,\lambda)(a_j-a_i) - (a,\lambda)(b_j-b_i)\right) =$$

$$\sum_{(i,j)}\left((a_j(b,\lambda) - b_j(a,\lambda)) - (a_i(b,\lambda) - b_i(a,\lambda))\right) = 0$$

Where (i, j) are the indices corresponding to the non-zero components of the vector λ and μ respectively.

Hence, for $\forall \mu(\lambda) \in \Lambda^{\ell}$ $I(\mu(\lambda)) = I(\lambda)$. In that case $\mu(\lambda)$ vector is the optimal solution for the problem (1).

Statement Has Been Proved. Statement 4. If condition (5) is appropriate for the components $\varphi(\lambda)$, then $\mu(\lambda)$ will be the only optimal solution to the problem (1).

Proof. Suppose that, $\mu(\lambda)$ and $\varsigma(\lambda)$ solutions for the problem (1), that is $I(\lambda) = I(\mu(\lambda)) = I(\varsigma(\lambda))$. In that case, according to Statement 1

$$(\varphi(\lambda), \mu(\lambda)) = 0,$$
$$(\varphi(\lambda), \varsigma(\lambda)) = 0$$

From these we get the following

$$(\varphi(\lambda), \mu(\lambda)) = (\varphi(\lambda), \varsigma(\lambda)) \Rightarrow (\varphi(\lambda), \mu(\lambda)) - (\varphi(\lambda), \varsigma(\lambda)) = 0$$

Furthermore, since $\mu(\lambda)$ and $\varsigma(\lambda)$ are optimal solutions, according to (3) we have

$$(\varphi(\lambda), \eta(\lambda)) = \max_{\eta\in\Lambda^{\ell}}(\varphi(\lambda), \eta)$$
$$(\varphi(\lambda), \varsigma(\lambda)) = \max_{\eta\in\Lambda^{\ell}}(\varphi(\lambda), \eta)$$

Conditionally (5) is appropriate for the $\varphi(\lambda)$ components. In that case

$$\max_{\eta\in\Lambda^{\ell}}(\varphi(\lambda), \eta) = \sum_{i=1}^{\ell}\varphi_{j_i}(\lambda)$$

It follows that $\mu(\lambda) = \varsigma(\lambda)$.

Statement Has Been Proved. We now examine the case where condition (6) is satisfied for $\varphi(\lambda)$.

Based on (6), we can divide the initial feature space into 3 groups, i.e.

$$F_{gr} = \{j_1, j_2, \ldots, j_{\ell-t}\};$$
$$S_{gr} = \{j_{\ell-t+1}, j_{\ell-t+2}, \ldots, j_{\ell+l}\};$$
$$T_{gr} = \{j_{\ell+l+1}, j_{\ell+l+2}, \ldots, j_N\}.$$

We define the vector components $\mu(\lambda)$ on the basis of $\varphi(\lambda)$ as follows:

$$\mu^i = \begin{cases} 1, & \text{if } i \in F_{gr} \\ 0 \text{ or } 1, & \text{if } i \in S_{gr} \\ 0, & \text{if } i \in T_{gr} \end{cases} \quad (7)$$

We denote ℓ *informative* $\mu(\lambda)$ vector sets for components of which (7) condition is appropriate as follows

$$\Lambda^\ell(\mu(\lambda)) = \begin{Bmatrix} \mu(\lambda) \in \Lambda^\ell : \mu^i = 1, \ i \in F_{gr}; \\ \mu^j \in \{0, 1\}, \ j \in S_{gr}; \ \mu^k = 0, \ k \in T_{gr} \end{Bmatrix}$$

Then for $\forall \mu(\lambda) \in \Lambda^\ell(\mu(\lambda))$

$$(\varphi(\lambda), \mu(\lambda)) = \max_{\eta \in \Lambda^\ell}(\varphi(\lambda), \eta) \quad (8)$$

Based on the above, the following statement is true for the problem (1).

Statement 5. If (6) for $\varphi(\lambda)$ and $I(\lambda) = I(\mu(\lambda))$ for any $\mu(\lambda) \in \Lambda^\ell(\mu(\lambda))$, then $\Lambda^\ell(\mu(\lambda))$ An arbitrary vector belonging to the set (1) will be the optimal solution for the problem.

Proof. Suppose that (6) is valid for $\varphi(\lambda)$ and that the condition $I(\lambda) = I(\mu(\lambda))$ is satisfied for any $\mu(\lambda) \in \Lambda^\ell(\mu(\lambda))$. Then according to Statement 1 $(\varphi(\lambda), \mu(\lambda)) = 0$ and according to Statement 2 it follows that $\mu(\lambda)$ is optimal for the problem (1).

Now let $\forall \varsigma(\lambda) \in \Lambda^\ell(\mu(\lambda))$, where $\varsigma(\lambda) \neq \mu(\lambda)$. Then according to (8) $(\varphi(\lambda), \varsigma(\lambda)) = \max_{\eta \in \Lambda^\ell}(\varphi(\lambda), \eta) = 0$. It follows that $I(\lambda) = I(\varsigma(\lambda))$ according to Statement 1 and according to Statement 2 $\varsigma(\lambda)$ is optimal solution for (1).

Statement Has Been Proved. It is obvious that if for $\varphi(\lambda)$ condition (6) is true, then the optimal solutions of the problem (1) equals to *card* $\left(\Lambda^\ell(\mu(\lambda))\right) = C_{t+l}^t$.

2.2 Algorithm

Here is the algorithm of the proposed method:

1-step. Input parameters: ℓ – required feature space measurement; N – the initial feature space measurement; $a, b - N-$ dimensional vectors.

2-step. Set $\lambda = \left(\underbrace{1, 1, ..., 1}_{\ell}, 0, 0, ..., 0\right)$;

3-step. Calculate the value $I(\lambda) = \frac{(a,\lambda)}{(b,\lambda)}$;
4-step. Calculate the value $\varphi(\lambda) = a(b, \lambda) - b(a, \lambda)$;
5-step. $\varphi_1(\lambda), \varphi_2(\lambda), ..., \varphi_N(\lambda)$ are arranged in descending order: if for $\varphi_i(\lambda)$ $\left(i = \overline{1, N}\right)$ (4) is true, then any $\mu(\lambda) \in \Lambda^\ell$ vector is informative and the process ends;
6-step. If (5) is true for $\mu(\lambda) \in \Lambda^\ell$ vector, if (6) is true for $\forall\mu(\lambda) \in \Lambda^\ell(\mu(\lambda))$ then calculate the value $I(\mu(\lambda))$;
7-step. Check the condition $I(\mu(\lambda)) > I(\lambda)$. If it's true, then set $\lambda = \mu(\lambda)$, $I(\lambda) = I(\mu(\lambda))$ and return to step 4. Otherwise the process ends and $\mu(\lambda)$ – optimal solution;
8-step. Output parameters:$\mu(\lambda)$, $I(\mu(\lambda))$.

It should be noted that ℓ *informative* systems formed by this method are optimal and reliable relative to a given heuristic criterion, and it is empirically proven that the number of iterations in the process of forming perfect systems is at most 2–3.

2.3 Example of Comparison

Below we consider the comparative analysis of ℓ *informative* systems formed by the "Complete Selection" and the proposed method in the above examples.

1. $N = 8$, $a = (6, 7, 5, 6, 4, 4.2, 4.1, 3)$ and $b = (5, 6, 6, 7.2, 5, 5.5, 5.4, 4)$ (Tables 1 and 2);
2. $N = 8$, $a' = (6, 7, 5, 7.3, 4, 4.2, 4.1, 3)$ and $b' = (5, 6, 6, 9, 5, 5.5, 5.4, 4)$ (Tables 3 and 4).

Table 1. Here is a detailed description of how the algorithm works at $\ell = 4$:

Feature	1	2	3	4	5	6	7	8	$I(*)$
a	6	7	5	6	4	4.2	4.1	3	
b	5	66	6	7.2	5	5.5	5.4	4	
λ	1	1	1	1	0	0	0	0	0.9917
$\varphi(\lambda)$	25.2	25.4	−23	−27.6	−23.2	−30.36	−30.38	−23.4	
$\mu(\lambda)$	1	1	1	0	1	0	0	0	1
$\varphi(\mu(\lambda))$	22	22	−22	−26.4	−22	−28.6	−28.6	−22	
$\lambda = \mu(\lambda)$	1	1	1	0	1	0	0	0	1
$\mu(\lambda)$	1	1	1	0	0	0	0	1	1
$\mu(\lambda)$	1	1	0	0	1	0	0	1	1

Table 2. Here is a detailed description of how the algorithm works at $\ell = 4$:

ℓ	Proposed	$I(*)$	Complete selection	$I(\lambda)$
1	1	1.2	1	1.2
2	1, 2	1.1818	1, 2	1.1818
3	1, 2, 8	1.0667	1, 2, 8	1.0667
4	1, 2, 3, 5; 1, 2, 3, 8; 1, 2, 4, 8	1	1, 2, 3, 5; 1, 2, 3, 8; 1, 2, 4, 8	1
5	1, 2, 3, 5, 8	0.9615	1, 2, 3, 5, 8	0.9615
6	1, 2, 3, 4, 5, 8	0.9337	1, 2, 3, 4, 5, 8	0.9337
7	1, 2, 3, 4, 5, 7	0.9096	1, 2, 3, 4, 5, 7	0.9096

Table 3. In this example we see how the algorithm works for $\ell = 5$:

Feature	1	2	3	4	5	6	7	8	$I(*)$
a	6	7	5	7.3	4	4.2	4.1	3	
b	5	66	6	9	5	5.5	5.4	4	
λ	1	1	1	1	1	0	0	0	0.9452
$\varphi(\lambda)$	39.5	41.2	−20.8	−37.4	−22.5	−30.95	−31.12	−24.2	
$\mu(\lambda)$	1	1	1	0	1	0	0	1	0.9615
$\varphi(\mu(\lambda))$	31	32	−20	−35.2	−21	−28.3	−28.4	−22	
$\lambda = \mu(\lambda)$	1	1	1	0	1	0	0	1	0.9615

Table 4. In this example we see how the algorithm works for $\ell = 5$:

ℓ	Proposed	$I(*)$	Complete selection	$I(\lambda)$
1	1	1.2	1	1.2
2	1, 2	1.1818	1, 2	1.1818
3	1, 2, 8	1.0667	1, 2, 8	1.0667
4	1, 2, 3, 5; 1, 2, 3, 8; 1, 2, 5, 8	1	1, 2, 3, 5; 1, 2, 3, 8; 1, 2, 5, 8	1
5	1, 2, 3, 5, 8	0.9615	1, 2, 3, 5, 8	0.9615
6	1, 2, 3, 5, 6, 8	0.927	1, 2, 3, 5, 6, 8	0.927
7	1, 2, 3, 5, 6, 7, 8	0.9024	1, 2, 3, 5, 6, 7, 8	0.9024

3 Conclusion

The paper discusses the issue of reducing the size of the space under study using a given study sample, which is one of the important problems in pattern recognition. To solve this problem, the objective function of the generalized Fisher-type criteria constructed by the authors using the Fisher criterion is visually formed and the problem of optimization is constructed. An excellent method and algorithm based on mathematical approaches to solving the optimization problem formed by the authors have been developed.

The theoretical and practical results obtained were used to normalize the problems of the initial processing of the data and to make the second object of the research object convenient for processing the space dimension under consideration.

The article develops methods and algorithms for solving this optimization problem, its reliability is tested experimentally in model problems of comparative analysis with the results of the full selection method.

References

1. Fazilov, Sh.Kh., Mirzaev, N.M., Mirzaeva, G.R.: Modified recognition algorithms based on the construction of models of elementary transformations. Procedia Comput. Sci. **150**, 671–678 (2019)
2. Fazilov, Sh.Kh., Mirzaev, N.M., Radjabov, S.S., Mirzaev, G.R.: Determination of representative features when building an extreme recognition algorithm. J. Phys. Conf. Ser. **1260**, 1–8 (2019)
3. Mirzaev, N.M., Khaydarova, M.Yu., Mirzaeva, G.R., Ibragimova, S.N.: Models of recognition operators defined in the space of large dimension attributes. J. Phys. Conf. Ser. **1260**, 1–8 (2019)
4. Fazilov, S., Lutfullaev, R.A., Mirzaev, N.M., Mukhamadiev, A.: Statistical approach to building a model of recognition operators under conditions of high dimensionality of a feature space. J. Phys. Conf. Ser. **1333**, 1–7 (2019)
5. Fazilov, Sh.Kh., Mirzaev, N.M., Mirzaev, G.R., Tashmetov, Sh.E.: Construction of recognition algorithms based on the two-dimensional functions. In: Communications in Computer and Information Science, Singapore, vol. 1035. pp. 474–483 (2019)
6. Fazilov, S., Mirzaev, N., Mirzaev, O., Mirzaeva, G., Ibragimova, S., Rustamov, B.: Feature extraction model in systems of face images for person identification. In: Proceedings of the 9th International Conference Advanced Computer Information Technologies (ACIT 2019), Ceske Budejovice, Czech Republic, 5–7 June 2019 (2019). https://doi.org/10.1109/ACITT.2019.8780089
7. Fazilov, S., Mirzaev, O., Saliev, E., Khaydarova, M., Ibragimova, S., Mirzaev, N.: Model of recognition algorithms for objects specified as images. In: Proceedings of the 9th International Conference Advanced Computer Information Technologies (ACIT 2019), Ceske Budejovice, Czech Republic, 5–7 June 2019 (2019). https://doi.org/10.1109/ACITT.2019.8779943
8. Mirzaev, N, Saliev, E.: Recognition algorithms based on radial functions. In: Proceedings of the 3rd Russian-Pacific Conference on Computer Technology and Applications, Vladivostok, Russky Island, Russia, 18–25 August 2018, pp. 1–6. FEFU, Vladivostok (2018)
9. Fazilov, Sh.Kh., Mirzaev, N.M., Radjabov, S.S., Mirzaev, O.N.: Determining of parameters in the construction of recognition operators in conditions of features correlations. In: Belim, S., Kononov, A., Kovalenko, Y. (eds.) Proceedings of the School-Seminar on Optimization Problems and their Applications (OPTA-SCL 2018), Omsk, Russia, 8–14 July 2018. CEUR-WS, 18 May 2018. https://ceur-ws.org/Vol-2098/paper10.pdf

10. Tukhtasinov, M.T., Mirzaev, N., Narzulloev, O.M.: Face recognition on the base of local directional patterns. In: 2016 Dynamics of Systems, Mechanisms and Machines Published, pp. 1–5 (2016)
11. Kamilov, M., Fazilov, S., Mirzaev, N., Radjabov, S.: Estimates calculations algorithms in condition of huge dimensions of features' space. In: Proceedings of the 4th International Conference on "Problems of Cybernetics and Informatics" (PCI 2012), Baku, vol. I. pp. 184–187 (2012)
12. Mirzaev, N.M.: About one model of image recognition. In: Proceedings of the First Russia and Pacific Conference on Computer Technology and Applications, Vladivostok, pp. 394–398 (2010)
13. Nishanov, A.Kh., Djurayev, G., Khasanova, M.A.: Improved algorithms for calculating evaluations in processing medical data. Compusoft Int. J. Adv. Comput. Technol. **8**(6), 3158–3165 (2019)
14. Kamilov, M., Nishanov, A., Beglerbekov, R.: Modified stages of algorithms for computing estimates in the space of informative features. Int. J. Innov. Technol. Explor. Eng. **8**(6S), 714–717 (2019)
15. Nishanov, A., Avazov, E., Akbaraliyev, B.: Partial selection method and algorithm for determining graph-based traffic routes in a real-time environment. Int. J. Innov. Technol. Explor. Eng. **8**(6S), 696–698 (2019)
16. Nishanov, A., Turakulov, K., Turakhanov, K.: A decision rule for identification of eye pathologies. Biomed. Eng. **33**(4), 178–179 (1999)
17. Nishanov, A.Kh., Turakulov, Kh.A., Turakhanov, Kh.V.: A decisive rule in classifying diseases of the visual system. Med. Tekh. (4), 16–18 (1999)
18. Nishanov, A., Ruzibaev, O., Tran, N.: Modification of decision rules 'ball Apolonia' the problem of classification. In: 2016 International Conference on Information Science and Communications Technologies, ICISCT 2016 (2016). https://doi.org/10.1109/ICISCT.2016.7777382
19. Nishanov, A.Kh., Djurayev, G., Khasanova, M.A.: Classification and feature selection in medical data preprocessing. Compusoft Int. J. Adv. Comput. Technol. **9**(6), 3725–3732 (2020)
20. Nishanov, A.Kh., Akbaraliev, B.B., Juraev, G.P., Khasanova, M.A., Maksudova, M.Kh., Umarova, Z.F.: The algorithm for selection of symptom complex of ischaemic heart diseases based on flexible search. J. Cardiovasc. Dis. Res. **11**(2), 218–223 (2020)
21. Nishanov, A.Kh., kbaraliev, B.B., Samandarov, B.S., Akhmedov, O.K., Tajibaev, Sh.Kh.: An algorithm for classification, localization and selection of informative features in the space of politypic data. Webology **17**(1), 341–364 (2020)

The Neural Network Based Decision Support System Using Aggregated Opinions of Public Observers

Elchin Aliyev, Ramin Rzayev(✉), and Khanmurad Abdullayev

Institute of Control Systems of ANAS, Vahabzadeh str. 9, AZ1141 Baku, Azerbaijan
{elchin.aliyev,khanmurad.abdullayev}@sinam.net,
raminrza@yahoo.com

Abstract. Creation of information technologies becomes especially actual when the accessible information is interpreted as a major and unique resource, a stock, a reserve, as an element of economic potential of the business entity, which should be used effectively for achievement of a competitive position in the market. Thereupon there is necessary a decision of some actual scientific problems: formation of the major concepts, development of theoretical bases of data resource management, creations of typical models of information systems for the enterprises, corporations and regional economy. In the framework of e-Governance the conception of decision making support system is offered on the base of the aggregated opinions of public observers. All estimations of these recipients are realized by fuzzy inference system and aggregated by structured neural network model. The neural network model combining two types three-layer feedforward neural networks, allows rather quickly reacting to possible changes in a conjuncture of recipients' priorities.

Keywords: Decision making support system · Recipient · Neural network · Fuzzy set · Fuzzy utility function

1 Introduction

Construction of decision-making information support system requires the integrated approach, using of achievement of different sciences, synthesis of various concepts and theories. In evolutionary aspect of development of the information society, the problems of the information were analysed in [1–3]. Theoretical bases of data resource management at various times were considered by many researchers, for example in [4–7]. However, the important stage in formation of modern understanding of information economy was introduction of concepts «an information resource» and «the integrated information management» [8, 9]. According to [9] the central moment of information management is that circumstance that the information should be considered as the major economic resource, as determinative of competitiveness, stability and safety of economic system. Economic forms of objects of information interactions, antagonisms and the tendency of development of relations between their subjects, research of the forms and substance

© The Author(s), under exclusive license to Springer Nature Switzerland AG 2021
R. A. Aliev et al. (Eds.): WCIS 2020, AISC 1323, pp. 113–120, 2021.
https://doi.org/10.1007/978-3-030-68004-6_14

of mechanisms and outer effects from implementation of the new information technology in economy requires scientific understanding. Questions on modes of representation of the scientific and technical information in various information systems, including in knowledge domain of the economic profile are insufficiently considered. Implementation and using of the modern information technology and mechanisms of application of a database and a knowledge base in the field of management of economic systems requires the further development. Necessity of new understanding is dictated to that sufficiently wide application in economy of intellectual information systems and computer networks, development of all types of information have led to origin of fundamental new aspects of the organisation of economic activities at all levels of economy. Thus, variety and heterogeneity of information resources, necessity of the scientific labour organization by search, gathering and processing of information, presence of information noise, necessity of the filtration of huge information collections, on the one hand, and social effects from wide implementation of the information technology in all line of business of legal and natural persons, on the other hand, are the basic preconditions for carrying out of researches on the course of further development of the information society.

In frameworks of e-Governance the various information systems capable to synthesise the decision more effective than the person are actively developed. Thus, used methods are based on knowledge and practice of the persons responsible for decision-making. However, procedures of replenishment of knowledge and systems of preferences demand permanent participation of public observers (recipients), because made decisions concerned the interests of them. Recipients, as a rule, are well informed on the preferences and intuitively estimate degrees of satisfaction (utility) from offered decisions. Their estimations are capable to provide purposeful activity of E-Governance and adequacy of final decision-making. In this case, the dissatisfaction of the recipient with the synthesised decision becomes a principal cause for updating of the initial data or even for revision of problem formulation. For example, there can be the new alternatives demanding introduction of new criteria. Therefore, decision-making process is dynamic, i.e. means the multiplex (iterative) decision of the assigned task for short intervals of time, and, hence, demands to use of effective information technology of support based, for example, on using fuzzy logic and multilayer neural networks.

Thus, closing system on recipients it is possible to achieve recurrence of decision-making process, to enter there a training element and, by that, to synthesise the adapted decisions. Conversationally there is an exchange of information between two subjects: informants and recipients, which permanent change of their roles, i.e. giving out and accepting the message.

2 Problem Definition

Information technologies used in economy do not provide full and qualitative satisfaction of information needs of experts in the field of the scientific management and the decision of economic problems under market uncertainty. Moreover, so far there is no rational technique of information resources management in various economic systems. Therefore, for the economic systems functioning under uncertainty and market transformation of economy conditions, one of priority directions is improvement of information mechanisms of management and development of regional socio-economic systems on the basis of involvement of the modern information technologies. Thus, for further development and democratisation of the information society it is necessary to develop the structure of recipients-closed autonomous adaptive information decision-making system integrated into e-Governance and through information portals providing public regulation. Thereupon, it is considered following problems: 1) to generate system of the base concepts defining the economic content of information mechanisms of management; 2) to develop the structure and functions of data resource management mechanism and to formulate main principles of its development; 3) to define directions of improvement of toolbox of the data resource management process; 4) to disclose and describe essence of an information product as the base resource; 5) to develop conceptual approaches to the creation of data resource management mechanism and to design of information decision-making support system at the level enterprise, inter-branch complex and region.

3 Fuzzy Utility Function

Among methods and approaches to decision-making the greatest interest is represented by those, which provide capability to take into consideration multi-criterion problems and uncertainty, and allow to carry out a choice of decisions from set of alternatives of the various nature under the criteria having different types of scales of measurement. Now those are methods of the utility theory, the analysis of hierarchy and the fuzzy sets theory, which have applied in systems of computer and information support. This choice is specified by that these methods at the maximal degree satisfy meet the requirements of universality, the account multi-criterion choice under uncertainty from discrete or continuous set of alternatives, simplicity of preparation and processing of the expert information. The method of expected utility was widespread among group of axiomatic methods of decision-making under in the risk and uncertainty conditions. The basic idea of this approach consists in calculating quantitative estimations of utility of the possible outcomes, which are consequences of decision-making process. Further, based on these estimations it is possible to choose the best outcome.

Traditional approaches in decision-making basically are based on the preferences of a group of persons responsible for development of the decision. However, in the conditions of dynamically developing economy and the competitive environment the opinions of end users (recipients) become dominating. Therefore, for calculation of quantitative estimations of outcomes it is offered to use the information on preferences of the recipients providing public regulation of decision-making process. Thus, it is necessary to mean that, first, recipients possess all completeness of the information concerning their consumer decisions, and, secondly, they are capable to range all conceivable outcomes on the basis of ability of each outcome to provide required utility. The recipient always tries to maximise the level of its satisfaction or, as economists identify, utility, which is defined as individual perception of satisfaction from any decision offered it.

From the mathematical point of view, the behavior of the recipient is expressed as choice of some point from «outcome space». On this space there is continuous real function $U(\cdot)$ named utility function, for which it is satisfied inequality $U(X) \geq U(Y)$ at preference of the outcome X over Y $(X \succ Y)$ [10]. Outcomes are quantitatively estimated by means of this function: to each outcome X_k value $U(X_k)$ is put in conformity.

Thus, an optimal outcome and optimal procedure of decision-making can be found by maximization of expected utility on all space of possible outcomes. It assumes presence of the adequate mathematical model connecting parameters of optimization (alternative scenarios of action) with variables of objective utility function. Formation of the utility function is difficult procedure. In various publications it is offered various its formalisms. However, any of them cannot be considered as universal – for each case it is necessary to create separate utility function. In the given paper for formalisation of utility function of consequences from decision-making it is offered to use the mechanism of fuzzy inference. At an estimation of outcomes in each specific case the recipient leans against on its subjective judgements and mentally forms own preference scale. Even if recipient estimates quantitative characteristics of outcomes, its estimation all the same has the "fuzzy" colouring since it, as a rule, uses qualitative categories. Therefore, believing that adopted decision is composite outcome every its component we will present in the form of a linguistic variable (LV) with values in the form of fuzzy terms-sets [11]. To unify the approach, numerical characteristics (components) of each outcome we will consider also fuzzy. Moreover, the concept utility is sooner a qualitative category than quantitative and, therefore, as utility criterion it is possible to use also the LV to possess fuzzy values. Functional dependence for utility we will define by creation of fuzzy implicative rules where LV "utility" we will consider as endogenous variable, and terms of correspondent linguistic outcomes we will consider as exogenous variables. Final problem at this stage is construction of family of fuzzy levels of utility from decision-making subject to consumer preferences. Defuzzified values of these levels we will consider as conventional values of desired utility function.

Now, let $C = \{X = (A_{1j}, A_{2j}, \ldots, A_{nj},)\}$ is the possible outcome space, where A_{kj} ($k = 1 \div n$); $j = 1, 2, \ldots$) is one of fuzzy values of LV x_k. Then, in consideration of capability of approximation the continuous functions by fuzzy implicative rules type of "If …, then …" the functional dependence of outcomes utility in terms of recipients one can represent as following set of fuzzy implicative rules:

$$\text{If}\, x_1 = A_{1i} \,\text{and}\, x_2 = A_{1j} \,\text{and}\, \ldots \,\text{and}\, x_n = A_{nr},\ \text{then}\ U = u_p \quad (1)$$

where x_k ($k = 1 \div n$) is LV indicating k-th outcome and possessing the fuzzy value (for example, A_{1i}, A_{1j}, etc.); U is LV "*utility*", which subject to outcomes possesses the fuzzy term-sets u_p ($p = 1, 2, \ldots$).

Adequacy of offered fuzzy model for utility function (see Fig. 2) depends on a correct (optimal) choice of rules (1) and the parameters of membership functions describing fuzzy values of LVs. If earlier this part of research was carried out exclusively by experts of problem domain, then in the conditions of public control over decision-making utility function should reflect system of preferences of recipients. Particularly, to realise the process of structural and parametrical optimisation of utility function (1) one can take advantage of multi-layered feedforward neural network (NN).

4 NN Model for Aggregation of Recipient Assessments

Figure 1 shows the multi-input and multi-output three-layer feedforward NN, which after appropriate training can act as necessary signal converter. The NN converts inputs in compliance with weights of its connections. In process of training, these weights are adjusted so that NN induced desirable outputs. Process starts with arbitrary choice of weights and, therefore, in the first stage outputs of NN will differ substantially from the desired values. Comparing the current outputs with corresponding desired values the NN adjusts connection weights so that to reduce this difference. Actually, it develops the basis of error backpropagation algorithm for training of the NN.

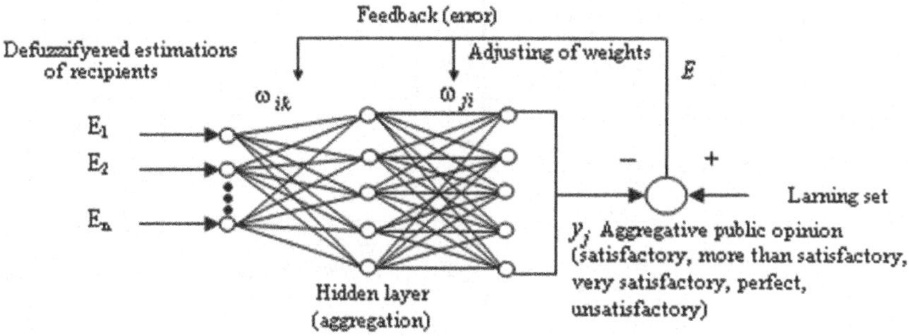

Fig. 1. Three-layer feedforward NN.

Owing to interpolation properties and ability to training the given NN is enough flexible: it can recreate, change its functions, and adequately react to the unrecorded conditions. Nevertheless, the offered NN can react only to static input-output communication in given moment of time. Therefore, it is necessary additional process, which for certain interval of time could take into consideration dynamic connections between changing in time input signals and the row of output samples. In other words, for storage of the calculated signals corresponding to available input samples additional memory nodes are necessary. However, it will make the NN very bulky and considerably will decelerate the process of its training.

For maintenance of public control over the decision-making process within the limits of neuro-fuzzy technology, it is adapted structured NN model (Fig. 2), which determines a row of the generalised estimations of recipients concerning made decisions. This model combines two types of three-layer NNs: *causal* and *instinctive.* Directly incoming from defuzzifier inputs and total outputs of NN model vary on interval [0, 1] (although in the elementary case they can be binary). In particular, assumed that gradation for output is uniform: [0, 0.2) – UNSATISFACTORY; [0.2, 0.4) – SATISFACTORY; [0.4, 0.6) – MORE THAN SATISFACTORY; [0.6, 0.8) – VERY SATISFACTORY; [0.8, 1] – PERFECT, it is possible to estimate the made decision (or alternative outcome) by corresponding value of the output of the NN model. The causal NN defines conformity of a multicomponent input from defuzzifier and induces the aggregated decision of recipients, for example, number 0.47, which indicate that the made decision is more than satisfactory.

In turn, the instinctive NN checks outputs of defuzzifier on conformity with the row of available estimations of the recipients, which have been declared in the certain period of time. For example, while defuzzifier induces "known" for the instinctive network signals the estimation of the corresponding outcome remains and it is necessary sequentially to continue to correct the decision. Therefore, the instinctive NN is significance in the

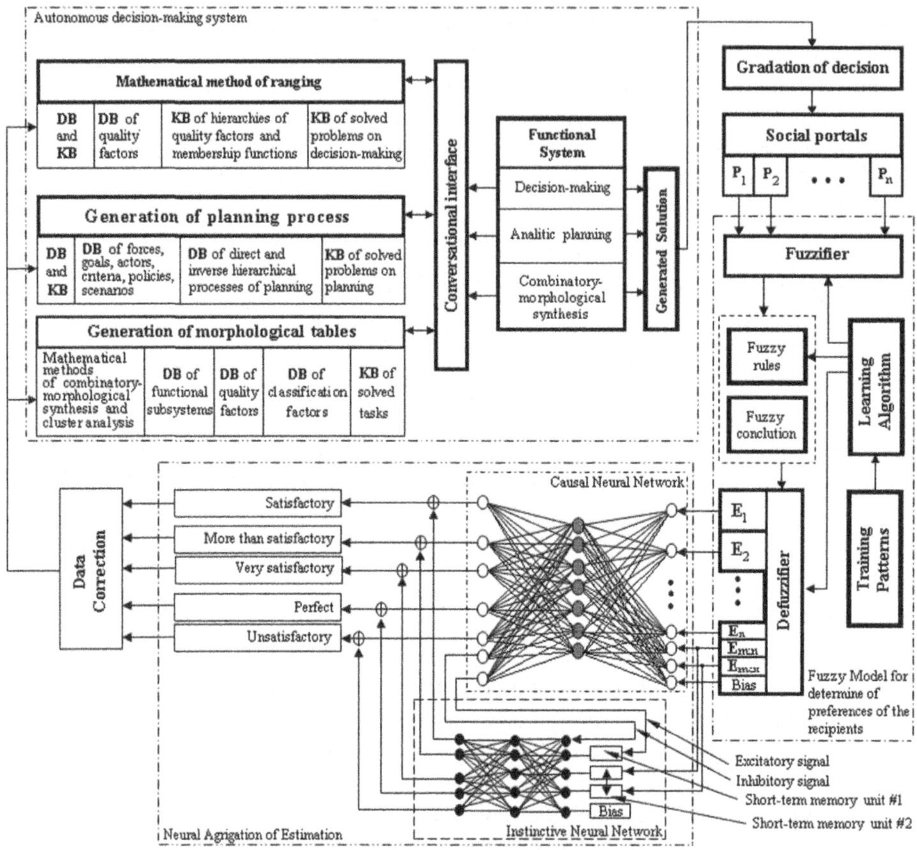

Fig. 2. The structure of autonomous adaptive decision-making system.

process of development of the comprehensive decision. The instinctive network has three short memory nodes, two of which possess mutually inhibitory connections. The short memory node keeps excitatory or inhibitory condition for the certain period.

As is shown in Fig. 2, some output neurons of causal NN are connected with the instinctive network by connections, through which excitatory and inhibitory signals are transmitted. When defuzzifier induces "known" signals, the inhibitory signal from the causal NN to the instinctive network is dumped. In this case, excitatory signal is transmitted from the causal NN through 2nd short memory node to the instinctive network and activates the instinctive network forcing it to take over control for the certain period. The instinctive NN starts and in own manner functions under the habitual (trained) scenarios: instantly processes known signals and induces the aggregated conclusion of recipients. Thus, when defuzzifier induces "unknown" signals the system exploits a causal neural network, which blocking work of the instinctive network combines opinions of recipients in the aggregated conclusion. Otherwise, if calculated by defuzzifier signals will be from the interval [$E_{\min}$, $E_{\max}$] where training samples have been chosen, then system will have closed up to the instinctive neural network, which will much fast induce the aggregated result. In essence, the causal NN is an extrapolator and the instinctive NN is an interpolator.

To realize given procedure of aggregation it would be possible to be limited only by single NN simultaneously combining in itself extrapolator and interpolator. However, in this case to obtain of the adequate aggregated estimations of recipients it would be required to use the fair-sized network (i.e. to increase number nonlinear neurons in the hidden layer) causing presence of rather more numbers of parameters (weight of synoptic connections and thresholds of nonlinear neurons in the hidden layer). For required adjustment of these parameters should to expand set of training samples considerably. Naturally, all it will essentially slow down procedure of NN training, and consequently decagon-making process will be ineffective. Within the framework of the offered approach it is possible, first, to simplify extremely configurations of neural networks by using of necessary and sufficient critical number of nonlinear neurons from their hidden layers, and, secondly, without additional timetable for training operatively to response to "known" outputs of defuzzifier. As a result, in comparison quickly it is possible to obtain the adequate aggregated estimation of recipients for the subsequent correction of synthesized decisions. For more effective training for correct behavior it is necessary to consider training samples in a combination. It will provide to NN system of aggregation (NNA) "reasonable" functioning on the basis of the minimum number of training samples. NNA can contain comparatively small number of neurons since it combines two types of neural networks. Otherwise, the total amount of combinations of input binary signals required for full-weight adjustment of NNA will compose the huge number 2^{n+2}. It will essentially slow down the training process, and finally it will reduce the information collapse of system in whole.

As is known, the basic advantage of the NN is that for its training is enough to select minimum of typical training samples. Therefore, NNA can induce adequate to various conditions and environments signals by skills acquired after typical training. Figure 3 shows the scheme of three-phase identification of NN model of signals aggregation system (SAS) providing use of training and test samples. The first is used directly for training, and the second verify the quality of training. On the first phase the training sample required for adjustment of parameters of NN model of SAS (Fig. 3) is formed.

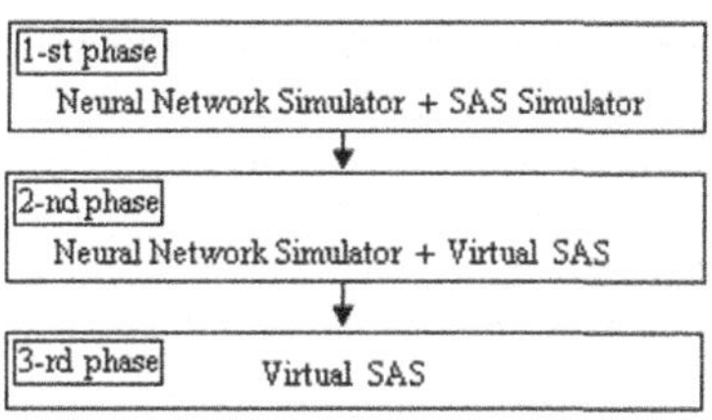

Fig. 3. Three-phase identification of NN model of SAS

Adequacy of NN model is verified by comparison of a root-mean-square error $E = (1/2)\sum_{j=1}^{s} (y_j - y_j^{\text{net}})^2$, where s is the number of carried out experiments with an admissible identification error e. If in the process of functioning, SAS inadequately responses to the unknown vector-signal from defuzzifier, then this signal together with the corresponding desirable output is added into the training sample. On the second phase the adequacy of functioning SAS is verified by test sample. In case of successful testing, already on the third phase optimum values for parameters of NN model are established. After loading the data of these parameters, the system begins to function independently. If any infringements are not observed, then the training is completed.

5 Conclusion

To provide continuous dialogue between subjects of process of development of the adapted decision, one of which use formal language and others use natural language, it is offered to use the automatic compiler on the basis of fuzzy and NN models. Fuzzy model allows to formalize subjective arguments of recipients and to give them point estimations. NN model processing this data induces the aggregated opinion of recipients. Both models are supplied by training elements that allows them to response to possible changes in the conjuncture of priorities of recipients. The offered approach to training of the NN can essentially raise the effectiveness of system as a whole since it provides training by bounded set of behavioral samples, which can be quickly and easily modified for various conditions and problems. Besides, the decomposition structure of NN model allows to train subnetwork independently.

References

1. Bell, D.: The third technological revolution and its possible socioeconomic consequences. Dissent **36**(2), 164–176 (1989)
2. Moiseyev, N.: Information society as a contemporary history stage. PC World **4**, 70–73 (1998). (in Russian)
3. Ursul, A.D.: Formations of an information society and transition to stable development. Inf. Probl. **4**, 13–20 (1997). (in Russian)
4. Hromov, G.P.: National Information Resources: Commercial Operation Problems. Nauka, Moscow (1984). (in Russian)
5. Blumendau, D.: Information and Information Service. Nauka, Moscow (1989). (in Russian)
6. Glushkov, V.M.: Bases of the Paperless Information. Nauka, Moscow (1982). (in Russian)
7. Klause, D.: Knowledge as the new paradigm of management. Probl. Theory Pract. Manag. **2**, 68–73 (1998). (in Russian)
8. Vogel, E.: Information's Managements. Berufliche Anforderungen und Konsequenz fur die Ausbildung Universitatsverlag, Konstanz (1992)
9. Herget, J., Schwuchov W.: Information's controlling. Konstanz (1995)
10. Intrilligator, M.: Mathematical Methods of Optimization and the Economics. Progress, Moscow (in Russian)
11. Zadeh, L.: The Concept of a Linguistic Variable and its Application to Approximate Reasoning. Elsevier, New York (1974)

Multi-model Virtual Analysers of Parameters of Technological Processes

A. N. Yusupbekov, U. A. Ruziev, and M. K. Shodiev(✉)

Tashkent State Technical University, 100095 Tashkent, Republic of Uzbekistan
m.k.shodiyev@gmail.com

Abstract. The article discusses program analyzers of difficultly measured or unmeasured parameters of technological processes. Mathematical models of virtual analyzers have been built using regression methods. Predictive models focused on the use of computer technology, as well as the application of the theory of indefinite sets in the creation of virtual analyzers are considered. The functions and methods of developing software analyzers are considered. Based on the results obtained, it is shown that the most suitable model is from a set of trained models based on a neural network. The ensemble methods are proposed that ensure the accuracy of forecasting virtual analyzers, allowing the selection of the optimal model for effective forecasting of product quality based on data analysis. The proposed virtual analyzer provides data preprocessing, automation of selection and maintenance, and also minimizes human intervention. The analyzer developed using machine learning methods is designed to quickly predict the state of the technological process under study and the quality of the final product.

Keywords: Software analyser · Majority voting · Density · Volumetric flow · Regulation · Predictive model · Variance · Standard deviation · Ensemble methods

1 Introduction

The computerization of control systems and a large number of measuring instruments in industry have led to the need to register and store a large amount of information. Although this information is currently circulating in low-level industrial control and monitoring systems, the hidden information contained in this data is outside the scope of these programs. However, gaining access to this hidden information is quite difficult, and its access requires advanced tools, one of which is virtual analysers, which have recently been the focus of research. Virtual analysers offer the practical benefits of a new generation at a relatively low cost to build and operate, as the methods and techniques required to develop these analysers are now readily available.

Most industrial enterprises are equipped with a huge number of sensors and primary measuring transducers. The main purpose of the latter is to deliver data for monitoring and process control. However, about two decades ago, researchers began using large amounts of data that are measured and stored in the manufacturing industry, creating

© The Author(s), under exclusive license to Springer Nature Switzerland AG 2021
R. A. Aliev et al. (Eds.): WCIS 2020, AISC 1323, pp. 121–126, 2021.
https://doi.org/10.1007/978-3-030-68004-6_15

predictive models called virtual analysers. Generally speaking, they can be thought of as simple predictive models that are trained using existing datasets and then used to predict the input data [1–5].

2 Machine Learning

Machine learning provides tools for developing and automatically maintaining virtual analysers. Developed using machine learning methods, these analysers are designed to quickly predict the state of the process under study and the quality of the final product.

A prerequisite for supervised learning is the availability of data sets S^{tr}, which consist of recorded and stored data from information systems of an automated process control system (APCS). This dataset has the following features [6]:

$$S^{tr} = \{(X^{tr}, Y^{tr})\} \tag{1}$$

where: $X^{tr} = (x_1, ...x_n)^T$, $Y^{tr} = (y_1, ...y_n)^T$, $X^{tr} \in R^{m \times n}$ – n input samples of the constructed matrix, each row of the matrix is a vector x_i, corresponding to one input sample, consisting of measured variables, i.e., $x_i = (x_{i,1}, ...x_{i,m})$.

Each sample of input x_i can be interpreted as a point of the m-dimensional input matrix of the space χ. Typically, a q-dimensional target value is given for each input sample $y_i \in R^q$. Since the purpose of the abstracting thesis is a virtual analysis of product quality control, the target area is limited to a one-dimensional output space, that is $y_i \in R^1$.

The target variable is generated by a hidden function φ mapping the entry space to the exit space:

$$\varphi = X \to Y. \tag{2}$$

Since the objective function is unknown, its detection is a function of the learning algorithm L, while the set of stored data S^{tr} and the prediction function are taken into account to find an approximation to the function.

The task of the learning algorithm L is to find an approximation to the dataset function S^{tr} and the function of prediction f^{init}:

$$f^{tr} \leftarrow L\{(S^{tr}, f^{init})\} \tag{3}$$

The object prediction model is a simple neural network regression model, either a principal component regression model with a fixed number of principal components, or a multilayer perceptron with a given number of hidden blocks and arbitrary initialized weight functions. As a result of training, a trained predictive model $f^{tr.}$ is created. The generated model displays the input samples in the output space, in other words, it approximates the predictive $y_i^{pred.} = f^{tr}(x_i)$ and objective function φ.

Assessment and systematization of learning outcomes is performed using the error function e (f, y), which allows you to calculate the difference between the target value y and the predicted value for a given input sample x. In case of noise:

$$y = \varphi(x) + e \tag{4}$$

Based on this, that the error function is a normal random variable with a mean value of zero, it follows that the optimal error function excluding the influence of random noise is the root-mean-square error:

$$e(f(X), y) = \frac{1}{n}\sum_{i=1}^{n}(f(x_i) - y_i)^2 \tag{5}$$

where n is the number of data samples.

In the presence of an error function, f^{opt} the function of minimizing $e(f(X), y)$ of the error function of the learning process is found as follows:

$$f^{opt} = \arg\min_{f} e(f(X), y) \tag{6}$$

In practice, when developing virtual analyzers, the best results are achieved by obtaining a model that works optimally with samples. However, at the stage of model development, as a rule, samples are not available, they can be obtained in the future. Due to the impossibility of calculating models on future data samples, existing information will have to be used. This separates the training data f^{tr} into training data and validation data. The prediction model is first trained on new training data and then validated against the validation data that evaluates prediction performance. Since this approach cannot be used to train new data from the training dataset, it was proposed to limit the size of the data collection by a number of factors and to solve the problem of loss of validation data.

Here is a way to solve the above problems: K-fold cross-validation. This method divides training data S^{tr} into periodic training data and test data. Here the size of the division depends on the number K. For K-fold cross validation, each function test set S^{tr} was equal $n\frac{1}{k}$ and the remaining $n\left(1 - \frac{1}{k}\right)$ samples were used as training data. As a result, we get k one-time training and testing, therefore, k-models. The main advantage of this method is that it is guaranteed that all available samples will be used to train and validate the model [7].

The availability of k-trained models based on neural networks allows you to create a multi-model virtual analyser. In this case, the predictive model of the virtual analyser consists of a set of trained models based on a neural network, and the output values of the models are generalized in the virtual analyser model. In the construction of multi-model virtual analyser is must be given special attention to the selection of models.

The generalized error of the predictive model can be divided into two components: standard deviation and variance. The fixed form for the squared error function is written as follows for various training sets:

$$\begin{aligned} E_S\left[\left(y^{test} - f_{(k)}\left(X^{test}\right)\right)^2\right] &= \left(E_S\left[y^{test} - f_{(k)}\left(X^{test}\right)\right]\right)^2 + E_S\left[\left(y^{test} - E_S\left[f_{(k)}\left(X^{test}\right)\right]\right)^2\right] \\ &= \text{bias}\left(f_{(k)}\right)^2 + \text{variance}\left(f_{(k)}\right) \\ f_{(k)} &\leftarrow L\left(S_i \in S^{tr}, f^{init}\right). \end{aligned} \tag{7}$$

The expected value E_S in expression (7) is plotted from the set of continuous-scale random training data set S_i. If the $f(X^{mecm}) - X^{mecm}$ and y^{mecm} inputs are the correct target for the squared error data, then the predictive model data is accepted.

The separation of errors is carried out by dividing the generalized error into two components that balance each other. The bias component describes the expected error for all trained $f_{(k)}$ predictive models. Finding the optimal predictive model means finding a balance between bias and variance error.

Application of ensemble model methods, which include a set of models involved in the development of a virtual analyzer model. There are two important tasks for ensemble modeling. The first is to combine sets of different ensemble methods, and the second is to combine models to produce final predictions, or to combine model predictions.

The ensemble method consists of p set of models:

$$F = \{f_{(f)}\}_{k=1}^{p} \quad (8)$$

At the same time, it is limited to predictive combinations and does not imply model-level combinations. There are several approaches to combining the predictive level. Since there is a test data set consisting of the X^{value} of the input and the y^{value} of the target variables:

$$S^{value} = \{X^{value}, y^{value}\} = \{(x_j, y_j)\}_{j=1}^{n^{3value}} \quad (9)$$

where n^{value} is the number of test samples.

When choosing models, the following approaches are used:

- choosing the best model: $f = \arg\min_i e(f_{(i)}(X^{value}), y^{value})$;
- mean value of forecasting models: $f = \frac{1}{p}\sum_i f_{(i)}$;
- weight function comparison $f = \sum_i \omega_i f_{(i)}$ c $\sum_i \omega_i = 1, \forall i : \omega_i > 0$;
- sorting the optimal function g by value $S^{value} : f = g(F)$, where $g : F \to y$.

Analysis shows that the mean value method is preferable to the best model selection method. On this basis, it was determined that the square error of the ensemble $(f(x) - y)^2$ decreases with an increase in the volume of the ensemble data:

$$\left(f(x) - y^2\right) = \frac{1}{p}\left(\frac{1}{p}\left(\sum_{k=1}^{p}(y - f_{(k)}(x))^2\right)\right) \quad (10)$$

The above equation shows that the approximate ensemble error can be reduced simply by adding new models to the ensemble. This seems to be a very satisfactory result, but unfortunately this equation only applies if the squared error of the prediction of the ensemble members is statistically independent, which becomes more difficult to achieve as the ensemble grows gradually.

In decomposition, covariance-displacement-variation is extended to ensembles in the form of decomposition. For a combination of values, the variance components can then be divided into the variance and covariance components. In the case of the average combination of predictions, the predictor model f is achieved by replacing the average combination $\bar{f} = \frac{1}{p}\sum_k f_{(k)}$, which leads to the following form of the variance term:

$$\begin{aligned} variance(\bar{f}) &= E\left[\left(\bar{f} - E\left[\bar{f}\right]\right)^2\right] = E\left[\left(\frac{1}{p}\sum_k f_{(k)} - E\left[\frac{1}{p}\sum_k f_{(k)}\right]\right)^2\right] \\ &= \frac{1}{p^2}E\left[\sum_k \sum_{l \neq k} \left(f_{(k)} - E\left[f_{(k)}\right]\right)\left(f_{(l)} - E\left[f_{(l)}\right]\right)\right] + \frac{1}{p^2}E\left[\sum_k \left(f_{(l)} - E\left[f_{(k)}\right]\right)^2\right] \\ &= \frac{1}{p^2}covariance(\bar{f}) + \frac{1}{p^2}variance(\bar{f}). \end{aligned} \quad (11)$$

In the above equation, E is calculated for a different set of training data and for different initializations of the prediction model. It is important to note that since the covariance portion of Eq. 11 can be negative, it can reduce the generalization error of the model. It also shows that the squared error of the ensemble depends on the variance, bias, and interdependence of the ensemble members, that is, the covariance.

As shown in Eq. 11, it can theoretically be seen that ensemble methods reduce the squared prediction error by decreasing the variance component of the theoretical bias expansion.

Based on the work carried out, an updated methodology for the development of a virtual analyzer is proposed, which provides most of the development of a virtual analyzer, especially at the stage of data preprocessing, automation of selection and maintenance, and also minimizes human intervention. The stages of development, operation and maintenance of the virtual analyzer are shown in Fig. 1.

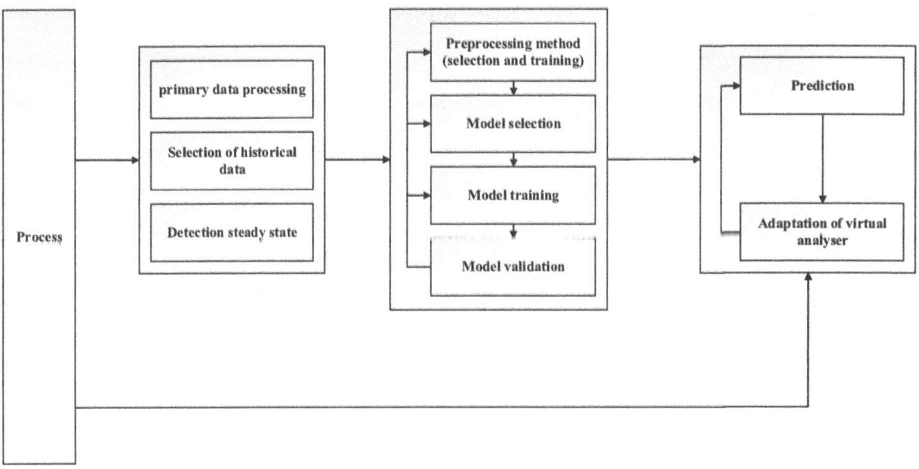

Fig. 1. Scheme of development and operation of a virtual analyzer.

3 Conclusion

Based on the results obtained, it is shown that the most appropriate is the ensemble model. The ensemble model does not benefit participants who make the same predictions, i.e.

the result of an ensemble that combines ten models that predict the same values is equal to the performance of one model.

The proposed virtual analyzer can be used in various areas of the manufacturing industry to control the quality of products. Virtual analyzers are the most affordable tool for monitoring process health.

References

1. Zade, L.A.: Ponyatie lingvisticheskoy peremennoy i ego primenenie k prinyatiyu priblijenny'h resheniy, Mir, Moscow (1976)
2. Takagi, T., Sugeno, M.: Fuzzy identification of systems and its applications to modeling and control. IEEE Trans. Syst. Man Cybern. **15**(116), 116–132 (1985)
3. Glazunov, V.F., Tarary'kin, S.V., Spichkov, Y.P.: O racional'nom postroenii datchika natyajeniya tkani v potochny'h liniyah tekstil'noy promy'shlennosti. Izv. vuzov. Tehnologiya tekstil'noy promy'shlennosti, vol. 1, pp. 78–82 (1981)
4. Glazunov, V.F., Kulenko, M.S.: Reducirovanie matematicheskih modeley petleobrazovateley v sisteme upravleniya natyajeniem tkani. Izv. vuzov. Tehnologiya tekstil'noy promy'shlennosti, vol. 6, pp. 101–104 (2008)
5. Ruziev, U.A., Shodiev, M.K.: Virtual analyser of quality of liquid products. Spec. Issue Int. Sci. Tech. J. «Chem. Technol. Control Manag.» 4–5 (2018)
6. Yusupbekov, A.N., Ruziev, U.A., Shodiev, M.K.: Identifikacionny'y podhod razrabotki virtual'ny'h analizatorov i ih algoritmov. Jurnal Himicheskaya tehnologiya. Kontrol' i upravlenie, vol. 2, pp. 74–80 (2019)
7. Ruziev, U.A., Shodiev, M.K.: Identification approach to creating software analyzers and their algorithms. Int. J. Adv. Res. Sci. Eng. Technol. **6**(7), 10167–10172 (2019). www.ijarset.com

Fuzzy Assessment of Technological Risks in the Main Oil Pipeline

H. G. Ismayilova[1], Z. I. Farzalizada[1](✉), J. R. Damirova[1], Y. Z. Alakbarov[2], and M. E. Shahlarli[3]

[1] Azerbaijan State Oil and Industry University, Azadlıq 20, AZ1010 Baku, Azerbaijan
ismayilova.hecer@bk.ru, zivar.farzalizade@mail.ru, cavida.damirova@asoiu.edu.az

[2] Geotechnical Problems of Oil and Gas and Chemistry SRI, Dilara Aliyeva, AZ1010 Baku, Azerbaijan
alakbarovsanz@mail.ru

[3] Oil and Gas Research and Design Institute, H. Zardabi 88a, AZ1012 Baku, Azerbaijan
mansursahlarli1994@gmail.com

Abstract. It is known that the amount of expected environmental damage with adverse effects is defined as the sum of the individual expected losses for different components of the environment. Estimation of damage caused by oil spills during tragic oil pipeline accidents is based on the existing normative documents and methodological guidance. The basis for developing priority measures to increase the safety of the main pipelines, which are considered a potential source of threat, is the risk assessment in the pipeline. Accidents are mainly associated with oil spills, which have a negative impact on the environment. Therefore, any section of the main oil pipelines should be assessed with certain risk parameters as it can cause accident.

Keywords: Risk · Oil leakage · Corrosion · Environment · Oil pipeline · Fuzzy analysis

1 Introduction

Accident risk assessment of main oil pipelines involves [1–3]:

- Forecasting of oil leakage rates in the pipeline and estimation of the volume of spilled and lost oil (technological risk);
- Evaluation of the consequences of oil spills for various components of the natural environment;
- Carrying out the hazardous rank of the pipeline route based on risk assessment and prioritizing safety measures for ranging.

The oil pipeline leak prediction is generally based on the combined effects of the groups shown in Table 1. The relative impact of each group of factors on the intensity

© The Author(s), under exclusive license to Springer Nature Switzerland AG 2021
R. A. Aliev et al. (Eds.): WCIS 2020, AISC 1323, pp. 127–131, 2021.
https://doi.org/10.1007/978-3-030-68004-6_16

of accident changes in the pipelines is considered with weight coefficient. Each of these groups, based on statistical data on leakage statistics (influencing factors) in relation to accidents on main oil pipelines, has a specific weighting coefficient ("contribution"). It should be noted that weight coefficient are preliminary and can be determined as a rule, taking into account the opinion of specialists (experts).

Table 1. Total statistics on accident stopping frequencies by factors (weight coefficients)

№	Name of groups	Shares of group factors
1	External anthropogenic influences (F1)	0,20
2	Corrosion (F2)	0,10
3	Quality of pipe production (F3)	0,05
4	Quality of construction and installation works (F4)	0,10
5	Constructive technological factors (F5)	0,10
6	Natural effects (F6)	0,10
7	Exploitation factors (F7)	0,05
8	Defects in pipe and weld seams (F8)	0,30

2 Solution and Discussion of the Problem

Taking into account of weight coefficients in order the assesment of effective parameters, the following ranking was held on a 5-point scale (Table 2). Then, the results of evaluating the factors that caused oil leakage from the pipelines to the environment by individual experts were examined. The research (opinions) conducted by three different experts on these factors are shown in Table 3. As can be seen in Table 3, linguistic assessments ("very low", "low", "medium", "high" and "very high") given by experts differ from each other and from the scale adopted so far. So if expert 1 exaggerates factor F_8 (defects in pipes and welds) as the most influential factor, experts 2 and 3 focus on factors F_7 (operational factors) and F_2 (corrosion), respectively. If expert 1 chooses F_3 (pipe production quality) and F_7 (exploitation factors) as the least effective factor, expert 2 accepts factors F_3, F_4, F_5, F_6 and 3rd expert accepts factors F_4 and F_5.

Aggregation of expert opinions on the above 8 factors (Table 3) was carried out for oil leaks. Initially, the linguistic values shown in Table 3 were converted to fuzzy numbers on the scale in Fig. 1. The final vector was determined based on the aggregation of fuzzy numerical vectors.

Table 2. Rank classification according to the weight coefficient of the factors by groups

Weight coefficient change range	Linguistic estimates	Rank
Less than 0,07	"very low"	1
0,7–0,13	"low"	2
0,14–0,20	"medium"	3
0,21–0,27	"high"	4
More than 0,28	"very high"	5

The values of these final vectors according to factors are given in Table 3. These prices are the general price vector of the group of experts.

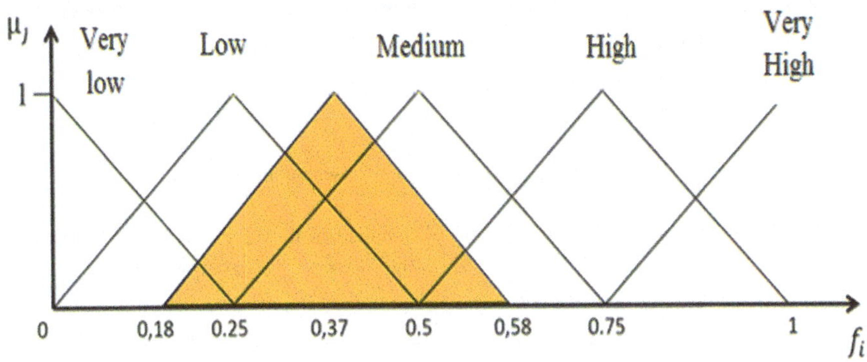

Fig. 1. Fuzzy numbers scale

Then, according to the method [4], the distance to the highest value vector (in other words, to the "very high" risk factor) of this vector and the sum vectors of the individual experts were calculated. The distances obtained are shown in Table 4. As can be seen from Table 4, these distances were 5.90, 6.39, 5.92 and 5.90, respectively, for the generalized vector of individual experts and the expert group. If we take into account that the distance between the "very weak" indicator and the "very high" indicator is 13.75 and convert the above-mentioned distances to the scale [0, 1], we get the following: 42, 46, 43, 43%.

These estimates indicate the percentage of accident risks far from the "very high" risk, taking into account the opinion of individual experts and expert groups.

Table 3. Possible risk factors for infield pipeline failures

Values of factors and risk factors	1st expert score	2nd expert score	3rd expert score	Fuzzy numerical final estimate
External anthropogenic influences – F_1	medium	medium	low	0.17, 0.42, 0.75
Corrosion – F_2	low	low	very high	0.25, 0.5, 0.67
Quality of pipe production – F_3	very low	very low	low	0, 0.17, 0.42
Quality of construction and installation works– F_4	low	very low	very low	0, 0.17, 0.42
Constructive-technological factors – F_5	low	very low	very low	0, 0.08, 0.33
Natural influence – F_6	high	very low	medium	0.25, 0.41, 0.67
Exploitation factors – F_7	very low	very high	low	0.25, 0.42, 0.58
Defects in pipes and weld seams – F8	very high	high	high	0.58, 0.75, 1

Table 4. Results of fuzzy assessment

Values of factors and risk factors	Sum values of fuzzy vectors	Distance from "very high" risk	Diverge "too high" risk %
1st expert score	0.18, 0.37, 0.59	5.90	42
2nd expert score	0.18, 0.31, 0.53	6.39	46
3rd expert score	0.18, 0.37, 0.59	5.92	43
Fuzzy numerical final estimate	0.18, 0.36, 0.58	5.90	43

3 Conclusion

Finally, the final assessment was based on the final vector calculated for individual expert meetings and the expert group. The results were [0.18, 0.37, 0.59], [0.18, 0.31, 0.53], [0.18, 0.37, 0.59] and [0.18, 0.36, 0.58], respectively (Table 4). As can be seen, these indicators do not differ much from each other and can be assessed as "weak" by 0.52° and "moderate" by 0.48° on the scale [0,1].

Therefore, despite the diversity of initial assessments by experts, the level of technological risk resulting from oil spills for oil pipelines is not very high, mainly can be assessed "weak" risks.

References

1. Mazur, I.I., Ivansov, O.M., Moldavanov, O.I.: Constructive reliability and environmental safety of pipelines. Nedra (1990)
2. Kent Muhibauer, W.: Pipeline Risk Managament Manual. Guff Publishing Company (1992)
3. Khodzhaeva, G.K.: Risk Assessment of the accidents in oil pipeline systems in terms of geodynamic processes. In: Monograph, p. 132, Nizhnevartovsk (2016)
4. Babanli, M.B.: Synthesis of new materials by using fuzzy big date concepts. Procedia Comput. Sci. **120**, 104–111 (2017). https://doi.org/10.1016/J.PROCS.2017.11.216

The Efficiency of the FSV Model in an Intelligent Search Engine

Bahodir Boltayevich Muminov and Ulugbek Bakhrom Bekmurodov(✉)

Tashkent University of Information Technologies named after Muhammad al-Khwarizmi, Street A.Temur 108-a, Tashkent, Uzbekistan
Ulugbek_bekmurodov@mail.Ru

Abstract. It is enough to have 3 basic stages of the modules in the SPD of a diversified corporate network: (F) - the method of submitting the request, i.e. the method of forming the expression of the information needs of the system user (S) - the function of the correspondence of the electronic resource to the request the degree of compliance with the request and the found electronic resource; (V) - method of presenting electronic resources. Combining these three stages for models, methods and software modules of the AML, we call FSV technology (FSV platform, FSV Framework). FSV technology is an instrumental software platform based on a client-server architecture, integration and modification of models, methods and algorithms of AML in the information environment of corporate networks. The following architecture has been developed for the FSV technology proposed for the search index in data retrieval systems.

Keywords: Intellectual search · FSV technology · Information resource · Software · Process · Intellectual analysis · Data processing · Fuzzy and stochastic information media · Search module · Program module · Data search and processing.ont · Style · Formatting

1 Introduction

In the world to meet the needs of the population in electronic resources, information is given great attention to the creation of intelligent data search systems in corporate networks of information and resource centers. Technologies of using information resources, the number of which is increasing every day, are one of the important issues of today. In this regard, including special attention is paid to improving the technology of searching and processing data based on semantic and fuzzy rules in corporate networks, database design and logical knowledge [1, 2].

The world conducts research work aimed at creating data retrieval systems, database design and semantic knowledge, creating and improving software modules and algorithms for mining, search, data storagein corporate networks. In this regard, the important tasks are the design, development of models, methods for searching and processing data, creating software for building data retrieval systems in fuzzy and stochastic information environments of corporate networks.

© The Author(s), under exclusive license to Springer Nature Switzerland AG 2021
R. A. Aliev et al. (Eds.): WCIS 2020, AISC 1323, pp. 132–140, 2021.
https://doi.org/10.1007/978-3-030-68004-6_17

2 Main Chapter

Modern technologies that do not replace many years of experience and highly qualified specialists, are designed to significantly improve the quality and efficiency of its work. This was discussed at the international forum held in Tashkent and dedicated to modern intellectual systems.

The event was attended by representatives of the Ministry for the Development of Information Technologies and Communications of the Republic of Uzbekistan, the Center for the Development and Implementation of Computer and Information Technologies "Uzinfocom", domestic and foreign IT specialists, developers of intelligent systems and technologies, manufacturers of server and network equipment and software.

As it was noted, the modern IT industry all over the world is going through a new stage of its development. The catalyst for this process is the latest technology that can solve problems that have traditionally been considered creative. The forum discussed trends in the development of modern intellectual systems, their implementation in various sectors of the economy, the health sector [3, 5].

Studies have shown that issues related to information retrieval, the structure and properties of data, the formation and processing of queries, the creation of a semantic core, the identification of electronic resources rating, the development of models and methods for intelligent search, the presentation and processing of data in fuzzy and stochastic information environments of corporate information biotech systems are not well understood.

A "search module" is a subsystem that, upon request, provides the most current ordered set of electronic resources, equipped with a user-friendly interface, including a database and statistical data.

The search module in the corporate network consists of two main elements that are independent of each other: an indexing element and a search element. The user sees only the capabilities of the search item and uses it. The indexing element is used to create the appropriate form for the effective presentation of information when searching for the necessary data in the data retrieval system (DPS).

Suppose that in the database are consistently given and are in the segment V = [1, n], with DosID identifiers with the set V, which has n electronic resources.

Proposition 1. A quote (or reference) is a pair of ordinal electronic resources. $(i, j) \in V^2$. i - incoming link and j outbound link electronic resource.

In the set V between the electronic resources from all the links we form the E set, the graph directed to the links can be called reference graphs $G = (V, E)$.

Proposition 2. For example, $G = (V, E)$ in this case the graph is the final set of vertices, $E = V \cdot V$ and $i \in V$, denote $I(i)$ the incoming set of links, and $O(i)$ the outgoing set of links denote as, then:

$$I(i) = \{e \in E | e = (i, j), j \in V\},$$
$$O(i) = \{e \in E | e = (i, j), j \in V\}.$$

Proposition 3. If $O(i)$ and $I(i)$ do not exist and, then they are denoted as in this case $\{\varnothing\}$, the rating of the electronic resource $i \in V$ equals 0.

Proposition 4. No electronic resource can provide its inbound links $I(i)$ and outbound links $O(i)$ to these links. The rating of incoming links I_r and outgoing links O_r is determined as follows:

$$I_r = \frac{\sum_{i=1}^{|I(i)|} Doc_r^i}{|I(i)| \sum_{count(I(i))} I(i)}, \quad O_r = \frac{|O(i)|}{\sum_{count(O(i))} O(i)} \tag{1}$$

Here $count(I(i))$ - is the number of incoming, $count(O(i))$ - the number of all outgoing links.

Using the expression given above (1), we determine the method of preliminary rating of an electronic resource using (2) an expression.

$$D_r = (I_r + O_r)/2 \tag{2}$$

(2) the expression allows you to define electronic resources with a high rating and select a set that is recognized as relevant based on the identification of the rating of electronic resources in corporate information library systems (ICBS).

Suppose that $q = \{q_i\}$ a lot of requests and given to q_i the corresponding to a variety of electronic resources

$$R^i = \left\{r_j^i\right\}, j = 1 \ldots M_d, j = 1 \ldots N_q \tag{3}$$

Here M_d – is the number of electronic resources viewed in the system, is the number of requests in the system. The method of logical-semantic search for a couple of queries, we will write as follows.

$$f(q_i, q_j) = \langle q_i | q_j \rangle = \frac{\left|R^j \cap R^i\right|}{\left|R^j\right|} \tag{4}$$

Here, $\langle a \mid b \rangle$ - denotes the relation and similarity of a with respect to b. $\left|R^j \cap R^i\right| \leq \left|R^j\right|$ - the condition must be relevant.

Rule 1. If $\langle q_i \mid q_j \rangle = 0$, then in this case q_i - is not semantically related to q_j, and is also considered not similar to each other.
Rule 2. If $\langle q_i \mid q_j \rangle = 1$, then in this case it is considered that q_i, q_j is completely semantically related and the similarity 1 is considered (q_i exactly similar).
Rule 3. If the expression (4) is equal $i = j$, then in this case $\langle q_i \mid q_j \rangle = 1$ it is considered semantically related, the similarity is 1 (it is exactly the same).
Rule 4. If $0 \leq \langle q_i | q_j \rangle \leq 1$, then q_i in this case, it is considered that semantically equal q_j and the similarity is calculated by the expression (4).

The method of logical-semantic search (4) for all queries, we write as follows:

$$f(q_1, q_2, \ldots, q_{N_q}) = \sum_{i=1}^{N_q} \sum_{j=1}^{N_q} \langle q_i | q_j \rangle = \sum_{i=1}^{N_q} \sum_{j=1}^{N_q} \frac{\left|R^j \cap R^i\right|}{\left|R^j\right|} \tag{5}$$

here $\sum$ is the union of the elements of the set. (5) the expression allows you to create a base of semantic knowledge in the data retrieval system (DPS).

Creating models, methods, and software for searching and processing data in fuzzy information environments, using the theory of fuzzy sets and using the linguistic variable model, defines the method for developing fuzzy queries in the data retrieval system (PDS) as follows:

1) β – defined linguistic variable;
2) $T(\beta)$ – we define a linguistically similar variable of a fuzzy set of terms;
3) X – for fuzzy sets of terms, we determine the set of values;
4) $G(\beta)$ – defined in the set of fuzzy terms, characterized by logical and modifier outputs;
5) μ_x – X defines the membership function of the set.

When designing a parametric membership function, the set μ_x – X is the linguistic variable of the corresponding object characteristic, the parametric design.

β-method of the membership function is an important element in the intellectualization capabilities of the data retrieval system (PDS). μ_x – in the membership function, enter 4 parameters and give the following definition:

Proposition 5. If *a, b, c, d* parameters are taken from X - the set of values, then in this case the function is considered to be $\mu_x(a,\ b,\ c,\ d)$ the membership function of the set of values - X.

Proposition 6. In the membership function $\mu_x(a,\ b,\ c,\ d)$, the parameters *a, b, c, d* among themselves must be in ascending order. So one of the conditions $a < b$ or $b < c$ or $c < d$ must be constantly met.

Proposition 7. $T(\beta)$ – the parameters of the membership function for each term of the set of fuzzy terms corresponding to linguistic variables $T(\beta)$, at least one distinguishing parameter must be introduced.

It is recommended to use the choice of classes for the selection of membership functions corresponding to the linguistic variables of a fuzzy set in intelligent search and processing in a fuzzy information environment.

When intelligently searching for data in a fuzzy information environment of corporate information library systems, the main problem is the creation and management of the semantic core. We give the definition of a semantic core for intelligent search and data processing.

In a fuzzy information environment:

The semantic core is a set of relationships recognized as relevant, relevant in content and essence to normalized terms and electronic inquiry.

Based on this definition, 3 methods have been developed for creating the semantic core of data retrieval systems in a fuzzy information environment. These include:

- Automatic construction of the semantic core as a result of search and processing.
- Creating a semantic core by processing the citation of links, data in electronic resources.
- Creating a semantic core with the help of expert groups. This method is carried out in version 2.

a) Creation of a semantic core by an expert by creating queries corresponding to an electronic resource.
b) Expert processing of the existing semantic core, cancellation, modification, creation by it of the relationship of the request with the electronic resource, if necessary.

A model of the semantic core, based on the implementation of the control of the semantic core and 3 algorithms of methods for creating a semantic core, has been developed.

To determine the relevance of the semantic core model 4 properties have been added to the semantic core. If in property 1 there is a semantic link of the electronic resource and the query is 1, otherwise it takes the value 0, property 2 activates N - the total number of links, property 3 - m - the number of link-references.

In a fuzzy information environment for intelligent search and data processing in the development of models and rules for fuzzy search, the core of the knowledge base will form a fuzzy query q in a variety of fuzzy queries q* (6) to form the rule condition as follows.

$$f : q \to q^* = \{q_1^*, q_2^*, \ldots, q_n^*\} \tag{6}$$

in this case, $q_i^* - q_{i1}^*$ the term consists of a fuzzy pair of terms q_{i2}^* and can be defined as $\langle q_{i1}^* | q_{i2}^* \rangle$, and, $i = 1 \ldots n, n \leq 2 \cdot |q_{i2}^*|, |q_{i2}^*| - q_{i2}^*$ is the number of fuzzy terms.

Conditions of the rule for the logical action "And" and q_1^* a pair of terms (6) we write as follows.

$$(q_1^* = q) \text{ И } (q_{11}^* = d) \text{ И } (q_{12}^* = t^*) \text{ И } (a_1 = \mu_{1,x_1}(q_{12}^*)) \tag{7}$$

here $(q_1^* = q)$ - q_1^* - q – is the pairing or non-pairing of the terms that make up the query, $(q_{11}^* = d)$ - the semantic relation of the electronic resource d and term, q_{11}^*, $(q_{12}^* = t^*)$ - the presence or absence in the set of fuzzy terms t^* of a database of fuzzy data q_{12}^*, $(a = \mu_x(q_{12}^*))$ - $(q_{11}^* = d)$, the weight a and the membership or non-membership of the values of membership functions of fuzzy terms q_{12}^* (Fig. 1).

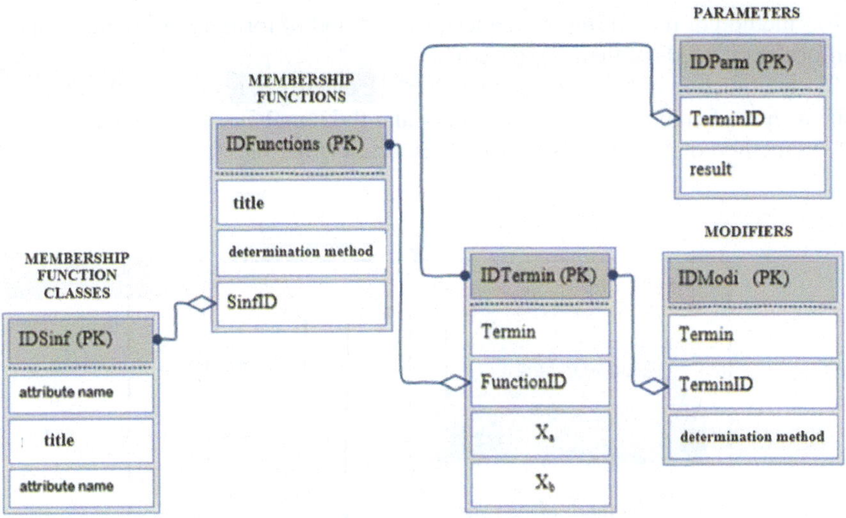

Fig. 1. IDEF1x core model knowledge base.

Creation and implementation of software for searching and processing data in the corporate network, is devoted to the development of technology and architecture of FSV, composite modules and models, for the effective operation of search modules in corporate information-library environments it is necessary to have several basic steps (Fig. 2).

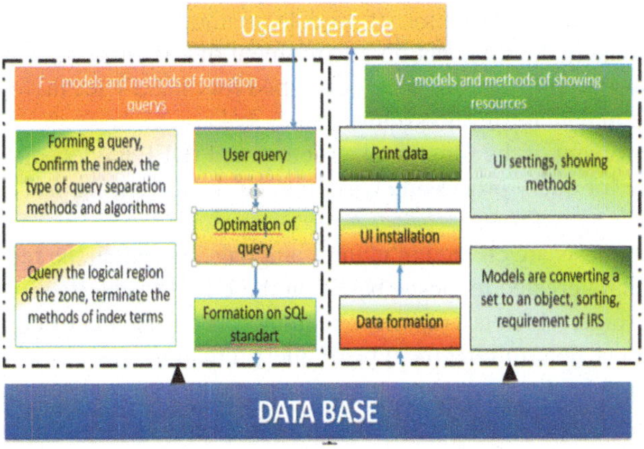

Fig. 2. Architecture FSV technology

It is enough to have 3 basic stages of the modules of the search box in the data retrieval system of a multi-profile corporate network. These include:

1. (F) - method of presenting the request, i.e. method of forming the expression of the information needs of the user system.
2. (S) - the function of the correspondence of the electronic resource to the request, i.e. the degree of compliance with the query and the found electronic resource.
3. (V) - method of presenting electronic resources (Fig. 3).

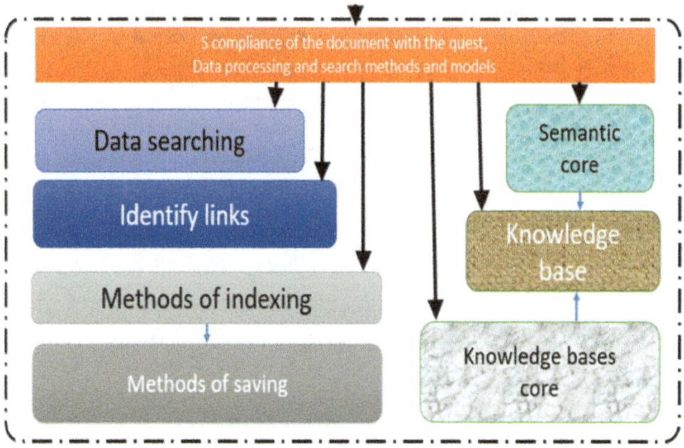

Fig. 3. Architecture FSV technology.

Combining these three stages for models, methods and software modules of the AML, we call FSV technology (FSV platform, FSV Framework).

FSV technology is an instrumental software platform based on a client-server architecture, integration and modification of models, methods and algorithms for searching and processing data in the information environment of corporate networks.

The following architecture has been developed for the FSV technology proposed for the search index in data retrieval systems.

In the development of the functional structure and IDEF models of the instrumental software module for query processing corporate networks request processing is divided into 2 stages.

At stage I, requests are created in natural language, based on the request processing methods, artificially created requests, based on the UI (User interface) methods.

The second stage, which is a recurrent sequence, is carried out by clearing additional and auxiliary terms, determining the type of query, creating indices of logical and fuzzy terms, of the zonal field.

When creating IDEF models and functional structure of a software search and presentation software module.

In the corporate network, an IDEF0 model was developed based on the methods, data retrieval system models in fuzzy and stochastic environments, traditional search, and S-ER matching process. Presentation V includes the processes of receiving a set of electronic resources, choosing a presentation method, determining the type of UI, providing constraints and filters and representing the set based on them, determining and controlling calls in the user's electronic resources.

FSV technology is integrated into such systems as: "ARMAT++" - an automated system of information resource centers, "Mobile-Library" - mobile software on library resources, "vLibrary" - a virtual library of written cultural heritage of Turkic peoples of Central Asia of the XIX century (Kazakhstan).

For the experiment in MARC21 format, you can use the ISO2709 data exchange standard when exchanging data in corporate information library systems. Therefore, more than 50,000 titles of various electronic resources and full texts were chosen. For comparison and analysis results, such parameters as "Speed", "Number of found data", "Average relevance value" were selected and their results are given in the following figures.

The results of the experiment show that high speed provides the user with, firstly, superfluous information, secondly, it is very difficult to satisfy the information needs, i.e. need to analyze a lot of data. Although the search takes more time, but finding the necessary data indicates the effectiveness of search systems, since the analysis and limitation of redundant information saves 25%, sometimes up to 50% of the time of users.

3 Conclusion Remarks

As a result of the scientific and practical work "Models, methods of searching and processing data in fuzzy and stochastic information environments" the following results were obtained and the following main conclusions were made:

For the evaluation of data retrieval systems, compliance with the criteria of accuracy and integrity, reducing search time, a logical-semantic search model was developed, which increased the search efficiency by 5%. It can serve to study the issues of indexing sets of predefined ERs and queries, building their semantic links, optimizing search speed based on criteria and mathematical methods for searching and processing data in multi-disciplinary and multi-stage information resources [9].

The algorithm and software module for determining the relationship between the parameters of parametric membership functions, the choice of fuzzy terms corresponding to the classes of membership functions in the methods of intelligent search and data processing have been developed. Improved processing of fuzzy requests by 22%.

Developed algorithms, software module and 3 ways on creation of a semantic core for data retrieval systems in a fuzzy information environment of corporate information library systems, as well as increased the amount of data matching the query, on 10%.

The architecture, technologies FSV-software for creating a system for searching and processing data in corporate information library systems has been created. The results of introducing FSV technology into the ARMAT++ corporate information library system make it possible to increase the effectiveness of the "Number of found data" by 7%, the Average relevance value by 10%, and the Search by 5%.

References

1. Anagnostopoulos, A., Andrei, Z., Kunal, P.: Effective search engine model. In: Proceedings of CIKM, pp. 208–217 (2006)

2. Arlitsch, K., Obrien, P., Rossmann, B.: Managing search engine optimization. J. Libr. Adm. **53**(2–3), 177–188 (2013)
3. Azevedo, C., Iacob, M., Almeida, J., Van, S., Pires, L., Guizzardi, G.: ArchiMate. Inf. Syst. **54**, 235–262 (2015)
4. Bhat, N.A., Ganaie, Sh.A.: E-resources. Y.S. Parmar University of Horticulture and Forestry. Collection Building **35**(1), 16–21 (2016)
5. Brock, E.O.: Declarative semantics of transactions in ORM. Inf. Syst. **60**, 85–94 (2016)
6. Cai, F., Rijke, M.: Learning from homologous queries and semantically query auto completion. Inf. Process. Manag. **52**(4), 628–643 (2016)
7. Cheang, B., Chu, S.: OR/MS journals evaluation based on a refined. Scientometrics A **100**(2), 339–361 (2014)
8. Muminov, B.B.: Methods of information retrieval and cash algorithms. ITPA, pp. 194–196 (2014)
9. Muminov, B.B.: The model of logical semantic searching. In: Proceedings of the Uzbek-Japan Symposium on Ecotechnologies, Tashkent, pp. 72–77 (2016)

Neural-Network Based Approach to Mechanism of Liquid Extraction

A. A. Hasanov(✉)

Azerbaijan State Oil and Industry University, Azadlig Avenue 20, Nasimi, Baku, Azerbaijan
alakbar48-48@mail.ru

Abstract. The process of mass transfer is studied in the system of liquid-liquid in this article. It was found that in such systems, mass transfer can occur not only through diffusion, but also because of movements called spontaneous surfaces as a result of convection (SSC). It turned out that the explanation of the SSC involving effects of Marongoni and instability inconsistent with experience and requires other physical and chemical prerequisites. In this regard, we have put forward the following hypotheses:

- at the time of occurrence of the SSC, the defining resistance is the free energy of the interfacial surface, while the diffusion resistance in the phases has almost no effect;
- SSC is manifested in conditions of excess of the free energy of the molecules of the extract over the free energy of the surface.

Keywords: Mass transfer diffusion · Parameter · Convection · Turbulence · Concentration · Effect

1 Introduction

During extraction in a liquid-liquid system, mass transfer can occur not only by diffusion, but also by spontaneously passing movements [1, 2] called, spontaneous surface convection (SSC). In the ongoing studies on interfacial turbulence, the main idea is to take into account the macrokinetic parameters [3].

At the same time, two theoretical concepts that explain the occurrence of SSC by the uneven mass transfer at various points on the surface [4] are most common. According to the first concept, the cause of the occurrence of the SSC is that local changes in the interfacial tension σ caused by concentration or temperature gradient forces give rise to forces spreading to the adjacent fluid layers, setting them in motion (due to the Marangoni effect). According to the second concept, the cause of occurrence of SSC is related with density fluctuations, changes can be caused by both concentration changes and thermal effects of mass transfer. Consider each of the proposed concepts.

Lewis and Pratt, who first reported spontaneous surface convection, explained its presence by the release of heat when a solute (extract) was transferred from a drop

© The Author(s), under exclusive license to Springer Nature Switzerland AG 2021
R. A. Aliev et al. (Eds.): WCIS 2020, AISC 1323, pp. 141–146, 2021.
https://doi.org/10.1007/978-3-030-68004-6_18

(raffiant) to the surrounding liquid (extractant). The release of heat, according to the authors, leads to a local change in σ and to a debelance of forces in the drop. This explanation was based on experiments with uranyl nitrate, the transition of which from organic solvent to water did not cause SSC.

The SSC appeared during the movement to the opposite direction. However, if a similar explanation was sufficient, then according it, the SSC would have arisen only with mass transfer in one direction. The described phenomenon can be observed with mass transfer in both directions. The appearance of the SSC during mass transfer from the organic phase to the aqueous and in the opposite direction was found in the extraction systems acetic acid - ethyl acetate - water, acetone - ethyl acetate – water and formic acid - butyl acetate – water [5]. In contrast to Lewis and Pratt, we assume that SSC is associated with local σ differences as a result of local changes in the concentration at the phase boundary. According to this hypothesis, SSC occurs only where σ is sufficiently lowered by the extract.

However, such extraction systems are known, which the SSC is found, despite the fact that the extract has a slight surface activity. Also known systems where SSC is absent, although the extract significantly changes σ. The process of mass transfer in the system liquid - liquid is studied.

It has been found that in such systems mass transfer can occur not only by diffusion, but also due to movements called spontaneous surface convection (SSC). It was found that the explanation of the SSC with the attraction of the Marangoni effects and instability does not correspond to the experience and requires different physical and chemical prerequisites. In this regard, we made the following hypotheses:

- at the time of occurrence of the SSC, the defining resistance is the free energy of the interfacial surface, while the diffusion resistance in the phases has almost no effect.
- SSC is manifested in conditions of excess of the free energy of the molecules of the extract over the free energy of the surface.

2 Experimental Part

We carried out a liquid in the system butyl glycol - water - isopropyl ether, acetic acid - water - isopropyl ether. The experimental data we obtained (Table 1) proves that a change in $\frac{d\sigma}{dC}$ with the same C_0 or a change in C_0 $\frac{d\sigma}{dC}$ is not associated with the occurrence of SSC. We also believe that the occurrence of the SSC may be due to local changes in σ, caused by local changes in concentration and temperature. However, this view is contrary to experience.

In Table 2 below, we cited experiments with a surfactant-inactive substance. Experiments show that the appearance of the SSC can only be attributed to the release of heat (ΔE) during extraction.

However, in the system with acetone, where SSC is observed, the value (ΔE) is less, for example, in the system in acetic acid, where SSC is not detected. Thus, the effect of local variation of σ during mass transfer cannot be the only cause of SSC.

It was evidenced by experiments in which the SSC was seen when one liquid was dissolved in another.

Table 1. About the role of effect in the process of occurrence of the SSC

Extraction system	C_0, mol/l	$\frac{d\sigma}{dc}$	$C_0 \frac{d\sigma}{dC}$, erg/sm^2	Presence or absence
butyl glycol from water to isopropyl ether	0,22	7,7	2,34	–
	0,49	6,28	3,75	–
	0,88	4,81	4,53	+
acetic acid from water to isopropyl ether	0,30	5,2	1,6	–
	0,95	3,7	3,5	–
	1,30	3,3	4,3	+
acetone from water to isopropyl ether	0,10	77	7,7	–
	0,20	29	5,8	+
	0,40	22	8,8	+

Despite the fact that the use of the Marongoni effect as the primary cause of the SSC is incorrect, it became widespread based on the said Stirling and Scriven effect [4], in which the laws of boundary dynamics (Marangoni effect) and hydrodynamics (equations of Navies - Stokes) and diffusive transport (Fick's laws).

It should be noted that the theory of hydrodynamic instability in some cases does not even withstand a qualitative test. The separate and simultaneous movement of phenol and propionic acid between carbon tetrachloride and water described in [6]. In accordance with, the system under study should be unstable only when passing from the aqueous phase to the organic one. In fact, when transmitting in the opposite direction, the system is not only unstable, but also the intensity of the SSC even increases. Numerous experimental facts that contradict the theory are also given in [7].

The discrepancy that the mass transfer coefficients in the SSC area does not depend on the intensity of the mixing of the phases, the "external" hydrodynamic conditions of the theory to experimental data, in our opinion, in fact the proposes that the basic premises of the theory are physically unjustified.

By using Neural Network based approach and Matlab environment the role of effect in the process of occurrence of the SSC is learned. The basic parameters of the created neural network are (Fig. 1): 3 input, 1 output, generated rule-7, Squash factor of 2.0 (Only find clusters that are far from each other), Accept ratio 0.8 (Only accept data points with a strong potential for being cluster centers), Reject ratio of 0.7 (Reject data points if they do not have a strong potential for being cluster centers), epoch = 500, optimization method –hybrid.

For example, the authors of the theory [4], in addition to attracting the Marangoni effect, use the laws of "external" hydrodynamics to describe the SSC. However, the cited experiments were showed.

According to the second concept of the causes of the occurrence of the SSC is often indicated [8, 9] to the occurrence of density fluctuations during mass transfer, leading to a deviation of fluid motion. This explanation is criticized in. Thus, Olander and Reddy [10], who studied the extraction of HNO_3 in the system of water - solutions of tributyl

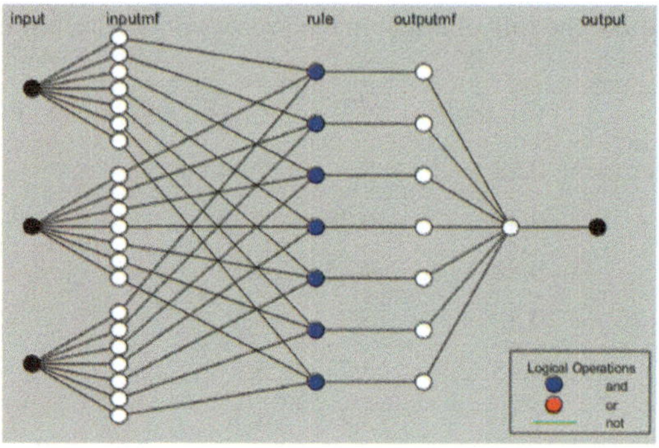

Fig. 1. Structure of the neural network

phosphate in n-hexane and CCl_4, came to the conclusion that density instability cannot be the cause of the formation of SSC. In this system, the aqueous phase is above the organic one and, when HNO_3 passes from water to the organic phase, the density of the latter on the surface of phase separation would be higher than in the rest of the phase volume. Since the organic phase is located below the interface, the mass transfer coefficient of the aqueous phase into the organic phase would increase if the cause of the formation of the SSC was instability of density, which does not consent with experience.The following result (Fig. 2) is determined by using the given initial data fragment in Table 1:

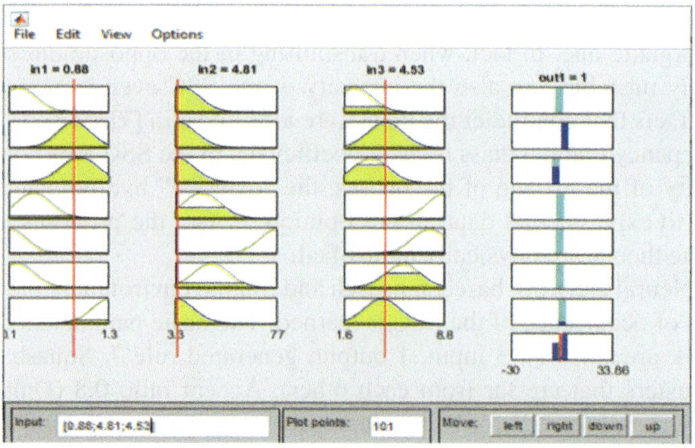

Fig. 2. The role of effect in the process of occurrence of the SSC

Analysis of the considered case, as well as other numerous examples (Table 2) indicates the impossibility of explaining the experimental facts from the standpoint of this hypothesis.

Table 2. SSC density instability

Extractional	Density			Transfer direction	Presence or absence of SPK
	Raffinate	Extractant	Extract		
butyl alcohol - water - isopropyl ether	1000,0	*725,0*	*820*	↓	+
acetic acid - water - isopropyl ether	1000,0	*725,0*	*890*	↓	+
acetone - water - isopropyl ether	1000,0	*725,0*	*740*	↓	+

3 Conclusions and Disscussion

Thus, the explanation of the SSC with the attraction of Marangoni effects and instability does not correspond to the experience and requires the introduction of other physicochemical prerequisites. In connection with the above, we made the following hypotheses:

- at the time of occurrence of the SSC, the defining resistance is the free energy of the interfacial surface, while the diffusion resistances in the phases have almost no effect;
- SSC will appear in conditions of excess of the free energy of the molecules of the extract M over the free energy of the interfacial surface occupied by 1 mol of extract N, under conditions of M/N ≥ 1.

Explicitly, the expression for M/N can be represented as follows:

$$\frac{M}{H} = \frac{(1 - \frac{C_2}{C_1 \kappa_\rho})}{\sigma s_0} \geq 1$$

Where C_1, C_2 - the concentration of the extract, respectively, in the phases of the raffinate and extractant at the time of termination of the SSC; K_ρ - distribution coefficient; R is the gas constant, T is the absolute temperature; S_0 - is the surface occupied in the condensed monomolecular layer of 1 mol of the extract.

Noting the features of the proposed criterion, we note that:

- Inequality M/N allows, based on the physicochemical parameters of the extraction systems (σ, S_0, T, $K\rho$), to calculate the conditions for the occurrence of SSC in them;
- In contrast to the considered theoretical concepts of the emergence of the SSC, the above inequality is derived without regard to the Marangoni effects. It has a fundamental importance, since it allows one to predict the occurrence of SSC, even while extracting surface - inactive substances;

- the criterion M/N $\geq$ 1 allows to explain from a unified standpoint the occurrence of the SSC not only during the "pure" extraction mass transfer, but also during the extraction complicated by a chemical reaction;
- the criterion M/N $\geq$ 1 allows for practically important cases to predict the concentration of the extract that can cause SSC in the presence of surface - active substances in the system.

In conclusion, we note that the validity of the proposed criterion is confirmed on a large amount of experimental material:

1) Extraction in the presence of surfactants;
2) during the extraction, complicated by a chemical reaction.

References

1. Treybal, R.: Liquid Extraction. Publishing House Chemistry (1996)
2. Braunstein, B.I., Shchegolev, V.V.: Hydrodynamics, mass and heat exchange in column apparatuses. Chemistry, Leningrad (1988)
3. Kafarov, V.V.: Base of mass transfer. Issue High School (1962)
4. Ditnerskiy, Y.I.: Processes and apparatus of chemical technology. Chemistry (1995)
5. Honji, A., Ishii, H., Mochizuki, A., Saho, N., Tsuyama, T., Isogami, H.: Treatment method of organic compounds included in waste water, a treatment apparatus of organic compounds included in waste water, a treatment system of organic compounds included in waste water and a bitumen collecting system. United States Patent 201001176039, Japan, 15 July 2010
6. Liquid Extraction (Thexry and Practic): Under reduction of A.Q. Kasatkin – M. State scientific technical issue of chemical literature (1958)
7. Qian, D., McLaughlin, J.B., Sankaranayanan, K., Sundaresan, S., Kontomaris, K.: Simulation of bubble breakup dynamics in homogeneous turbulence. Chem. Eng. Comm. **193**, 1038 (2006)
8. Altunbas, A., Kelbaliyev, G., Ceylan, K.: Eddy diffusivity of particles in turbulent flow in rough cannels. J. Aerosol. Sci. **33**(7), 1075 (2002)
9. Martinez-Bazan, C., Montanes, J., Lasheras, J.C.: On the break-up of air bubble injected into fully developed turbulent flow. Part 1: break-up frequency. J. Fliud Mech. **401**, 157 (1999)
10. Sliencher, C.A.: Maximum stable drop size in turbulent flow. AIChE J. **8**(4), 471 (2004)

Predicting the Performance of Marital Gas Pipelines with Different Coatings

Iravanli Konul Babek and Habibov Ibrahim Abulfas(✉)

Azerbaijan State Oil and Industry University, Baku, Azerbaijan
konul.iravanli@gmail.com, h.ibo@mail.ru

Abstract. Protection of gas pipelines from corrosive effects is an urgent problem of the oil industry in Azerbaijan. Currently, in order to protect gas pipelines from corrosive effects, various methods are used, among which the insulation of pipeline surfaces using polymer coatings holds a special place. In this paper, the goal is to develop a program for predicting reliability indicators (water permeability and adhesive strength of the coating) of gas pipelines.

To construct a γ-dimensional dynamic model of the gas pipeline functioning, we used the technique of reconstructing the attractor of the attached system according to the one-dimensional "projection" of the phase trajectory of the system using the influence of the phase space into the reduced space of dimension $m < n$. The choice of the dimension of the nesting space was made taking into account the false nearest neighbors and false neighbors on the attractor folds.

The quality of the coating was: Oil insulating bitumen (BNI), a composition based on bitumen filled with phenol-formaldehyde oligomer and butyl rubber (BNI + FFO + BK), also a composition based on bitumen filled with epoxy resin and copper nanoparticles (BNI + ED-20 + Cu). It was found that the use of nano-coating differs sharply in controlled properties from other

Keywords: Reconstruction of the attractor · Nesting space · False neighbors in proximity to the distance and on the folds of the attractor · Phase trajectory and its one-dimensional "projection" in the form of scalar time series · Converging neighbors

1 Introduction

Analysis of the results of studies to assess the state of gas pipelines showed that they all have basically exhausted their resources. Gas pipelines pass sections with different reliefs: flat steppe terrain, cut by beams, ditches, ravines, with rivers and canals through which there are air passages. In some places, they cross arable land, sown with various crops, orchards, vineyards and bushes, the roots of which in some cases damage the insulating coating of gas pipelines.

The soils on these routes are diverse and are represented by steps, sands, dense loams, heavy clays, gravel, which are moistened and saline to some extent.

To assess the state of the gas pipelines, studies were carried out according to the following criteria: the potential difference of pipes along the pits and the specific resistance

© The Author(s), under exclusive license to Springer Nature Switzerland AG 2021
R. A. Aliev et al. (Eds.): WCIS 2020, AISC 1323, pp. 147–154, 2021.
https://doi.org/10.1007/978-3-030-68004-6_19

of the soil. In addition, ultrasonic measurement of pipe wall thickness and assessment of the state of corrosion protection of gas pipelines were carried out.

Ultrasonic measurement of the pipe wall thickness was carried out by two methods: in places where there was no corrosion at 4 points along the pipe perimeter and in places where corrosion was present in several measurements in places of the deepest caverns and ulcers.

Analysis of soil resistivity along 3 routes of the gas pipeline showed that it varies within 0.63 … 44.96 × 10. Seasonal climate fluctuations vary from −5 to 46 °C.

In this work, based on the time series presented in Fig. 1 and Fig. 2, changes in the indicators of water permeability (B) and adhesion strength (A) for coatings of three types (BNI; BNI + FFS + BK; nanonized BNI + ED-20) are constructed for each indicator and type coverage (by months) dynamic model displays, which are used to obtain predictive estimates of indicators.

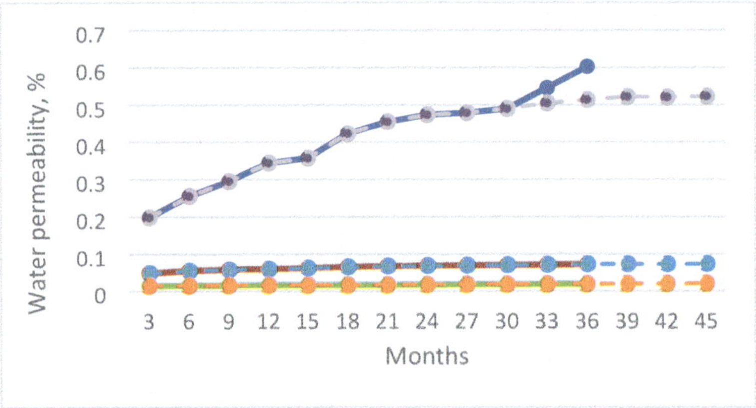

Fig. 1. Changes in coating water permeability, in %

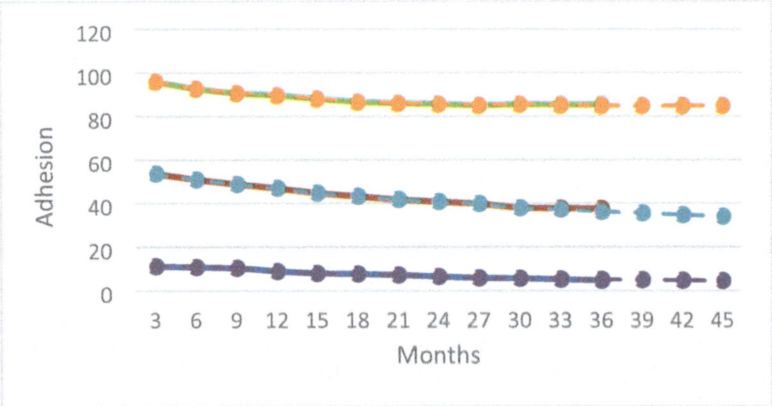

Fig. 2. Changes in the adhesive strength of the coating, in MPa

2 Construction of a Dynamic Model for a Time Series

Consider a dynamical system (DS) represented in the form of a system of ordinary differential equations of the first order

$$\frac{d\vec{x}}{dt} = \vec{F}(\vec{x}), \quad \vec{x} \in R^n \tag{1}$$

where n - is determined by the number of equations; R^n - Euclidean space of dimension. The phase portrait of the dynamical system (1) can be reconstructed from the time series a_i (which is a discretized time Δt dependence of one of the "observed" variables DS $x_j(t) : a_i = a(i\Delta t) = x_j(i\Delta t), i = 1, \ldots, N)$ if the same series a_i is used as the missing coordinates of the state vector, taken with a certain delay [1].

The reconstruction technique consists in the choice of delay τ, dimension of the nesting space m and in the formation of an array of vectors $\vec{z}(i\Delta t)$ [2–4].

The behavior of the vector will be modeled using a mapping $\vec{z}_{i+1} = G(\vec{z}_i)$. This mapping is reduced to a simpler form: $x_j(t_{i+1}) = x_{j+1}(t_i)$ $\left(j = \overline{1, m-1}\right)$, $x_{m,}(t_{i+1}) = G(x_1(t_1), x_2(t_i), \ldots, x_m(t_i))$ which is equivalent to the mapping.

$$a_{i+m} = G(a_i, a_{i+1}, \ldots, a_{i+m-1}) \tag{2}$$

We choose the function G in the form of an order polynomial K:

$$G(x_1, x_2, \ldots, x_m) = \sum_{l_1, l_2, \ldots, l_m=0}^{K} c_{l_1, l_2, \ldots, l_m} \prod_{i=1}^{m} x_i^{l_i}, \sum_{i=1}^{m} l_i \le K \tag{3}$$

To build the model (1.3), you need to determine the values of the M coefficients: $M = (m+K)!/(m!K!)$. These values are chosen where for the equation so that the root-mean-square error of the approximation is minimal (least squares method): $\varepsilon_{train}^2 = \frac{1}{N_{train}-m} \sum_{i=1}^{N_{train}-m} \left[a_{i+m} - G(a_i, a_{i+1}, \ldots a_{i+m-1})\right]^2 \to \min$.

The degree K of polynomial (3) is selected by enumeration K from 1 to $K_{\max}$ according to the criterion

$$\max(\varepsilon_{train}, \varepsilon_{test}) \to \min_{K=1, K_{\max}} \tag{4}$$

3 Choosing the Dimension of the Embedding Space

When approximating the mapping (2) at some point $(a_i, a_{i+1}, \ldots a_{i+m-1})$, the values at the nearest neighboring points $\vec{z}_i$ and $G(\vec{z}_i)$ are used. It is assumed that for close pairs of points $\{\vec{z}_0, \vec{z}_k\}$ in the reconstructed space, their preimages in the original space $\{\vec{x}_0, \vec{x}_k\}$ are also close. However, the properties of the mapping $\vec{z}(t) = \Lambda_m(\vec{x}(t)) = \{a(t), a(t+\tau), \ldots, a(t+(m-1)\tau)\}$ are such that this assumption is sometimes violated. False neighbors appear in the reconstruction. They can be conditionally divided into two types [5], which are sometimes called False Nearest Neighbors (FNN) and False

Neighbors on Folds (FNF). They correspond to two types of distortions: FNN occurs when reconstruction window lengths are too small, and FNF - when too large.

For attractors of small dimensions, there is usually an interval of values of the parameters m and τ, at which the FNN neighbors have already disappeared, and the FNF neighbors have not yet appeared. For attractors of large dimension, both types of false neighbors may even coexist. It is because of this effect that reconstruction may become generally inapplicable for dynamic methods of time series processing [5].

There is a simple method for an integer upper bound for the attractor dimension with allowance for false nearest neighbors (FNN), the so-called "false neighbor method" [4]. It is based on checking the property that the phase trajectory reconstructed in a space of sufficient dimension should not have self-intersections.

One of the variants of the algorithm is as follows [4]. For a trial dimension for each vector $\vec{z}_k = (s_k, s_{k+1}, \ldots s_{k+m-1}), (k = m+1, \ldots, N-m+1)$ reconstructed according to model usually look for the nearest neighbor $\vec{z}^{NN}(k) = d(s_k^{NN}, s_{k+1}^{NN}, \ldots s_{k+m-1}^{NN})$.

Increasing m by 1, we determine which of the neighbors turned out to be false and which ones were true (the definition of false neighbors is given below).

We calculate

$$\lambda = \frac{\text{Number of false neighbors } \vec{z}^{\text{NN}}(\text{k})}{\text{Total number of recovered vectors } \vec{z}(\text{k})} \tag{5}$$

Let us postpone this number depending on the plane (). If with increasing this relative number of self-intersections decreases to zero at a certain value, then the latter is the estimate of the dimension of the space in which the embedding of the trajectory of the process being modeled is achieved. In practice, starting with some "correct" dimension, the number of false neighbors becomes sufficiently small (but not strictly zero due to noise, etc.) and does not decrease further. This value is taken as an estimate of the dimension of the model.

We will determine false neighbors according to the following criteria [6, 7]:

Criterion 1. Let be the nearest neighbor to the vector $\vec{z}_k = (s_k, s_{k+1}, \ldots s_{k+m-1})$ in Euclidean space R^m, $\vec{z}(k)$ and $\vec{z}^{NN}(k)$ the distance $R_m(k)$ between the vectors and is determined in the usual Euclidean metric. Then the distance between $\vec{z}(k)$ and $\vec{z}^{NN}(k)$ the vectors will be $R_m(k) = \sum_{l=0}^{m-1} \left(s_{k+l} - s_{k+l}^{NN}\right)^2$.

In space R^{m+1}, vectors $\vec{z}(k)$ and $\vec{z}^{NN}(k)$ have the same components as in R^m, the first m components, and their $(m+1)$ components will be respectively s_{k+m} and s_{k+m}^{NN}. Wherein $R_{m+1}^2(k) = R_m^2(k) + \left(s_{k+m} - s_{k+m}^{NN}\right)^2$. We will consider points $\vec{z}(k)$ and $\vec{z}^{NN}(k)$ as false neighbors if

$$\frac{\left|s_{k+m} - s_{k+m}^{NN}\right|}{R_m(k)} > \tilde{R} \tag{6}$$

where $10 \leq \tilde{R} \leq 50$. As indicated in [7], some caution must be exercised when using this criterion. Namely, when applied to data obtained with a high-dimensional random number generator, it will show that a set of observations can be nested in a low-dimensional space. With an increase in the number of analyzed points, the nesting dimension determined with its help will increase. In this regard, with a uniform distribution of objects (points) in d-dimensional space with a fixed number of points, the following problem arises: the points must move further and further from each other, since it increases and most of the given d points will be separated by a large distance. With an infinitely large amount of data, such a problem does not arise, and with a finite amount of data, due to a decrease in the defined dimension of the nesting (as mentioned above), by chance all points will have "nearest neighbors", which will not allow them to be moved too far in space. In this case, the fact that the points are the nearest neighbors does not at all mean that they are close to each other in the sense of a distance in a space of dimension equal to the approximate dimension of the attractor R_A. If the closest $\vec{y}(k)$ neighbor to is actually close to it $R_d(k) \approx R_A$, so, $R_{d+1}(k) \approx 2R_d(k)$ then. This means that the distance between the nearest neighbors will expand to extreme values of the distances on the attractor, i.e. such neighbors may turn out to be false nearest neighbors.

Therefore, in practice, an additional criterion is used to test the falsity of the nearest neighbors.

Criterion 2. Points $\vec{z}(k)$ and $\vec{z}^{NN}(k)$ are false neighbors if

$$\frac{R_{m+1}(k)}{R_A} \geq 2 \tag{7}$$

where, $R_A^2 = \frac{1}{N-2m-1} \sum_{k=m+2}^{N-m} (s_k - \bar{s})^2$, $\bar{s} = \frac{1}{N-2m-1} \sum_{k=m+2}^{N-m} s_k$, N - the volume of the data sample.

If the nearest neighbors by at least one of these two criteria turn out to be false, we will consider them as false neighbors.

False nearest neighbors (FNN) arise when the dimension m of the embedding is very small, then two different points lying on the attractor, after being projected into the embedding space, can coincide, thereby forming points of the FNN type. This effect is further enhanced when, moreover, the delay time is short τ [7–9]. For high-dimensional systems, the delay method is applicable for system reconstruction, albeit with great difficulties, while FNNs always exist and can coexist with FNFs [9, 10]. In the presence of FNF, the dimension of the embedding space is determined using the correlation integral [10].

For the total number N_2 of pairs in the $(\vec{z}_i, \vec{z}_j)$ embedding space R^m, the correlation integral $C(\varepsilon)$ at a given scale ε is determined by the formula [10]

$$C(\varepsilon) = \lim_{N\to\infty} \frac{1}{N_2} \sum_{i,j(i+j)} \theta\left(\varepsilon - \left\|\vec{z}_i - \vec{z}_j\right\|\right), \tag{8}$$

where $\theta(x)$ - is the Heaviside function: at $\theta(x) = 1$ $x > 0$ and $\theta(x) = 0$ $x \leq 0$; $N_2 = C^2_{N-(m-1)} = \frac{1}{2}(N-m)(N-m+1)$.

The correlation dimension is defined as the limit equality

$$\upsilon = \lim_{\varepsilon\to 0} \lim_{N\to\infty} \frac{\ln C(\varepsilon, N)}{\ln \varepsilon}, \tag{9}$$

$$C(\varepsilon, N) = N_2^{-1} \sum_{i,j(i \pm j)} \theta\left(\varepsilon - \left\|\vec{z}_i - \vec{z}_j\right\|\right), \tag{10}$$

In numerical N experiments, it is always finite, $\varepsilon \to 0$ and $N \to \infty$ both limits are meaningless; therefore, the rati $\ln C(\varepsilon, N)/\ln \varepsilon$ o is considered at not too small values ε. To calculate the dimension m of the embedding using the correlation dimension υ, proceed as follows [9]. An arbitrary small value of the dimension of the nesting space m (for example, $m = 1$) is selected and calculated by the slope υ of the linear plot of the graph from $\ln C(\varepsilon, N)$ to $\ln \varepsilon$. Then it is increased m by one and determined again υ. Thus, the dependence of the results of calculating the correlation dimension on the choice of the m dimension of the embedding space is analyzed. This technique allows us to conclude that there is a low-dimensional dynamics ($\upsilon < 4$): in its presence $\upsilon(m)$, the dependence quickly reaches saturation and with a further increase the value does not change within the limits of the calculation accuracy. In this case, the linear section $\ln C(\varepsilon, N)$ of the dependence $\ln \varepsilon$ on becomes parallel to the axis (with an inclination angle equal to zero), which is reflected in the graph $(\ln C(\varepsilon, N), \ln \varepsilon)$ in the form of a plateau. The value $m = m^*$ corresponding to this plateau is taken as the true value of the dimension of the nesting space.

If there is no low-dimensional dynamics, then υ it increases with growth m. In this case, for sufficiently large m, saturation is possible, which is caused by fundamental restrictions on the value of the correlation dimension, associated with the finite length of the analyzed time series N_{t0t}:

$$\upsilon\text{max} \approx \frac{2 \ln N_{t0t}}{\ln(1/r)}, \quad r = \frac{\varepsilon}{\varepsilon\text{max}}, \tag{11}$$

where ε_{max} is the value ε at which $C = 1$.

4 Direct and Iterative Forecasting Methods Based on Dynamic Model Mapping

The direct forecasting method consists in the fact that for forecasting sampling intervals T ahead (with institution T) according to model (1.2), the predicted values $a(t)$ at time points are directly approximated with a step of 1, using $t_{N+1}, t_{N+2}, \ldots, t_{N+T}$ the same mapping G constructed according to criterion (1.4). In this case, already found approximations $a(t_{N+1}), a(t_{N+2}), \ldots, a(t_{N+T})$, etc. are used as new known values $a(t)$ of the observable.

In the iterative forecasting method, the predicted value $a^{np}(t_{N+1})$ is calculated $a^{np}(t_{N+1})$ as in the direct forecasting method. Subsequent predicted values are constructed differently.

To build a forecast at a t_{N+2} point, considering it as a known value, we split the interval $[1, N+1]$ again into a training (training) $[1, N_{train}^H]$ sample and a test (validation) $[N_{train}^H + 1, N+1]$ sample. We repeat stages 1 and 2, as a result we find the "correct" m^H and K^H ("H" means new) and build a forecast $a^{np}(t_{N+2}) = G^H(a_{N-m+2}, a_{N-m+3}, \ldots, a_{N+1})$. Subsequent forecasts in points $t_{N+3}, \ldots, t_{N+T}$ are constructed similarly. In [2] it is shown that the forecast accuracy using the iterative method is somewhat higher $a^{np}(t_{N+2}) = G^H(a_{N-m+2}, a_{N-m+3}, \ldots, a_{N+1})$.

5 Numerical Implementation of the Method and Discussion of Results

The dimension of the embedding space *m* was determined using criteria 1 and 2, which determine false neighbors of the FNN type, as well as using the correlation integral, which determines false neighbors of the FNF type.

Numerical value of the smallest dimension of the embedding space, found using criteria 1 and 2 and using the correlation integral $m = 2$.

When constructing a dynamic model for a time series with $m = 2.$, polynomials of the form (9) with were tested. (1) model (9) with K = 2 turned out to be the best according to criterion (11). The predicted values for such a model were constructed using the iterative forecast method.

Table 1 shows the initial data (for the last 6 months) and predicted (retro and future) values of the water permeability and adhesive strength of the tested materials.

Table 1. Baseline data for the last 6 months and projected values water permeability and adhesive strength of the test materials

Coating materials	Changes in the water permeability of the coating, in % by month				
	Retro forecast		Future		
	33	36	39	42	45
Insulating oil bitumen (BNI)	0.507	0.516	0.524	0.523	0.525
BNI + FFO + BK	0.074	0.750	0.0753	0.756	0.0759
Nanonized BNI + ED-20 + Cu	0.0218	0.0221	0.0 + 222	0.0223	0.0224
BNI	Changes in the adhesion strength of the coating, in MPa, by months				
	5.61	5.46	5.26	5.03	4.91
BNI + FFO + BK	37.90	37.80	37.73	37.60	37.48
Nanonized BNI + ED-20 + Cu	85.68	85.62	85.48	85.40	85.31

Thus, the following conclusions can be drawn:

1. Relative changes in the predicted values for the 45th month and the observed values for the 36th month as a percentage for a water permeability of 2.1; 1.2; 7.7 and for adhesive strength 2; 9.9; 0.6.
2. Compared with conventional (BNI) coating material, the use of nanotechnology (BNI + EV-20 + Med) gives better results than others
3. This material is recommended for the application of anti-corrosion coating for main gas pipelines

4. The developed software product can be used to analyze and predict the behavior of pipeline systems.

References

1. Takens, F.: Detecting strange attractors in turbulence. In: Rand, D., Young, L.-S. (eds.) Dynamical Systems and Turbulence, Warwick 1980. LNM, vol. 898, pp. 366–381. Springer, Heidelberg (1981). https://doi.org/10.1007/BFb0091924
2. Casgagli, M.: Nonlinear prediction of chaotic time series. Physica D **35**, 335–356 (1989)
3. Bezruchko, B.P., Smirnov, D.A.: Construction of model mappings from chaotic time series: a tutorial, 38 p. The Educational Center – College, Saratov State (2000). (in Russian)
4. Bezruchko, B.P., Smirnov, D.A.: Mathematical modeling and chaotic time series, 319 p. The Educational Center – College, Saratov State (2005). (in Russian)
5. Potapov, A., Kurths, J.: Correlation integral as a tool for distinguishing between dynamics and statistics in time series data. Physica D **120**, 369–385 (1998)
6. Malinechkiy, G.G., Potapov, A.B.: Modern problems of nonlinear dynamics. M.: Editorial, 336 p. (2000)
7. Schreiber, T.: Determination of the noise level of chaotic time series. Phys. Rev. E **48**, RB–R-16 (1993)
8. Abarbanel, H.D.I., Brovn, R., Sidorovich, I.I., Tsimring, L.: The analysis of observed chaotic data in physical systems. Revives Modern Phys. **65**(4), 1331–1392 (1993)
9. Pavlov A.N.: Complex Signal Analysis Techniques, 120 p. Scientific Book, Saratov (2008). (in Russian)
10. Shuster, G.: Deterministic Chaos: An Introduction. M: World, 240 p. (1988). (in Russian)

Enhancing Teaching Approach with 3D Primitives in Virtual and Augmented Reality

F. M. Nuraliev[1(✉)], U. E. Giyosov[1], and Yoshihiro Okada[2]

[1] Tashkent University of Information Technologies named after Muhammad al-Khwarizmi, Amir Temur, 108, 100084 Tashkent, Republic of Uzbekistan
f.nuraliev@tuit.uz, bek99989@gmail.com

[2] Graduate School of Information Science and Electrical Engineering, Kyushu University, Motooka 744, Nishi-Ku, Fukuoka 819-0395, Japan
okada@inf.kyushu-u.ac.jp

Abstract. Today, virtual reality (VR) and augmented reality (AR) are widely used in various fields, especially in computer systems, engineering and medicine. However, it remains one of the main barriers to education due to the high cost of conducting experimental research. This interactive virtual learning is very convenient and effective. Nowadays, it offers great opportunities to improve the quality of education by creating new methods and techniques. Traditionalist methods of substance introduction (fixed video, sound, contents) need personalization and communication. In addition to create a private virtual education environment which allows the students and researchers to get enough learning that is interactive and more visual, where students can learn through experimentation and interactions in the virtual world. There is a requirement for MOOC to grasp propelled learning strategies to give all the more captivating experience to its students. Finally, we create internal avatar model while using in virtual reality (VR) and augmented reality (AR) interfaces.

Keywords: Virtual reality for education · Virtual reality (VR) · Augmented reality (AR) · Passive and active learning activity · 3D primitives · 3D avatars

1 Introduction

The establishment of virtual universities abroad is recognized as a modern innovative pedagogical technology. The number of supporters of virtual education is growing. In countries where the system is in place, such as the United States, the United Kingdom, Germany, Korea, and Japan, there are several million higher education institutions a year. This is because distance learning is both more convenient and cheaper for users than traditional methods of learning in educational institutions, and most importantly, it allows users to learn at a time that suits them.

General methodological principles for the development of software systems for creating models of the organization of the virtual 3D university educational process. Today is the beginning of the application of virtual reality technologies and principles for the

© The Author(s), under exclusive license to Springer Nature Switzerland AG 2021
R. A. Aliev et al. (Eds.): WCIS 2020, AISC 1323, pp. 155–163, 2021.
https://doi.org/10.1007/978-3-030-68004-6_20

development of virtual reality applications. This methodological principle approaches the software in three areas: passive virtual being monitored by a non-human autonomous graphic image with sound, scripted virtual being presented to the user in a limited number of scenarios, image, sound selection, interactive virtual being tracking is that the user can control the virtual environment on the basis of the laws of the world, created using a special device that can perform its function. In many IT organizations, the development of virtual reality applications is a collaborative effort involving project engineers, managers, and other stakeholders [3].

Today, a number of systems for evaluating virtual education in foreign countries (Codeingame.com, vAcademia.com, fun-mooc.fr, rwaq.com, VirBela.org etc.) [1, 2].

With regards to AR and VR, late innovative advances have permitted augmented and virtual encounters to be conceivable on handheld cell phones or minimal effort head-mount-shows (HMD). This implies it is feasible for additional individuals to encounter fundamental AR and VR content utilizing their own gadgets. Be that as it may, figuring execution and capacities of clients' gadgets change dependent upon a few conjectured taking care of variables including multifaceted nature of computerized content and concentrated sensors.

The paper is organized out as follows. Section 2 talks about Augmented and Virtual reality (AR/VR) in e-learning, Sect. 3 How to implement graphics field, Sect. 4 Implication of 3D modeling by software.

2 Augmented and Virtual Reality (AR/VR) in E-Learning

Augmented Reality alludes to an idea where genuine world is upgraded by mixing it with virtual world [10]. This present reality static items are powerfully transformed with setting delicate virtual data like video, sound, or a realistic overlay to improve client's experience [4, 10].

2.1 Using AR in Education

These days, a typical case of AR is seen by method of applications like Nokia City Lens (Windows App) or Across Air (iOS application) or Google Goggles (android application) which use GPS and compass of cell phone to help client in route around a region. When the client holds the cell phone upstanding with camera in an area, it begins to show insights regarding eateries, inns, milestones and other geo-labeled sections. Coasting picture inflatables or symbols are shown on the versatile screen which on clicking show increasingly nitty gritty data about the picked object. AR applications take into account more up to date approaches to investigate an item and its related space by empowering better commitment and client cooperation.

Augmented Reality is additionally changing the learning condition by upgrading student's genuine condition with virtual data, in different interactive media positions like designs, video and sound and so on [4]. The overlaid data is covered up under the prompts which when examined with AR empowered gadgets breath life into static condition and gives better learning experience to understudies.

Three systems have accepted a key activity in e-learning. To consider the need of an individual understudy and reinforce individualization another learning methodology rose, titled Adaptive Learning or Intelligent Tutoring. Furthermore, if understudies are getting some answers concerning un-recognizable structures like close planetary framework, photosynthesis process, nuclear reactions or bona fide goals; it is obvious that understanding these thoughts without having the choice to see them really is an anticipation in method of expansive learning. Augmented reality makes virtual things which when blended in with real world, give a striking learning experience to learners. (It is described and shown Fig. 1).

Augmented Reality insinuates a thought wherein veritable world is redesigned by blending it in with virtual world. This current reality static articles are effectively changed with setting fragile virtual information like video, sound, or a sensible overlay to improve customer's experience.

Augmented reality is a system that:

- Combines virtual and real
- Interacts in real time
- Works in three-dimensional space

Virtual reality - a completely digital alternative to reality is possible to plunge into digital space with the full illusion of presence, but it is divorced from the real world, in contract to augmented reality.

2.2 Virtual Reality (VR) for Education

Augmented reality innovation (VR) and communication by 3D geometric model could stop the uninvolved realizing which is followed in the conventional technique of training. They additionally lead to advantageous correspondence between different members in training process [11].

Virtual reality frameworks included time and area and great nature of comprise of PCs, clients, equipment and programming. They isolated into two primary classes: Immersive Virtual and Non immersive. In addition to, evaluation of e-learning increased year by year. (It is described and shown Fig. 2).

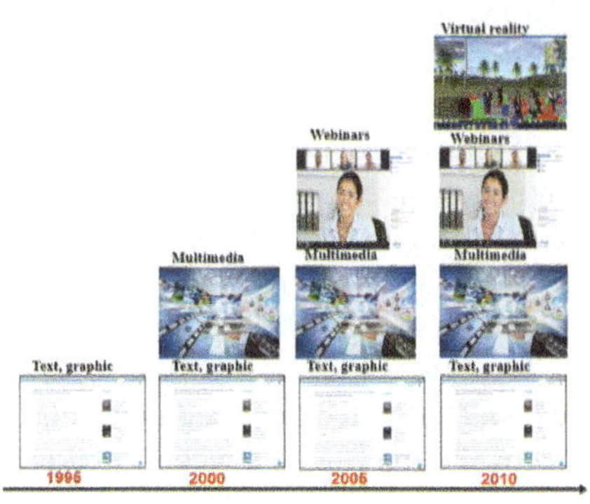

Fig. 1. E-learning development tools.

2.3 The Need for Virtual Reality in Education

The following reasons support virtual reality in education:

Gives new structures and techniques for perception, drawing on the qualities of visual portrayals, and gives an elective strategy for speaking to the material. VR can likewise more precisely show certain highlights, forms than different strategies, taking into consideration extraordinary intensive investigation of an item, perception from a significant distance, just as perception and research of territories and occasions that are not accessible by different methods.

Propel and support dynamic investment and communication with respect to understudies, instead of lack of involvement. A few kinds of virtual reality, for example, community computer virtual reality utilizing content contribution with virtual universes, support or require cooperation and make a social climate.

Virtual reality permits the understudy to experience understanding over a wide timeframe. This permits individuals with inabilities to take an interest in an analysis or learning condition and conquers all language boundaries. On account of content access, it gives equivalent chances to correspondence with understudies in different societies, permitting the understudy to assume the job of man in various societies. Potential favorable circumstances of utilizing VR in instruction and preparing: representation and renewal, an elective strategy for introducing material; learning in conditions that are inconceivable or hard to involvement with reality; expanded inspiration; advancement of collaboration; flexibility, giving the chance to figuring out how to adjust to the qualities and requirements of the understudy; and appraisal and assessment, offering extraordinary potential as a device for assessment because of the simple checking and recording of meetings in a virtual situation [1].

3 How to Implement Graphics Field

Forcing classes on any field is perilous, however most illustrations experts would concede to the accompanying significant zones of PC designs:

- Modeling manages the numerical determination of shape and appearance properties in a manner that can be put away on the PC. For instance, an espresso cup may be depicted as a lot of requested 3D focuses alongside some addition rule to interface the focuses and a reflection model that portrays how light collaborates with the cup.
- Rendering is a term acquired from craftsmanship and manages the formation of concealed pictures from 3D PC models.
- Animation is a method to make a deception of movement through successions of pictures. Activity utilizes demonstrating and rendering however includes the key issue of development after some time, which isn't normally managed in fundamental displaying also, rendering. There are numerous different zones that include PC designs, and whether they are center illustrations regions involves supposition. These will all be at any rate addressed in the content. Such related regions incorporate the accompanying:
- User collaboration manages the interface between input gadgets, for example, mice and tablets, the application, criticism to the client in symbolism, and other tactile criticism. Truly, this zone is related with illustrations to a great extent since designs

scientists had probably the most punctual access to the info/yield gadgets that are presently omnipresent.
- Virtual reality attempts to drench the client into a 3D virtual world. This regularly requires at any rate sound system illustrations and reaction to head movement. For genuine computer-generated simulation, sound and power input ought to be given too. Since this zone requires propelled 3D designs and propelled show innovation, it is regularly firmly connected with illustrations.
- Visualization attempts to give clients knowledge into complex data through visual showcase. Regularly there are realistic issues to be tended to in a perception issue.
- Image handling manages the control of 2D pictures and is utilized in both the fields of illustrations and vision.
- 3D examining utilizes extend discovering innovation to make estimated 3D models. Such models are valuable for making rich visual symbolism, and the preparing of such models regularly requires designs algorithms.

Virtual objects (artifacts):

- What is possible only in Virtually
- You can simulate the environment
- You can simulate events
- You can simulate joint actions

Further training or retraining of teachers involved in virtual learning can be achieved by increasing the level of their training. To solve these problems, it is necessary, first of all, to revise the existing regulatory legal acts on virtual education and develop standards and rules necessary for the implementation of virtual education. Also, it is necessary to analyze the current technical condition of existing computer networks, increase their speed and take measures to strengthen their material and technical base.

3D Auditorium-Trainers who address in the 3D Auditorium will have the option to stack explicit introductions from VLE and even transfer slides. Homeroom/Meeting Room-These rooms can be utilized as a gathering territory for venture accomplices, or as a study hall for a little gathering of understudies. We will survey the model of the college, a few understudies pick their subject over the span of the learning plan. Presently understudies can encounter the themes they are learning. The utilization of augmented reality innovation has been appeared to build understudy cooperation and consideration, while vivid and intelligent situations urge understudies to become dynamic students.

4 Implication of 3D Modeling by Software

We The user operates the application with the Body-Hand user interface, using head movement, thumb tracking, pinch gloves, and a menu system to perform the following object manipulation tasks: Constructive Solid Geometry allows the construction of complex 3D graphical shapes using: Object selection – the user can point at objects and select them, placing them into one of several clipboards. Object transform–perform translate, rotate, and scale operations, in a variety of different ways. Create primitives – 3D primitives can be created in the virtual world, from infinite planes as the simplest, to complex graphical models such as a water heater. Combine primitives – previously constructed and manipulated primitives may be combined together using Constructive Solid Geometry (CSG) operations to produce higher level graphical objects. only a small number of primitives. Solid shapes contain an inside and outside, and using set operations, it is possible to perform 3D union, intersection and difference operations (see Fig. 2, Fig. 3, Fig. 4). By combining these operations, it is possible to produce any kind of object required. We create internal avatar model while using in virtual reality (VR) and augmented reality (AR) interfaces.

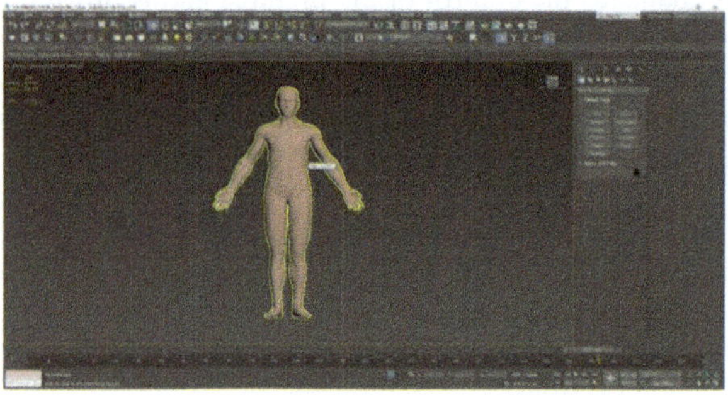

Fig. 2. Modeling using the Body Cylinder.

Body avatar is a Kinect-based interactive system that allows users without professional skills to create freeform 3D avatars using body gestures. Unlike existing gestures-based 3D modeling tools, Body avatar centers around a is first person "you the avatar" metaphor, where the user treats their own body as a physical proxy of the virtual avatar. Based on an intuitive body-centric mapping, the user performs gestures to their own body as if wanting to modify it, which in turn results in corresponding modifications to the avatar. Body avatar provides an intuitive, immersive and playful creation experience for the user.

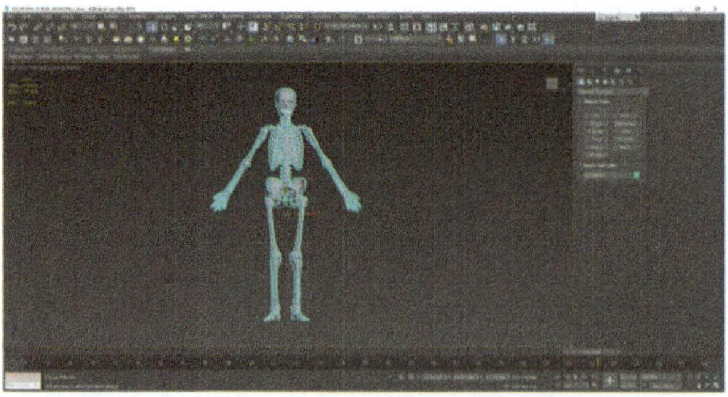

Fig. 3. Human skeleton - human skeleton drawn through the 3D Max program.

To do this we model the 3D Max program by using the Box element in the Geometry section. The most VR/AR applications in Education are useful as training systems for medical/dental students. So, we use teaching and learning approach that subjects of English, Russian and Korean classroom with 3D primitives with VR/AR are useful for. In addition to, educators can experiences hairdresser laboratory room's project, cooking room's, mechanical experience laboratory room's. (It is described and shown Fig. 5).

Fig. 4. This is the case with the Elements of Editing Poly by using the Elements of the Geometry section of the male teacher's clothing.

Now we will see some experimental result of enhanced teaching and learning Central Asian one of best University Tashkent University of Information Technology named after Muhammad Al-Khwarizmi.

Fig. 5. Some of the steps of the construction works process that presented at the BC/VR Software.

Testing models created with virtual reality applications by using 3D primitives. (a) the main view of the educational process in English classroom, (b) the main view of the educational process in Russian classroom, (c) view of hairdresser laboratory room's project, (d) view of cooking room's, (e) view of mechanical experience laboratory room's, (f) the main view of the educational process in Korean room [8, 9].

5 Conclusion

This paper presents new way of review in this example model of University and there are students choose their subjects where they learning his timetable. Use of virtual reality technology has been shown to increase student engagement and focus, while the immersive and interactive environment encourages the students to become active learners. The educational software helped to take the Formal Language subject from classrooms to the computer labs, making the teaching or learning process more interesting and pleasant to the students, facilitating the teachers work during the evaluation performance too.

References

1. Thomas, B.H., Plekarski, W.: Unifying augmented reality and virtual reality user interfaces. https://www.researchgate.net/publication/228588353
2. Jantjies, M., Maart, R., Moodley, T.: Experiential learning through Virtual and augmented reality in higher education (2019). https://www.researchgate.net/publication/331423337
3. Wu, H.K., Lee, S.W.Y., Chang, H.Y., Liang, J.C.: Current status, opportunities and challenges of augmented reality in education. Comput. Educ. **62**, 41–49 (2013)
4. Epson. Cue Educate. Innovate. Explore, http://www.cue.org/sites/cue.org/files/images/121013_OnCUE_W13_lorez-1.pdf. Accessed June 2015
5. Wikipedia contributors, "Augmented reality." Wikipedia, The Free Encyclopedia. Wikipedia, The Free Encyclopedia, 16 Aug. 2015. Web. 17 Aug. 2015. McGraw-Hill, "McGraw-Hill Launches First-Ever Adaptive Learning Ebook in Canada". http://www.digitalbookworld.com/2013/mcgraw-hill-launches-first-ever-adaptive-learning-ebook-in-canada/. Accessed 26 Feb 2013
6. Sampaio, A.Z., Ferreira, M.M., Rosario, D.P., Martins, O.P.: 3D and VR models in civil engineering education: construction, rehabilitation and maintenance. Autom. Constr. **19**(7), 819–828 (2010). https://doi.org/10.1016/j.autcon.2010.05.006
7. Marschner, S., Shirley, P.: Fundamentals of Computer Graphics, 4th edn. Cornell University (2016)
8. Nuraliev, F.M., Giyosov, U.E.: On the design of virtual reality environments in education. In: 8 th International Conference on advanced info telecommunications, ICAIT 2019, 27–28 February 2019, pp. 4–14. Saint-Petersburg (2019). http://apino.spbgut.ru/old/apino8
9. Nuraliev, F.M., Giyosov, U.E.: Integration of virtual reality and 3D modeling use of environments in education. In: International Conference On Information Science And Communications Technologies ICISCT 2019 Applications, Trends And Opportunities, 4–6 November 2019, Tashkent Uzbekistan (2019). https://ieeexplore.ieee.org/xpl/conhome/8982113/proceeding

10. Chauhan, J., Taneja, S., Goel, A.: Enhancing MOOC with augmented reality, adaptive learning and gamification. In: 2015 IEEE 3rd International Conference on MOOCs, Innovation and Technology in Educa-tion (MITE) (2015). https://www.researchgate.net/publication/282798098
11. Talley, S., Martinez, D.H.: Tools for schools: School reform models supported by the National Institute on the Education of At- Risk Students. US Department of Education, Office of Educational Research and Improvement, Washington, DC (1998). [verified 31 Oct 2004] http://www.ed.gov/pubs/ToolsforSchools/title.html

Modelling of Gas Lift Dynamic Processes Using ANFIS and ANN

I. Salahaddin Yusifov[1], B. Allahverdi Hasanov[2], S. Rza Safarov[1](✉), and O. Azim Logmanli[1]

[1] Azerbaijan State Oil and Industry University, Azadliq Avenue 1010, Baku, Azerbaijan
siyusifov@yahoo.com, rzasafarov@mail.ru, ezim1497@gmail.com
[2] Azerbaijan National Academy of Sciences, Institute of Control Systems, Baku, Azerbaijan
hesenli_ab@mail.ru

Abstract. Gas lift optimization and prediction parameters is becoming more important now a day in petroleum industry. The objective of this work has been an estimation of potentiality of Adaptive Neuro-Fuzzy Inference Systems and Artificial Neural Networks in gas lift operations optimization. The estimation of two main parameters of the process, i.e. the rate of optimal gas injection and the depth of optimal injection. A set of nine input parameters having a basic influence on the gas lift optimization. In this paper, two models are presented for optimization and predicting of both the optimal gas injection depth and gas injection rate, using these methodologies. The results obtained utilizing the two methods show that while both are suitably tools to prediction injection depth and injection rate, most of the time ANN has lower error than ANFIS.

Keywords: Gas lift · Artificial neural network · Adaptive Neuro-Fuzzy Inference System · Gas injection rate · Gas injection depth

1 Introduction

Gas lift is one of methods of artificial lift. It uses a high-pressure gas source to supplement a gas formation for well fluids lifting. If the well has an insufficient reservoir pressure then gas lift is used. Injected gas diminishes the density of fluid subsequently the hydrostatic column weight is decreased. It causes backpressure to be lower that permits the reservoir to product more fluid to the surface [1].

In gas lift processing it is critical to select the proper gas injection pressure. Several factors can influence it. However, one of them should be taken into account. Any compromise with it can cause gas injection to be less efficient. At the injection depth, the pressure of injected gas must be greater than pressure of flowing. Another factor is the rate of the gas injected into a well and one must choose it correctly. Injecting gas only decreases the density of fluid above the gas injection depth. Then it is important how much gas at which depth should be injected. At the lower depth, a smaller volume is more efficient than a higher volume at the upper part of the column. Two problems in gas lift processing are most important: defining of the optimal injection point and finding the

© The Author(s), under exclusive license to Springer Nature Switzerland AG 2021
R. A. Aliev et al. (Eds.): WCIS 2020, AISC 1323, pp. 164–169, 2021.
https://doi.org/10.1007/978-3-030-68004-6_21

rate of optimal gas injection. In this paper, data for 40 gas lift wells are used. For each of them, the following data set are collected to be used in the research: Bottom Hole Flowing Pressure (BHFP), Bottom Hole Static Pressure (BHSP), Well Head Flowing Pressure (WHFP), Productivity Index (PI), Tubing Size, Well-Bore Size, Oil Production Rate (Qoil), Well Depth, Water-Cut Percent (WC%), the Depth of Injection and the Rate of Injection (Qgasinj). The two last parameters are output data; others are input data [2]. In this research, MATLAB 8.4 is used for the predicting process.

2 Method

The basic difference between ANFIS and Mamdani FIS emanates at the final of the third step of the Mamdani FIS. In the model of Takagi-Sugeno, an output of each rule is crisp. The general output is a weighted average of the outputs of single rules. In ANFIS, each rule's output can be considered as a linear combination made of input variables adding a constant term or it is only a constant term. The final output can be considered as the weighted average from the output of each rule.

This structure of a neural network consists of 5 layers together with the input layer (Layer 0):

1. The input layer (Layer 0) consists of n (the number of the system inputs) nodes.
2. In the fuzzification layer (Layer 1) each node means a membership value for a linguistic term in the form of a Gaussian function with the given mean:

$$\mu_{A_i}(x) = \frac{1}{1 + \left[\left(\frac{x - c_i}{a_i}\right)^2 \right]^2},$$

where a_i, b_i, c_i are the Gaussian function parameters. They are parameters adaptive. While learning, we use the back-propagation algorithm for adaptation. The linguistic term A_i of the membership function changes as the values of the parameters change.

3. Each node in Layer 2 provides the rule's strength by means of a multiplication.
4. The normalization layer (Layer 3) normalizes the all rules strength in accordance with the following equation:

$$\bar{w_i} = \frac{w_i}{\sum_{i=1}^{R} W_i},$$

where $\bar{w_i}$ is the ith rule's firing strength computed in Layer 2. Node i computes the ratio of the firing strength of the ith rule to the sum of firing strengths of all rules. In Layer 3, there are p^n nodes.

5. Every node in Layer 4 (the adaptive nodes layer) calculates function that is linear. We use the error function of the multi-layer feed-forward neural network to adapt coefficients of this function.
6. The function of the output layer Layer 5 is summing of net outputs of the Layer 4 nodes. This output can be calculated as.

$$\sum_i \bar{w}_i f_i = \frac{\sum_i w_i f_i}{\sum_i w_i}$$

A hybrid-learning algorithm is used in ANFIS to train the network. In layer 1, a back-propagation algorithm is used for its parameters. We use a variation of least squares approximation for training of Layer 4 parameters. We applied the hybrid algorithm to this architecture. Functional signals move forward up to Layer 4 in the forward pass of the hybrid-learning algorithm. The least squares evaluation was used to identify the consequent parameters. The initial parameters are updated by the gradient descent in the last backward [3].

3 Results

As it was above mentioned, one of the main purposes of this research is comparing the performances of ANFIS and ANN to model gas lift operations. The ANFIS method builds a fuzzy inference system (FIS) through the set of input/output data. The FIS membership function parameters can be adjusted (tuned) by means of either a back propagation algorithm in combination with a least squares type of method or a back propagation algorithm only. It allows a fuzzy system to learn from its modeling data. In the ANFIS structure, the membership function's shape is essential. A membership function (MF) shows how the input space can be mapped to a membership grade changed in the interval (0. 1).

The given ANFIS is a neural network, which has four layers. It works on the operating manner of a FIS. In layers 1 and 4, the linguistic nodes represent linguistic variables respectively for the input and output. In layer 2, nodes are term nodes. They act as membership functions of variables of input.

Nine feature variables, bottom hole flowing pressure, tubing size, well head flowing pressure, bottom hole static pressure, productivity index (PI), well-bore size, oil production rate, well depth, water-cut percent (WC%) are chosen as ANFIS inputs. We assigned two membership functions (Mfs) for each linguistic variable while the membership function of the generalized bell shape is adopted.

The process of ANFIS modeling begins by obtaining the input-output data set and dividing this set into data sets of testing, training and checking. Training data forms sets of input vectors and output vectors. In order to make the data suitable to the process of training those data are normalized.

One uses the data set of training to determine the initial parameters of the Mfs by equally placing each of them. One determines a threshold value for the error signal between the desired and actual output. We find the consequent parameters by means of the least-squares method. Subsequently we find an error signal for each data pair. If the error signal is larger than the threshold, then we update the initial parameters by means of the method of gradient decent.

Finally, we prepare a comparison Table 1 to show the comparison between ANFIS and ANN.

A neural network (NN) is a big processor distributed in parallel. It is built of units of simple processing. NN has a natural tendency to store experiential data and make the knowledge suitable for use. NN is like the brain for two reasons:

1. Network acquire knowledge from its environment during a process of learning.
2. In order to store the acquired knowledge interneuron connection strength (by other words, synaptic weights) are used.

When we talk about NN, we usually mean Artificial Neural Network (ANN). ANN processes information as a brain (biologically nervous systems) and in manner it processes information. The essential base of this pattern is the biologically nervous systems structure. In this structure a big number of elements interconnected for processing operate together for problem solving. These elements are known as neurons. Like a human being ANNs learn by examples. The name of the operation used for a learning process is a learning algorithm. This algorithm tunes the network synaptic weights to originate design objective that is required [4].

Fundamental processing elements of the artificial neural network are neurons (artificial), like a human brain. An ANN is built of neurons interconnected each other. They are formed in several layers.

- for the network, raw information is represented by the input layer.
- Hidden layers. Their number can be one or more. They are used to transduce the inputs into the thing that output layers are in need of.
- An output layer. This layer sends the final result (the output) to the outside world.

Values of the weights of connections between the neurons thresholds and the neurons represent the knowledge of a neural network. The determination of those values is the learning process.

One uses a learning algorithm to tune the weights of the bonds in a supervised learning process in such a way when the error between supervised outputs and the output of the model is reduced. The NN computes the derivative's value of the error of the weights to carry out this process.

One of the general learning supervised algorithms for network training is the back propagation algorithm. Firstly the weight matrix is initiated by means of some little random values. Then we apply input data for this network. Then we compute the output. This stage is known as feed forward. Each neuron's error is computed in the direction of backward in the next stage (so called back propagation). I.e. errors are computed from the last layer and then the layer (so called hidden) just before the output (last) layer and it goes on with this rule. Subsequently the system uses those errors to change each neuron weight. So those errors get smaller in such a way. That process repeats until we get the minimal error [4].

4 Discussion

In this paper, we use a back-propagation algorithm during the ANN training. This algorithm is used for problems of prediction which are associated with input-output pairs. Suppose that there are many parameters in a system. They describe the given system. Our task is to build a system of interconnected elements where we want to know the relationship of the given parameter to other parameters.

In this article we consider MATLAB NN Toolbox (NNTOOL) for purposes of ANN buildup and training of network. The depth of gas injection and the rate of gas injection is the output of the ANN. We carry out the data representation for ANN after creating the input and output vectors. We use a network consisting of three (1 input, 1 hidden and 1 output) layers to train the ANN. Two stopping conditions are used while training. Those are zero (0) error conditions between the real and predicted values of gas injection depth and gas injection rate, and data over fitting.

The next step after defining the outputs, inputs, and the hidden neurons for the NN is training the given model. We used 70% of our dataset to model training when we use only 20% for validation of the training and 10% to testing of the model. Suppose R value is a relation between the targets and the outputs. In case of a precise linear regression between them $R = 1$. When $R \approx 0$, then the relation between them is not linear.

Therefore, this research concentrated on predicting of the depth of the gas injection and the rate of the gas injection of gas lift wells based on ANFIS and ANN. These parameters plays a significant role in gas lift operations.

5 Conclusion

The models were developed in MATLAB software using ANFIS and NN toolboxes. While both methods were quite capable of predicting, the ANN procedure had better performance than the ANFIS method. A comparison between ANN and ANFIS results is given below (Table 1).

Table 1. The comparison results gotten from the ANFIS and ANN models

Wells №	Real optimum depth of injection (meter)	Real optimum gas injection rate (m^3/day)	Optimum depth of injection with ANFIS (meter)	Optimum gas injection rate with ANFIS (m^3/day)	Optimum depth of injection with ANN (meter)	Optimum gas injection rate with ANN (m^3/day)
I test	3307	82320	3370	87366	3272	83723
II test	3322	33040	2834	30525	3319	33881
III test	3527	43960	2407	48442	3192	46463
IV test	3765	41720	2572	50968	3736	42837

References

1. Amao, M.: Artificial Lift Methods. https://pdfs.semanticscholar.org/e2bd/3fcb811feedb505d29caf9cbb21b2110766b.pdf
2. Yusifov, S.I.: Optimization of the dynamic working mode of gas lift wells. In: 9th World Conference on Intelligent Systems for Industrial Automation, Tashkent, Uzbekistan (2016)
3. Jang, R.J.: ANFIS: adaptive-network-based fuzzy inference system. IEEE Trans. Syst. Man Cybern. **23**(3), 665–684 (1993)
4. Arbib, M.A.: The Handbook of Brain Theory and Neural Networks, 2nd edn. MIT Press, Cambridge (2003)

Recognition and Classification of Pollen Grains Based on the Use of Statistical, Dynamic Image Characteristics, and Unique Properties of Neural Networks

Jumanov Isroil Ibragimovich(✉), Djumanov Olim Isroilovich, and Safarov Rustam Abdullayevich

Samarkand State University, Samarkand, Uzbekistan
Isroil.jumanov2019@gmail.com, o.djumanov2019@gmail.com, rustammix.rs@gmail.com

Abstract. Constructive approaches, principles, and methods of identification, recognition, and classification of pollen grains have been developed based on dynamic models, neural networks (NN), and mechanisms for extracting statistical, dynamic, specific information characteristics, preliminary image processing, and setting variables. Pollen recognition models are proposed, which rely on mechanisms for checking the similarity of micro-objects, highlighting the outline, dividing them into segments (classes), using specific features that differ from other classes, extracting the desired properties, characteristics, knowledge, performing image identification based on a set of numerical characteristics, database and knowledge base rules [1, 2]. The traditional computational schemes of multilayer NN, Kohonen NN, radial - basic network, which is optimized on the basis of network learning mechanisms with the variable setting, control of identification error along the boundaries of permissible values, taking into account the stationary, quasi-stationary and nonstationary behavior of the dynamics of the image contour, are investigated. Generalized algorithms for identifying images of pollen grains based on combined models and mechanisms of NN are proposed. The efficiency of the algorithms was investigated according to the criteria of the mean square error and the speed of information processing with combined dynamic and neural network identification models, linear, nonlinear, and compositional connections of neurons between network layers [3]. A software package for visualization, recognition, and classification of images of pollen grains has been developed, the implementations of which have been tested taking into account the conditions of a priori insufficiency, parametric uncertainty, and nonstationarity of processes.

Keywords: Image · Pollen grains · Identification · Information processing · Error · Control mechanisms · Optimization · Software package

1 Introduction

Since the 90s, technologies for recognizing faces, license plates, speech, fingerprints, machine vision, and image processing of micro-objects have been improved and their

© The Author(s), under exclusive license to Springer Nature Switzerland AG 2021
R. A. Aliev et al. (Eds.): WCIS 2020, AISC 1323, pp. 170–179, 2021.
https://doi.org/10.1007/978-3-030-68004-6_22

field of application has significantly expanded. In 2003, systems for intellectual analysis and information processing were created, aimed at solving problems of image visualization, recognition, and classification of micro-objects by leading experts from England, France, Germany, Italy, Spain, Austria, New Zealand, etc. One of these tasks is the identification and counting of pollen grains in air samples. To predict and study air pollution, the German Weather Service (DWD), MeteoSwiss, and the Institute for Computer Science at the University of Freiburg have developed computer database technologies and automatic identification of pollen grain images. Nineteen European countries are united in the European EAN/EPI Aero-Allergy Network collecting information on pollen. In all, 188 pollen organizations distributed across Europe send weekly pollen information to national networks, which forward it to the European Coordinating Center. In existing publications, it is noted that the standards used are insufficient for their acceptance as the basis for research on the creation of an automated system for visualization, recognition, classification, and accounting for images of micro-objects [4]. There is an urgent need to research and develop a technology for automatic classification and counting of micro-objects from a slide based on intelligent programs that determine their class, type, genus, type of images obtained by optical microscopes; methods, models, algorithms, software package (SP) for recognition, classification of images of micro-objects, eliminating the existing shortcomings in the known methods and providing the required accuracy of data processing with low time and material costs.

2 Main Part

2.1 Main Features, Principles, and Characteristics of the Technology for Identifying Images of Pollen Grains

Pollen recognition methods are based on the following mechanisms: checking the similarity of micro-objects of the same type; selection of the contour, dividing them into segments (classes); using specific features that differ from other classes; extraction of the required properties, characteristics, knowledge; performing image identification based on a set of numerical characteristics generated in databases (DB) and rules from the knowledge base (KB).

The stage has been investigated - the technology of image input is carried out by means of capture from an optical device (photo, video camera, microscope) or reading. In the process of reading and converting to a digital array, the image undergoes various distortions and noise due to the presence of interference, blur, which require the use of mechanisms for their improvement and restoration [4, 5].

A modified technology for preliminary processing of the pollen image is proposed, which is aimed at improving the quality, cutting off unnecessary non-stationary and non-typical parts, converting to a grayscale image, leveling lighting, smoothing and filtering Filtering by technology is carried out by image convolution, masking, application of the Gaussian filter, which also represents blur operations. Filtering and image convolution is associated with finding the desired point and the direction of the optimization gradient, highlighting local maxima, reducing the width of the stripes to one pixel, linking, cropping at the threshold, determining the lower and upper boundaries. The threshold is used when initializing and continuing curves. The edge enhancement is based on the

Canny detector, which has the advantages of good detection, minimal chance of missing real difference in brightness, and false detection of edge (output signal to noise ratio).

The proposed mechanism for recognizing micro-objects based on neural networks (NN), which is aimed at checking the robustness of an image using a template or a reference model. The mechanism includes operations for selecting and using a template, various faces, context, characteristics, properties, structure, and their specific features.

The developed technology assumes that visual information is presented in the form of curves, diagrams, samples, images, which during identification are determined by the following designations:

$X = \{X_i\}$ - set of vectors of values of variables $X_i = <x_{ij}> = <x_{i1}, ..., x_{iN}>$, where $i = 1, ..., M$; M - power $X, j = 1, ..., N = 5$; $j = 1$ corresponds to the value of the variable P_1 - the background component of the image, $j = 2 - P_2, j = 3 - P_3$, $j = 4 - P_4, j = 5$; $A = \{A_p\}$ is the set of vectors of values of the reference variables A_p, $p = 1, ...N_p = 6$. Moreover, $p = 1$ corresponds to parameter $P_1, p = 2 - P_2, p = 3 - P_3$, $p = 4 - P_4, p = 5 - P_5, p = 6 - P_6$; $A = \{A_{pg}\}$ - a set of A_{pg} gradations for each of the A_p variables; $A_{pg} = \frac{1}{N_g} \sum_{i=1}^{N_g} x_{ig}^p$ - standard g grade A_p; N_g - the number of retrospective gradations g, $g = 1, ..., 7$. A_{pg} - is presented as the arithmetic mean of retrospective $\{X_i\}$ data by g gradations for each $p : X_i = \{< x_{ij} >\} = \{< x_{i1}, ..., x_{iN} >\}$.

The texture of the image of a micro-object, its features are reflected on the basis of special E matrices of entries, which cover, reflect useful and hiding characteristics.

Each matrix describes textures in the corresponding direction from one edge of the image to the other. A vector of texture features is compiled, the characteristics of E matrices are calculated, and considered as a covering pixel.

When the Haralik texture features are used, a window with a size of I pixels is formed for the θ_h image. A set of texture features is calculated for each window.

The θ_h value is generally chosen to be seven pixels. The algorithm for the implementation of this technology is considered in two stages.

Stage 1. Several directions are selected $\alpha = \{\alpha_X\}$; $x = \overline{1, X}$; $\alpha_x \in [0, 2\pi)$, α_x- the angle describing the direction corresponding to the matrix of occurrence. Matrices of occurrences for directions $\alpha = \{0, \frac{\pi}{4}, \frac{\pi}{2}, \frac{3\pi}{4}\}$ are calculated.

Stage 2. Implementation of the algorithm, which consists of the following steps.

Step 1. Size θ_{alp} of the neighborhood of each pixel in image I_h is selected, $\theta_{alp} = 3$ is the optimal choice for image I_h of size $\theta_h = 7$. A larger window will lead to a significant degeneration of the entry matrices as it too often goes beyond the image.

Step 2. Entry matrices $E = \{E_x\}$; $x \in \overline{1, X}$ of size $B * B$ elements are filled with zeros; B - the number of levels of quantization of spectral brightness in image I_h.

Step 3. For each pair ((«pixel», «matrix of entries») $(I_h(m_h, n_h), E_x)$; $m_h = \overline{1, \theta_h}$; $n_h = \overline{1, \theta_h}$; $x = \overline{1, X}$, the following operations are performed: a $\theta_{alp} * \theta_{alp}$ pixel window is "superimposed" on pixel $I_h(m_h, n_h)$; in the α_x direction in this window are the coordinates of the extreme pixel $(m_{h\,x}, n_{h\,x})$ for $\alpha_x = 0$;

$$\theta_{alp} = 3: \ m_{hx} = m_h; \ n_{hx} = n_h + 1; \ \alpha_x = \frac{\pi}{4};$$

$$\theta_{alp} = 3: \ m_{hx} = m_h - 1; \ n_{hx} = n_h + 1; \ \alpha_x = \frac{\pi}{2}; \ \theta_{alp} = 3: \ m_{hx} = m_h - 1; \ n_{hx} = n_h; \ \alpha_x = \frac{3\pi}{4};$$

$$\theta_{alp} = 3: \ m_{h\,x} = m_h - 1; n_{h\,x} = n_h - 1;$$

- if the (m_{hx}, n_{hx}) coordinates of the extreme pixel in the direction α_x go beyond the limits of the I_h image, then the transition to the next pair is performed ("pixel", "matrix of entries");
- the following equality is checked

$$E_x(I_h(m_h, n_h), I_h(m_{h\,x}, n_{h\,x})) = E_x(I_h(m_h, n_h), I_h(m_{h\,x}, n_{h\,x})) + 1.$$

Step 4. Matrices of E entries are scaled by division to the sum of their elements equal to one. For what, a transformation of the form is performed.

$$E_x(I_h(m_h, n_h), I_h(m_{h\,x}, n_{h\,x})) = \frac{E_x(I_h(m_h, n_h), I_h(m_{h\,x}, n_{h\,x}))}{\sum_{i=0}^{B}\sum_{j=0}^{B} E_x(i,j)}; \; x = \overline{1, X}.$$

Step 5. For each matrix E_x, the following characteristics $E_x(k)$ are calculated in the form: $E_x(k) = \sum_{i=1}^{B}\sum_{j=1}^{B} E_x(i,j); \; i+j = k$. Average μ_X - by rows and μ_X - by columns $\mu_{Xw} = \sum_{i=1}^{B} i \sum_{j=1}^{B} E_x(i,j); \mu_{Xh} = \sum_{j=1}^{B} j \sum_{i=1}^{B} E_x(i,j)$. Variance σ_X by rows and σ_X by columns $\sigma_{Xw} = \sum_{i=1}^{B} (i - \mu_{x\;w})^2 \sum_{j=1}^{B} E_x(i,j); \sigma_{Xh} = \sum_{j=1}^{B} (j - \mu_{x\;h})^2 \sum_{i=1}^{B} E_x(i,j)$.

Step 6. For each E_x matrix, the following characteristics are calculated: $e_{x1} = \sum_{i=1}^{B}\sum_{j=1}^{B} (E_x(i,j))^2; \; e_{x2} = \sum_{n=0}^{B-1} n^2 \sum_{i=1}^{B} \sum_{j=1}^{B} (E_x(i,j))^2; \; e_{x3} = \frac{\sum_{i=1}^{B}\sum_{j=1}^{B} ijE_x(i,j) - \mu_{x\;w}\mu_{x\;h}}{\sigma_{X\;w}\sigma_{X\;h}}; \; e_{x4} = \sum_{i=1}^{B}\sum_{j=1}^{B} (i - \mu_{x\;w})^2 E_x(i,j); \; e_{x5} = \sum_{i=1}^{B}\sum_{j=1}^{B} \frac{1}{1+(i-j)^2} E_x(i,j);$ $e_{x6} = \sum_{i=2}^{2B} iE_x(i); \; e_{x7} = -\sum_{i=2}^{2B} (i - e_{x\;6})^2 E_x(i); \; e_{x8} = -\sum_{i=1}^{B}\sum_{j=1}^{B} E_x(i,j)\log(E_x(i,j));$ $e_{x9} = -\sum_{i=2}^{2B} E_x(i)\log(E_x(i)+); \; E_x(i) = 0.$

Pixel $I_h(m_h, n_h)$ composing with distance characteristic $I\;d_{mn} = \{e_{xy}\}, \; x = \overline{1, X}, \; y = \overline{1, 9}$; m and n - coordinates.

The efficiency of the generalized algorithm for identifying images of pollen grains increases on the basis of the synthesis of parallel computing mechanisms, computational training schemes for a radial-basis neural network (RBNN), Kohonen's NN, a three-layer one, as a result of which the influence of a significant amount of calculations of texture features according to Haralik is eliminated.

2.2 Generalized Algorithms for Identification, Recognition, and Classification of Images of Micro-objects

The large dimension of problems, limited a priori information, parametric uncertainty, non-stationary dynamics of the image contour, a significant amount of information require consideration of problems of identification of micro-objects based on a new approach aimed at constructing mechanisms for regulating variable computational schemes of structural components, learning algorithms for multilayer NN, RBNN, Kohonen NN with clustering, as well as using the properties of self-approximation, self-learning, self-organization of networks.

Network learning mechanisms are associated with dividing the N-dimensional area of the image into L sections (segments), respectively, into clusters of representing input information for recognizing a micro-object. Each cluster is associated with a vector of variables of dimension M at the output of the NN, the elements of each of which are the approximated values of the points of the contour curve [6, 7].

The input layer of the network has $N_{1,l}(l = 1, 2, ..., L)$ neurons, each of which is represented as a product with a nonlinear neuron f_l. All the neurons in the intermediate layer $N_{2,m}(m = 1, 2, ..., M)$ are connected to the neurons in the output layer. The values of nonlinear neurons in this layer are given as.

$y_{1,l}^{(f)} = y_{1,l} \cdot f_l\left(\sum_{n=1}^{N} x_n w_{0,n;1,l}^{(f)}\right)$, $l = 1, 2, ..., L$, where f - nonlinear neuron element.

The values of the neurons of the output layer are defined as:

$$y_m = \sum_{l=1}^{L} y_{1,l}\left[w_{1,l;2,m} + w_{1,l;2,m}^{(f)} \cdot f_l\left(\sum_{n=1}^{N} x_n w_{0,n;1,l}^{(f)}\right)\right].$$

The weights of connections between nonlinear neurons of the intermediate layer and the output layer will be equal $w_{1,l;2,m}^{(f)} = 1,\ l = 1, 2, ..., L,\ m = 1, 2, ..., M$.

The values of the output neurons are set as $y_m = w_{1,\tau;2,m} + f_l\left(\sum_{n=1}^{N} x_n w_{0,n;1,\tau}^{(f)}\right)$.

A generalized algorithm for adaptive identification of images of micro-objects based on the Kohonen NN with a clustering mechanism is constructed, the features of which are considered in two stages.

Stage 1. Launching the self-organization mechanism of the neural network, in which the weights $w_{0,h;1,i}$ of neurons, the center of the corresponding cluster, are determined, the search for clusters is carried out. The L vector is divided into a smaller number of clusters, the centers of which correspond to the weight values of neurons.

Stage 2. The weight coefficients $w^{(f)}{}_{0,n;1,\phi}$and $w_{1,\phi;2,m}$of neurons are in accordance with the requirement of the network with backpropagation of errors in the form:

$$w_{1,\tau;2,m}(t+1) = w_{1,\tau;2,m}(t) + \eta(d_m - y_m);$$

$$w_{0,n;1,\tau}^{(f)}(t+1) = w_{0,n;1,\tau}^{(f)}(t) + \eta x_n f' \sum_{m=1}^{M}(d_m - y_m),\ 0 < \eta \le 1,$$

where d_m - m-th output of the Kohonen neural network (KNN) with clustering; c- learning rate coefficient; f' is a derivative of the activation function of a nonlinear neuron associated with neuron $N_{1,\phi}$.

The mechanism of the modified learning KNN performs the functions of calculating the distance between clustering objects, using statistical, dynamic, specific properties of images in the form $dst(x_i, x_j) = \sum_{n=1}^{N} \alpha^{(n)} d_{i,j}^{(n)} \delta_{i,j}^{(n)} / \sum_{n=1}^{N} \delta_{i,j}^{(n)}$, where x_i-th vector of the training set; δ^n - the influence of the n-th factor in the training set; $d_{i,j}^{(n)}$ is the distance between the n-th elements of the vectors i and j. If there is a gap of the n-th element for one of the compared vectors i or j, then the distance $d_{i,j}^{(n)} = 0$.

Now the generalized algorithm for identifying images of micro-objects based on neural networks includes mechanisms for filtering «noise». This version of the mechanism uses information about the nearest common neighbors (NCN), estimates of the degree of similarity of the input and reference vectors. The vector element with the lowest similarity coefficient value is considered to be susceptible to "noise" and is removed.

The mechanism algorithm includes the following steps.

Step 1. The choice of K values is the distance of the nearest neighbors; R - the minimum distance of the nearest common neighbors for two vectors, which determines the belonging to the R*, K, > R* > R clusters.

Step 2. Combining vectors into clusters according to the principle of cross-neighborhood. In this case, it is considered that two vectors are included in one cluster if between them it is possible to build a chain of neighbors, each of which will have an S indicator greater or equal to R*.

Step 3. Adding to clusters S, greater or equal value of R on the principle of maximization. If so, then the vector is added to the S cluster that has the maximum magnitude. If a vector has an equal distance of neighborhood with vectors from two, more clusters, then such a cluster is selected, in which the object is closer to the given vector in distance.

Step 4. The center of the clusters is determined, the execution of the procedure for this algorithm is specified in the following steps.

Step 4.1. Selection of selective A subsets and their values B.

Step 4.2. Selection of the largest S from $C_{\max}$ clusters by the following partition.

$$C_{\max} = \max_{p=1...S} \frac{1}{|C_s|} \sum_{p=1}^{|C_S|} \left(M_S, c_h^{(s)}\right), \quad \mu_s = \frac{1}{|C_s|} \sum^{|C_S|} c_h^{(s)}$$

where μ_s - the center of cluster C_s; $c_h^{(s)}$ - h-th vector of the C_s cluster.

Step 4.3. Formation of $C^1{}_{\max}$, $C^2{}_{\max}$, ..., $C^A{}_{\max}$ centroids.

Random selection from a set of $C_{\max}$, its initialization for pairs of clusters $M = \left\{\left(\mu_{\max}^{a,1}, \mu_{\max}^{a,2}\right)\right\}$, $a = 1, 2, ..., A$. Assigning each vector to one of the $C_{\max}^{a,1}$, $C_{\max}^{a,2}$ clusters. Determining the proximity of an element to the centroid. Recalculation of pairs of clusters M.

Step 4.4. Condition check.

$$\sum_{s=1}^{S} dst(m_s(t-1), m_s(t)) < \varepsilon,$$

where $\varepsilon \geq 0$ is the maximum allowable value of the weighted centroid deviation at two successive iterations; $\mu_s(t-1)$ - coordinates of the s-th centroid, obtained at the previous iteration; $\mu_s(t)$ - coordinates of the s-th centroid obtained at the current iteration [8].

Step 4.5. The choice of the coordinates of the centers of two clusters ${C^1}_{max}$ and ${C^2}_{max}$, formed from the cluster C_{max} so that the distance from the center of the cluster to all objects belonging to it was minimal.

Step 4.6. Saving objects C_{max} in two new clusters, increasing the counter S. If $S < L$, then go to step 2, otherwise set the coordinates of the corresponding clusters.

It has been found that a larger value of $R*$ and a small value of R under the condition of a $K >> R$ lead to the formation of a number of small clusters with a core consisting of several vectors.

To modify the algorithm of the considered mechanism, the principle of minimizing the total distance to the center of the combined clusters was implemented in the form.

$$\mu^* = \frac{1}{|C_i| + |C_j|}\left(\sum_{g=1}^{|C_i|} c_g + \sum_{h=1}^{|C_i|} c_h\right).$$

Figures 1, 2 show graphical illustrations of the efficiency of the generalized algorithm for identifying micro-objects according to the criterion of the mean square error and the speed of information processing (t sec), depending on the size of the training sets submitted for processing. The results of identification of micro-objects based on a three-layer NN, KNN with automatic clustering, RBNN are compared. Implementation of computational schemes for the recognition and classification of images of micro-objects are investigated for the values of the parameters: $R = 30,\ K = 50$.

From a large sample of initial information, 500 vectors were randomly taken that formed a control sample training (CS), the remaining 3677 vectors formed a reference sample (RS). The following types of training test samples (TS) were formed: with pseudo omissions of elements in the amount of 10%; categorical selections (CTS) with missing elements; TS with gaps in continuous data (ConTV); TS of mixed type (MTS). The procedures for the restoration of TS gaps with the receipt of CTS, ConTS and MTS have been tested Table 1. The values of the indicator of the identification rate of micro-objects are given [9–11].

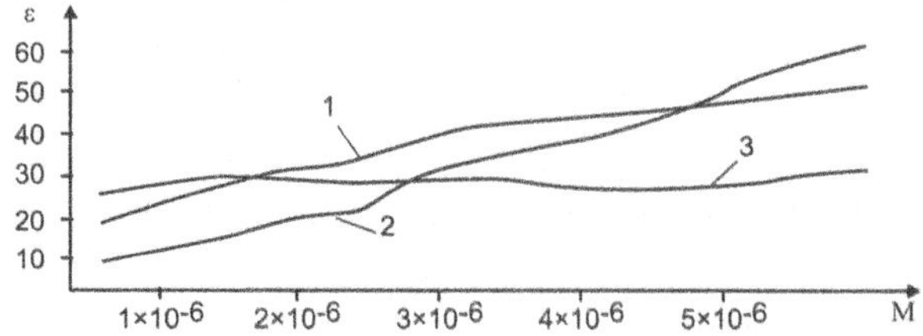

Fig. 1. Root mean square error identification of micro-objects based on three-layer NN – graph 1, Kohonen NN – graph 2, RBNN – graph 3.

The influences of the components caused by "noise", "smear"s and other types of bursts were investigated when recognizing micro-objects based on neural network identifiers. For this, pseudo-components are reflected in the image contour, evenly distributed over the range of change in point values.

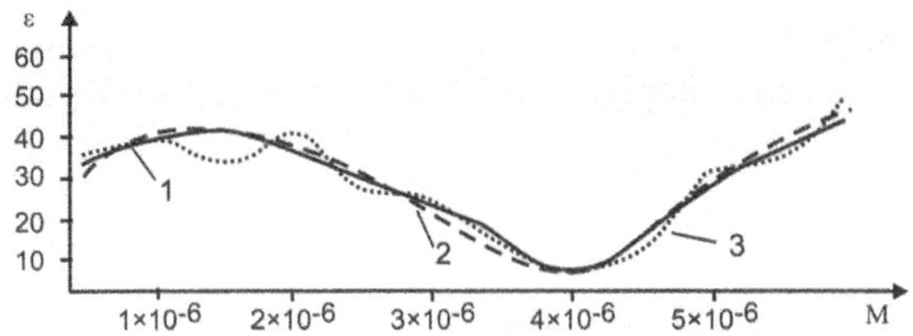

Fig. 2. Information processing rates of a three-layer NN – graph 1, Kohonen NN – graph 2, RBNN – graph 3.

Table 1. The values of the speed of identification of micro-objects based on the NN.

Networks	Test samples				Test sample with «noise»			
	CTS	ConTS	MTS	RS	*CTS	*ConTS	*MTS	*RS
Three-layer	205	241	164	122	191	233	235	162
RBNN	77	76	76	45	79	80	79	49
KNN	73	88	71	43	79	77	84	52
KNN with feature map	42	43	42	42	43	44	43	44

The following advantages of the mechanisms have been established: the speed of information processing during identification, recognition and classification of images of micro-objects based on the use of KNN rises from 1.7 to 5.4 times, elements with gaps are restored to 98%, the iteration of the clustering algorithm with a network learning map decreases 3–4 times, and when three-layer NN 6–8 times; KNN and RBNN, when working with reference images, are compatible with the difference in the learning rate, instability of the KNN learning rate is determined by the mechanism for choosing cluster centers, using the property of self-organization, self-learning, self-approximation of the network; KNN with automatic clustering learns 1.65 times faster and shows good speed results than a three-layer NN.

Table 2. Values of the mean square error of identification of micro-objects based on the NN.

Networks	Test samples				Test sample with «noise»			
	CTS	ConTS	MTS	RS	*CTS	*ConTS	*MTS	*RS
Three-layer	0,021	0,031	0,025	0,020	0,030	0,035	0,034	0,047
RBNN	0,024	0,023	0,027	0,022	0,034	0,041	0,048	0,056
KNN	0,027	0,028	0,031	0,024	0,036	0,042	0,048	0,059
KNN with feature map	0,023	0,025	0,028	0,021	0,036	0,039	0,038	0,051

Table 2 gives the values of the mean square error of identification of micro-objects based on the NN, obtained under real conditions of testing identification algorithms.

3 Conclusion

Constructive approaches, principles, methods of constructing technologies for identification, recognition, clustering of images of micro-objects are proposed, which are recommended for use in systems of medical diagnostics, environmental monitoring, selection, in the management of industrial and technological complexes.

The methodology for the improvement and development of computer vision systems for images of micro-objects is investigated, their main features and characteristics are highlighted, a range of models and algorithms is determined, a software package for the identification, recognition, and classification of images of pollen grains is developed. The efficiency of the implemented algorithms for the identification of micro-objects based on NN of various topologies is investigated with optimization mechanisms according to the criteria of the minimum mean-square error and information processing speed. It was found that the information processing speed of the algorithm based on the KNN with automatic clustering is several times higher than the KNN without the clustering mechanism, the three-layer NN, and RBNN. The superiority in the learning speed of the KNN with automatic clustering is provided due to its simplified architecture, the use of self-organizing network properties.

References

1. Rendondo, R., Marcos V.: Pollen segmentation and feature evaluation for automatic classification in bright-field microscopy. Comput. Electron. Agric. **110**, 56–69 (2017). https://www.sciencedirect.com/science/journal/01681699/
2. Chica, M.: Standard methods for inexpensive pollen loads authentication by means of computer vision and machine learning. https://www.arxiv.org/ftp/arxiv/papers/1511/1511.04320.pdf (2017)
3. Chica, M., Campoy, P.: Discernment of bee pollen loads using computer vision and oneclass classification techniques. J. Food Eng. **112**, 50–59 (2012)
4. Xanjina, N.Y., Zamyatina, Y.B.: Ispolzovaniye klassicheskix metodov i neyronnix setey dlya raspoznavaniya pilsevix zeren.Vestnik Permskogo universiteta. Matematika. Mexanika. Informatika. **4**(23), 111–119 (2013)
5. Chernix, A.S., Zamyatina, Y.B.: Issledovaniye vozmojnosti primeneniya ryada klassicheskix metodov dlya raspoznavaniya pilsevix zeren. Perm: Izd-vo Perm. gos. nas. issled. un-ta, (2012)
6. Hecht-Nielson, R.: Counterpropagation networks. In: Proceedings of the IEEE First International Conference on Neural Networks, New York, pp. 19–32 (1987)
7. Tarxov, D.A.: Neyronniye seti. Modeli i algoritmi. Kn. 18. - M.: Radiotexnika (2005)
8. Jumanov, I.I.: Optimizasiya obrabotki izobrajeniy mikroobyektov na osnove rekurrentnogo obucheniya neyronnoy seti i implikativnogo otbora informativnix priznakov.Problemi informatiki i energetiki **4**, 12 (2016)
9. Djumanov, O.I.:Adaptivnaya neyrosetevaya sistema vizualizasii izobrajeniy, raspoznavaniya i klassifikasii mikroobyektov.Vestnik SibGUTI. **2**, 76–86 (2008). ISSN 1998-6920

10. Jumanov, I.I.: Optimizasiya obrabotki dannix nestasionarnix obyektov na osnove nechetkix modeley identifikasii s nastroykoy parametrov. Jurnal Vestnik TUIT, **1**(41), 34–47 (2017)
11. Djumanov, O.I.: Neyrosetevaya sistema adaptivnoy obrabotki informasii nestasionarnoy prirodi pri upravlenii Vuzom. IN: Sb. mat. X mejd. nauch.-metod. konf. Vuzov i fakultetov telekommunikasiy, Rostov-na-Donu, pp. 81–85 (2008)

Fuzzy Type-2 Decision Making Method on Project Selection

Kamala Aliyeva(✉)

Azerbaijan State Oil and Industry University, 20 Azadlig Avenue, AZ1010 Baku, Azerbaijan
kamalann64@gmail.com

Abstract. Selection of appropriate project become a real challenge for multi-criteria group decisions if the group of decision-makers indicate their precedencies depending on the character of alternatives and various attributes in relation to their information of them. The goal of the project selection process is to analyze the vitality of a project and to approve or reject project suggestions on base of determined criteria. This decision is very complex because they need the determination, consideration and analysis of different tangible and intangible attributes. This article represents a multi-criteria group technique to decision-making for the problem of choosing a project in a fuzzy environment of the second type. A step-by-step procedure is used to rank and evaluate alternatives in the offered technique, and the best decision is selected taking into account both beneficial and unfavorable criteria. We use four sets of criteria weights to analyze the robustness of the used technique.

Keywords: Fuzzy numbers · Type-2 fuzzy sets · Multicriteria decision making · Uncertainty

1 Introduction

The problem of choosing a project is a very important point in making management decisions. The basic goal of the project selection task is to choose a portfolio of projects that meets the goals set without troubling the accessible capital resources. When choosing a project, two questions are of particular importance. They are critical factors and goals to consider, as well as methods that can be used to compare and select projects. In the first question, decision makers looked at different attributes such as raw materials availability, equipment, human resources, etc. Various goals are considered, such as minimizing costs and maximizing profits. Extended portfolio selection model with project divisibility and interdependence has been proposed by Lee et al. [1]. In most of the studies devoted to the problem of project selection, parameters were considered as deterministic. However, real problems are faced with imperfect and uncertain information. Based on the reliability criterion, Huang [2] included random fuzzy uncertainty in the choice of the project by integrating a genetic algorithm with random fuzzy modeling. Karlsson et al. [3] represented a technique for evaluating R&D project options by fuzzy mixed integer programming model for choosing an optimal R&D portfolio. Huang and Zhao [4] offered

© The Author(s), under exclusive license to Springer Nature Switzerland AG 2021
R. A. Aliev et al. (Eds.): WCIS 2020, AISC 1323, pp. 180–185, 2021.
https://doi.org/10.1007/978-3-030-68004-6_23

project selection and structuring, taking into account the uncertainty of net profit and investment costs. Rafiee and Kianfar [5] suggested a decision tree method to solve a multi-period project selection problem using the real options pricing technique. Perez and Gomez [6] suggested a sample of a multipurpose project portfolio with fuzzy constraints. The fuzzy numbers in the constraints were ordered and a comparing method on base of confidence levels was applied. Wang and Hwang [7] offered a method for selecting an R&D portfolio to overcome R&D uncertainty and flexibility of information. Wey [8] suggested the uncertainty of the accessible budget, the chances of success and the efficient allocation of the project team when choosing urban renewal projects. The main contribution of this paper is selection of the project by using fuzzy type-2 decision making method [9]. This article is organized as follows: Important definitions of type-2 fuzzy sets introduced in Sect. 2. In Sect. 3, the results of approache are determined from analysis.

2 Preliminaries

Definition 1 [9]. A type-2 fuzzy set, signed by $\tilde{A}$, is described by a type-2 membership function $\xi_{\tilde{A}}(x, u)$, where $x \in X$ and $\xi \in J_x \subseteq [0, 1]$, that is, $\tilde{A} = \{(x, u), \xi_{\tilde{A}}(x, u) | \forall x \in X, \forall u \in J_x \subseteq [0, 1]\}$ in which $0 \leq \xi_{\tilde{A}}(x, u) \leq 1$.

$\tilde{A}$ can also be expressed as $\tilde{A} = \int_{x \in X} \int_{u \in J_x} \frac{\xi_{\tilde{A}}(x,u)}{(x,u)}$, $J_x \subseteq [0, 1]$ where $\int \int$ denoted the union over all permissible x and u. For a universe of discrete discourse, $\int$ is replaced by $\sum$.

Definition 2 [9]. The domain of secondary membership function is called the primary membership of x. In the representation of $\tilde{A}$, J_x is the primary membership of x, where $J_x \subseteq [0, 1], \forall x \in X$.

Definition 3. [9]. Uncertainty in the primary membership of type-2 fuzzy sets, $\tilde{A}$ consists of a bounded area and it is called the footprint of uncertainty (FOU). It is the union of all primary membership, that is, $FOU\left(\tilde{A}\right) = \cup J_x, \forall x \in X$.

Definition 4. Let X be the universe of discourse. A triangular type-2 fuzzy sets, signed as $\tilde{A} = \{(a, c, b; \xi(c)), (a', c', b'; b')\}$, is characterized by piecewise linear function $\xi_{\tilde{A}}(u)$. It is defined as next function:

$$\xi_{\tilde{A}}(u) = \begin{cases} 0, & 0 \leq u \leq a, \ b \leq u \leq 1, \\ \frac{\xi(c)(u-a)}{(c-a)} & a \leq u \leq c, \\ \frac{\xi(c)(u-b)}{(c-b)} & c \leq u \leq b, \end{cases}$$

where $a < c < b$ and $\xi(c)$ is the maximum value of the secondary membership $\xi(u)$, that is $\max(\xi(u)) = \xi(c)$ and u is primary membership function, and characterized by a linear function $u_{\tilde{A}}(x)$. It is determined as follows:

$$u_{\tilde{A}}(x) = \begin{cases} 0, & 0 \leq x \leq a', \ b' \leq x \leq 1, \\ \frac{b(x-a')}{(c'-a')}, & a' \leq x \leq c', \\ \frac{b(x-b')}{(c'-b')}, & c' \leq x \leq b', \end{cases}$$

where $0 < a' < c' < b' < 1$ and b is the maximum value of the primary membership u, that is, $\max(u) = b$.

3 Statement of the Problem

Suppose that there are four possible projects A_1, A_2, A_3, A_4 for selection with four criteria's: C_1 - Technical performance, C_2 - Market potential, C_3 - Policy environment and C_4 - Investment risk.

Table 1. Linguistic values of type-2 fuzzy sets for linguistic terms.

Linguistic terms	Linguistic values of type-2 fuzzy sets	
	Secondary membership function	Secondary membership function
Extremely Poor	(0,0,0;0)	(0,0,0;0)
Very Poor	(0.1,0.2,0.3;0.3)	(0.1,0.2,0.3;0.3)
Poor	(0.2,0.3,0.4;0.4)	(0.2,0.3,0.4;0.4)
Slightly Poor	(0.25,35,0.45;0.45)	(0.25,35,0.45;0.45)
Fair	(0.3,0.4,0.5;0.5)	(0.3,0.4,0.5;0)
Slightly Good	(0.4,0.5,0.6;0.6)	(0.45,0.55,0.65;0.65)
Good	(0.5,0.6,0.7;0.7)	(0.55,0.65,0.75;0.75)
Very Good	(0.6,0.7,0.8;0.8)	(0.65,0.75,0.85;0.85)
Extremely Good	(0.7,0.8,0.9;0.9)	(0.7,0.8,0.9;0.9)

4 Solution of the Problem

The detailed steps involved in the presented method to find the best alternative are listed down:

Step-1. The four possible alternatives should be evaluated according to the four attributes appropriating to the linguistic meanings of type 2 fuzzy sets for linguistic terms, represented in Table 1. Weights of criteria based on expert's opinion and values of alternatives are represented in Table 2.

Step-2. By using model (1) we determine the ranking weight value W_j for criterion $\tilde{C}_j(j = 1, 2, 3, 4)$:

$$R\left(\tilde{A}_i\right) = \frac{1}{2}\left\{\frac{1}{3}\frac{c_i^3(b_i - a_i) + c_i\left(b_i^3 - a_i^3\right) - a_i b_i\left(b_i^2 - a_i^2\right)}{c_i^2(b_i - a_i) + c_i\left(b_i^2 - a_i^2\right) - a_i b_i(b_i - a_i)} + \frac{b_i\left(a' + 2b_i' + c_i'\right)}{2}\right\} \quad (1)$$

Weights are normalized by using the next formula:

$$W_j = \frac{R\left(w_j\right)}{\sum_{j-1}^{n} R\left(w_j\right)}. \quad (2)$$

Table 2. Weights of criteria's and values of alternatives

	$\tilde{C}_1$	$\tilde{C}_2$	$\tilde{C}_3$	$\tilde{C}_4$
$\tilde{A}_1$	(0.2,0.3,0.4,0.4; 0.25,35,0.45;0.45)	(0.4,0.5,0.6,0.6; 0.55,0.65,0.75;0.75)	(0.3,0.4,0.5,0.5; 0.55,0.65,0.75;0.75)	(0.5,0.6,0.7,0.7; 0.55,0.65,0.75,0.75)
$\tilde{A}_2$	(0.6,0.7,0.8,0.8; 0.7,0.8,0.9,0.9)	(0.2,0.3,0.4,0.4; 0.25,35,0.45;0.45)	(0.5,0.6,0.7,0.7; 0.55,0.65,0.75,0.75)	(0.3,0.4,0.5,0.5; 0.55,0.65,0.75;0.75)
$\tilde{A}_3$	(0.3,0.4,0.5,0.5; 0.55,0.65,0.75;0.75)	(0.4,0.5,0.6,0.6; 0.55,0.65,0.75;0.75)	(0.6,0.7,0.8,0.8; 0.7,0.8,0.9,0.9)	(0.5,0.6,0.7,0.7; 0.55,0.65,0.75,0.75)
$\tilde{A}_4$	(0.5,0.6,0.7,0.7; 0.55,0.65,0.75,0.75)	(0.6,0.7,0.8,0.8; 0.7,0.8,0.9,0.9)	(0.3,0.4,0.5,0.5; 0.55,0.65,0.75;0.75)	(0.4,0.5,0.6,0.6; 0.55,0.65,0.75;0.75)
W_j	(0.2,0.3,0.4,0.4; 0.25,35,0.45;0.45)	(0.4,0.5,0.6,0.6; 0.55,0.65,0.75;0.75)	(0.3,0.4,0.5,0.5; 0.55,0.65,0.75;0.75)	(0.5,0.6,0.7,0.7; 0.55,0.65,0.75,0.75)

The obtained results:

$$R\left(\tilde{A}_1\right) = 0.20,\ R\left(\tilde{A}_3\right) = 0.40,\ R\left(\tilde{A}_4\right) = 0.54,\ R\left(\tilde{A}_2\right) = 0.50,$$

$$\sum R(w_j)0.20 + 0.50 + 0.40 + 0.54 = 1.64,$$

$$w_1 = \frac{0.20}{1.64} = 0.12,\ w_2 = \frac{0.50}{1.64} = 0.30,\ w_3 = \frac{0.40}{1.64} = 0.25,\ w_4 = \frac{0.54}{1.64} = 0.33,$$

Step-3: For all alternatives $\widetilde{A}_i (j = 1,\ 2,\ 3,\ 4)$ weighted ranking value is determined by using Eq. (3):

$$R_{w_j}\left(\tilde{A}_i\right) = \sum_{j=1}^{n} w_j R(S_{ij}) \tag{3}$$

The results:

$$R_w\left(\tilde{A}_{11}^L\right) = 0.11,\quad R_w\left(\tilde{A}_{11}^U\right) = 0.126,$$

$$R_w\left(\tilde{A}_{12}^L\right) = 0.45,\quad R_w\left(\tilde{A}_{12}^U\right) = 0.59,$$

$$R_w\left(\tilde{A}_{13}^L\right) = 0.3,\quad R_w\left(\tilde{A}_{13}^U\right) = 0.49,$$

$$R_w\left(\tilde{A}_{14}^L\right) = 0.59,\quad R_w\left(\tilde{A}_{14}^U\right) = 0.64.$$

$$Rank\left(A_1^L\right) = \frac{1}{n(n-1)} \cdot \left(\sum P\left(A_1^L\right) + \frac{n}{2} - 1\right) = \frac{1}{12}(1.45 + 1) = 0.21,$$

$$Rank\left(A_1^U\right) = \frac{1}{n(n-1)} \cdot \left(\sum P\left(A_1^U\right) + \frac{n}{2} - 1\right) = 0.24,$$

$$Rank(A_1) = \frac{Rank(A_1^L) + Rank(A_1^U)}{2} = \frac{0.21 + 0.24}{2} = 0.225.$$

We apply all these calculations for other alternatives:

$$Rank(A_2) = \frac{Rank(A_2^L) + Rank(A_2^U)}{2} = \frac{0.20 + 0.23}{2} = 0.215,$$

$$Rank(A_3) = \frac{Rank(A_3^L) + Rank(A_3^U)}{2} = \frac{0.23 + 0.26}{2} = 0.245,$$

$$Rank(A_4) = \frac{Rank(A_4^L) + Rank(A_4^U)}{2} = \frac{0.22 + 0.26}{2} = 0.24.$$

Step-4: In last step we rank the alternatives and choose best alternative with weighted ranking values $R_{w_j}(\widetilde{A}_i)$. The results:

$$Rank(A_1) = 0.225,$$

$$Rank(A_2) = 0.215,$$

$$Rank(A_3) = 0.245,$$

$$Rank(A_4) = 0.24,$$

$$A_3 > A_4 > A_1 > A_2.$$

This ranking represents that alternative A_3 is best and A_2 is the worst alternatives in project selection.

5 Conclusion

In this work, is used the method of ranking triangular type-2 fuzzy sets. In the proposed technique, was used the traces of uncertainty as a set of the primary membership function and the centroid of the secondary membership function. This represented that the technique that is used in paper provides four acceptable characteristics of the fuzzy value, so it is more expedient than the presenting selecting technique. The offered ranking method is applied to a multi-attribute decision making problem for determining the best alternative for various attributes. As a conclusion, in this work, it was possible to conclude that the most famous alternatives associated with the project is A_3.

References

1. Li, X., Fang, S.C., Guo, X., et al.: An extended model for project portfolio selection with project divisibility and interdependency. J. Syst. Sci. Syst. Eng. **25**, 119–138 (2016). https://doi.org/10.1007/s11518-015-5281-1
2. Huang, X.: Optimal project selection with random fuzzy parameters. Int. J. Prod. Econ. **106**(2), 513–522 (2007). https://doi.org/10.1016/j.ijpe.2006.06.011
3. Carlsson, C., Fullér, R., Heikkilä, M., Majlender, P.: A fuzzy approach to R&D project portfolio selection. Int. J. Approximate Reasoning **44**(2), 93–105 (2007). https://doi.org/10.1007/3-540-34783-6_2
4. Huang, X., Zhao, T.: Project selection and scheduling with uncertain net income and investment cost. Appl. Math. Comput. **15**(247), 61–71 (2014). https://doi.org/10.1016/j.amc.2014.08.082
5. Rafiee, M., Kianfar, F.: A scenario tree approach to multi-period project selection problem using real-option valuation method. Int. J. Adv. Manuf. Technol. **56**(1), 411–420 (2011). https://doi.org/10.1007/s00170-011-3177-x
6. Perez, F., Gomez, T.: Multiobjective project portfolio selection with fuzzy constraints. Ann. Oper. Res. **245**(1–2), 7–29 (2016). https://doi.org/10.1007/s10479-014-1556-z
7. Wang, J., Hwang, W.L.: A fuzzy set approach for R&D portfolio selection using a real options valuation model. Omega **35**(3), 247–257 (2007). https://doi.org/10.1016/j.omega.2005.06.002
8. Wey, WM.: A multiobjective optimization model for urban renewal projects selection with uncertainty considerations. In: Fourth International Conference on Natural Computation. ICNC 2008, 18 October 2008, vol. 6, pp. 423–427. IEEE (2008). https://doi.org/10.1109/icnc.2008.554
9. Mendel, J.M., John, R.I.B.: Type-2 fuzzy sets made simple. IEEE Trans. Fuzzy Systems **10**, 117–127 (2002). https://doi.org/10.1109/91.995115

IoT and Intelligent Wireless Sensor Network for Remote Monitoring Systems of Solar Power Stations

Ilkhom Siddikov(✉), Khalim Khujamatov, Doston Khasanov, and Ernazar Reypnazarov

Tashkent University of Information Technology named after Muhammad al-Khwarizmi, Tashkent 100200, Uzbekistan
ikhsiddikov@mail.ru, kh.khujamatov@gmail.com, dhasanov0992@gmail.com, reypnazar0vernazar@gmail.com

Abstract. Nowadays, the widespread use of the Internet of Things in real life has paved the way for the emergence of systems such as Smart Homes, Smart Cities, and Smart Industries. They are called smart because of their ability to work autonomously. That is, smart systems have the ability to self-control and manage by sensing various physical influences and changes, and making decisions when necessary. Today, such systems detect physical processes and changes using special wireless sensors.

In this paper was explored smart wireless sensor networks (WSN) that can be used in the monitoring systems of solar power station (SPS). Six main types of sensors (such as voltage, current, temperature and humidity, illumination, solar radiation, and dust) were obtained for this monitoring system. Based on the structure and working principle of each sensor, the method of connecting to the wireless network module was illustrated. WSN for monitoring system of SPS was built, on the AnyLogic software, a simulation was created and results were obtained on its characteristics.

Keywords: Smart systems · IoT · Wireless sensors · Sensor nodes · Remote monitoring · SPS · WSN · AnyLogic

1 Introduction

With the development of wired and wireless network technologies, home appliances, industrial devices and various sensor devices connected to the Internet are increasing. Such systems collectively can be called the Internet of Things (IoT). In essence, IoT is an information exchange environment in which tools used in everyday life or in industry are connected to wired and wireless networks. In the last time, this technology has been used not only in the field of household electronics and application, but also in other fields such as Healthcare, Smart City, Smart Home, Smart Car, Energy System and Industrial Safety [1]. Besides, the applying of this technology in various power generation systems, as well as their remote control and management, has been proven by many scientists to be

© The Author(s), under exclusive license to Springer Nature Switzerland AG 2021
R. A. Aliev et al. (Eds.): WCIS 2020, AISC 1323, pp. 186–195, 2021.
https://doi.org/10.1007/978-3-030-68004-6_24

highly effective. One such power generation system is solar power station (SPS) based on photovoltaic panels. Solar energy is becoming a potential solution for sustainable energy supply in the future. Therefore, the creation of renewable solar power station for various industries is performing intensively [2]. However, there is growing privation to monitor real-time data received from solar photovoltaic plants in order to optimize the overall performance of solar power stations, increase their stability, and generate electrical energy.

The presence of large amounts of voltage in power plants and its sufficient risk to human life is another important reason for their remote monitoring and control. For the monitoring system, the use of digital technologies, more modern computing equipment and IoT technologies is giving their results. The ability of solar power stations to generate electrical energy is directly related to sunlight, temperature, dust on the panel surface, the amount of light in the environment, and other similar natural factors.

The paper uses IoT and smart wireless sensor approaches in the development of a remote monitoring system for a solar power station. Remote monitoring systems overcome the shortcomings of traditional wired networks [3], simplify the process of measuring and controlling parameters, and in such IoT-based systems, web monitoring plays an important role in the development of smart decisions [4]. The decentralized architecture and installation flexibility of remote monitoring and control systems are proposed as the optimal solution for its use in various fields. One of the most important issues for solar panels is the maintenance process, which includes the detection of system faults and malfunctions. It also allows for comprehensive use of equipping, electrical technology and computer software together [5].

In general, remote monitoring systems should be able to remotely receive, analyze, transmit, manage, and feedback the information remotely [6] using the advanced direction in communication technologies and other fields. However, such systems should include automatic diagnostic methods for solar power plants. Nowadays, widespread monitoring of the solar power system causes some problems, such as autonomous working mode and real-time [7]. These problems can be prevented remotely by monitoring and controlling environmental information via smart wireless sensor networks.

2 Configuration of Intelligent WSN

Current developments related to WSN include new features that provide a certain intelligence for this type of network. It is one of the technologies required and used for Smart Home, Smart City or Smart Communication.

To add a smart function to WSNs, it is necessary to propose intelligent algorithms that can manage all restrictions and requirements together and make decisions on optimizing WSN performance [8]. Restrictions and requirements include data received from WSNs (battery status, transmitted data, etc.), delays, channel or security constraints (delay limit, counter node positions, adverse propagation conditions) and the intended performance of the WSN (minimum power consumption, service lifetime, and data transmission coverage) and routing [9, 10].

By parameters to be configured (see Fig. 1), the algorithm can work dynamically on the following parameters: first, an algorithm that allows for the strategic placement of sensors for the position of the sensors; second, an algorithm that detects whether sensors transmit signals in conjunction with other sensors (using beamforming) or individually (without beamforming). That is, one node can transmit data individually, with another node, or with several other nodes [11]. Different methods can be selected depending on the type and status of the network being built to implement a smart algorithm for WSNs [8]. This research was carried out on the second type of Optimization Algorithm for intelligent WSN named «Parameters to be configured».

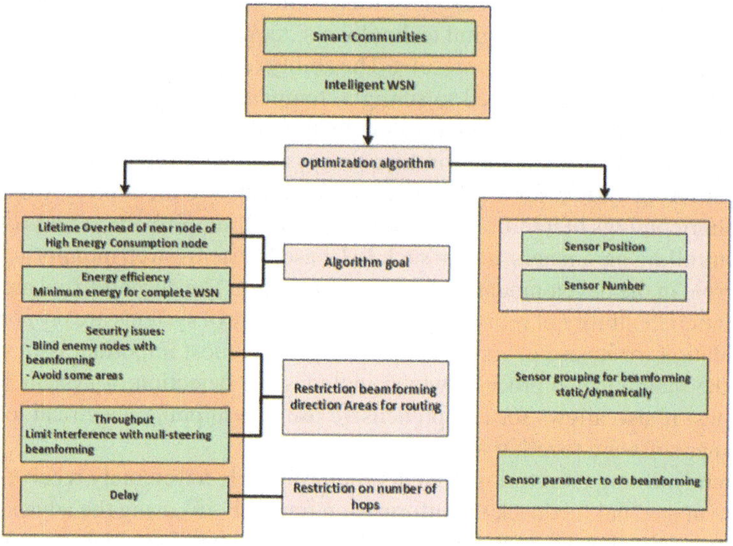

Fig. 1. Construction of an optimization algorithm for the intelligent function of WSNs.

3 The Principle of Construction of Intelligent Wireless Sensor Networks

When developing a system for remote monitoring of solar energy sources of Telecommunication systems through smart wireless sensor networks, required to pay attention to all the parameters [12, 13]. Therefore, eight indicators have been identified as key parameters for remote monitoring of SPS in telecommunications systems:

- Power generated by SPS;
- The level of radiation of the sun;
- The amount of illumination of the environment;
- Temperature of environment and station equipment;
- Ambient humidity;
- The amount of dust on the panel surface;
- Voltage at SPS output;
- Current at SPS output.

These indicators determine the main characteristics of the system. For this reason, six sensors were selected for monitoring of SPS. These are current, voltage, temperature & humidity, illumination, radiation, and dust sensors.

Today, high-precision, reliable, and efficient multifunctional sensor devices are widespread. Therefore, it's so important to design a high-speed system that transmits data over a network.

Bluetooth, ZigBee, Wi-Fi, GSM, Sigfox and LoRa are the most commonly used wireless technologies in WSNs. In this paper was used ZigBee (IEEE 802.15.14) wireless technology [14, 15]. This technology allows exchanging relatively small amounts of data over long distances with low energy consumption. However, it is a widely used standard in monitoring systems today [16]. ZigBee technology can be configured based on a star, cluster tree or mesh topology [17, 18].

We want to propose the structure of a wireless sensor node device designed based on the construction concept of SPS monitoring system. The device is based on sensors, AVR microcontroller computing system and ZigBee wireless network module (see Fig. 2).

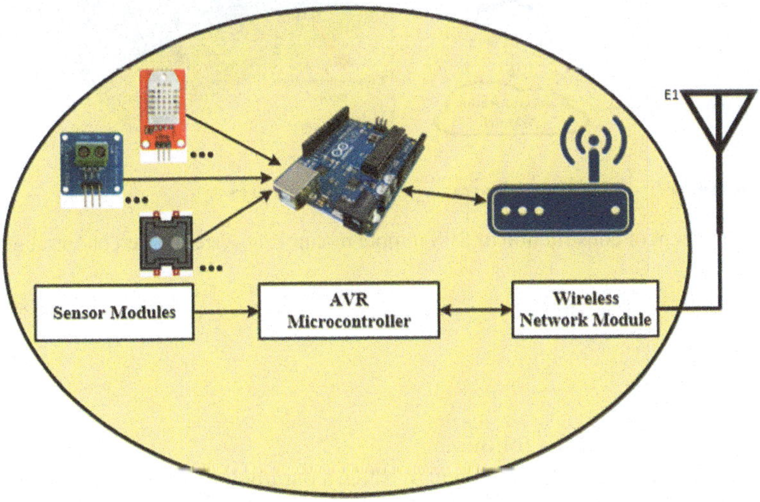

Fig. 2. The structure of the end node device of a wireless sensor network.

This device represents a single end node setup for WSN. There can be many such wireless sensor nods for a monitoring system.

4 Construction Concept of SPS Remote Monitoring System

Typically, a SPS monitoring system is performed on hundreds of solar panels located over a large area [19]. An IoT-based smart wireless sensor network solution was chosen for SPS because the transmission of data between monitoring equipment to the control room of the monitoring center via wired network solutions requires more time, labour and cost [20, 21].

There are several important factors that need to be tested when evaluating wireless network technologies used in SPS monitoring. Key factors include: monitoring coverage, topology, data requirements, network security, and energy necessity [22]. The importance and value of the various factors of wireless networks depends on a special system on which wireless technology is installed for remote monitoring of SPS [23, 24].

Consequently, the general concept of SPS monitoring system based on intelligent SST can be divided into 3 areas (see Fig. 3):

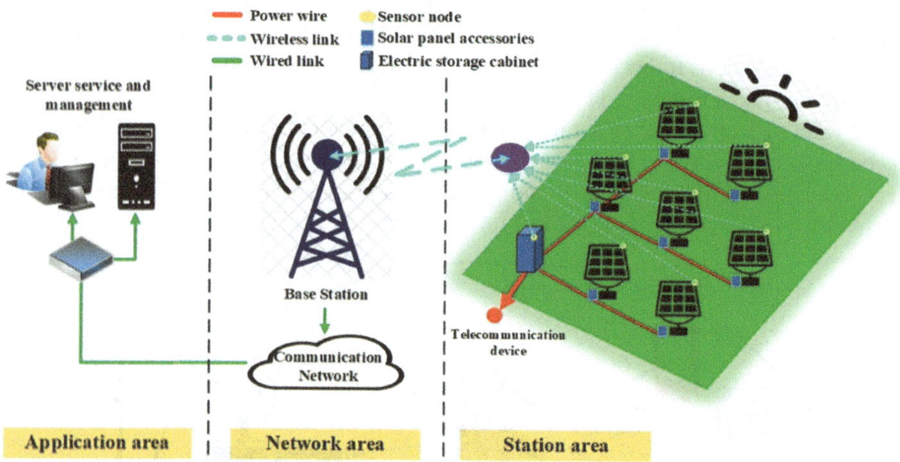

Fig. 3. The concept of construction of SPS remote monitoring system based on intelligent WSN.

- Station area;
- Network area;
- Application area.

They can be described as follows.

The setting required for a intelligent WSN is a parameter for monitoring devices that need to be installed in different locations along the area where the panels are installed. The number of monitoring devices required across the SPS installation area depends on the area of the station and the accuracy of the required measurement details.

The station-area parameters of this monitoring system are detected by the WSN and transmitted to the network level. The station-area wireless sensor network structure of the monitoring system under research is illustrated in Fig. 4. This wireless network is based on a combination of star and mesh topologies. The sensor network consists of 3 nodes: Sensor node, Router node and Coordinator node [25, 26]. In this case, the number of sensor nodes and routers can be selected relative to the size of the station area. The coordinator node is one for the specified area.

Sensor nodes measure station-area parameters such as current, voltage, panel temperature, ambient temperature and humidity, device temperature, illumination, and solar radiation. The measured parameters are transmitted to the router to which the sensor node is connected. The router transmits the received parameters to the Coordinator using the "Link State" algorithm. At the same time, the Coordinator node performs tasks such

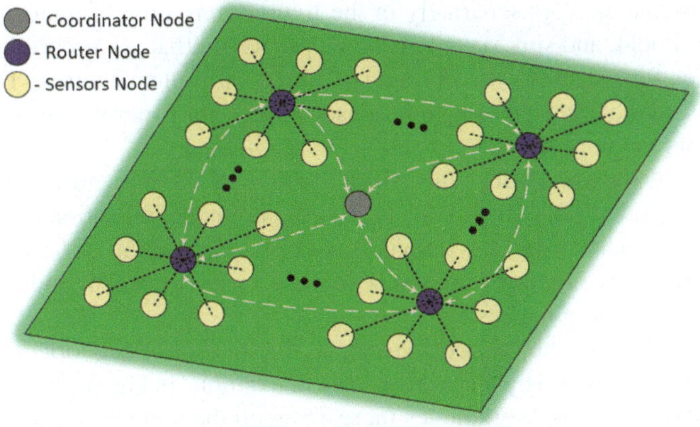

Fig. 4. Station-area wireless sensor network.

as managing the wireless sensor network at the station area, assigning IP addresses to them, and selecting routing directions. Data received at the network area is sent to the application level for viewing, storage and analysis over the local or Internet network.

At the application level, solar panels monitoring, fault monitoring, maintenance, inspection planning, parameter analysis and system status prediction are done, and various decisions are made based on them.

5 Simulation and Results

A network simulation is created in AnyLogic to obtain and analyze various characteristics of the wireless sensor network at the station level illustrated and described above (see Fig. 5). This network simulation is configured for a single Coordinator, four routers, as well as multiple sensor nodes.

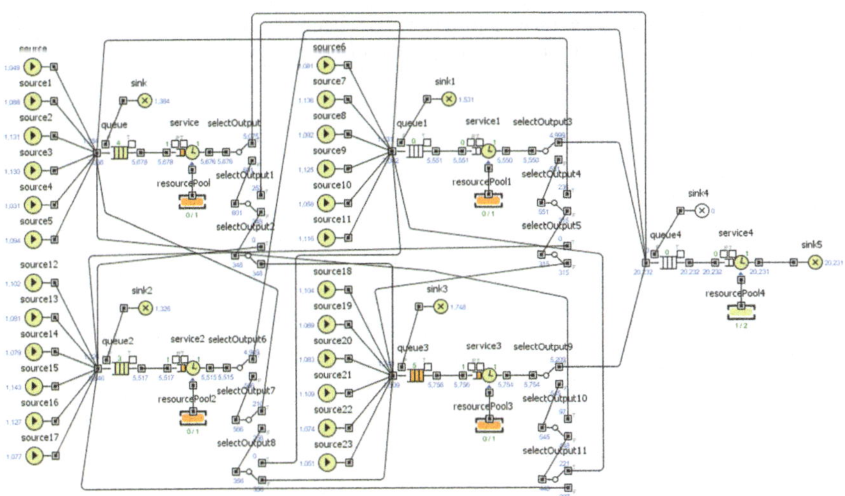

Fig. 5. Simulation model of Intelligent WSN.

The section that appears separately in the figure (consisting of queue4, sink4, service4, resourcePool4, and sink5) contains node properties that act as coordinators. Similarly, the other four sections are configured based on router features. The "Source" sections represent the corresponding end sensor nodes. They transmit the collected data from the sensors to the network.

As a result of the simulation, the characteristics of the utilization dependence of each router (Util_1, Util_2, Util_3, Util_4), coordinator node utilization (Util_5), packet queuing time (w), packet time in system (U) and packet loss in the system (Packet Loss) on the packet intensity coming from the sensors (λ) were obtained, and their graphics are given in the following figures.

According to the proposed network structure, the state of the network can be assessed in terms of data transmission by these graphics. For example, in Fig. 6a, it can be observed that the utilization of the network nodes increases with the number of transmissions per second of the sensor nodes. Or, as shown in Fig. 7a, the loss of packets in the network increases as the number of transmissions per second from the sensor nodes (Sources) increases. In this case, the high number of the packet loss corresponds to a transmission of 6–12 times per second. This means that for the proposed intelligent WSN structure, setting the number of transmissions of the sensor nodes to a maximum of 6 times per second provides the highest efficiency in the network.

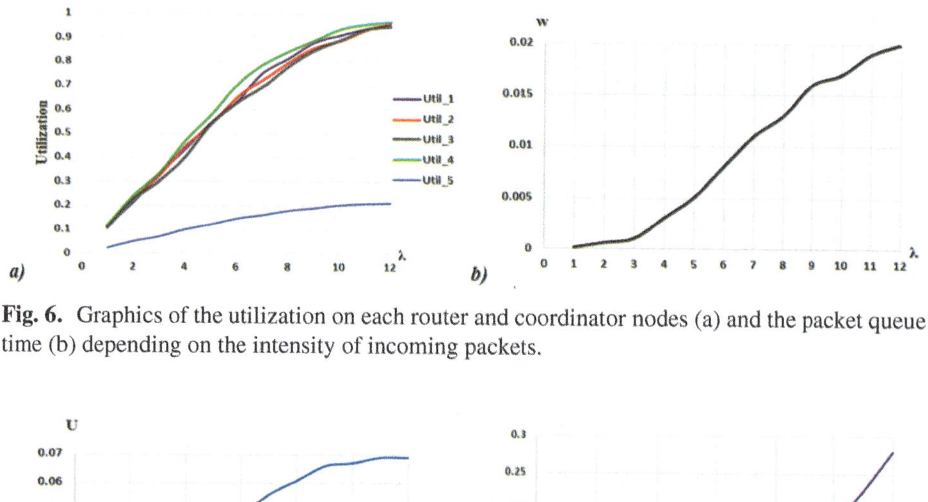

Fig. 6. Graphics of the utilization on each router and coordinator nodes (a) and the packet queue time (b) depending on the intensity of incoming packets.

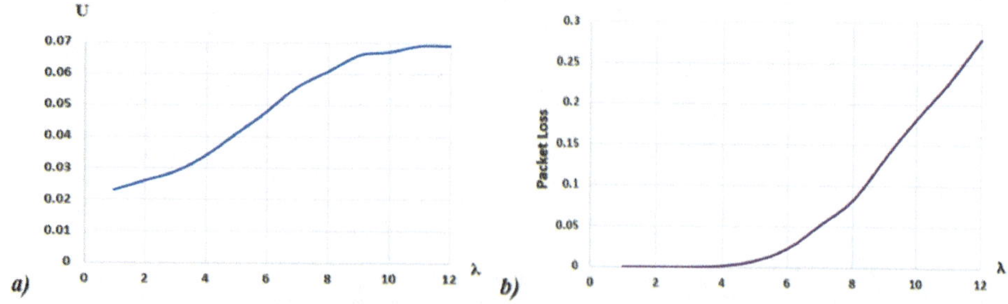

Fig. 7. Graphics of the packet time in the system (a) and the packet loss in the system (b) depending on the intensity of incoming packets.

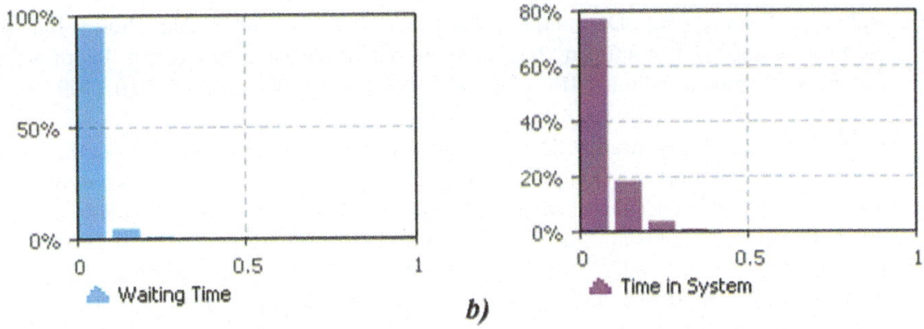

a) *b)*

Fig. 8. Histogram of distribution of packet queuing time (a) and packet time in the system (b).

6 Conclusion

As mentioned above, an optimization algorithm is required to add the concept of intelligence to wireless sensor networks. The optimization algorithm is divided into two types: Optimization parameters and Configuration parameters.

In this research work, the concept of building an intelligent wireless sensor network based on the second parameter, namely the Configuration parameters, was proposed for the SPS monitoring system. In the paper, a network structure was created based on the selection of the number of sensors by area size, the placement of sensor nodes across the field, and their data transmission parameters. However, the «Link State» routing algorithm is selected for the Station-level wireless sensor network according to the Optimization Parameters.

References

1. Trung, P.M., Mariappan, V., Cha, J.S.: A study on IoT/LPWA-based low power solar panel monitoring system for smart city. J. Korea Inst. Intell. Transp. Syst. **18**(1), 74–82 (2019). https://doi.org/10.12815/kits.2019.18.1.74
2. Padmanathan, K., Govindarajan, G., Ramachandaramurthy, V.K., Sudar, U.S., Jeevarathinam, B.: Integrating solar photovoltaic energy conversion systems into industrial and commercial electrical energy utilization—asurvey. J. Ind. Inf. Integr. **10**, 39–54 (2018). https://doi.org/10.1016/j.jii.2018.01.003
3. Serpen, G., Li, J., Liu, L.: AI-WSN: adaptive and intelligent wireless sensor network. Procedia Comput. Sci. **20**, 406–413 (2013). https://doi.org/10.1016/j.procs.2013.09.294
4. Lif, S.: Intelligent wireless sensor nodes in water monitoring systems. In: IEEE Workshop on Environmental, Energy, and Structural Monitoring Systems Proceedings, pp. 192–197 (2014). https://doi.org/10.1109/eesms.2014.6923290
5. Elov, J., Khamdamov, U., Makhmanov, O.: The data structure and information model for the information system of monitoring the educational process in the medical field. In: International Conference on Information Science and Communications Technologies ICISCT 2019, Tashkent, Uzbekistan (2019). https://doi.org/10.1109/icisct47635.2019.9012053
6. Wang, X., Li, J., Gao, X.: Research on remote intelligent monitoring system. In: International Conference on Uncertainty Reasoning and Knowledge Engineering, pp. 52–55 (2011)

7. Adhya, S., Saha, D., Das, D., Jana, J., Saha, H.: An IoT based smart solar photovoltaic remote monitoring and control unit. In: 2nd International Conference on Control, Instrumentation, Energy & Communication (CIEC), pp. 432–436 (2016). https://doi.org/10.1109/ciec.2016.7513793
8. Khujamatov, Kh.E., Khasanov, D.T., Reypnazarov, E.N.: Research and modelling adaptive management of hybrid power supply systems for object telecommunications based on IoT. In: International Conference on Information Science and Communications Technologies ICISCT 2019, Tashkent, Uzbekistan (2019). https://doi.org/10.1109/icisct47635.2019.9011831
9. Luna, F., Valenzuela-Valdes, J.F., Sendra, S., Padilla, P.: Intelligent wireless sensor network deployment for smart communities. IEEE Commun. Mag. **56**(8), 176–182 (2018)
10. Khamdamov, U.: Algorithms for multithreaded parallel processing of signals on heterogeneous digital telecommunication systems. In: International Conference on Information Science and Communications Technologies ICISCT 2019, Tashkent, Uzbekistan (2019)
11. Davronbekov, D., Isroilov, J., Akhmedov, B.: Principle of organizing database identification on mobile devices by IMEI. In: International Conference on Information Science and Communications Technologies ICISCT 2019, Tashkent, Uzbekistan (2019)
12. Matyokubov, U.K., Davronbekov, D.A.: The impact of mobile communication power supply systems on communication reliability and viability and their solutions. Int. J. Adv. Sci. Technol. **29**(5), 3374–3385 (2020)
13. Muradova, A., Khujamatov, Kh.: Results of calculations of parameters of reliability of restored devices of the multiservice communication network. In: International Conference on Information Science and Communications Technologies ICISCT 2019, Tashkent, Uzbekistan (2019)
14. Wei-zhao, Zh., Sheng-qian, M.A., Man-hong, F., Jin-long, L.: Design of temperature and humidity intelligent monitoring system. In: International Conference on Electrical and Control Engineering, pp. 782–785 (2010). https://doi.org/10.1109/icece.2010.200
15. Siddikov, I.Kh., Sattarov, Kh.A., Khujamatov, Kh.E.: Modeling of the transformation elements of power sources control. In: International Conference on Information Science and Communications Technologies (ICISCT) Applications, Trends and Opportunities, Tashkent, Uzbekistan, 2–4 November 2017
16. Davronbekov, D.A., Aliev, U.T., Isroilov, J.D., Alimdjanov, X.F.: Power providing methods for wireless sensors. In: International Conference on Information Science and Communications Technologies. Applications, Trends and Opportunities, ICISCT (2019). https://doi.org/10.1109/icisct47635.2019.9011850
17. Khujamatov, Kh., Reypnazarov, E., Khasanov, D., Akhmedov N.: Networking and computing in Internet of Things and cyber-physical systems. In: The 14th IEEE International Conference Application of Information and Communication Technologies, Tashkent, Uzbekistan, 07–09 October 2020
18. Khujamatov, Kh., Reypnazarov, E., Akhmedov, N., Khasanov, D.: Blockchain for 5G healthcare architecture. In: International Conference on Information Science and Communications Technologies ICISCT 2020, Tashkent, Uzbekistan (2020)
19. Khujamatov, Kh.E., Khasanov, D.T., Reypnazarov, E.N.: Modeling and research of automatic sun tracking system on the bases of IoT and Arduino UNO. In: International Conference on Information Science and Communications Technologies ICISCT 2019, Tashkent, Uzbekistan (2019). https://doi.org/10.1109/icisct47635.2019.9011913
20. Khujamatov, Kh., Reypnazarov, E., Akhmedov, N., Khasanov, D.: IoT based centralized double stage education. In: International Conference on Information Science and Communications Technologies ICISCT 2020, Tashkent, Uzbekistan (2020)

21. Siddikov, I.Kh., Sattarov, Kh.A., Khujamatov, Kh.E., Dekhkonov, O.R.: Modeling the processes in magnetic circuits of electromagnetic transdusers. In: International Conference on Information Science and Communications Technologies ICISCT 2016, Tashkent, Uzbekistan, 2–4 November 2016
22. Khujamatov, Kh., Reypnazarov, E., Khasanov, D., Akhmedov, N.: Industry digitalization concepts with 5G-based IoT. In: International Conference on Information Science and Communications Technologies ICISCT 2020, Tashkent, Uzbekistan (2020)
23. Siddikov, I.Kh., Sattarov, Kh.A, Khujamatov, Kh.E: Modeling and research circuits of intelligent sensors and measurement systems with distributed parameters and values. "Chemical technology control and management". Int. Sci. Tech. J. **2018**(3), 50–55 (2018)
24. Siddikov, I.Kh., Khujamatov, Kh.E., Khasanov, D.T., Reypnazarov, E.N.: Modeling of monitoring systems of solar power stations for telecommunication facilities based on wireless nets. Int. Sci. Tech. J. Chem. Technol. Control Manage. **2**(93), 20–28 (2020)
25. Khujamatov, Kh., Khalil, A., Reypnazarov, E., Khasanov, D.: Markov chain based modeling bandwith states of the wireless sensor networks of monitoring system. Int. J. Adv. Sci. Technol. **29**(4), 4889–4903 (2020)
26. Arsheen, S., Wahid, A., Ahmad, Kh., Khujamatov, Kh.: Flying Ad hoc network expedited by DTN scenario: reliable and cost-effective MAC protocols perspective. In: The 14th IEEE International Conference Application of Information and Communication Technologies, Tashkent, Uzbekistan, 07–09 October 2020

Path Planning Collision Free for Warehouse Mobile Robot

A. B. Sultanova[1,2(✉)] and M. Y. Abdullayeva[2]

[1] Azerbaijan State Oil and Industry University, Azadlig Avenue 20, 1010 Baku, AZ, Azerbaijan
saxira@mail.ru

[2] Institute of Control Systems of ANAS, Baku, Azerbaijan
mayaabdullayeva@hotmail.com

Abstract. In this paper, we propose a new method for planning and optimizing the trajectory of warehouse robots using an improved algorithm. The paper reflects algorithms for planning the trajectory of collision less movement of a warehouse mobile robot and describing the functioning of the system. The main goal of many scientific research in recent decades is to replace people with automatic systems and robots in order to improve the efficiency of modern technological processes and various industries. Thus, the problem of creating mobile robotic systems that can act independently and perform the task is urgent. At the same time, an important role is played by the issue of creating an intelligent control system that allows the robot to independently perform the task in a closed environment with minimal human participation.

Keywords: Automation · Mobile robot · Control algorithm · Warehouse · Path planning · Motion planning · Uncertainty

1 Introduction

In accordance with the principles of supply chain management, modern companies strive to achieve large volumes of production and distribution, using minimal resources throughout the logistics chain, which must be delivered in the shortest possible time. An automatic controlled vehicle is widely used for this purpose.

Starting in the late 1950s, cars operated by traditional manual operations were replaced by cars with automatic controls [1, 2] These machines were first introduced to provide faster delivery of goods and products to warehouses and production areas. Automation of cargo transportation in the warehouse system is necessary to increase labor productivity and reduce costs for modern logistics companies. The use of robots, which are the main element of automatic warehouse and logistics systems, allows you to plan, optimize tasks, increase productivity and reduce costs. Using mobile robots to select shelves inside the warehouse, load and unload cargo, and plan the optimal placement of goods on warehouse shelves saves time and money for the warehouse business.

Mobile warehouse robots are independent wheeled mobile robots that can move shelves inside a warehouse from a storage area to a consolidation area [3, 5]. The main

© The Author(s), under exclusive license to Springer Nature Switzerland AG 2021
R. A. Aliev et al. (Eds.): WCIS 2020, AISC 1323, pp. 196–200, 2021.
https://doi.org/10.1007/978-3-030-68004-6_25

task of the mobile warehouse robot is to load, unload, deliver cargo to the desired area, place and reverse movement inside the warehouse to facilitate the operation of picking orders [3, 6].

One of the main challenges in this area is the provision of navigation capabilities inside warehouses [2, 4]. Various methods are used for this purpose. Some use markings on the floor, walls or installation of special markings (wireless or reflective) on the shelves when preparing the shed, mainly systems based on intelligent systems are used. For example, image recognition, computer vision systems, and so on [4–7].

2 Method

In order for robots to move freely around the warehouse, you need to plan the trajectories of mobile robots. Trajectory planning refers to building a trajectory from the starting point to the end point where the robot should move. The movement must be set so that the robot does not encounter any obstacles along the path of movement. A possible point (for example, Q) in the robot's consolidation zone may be the starting point. The target (destination point) point can be any shelf (rack B) that depends on the customer's order.

The pure pursuit algorithm was originally developed as a method for calculating the arc required to return the robot to its trajectory. This first application of the method came with the Terragator, a six-wheeled robot with skids that was used for outdoor experiments in the early 80's. The standard references to the original conclusions of this work refer to Wallace [9].

There is a set of path planning methods, obstacle avoidance algorithms, and pure pursuit algorithms. Testing all these algorithms showed that the pure pursuit method was the most reliable and reliable method. We have modified this method. The method we proposed allows you to work with group work in a closed and undefined environment.

2.1 Current Situation and Warehouse Mobile Robot

There is no fully robotic warehouse in the world today. Professional and systematic research on road planning for the acquisition of warehouse and logistics robots is relatively small. It is necessary to design complex facilities on the basis ofuolual and fuzzy control systems developed to ensure complete independence of logistic buildings. Smart warehouse is a new automated warehouse built on the basis of artificial intelligence (AI) technology. Intelligent warehouse has high automation, low labor cost and high efficiency.

The logistics warehouse robot plays a key role in the process of creating an intelligent warehouse. Thus, a logistics warehouse robot is a robot that is widely used in the field of warehouse and logistics technologies. This type of robot can independently perform tasks such as placing, transporting goods, and sorting goods.

Currently, warehouse and logistics robots are mainly divided into two categories. The first category includes the Kiva robotic system provided by Amazon, and the other category includes robots that perform the function of receiving and delivering goods provided by Fetch. Despite the fact that research on logistics robotics technologies starts late, the pace of development of these companies is higher.

3 Results and Discussions

The warehouse (xy) is described in a two-dimensional coordinate plane. The warehouse has walls, shelves and other processing equipment as a static barrier. The task of planning the trajectory of a logistics warehouse robot is to determine its position after existing positions in the working environment, preventing the robot from colliding with other robots and obstacles (of a static nature), to determine the length of the route and the time to reach the goal. The issue of planning the trajectory of a logistics warehouse robot is formed using the following principles.

The starting position (G_K) of the robots and the target point (A_K, B_K, C_K) are given in a two-dimensional working environment. 2-dimensional workspace each warehouse robot is oriented in different directions and in different positions to avoid obstacles:

$$A_k = \left(a_k^1 \ldots, a_k^d \ldots, a_k^n\right);\ B_k = \left(b_k^1 \ldots, b_k^d \ldots, b_k^m\right);\ C_k = \left(c_k^1 \ldots, c_k^d \ldots, c_k^l\right)$$

where a_k^n, (k = 1,2, .., n);b_k^m, (k = 1,2, .., M); c_k^l, (k = 1,2, .., l).

Static barriers in the warehouse- walls, additional robots do not face each other. In order to better understand how the system works, we will describe the interaction of the algorithm and its components:

1. The robots were in standby mode
2. The control system schedule includes information about the delivery of X cargo to the destination.
3. The control system analyzes the information received and fixes the location of the cargo.
4. The control system determines the route of movement of the robot. In this case, he receives information about the current state of other robots in order to avoid collisions.
5. The warehouse robot is instructed to follow the specified route.

The distance of the Robot's path is calculated using the following equation

$$GA = \sqrt{(a_x - nx_i)^2 + (a_y - ny_i)^2}$$

a_x , a_y -coordinates of current point of the robot, nx_i, ny_i– coordinates of neighbouring points of the robot. In the absence of the task of the warehouse robot is in the mode of waiting (x_0, y_0). In this case, the information about the trajectory and the current position of the robot will be recorded in memory. This information takes the form of a two-dimensional matrix, called a coordinate matrix, which contains the coordinates of all points of the movement. The coordinate array is given in the table below (see Table 1).

Table 1. Defining a set of waypoints for the desired robot path. (The coordinate array)

	Robot1		Robot2		Robot3	
	Towards the goal	Back to	Towards the goal	Back to	Towards the goal	Back to
1	(x, y)	(x,y)	(x,y)	(x,y)	(x,y)	(x,y)
1	(0 0)	(4.5 12.0)	(0 0)	(8.0 10.0)	(0 0)	(11. 11.0)
2	(1.0 2.0)	(4.0 10.0)	(2.0 1.5)	(7.0 9.0)	(2.0 1.0)	(10.0 6.0)
3	(1.5 3.0)	(3.5 9.0)	(3.0 2.0)	(5.0 6.0)	(5.0 2.0)	(7.0 4.0)
4	(2.0 5.0)	(3.0 7.0)	(4.0 4.0)	(4.0 4.0)	(7.0 4.0)	(5.0 2.0)
5	(3..0 7.0)	(2.0 5.0)	(5.0 6.0)	(3.0 2.0)	(10.0 6.0)	(2.0 1.0)
6	(3.5 9.0)	(1.5 3.0)	(7.0 9.0)	(2.0 1.5)	(11.0 11.0)	(0 0)
7	(4.0 10.0)	(1.0 2.0)	(8.0 10.0)	(0 0)	(12.0 14.0)	–
8	(4.5 12.0)	(0 0)	(9.0 14.0)	–	–	–
9	(5.0 14.0)	–	–	–	–	–

The simulation was conducted with the help of the MATLAB program package and the following results were obtained (Table 2):

Table 2. Possible ways of warehouse robots

Case	Number of points passed	The robot traveled (meter)
Robot 1	9	17
Robot 2	8	19.2
Robot 3	7	22.3

4 Conclusion

The geometric controller is widely used to solve the problem of tracking the trajectory of movement in an Autonomous vehicle. This is a new technique that is being developed here.

In this article, we use the pure pursuit geometric controller as the base controller.

One of the ways to control the movement of a group of robots is presented.

The article implements an algorithm for planning the trajectory of several warehouse robots located in a limited space, and this algorithm, acting on a known coordinate, sets the task of achieving the goal.

The simulation was upgraded in the MATLAB environment and the experiment was conducted with 3 warehouse robots.

References

1. Chernigovsky, A.S., Tsarev, R.Y., Kapulin, D.V.: Scheduling algorithms for automatic control systems for technological processes. IOP Conf. Ser. J. Phys. Conf Ser. **803**, 012028 (2017). https://doi.org/10.1088/1742-6596/803/1/012028
2. Chang, C.F., Lin, T.Y., Tai, C.L., et al.: Advanced information of parity bits for decoding short linear block codes using the A* algorithm. IEEE Trans. Commun. **61**(4), 1201–1221 (2013). https://doi.org/10.1109/TCOMM.2013.020813.120216
3. Kapulin, D.V., Chemidov, I.V., Kazantsev, M.A.: The design of the automated control system for warehouse equipment under radio-electronic manufacturing. IOP Conf. Ser. J. Phys. Conf. Ser. **803**, 012064 (2017). https://doi.org/10.1088/1742-6596/803/1/012064
4. Tuncer, A., Yildirim, M.: Dynamic path planning of mobile robots with improved genetic algorithm. Comput. Electr. Eng. **38**(6), 1564–1572 (2012). https://doi.org/10.1016/j.compeleceng.2012.06.016
5. Jose, K., Pratihar, D.K.: Task allocation and collision-free path planning of centralized multi-robots system for industrial plant inspection using heuristic methods. Robot. Auton. Syst. **80**, 34–42 (2016). https://doi.org/10.1016/j.robot.2016.02.003
6. Vivaldini, K.C.T., Galdames, J.P.M., Bueno, T.S., Araujo. R.C., Sobral, R.M., Becker, M., Caurin, G.A.P.: Robotic forklifts for intelligent warehouses: routing, path planning, and auto-localization. In: IEEE International Conference on Industrial Technology, Vina del Mar, Chile (2010). https://doi.org/10.1109/ICIT.2010.5472487
7. Hamberg, R., Verriet, J.: Automation in Warehouse Development. Springer- Verlag London Limited (2012). https://doi.org/10.1007/978-0-85729-968-0
8. Li, C., Bodkin, B., Lancaster, J.: Programming Khepera II robot for autonomous navigation and exploration using the hybrid architecture. In: Proceedings of 47th Annual South East Regional Conference, ACM, p. 31 (2009)
9. Wallace, R., et al.: First Results in Robot Road-Following, repon witbio CMU-RI-TR-8&4

Simulation Model of the Multi-position Mechatronic Module Functioning of an Intelligent Robot

Kh. N. Nazarov[1(✉)], N. R. Matyokubov[2], and T. O. Rakhimov[1]

[1] Department of Electronics and Automation, Tashkent State Technical University named after Islom Karimov, 2, University Street, 100095 Tashkent, Uzbekistan
{nazarov_hayriddin,rahimov_timur}@bk.ru

[2] Tashkent State Technical University named after Islam Karimov, 2, University Street, 100095 Tashkent, Uzbekistan
nm85@mail.ru

Abstract. The simulation concept construction of models for the multi-position mechatronic module functioning of an intelligent robot is considered. A simulation model of the functioning of a multi-position mechatronic module of linear motion is constructed and formally described. With the help of the reachability graph, possible options for the functioning of the network are described, and the absence of dead-end marking is determined. The constructed Petri net and the MATLAB package reflect the structure of the multi-position mechatronic module and its output horizontal and vertical movements, which are recorded using sensors. The conditions for starting composite electromagnets and fixing couplings are described. Individual positions are used to simulate the control processes of the mechatronic module, taking into account the information of the position sensors. The process of changing the state of an electromagnet and the conditions for switching on clutches using a set of transitions are considered.

Keywords: Simulation model · Mechatronic module · Intelligent robot · Petri net · MATLAB package · Position sensor · Electromagnet · Coupling

1 Introduction

Currently, the most important and integral stage in the design of modern multi-position mechatronic devices and systems of intelligent robots is simulation. Due to the fact that a mechatronic system is a synergistic combination of mechanical, electrical and computer parts, modeling tools should allow joint modeling of these parts on a single methodological basis, making it possible to build and explore multidimensional models [1, 2].

As a result of simulation, the functional requirements and technical parameters of the system can be set many times and, if necessary, will be adjusted. This saves time and money. Therefore, the issues of creating various mathematical models of the mechatronic system of intelligent robots are always relevant.

© The Author(s), under exclusive license to Springer Nature Switzerland AG 2021
R. A. Aliev et al. (Eds.): WCIS 2020, AISC 1323, pp. 201–206, 2021.
https://doi.org/10.1007/978-3-030-68004-6_26

A multi-position mechatronic module (MPM) of linear motion is modeled. Designed for the implementation of horizontal and rotational movements of the manipulator of an intelligent robot. The MMM includes a block of electromagnets with moving elements, a block of couplings, in particular electromagnetic, which, depending on their functional purpose, are used for working and fixing [2, 3].

2 Research Methodology

As an object of modeling, take the developed robot based on the MPM, which has four degrees of mobility and provides independent fixed displacements of the corresponding degrees of mobility [4, 5].

Consider the principle of operation of the MPM, the first output of the MPM activates the horizontal movement along the x coordinate. The second output implements movement around the $z-$axis ($z-$ output). The third and fourth outputs implement movements around the $x-$axis; the MPM output provides opening and closing of the robot's gripper ($a-$ output) [6, 7].

The movement of the robot arm horizontally is limited by the coordinates x_1 and x_2, around the $z-$axis by the coordinates, α_1 and α_2; around the $x - \beta_1$ and β_2 axis with coordinates, around the $x-$axis with γ_1 and γ_2 coordinates, then when these coordinates are reached, sensors a, b, c, d are triggered, respectively.

The simulation models of the outputs 1, 2, 3, 4 MPM of the robot are identical, they reflect the functioning of the degrees of mobility of the robot and are built using the MATLAB program.

In the model of the first four MPM outputs, positions 1, 2 describe the conditions for starting to the right and left or down and up, respectively. The presence of markers in positions 3 and 4 indicates the closed contacts of $x_2(a_1, \beta_2, \gamma_2)$, and in positions 5 and 6—contacts $x_1(a_2, \beta_1, \gamma_2)$. The marker at position 7 denotes the closedness of either $x_2(a_1, \beta_2, \gamma_1)$ or $x_1(a_2, \beta_1, \gamma_2)$. Position 8 simulates stopping the engine, and positions 9, 10—movements down or up (right or left), respectively. These positions are assigned a marker delay time equal to the travel time between the extreme positions [8].

3 Analysis and Results

Let's consider the principle of functioning of one of the MPM outputs. For example, the markers are at positions 5, 6, 8 (Fig. 1). Downward triggering can occur only after the marker appears in position 1. When the t_1 transition is triggered, the marker appears in position 9, which means the gripper (or degrees of movement) moves down after the t_3 transition and t_5 transition are triggered, the markers arrive at positions 3, 4, 8. The resulting marking is the initial position for launching upward, 4, 6, 7 were introduced to simulate the actual control processes, and positions 3, 4, 5—to simulate the manipulator control process.

The MPM simulation model is based on a Petri net, the mathematical description of which is:

$$N = (P, T, E, H, \mu_0) \tag{1}$$

where $P = \{p_1, p_2, p_3, p_4, p_5, p_6, p_7, p_8, p_9, p_{10}\}$ – is a finite non-empty set of positions;
$T = \{t_1, t_2, t_3, t_4, t_5, t_6, t_7, t_8, t_9, t_{10}\}$; – a finite non-empty set of transitions;
$\mu_0 = (1, 0, 0, 0, 0, 1, 0, 1, 0, 0)$– many markers.

$$F : P * T \to \{0, 1, 2 \ldots\};$$

$H = T * P \to \{0, 1, 2 \ldots\}$ – functions of input and output incidents;

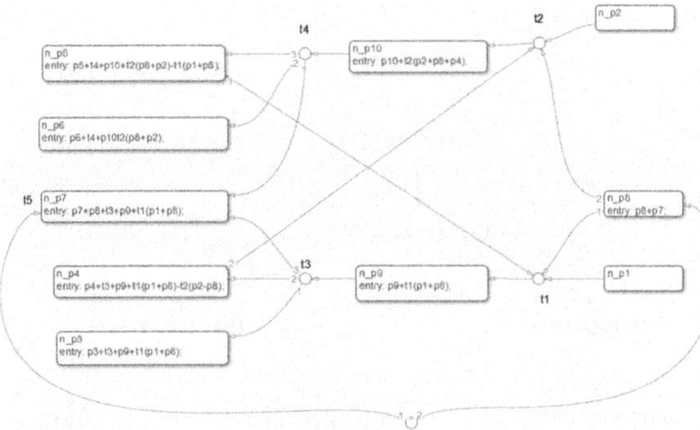

Fig. 1. Simulation model of four MPM outputs.

Functions F and H are given by matrices:

$F =$ (2)

	t_1	t_2	t_3	t_4	t_5
p_1	1	0	0	0	0
p_2	0	1	0	0	0
p_3	0	0	0	0	0
p_4	0	1	0	0	0
p_5	0	0	0	0	0
p_6	1	0	0	0	0
p_7	0	0	0	0	1
p_8	1	1	0	0	0
p_9	0	0	1	0	0
p_{10}	0	0	0	1	0

$$H = \begin{array}{c|cccccccccc|} & p_1 & p_2 & p_3 & p_4 & p_5 & p_6 & p_7 & p_8 & p_9 & p_{10} \\ \hline t_1 & 0 & 0 & 0 & 0 & 0 & 0 & 0 & 0 & 1 & 0 \\ t_2 & 0 & 0 & 0 & 0 & 0 & 0 & 0 & 0 & 0 & 1 \\ t_3 & 0 & 0 & 1 & 1 & 0 & 0 & 1 & 0 & 0 & 0 \\ t_4 & 0 & 0 & 0 & 0 & 1 & 1 & 1 & 0 & 0 & 0 \\ t_5 & 0 & 0 & 0 & 0 & 0 & 0 & 0 & 1 & 0 & 0 \end{array} \quad (3)$$

A fragment of the reachability graph for a given network is shown in Fig. 2.

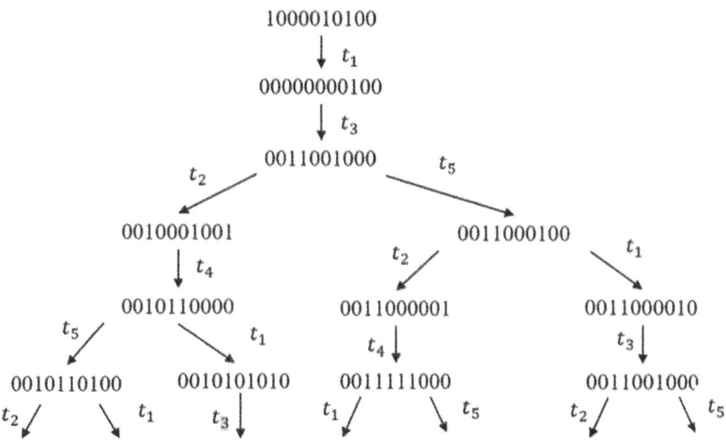

Fig. 2. Fragment of the reachable graph.

The reachability graph is used to describe possible options for the operation of the network. With regard to this network, based on the reachability graph, we can say that there is no dead-end marking [3, 9].

Let's consider a Petri net, which describes in more details the operation of a multi-position mechatronic module (Fig. 3).

It consists of two degrees of freedom, one degree of freedom is realized with the help of a rod (1), which makes horizontal movement to the left (right), the second - with the help of a flexible rod (2) (or a cable), which carries out a vertical movement down (up), position rods are fixed vertically and horizontally using position sensors in the coordinates y_1, y_2 and x_1, x_2, respectively.

The MPM simulation model with two degrees of freedom was built using the MATLAB package (Fig. 4). In this case, positions 1, 2 describe the conditions for starting the electromagnets of the MPM, positions 3, 4 and 5, 6 reflect the states of these electromagnets. Positions 7, 8 and 9, 10 reflect the states of the couplings, which are rigidly connected to the moving element, positions 11 and 12 are introduced to simulate the control processes of electromagnetic clutches, positions 13 and 14 reflect the state of the output rod, positions 15, 16, 17 are used to simulate motor control processes taking into account information from position sensors, position 18 simulates a stem stop. Transitions t_1 reflects the process of changing the state of the electromagnet; transition t_2 reflects the clutch engagement condition; t_3 means the transition of the coupling from one state to another, and t_4– the transition of the stem from one position to another [5, 8, 9].

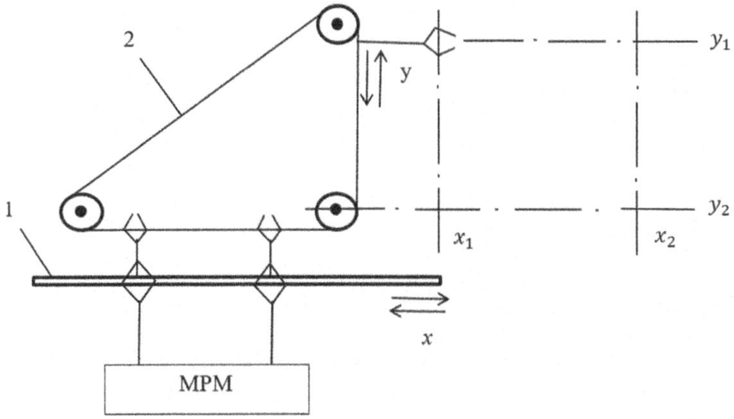

Fig. 3. MPM with two degrees of freedom.

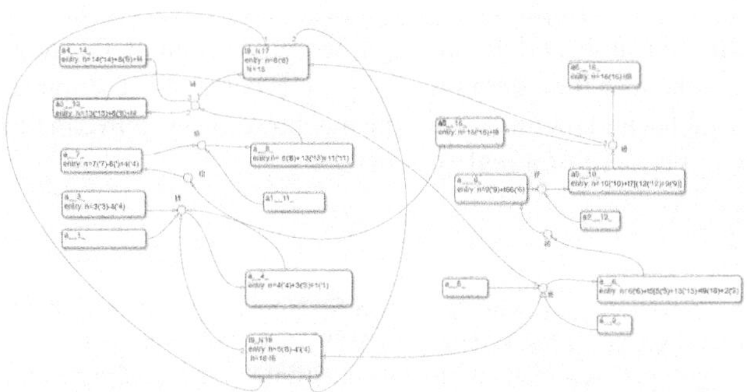

Fig. 4. Simulation model of MMM functioning.

The presence of markers in positions 13 and 14 means the sensor is triggered in position x_1, and in positions 15 and 16 in position x_2. The marker in position 17 denotes the triggering of the sensor in the position either x_1 or x_2, and in position 4.5– movement to the left and corresponds to the marker delay time.

Let the markers be in positions 15, 16, 17. The start of the MPM electromagnets can occur only after the appearance of the marker at position 1. When the transition t_1 is triggered, the marker appears at position 4, which means the moving element moves to the left. After the transitions t_2, t_3, t_4 are triggered, the markers arrive at positions 13, 14, 17. In this case, the resulting marking is the starting point for the movement of the stem to the right. Thus, positions $\bar{1}, \bar{2}, \bar{3}, \bar{4}, \bar{7}, \bar{8}, \overline{13}, \overline{14}$ simulate the movement of the output rod downward, and positions $\bar{2}, \bar{5}, \bar{6}, \bar{9}, \overline{10}, \overline{15}, \overline{16}$, – up.

The constructed MPM model is formally described as follows:

$$\begin{aligned}
&P = \{p_1, p_2, \ldots, p_{18}\}, T = \{t_1, t_2, \ldots, t_9\}.\\
&I(t_1) = \{p_1, p_3, p_{15}, p_{18}\}.O(t_1) = \{p_u\}.\\
&I(t_1) = \{p_u\}, O(t_2) = \{p_1\}, I(t_3) = \{p_1, p_{11}\}.\\
&O(t_3) = \{p_8\}, I(t_4) = \{p_8\}, O(t_4) = \{p_{13}, p_{14}, p_{17}\}.\\
&I(t_5) = (t_5) = \{p_2, p_5, p_{18}\}, O(t_5) = \{p_6\}, I_6 = \{p_6\}.\\
&I(t_6) = \{p_6\}, I(t_7) = \{p_9, p_{12}\}, O(t_7) = \{p_{10}\}.\\
&I(t_8) = \{p_{10}\}, O(t_8) = \{p_{15}, p_{16}, p_{17}\}.\\
&I(t_9) = \{p_{10}\}, O(t_9) = \{p_{18}\}.
\end{aligned} \tag{4}$$

4 Conclusion

Thus, the proposed simulation model based on the MATLAB program makes it possible to analyze the functioning of multi-position mechatronic modules and implements the modular principle of modeling a multi-position mechatronic system of an intelligent robot manipulator, taking into account its physically heterogeneous components and their interrelationships when they function in time. The presented simulation model makes it possible to quite simply change the description of the region of motion of the degrees of mobility of robots and require significantly less financial and time costs, since modeling is much cheaper than creating real prototypes and is completely controllable, since all input variables and parameters of the mechatronic system can be predetermined, which is usually not typical of a real system.

References

1. Glazunov, V.A., Vorobev, Ye. I., Gavryushin, S.S.: New mechanisms in modern robotics, Pod redaksiey – M.: TEXANOSFERA, 316 p. ISNB 978–5–94836–537–4 (2018). (in Russian)
2. Borisov, I.I., Kolyubin, S.A.: Simulation of mechatronic systems, SPb: Universitet ITMO, 103 p. (2020). (in Russian)
3. Musalimov, V.M., Zamoruev, G.B., Kalapishina, I.I., Perechesova, A.D., Nujdin, K.A.: Modeling of mechatronic systems in the MATLAB environment (Citylink/Simmechanics): a textbook for higher educational institutions, SPb: NIU ITMO, 114 p. (2013). (in Russian)
4. Nazarov, K.: Intelligent mechatronic modules for linear motion of robotic systems. Mechatronic Autom. Control **4**, 26–31 (2005). (in Russian)
5. Nazarov, Kh.N., Xasanov, P.F.: Industrial robot. Patent, № 1598380, N02K33/02, BI № 23 (1996) (in Russian)
6. Yusupbekov, A.N., Nazarov, K., Matyokubov, N.R., Rakhimov, T.O.: Conceptual bases of modeling multi-coordinate mechatronic robot modules, chemical technology control and management. Int. Sci. Tech. J. **4**(94), 10–15 (2020)
7. Nazarov, K., Rakhimov, T.O.: Principle of construction of intellectual multicordinate mechatrone module with power moment sensitivity. J. Modern Tech. Engineer. **5**(2), 181–187 (2020)
8. Pelz, G., Waddington, R.: Mechatronic Systems. Wiley Online Library (2004)
9. Seok, S., Wang, A., Chuah, M.Y., Otten, D., Lang, J., Kim, S.: Design principles for highly efficient quadrupeds and implementation on the MIT cheetah robot. In: 2013 IEEE International Conference Robotics Automation, pp. 3307–3312 (2013)

Decision-Making in the Production of Hard-To-Recover Oil Reserves Under Uncertainty

G. M. Efendiyev[1](✉), R. H. Kuliyev[2], M. K. Karazhanova[3], I. A. Piriverdiyev[1], D. A. Akhmetov[3], and L. B. Zhetekova[3]

[1] Institute of Oil and Gas of Azerbaijan National Academy of Sciences, 9, F.Amirov Street, Baku 1000, Azerbaijan
galib_2000@yahoo.com, igorbaku@yandex.ru

[2] Institute of Geology and Geophysics of Azerbaijan National Academy of Sciences, H.Javid Avenue, 119, Baku 1073, Azerbaijan
ruslankuliyev@gmail.com

[3] Yessenov University, 32 Microregion, Main Building, 130000 Aktau, Republic of Kazakhstan
mikado_70@inbox.ru, aldee@list.ru, zhetekova81@mail.ru

Abstract. The problem of decision-making in oilfield practice has been in the focus of specialists' attention for many years. At the same time, one has to make decisions when applying various technologies under conditions of uncertainty. The use of traditional methods does not allow fully studying the conditions for making decisions and ensure the effectiveness of the methods used. When using hard-to-recover oil recovery methods, the formulation of the task is to achieve maximum production with a minimum volume of injected steam (or gas). It is possible to achieve success in such a situation using the basic provisions of the theory of fuzzy sets. Based on this, an attempt is made in this paper to show the possibility of making decisions using approaches known from the theory of fuzzy sets on the example of two methods of oil extraction. The manuscript provides a brief overview of one of the methods of oil production, which showed the difficulties arising in its application and allowed to justify the formulation of the problem. The results of the analysis of information on the implementation of oil production methods and decision-making using an approach based on the provisions of the theory of fuzzy sets are presented.

Keywords: Decision-making · Uncertainty · Optimization of oil production · Membership function · Oil production

1 Introduction

The problem of decision-making in oilfield practice has been in the focus of specialists' attention for many years. At the same time, one has to make decisions when applying various technologies under conditions of uncertainty. The use of traditional methods does not allow to fully study the conditions for making decisions and ensure the effectiveness of the methods used. The use of traditional methods in the design of field development is

© The Author(s), under exclusive license to Springer Nature Switzerland AG 2021
R. A. Aliev et al. (Eds.): WCIS 2020, AISC 1323, pp. 207–214, 2021.
https://doi.org/10.1007/978-3-030-68004-6_27

associated with two problems. The first of them is that the decision-making process requires a comprehensive assessment and classification of hard-to-recover reserves, while the existing classifications do not take into account the "blurred" boundaries between certain categories of oils and conditions of their occurrence. The second problem is that traditional methods do not take into account uncertainty, in particular, multicriteria and unclear (fuzzy) formulation of goals and constraints when making decisions. For example, when applying hard-to-recover oil recovery methods (such as gas-lift and thermal methods), the formulation is to achieve maximum production with a minimum volume of injected steam (or gas). It is possible to achieve success in such a situation using the basic provisions of the theory of fuzzy sets. Based on this, an attempt is made in this paper to show the possibility of making decisions using approaches known from the theory of fuzzy sets on the example of two methods of oil extraction. The following brief overview of the application of one of the oil recovery methods shows the difficulties encountered in its application.

2 A Brief Overview of Optimization Methods for Gas-Lift Production

This section of the manuscript provides an overview of the methods and technologies developed over the past two decades to solve the problems of optimizing gas-lift wells. The analyses of well data has been carried out over a wide range, from one well up to multidimensional schemes of optimization in real time and covering all wells in the field. A number of works are devoted to modeling, the development of computational schemes and computer programs. The computer system developed by the authors [1] is a toolkit for making a reasonable optimal decision on this problem. Operational control tasks is arisen while operating a gas-lift complex. In this regard, the computing system proposed in [1] allows developing the optimized decisions of operator including the programs of the detailed modeling of process in the contour of operational control in real time. Detailed computer modeling of processes with continuous and periodic gas-lifts allowed analyzing the dynamics of the output parameters of the processes with a wide variation in the initial and adjustable parameters. Such an analysis allowed the authors clarifying the model. The created software package allows solving the problem of choosing technical solutions while equipping gas-lift field gas-liquid elevators, as well as calculating the operating modes of a gas-lift well during its operation. Work [2], which is a comprehensive approach using dip curves for a reservoir together with network and technological models [3] is also devoted to modeling using an integrated approach. An integrated model of asset provides more accurate conditions of processing imposed by communications in the real world. The iterative offline-online scheme proposed by Rashid [4] has been used to optimize gas-lift. In a situation where it is necessary to make a choice between modes with a minimum specific gas flow rate or a maximum liquid flow rate or some average between these modes, one of the criteria of decision making under uncertainty is used. So, the Hurwitz criterion, which takes into account the heuristic coefficient of success has been used in [5]. In this case, the optimal mode is selected according to the group characteristics based on the condition [5]:

$$Q_{opt.l.} = \alpha \cdot Q_{\max .l.} + (1 - \alpha) \cdot Q_{\min .s.f.},$$

where α is heuristic coefficient of success; $Q_{max.\ l.}$ corresponds to the maximum value of fluid selection; $Q_{min.\ s.f.}$ is fluid selection corresponding to minimum specific gas flow rate. When the increase of flow rates of fluid and oil are equal $\alpha = 1$, in accordance with an equation, the optimal mode coincides with the mode of the maximum flow rate. In case of absence of increments of oil production due to increase of fluid production ($\alpha = 0$), the optimal mode provides the minimum of specific flow rate. In this case, the task of optimization is two-criteria, that is, the optimal mode should ensure maximum flow rate and minimum specific gas flow. In accordance with this, one or another method of making the optimal decision should be chosen. At the same time, the proposed criterion is generalizing, that is, allowing the reduction of the two marked criteria to one and thereby the task is reduced to a single-criterion. An overview of methods and techniques developed to solve the problem of continuous optimization of gas-lift over the past two decades has been presented in [6]. They range from an isolated analysis of one well to a multidimensional scheme of real-time covering all the wells in the considered field. Although some of the considered methods are clearly limited due to the fact that they neglect the processing of results of the observed effects of interdependent wells with common flow lines, other methods are limited due to the efficiency and quality of the solution obtained while working with large-scale networks consisting of hundreds of hard-to-reach wells. The purpose of the present work, as noted in [6] is to create an idea of the developed approaches reflected in studies accumulated to present date and to highlight the remaining problems of interest in the future. The work dedicated to the necessity to use an optimization device [7] while considering fields with many interrelated gas-lift wells and components of an object is noted. A strict multiphase solving device of flow network has been used to solve pressure and flow problems throughout the network in combination with a sequential quadratic programming device for a limited distribution problem. The advantages of a simultaneous solution to obtain accurate results are shown [6, 7], but the corresponding calculations are not presented. In [8], a network system has been defined to completely consider the interaction between the producing and gas lift systems at the same time for a complex offshore of operation on gas-lift to increase production and reduce costs. A kind of gas lift curves has been constructed for varying well-head pressure, which, according to the author, is a fairly good description of the considered process. The curves have been assumed to be piecewise linear, and the method of sequential linear programming has been used for optimization. Vazquez et al. [9] proposed an approach combining the genetic algorithm with the Taboo search method. The latter performs a local search and saves a set of search points marked as viable, known as Tabu (taboo) steps. These studies are moving in the direction where improvement in objective function is most likely. The method has been tested on a 25-well system and 5% increase of production has been reported compared to the initial state. Other methods of the mechanized production have also been considered, the scheme can be equally applied to the gas-lift scenario. The main drawback of the approach was great difficulties and the associated high cost of evaluating the function due to the global stochastic scheme.

The results considered in this brief review demonstrate significant difficulties in the decision-making conditions for oil production in the indicated way, which necessitates the search and application of methods to avoid the difficulties associated with the use of traditional methods.

3 Problem Statement

The essence of the gas-lift method of operation is to increase the production of the well supplying it with the necessary amount of compressed gaseous working agent. As the review of studies shows, the gas-lift method of operation is one of the oldest methods of mechanized oil production. As can be seen from the review, traditional technologies developed in the last century do not meet all the requirements associated with high pressure, high efficiency and safety. In this regard, great success has been achieved in the creation of new equipment capable of working in difficult conditions. Decision-making problems are posed using two criteria, which requires the use of modern methods that take into account this circumstance introducing fuzziness in the studies performed in recent years in solving various problems of oil-field practice [6–10]. According to it, the task of choosing optimal modes based on the analysis of field data using the theory of fuzzy sets has been set.

4 Results of Research

Data have been collected to solve this problem on operation of wells by the gas-lift method, which have been subjected to statistical processing. Initially, the dynamics of oil production and gas flow have been analyzed. A method known as the "moving average method" has been applied in analyzing the dynamics. The moving average method is based on the property of the average to repay random deviations from the general regularity. As a result, the calculation of the average somehow slides from the beginning of a series of dynamics to its end. According to this method, a certain number of points is selected, which can be changed or clarified while building dependencies. In this case, it is chosen equal to 3. First, the first three points are averaged over the abscissa and ordinate, then, the next three, etc. starting from the second. An averaged series has been obtained as a result. The dependences of gas flow and oil production on time are constructed according to these data. The main characteristics - these are graphical dependencies of the specific gas flow and production from the total gas flow rate are often used in practice using gas-lift method (see Fig. 1).

Here, curve Q_{oil} characterizes the dependence of the flow rate on volumetric flow of gas, and R curve - the specific flow rate of working agent on volumetric flow of gas. The optimal modes are determined in practice based on it. In this case, the correspondence of the optimal mode to the minimum value of the specific gas flow (coordinates of point A) is taken into attention. This point is the tangent point to the curve Q_{oil} passing through the beginning of coordinates. Along with this, the mentioned analysis is significantly complicated by the presence of uncertainties. The mentioned uncertainty is arisen due to a fuzzy formulation of the purpose and limitations while making decisions in this case. The task is to obtain the maximum amount of oil with a minimum amount of

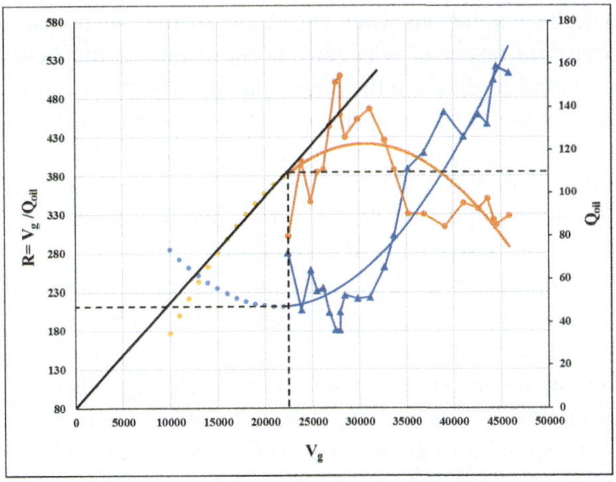

Fig. 1. Graphic dependencies of specific gas and well flow on total gas flow.

Table 1. The results of the analysis of mining data for decision making in fuzzy conditions.

Month	V_g	Qoil	μ V_g	μ Qoil	μ D
25	22433,33	135	0,90	0,57	0,57
29	23833,33	115	0,84	0,34	0,34
21,67	26833,33	131,33	0,73	0,52	0,52
…	…	…	…	…	…
18	**27500**	**151,33**	**0,71**	**0,83**	**0,71**
18,67	28000	154,67	0,69	0,89	0,69
…	…	…	…	…	…

gas injected into the well in this case. As a usual, it is impossible to provide both of the above mentioned conditions at the same time while solving such problems, and in this case, methods are used to make a compromise solution. Such an approach is the approach known from the theory of fuzzy sets. According to this, it is necessary to evaluate the membership function for each of the selected criteria. In this regard, appropriate calculations have been performed taking into account the statements set in [11].

Further, the smallest of this value pair is found on these values which represents the values of membership functions of the set of solutions (data are partially presented in Table 1). The maximum value of this membership function corresponds to the optimal solution. The calculation results are shown in the table where the optimal solution is highlighted.

Similarly, a search for solutions when choosing an effective option using the method of steam injection into a reservoir was carried out at one of the fields in Kazakhstan. Theoretical and practical research in the development of high-viscosity oil deposits and the experience accumulated in Kazakhstan [12, 13] show that at the initial stage, when the water cut level is low, the most promising areas for the development of such deposits are thermal stimulation methods. This technology has been successfully applied for over 20 years at the Karazhanbas field. Therefore, the problem of decision-making in the implementation of the above method is relevant.

The challenge in this case is to obtain the maximum volume of oil with the minimum volume of steam injected into the well. Usually, when solving such problems, it is impossible to provide both noted conditions at the same time, and in this case, methods are used that allow making a compromise solution. Accordingly, it is necessary to evaluate the membership function for each of the selected criteria, as in the previous case. In this regard, the corresponding calculations were carried out taking into account the provisions set forth in the work [6]. As can be seen from the noted data and calculation results (Table 2), the best result was observed in 2011, when 116,704 tons of oil were produced with an amount of 207,684 tons of injected steam.

Table 2. Production indicators under thermal steam impact and corresponding membership functions.

Date	Steam injection	Oil production by steam injection	μ_1	μ_2	μ_3
01.05.2010	263,176	90,255	0,614596	0,152333	0,152
01.06.2010	262,597	91,676	0,718065	0,170547	0,170547
01.07.2010	268,971	97,007	0,696624	0,256613	0,256613
...	...	...	...	...	...
01.05.2011	**207,684**	**116,704**	**0,927801**	**0,976167**	**0,9278**
01.06.2011	220,906	106,395	0,872623	0,500281	0,500281
...	...	...	...	...	...

Graphically, changes in membership functions and the optimal decision corresponding to their intersection point on the graph are shown in Fig. 2.

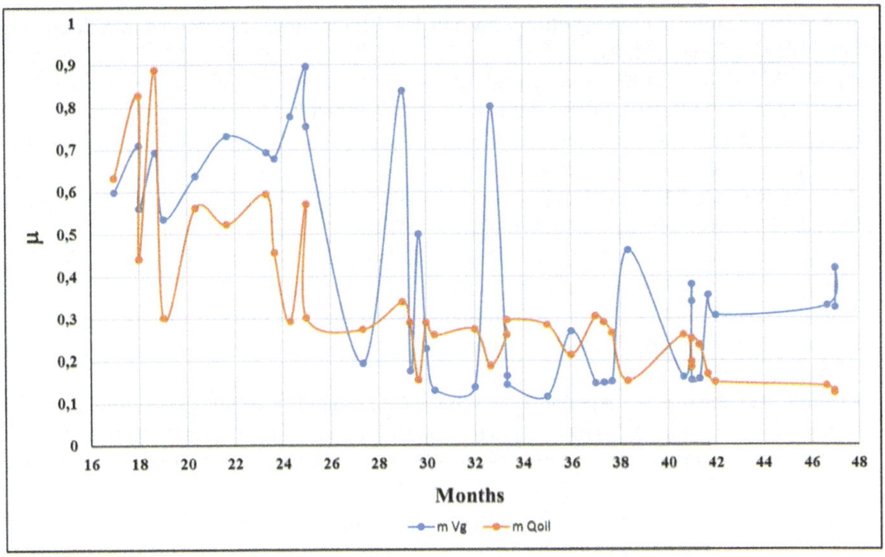

Fig. 2. Change of membership functions of sets of goals and constraints in time.

5 Conclusion

The performed analysis makes it possible to substantiate the need for a detailed comprehensive analysis of technical and technological factors and indicators together with the characteristics of geological conditions, which allows making decisions on assessing the required volume of injected steam. In this case, it should be noted the uncertainty of the processes occurring in the reservoir sys-tem when stimulating it. Thus, it is necessary to make decisions in the presence of several criteria. The use of methods known from the theory of fuzzy sets and intended for decision-making under these conditions makes it possible to improve the efficiency of various methods, in particular, in the gas-lift method and the thermal steam treatment method we have considered by increasing oil production with minimal volumes of injected gas (steam). The possibility of making decisions in conditions with a fuzzy formulation of goals and limitations is shown and implemented, which makes it possible to increase the efficiency of one or another method of stimulating the reservoir with the implementation of the gas-lift method and thermal steam treatment as an example by increasing oil production with minimal volumes of injected gas (steam).

References

1. Barashkin, R.L., Samarin, I.V.: Modeling of modes of operation of gas-lift well. In: Proceedings of Tomsk Polytechnic University, Natural Sciences, vol. 6, no. 309, pp. 42–46 (2006). (in Russian)
2. Gutierrez, F., Hallquist, A., Shippen, M., Rashid, K.: A new approach to gas lift optimization using an integrated asset model. In: SPE International Petroleum Technology Conference (IPTC 2007), Dubai, UAE, pp. 1371–1380 (2007). https://doi.org/10.2523/IPTC-11594-MS

3. Schlumberger Information Solutions, Avocet-IAM Integrated Asset Modeler. Schlumberger SIS. Calgary, Canada (2007)
4. Rashid, K.: Optimal allocation procedure for gas-lift optimization. Ind. Eng. Chem. Res. **5**(49), 2286–2294 (2010). https://doi.org/10.1021/ie900867r
5. Buy, D.C.: Improving the efficiency of technologies of raising products in gas-lift wells by controlling the flow structure. PhD thesis, p. 148 (2014). (in Russian)
6. Bellman, R.E., Zadeh, L.A.: Decision-making in a fuzzy environment. Manage Sci. **17**, 141–164 (1970). https://doi.org/10.1287/mnsc.17.4.B141
7. Rashid, K., Bailey, W. Couet, B.: A survey of methods for gas-lift optimization. Model. Simul. Eng. June (2012). https://doi.org/10.1155/2012/516807
8. Dutta-Roy, K., Barua, S., Heiba, A.: Computer-aided gas field planning and optimization. In: Proceedings of the SPE Production Operations Symposium, Oklahoma, USA, pp. 511–515 (1997)
9. Vazquez, M., Suarez, A., Aponte, H., Ocanto, L., Fernandes, J.: Global optimization of oil production systems, a unified operational view. In: Proceedings of the SPE Annual Technical Conference & Exhibition (ATCE 01), New Orleans, USA (2001). https://doi.org/10.2118/71561-MS
10. Efendiyev, G.M., Mammadov, P.Z., Piriverdiyev, I.A., Sarbopeyeva, M.D.: Selection of the best combination of bit types and technological parameters during drilling taking into account uncertainty. In: 9th International Conference on Theory and Application of Soft Computing, Computing with Words and Perception (ICSCCW-2017), Budapest, Hungary, pp. 67–74 (2017). https://doi.org/10.1016/j.procs.2017.11.211
11. Abraham, K.: High prices, instability keep activity high. World Oil **9(227)** (2006)
12. Zhao, X., Gao, P., Wu, Y.: Superheated steam injection in North Kenkiak shallow heavy oil reservoir. J. Geol. Min. Res. **5**(6), 136–146 (2013). https://doi.org/10.5897/JGMR2013.0173
13. Esetov, Z., Turdiyev, M.F., Kemalov, A.F., Abdrafikova, I.M.: Thermal-steam cyclic processing technology of development objects in Karazhanbas Kazakhstan field. Indian J. Sci. Technol. **9**(18), 1–7 (2016). https://doi.org/10.17485/ijst/2016/v9i18/93749

Automatic Recognition of Uzbek Speech Based on Integrated Neural Networks

Muhammadjon Musaev(✉), Ilyos Khujayorov, and Mannon Ochilov

Tashkent University of Information Technologies named after Muhammad Al-Khwarizmi, Prospect Amir Temur-108, Tashkent City 100200, Uzbekistan
mm.musaev@rambler.ru

Abstract. The article deals with the approach based on integrated models of speech recognition in the Uzbek language. As well as, the advantages of integral models in speech recognition, the stages of their formation and the functions of their components are pointed out. Works on training and testing of a collection of words consisting of names of geographical places in the Republic, terms used in urban planning cadaster, names of constituent elements of urban planning documents were carried out on the basis of the architecture of the speech recognition system based on the proposed integrated model. As a result of testing, the eval uation metric for the number of incorrectly recognized characters was CER = 17%.

Keywords: ANN · End-to-End · RNN · LSTM · BRNN CTC · Dataset · MFCC · CNN · Neural network · Automated speech recognition · Hybrid systems · Integrated systems

1 Introduction

One of the relevant areas of application of artificial intelligence systems is the recognition of human speech. Many examples of programs that use automatic speech analysis tools can be mentioned: voice control programs, speech-text systems, speech help systems, language training and linguistic rehabilitation programs, voice search, mobile speech dictionaries, and tools for helping people with disabilities (deaf, blind, dumb), smart home systems, biometrics, scientific research.

In our republic, little attention has been paid to the problem of speech recognition programs for automatic recognition of exactly Uzbek speech have not yet been created.

The main feature of the speech signal is that it varies greatly in many parameters: duration, pace, pitch, the variability of the human voice path, the emotional states of the speaker, the difference in voices of people of different ages and gender.

The features of the process of creating effective speech processing algorithms are:

Finding forms of statistical averaging with an infinite number of variations of speech signals;

The need for a quick search in databases and knowledge bases of artificial intelligence systems requires high-speed data processing tools;

© The Author(s), under exclusive license to Springer Nature Switzerland AG 2021
R. A. Aliev et al. (Eds.): WCIS 2020, AISC 1323, pp. 215–223, 2021.
https://doi.org/10.1007/978-3-030-68004-6_28

Construction of hierarchical means of segmentation and parametric representation of speech components: sounds, syllables, words and phrases;

To solve speech recognition problems, flexible configuration architectures for the actual task are required.

All these qualities are possessed by neural networks, which also have the ability to control their actions with subsequent correction. Another very important feature of neural networks is the ability to translate memorized images into a new parametric space with a much smaller dimension.

In this paper, the approach to the implementation of word recognition in the Uzbek language on the basis of integrated models is considered. However, some theoretical information is provided about the architectures of neural networks used in the integrated model and the results of preliminary experimental researches based on them.

2 End-to-End Speech Recognition Systems

Nowadays, there are effective standard models of speech recognition that consist of many different parts and provide good results in speech recognition. Figure 1 depicts its structure.

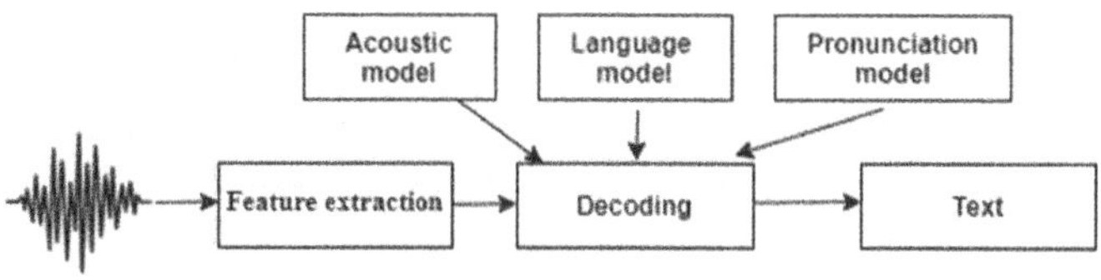

Fig. 1. Conventional automatic speech recognition pipeline

These models contain many components such as acoustic, lexical, language, and pronunciation models. These components involve separate training. In addition, this process leads to various complications. For instance, the process of teaching the acoustic model, the complexity of integrating all the components to perform decoding, the large amount of memory required to store language models, and the above components are optimized separately for different purposes. This causes to contradictions in optimization.

In addition, the use of hidden Markov models (HMM), which are widely used for acoustic modelling, has disadvantages: the assumption is that the observations are independent random variables and the "weak models" - Markov models of the first order - are used.

In recent years, in order to avoid from such complications, another approach has begun to be applied in speech recognition systems. This approach performs the process of distinguishing characters from speech signals, in a single model that integrates teaching and testing. Typically, such models are called integral models (end-to-end). The benefits of integrated models over standard models are that their implementation is simple, high efficiency in terms of speed and recognition accuracy, and their requirements of low memory [5].

In many works, it has been shown that the use of neural networks at each step of the script of a standard speech recognition system improves the quality of its work. So, for example, in [1–4] methods of feature extraction are presented, in the work [6] language models were trained using recurrent neural networks (RNN), in [8, 9] deep neural networks showed high results for the constructing acoustic models.

Therefore, the idea of using neural networks at all stages of speech recognition emerged. For that reason, another approach called the end-to-end model has begun to be used in speech recognition systems. These end-to-end systems consist of the following parts: an coder that divides the incoming ongoing speech into a vector of sequential features; a balancer that makes compatibility between languages and sequence of objects; a decoder that performs the decoding of the final identification result.

All of the above parts of this end-to-end system work as a single system. You can see that the integrated speech recognition system greatly simplifies the complexity of traditional speech recognition. There is no need to manually label information; in a neural network, you can automatically learn the language or pronunciation information, as shown in Fig. 2.

Typically, the structure of integrated systems of speech recognition consists of a convolution layer designed to separate important characters from incoming data and a recurrent layer that predicts the distribution of labels for each frame, as well as transcription layers that predict the sequence of final labels in each frame.

The Convolution Neural Network (CNN) is commonly used due to the spectrogram images generated from these coefficients or the Mel Cepstral coefficient (MFCC) as the login information in the convolution layer. This is due to the fact that, CNN is most effective at separating the local spatial characteristics of a speech signal.

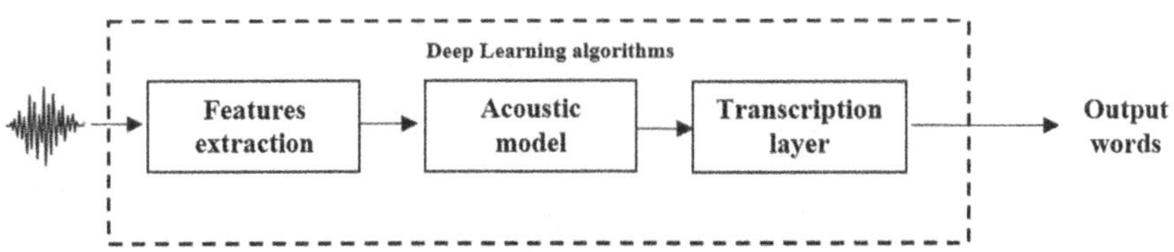

Fig. 2. End-to-end automatic speech recognition pipeline

The recurrent layer is corresponds to the separation of time-dependent characteristics of speech and in turn it performs acoustic modeling from the character vector generated by the convolution layer. At this level, various RNN architectures are applied.

The standard recurrent network for a given incoming sequence $x = (x_1, ..., x_T)$ calculates the output vector sequence $y = (y_1, ..., y_T)$ and the hidden vector sequence $h = (h_1, ..., h_T)$ based on the iteration to $t = 1, ..., T$. In this h_t and y_t is expressed as follows.

$$h_t = \sigma(W_{xh}x_t + W_{hh}h_{t-1} + b_h)$$

$$y_t = soft\max(W_{hy}h_t + b_y)$$

where, W-weight means a matrix (for example W_{xh}- a hidden incoming weight matrix), b-bias vector.

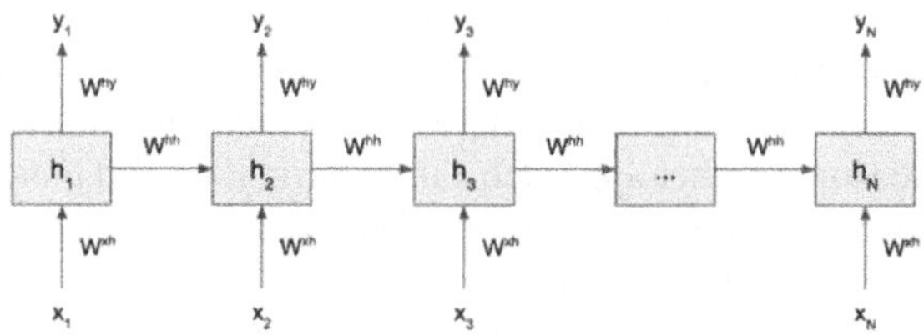

Fig. 3. Recurrent neural network

Many variants of RNN cells are used today. To name a few, from the most basic to more complex: RNN, GRU and LSTM. Among them, LSTM is considered the most widely used.

LSTM network consisting of special elements called memory blocks. An example of the LSTM network memory block is shown in Fig. 4. In the figure x_t- the input vector at the moment of time t, h_t- the output vector.

The LSTM network cell is considered as a complex element of the network that stores data for a long time. Gates determine when input data is important and remember it and when to send data to output.

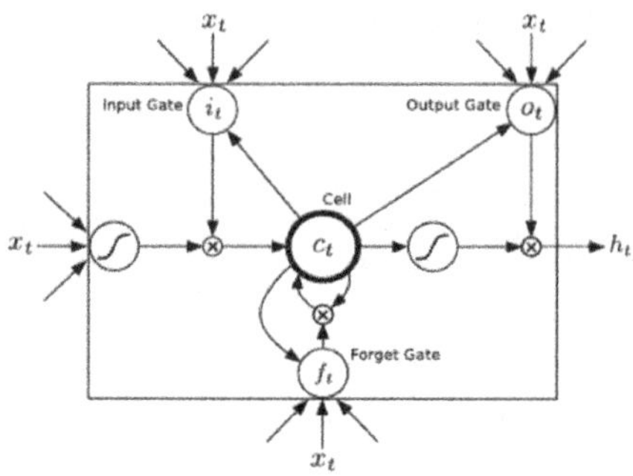

Fig. 4. An example of the LSTM network memory block

The LSTM network works based on the following expressions:

$$i_t = \sigma(W_{xi}x_t + W_{hi}h_{t-1} + W_{ci}c_{t-1} + b_i)$$

$$f_t = \sigma(W_{xt}x_t + W_{ht}h_{t-1} + W_{ct}c_{t-1} + b_t)$$

$$c_t = f_t c_{t-1} + i_t \tanh(W_{xc}x_t + W_{hc}h_{t-1} + b_c)$$

$$o_t = \sigma(W_{xo}x_t + W_{ho}h_{t-1} + W_{co}c_t + b_o)$$

$$h_t = o_t \tanh(c_t)$$

where, σ- the sigmoid function, *i, f, o,* and *c* are the input, forget, output gates and state cells, respectively, all of them have the same size with the hidden vector *h* [10].

In the transcription layer come the methods that perform marking of sequences and the methods that optimize the probability distribution of a variable output sequence based on long incoming sequences[11]. One of these methods is the Connectionist Temporal Classification (CTC).

CTC has been proposed for marking the sequence data in a single network architecture that does not require initial segmentation and processing. The basic idea of the CTC is to interpret network outputs as a conditional probability distribution across all possible output tags of a sequence. With this distribution, it is possible to obtain an objective function that maximizes the probability of a sequence of correct labels.

Training an acoustic model using CTC as a loss function is a comprehensive training that does not require prior alignment of data, but only requires training in the input and output sequences. Thus, there is no need to align and mark the data one by one, and the probability of directly predicting the CTC of the output sequence does not need external post-processing.

At the stage of neural network training, where CTC - function is used as a loss function, the neural network output layer contains one block for each character of the output sequence (in our case, Uzbek letters) and one more block for additional character "blank" ("space" for our specific task) corresponding to the silence on the audio track.

The output vector w_m is normalized using the softmax function, which can be interpreted as the probability of the appearance of a symbol with index *k* at time moment m, where w_m^k- is the *k*-th element of w_m :

$$P(k, m) = \frac{e^{w_m^k}}{\sum_{k'=0}^{|w_m|} w_m^{k'}}. \tag{1}$$

Let, α-be a sequence of "blank" indexes and symbols of some length *T*.

The probability of occurrence of a certain sequence of characters can be represented as the product of the probabilities of occurrence of characters at each moment of time:

$$P(\alpha) = \prod_t^T P(\alpha_t). \tag{2}$$

For a given sequence of w_m there are as many possible alignments as there are ways to place "blanks" between characters.

For example, alignment of the sequence (a, "blanks", b, c, "blanks", "blanks") and ("blanks", "blanks", a, "blanks", b, c) correspond to the sequence (a, b, c). Thus, when a series of repeating characters comes in succession, they are replaced by one: (a, a, a, b, b, c) and (a, "blanks", b, "blanks", c) correspond to (a, b, c).

Based on the foregoing, the total probability of the output sequence *w* is equal to the sum of the probabilities of all possible corresponding alignments, where *B* is the operator that removes repetitions and omissions, and B^{-1} is the operator inverse to B :

$$P(w) = \sum_{\alpha \in B^{-1}(w)} P(\alpha) \tag{3}$$

This expression allows the neural network to train on non-segmented data. In other words, by training the neural network without the exact location of the labels, we sum over all the locations where they may appear. At the decoding stage, the method of finding the best alignment of the output sequence is used.

This method is based on the assumption that the most probable alignment corresponds to the most probable output sequence. In the predicted acoustic model of the matrix, each column will contain the distribution of probabilities by alphabet characters. The higher the value of the matrix element, the more the acoustic model is confident in the symbol corresponding to the matrix row at a certain time step.

The result of the prediction is calculated using the maximum decoding (greedy search). The most likely result from the predicted matrix in Fig. 6 will be:

$$h(x) = \arg\max p(l|x), \text{ when } l \in A \leq T. \tag{4}$$

where, $p(l|x)$ for each time step t is calculated by the following formula:

$$p(l|x) = \sum \frac{\alpha_t(s)\beta_t(s)}{y^t_{l'_s}}, \tag{5}$$

where, $y^t_{l'_s}$ is the probability of the "blank" character at the last time step.

But, since the number of possible results l for enumeration is too high, we will use approximate solutions. The obvious solution is to choose a path assembled from elements with a maximum probability at each step t in the prediction matrix.

But, since the number of possible results l for enumeration is too high, we will use approximate solutions. The obvious solution is to choose a path assembled from elements with a maximum probability at each step t in the prediction matrix.

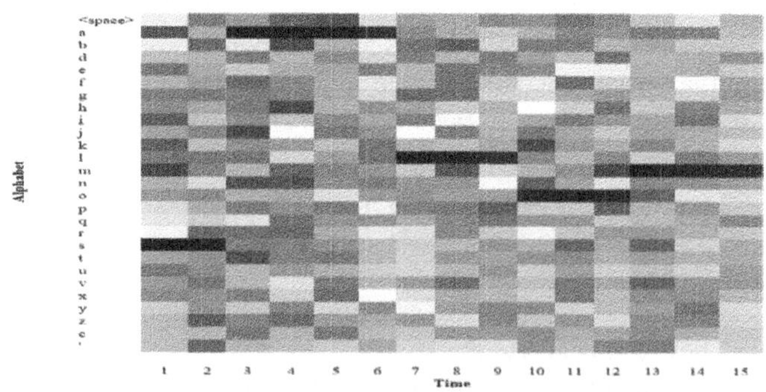

Fig. 5. The predicted matrix of probabilities

An illustration of the solution is shown in Fig. 6.

$$h(x) = B(\arg\max p(\pi|x)), \ when \ \pi \in N^T. \tag{6}$$

But such a solution does not always give the most probable result due to the fact that several ways can correspond to the same result and the probability of some result may turn out to be greater than the probability of an obvious solution.

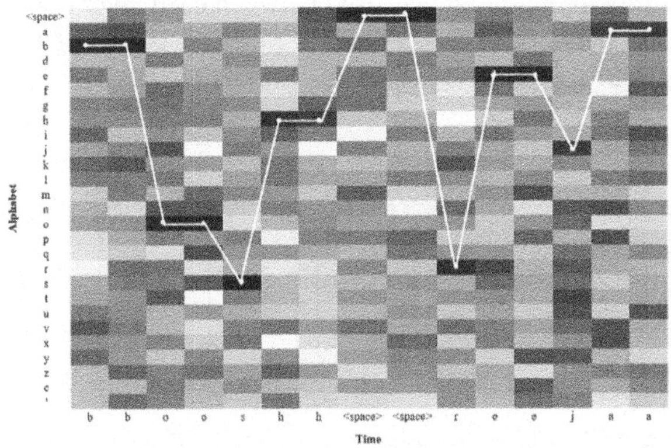

Fig. 6. Naive decoding solution

3 Experiments

During the research were used spectrograms formed from Mel Cepstral coefficients as incoming data for automatic recognition of Uzbek words based on an integral system and the LSTM network for acoustic modeling and the CTC Loss functions for phoneme-based word prediction.We will consider the parameters of the system architecture, the data set used in training the system, and the test results.The experiments were carried out in 100 words which were include the names of geographical places in the country, terms used in the urban cadaster, the names of the constituent elements of urban planning documents.

Dataset. The audio files on the dataset are in wav format, and digitization parameters are single-channel, 16-bit, 16 kHz. In the creation of the dataset narration were participated 50 people a total of 20 men and 30 women. Each speaker created audio files based on the pronunciation of these words. The total number of audio files is 3500. The process divided for 80% of the Dataset is for training, 10% for network validation and 10% for testing.

Network Architecture. The network architecture of the system, based on integrated models of automatic recognition of Uzbek words based on phonemes, consists of a convolution layer, recurrent layer, a fully connected layer and a CTC layer that performs decoding at the end of the network.Mel spectrogram that taken from a speech fragment every 20ms to the network architecture comes as characters of the vector values in the dimension 40×1 . At the beginning of the network there was located three consecutive one-dimensional convolution layers. There are 256 filters in each convolution layer, and the dimension of the filter is 5 × 1. Then, there comes the recurrent layer. The type of recurrent network is LSTM, with layer size = 256 and batch size = 32. Two fully connected layers were placed after this recurrent layer.

The first has 256 units and the second has 27 units that was used softmax activation function. A 20% dropout operation was used between each layer to prevent the problem of overfitting in the network. The CTC Loss function is implemented to obtain the predicted sequences in the last layer of the network (Fig. 7).

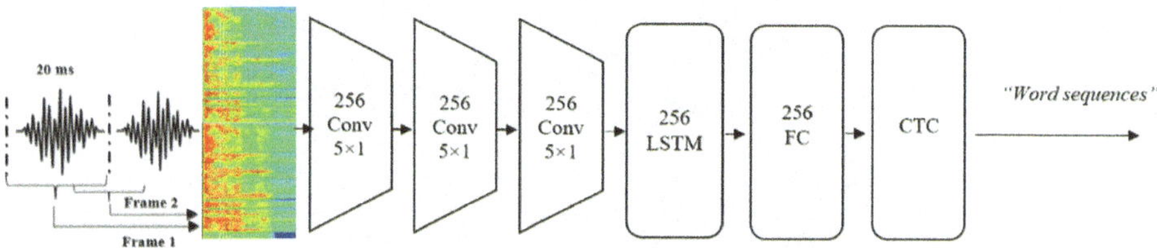

Fig. 7. Proposed network architecture

Testing. The metrics such as the Word Error Rate (WER), the Phoneme Error Rate (PER), or the Character Error Rate (CER) are typically used in continuous speech recognition systems for evaluating and measuring the work quality. These metrics are calculated using the Levenshtein distance [12] between given and obtained sequence of words, phonemes, or characters:

$$WER/PER/CER = \frac{D + S + I}{N} \times 100\% \tag{7}$$

where, N - the total number of words, phonemes or characters in a given sequence, D - the number of deletions, S - the number of substitutions, and I - the number of inserts.

When evaluating the test result, the evaluation metric for the number of characters unrecognized during the testing process was CER = 17%.

The some examples of the recognition results based on the proposed model of automatic word recognition system on a basis of integrated systems are given in Fig. 8.

Ground truth: Tayanch reja
Predictions: **Tayach reja**
Ground truth: Magistral
Predictions: **Magistral**
Ground truth: Bosh reja
Predictions: **Bosh reja**
Ground truth: Tasdiqlangan
Predictions: **Tasdiqlangan**
Ground truth: Tasdiqlanmagan
Predictions: **Tasdiqlanmagan**
Ground truth: Muhandislik tayyorgarligi
Predictions: **Muqandislik tayorgarlig**

Ground truth: Rejalashtirilgan
Predictions: **Rejalashtirilgan**
Ground truth: Qizil chiziqlar
Predictions: **Qizilchizilar**
Ground truth: Respublika
Predictions: **Respublika**
Ground truth: Viloyat
Predictions: **Viloyat**
Ground truth: Tuman
Predictions: **Tuman**
Ground truth: Shahar
Predictions: **Shahar**

Fig. 8. Tested words and the results of their recognition

By the help of using these words the geographic information system of the State Urban Cadaster of the Republic of Uzbekistan and the names of the regional areas, urban planning regulations, objects of town-planning documents can be found based on voice search.

4 Conclusion

Several modules are integrated into a single network in order to co-teaching in integrated systems of automatic speech recognition. This in turn, does not require the development of many modules in order to perform case descriptions between modules. The convolution and recurrent layers used in the integrated system architecture distinguish the time-dependent local characters of the speech signal. This improves the efficiency of speech recognition.

References

1. Gulmezoglu, M.B., et al.: A novel approach to isolated word recognition. IEEE Trans. Speech Audio Process. **7**(6), 620–628 (1999)
2. Zhang, J., Xiao, S., Zhang, H., Jiang, L.: Isolated word recognition with audio derivation and CNN. In: Proceedings International Conference on Tools with Artificial Intelligence, pp. 336–341, ICTAI (2018)
3. Polap, D., Woźniak, M.: Image approach to voice recognition. In: IEEE Symposium Series on Computational Intelligence, SSCI 2017 – Proceedings (2017)
4. Musaev, M., Khujayorov, I., Ochilov, M.: Image approach to speech recognition on CNN. In: Proceedings of the 2019 3rd International Symposium on Computer Science and Intelligent Control, pp. 1–6 (2019). https://doi.org/10.1145/3386164.3389100.
5. Markovnikov, N.M., Kipyatkova, I.S.: Analytical review of integral speech recognition systems. Tr. SPIIRAN **58**, 77–110 (2018). https://doi.org/10.15622/sp.58.4
6. Mikolov, T., et al.: Recurrent neural network based language model. Interspeech **2**, 1045–1048 (2010)
7. Rao K., Peng F., Sak H., Beaufays F.: Grapheme-to-phoneme conversion using long short-term memory recurrsent neural networks. In: 2015 IEEE International Conference on Acoustics, Speech and Signal Processing (ICASSP), pp. 4225–4229 (2015)
8. Graves, A., Schmidhuber, J.: Framewise phoneme classification with bidirectional LSTM and other neural network architectures. Neural Netw. **18**(5–6), 602–610 (2005)
9. Lu, L., Zhang, X., Cho, K., Renals, S.: A study of the recurrent neural network encoder-decoder for large vocabulary speech recognition. In: INTERSPEECH (2015)
10. Hochreiter, S., Schmidhuber, J.: Long short-term memory. Neural Comput. **9**(8), 1735–1780 (1997)
11. Wang, S., Li, G.: Overview of end-to-end speech recognition. IOP Conf. Ser. J. Phys. Conf. Ser. **1187**, 052068 (2019). https://doi.org/10.1088/1742-6596/1187/5/052068
12. Levenshtein, V.I.: Binary codes capable of correcting deletions, insertions, and reversals. Soviet Phys. Doklady. **10**, 707–710 (1996)

Software Selection Problem by Using Single Valued Neutrosophic (SVN)-Based TOPSIS Method

Nihad Mehdiyev(✉)

Department of Instrument-Making Engineering, Azerbaijan State Oil and Industry University, 20 Azadlig Avenue, 1010 Baku, Azerbaijan
nihadmehdi@gmail.com

Abstract. The single-valued neutrosophic set (SVNS) regarded as a seductive tool for dealing uncertain and incomprehensible information. With this consideration, different from the most current distance-based method for order preference by TOPSIS technique, this approach offer correlation-based TOPSIS method for addressing the single-valued neutrosophic (SVN) multiple attribute decision making (MADM) problems. The weighted correlation coefficient of the SVN is presented in order to emphasize the importance of the factors. The concept index on the basis of correlation is further proposed for the installation of the focal framework of the TOPSIS, which is denoted the TOPSIS method on the base of SVN correlation and in this paper these approach is used for selection of software in enterprise.

Keywords: Fuzzy numbers · Multicriteria decision making · Uncertainty · SVN · TOPSIS

1 Introduction

The choice of enterprise software can be linked to a wide range of needs and products, and it is necessary to focus on the key attributes that affect the selection process [1, 2]. Modeling is a very important tool for adopting new systems and assessments, monitoring systems. Currently, the market offers several software packages for modeling [3]. Thus, a formulated and standardized type of modeling software selection is needed. Many companies select a toolbox without establishing formal valuation factors. It is very significant to identify ways to consequentially and objectively evaluate the use of the tool [4]. Some authors [5] described a comprehensive methodology choosing software discrete modeling for large companies. Dorado et al. [6] offered the AHP methodology for selecting of software in education. The results of these works showed that a lot of general selection techniques and the AHP method are widely used as a decision-making tool in software selection. Hlupic et al. [7] suggested a detailed structure for estimating software where the size is more than 300 criterias. Fuzzy techniques permit decision makers to use linguistic terms to estimate factor in a simple and intuitive way. This refines the software selection process by accommodating the indeterminacy that

© The Author(s), under exclusive license to Springer Nature Switzerland AG 2021
R. A. Aliev et al. (Eds.): WCIS 2020, AISC 1323, pp. 224–228, 2021.
https://doi.org/10.1007/978-3-030-68004-6_29

take place in decision making. The use of fuzzy methodology in the software selection process can also correct the uncertainties inherent in peer review. This article uses the TOPSIS methodology based on SVN correlation as the MCDM method for software selection. The SVN correlation coefficient is identically to the Pearson's coefficient between real sets, which represents toughness of the relationship between a single-valued neutrosophic set and describes whether they are negative or positive correlation. This article is formulated as next: Sect. 2 described proposed methods. In Sect. 3, techniques that are used in this paper tested by using sensitivity analysis. In Sect. 4 results of this paper are represented.

2 Preliminaries

The single-valued neutrosophic set (SVNS) determined as a tempting implementation for dealing with uncertain and inexplicable information.

Definition 1 [8]. Let y is element in a finite set Y. Single-valued neutrosophic set S in Y can be denoted as in (1).

$$S = \{\langle y, T_s(y), I_s(y), F_s(y)\rangle | y \in Y\} \tag{1}$$

There are next conditions of $T_s(y)$ - truth membership function, $I_s(y)$ - indeterminacy-membership function and $F_s(y)$ - falsity-membership function:

$$0 \leq T_s(y), I_s(y), F_s(y) \leq 1, \quad 0 \leq T_s(y) + I_s(y) + F_s(y) \leq 3 \tag{2}$$

The triplet $(T_s(y), I_s(y), F_s(y))$ is denoted a single-valued neutrosophic number and represented as $y = \left(T_y, I_y, F_y\right)$ for convenes.

Definition 2 [9]. Let Y is nonempty fixed set, $P = (p_1, p_2, \cdots, p_n)$ and $S = (s_1, s_2, \cdots, s_n)$ are two SVNS in Y.The SVN coefficient of determination $K(P, S)$ between P and S determined by using next formula:

$$K(P, S) = \frac{\sum_{i=1}^{n} (T_{p_i} T_{s_i} + I_{p_i} I_{s_i} + T_{p_i} T_{s_i})}{\sum_{i=1}^{n} \left[(T_{p_i})^2 + (I_{p_i})^2 + (F_{p_i})^2\right] \sqrt{\left[(T_{s_i})^2 + (I_{s_i})^2 + (F_{s_i})^2\right]}} \tag{3}$$

Definition 3. Let Y is nonempty fixed set, $P = (p_1, p_2, \cdots, p_n)$ and $S = (s_1, s_2, \cdots, s_n)$ are two SVNs in Y. The SVN coefficient of determination $r(P, S)$ between P and S represented by the next formula:

$$r(P, S) = \frac{1}{3}(r_T(P, S) + r_I(P, S) + r_F(P, S)) \tag{4}$$

where

$$r_T(P, S) = \frac{\sum_{i=1}^{n} \left(T_{p_i} - \overline{T}_p\right)\left(T_{s_i} - \overline{T}_s\right)}{\sqrt{\sum_{i=1}^{n} \left(T_{p_i} - \overline{T}_p\right)^2 \sqrt{\sum_{i=1}^{n} \left(T_{q_i} - \overline{T}_q\right)^2}}} \tag{5}$$

$$r_I(P,S) = \frac{\sum_{i=1}^{n}\left(I_{p_i} - \overline{I}_p\right)\left(I_{s_i} - \overline{I}_s\right)}{\sqrt{\sum_{i=1}^{n}\left(I_{p_i} - \overline{I}_p\right)^2}\sqrt{\sum_{i=1}^{n}\left(I_{q_i} - \overline{I}_q\right)^2}} \tag{6}$$

$$r_F(P,S) = \frac{\sum_{i=1}^{n}\left(F_{p_i} - \overline{F}_p\right)\left(F_{s_i} - \overline{F}_s\right)}{\sqrt{\sum_{i=1}^{n}\left(F_{p_i} - \overline{F}_p\right)^2}\sqrt{\sum_{i=1}^{n}\left(F_{q_i} - \overline{F}_q\right)^2}} \tag{7}$$

Among Eqs. (4)–(7) suitable means of degree of truth, uncertainty and belonging to falsity membership in conformity with p_i and s_i are given by

$$\overline{T}_p = \sum\nolimits_{i=1}^{n} {}^{T_{p_i}}/_{n},\ \overline{T}_s = \sum\nolimits_{i=1}^{n} {}^{T_{s_i}}/_{n},\ \overline{I}_p = \sum\nolimits_{i=1}^{n} {}^{I_{p_i}}/_{n},\ \overline{I}_s = \sum\nolimits_{i=1}^{n} {}^{I_{s_i}}/_{n},$$
$$\overline{F}_p = \sum\nolimits_{i=1}^{n} {}^{F_{p_i}}/_{n},\ and\ \overline{F}_s = \sum\nolimits_{i=1}^{n} {}^{F_{s_i}}/_{n}.$$

3 Statement of the Problem

Suppose that an MCDM problem involves 4 criteria's - C_1, C_2, C_3, C_4 and 3 alternatives - A_1, A_2, A_3. Basic criteria's are C_1 - vendor, C_2 - functionality, C_3 - reliability, C_4 - portability.

Table 1. The single-valued neutrosophic decision matrix.

	C1	C2	C3	C4
A1	(0.4,0.3,0.2)	(0.7,0.4,0.2)	(0.6,0.3,0.1)	(0.4,0.3,0.2)
A2	(0.6,0.4,0.2)	(0.5,0.3,0.1)	(0.7,0.4,0.2)	(0.5,0.3,0.1)
A3	(0.7,0.5,0.3)	(0.5,0.3,0.2)	(0.5,0.3,0.2)	(0.4,0.2,0.1)

The detail steps associated with the presented method for finding the best alternative are listed below.

4 Solution of the Problem

Step-1. The evaluation information represented by SVNN decision makers is described in Table 1.

Step-2. Employing Eqs. (8) and (9) we compute the SVN PID $y^+ = \{y_1^+, y_2^+, \cdots, y_n^+\}$ and $y^- = \{y_1^-, y_2^-, \cdots, y_n^-\}$:

$$y_j^+ = \left(T_j^+, I_j^+, F_j^+\right) = \begin{cases} \max\{T_{ij}\}, \min\{I_{ij}\}, \min\limits_{i}\{T_{ij}\},\ for\ benefit\ attribute\ C_j \\ \min\{T_{ij}\}, \max\{I_{ij}\}, \max\{T_{ij}\},\ for\ benefit\ attribute\ C_j \end{cases} \tag{8}$$

$$y_j^- = \left(T_j^-, I_j^-, F_j^-\right) = \begin{cases} \min\{T_{ij}\}, \max\{I_{ij}\}, \max\{T_{ij}\}, & \textit{for benefit attribute } C_j \\ \qquad\qquad\qquad\quad i & \\ \max\{T_{ij}\}, \min\{I_{ij}\}, \min\{T_{ij}\}, & \textit{for benefit attribute } C_j \end{cases} \tag{9}$$

$$p^+ = \{(0.7, 0.3, 0.2), (0.7, 0.3, 0.1), (0.7, 0.3, 0.1), (0.5, 0.2, 0.1)\}$$

$$p^- = \{(0.4, 0.5, 0.3), (0.5, 0.4, 0.2), (0.5, 0.4, 0.2), (0.4, 0.3, 0.2)\}$$

Step-3: Optimal weighting vector of the attributes is $w = (0.38,\ 0.25,\ 0.22,\ 0.15)$. Weighted SVN correlation coefficients $r^w(Y_i, y^+)$ and $r^w(Y_i, y^-)$ for each $A_i \in A$ are yielded by using (10), which are presented in Table 2.

$$r_N^w(P, S) = \frac{\sum_{i=1}^n w_i\left(F_{p_i} - \overline{F}_p\right)\left(F_{s_i} - \overline{F}_s\right)}{\sqrt{\sum_{i=1}^n w_i\left(F_{p_i} - \overline{F}_p\right)^2}\sqrt{\sum_{i=1}^n w_i\left(F_{s_i} - \overline{F}_s\right)^2}} \tag{10}$$

Table 2. Computational result

	$r^w(Y_i, y^+)$	$r^w(Y_i, y^-)$	$\xi(Y_i)$
A1	−0.51	0.43	−0.71
A2	−0.38	0.47	−0.63
A3	0.27	0.31	0

Step 4. The SVN regression based comprehensive index $\xi(Y_i)$ for each $A_i \in A$ is determined by using (11):

$$\xi(Y_i) = \frac{1 + r^w(Y_i, y^+)}{1 + \max_i\{r^w(Y_i, y^+)\}} - \frac{1 + r^w(Y_i, y^-)}{1 + \min_i\{r^w(Y_i, y^-)\}} \tag{11}$$

For each alternative index $\xi(Y_i)$ determined as below:

$$\xi(Y_1) = \frac{1 + (-0.51)}{1 + 0.27} - \frac{1 + 0.43}{1 + 0.31} = -0.71,$$

$$\xi(Y_2) = \frac{1 + (-0.38)}{1 + 0.27} - \frac{1 + 0.47}{1 + 0.31} = -0.63,$$

$$\xi(Y_1) = \frac{1 + 0.27}{1 + 0.27} - \frac{1 + 0.31}{1 + 0.31} = 0.$$

Step 5. By using Table 2, we determine that $\xi(Y_i)$ for all $A_i \in A$ are determined as follows:

$$\xi(Y_1) = -0.71, \xi(\mathrm{Y}_2) = -0.63, \xi(\mathrm{Y}_3) = 0.$$

The lager $\xi(Y_i)$ the better alternative A_i.Therefore, the ranking orders of the alternatives are $A_3 > A_2 > A_1$.

5 Conclusion

The correlation coefficients measure of SVNS is used for dealing with software selection problem. By measuring the correlation coefficients between each alternative and ideal alternatives, the ranking order of all alternatives can be obtained. In conclusion, this work was able to conclude that the most prominent alternatives related to software A_3 with the highest SVN correlation-based comprehensive index compare to the others alternatives.

References

1. Bresnahan, J.: Mission impossible. CIO Mag., 15 (1996)
2. Franch, X.: On the lightweight use of goal-oriented models for software package selection. Active Flow Comb. Control **2018**, 551–566 (2005). https://doi.org/10.1007/11431855_38
3. Gupta, A., Varna, R., Singh, K.: Smart sin selector. a software for simulation software selection. Int. J. Eng. (DE) **3**(3), 175–185 (2009)
4. Chikofsk, E.J., Martin, D.E., Chang, H.: Assess* the state of tools asmssment. IEEE Softw. **9**(3), 18–21 (1992)
5. Tewoldeberhan, T.W., Verbraeck, A., Hupic, V.: Implementing a discrete-event simulation software selection methodology for supporting decision making at Accenture. J. Oper. Res. Soc. **61**(10), 1446–1453 (2010)
6. Dorado, R., Gómez-Moreno, A., Torres-Jiménez, E., López-Alba, E.: An AHP application to select software for engineering education. Comput. Appl. Eng. Educ. **22**(2), 200–208 (2011). https://doi.org/10.1002/cae.20546
7. Hupic, V., Irani, Z., Paul, R.: Evaluation framework for simulation software. Int. J. Adv. Manuf. Technol. **15**(5), 366–382 (1999)
8. Ye, J.: Single-valued neutrosophic clustering algorithms based on similarity measures. J. Classif. **34**, 148–162 (2017). https://doi.org/10.1007/s00357-017-9225-y
9. Ye, J.: Multicriteria decision-making method using the correlation coefficient under single-valued neutrosophic environment. Int. J. Gen Syst **42**(4), 386–394 (2013). https://doi.org/10.1080/03081079.2012.761609

Neural Network Model for Adaptive Control of Nonlinear Dynamic Object

N. R. Yusupbekov[1(✉)], D. P. Mukhitdinov[1], and O. U. Sattarov[2]

[1] Tashkent State Technical University, Tashkent 100095, Uzbekistan
dodabek@mail.ru, muxitdinov48@mail.ru
[2] Navoi State Mining Institute, Navoi 706800, Uzbekistan
olim81@bk.ru

Abstract. The article considers the possibility of constructing artificial neural networks based on the theory of fuzzy logic. The possibility of replacing linear controllers with neural network ones is investigated, taking into account the choice of neural network architecture, as well as identifying factors that determine the quality of simulation of the original controller. To cover the entire frequency band that is significant for this system, the control system and its components are studied in discrete time with a sufficiently small and constant sampling step, which makes it possible to model the SAS itself and neural networks on a computer. Methods for connecting the NSR are considered. A series of computer experiments were conducted to compare the results of the formation of the input vector of the NSR obtained from different variants. The experiments were carried out under stochastic "setpoint" conditions, taking into account the effect of random interference on the measured output values of the control object at the value of the forming filter "setpoint". The obtained results were tested in the construction of a neural network for controlling the phosphoric acid production process.

Keywords: Control object · Dynamic process · Control action or "setpoint" · Closed loop · Stationary mode · Neural network controller · Neural network identifier · Neural networks · Optimal neuro controller · Daptation method

1 Introduction

Classical mathematical models assume the existence of an exact (analytical) mathematical model. However, the technological process may be affected by uncertain parameters that cannot be taken into account by the developed models.

Therefore, the development of mathematical models that take into account all the factors affecting the control object is an unrealizable task, so it is appropriate to talk about the uncertainty of the control object model, and since the model replaces the control object, we can talk about the uncertainty of the object.

In order for the quality of the designed technical system to meet the required indicators, it must synthesize a process control system. The synthesis of the control system consists in the analysis of the control object, development of the system structure, selection of the controller and adjustment of the system parameters.

© The Author(s), under exclusive license to Springer Nature Switzerland AG 2021
R. A. Aliev et al. (Eds.): WCIS 2020, AISC 1323, pp. 229–236, 2021.
https://doi.org/10.1007/978-3-030-68004-6_30

Since in such conditions classical methods of control theory give poor results or are not applicable at all, there is a need to develop methods for analyzing and synthesizing control systems with uncertain mathematical models.

A possible way out of this situation may be PID regulation, which assumes that the object is affected by the disturbing effect of f and that there is information about the object's state that is not available for measurement. Of course, it is assumed that the object state measurements are performed with an error of N, and the controller develops control actions aimed at minimizing misalignment in the control system in case of possible adverse combinations of uncertainty factors (f and N).

The behavior of control systems in situations involving uncertainties, such as fluctuations in the quality of the raw material flow, environmental influences, economic factors, etc., makes it much more difficult to predict the behavior of complex systems. At the same time, there are situations when factors that do not have a significant impact on the behavior of the system, over time, begin to play a key role. Thus, operational management of large systems based on standard solutions is difficult in real time.

To solve such problems, it is advisable to use models and algorithms for building intelligent automatic control systems that allow you to train, adapt or configure controllers by storing and analyzing information about the object's behavior, its control system, and external influences.

Artificial neural networks (NS), based on the theory of fuzzy logic, are one of the modern approaches that allow implementing nonlinear control in technical systems. The neural network has the ability to "learn", which allows you to use the experience of an expert to teach the neural network the art of adjusting the coefficients of the regulator.

It follows from the above that it is necessary to replace linear controllers with neural network ones, taking into account the choice of neural network architecture, as well as identifying factors that determine the quality of simulation of the original controller.

A neural network can perform nonlinear mapping of a set of input signals to a set of output signals, while the NS has no restrictions on the linearity of the system, it is effective under the influence of disturbing influences after completing the training, it provides real-time management and, in addition, neural network SYSTEMS not only implement adaptive management methods, but also offer their own algorithmic approaches to a number of tasks.

2 Main Part

An automatic control system with feedback can be represented as the following scheme [1–3].

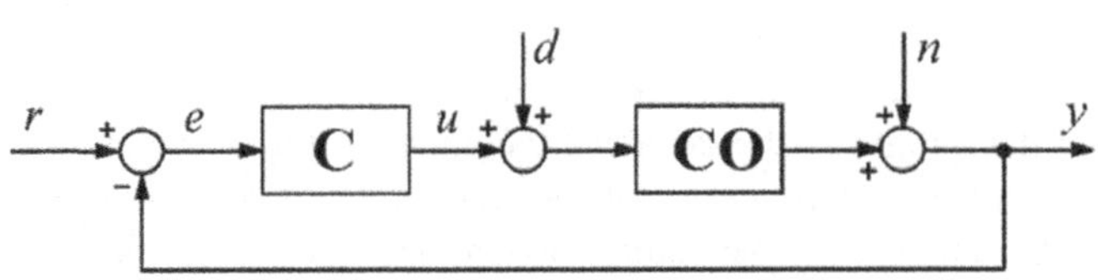

Fig. 1. Block diagram of the SAC technological process parameters.

Where: C - controller, CO - control object, r - control action or "setpoint", e - mismatch signal (or error), u - output value of the controller, y - adjustable value.

It is assumed that the value measured at the output of the system corresponds to the state of the object, but there is an additive interference with certain constant properties in the observation channel. The parameters of the control object are unchanged over time. (It is assumed that the linear controller maintains the state of the object close enough to the specified value $r(t)$).

Replace the controller in Fig. 1. SAC with a neural network (NSR). To compare the characteristics, we use a PID controller, as belonging to the most common type, providing stability and a certain satisfactory quality of control.

With some degree of accuracy, it can be assumed that a simulator that functions with sufficient accuracy will control the system as well. The quality of training for an artificial neural network [4, 5] depends on the choice of architecture, training data, and parameters of the training method.

Will see at how these factors affect the solution of the problem of simulating the original controller.

To cover the entire frequency band that is significant for this system, the control system and its components are studied in discrete time with a sufficiently small and constant sampling step, which makes it possible to model the SAR itself and neural networks on a computer. The operation of the linear controller for process control, shown in Fig. 1, can be described by some function

$$u_k = f(e_k, s) \tag{1}$$

Where: s - is the internal state of the controller, which changes its value at each clock cycle according to a certain rule.

The task of synthesizing a neural network controller $N^p(i_k)$ based on the standard deviation of the criterion $f(e_k, s)$ can be formulated as:

$$\sum_k (f(e_k, s) - N^p(i_k))^2 \rightarrow \min_{\forall i_k \varepsilon I} \tag{2}$$

where: - available information about the management system at the time t_k; e_k - control error, I - range of possible values t_k for the controller.

To form a vector, it is necessary to simulate the original controller with a neural network. To solve this problem, it is necessary to develop a method for its formation.

The task of neural network controller synthesis is to select the architecture of the neural network controller that includes the vector of information about the system, collecting training data, training the neural network controller, and establishing the quality of simulation and operation of the neural network controller.

In order for the controller's neural network to successfully simulate, it must also have the dynamic properties inherent in the PID controller [1, 3, 6].

The task of teaching NSR and the task of synthesizing a neural network model of an object (NSMO) have a certain similarity: both of them imitate a certain linear system, however, there are two significant differences:

- The NSMO can be implemented as an independent model that reflects the most significant characteristics of the object, or as an object-dependent model that predicts its state, while the task of the NSR is to completely replace the original controller;
- the purpose of the NSMO is to repeat the behavior of the control object, and the neural network controller must first be trained, like the original linear controller, and then, in accordance with the minimization of the RMS error, it can change its properties.

In the synthesis of NSMO, the problem of choosing between static and dynamic neural network architectures may arise [6, 7]. The positive side of dynamic architecture is the ability to simulate the properties of a linear dynamic system using an Autonomous model. However, when using feedbacks in the NSR, it has certain disadvantages associated with the complexity of analyzing the stability of the system due to the formation of an additional feedback loop with the neural network, the complexity of the learning algorithm in the loop, and increasing resource requirements [8, 9].

There may be some variations in the formation of the input value vector for the NSR. When checking a connection option similar to the original controller, only the error value at the current time is sent to the input. Studies have shown that with this option of connecting a static neural network, the neural network model is not able to provide a proper simulation of the properties of a dynamic system. To overcome the limitation of static display, it is necessary to artificially organize the memory with the repetition of previous inputs and investigate the behavior of the NSR at different values of memory capacity within.

The methods of connecting the NSR with inputs e_k, Δ_{ek} i $r_{k,}e_k$ are also subject to consideration.

In the first method of connecting the NSR, the neural network is provided with information about dynamic changes using a discrete derivative of the control error signal. In the second version, the concepts of perturbation $r(t)$ and deviation control are combined $e(t)$.

A series of computer experiments were conducted to compare the results of the formation of the input NSR vector obtained from different variants. The experiments were carried out under stochastic "setpoint" conditions, taking into account the effect of random interference on the measured output values of the control object at the value of the forming filter "setpoint" $R^*(z) = \frac{0.625z}{z-0.779}$ and interference $N^*(z) = 0.4$.

The object was managed by a PID controller $P^*(z) = \frac{z}{z-0.5}$.

$$C^*(z) = 0.4 + 0.5\frac{z}{z-1} + 0.05\frac{z^2 - 2z + 1}{z(z-1)}.$$

The neural network of the controller with the architecture $N^p_{x,7,3,1}$ (x dimension of the input vector) was trained on a sample length of 200 for 400 epochs, and the control sample had a length of 500. The minimum values of the RMS playback error were used as the learning criterion. The resulting numerical values obtained as a result of the experiment are shown in column A of Table 1.

Analyzing the results of calculating the RMS error, we can see that with increasing memory of previous inputs, the simulation error decreases slightly, and the results obtained under the conditions presented in the sixth and second rows of table 1 have

Table 1. Comparison of the accuracy of simulation of the PID controller outside the control loop with different methods of forming the input vector of the NSR and different parameters of the setpoint and interference.

Experiment		A	B	V	G
Setpoint		Stochastic $R^*(z)$		$\sin(t)$	0
Interference		$N^*(z) = 0.4$	$N^*(z) = 0.7$	$N^*(z) = 0.4$	$N^*(z) = 0.4$
Experimental results					
№	Input vector i_k	Mean-square error in the control sample			
1	e_k	0,3228	0,2906	0,1290	0,0027
2	e_k , e_{k-1}	0,2993	0,2710	0,1291	0,0022
3	$e_k \dots e_{k-2}$	0,2709	0,2459	0,1270	0,0016
4	$e_k \dots e_{k-3}$	0,2482	0,2284	0,1240	0,0013
5	$e_k \dots e_{k-4}$	0,2385	0,2163	0,1188	0,0013
6	e_k , Δe_k	0,2980	0,2738	0,1293	0,0023
7	r_k, e_k	0,0388	0,039	0,0039	0,0040

very similar values, which indicates the information equivalence of these options for the neural network controller.

The results of the comparison are shown in Fig. 2.

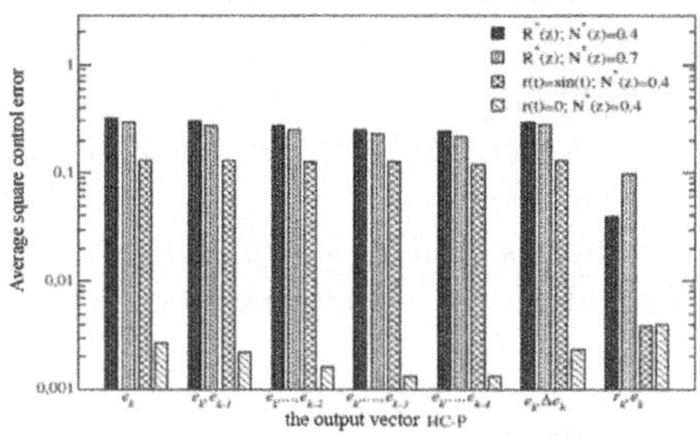

Fig. 2. Mean-square error control in the case of different test signals and input vectors of the NSR

With the input vector r_k, e_k (with an error level less than almost 10 times compared to other experiments), the highest quality of simulation was achieved.

However, this advantage, as can be seen from the results of the experiment given in column B, is lost with increasing power of disturbing effects, noting the justification for using the input vector in certain variants $e_k \dots e_{k-d}$.

Experiments conducted with pre-training of the NSR with variations in the "setpoint" form indicate the best quality of the input vector formation with combined perturbation and deviation control: (line 6). In addition, experiments show the independence of the final error level at the periodic setpoint from its form.When solving the stabilization problem with a constant "setpoint" value (for example, 0), the lowest result was obtained (column G of Table 2) for a neuroregulator with an input vector r_k, e_k from all variants with a repetition of past error values. Obviously, the input r_k without the influence of the "setpoint" negatively affects the neural network, manifesting itself as an unnecessary malicious element.

Evaluating the results obtained, we can draw the following conclusion on the variation of the input vector of the neural network controller relative to the quality of simulation of the original PID controller:

- when solving the stabilization problem with a constant "setpoint" value, use the input vector with the error and its first difference e_k , Δe_k If this increases the memory capacity of past values, then even if there are external disturbances, the quality of simulation will increase.
- if the setpoint value varies, a combined perturbation and deflection control is desirable r_k, e_k.
- if you use a combination of the previous two, when the "setpoint" keeps its value low for a certain period of time, you must use a combination of a set of inputs: r_k, $e_k \ldots e_{k-d}$.

To evaluate the quality of simulation of the original controller and the influence of the number of hidden layers and neurons in them on the learning process, we used several neural networks with inputs and the number of hidden layers from 0 to 3, trained them on the same sample with a length of 800 samples and a control sample with a length of 200 samples. For the unit of measurement of the duration of batch training of NSR, "epochs" were taken, during which the error reached the set value (in our experiment –0.005)). If the specified error value was exceeded for 3000 epochs, the training was completed with the last satisfactory result. The trained controller was connected to the control circuit.

Under deterministic conditions, a step-by-step "setpoint" was applied to the system input. The quality of management was evaluated by the standard error of control.

Table 2. Influence of the internal structure of the NSR on training and simulation quality.

N^0	Architecture NN-R	Training error	Control error	Computing time, *c*
1	$N^p_{r+e,1}$	0,0066	0,0058	57,0
2	$N^p_{r+e,5,1}$	0,0065	0,0061	76,2
3	$N^p_{r+e,20,1}$	0,0065	0,0060	132,1
4	$N^p_{r+e,7,5,1}$	0,0067	0,0061	120,2
5	$N^p_{r+e,20,10,1}$	0,0067	0,0063	306,6
6	$N^p_{r+e,7,9,5,1}$	0,0070	0,0062	194,4
7	$N^p_{r+e,15,20,10,1}$	0,0067	0,0064	571,1

Based on the results shown in Table 2, we can conclude that the internal structure of a multi-layer NSR with inputs almost does not affect the quality of simulation of a pid controller that controls a linear object, and the number of neurons in the neural network directly determines the computational complexity of the learning process.

The obtained results were tested in the construction of a neural network for controlling the production of phosphoric acid.

The root-mean-square error with the number of neurons in the hidden layer equal to 10 was $\Delta = 10^{-4}$. The selection method was used to determine the optimal number of neurons in the hidden layer (4).

The result of neural network modeling can be seen in Fig. 3.

During the simulation at this stage, a concentration vector with perturbations was applied to the input of the trained network. A solid line shows graphs of the dependence of distorted concentrations on the time spent in the reactor without the participation of a neural network.

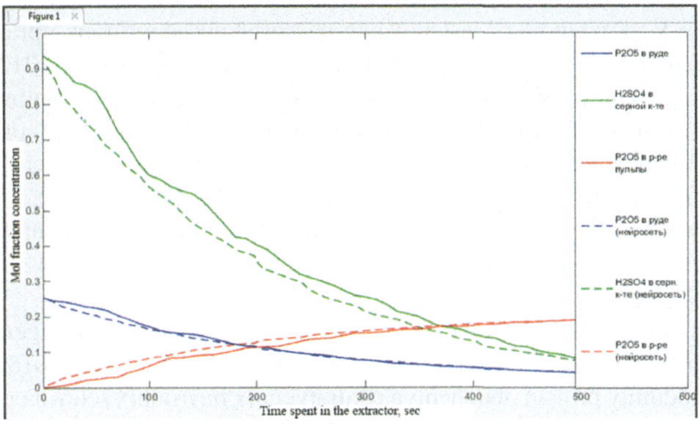

Fig. 3. Result of neural network modeling.

The dotted line shows the work of a trained neural network to stabilize perturbations. Thus, control using a neural network allows us to more reliably determine the optimal time spent in the reactor (even in the absence of intermediate laboratory analyses of changes in concentrations during the reaction).

3 Conclusion

The conducted experimental studies allow us to draw the following conclusion:

- if you complicate the internal architecture of the NSR, the calculation time increases, as well as the generalizing ability of the network is lost.

It is appropriate to note (if you know in advance about the nonlinearity of the object) that neural network approximation of an arbitrary function is possible only for networks with at least two layers of nonlinear neurons.

References

1. Muxitdinov D.P., Kadirov YO.B., Sattarov O.U., Boybutaev S.B.: Proporsionalno-integralno-differensiruyushie regulyatori v avtomatizirovannix sistemax upravleniya texnologicheskim protsessom, Nauchno - texnicheskiy i proizvodstvenniy jurnal «Gorniy vestnik Uzbekistana» № 4 (2014)
2. Yusupbekov, N.R., Muxitdinov D.P., Sattarov O.U., Boybutaev S.B.: Construction of a neural network using an approach to a genetic algorithm, Int. J. Adv. Res. Sci. Eng. Technol. **6**(6) (2019)
3. Eremenko, Yu.I., Poleshenko, D.A., Glushenko, A.I.: Issledovanie primenimosti neyrosetevogo optimizatora parametrov PID-regulyatora dlya upravleniya nagrevatelnim ob'ektom//Trudi III Vserossiyskoy nauchnoy konferensii molodix uchenix s mejdunarodnim uchastiem «Teoriya i praktika sistemnogo analiza», 21–24 maya
4. Eremenko, Yu.I., Poleshenko, D.A., Glushenko, A.I., Fomin, A.V.: Ob otsenke primenimosti razlichnix struktur neyronnoy seti v realizatsii neyrosetevogo optimizatora parametrov PI-regulyatora dlya upravleniya teplovimi ob'ektami//s Sistemi upravleniya i informatsionnie texnologii., № 3.2(57), S. 236–241 (2014)
5. Leonidovich, E.V.: Razrabotka i issledovanie neyrosetevix algoritmov upravleniya statsionarnimi i nestatsionarnimi ob'ektami»: atoref. dis. kan.tex. nauk-Moskva 2012
6. Terexov, V.A., Tyukin, I.: Sintez adaptivnix neyrosetevix regulyatorax nelineynix dinamicheskix ob'ektov. Institute for Physical and Chemical Research. (RIKEN), Japan (2017)
7. Nekotorie rezultati ispolzovaniya iskusstvennix neyronnix setey v kachestve regulyatora avtomaticheskoy sistemi regulirovaniya, Sabanin V. R., Dementev D. A., Kazmiruk I. Yu./ Sbornik trudov 26-y mejduna-rodnoy nauchno-texnicheskoy konferensii «Sovremennie texnologii v zadachax upravleniya, avtomatiki i obra-botki informatsii» g. Alushta, Krim, Rossiya 14–20 sentyabrya g. s. 60–61 (2017)
8. Kabulov, A.V., Normatov, I.H.: Kalandarov II Algorithmic model of management on the basis of algebra over functioning tables (FT). Sci. World Int. Sci. J **1**, 10–13 (2015)
9. Rikov, V.P. :Modulniy prinsip obucheniya iskusstvennix neyronnix setey na primere izvestnix neyrosetevix topologiy. Vestnik TGU, t. 19, vip. 2, S. 583–586 (2014)

Identification and Ranking of Key Factors for Pattern of Consumer Buying Decisions in Digital Marketing

Gunay Imanova(✉)

Azerbaijan State Oil and Industry University, 20 Azadlyg Avenue, Baku, Azerbaijan
gunayimanovaa@gmail.com

Abstract. An efficient marketing strategy and its practices should start from understanding the customers' needs and wants, and finding solutions to their problems through delivering value, better than competitors do. As the access to advanced digital technology and the internet increase, the use of e-commerce also increases considerably. In this study, we identify and rank several factors, such as effects of trendy and creative products; perceived risk of online purchases; brand names and quality; incentives for buying; and post-purchase services on consumer decision-making for online shopping. Then, we investigate the selection of optimal e-commerce among the given alternatives using fuzzy-TOPSIS method. The study is based on consistency driven approach for identification of weights of criteria, and distance between ideal positive and ideal negative solution under Z-number. Numerical case is provided to check the validity of the considered method. Moreover, effect of "trendy and creative products" is found to be the most important criterion, which is followed by "brand name and quality", "incentives for buying", "perceived risk of online shopping" and "post-purchase services", respectively, for the customers that are shopping online.

Keywords: e-commerce · Digital marketing · Consumer buying decision · Z-valued matrix · Consistency of pairwise comparison matrix · Fuzzy-TOPSIS

1 Introduction

Digital marketing is one of the most important and indispensable resources of our technology age. Furthermore, understanding consumer behavior and consumer decision making patterns in digital marketplaces have been the subject of researches for many years. Digital marketers should study the consumer behavior to understand how and what the customers are taking into account during the evaluation process of the products/services. Unless the marketers are informed about the consumer decision making stages, they couldn't influence the customers effectively and efficiently [1, 2].

Digitalization in the marketplace provides opportunities for the companies such as acting as a driver for globalization as it can serve many segments easily through broader customer access capability without any limitation for shelf space of products as in traditional brick-and-mortar stores [2].

© The Author(s), under exclusive license to Springer Nature Switzerland AG 2021
R. A. Aliev et al. (Eds.): WCIS 2020, AISC 1323, pp. 237–245, 2021.
https://doi.org/10.1007/978-3-030-68004-6_31

Consumers are changing their buying habits as the technology advances and the access rate to internet increases. Moreover, companies gain competitive advantage by capitalizing on the digital marketing technologies to understand their customers better [1]. Use of the technology provides better and interactive communication between customer and company to establish a long-term customer relationship. Besides, vast amount of accessible information about particular product/service and personalized promotions for customers, and cost efficiency for both parties are other drivers for digital marketing [3].

The main issue is to understand the consumer decision making pattern and the factors that are considered as most important by them. So, marketers can use this information to influence their decision-making phases. Companies can provide the desired product/service on the site and create satisfaction and customer loyalty eventually, and boost their profits accordingly. However, there are few actual researches based on dynamic factors that affect consumer decision patterns during online shopping, such as [4, 5]. Nevertheless, in most cases, the reliability associated with the values of the considered variables are tend to be ignored.

As a result, the aim of this study is to identify and rank the factors that are most influential on consumer buying behavior on digital marketplaces. In our research we also take into account the uncertainty related to the variables of interest and solve the MCDM problem accordingly.

The rest of the study involves the following sections. Section 2 represents preliminary information for further explanation that is followed by the statement of MCDM problem. The suggested solution method based on Z-matrix is presented in Sect. 3 to identify the optimal alternative. The next part, Sect. 4 represents an illustrative case study with the considered criteria and alternatives that implement the suggested method to consumer decision making problem for online purchases. At the end, Sect. 5 involves the conclusion notes for the study.

2 Preliminaries

Definition 1. Fuzzy Numbers [6]: Real-world involves complicated decision-making environment, and experts or DMs are usually imprecise or uncertain with the possibility of the given values, and the possible degrees of this impreciseness also remain indefinite for them. Fuzzy sets involve assigned degrees for each possible value of x, $\mu(x) \in [0, 1]$.

$x_0 \in R$, where $\mu_M(x_0) = 1$, for any $0 \leq \alpha \leq 1$, $A_\alpha = \left[x, \mu_{A_\alpha}(x) \geq \alpha\right]$ is a closed interval, where F(R).

Definition 2. Triangular Fuzzy Numbers [6]: Triangular fuzzy number, which is denoted by $\tilde{A}$ is a triplet, shown as (a_1, a_2, a_3), where, the membership is assigned using the formula below:

$$\mu_{\tilde{A}}(x) = \begin{cases} 0 & x \in (-\infty, a_1) \\ \frac{x-a_1}{a_2-a_1} & x \in [a_1, a_2] \\ \frac{c-x}{a_3-a_2} & x \in [a_2, a_3] \\ 0 & x \in (a_3, +\infty) \end{cases}$$

Definition 3. Z-Numbers [7]: Reliable data or information is needed for a sound decision. Z-number concept is based on this reality. Each Z-number consists of two parts A and B, or Z = (A, B). Here, A indicates the fuzzy restriction on the real values that a random variable X can hold; where B identifies the fuzzy value for the probabilistic measure of reliability on the part A. Here, both A and B are indicated in natural language (NL) and they are fuzzy numbers.

Definition 4. Z-Valued Matrix [8]: Z-valued matrix is developed for the elements of Z-numbers, and identifies partial reliability of the information based on values of the random variables, $X_{ij}, i, j = 1, \ldots, n$:

$$(Z_{ij} = (A_{ij}, B_{ij})) = \begin{pmatrix} Z_{11} = (A_{11}, B_{11})) \ldots Z_{1n} = (A_{1n}, B_{1n})) \\ \cdot \qquad \ldots \qquad \cdot \\ Z_{n1} = (A_{n1}, B_{n1})) \ldots Z_{nn} = (A_{nn}, B_{nn})) \end{pmatrix}$$

3 Statement of the Problem

In our study, every alternative e-commerce website, $A = \{a_1, a_2, a_3\}$, is characterized by 5 main criteria $C = \{c_1, c_2, c_3, c_4, c_5\}$. The main purpose of this research is to evaluate and prioritize the factors that have an impact on online buying decision of consumers in order to increase the company sales and provide a win-win situation for both parties. Identified criteria that must be feed are shown below with the supporting literatures. Researches [4, 5] identify the most influential factors for pattern of consumer decisions in online marketplaces. By integrating the mentioned research factors, we offer the following criteria:

C1: Creative and Innovative products: One of the most important factors that influences the consumer decision making process is related to how trendy or up to date the products are on online stores [5]. In order to trigger the need recognition stage of the customers marketers should use creative product advertisements [1] and sell trendy products on their website to attract customers. So, even if their prior purpose was to take a glance at the web, trendy products may convert the visitors into consumers. Furthermore, fashionable clothes with convenient prices trigger the fast consumerism [9].
C2: Perceived risk of online purchase: Perception of risk by consumers may obstruct or slow down the proliferation of online shopping [10]. Perceived technology risk is associated with the degree of consumer belief that she/he will suffer from poor personal information security or download problems. Seller/vendor perceived risk is related to the degree of the belief that the ordered products may be undelivered, or personal information may be unofficially used by the vendor, etc. Last two type of risks are related to consumer and product-based risks such as negative reactions from acquaintances about the online purchased products, and delivery of the inferior products [11]. Three-dimensional theory of the perceived risk is suggested in [12]. First dimension is related to expected product benefit failure risk that may arise when the consumer cannot get the expected benefit from the product purchased online. Second dimension is about the misuse or loss of the private personal information that consumer indicates during online

purchase. Functionally inefficiency risk is the third dimension and associated with the lost energy, time or money of the customers during online shopping [13]. As a result, online stores should provide an organized or user-friendly website with chatbot alike programs, secure purchase, and a fair business policy to increase its reputation [5].
C3: Brand name and Quality: As it is put forward in [4], consumers pay attention to the brand names besides, the status and quality associated with the name. Here, name guarantees the quality and the value to the consumer that creates the base for the customer relationship and the loyalty accordingly. So, brands and the associated quality serve as an important factor to influence the consumer decision on online sales also.
C4: Incentives for buying: This criterion involves the ways of appealing to customers. For online sales inducements such as free shipping, cash-on-delivery and discount coupons are usually leveraged. Several researches such as [5, 14] have shown that there is positive relationship between promotions and purchase rate. Furthermore, as there is no face to face contact as in usual stores, user-friendly sites with true information and smooth flow features provide better customer experience and satisfaction [15].
C5: Post-purchase service: Post-purchase services may involve cash-back and a fair return system [5] that may relieve the consumer's mind. Return system should be utilized for customer satisfaction and for high website traffic to increase the sales and build a strong customer relationship accordingly. Usually many websites offer free pick-ups and cash return [5]. Moreover, once the customers are satisfied, they will provide positive electronic word-of mouth (e-WOM) through digital platforms (through comments, likes, etc.) which is important for today's consumers as they are highly influenced by their acquaintances and peers rather than vendors [16].

4 Solution of the Problem

Solution method is based on consistency driven approach for determination of weights of the 5 criteria (Creative and Innovative products, Perceived risk of online purchase, Brand name and Quality, Incentives for buying and Post-purchase service) and identification of distance between ideal positive and negative solution under Z-number.

Step 1: We evaluate the consistency of DM's preferences using consistency-driven approach to Z-valued preferences. Weights of the criteria are assigned by a DM based on the literature survey and experiences (Table 1). The DM qualifies a certainty degree in linguistic terms on B part, based on the first part of evaluation, A, that is identified in triangular fuzzy numbers. Optimal Z-matrix is calculated and given in Table 2.
Step 2: Crisp maximum Eigenvalue and Eigenvectors are found.
Step 3: Ideal positive and negative solutions based on the weights of criteria are formulated using TOPSIS method. Ranking of the alternatives $\{a_1, a_2, a_3\}$ is found.

Table 1. Pairwise comparison matrix

	C1	C_2	C_3	C_4	C_5
C_1	(1, 1, 1), (0.8, 0.9, 1.0)	(2, 3, 4), (0.8, 0.9, 1.0)	(1, 2, 3), (0.8, 0.9, 1.0)	(1, 2, 3), (0.8, 0.9, 1.0)	(1, 2, 3), (0.8, 0.9, 1.0)
C2	(1/4, 1/3, 1/2), (0.8, 0.9, 1.0)	(1, 1, 1), (0.8, 0.9, 1.0)	(1/3, 1/2, 1/1), (0.8, 0.9, 1.0)	(1/3, 1/2, 1/1), (0.8, 0.9, 1.0)	(1, 2, 3), (0.8, 0.9, 1.0)
C3	(1/3, 1/2, 1/1), (0.8, 0.9, 1.0)	(1, 2, 3), (0.8, 0.9, 1.0)	(1, 1, 1), (0.8, 0.9, 1.0)	(2, 3, 4), (0.8, 0.9, 1.0)	(1, 2, 3), (0.8, 0.9, 1.0)
C_4	(1/3, 1/2, 1/1), (0.8, 0.9, 1.0)	(1, 2, 3), (0.8, 0.9, 1.0)	(1/4, 1/3, 1/2), (0.8, 0.9, 1.0)	(1, 1, 1), (0.8, 0.9, 1.0)	(1, 2, 3), (0.8, 0.9, 1.0)
C_5	(1/3, 1/2, 1/1), (0.8, 0.9, 1.0)	(1/3, 1/2, 1/1), (0.8, 0.9, 1.0)	(1/3, 1/2, 1/1), (0.8, 0.9, 1.0)	(1/3, 1/2, 1/1), (0.8, 0.9, 1.0)	(1, 1, 1), (0.8, 0.9, 1.0)

Table 2. Optimal Z-matrix

	C1	C_2	C_3	C_4	C_5
C_1	(0.997, 0.999, 0.999), (0.799, 0.903, 0.999)	(1.504, 2.793, 4.119), (0.938, 0.938, 0.999)	(0.880, 1.653, 2.986), (0.831, 0.895, 0.999)	(1.059, 2.270, 3.163), (0.955, 0.996, 0.996)	(1.003, 2.714, 2.714), (0.806, 0.906, 0.999)
C2	(0.258, 0.332, 0.501), (0.803, 0.897, 0.999)	(0.998, 0.999, 0.999), (0.804, 0.901, 0.999)	(0.238, 0.316, 0.835), (0.808, 0.974, 0.974)	(0.372, 0.635, 0.635), (0.796, 0.905, 0.999)	(0.499, 1.865, 2.969), (0.953, 0.986, 0.986)
C3	(0.331, 0.493, 0.999), (0.805, 0.897, 0.999)	(0.998, 2.001, 2.998), (0.790, 0.897, 0.999)	(1.001, 1.003, 1.003), (0.854, 0.854, 0.999)	(0.984, 2.672, 3.982), (0.844, 0.900, 0.999)	(0.993, 2.578 2.578), (0.801, 0.949, 0.949)
C_4	(0.330, 0.505, 0.994), (0.800, 0.894, 0.999)	(0.997, 1.993, 3.005), (0.812, 0.894, 0.999)	(0.244, 0.416, 0.416), (0.800, 0.896, 0.997)	(0.999, 1.000, 1.000), (0.797, 0.944, 0.944)	(0.728, 1.840, 2.841), (0.953, 0.953, 0.999)
C_5	(0.335, 0.500, 0.994), (0.803, 0.902, 0.998)	(0.331 0.500, 0.997), (0.804, 0.895, 0.998)	(0.327, 0.501, 0.998), (0.805, 0.899, 0.998)	(0.335, 0.492, 0.995), (0.792, 0.898, 0.994)	(0.996, 0.999, 0.999), (0.802, 0.951, 0.951)

The inconsistency index I for Z-number-valued PCM is indicated as it's given below:

$$I((Z_{ij})) = \max_{i<j<k} \min\left\{D\left(Z(1), \left(\frac{Z_{ik}}{Z_{ij}Z_{jk}}\right)\right), D\left(Z(1), \left(\frac{Z_{ij}Z_{jk}}{Z_{ik}}\right)\right)\right\}.$$

where:

- $Z(1)$ is the unit Z-number.
- $Z\_ik/(Z\_ij\ Z\_jk)$ and $(Z\_ij\ Z\_jk)/Z\ \_ik$ are the results of multiplication and division of Z numbers, $Z\_ik, Z\_ij, Z\_jk, i - 1, 2, 3 \ldots, j - 1, 2, 3 \ldots, k - 1, 2, 3$, and $i < j < k$.
- D is the distance between Z-numbers.

Inconsistency of original matrix = 0.521168530567112,
Inconsistency of optimal matrix = 0.0381041691329965.

Finding Crisp Maximum Eigenvalue and Eigenvectors. Here, maximum Eigenvalue is identified by solving the equation: $det(A - \lambda I) = 0$. Only medium values are involved in matrix A of fuzzy elements. Objective function DE is minimized, $CF(\lambda) = D^2(\lambda) - \varepsilon\lambda$, and λ is the parameter to optimize; $D(\lambda) = det(A - \lambda I)$ is the function dependent of λ, where ε is a constant (and might need to be adjusted for a concrete problem, in our example $\varepsilon = 100$). (DE search space dimension is 1).

A new DE objective function is constructed to identify Eigenvectors: $CF(X) = VectorDistance(AX, \lambda X)$. Here, parameters to search are (N − 1) elements of vector X, (except the first one that is set to 1), and DE search space dimension is N − 1.

Finding Fuzzy Eigenvalues. We use Fuzzy A matrix and construct a new DE objective function as follows: $CF(\lambda_L, \lambda_R) = FuzzyDistance(AX, (\lambda_L, \lambda_M, \lambda_R)X)$, where "FuzzyDistance" is stands for the sum of fuzzy distances between corresponding elements of vectors. Vector X compromises fuzzy singletons, from crisp Eigenvectors given in the first step. Only left and right components of fuzzy Eigenvalues are optimized (Medium component is taken from the first step). DE search space dimension is 2.

Finding Fuzzy Eigenvectors. Fuzzy eigenvalues and fuzzy matrix A are utilized to construct DE objective function as $CF(X) = FuzzyDistance(AX, \lambda X)$. Parameters to search are N elements of fuzzy vector X. Only left and right components are subject to search, and we select middle component from the first step. DE search space dimension is 2 * N.

STEP1
MAX LAMBDA = 5.2327959793008
Eigen Vector = 1 0.348061107771654 0.783295303148861 0.49527477460671 0.309811672843413
Correctness Check: A*X = 5.23278 1.82129 4.09881 2.59164 1.62113
Lambda*X = 5.23280 1.82133 4.09882 2.59167 1.62118
STEP2
Fuzzy LAMBDA = [3.04586406882296, 5.2327959793008, 7.18006224215461]
STEP3
Fuzzy LAMBDA = [3.04586406882296, 5.2327959793008, 7.18006224215461]
Fuzzy Eigen Vectors = [0.00000 1.00000 1.03093] [0.00000 0.34806 0.34806] [0.00000 0.78330 0.80819] [0.00000 0.49527 0.53220] [0.00000 0.30981 0.30981]
Correctness Check: A*X = [0.00000 5.23278 7.40214] [0.00000 1.82129 2.79910] [0.00000 4.09881 5.80288] [0.00000 2.59164 3.82125] [0.00000 1.62113 3.01805]
Lambda*X = [0.00000 5.23280 7.40214] [0.00000 1.82133 2.49910] [0.00000 4.09882 5.80288] [0.00000 2.59167 3.82125] [0.00000 1.62118 2.22447]

Criteria weights are found by normalizing eigenvectors using the following formula:

$$Z = \frac{X - min X}{max X - min X}$$

Normalized Fuzzy Eigenvectors =
[0.00000 0.10000 0.10000], [0.00000 0.05542 0.05304], [0.00000 0.68603 0.69112] [0.00000 0.26871 0.30840], [0.00000 0.00000 0.00000].

The calculated fuzzy criteria weights are indicated in Table 3, which presents the overall fuzzy decision matrix. Alternative e-commerce websites are evaluated according to identified criteria.

According to the calculation on normalized weights of criteria, effects of trendy and creative products (C_1) is found to be the most important criterion which is followed by brand name and quality (C_3), incentives for buying (C_4), perceived risk (C_2) and post-purchase services (C_5), respectively. C_1, C_3, C_4 and C_5 are benefit criteria, however, C2 is a cost criterion. Table 4 shows the weighted normalized fuzzy decision matrix computed by the following formula;

$$\tilde{V} = (\tilde{v}_{ij}), \text{ where } \tilde{v}_{ij} = \tilde{r}_{ij} \times w_j$$

Table 3. Fuzzy decision matrix.

	C_1-innovation	C_2-risk	C_3-brand	C_4-incentives	C_5-post-purchase
w	[0.00 1.00 1.00]	[0.0 0.06 0.05]	[0.00 0.69 0.70]	[0.00 0.27 0.31]	[0 0 0]
a_1	0.6, 0.7, 0.8	0.8, 0.9, 1	0.2, 0.3, 0.4	0.4, 0.5, 0.6	0.6, 0.7, 0.8
a_2	0.8, 0.9, 1	0.2, 0.3, 0.4	0.6, 0.7, 0.8	0.6, 0.7, 0.8	0.4, 0.5, 0.6
a_3	0.4, 0.5, 0.6	0.2, 0.3, 0.4	0.4, 0.5, 0.6	0.8, 0.9, 1	0.4, 0.5, 0.6

Table 4. Weighed normalized fuzzy decision matrix.

	C_1-innovation	C_2-risk	C_3-brand	C_4-incentives	C_5-post-purchase
a_1	0.0, 0.7, 0.8	0.0, 0.054, 0.05	0.0, 0.207, 0.28	0.0, 0.135, 0.186	0.0, 0.0, 0.0
a_2	0.0, 0.9, 1	0.0, 0.018, 0.02	0.0, 0.483, 0.56	0.0, 0.189, 0.248	0.0, 0.0, 0.0
a_3	0.0, 0.5, 0.6	0.0, 0.018, 0.02	0.0, 0.345, 0.248	0.0, 0.243, 0.31	0.0, 0.0, 0.0

For the next step we compute the Fuzzy Positive Ideal Solution (FPIS) and Fuzzy Negative Ideal Solution (FNIS) (Table 5).

$$A^* = \tilde{v}_1^*, \tilde{v}_2^*, \ldots, \tilde{v}_n^*, \text{ where } \tilde{v}_j^* = \max_i\{v_{ij}\}$$

$$A^- = \tilde{v}_1^-, \tilde{v}_2^-, \ldots, \tilde{v}_n^-, \text{ where } \tilde{v}_j^* = \min_i\{v_{ij}\}$$

Table 5. Fuzzy positive ideal and negative ideal solutions.

	C_1-innovation	C_2-risk	C_3-brand	C_4-incentives	C_5-post-purchase
a^*	0.0, 0.9, 1	0.0, 0.054, 0.05	0.0, 0.483, 0.56	0.0, 0.243, 0.31	0.0, 0.0, 0.0
a^-	0.0, 0.5, 0.6	0.0, 0.018, 0.02	0.0, 0.207, 0.28	0.0, 0.135, 0.186	0.0, 0.0, 0.0

Then, the distance from each alternative to the FPIS and to the FNIS is computed.

$$d_i^* = \sum_{j=1}^{n} d\left(\tilde{v}_{ij}, \tilde{v}_j^*\right),$$

$$d_i^- = \sum_{j=1}^{n} d\left(\tilde{v}_{ij}, \tilde{v}_j^-\right).$$

Finally, the closeness coefficient CC_i for each alternative is calculated for the ranking, which is shown in Table 6:

$$CC_i = \frac{d_i^-}{d_i^- + d_i^*}.$$

Table 6. Closeness coefficient of each alternative and ranking.

Alternatives	d_i^*	d_i^-	CCi	Ranking
a_1	0.485	0.19	0.28	2
a_2	0.074	0.607	0.89	1
a_3	0.543	0.177	0.25	3

As a consequence, according to the achieved results, second e-commerce business platform is the optimal one to be selected and used by the customers. Results are shown as; $a_2 > a_1 > a_3$.

5 Conclusion

As the technology improves, consumer behavior is subject to change in the light of digitalization. Digital marketing has become an inevitable part of marketing practices. Access to the advanced digital technology and the internet is constantly increasing. As a result, the use of e-commerce also increases. In this study, several factors such as effects of trendy and creative products; perceived risk of online purchases; brand names and quality; incentives for buying; and post-purchase services on consumer decision-making for online shopping are identified. Then, the selection of optimal e-commerce platform among the given alternatives is investigated. This research is based on consistency driven approach for identification of weights of criteria, and the distance between ideal positive and ideal negative solution under Z-number. Numerical example is given to check the validity of the considered method. Moreover, C_1-effects of trendy and creative products is found to be the most important criterion which is followed by brand name and quality (C_3), incentives for buying (C_4), perceived risk (C_2) and post-purchase services (C_5), respectively.

References

1. Kotler, P., Armstrong, G., Opresnik, O.M.: Principles of Marketing. Pearson, Harlow (2018)
2. Teo, T.S., Yeong, Y.D.: Assessing the consumer decision process in the digital marketplace. Omega **31**(5), 349–363 (2003). https://doi.org/10.1016/S0305-0483(03)00055-0
3. Durmaz, Y., Efendioglu, I.H.: Travel from traditional marketing to digital marketing. Glob. J. Manag. Bus. Res. **16**(2), 35–40 (2016)
4. Kumar, A., Mangla, S.K., Luthra, S., Rana, N.P., Dwivedi, Y.K.: Predicting changing pattern: building model for consumer decision making in digital market. J. Enterp. Inf. Manag. **31**(5), 674–703 (2018). https://doi.org/10.1108/JEIM-01-2018-0003
5. Singh, D.K., Kumar, A., Dash, M.K.: Using analytic hierarchy process to develop hierarchy structural model of consumer decision making in digital market. Asian Acad. Manag. J. **21**(1), 111 (2016)
6. Aliev, R.A., Aliev, R.R.: Soft Computing and Its Application. World Scientific (2001)
7. Zadeh, L.A.: A note on Z-numbers. Inf. Sci. **181**(14), 2923–2932 (2011)

8. Imanova, G.E., Imanova, G.: Digital marketing technologies selection under Z-environment. In: 14th International Conference on Theory and Applications of Fuzzy Systems and Soft Computing. Springer, Cham (2020)
9. Yang, S., Song, Y., Tong, S.: Sustainable retailing in the fashion industry: a systematic literature review. Sustainability **9**(7), 1266 (2017). https://doi.org/10.3390/su9071266
10. Solomon, M.R.: Consumer Behavior: Buying, Having, and Being, 10th edn. Pearson College Div (2012)
11. Benbasat, I., Barki, H.: Quo vadis TAM? J. Assoc. Inf. Syst. **8**(4), 7 (2007). https://aisel.aisnet.org/jais/vol8/iss4/7
12. Lim, N.: Consumers' perceived risk: sources versus consequences. Electron. Commer. Res. Appl. **2**(3), 216–228 (2003). https://doi.org/10.1016/S1567-4223(03)00025-5
13. Glover, S., Benbasat, I.: A comprehensive model of perceived risk of e-commerce transactions. Int. J. Electron. Commer. **15**(2), 47–78 (2010). https://doi.org/10.2753/JEC1086-4415150202
14. Fortes, N., Rita, P.: Privacy concerns and online purchasing behaviour: towards an integrated model. Eur. Res. Manag. Bus. Econ. **22**(3), 167–176 (2016). https://doi.org/10.1016/j.iedeen.2016.04.002
15. Ye, L.R., Zhang, H.H.: Sales promotion and purchasing intention: applying the technology acceptance model in consumer-to-consumer marketplaces. Int. J. Bus. Humanit. Technol. **4**(3), 1–5 (2014)
16. Moran, G., Muzellec, L., Nolan, E.: Consumer moments of truth in the digital context: How "search" and "e-word of mouth" can fuel consumer decision making. J. Advert. Res. **54**(2), 200–204 (2014). https://doi.org/10.2501/JAR-54-2-200-204

The Possibilities of Using Cloud Computing for Applications of Industrial Automation

N. R. Yusupbekov[1], Sh. M. Gulyamov[1], and N. B. Usmanova[2(✉)]

[1] Tashkent State Technical University, Tashkent, Uzbekistan
dodabek@mail.ru, shukhrat.gulyamov@mail.ru
[2] Tashkent University of Information Technologies, Tashkent, Uzbekistan
nargizausm@mail.ru

Abstract. With tremendous efforts in developing of computer networks and systems, cloud computing capabilities are in particular importance, providing access to common computing resources in an autonomous, dynamically scalable and verified mode: these are obvious advantages in speed, operability and efficiency.

Cloud computing is considered as a new paradigm shift in information technology industry. Cloud computing and its associated benefits in cost savings, increased efficiency, improved agility, opportunities for growth and innovation, tend to variety of implementation options for Industry 4.0. tasks. Having discussed the possibilities of using cloud technologies in industrial automation, the authors propose a methodical basis for the application of cloud computing in industrial environments, thus allowing to predict the behavior of the system, the situations occurred and determine critical processes taking into account rapid changes in the field of hardware and software of modern industries. The methodological foundations for using cloud computing in industrial automation are summarized and computational experiment for virtualization is demonstrated.

Keywords: Cloud computing · Industrial cloud · Industrial automation · Approach for applying cloud solutions

1 Introduction

With tremendous efforts in developing of computer networks and systems, cloud computing capabilities are in particular importance, providing access to common computing resources in an autonomous, dynamically scalable and verified mode: these are obvious advantages in speed, operability and efficiency.

The last decades demonstrated the rapid progress in computer hardware, software, internet, sensor networking, mobile communications, multimedia technologies, distributed and cloud computing systems, and related fields; it is worth to note, that cloud technologies representing the ever growing area of research, pose challenges for various applications, including networks and systems, while satisfying increasing demands for better quality of services and lower cost. From the other hand, the impact of Internet on the industry in the coming decade will be huge and will bring many benefits to individual

© The Author(s), under exclusive license to Springer Nature Switzerland AG 2021
R. A. Aliev et al. (Eds.): WCIS 2020, AISC 1323, pp. 246–253, 2021.
https://doi.org/10.1007/978-3-030-68004-6_32

users and companies. Among such advantages with respect to cloud capabilities are [1, 2]: accessibility, i.e. clouds are accessible to everyone, from anywhere; low cost due to reducing the cost of servicing the virtual infrastructure and paying for the actual use of resources; flexibility as unlimited computing resources (memory, processor, disks), through the use of virtualization systems; reliability - especially for those located in specially equipped data centers and a number of others. Along with this, there are a number of drawbacks, such as the need for a permanent stable connection to the network; there are restrictions on the software that can be deployed on clouds and provide to the user; the confidentiality of data stored in public clouds (which currently causes a lot of controversy, but in most cases, experts agree that it is not recommended to store the most valuable documents in a public cloud); the inability to recover information located in the cloud, etc.

Assumptions from experts and engineers working on Cloud Computing and benefiting cost savings, increased efficiency, improved agility, opportunities for growth and innovation, and others, stated in reports and related papers the necessity to imply for the companies and organizations the Cloud computing solutions, to scale and innovate at the rate of their competitors [3, 4]. Some statistics include: 83% of enterprise workloads will be in the cloud by the end of 2020, of which 41% of enterprise workloads will be run on public cloud platforms; public cloud services grew 18% in 2019, upwards of $215b, up from $180b from 2018.

Summarizing cloud computing capabilities, it can be noted that cloud computing is becoming increasingly popular in many industries, including the core business support functions [4–6].

In general, cloud computing implementation varies from industry to industry due to different levels of security and other requirements. Some of the solutions offered in the market with cloud services are shown in Fig. 1.

2 Methodical Aspects of Using Cloud Computing in Industry

Based on the trends of cloud computing evolution, the following main tendencies in the development of information technologies with respect to clouds can be noted: hybrid cloud and IT companies as a service broker; cloud-client architecture (the cloud serves as a management point, and a system or applications can be spread to many client devices); personal cloud (transition from devices to services); Software-defined anything (SDx) is a collective term describing the growing market demand for improved standards for programmable infrastructure and interaction with data centers; Web - scaling of information technologies. In addition, modern cloud technologies are not only used in ready-made network and server equipment, but also gradually penetrate the embedded cloud market and cause large-scale market restructuring [7].

The introduction of embedded systems leads to the placement of computer processors in such products as resource consumption meters, smart sensors, modules, cars, household appliances, etc. This allows to control the operation of devices, data collection and provision of interactive capabilities by connecting to a computer network [7, 8].

Along with this, in the near future the development of distributed high-performance systems will make it possible to achieve a significant breakthrough in many industries,

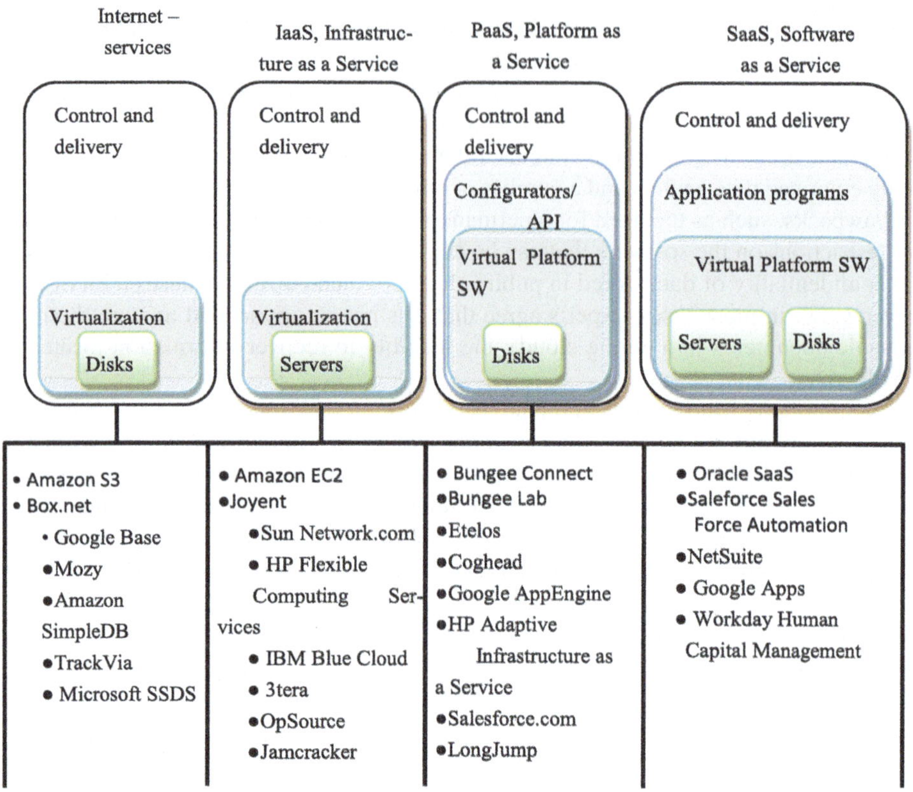

Fig. 1. Main types of services for cloud-based solutions

based on the results of numerical modeling in distributed environments, creating and promoting unified platforms for consolidating heterogeneous computing resources in a distributed environment based on cloud computing technologies.

Various enterprise management systems, customer relationship systems, and others can be hosted in the cloud, and industrial automation applications will require production data for their operation, the collection of which starts from the end devices (which may be sensors, smart meters, smart devices); such components can be connected to the controller, interfaces of various types and, depending on the functionality of these interfaces, process data directly at the source, filter the data, or save and consolidate data before sending it to the cloud [9–11]. From a technical standpoint, issues of proper resource configuration are important (for example, SaaS, IaaS services are often used to provide cloud computing, as well as to store and analyze data [12]). Thus, taking into account the existing requirements and capabilities for configuring resources in the cloud, we can consider a generalized configuration management scheme (Fig. 2), which can serve as a methodological basis for the interconnected functioning of the cloud computing environment, regardless of what model: SaaS, CaaS, PaaS, IaaS, NaaS, is implemented.

Computing resources are used to provide the necessary cloud service capabilities and support other system functions, such as resource abstraction and management, security, and monitoring. The peculiarities of providing access to resources in the cloud computing infrastructure substantiate the principles, models and essential characteristics of this technology when studying issues of ensuring the quality of the services provided [13].

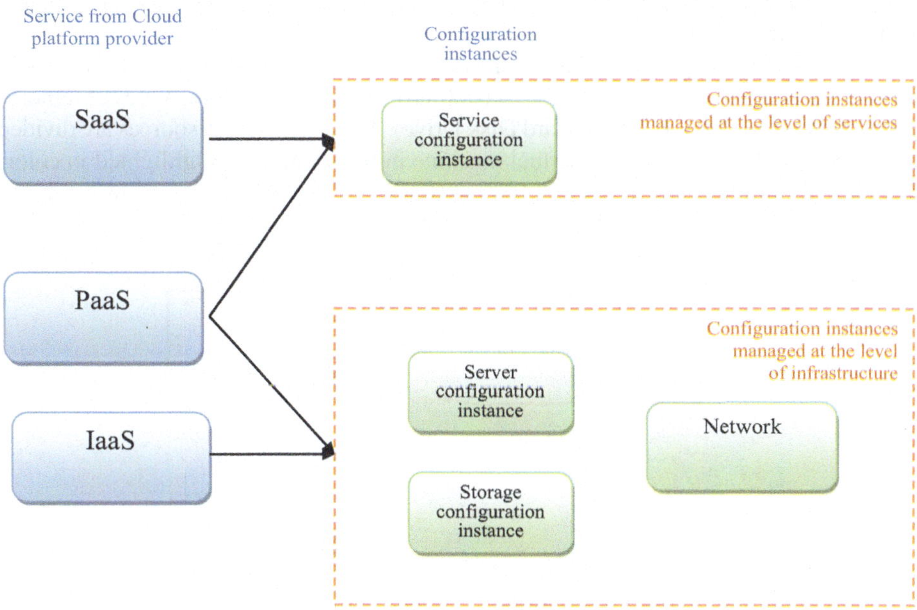

Fig. 2. Principles of configuring in cloud environment

3 Generalized Scenario of Virtualization in the Cloud Infrastructure

For industrial automation applications, two categories of virtualization can be specified: virtualization of platforms and virtualization of resources.

Platform virtualization refers to creation of software systems based on real software and hardware systems, which are both dependent and independent of them.

Platform virtualization can be divided into the following types:

Full emulation (simulation), a virtual machine virtualizes all hardware and keeps the guest operating system unchanged.

Partial emulation, the virtual machine virtualizes only the necessary amount of hardware so that it can run in isolation.

Partial virtualization (address space virtualization), when using this approach, the virtual machine simulates several instances of the hardware environment, in particular, the address space.

Paravirtualization, makes it possible not to simulate hardware, but instead, a special API (Application Programming Interface) is used to interact with the guest operating system.

Application-level virtualization: for an application, a container is created that has everything necessary for the operation of the application (registry files, configuration files, user and system objects).

Resource virtualization makes it possible to concentrate and simplify the management of resource groups.

The simulation is performed using the AnyLogic software environment. In the scenario there is a physical machine that has 20 GB of RAM (Random Access Memory) and a hard disk of 250 GB HDD (Hard Disk Drive). These memory resources are divided among 5 virtual machines VM (virtual machine in the scenario is highlighted in color) (Fig. 3).

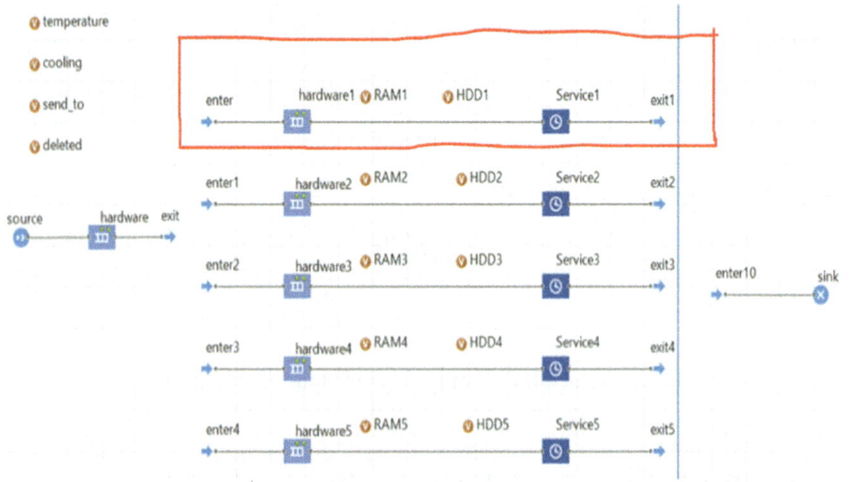

Fig. 3. Virtualization scenario in AnyLogic, creating virtual machine

For each VM there is 4 GB of RAM and 50 GB HDD and two kinds of variables for RAM and HDD, respectively. In the simulation scenario for the source, the value of the received requests is determined (random variable). The values at the source output are considered as the input number for the physical machine. In addition, there are some output codes that illustrate the VM usage algorithm associated with the size of the RAM and the size of the hard disk.

The study introduces the parameters of energy, temperature, heating and cooling. In the AnyLogic environment, the maximum number of agents is 50,000; in this regard, the maximum temperature value is defined as equal to 0.005. This means that if a physical machine serves each agent, the temperature rises to 0.005. If the system operates in cooling mode, the temperature value is reduced by - 0.1°. These conditions were introduced into Service and Sink (Fig. 4a), b)) and graphs were obtained demonstrating the operation of the modeling system. These are timeInSystemDistr, temp, RAM1_,

HDD1_, RAM2_, HDD2_, RAM3_, HDD3_, RAM4_, HDD4_, RAM5_ and HDD_5. Figure 5 shows the timeInSystemDistr and temp graphs. The graph of dependencies between timeInSystemDistr and the number (percentage) of agents illustrates how many (percentage) agents are in the system.

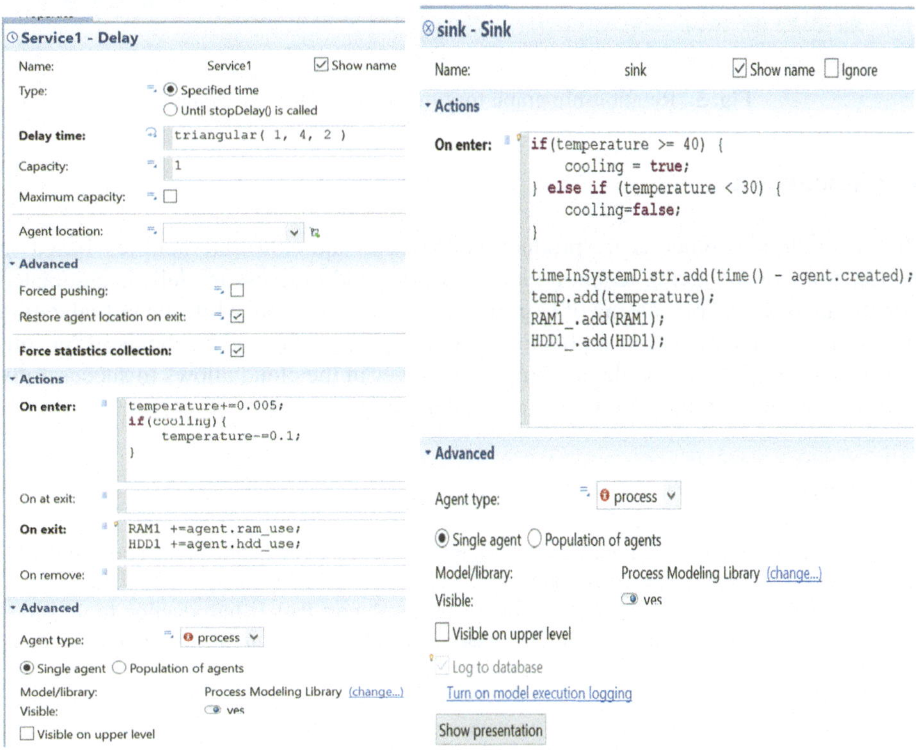

Fig. 4. Screenshots for Service (a) and for the Sink (b)

A graph of the relationship between temperature and the percentage of agents illustrates how many agents are at the indicated temperatures. For example, 16% of agents are at temperature of about 31°, while only 1% of agents are at 29°. This means that the cooling system operates properly.

The scenario of studying the properties of virtualization can be expanded in the implementation details when simulating the specific infrastructure solutions. Given that the cloud environment is based on the virtualization of all computing resources, such an approach can serve as the basis for determining the necessary parameters of a virtualized cloud infrastructure when planning and implementing cloud technologies [14].

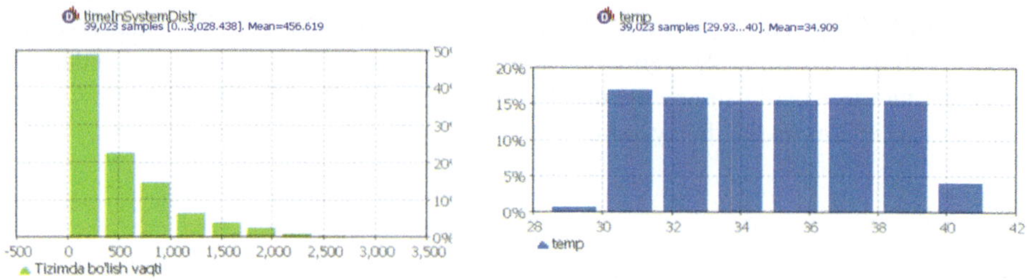

Fig. 5. Relationship graph for timeInSystemDistr and temp

4 Conclusion

The use of cloud computing for production processes and industrial facilities, based on the functions of managing technological processes and business models, is associated with the analysis and processing of large amounts of information, while the virtualization of most industrial automation applications requires proper rules and mechanisms for configuring resources, while scalable computing power in the cloud allows to successfully perform tasks that require significant computational resources.

References

1. The NIST Definition of Cloud Computing. https://csrc.nist.gov/publications/drafts/800-145/Draft-SP-800-145_cloud-definition.pdf
2. ITU-Focus Group on Cloud Computing Technical Report Part 1: Introduction to the cloud ecosystem: definitions, taxonomies, use cases and high-level requirements, FG P.1, February 2012
3. Cloud Computing: Current Trends. https://www.cloudreach.com/en/resources/blog/top-8-trends-in-cloud-computing-and-how-companies-can-prepare-for-2020/
4. Industries Aim to Evolve Cloud Computing Beyond Support Functions to More Strategic Uses, 25 May 2012. ID: G00234673. https://www.gartner.com/id=2027216
5. Kyriazis, D. (ed.): Cloud Computing SLAs: Exploitation of Research Results. European Commission Directorate General, Communications Networks, Content and Technology - Unit E2 - Software and Services, Cloud (2013)
6. Yusupbekov, N., Adilov, F., Ergashev, F.: Development and improvement of systems of automation and management of technological processes and manufactures. J. Autom. Mob. Robot. Intell. Syst. **3**, 53–57 (2017). https://doi.org/10.14313/JAMRIS_3-2017/28
7. Breivold, H.P.: Towards factories of the future: migration of industrial legacy automation systems in the cloud computing and Internet-of-Things. J. Enterp. Inf. Syst. **14**, 542–562 (2020). https://doi.org/10.1080/17517575.2018.1556814
8. Givehchi, O., Trsek, H., Jasperneite, J.: Cloud computing for industrial automation systems -a comprehensive overview. In: 2013 IEEE 18th Conference on Emerging Technologies & Factory Automation (ETFA), Cagliari, pp. 1–4 (2013). https://doi.org/10.1109/ETFA.2013.6648080
9. Velásquez, N., Estevez, E., Pesado, P.: Cloud computing, big data and the industry 4.0 reference architectures. J. Comput. Sci. Technol. **18**(03), e29 (2018). https://doi.org/10.24215/16666038.18.e29

10. Hegazy, T., Hefeeda, M.: Industrial automation as a cloud service. IEEE Trans. Parallel Distrib. Syst. **26**(10), 2750–2763 (2015). https://doi.org/10.1109/TPDS.2014.2359894
11. Li, G., Wang, H., Hardjawana, W.: New advancement in information technologies for industry 4.0. J. Enterp. Inf. Syst. **14**(4), 402–405 (2020). https://doi.org/10.1080/17517575.2019.1651399
12. Atobishi, Th., Gábor, S.Z., Podruzsik, S.: Cloud computing and big data in the context of industry 4.0: opportunities and challenges. In: IISES Annual Conference, Sevilla Conference: Proceedings of the IISES Annual Conference, Sevilla, Spain, March 2018 (2018). https://doi.org/10.20472/IAC.2018.035.004
13. Yusupbekov, N.R., Tarassov, V.B.: Integrated intelligent systems for industrial automation: the challenges of industry 4.0, information granulation and understanding agents. In: Proceedings of Tenth International Conference on Intelligent Systems for Industrial Automation – WCIS-2018, Tashkent, 25–26 October 2018, pp. 13–25 (2018).
14. Yusupbekov, N.R., Abdurasulov, F.R., Adilov, F.T., Ivanyan, A.I.: Application of cloud technologies for optimization of complex processes of industrial enterprises. In: 13th International Conference on Theory and Application of Fuzzy Systems and Soft Computing - ICAFS-2018, pp. 852–858. Springer (2018). https://doi.org/10.1007/978-3-030-04164-9_112

Application of Fuzzy AHP Method to Material Selection Problem

Mustafa Babanli(✉) and Tural Gojayev

Azerbaijan State Oil and Industry University, Azadlig Avenue 20, Nasimi, Baku, Azerbaijan
mustafababanli@yahoo.com, gojayev.tural@gmail.com

Abstract. Material selection is a significant problem attracting interest both theoretically and practically. Currently, many types of materials and alloys have been developed. In many alloys, certain characteristics are acceptable and meet the requirements, but some of them are unacceptable. Therefore, scientists have developed various selection methods due to all of these reasons. Multi-criteria decision making (MCDM), also implemented in material selection, is obtaining importance as a potential method for the analysis of real complex problems in terms of enabling the selection of the most optimal alternative on various criteria. There are various multi-criteria decision making methods. The Analytical Hierarchy Process (AHP) introduced by T.L. Saaty was recently popularized. Fuzzy AHP, which is an extension of AHP, plays a significant role when the information is under conditions of uncertainty and ambiguity. This paper aims to select the optimal alloy alternative of the austenitic type of stainless steels intended to use in the marine environment, using the Fuzzy AHP method based on the impact strength, brinell hardness, machinability index, and cost criteria.

Keywords: Multi-criteria decision · Fuzzy-AHP method · Material selection · Stainless steels

1 Introduction

Multi-Criteria Decision Making (MCDM) is known as one of the fastest growing fields in recent years. Because of its usefulness and attractiveness, it has played a significant role in different application fields. Multi-Criteria Decision Making includes a finite number of alternatives for a decision maker to choose and rank among these alternatives, and there is a finite number of criteria weighted by the decision maker according to their significance.

The Analytical Hierarchy Process (AHP) is a frequently used Multi-Criteria Decision Making method and was first offered by T.L. Saaty in 1980. The Analytical Hierarchy Process (AHP) works correctly if it does not take into account the conditions of uncertainty in comparative judgment. In AHP method the decision is made on net worth, and the ranking is fairly accurate. However, the choices or judgments of decision makers always include ambiguous statements and inconsistencies. Thus, AHP is not an appropriate method to get good results in situations of uncertainty. In this case, Fuzzy decision

© The Author(s), under exclusive license to Springer Nature Switzerland AG 2021
R. A. Aliev et al. (Eds.): WCIS 2020, AISC 1323, pp. 254–261, 2021.
https://doi.org/10.1007/978-3-030-68004-6_33

making plays a crucial role when the information is under conditions of uncertainty. The fuzzy concept was first introduced in 1965 by professor Lotfi A. Zadeh. Hence, better results can be obtained by using the Fuzzy Analytical Hierarchy Process (FAHP), which is an extension of AHP. Fuzzy concept can be used in to the pairwise comparison to resolve such vagueness and uncertainty issues. The Fuzzy AHP method allows decision makers for better determining their preferences in fuzzy environments.

In this article, the Fuzzy AHP method has been applied to select the appropriate material that is necessary for the success of a particular engineering activity. The efficiency of a material in a marine environment depends on convenient material selection, service parameters, corrosion control methods, type of the environment and design configuration. The necessity of choosing the appropriate material increases, in particular in the corrosive nature of the work environment. The materials chosen for each component in a system must meet all performance requirements, provide long system life with minimal or no maintenance, and be economical. The most crucial factors to take into account when choosing materials for using in the marine environment are; strength, hardness, machinability, high resistance to corrosion, biological fouling and thermal conductivity. Moreover, the magnitude and periodic nature of wave loads bring many problems in the marine environment [2–4, 9, 10]. The following studies have been implemented on Fuzzy AHP.

2 Preliminaries

Definition 1. Fuzzy Set. Firstly, let X be the universe set of the objects, then let x be the elements of the universe set of X. Secondly, let A be the subset of X which is used for membership function μ_A from X to {0, 1} can be specified as follows:

$$\mu_A(x) = [1 \text{ if and only if } x \in A \text{ and } 0 \text{ if and only if } x \notin A]$$

[0, 1] is the value set, where 1 specifies the membership and 0 shows non-membership. If the value set is in the range of [0, 1], then we can say A is fuzzy set. $\mu_A(x)$ is the extent of the membership of elements x in the fuzzy set A, $\mu_A : x \to [0, 1]$. If the value of the $\mu_A(x)$ is close to 1, it means x is attributing to the fuzzy set A [1].

$$A = [(X, \mu_A(x)), x \in X]$$

Definition 2. Fuzzy number, Triangular Fuzzy Number. Firstly, let y be the fuzzy number and let X be the universe set. If y is both normal and convex, then we can say that y is a fuzzy subset of the universe set X. There are 2 significant characteristics of Fuzzy numbers:

1) Membership function of them, that is $\boldsymbol{\mu_y(x)}$ is continuous from R to {0, 1}.
2) Their membership function ($\mu_y(x)$) is convex and normal that is there exists a number such as y_0 that enables $\mu_y(x) = 1$.

If a triangular fuzzy number y is expressed as (y_1, y_2, y_3), then the membership function of the triangular fuzzy number $\mu_y(x)$ can be indicated with the followings [7]:

$$\mu_y(x) = \begin{cases} \frac{x-y_1}{y_2-y_1} r, \text{If } y_1 \le x \le y_2 \\ \frac{x-y_3}{y_2-y_3} r, \text{If } y_2 \le x \le y_3 \\ 0, \text{otherwise} \end{cases}$$

3 Fuzzy AHP

The details of the fuzzy AHP method will be given in this section below.

Step 1. Constructing a fuzzy matrix based on pair-wise comparison.

Step 2. Calculate the fuzzy geometric mean $(\bar{r}_i)$ value by using the following formula: Let X $(X_1, X_2, \ldots. X_n)$ be the fuzzy number, then in order to multiply 2 (or more) fuzzy numbers following formula will be used: n – shows the number of criteria, $X_1 * X_2 = (l_1, m_1, u_1) * (l_2, m_2, u_2) = ((l_1 * l_2)^{\frac{1}{n}}, (m_1 * m_2)^{\frac{1}{n}}, (u_1 * u_2)^{\frac{1}{n}})$.

Step 3. Calculate the fuzzy weights $\bar{w}_i(\bar{w}_1, \bar{w}_2, \ldots, \bar{w}_n)$ with the following formula: $\bar{w}_i = \bar{r}_i * (\bar{r}_1 + \bar{r}_2 + \bar{r}_n)^{-1}$. $\bar{w}_i$ – indicates the fuzzy weights, $\bar{r}_i$ – geometric mean values.

Step 4. Calculate the de-fuzified crisp numerical values by using the following formula. In the formula l, m, u is the lower, middle, and upper values of fuzzy number. Centre of Area (COA) $= W_i = \frac{l+m+u}{3}$ [6].

Step 5. Calculate the consistency index and consistency ratio of the fuzzy matrix $\tilde{C}$ with the following formulas [11]. $CI = \frac{\lambda\max - n}{n-1}$ and $CR = \frac{CI}{RI}$

4 Application

An application of fuzzy AHP (proposed by Buckley - 1985) will be examined in this section by using geometric mean. Consistency will be calculated for fuzzy AHP as well. This Fuzzy AHP method problem will be solved for making decision based on the material properties of candidate Austenitic Stainless Steel.

4.1 Calculations for Criteria

Weighted sum value should be calculated for each raw of the matrix by using the following formula: $\sum V_1 = V_{11} + V_{12} + V_{13} + V_{14}$. For example, $V_1 = 2.7$. After calculating the weighted sum values (V), we should get the ratio of $\frac{V}{W}$ (W – is denoting the criteria weights). For example, $\frac{V1}{W1} = 4.8268$. This ratio has been solved for each of the values. $\lambda(\max) = \sum \frac{Vij}{Wij}/n = 4.201679601$ (n = 4) n – shows the number of compared elements. Consistency index: CI = 0.067226534. Consistency ratio: CR = 0.074696148. (If n = 4, then RI = 0.9). Therefore, 0.074696148 < 0.10 (matrix is reasonable consistent) (Tables 1, 2 and 3).

Table 1. Fuzzified pair-wise comparison matrix

Criterion	Impact Strength (J/m^2)	Brinell Hardness (kgf/mm^2)	Machinability index (%)	Estimated $ per lbs
Impact Strength (J/m^2)	(1 1 1)	(4 5 6)	(3 4 5)	(6 7 8)
Brinell Hardness (kgf/mm^2)	(1/6 1/5 1/4)	(1 1 1)	(1/3 1/2 1/1)	(2 3 4)
Machinability index (%)	(1/5 1/4 1/3)	(1 2 3)	(1 1 1)	(2 3 4)
Estimated $ per lbs	(1/8 1/7 1/6)	(1/4 1/3 1/2)	(1/4 1/3 1/2)	(1 1 1)

Table 2. Fuzzy geometric mean

Fuzzy geometric mean ($\bar{r}_i$)	Fuzzy weights ($\bar{w}_i$)	De-Fuzification	Weights
(2.91, 3.43, 3.93)	(0.42, 0.51, 0.86)	0.596666667	0.559375
(0.57, 0.74, 1)	(0.08, 0.11, 0.21)	0.133333333	0.125
(0.79, 2.21, 1.41)	(0.11, 0.32, 0.31)	0.246666667	0.23125
(0.29, 0.35, 0.45)	(0.04, 0.14, 0.09)	0.09	0.084375

Table 3. Calculating the consistency

Criterion	Impact Strength (J/m^2)	Brinell Hardness (kgf/mm^2)	Machinability index (%)	Estimated cost $ per lbs
Impact Strength (J/m^2)	0.559375	0.625	0.925	0.590625
Brinell Hardness (kgf/mm^2)	0.11	0.125	0.115625	0.253125
Machinability index (%)	0.13984375	0.25	0.23125	0.253125
Estimated cost $ per lbs	0.079910714	0.041666667	0.077083333	0.084375

4.2 Calculations for Impact Strength Criterion

Weighted sum value should be calculated for each raw of the matrix by using the following formula: $\sum V_1 = V_{11} + V_{12} + V_{13} + V_{14}$. For example, $V_1 = 1.5405405$.

After calculating the weighted sum values (V), we should get ratio of $\frac{V}{W}$. For example, $\frac{V1}{W1} = 2.630769231$. This ratio has been solved for each of the values. $\lambda(\max) = \sum \frac{Vij}{Wij}/n$ $= 4.324130037$ (n = 4). n – shows the number of compared elements. Consistency index formula: $CI = \frac{\lambda(max)-n}{n-1}$ then, CI = 0.108043346. Consistency ratio: CR = 0.120048162 (If n = 4, then RI = 0.9). Therefore, 0.120048162 < 0.10 (matrix is reasonable consistent) (Table 4).

Table 4. Fuzzy matrix for Impact Strength criterion

Impact Strength	200 (J/m^2)	195 (J/m^2)	195 (J/m^2)	110 (J/m^2)
200 (J/m^2)	(1 1 1)	(1 2 3)	(1 2 3)	(3 4 5)
195 (J/m^2)	(1/3 1/2 1/1)	(1 1 1)	(1 1 1)	(2 3 4)
195 (J/m^2)	(1/3 1/2 1/1)	(1 1 1)	(1 1 1)	(2 3 4)
110 (J/m^2)	(1/5 1/4 1/3)	(1/4 1/3 1/2)	(1/4 1/3 1/2)	(1 1 1)

4.3 Calculations for "Brinell Hardness" Criterion

Weighted sum value should be calculated for each raw of the matrix by using the following formula: $\sum V_1 = V_{11} + V_{12} + V_{13} + V_{14}$. For example, $V_1 = 2.41311477$. After calculating the weighted sum values (V), we should get ration of $\frac{V}{W}$. For example, $\frac{V1}{W1} = 4.088888947$. This ratio has been solved for each of the values. $\lambda(max) = \sum \frac{Vij}{Wij}/n = 4.10015548$ (n = 4). n – shows the number of compared elements. Consistency index: CI = 0.0333852. Consistency ratio: CR = 0.0370946 (If n = 4, then RI = 0.9). Therefore, 0.0370946 < 0.10 (matrix is reasonable consistent) (Table 5).

Table 5. Fuzzy matrix for Brinell Hardness criterion

Brinell Hardness	192 (kgf/mm^2)	207 (kgf/mm^2)	183 (kgf/mm^2)	207 (kgf/mm^2)
192 (kgf/mm^2)	(1 1 1)	(4 5 6)	(3 4 5)	(4 5 6)
207 (kgf/mm^2)	(1/6 1/5 1/4)	(1 1 1)	(1/4 1/3 1/2)	(1 1 1)
183 (kgf/mm^2)	(1/5 1/4 1/3)	(2 3 4)	(1 1 1)	(2 3 4)
207 (kgf/mm^2)	(1/6 1/5 1/4)	(1 1 1)	(1/4 1/3 1/2)	(1 1 1)

4.4 Calculations for Machinability Index Criterion

Consistency calculations is not needed for Machinability index criterion as all of the sub-criteria of machinability are equal to 1 (Table 6).

4.5 Calculations for Estimated Cost $ Per Lbs Criterion

Weighted sum value should be calculated for each raw of the matrix by using the following formula: $\sum V_1 = V_{11} + V_{12} + V_{13} + V_{14}$. For example, $V_1 = 2.459858703$.

Table 6. Fuzzy Matrix for Machinability Index criterion

Machinability index (%)	30%	30%	30%	30%
30%	(1 1 1)	(1 1 1)	(1 1 1)	(1 1 1)
30%	(1 1 1)	(1 1 1)	(1 1 1)	(1 1 1)
30%	(1 1 1)	(1 1 1)	(1 1 1)	(1 1 1)
30%	(1 1 1)	(1 1 1)	(1 1 1)	(1 1 1)
30%	(1 1 1)	(1 1 1)	(1 1 1)	(1 1 1)

After calculating the weighted sum values (V), we should get ratio of $\frac{V}{W}$. For example, $\frac{V1}{W1} = 4.255555557$. This ratio has been solved for each of the values. $\lambda(\max) = \sum \frac{Vij}{Wij}/n = 4.189953112$ (n = 4). n – shows the number of compared elements.

Consistency index formula: $CI = \frac{\lambda(\max) - n}{n-1}$ then, CI = 0.0633177 Consistency ratio formula: $CR = \frac{\frac{\lambda(\max)-n}{n-1}}{RI}$ then, CR = 0.070353 (If n = 4, then RI = 0.9). Therefore, 0.070353 < 0.10 (matrix is reasonable consistent) (Table 7).

Table 7. Fuzzy Matrix for Estimated Cost $ per lbs criterion

Estimated cost $	0.98 $	1.29 $	1.46 $	1.96 $
0.98 $	(1 1 1)	(4 5 6)	(3 4 5)	(4 5 6)
1.29 $	(1/6 1/5 1/4)	(1 1 1)	(1/4 1/3 1/2)	(6 7 8)
1.46 $	(1/5 1/4 1/3)	(2 3 4)	(1 1 1)	(2 3 4)
1.96 $	(1/6 1/5 1/4)	(1/8 1/7 1/6)	(1/4 1/3 1/2)	(1 1 1)

5 Global and Local Weights Calculations

Next step is the calculation of local and global weights. We have expressed the criteria weights with W (W_1, W_2, W_3, ..., W_n). Sub-criteria weights will be W_s (W_{S1}, W_{S2}, W_{S3}, ..., W_{SN}). Sub-criteria weights are the local weights at the same time. In order to calculate the global weights, criteria weights should be multiplied (W_n) by sub-criteria weights (W_{SN}). Let, global weights be W_G (W_{G1}, W_{G2}, ..., W_{GN}), then formula will be: $W_g = W_n * W_{SN}$. For instance, global weight of impact strength criterion is $W_{G1} = 0.559375 * 0.602561272 = 0.33706$. Rest of the global weights have been calculated based on the sub-criterion weights.

6 Decision Matrix

Next step is to form the decision matrix. Firstly, the global weights calculated in step 5, should be multiplied by criterion weights of the alternatives. After this we can add together all the outcomes (X_{ij}) in order to get priority values (P_n).

Table 8. Decision matrix

Decision matrix					
Weights	0.215247719	0.125	0.23125	0.084375	
Criteria	Criterion-1	Criterion -2	Criterion -3	Criterion -4	Priority values
A-1	0.188542938	0.00922125	0.013368563	0.004114969	0.215247719
A-2	0.053890188	0.00148625	0.013368563	0.001074094	0.069819094
A-3	0.053890188	0.003535	0.013368563	0.001623375	0.072417125
A-4	0.016579875	0.00138375	0.013368563	0.000306281	0.031638469

Let each outcomes be X_{ij}, and priority values P_n (P_1, P_2, P_3, ..., P_n), then formula will be, $P_n = X_{11} + X_{21} + X_{31} + X_{41}$. For instance, $P_1 = 0.215247719$. All of the results have been calculated and shown in Table 8.

7 Ranking the Alternatives

Finally, alternative 1 is the best ranked according to the priority value of 0.215247719. The value of 0.072417125 was determined as the 2nd, the value of 0.069819094 as the 3rd, and the value of 0.031638469 as the 4th alternative.

8 Discussion and Conclusion

The results of this study specify that the proposed Fuzzy AHP method can provide a reliable material selection for use in seawater. The Fuzzy AHP method is easy to apply and lets you to simultaneously deal with the qualitative and quantitative characteristics of materials. In this paper, the goal was to select the optimal alloy alternative based on the criteria of impact strength, brinell hardness, machinability and cost of the austenitic type of stainless steel intended for use in seawater by using the Fuzzy AHP method. Optimal alloy alternatives are shown according to the Unified Numbering System (UNS): S30400, S31600, S31703, S31254. As can be seen in Table 9, the most optimal alternative obtained as a result of ranking was selected 1st (S30400) with a value of 0.215247719. The value of 0.072417125 was determined as the 2nd, the value of 0.069819094 as the 3rd, and the value of 0.031638469 as the 4th alternative. Thus, the S30400 is the best ranked alternative, with an impact strength of 200 J/m2, a brinell hardness of 192 kgf/mm2, a machinability index of 30%, and a cost of $ 0.98 per lbs of material, austenitic stainless steel intended for use in the marine environment.

Table 9. Ranking the alternatives

Alternatives	Priority values	Ranking
Alternative-1	**0.215247719**	**1**
Alternative-2	0.069819094	3
Alternative-3	0.072417125	2
Alternative-4	0.031638469	4

References

1. Aliev, R.A., Fazlollahi, B., Aliev, R.R.: Soft Computing and Its Applications in Business and Economics. Springer, Heidelberg (2004). https://doi.org/10.1007/978-3-540-44429-9
2. Alvarez-Aramas, I.: Duplex stainless steels: brief history and some recent alloys. Mech. Eng. **1**(1), 51–57 (2008). https://doi.org/10.2174/1874477X108010051
3. Odwani, A., Al-Tabtabaei, M., Abdel-Nabi, A.: Performance of high chromium stainless steels and titanium alloys in Arabian Gulf seawater. Dealination **120**, 73–81 (1998). https://doi.org/10.1016/s0011-9164(98)00203-3
4. Shifler, D.A.: Understanding material interactions in marine environments to promote extended structural life. Corros. Sci. **47**, 2335–2352 (2005)
5. Hipel, K.W., Kilgour, D.M., Bashar, M.A.: Fuzzy preferences in multiple participant decision making. Scientia Iranica D **18**(3), 627–638 (2011)
6. Buckley, J.J.: Fuzzy hierarchical analysis. Fuzzy Sets Syst. **17**(1), 233–247 (1985). https://doi.org/10.1016/0165-0114(85)90090-9
7. Aliev, R.A., Kacprzyk, J., Pedrycz, W., Jamshidi, M., Sadikoglu, F.M. (eds.): ICAFS 2018. AISC, vol. 896. Springer, Cham (2019). https://doi.org/10.1007/978-3-030-04164-9
8. Babanli, M.B., et al.: Material selection methods: a review. In: Aliev, R.A., Kacprzyk, J., Pedrycz, W., Jamshidi, M., Sadikoglu, F.M. (eds.) ICAFS 2018. AISC, vol. 896, pp. 929–936. Springer, Cham (2019). https://doi.org/10.1007/978-3-030-04164-9_123
9. Reuben, R.L.: Materials in Marine Technology. Springer, London (1994). https://doi.org/10.1007/978-1-4471-2011-7
10. Ross, R. W.: New technology stainless steels and nickel alloys for marine applications in the year 2000 and beyond. In: IEEE Conference, pp. 1597–1605 (2000). https://doi.org/10.1109/oceans.2000
11. Saaty, T.L.,: The Analytic Hierarchy Process. McGraw-Hill International, New York (1980). https://doi.org/10.1002/0470011815.b2a4a002

Problems of Visualization and 3d Modeling in Chemistry: Analysis of Electronodonor Properties and Spatial Complex-Forming Structure of the Obtained Sulfocationite

S. M. Turabdzhanov[1], A. G. Eshimbetov[2,3], A. A. Nabiev[4], and L. S. Rakhimova[1](✉)

[1] Tashkent State Technical University, 100095 Tashkent, Republic of Uzbekistan
tur_sad@mail.ru, latofat.2011@mail.ru

[2] Institute of Bioorganic Chemistry, 100125 Tashkent, Uzbekistan
ealisherg@yahoo.com

[3] National University of Uzbekistan, 100174 Tashkent, Republic of Uzbekistan

[4] Tashkent Chemical Technological Institute, 100011 Tashkent, Republic of Uzbekistan
nabievximik@mail.ru

Abstract. This work is devoted to the study of 3D structures of the obtained sulfonic cation exchanger based on diphenyloxide and furfural using the Avogadro program. Electron-donating properties in modeling the chemical structure of the complexes showed that donor-acceptor bonds are formed due to the presence of an ionic bond of sulfonic groups, as well as between the atoms of diphenyloxide and the ring of furan. π-stacking interacts between aromatic rings when connecting intermediate metals.

Keywords: 3D modeling · Complex-forming · Distribution coefficient · Spatial structure · Cation exchanger

1 Introduction

Graphic design, visualization and 3D modeling of chemical compounds has become an integral part of modern chemistry in developed countries, where 3D modeling is widely used in teaching process of chemistry and in the study of reactivity, biological activity, spectral and other properties of compounds [1–5]. Many developed countries have their own software packages used in the field of In Silico [6] for computer modeling (simulation) of biological processes and for theoretical assessment of the biological activity of pharmacologically unexplored compounds. Prediction and screening of chemical compounds can reduce the costs of pharmacological research by tens and hundreds of times [6].

Today, there are simulation programs based on empirical (for example, Avogadro [7], VMD [8]), semi-empirical, ab initio of methods density functional theory (methods based Density Functional Theory–DFT). DFT methods have found the widest application

© The Author(s), under exclusive license to Springer Nature Switzerland AG 2021
R. A. Aliev et al. (Eds.): WCIS 2020, AISC 1323, pp. 262–271, 2021.
https://doi.org/10.1007/978-3-030-68004-6_34

for low molecular weight compounds [9–11]. For this, used software packages such as Gaussian [12], ORCA [13, 14], GAMESS [15, 16], Firefly [17] and others are used. The known scientist in quantum chemistry - J. Pople is the developer of the Gaussian program and quantum-chemical methods, who was awarded the Nobel Prize in Chemistry 1998. Perhaps this motivated as an incentive for programmer chemists to develop new software packages.

With the development of information technologies, computer programs used in chemistry have grown quantitatively and qualitatively, even android versions of some programs have appeared [18, 19]. In this work, using the example of Sulfonic-acid-type ion-exchanger obtained on the basis of diphenyloxide and furfural, the electron-donating ability was investigated by the theoretical method DFT/6-31G. And also, an analysis of the spatial structure of this sulfonic cation-exchanger was carried out to determine the simultaneous participation of the sulfonic group and the oxygen atom of the furan ring during complex-forming (Fig. 1).

2 Method of Calculation

For facilitate the calculation of the sulfonic cation exchanger, the initial geometry of the model monomer was built with a methyl (-CH_3) group in the "Head" and "Tail" parts using the Avogadro program [7].

The geometry of this monomer was completely optimized by the DFT/6-31G method using the ORCA 4.2 software package [13, 14]. The Becke method [20] modified by Lee, Young and Parr (B3LYP) [21] was chosen as the DFT method. The analysis of electrostatic potential surfaces was carried out using the Multiwfn [22] and VMD [8] software packages.

3 DFT Analysis

The fully optimized geometry (spatial structure) and charge distribution on the atoms of the model monomer is illustrated in Fig. 1.

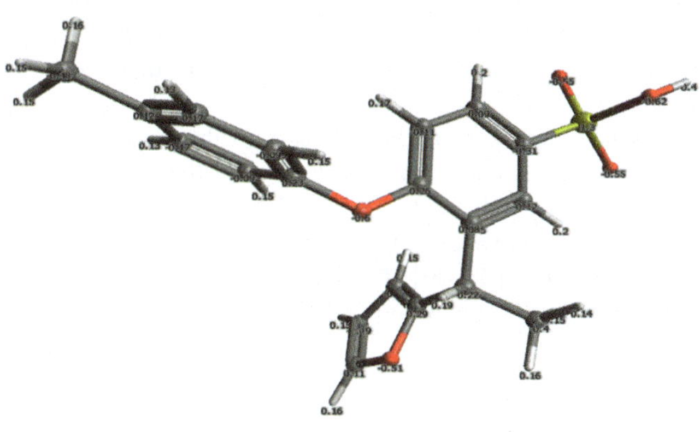

Fig. 1. Spatial structure and charge distribution on atoms of the model sulfonic cation exchange monomer. 3D visualization of the model monomer was done in the Avogadro program [7].

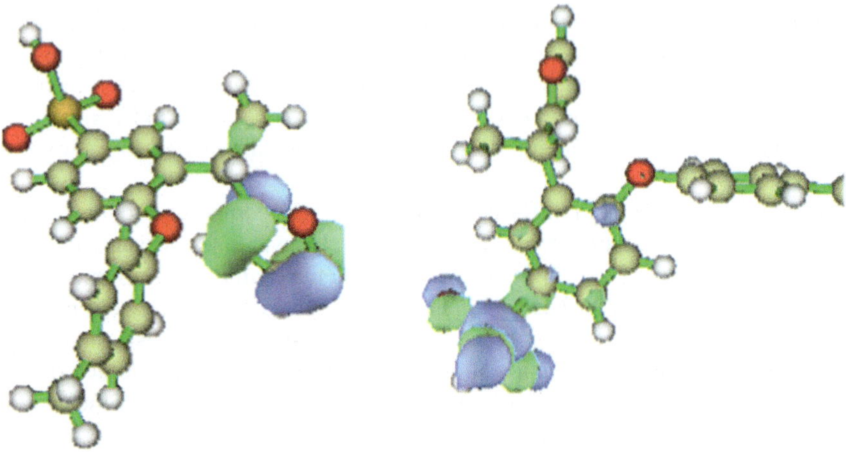

Fig. 2. Charge distribution in HOMO (left) and LUMO (right). 3D visualization of the model monomer obtained using the program Multiwfn [22].

From theoretical calculations, it is possible to extract information about the electronic structure of the compounds under study. Information about the electronic structure is extracted from parameters such as the distribution of the total charge on atoms and frontier orbital (Fig. 2), the energies of the frontier orbital, the electric dipole moment and the total energy of molecules, etc.

According to Koopmans' theorem, the energy of the highest occupied MO is equivalent to the ionization potentials ($-E_{HOMO} \approx IP$) chemical compounds [23–26]. Then, the energy of the lowest unoccupied MO is equivalent in affinity to an electron ($E_{LUMO} \approx IP$) molecule [23–26].

The distribution of the charge on atoms according to Mulliken [27] shows the localization of the largest negative charge on the oxygen atoms of the sulfonic group and diphenyloxide, as well as furan. The greatest positive charge is localized at the mobile atom H of the sulfonic group.

Based on the energy of the boundary orbitals, the following parameters of the electronic structure are determined: Energy Gap $|\Delta E| = E_{HOMO} - E_{LUMO}$); Chemical hardness, $\eta = (I - A)/2$ and Chemical softness, $s = 1/(2\eta)$); Electronegativity, $= (I + A)/2$); Chemical potential, $\mu_p = -(I + A)/2$ and Electrophilicity index, $\omega = \mu_p^2/2\eta$ [23–26] (Table 1).

High electric dipole moment, and relatively low η value, indicates instability (higher reactivity) of the monomer. An additional parameter of the electronic structure can be the electrostatic potential, which ESP is a useful guide to the reactive behavior of a molecule, especially in non-covalent interactions and indicates nucleophilic and electrophilic regions of molecules [28, 29]. The blue ESP regions represent a local surface maximum with positive values, which indicates a deficit in atoms.

Table 1. Quantum chemical parameter.

Parameter	SCE *	ASCE **
E_{HOMO}, eV	−6.06	−2.11
E_{LUMO}, eV	−2.13	1.85
$\lvert\Delta E\rvert = E_{HOMO} - E_{LUMO}$, (eV)	3.93	3.96
Ionization potentials, $I = -E_{HOMO}$, (eV)	6.06	2.11
Electron affinity, $A = -E_{LUMO}$, (eV)	2.13	1.85
Electronegativity, $\chi = (I + A)/2$ (eV)	4.09	1.98
Chemical hardness, $\eta = (I - A)/2$ (eV)	1.96	0.26
Chemical potential, $\mu_p = -(I + A)/2$ (eV)	−4.09	−1.98
Chemical softness, $s = 1/(2\eta)$ (eV^{-1})	0.25	1.92
Electrophilicity index, $w = \mu_p^2/2\eta$ (eV)	4.26	2.04
Dipole moment, μ (Debye)	6.31	19.9

** SCE - sulfonic cation exchanger, neutral model monomer*
*** ASCE-anionic form of model sulfonic cation exchanger*

At the same time, the red colored areas are local surface minima with negative values, which indicate electron-rich groups. The largest ESP maximum is 49.66 and is located in the immediate vicinity of the H group of NH. The largest positive charge on the hydrogen atom is the S-OH group (Fig. 1) and the largest ESP value near this atom (Fig. 3) shows the tendency of mobile hydrogen to hydrogen bonds, as well as to nucleophilic attack.

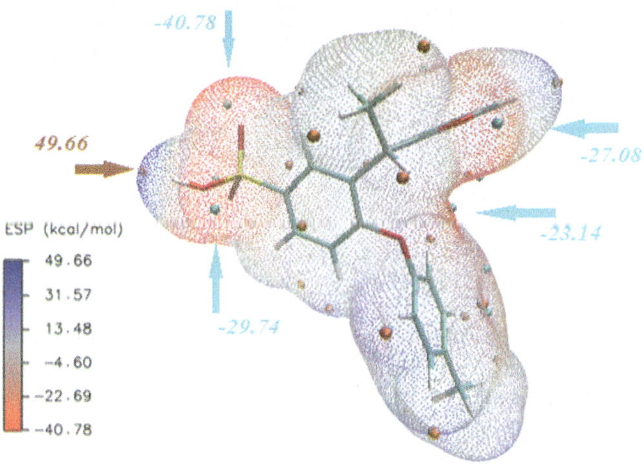

Fig. 3. The maximums (brown balls) and minimums (blue balls) of the electrostatic potential of the model monomer: determined using the programs Multiwfn [22] and VMD [8].

The smallest value of the minima of the ESP surface is in the immediate vicinity of the oxygen atoms of the sulfonic group. The next minima are localized near the oxygen atoms of the furan ring (−27.08) and diphenyloxide (−23.14). These values show that the oxygen atoms of the diphenyl oxide and the furan ring can also enter into intermolecular interactions as an electron donor. Analysis of the ESP and the charge distribution on the atoms of the model sulfonic cation exchanger shows a high mobility of the hydrogen atom of the OH group. During complex-forming, the anionic form of the sulfonic cation exchanger (ASCE) can easily form and, therefore, were also calculated the quantum-chemical parameters and charge distribution on the ASCE atoms (Table 1). The energy of HOMO ($E_{HOMO} = -2.11$ eV) ASCE is strongly increased relative to the neutral molecule ($E_{HOMO} = -06$ eV), which shows the high electron-donating ability of the anionic form. The quantum-chemical parameters μ, σ, η of the anionic form also strongly differ from the parameters of the neutral SCE and show instability (higher reactivity) of the neutral form.

As in the case of a neutral molecule, in the anionic form, the largest negative charge is localized on the oxygen atoms of the sulfonic group, then on the oxygen atom of diphenyloxide and the furan ring. The sulfonic group is the reaction center for the formation of ionic bonds with metals (Fig. 4).

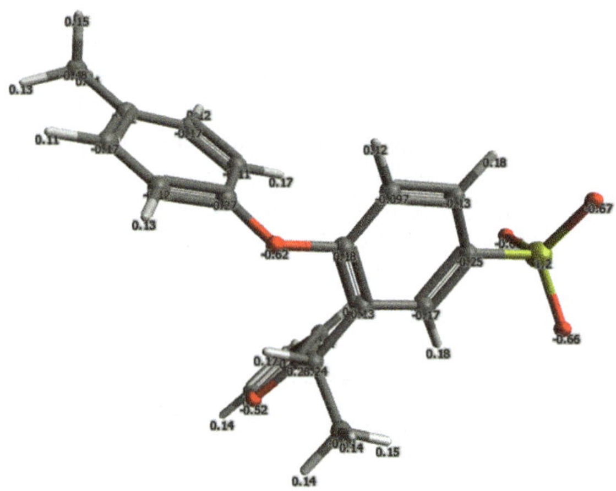

Fig. 4. Charge distribution on atoms of the anionic form of sulfonic cation exchanger. 3D visualization was done in the Avogadro program [7].

According to the results of X-ray structural analysis, the number of central ionic coordination in complexes with the participation of sulfonic groups is six. A single metal can also form an ionic bond with two sulfonic groups, filling in the remaining numbers in exchange for water molecules [30].

4 Research of Including Complex-Forming

The distribution coefficient is the most important indicator of the ratio of non-ionized compound between the concentration of the solution. For measurements of distribution coefficient (K_d), the pH of the aqueous phase is buffered to a specific value such that the pH is not significantly perturbed by the introduction of the compound [31]. In addition to the specific properties of the cation exchanger, K_d depends on the pH of the solution, temperature, type and concentration of competitive ions, and the presence in solutions.

The complexing properties of the synthesized cation exchange polymer [8] were investigated earlier by complexometric titration with the sodium salt of ethylenediaminetetraacetic acid (EDTA). Sorptional and selectivity properties of the cation exchanger, was carried out depending on the pH and ionic form of the cation exchanger from 0,1 N solutions of $NiSO_4$, $CuSO_4$, $CoCl_2$, leaving for 24 h. Experiments were carried out in ± 25 °C. Were used the obtained experimental data to calculate the distribution coefficient K_d. The distribution or partition coefficient, Kd, measured at equilibrium, is a useful concept that expresses the relative affinity for a sorbate in solution to sorba particular solid. The parameter has been used extensively in models to predict the behavior of contaminants in the environment. The batch method is often used to determine this coefficient in the laboratory and the results extrapolated to field situations. The K_d is defined as [32]:

$$K_d = \frac{(mass\ sorbed\,/\,mass\ of\ solid)}{mass\ in\ solution/volume\ of\ solution} \tag{1}$$

Or, alternatively,

$$K_d = \frac{(mass\ sorbed)}{(mass\ in\ solution)C_p} \tag{2}$$

C_p - mass concentration of particles per unit volume. Table 2 shows the obtained research data.

Table 2. Distribution coefficient depending on the pH environment in different forms of cation exchanger

Form of cation exchanger	pH	K_d, mg/g
Cu^{2+}/H-form	2,40	59
	2,37	125
	7,15	90
Cu^{2+}/Na-form	4,78	36
	4,41	125
	8,75	650
Co^{2+}/H-form	2,38	10,2
	2,34	1,68
	7,25	62,5
Co^{2+}/Na-form	6,90	9,14
	5,05	4,76
	7,84	44,7
Ni^{2+}/H-form	3,42	9,55
	2,31	12,38
	7,15	42,95
Ni^{2+}/Na-form	6,3	28
	5,24	36,2
	7,92	74,5

According to the literature data, the Kd value shows the selectivity of the cation exchanger for the ions of certain metals [33]. With an increase in the degree of neutralization and a change in the form of the cation exchanger, the sorption of the ions of the studied metals increases almost 2 times. With an increase in pH $= 9 \div 11$, the ions of the tested metals are sorbed in the form of an ammonia complex as follows:

$$2Me^{2+} + 2NH_4OH \rightarrow [Cu(NH_3)_4]^2 + OH \quad (3)$$

The data obtained indicate that the use of the cation exchanger is promising for the sorption of copper ions.

Depending on the simulated medium, stable complexes are formed in the reactions of sorption of d-metal ions with the obtained cation exchanger. Donor-acceptor interactions between the ligand and the central atom play an important role in the formation of complex compounds. The donor of an electron pair is usually a ligand. The acceptor is the central atom, which has unoccupied orbitals. This bond is strong and does not break when the complex dissolves (non-ionic), and it is called coordination [34] (Fig. 5).

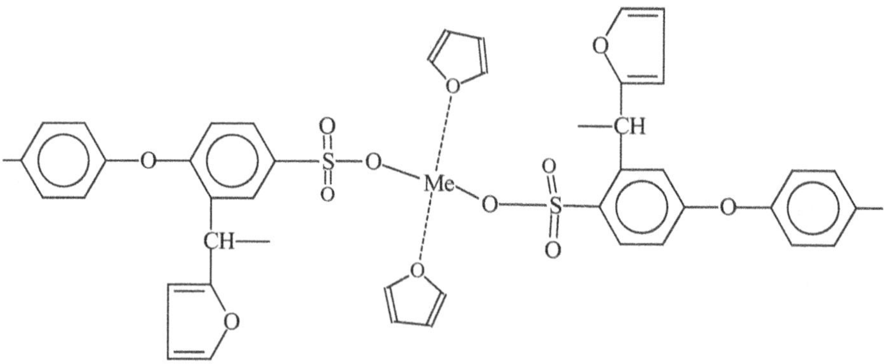

Or,

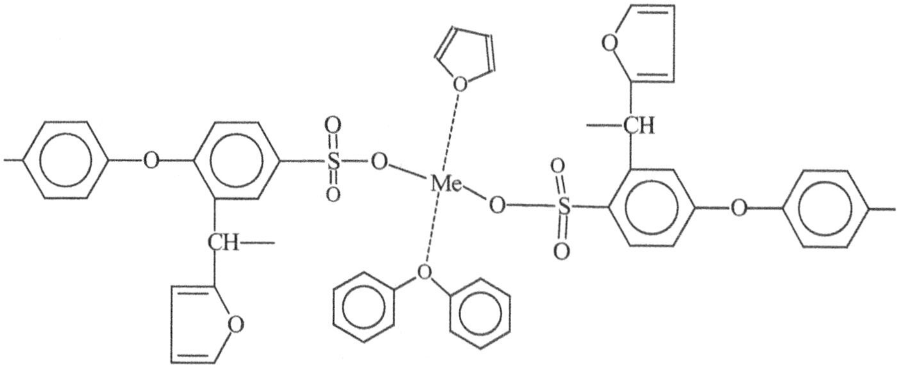

Fig. 5. Complex compounds of the cation exchanger with metals.

In this case, the chelate complex, metals as the central atoms forms a stable complex. Along with the general factors determining the stability of coordination compounds, such as the nature of the central ion and coordinated donor atoms of the ligand, in this case the number of atoms in the formed ring and the presence of multiple bonds in the bridge between donor atoms of the chelate ligand play an important role. In theory, the smaller the size chelate complex, the more stable it is. Me^{2+} cations increase as a complex $Cu^{2+} > Ni^{2+} > Co^{2+}$.

The reason can be explained by an increase in the atomic radius. Forms ionic and donor-acceptor bonds with the O atoms of the $-SO_3H$ compound and the furfural group obtained as a ligand. The reason for the impossibility of bond formation with the O atom in the renoxy group can be explained by the spatial barrier of the benzene ring (Fig. 6).

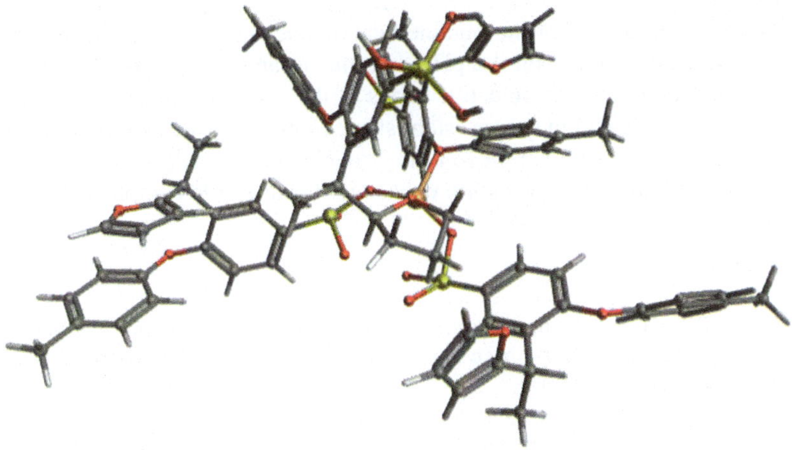

Fig. 6. 3D modeled copper complex in the Avogadro program [7]: Ionic bond of sulfonic group, donor-acceptor bond with oxygen atoms of diphenyloxide and furan. There is interaction π-stacking between aromatic rings.

Along with the σ-bonds, π-bonds are formed by the donor-acceptor mechanism. In this case, a metal ion serves as a donor, donating its paired d-electrons to the ligand, which has energetically favorable vacant orbitals.

5 Conclusion

In the modern world, there is an intensive change in the environment, an active penetration of scientific and technological progress into all spheres of life. Modeling has become one of the most promising methods of modern chemistry, which opens up a number of additional opportunities in acquaintance with the surrounding world. Modeling the complex-forming of d-metal ions by cation exchangers is a task in quantum chemistry. The extraction of copper, nickel and cobalt from industrial wastewaters and the development of their 3D structures allows predicting the physicochemical properties of ion-exchange polymers with desired properties with a high degree of probability. It is advisable to use chemical information on the structure and properties of ion-exchange polymers in order to determine the range of new molecules using computer computational technologies. To do this, it is necessary to solve the problem of the quantitative relationship between the structure and activity, the structure and properties of a molecule, and also to establish relationships that link the characteristic of a particular property with the known chemical properties of molecules - molecular geometry, dipole moment, charges on atoms, etc.

With the quantum-chemical calculations, information was obtained on the electronic structure of the sulfonic cation exchanger and a very high deficit of electrons near the H atom of the sulfonic group was shown. And also, the results of ESP showed the electron-donating ability of the oxygen atoms of the furan ring and diphenyloxide. However, due to steric hindrances, oxygen atoms cannot coordinate together with the sulfo group of one molecule.

The sorption properties of the obtained cation exchanger to copper, nickel, cobalt ions were studied depending on the pH-environment and ionic form. The distribution coefficients are found and the selective series of the action of the cation exchanger to the ions of the investigated metals is established; copper ions are especially promising.

The spatial structures of complexons are modeled. The obtained computational and mathematical data show that the cation exchanger forms stable complexes.

References

1. Esselman, B.J., Hill, N.J.: Integration of computational chemistry into the undergraduate organic chemistry laboratory curriculum. J. Chem. Educ. **93**, 932–936 (2016). https://doi.org/10.1021/acs.jchemed.5b00815
2. http://tccc.iesl.forth.gr/chemistry/scied_chem.html
3. https://www.acs.org/content/acs/en/careers/college-to-career/chemistry-careers/computational-chemistry.html
4. Miller, B.T., Singh, R.P., Schalk, V., Pevzner, Y., Sun, J., Miller, C.S., Boresch, S., Ichiye, T., Brooks, B.R., Woodcock, H.L.: Web-based computational chemistry education with CHARMMing I: lessons and tutorial. J. PLoS Comput. Biol. **7**, e1003719 (2014). https://doi.org/10.1371/journal.pcbi.1003719
5. Silva, F.D., Santos, K.L.B., Santos, L.D., Lobato, C.C., Costa, J.S., Lopes, G.A.C., Santos, C.B.R.: Computational chemistry programs as a facilitating tool in the teaching and learning process. Brit. J. Educ. **8**, 134–146 (2015)
6. Pagadala, N.S., Syed, Kh, Tuszynski, J.: Software for molecular docking: a review. J. Biophys. Rev. **9**, 91–102 (2017). https://doi.org/10.1007/s12551-016-0247-1
7. Hanwell, M.D., Curtis, D.E., Lonie, D.C., Vandermeersch, T., Zurek, E., Hutchison, G.R.: Avogadro: an advanced semantic chemical editor, visualization, and analysis platform. J. Cheminform. **4**, 4–17 (2012). https://doi.org/10.1186/1758-2946-4-17
8. Humphrey, W., Dalke, A., Schulten, K.: VMD: visual molecular dynamics. J. Molec. Graph. **14**, 33–38 (1996). https://doi.org/10.1016/0263-7855(96)00018-5
9. Jones, R.O.: Density functional theory: its origins, rise to prominence, and future. J. Rev. Mod. Phys. **87**, 897–923 (2015). https://doi.org/10.1103/revmodphys.87.897
10. Becke, A.D.: Perspective: fifty years of density-functional theory in chemical physics. J. Chem. Phys. **140**, 18A301 (2014). https://doi.org/10.1063/1.4869598
11. Kryachko, E.S., Ludena, E.V.: Density functional theory: foundations reviewed. J. Phys. Rep. **554**, 123–239 (2014). https://doi.org/10.1016/j.physrep.2014.06.002

12. www.gaussian.com
13. Neese, F.: ORCA Program system. J. Comput. Mol. Sci. **2**, 73–78 (2012). https://doi.org/10.1002/wcms.81
14. https://orcaforum.kofo.mpg.de/app.php/portal
15. https://www.msg.chem.iastate.edu/gamess/
16. http://www.cfs.dl.ac.uk/
17. http://classic.chem.msu.su/gran/gamess/index.html
18. Kholikova, Z.Z., Azimova, A.K, Yeshimbetov, A.G.: Android applications used in chemistry and their importance. In: Proceedings of the VI Republican Scientific-Practical Conference (2020)
19. https://play.google.com/store/search?q=Chem3D&c=apps&hl=en_US
20. Becke, A.D.: Density-functional thermochemistry. III. The role of exact exchange. J. Chem. Phys. 98, 5648–5652 (1998)
21. Lee, C., Yang, W., Parr, G.: Development of the Colle-Salvetti correlation-energy formula into a functional of the electron density. J. Phys. Rev. B. **37**, 785–789 (1988). https://doi.org/10.1103/physrevb.37.785
22. Lu, T., Chen, F.: Multiwfn: a multifunctional wavefunction analyzer. J. Comput. Chem. **33**, 580–592 (2012). https://doi.org/10.1002/jcc.22885
23. Rauk, A.: Orbital Interaction. Theory of Organic Chemistry. Wiley-Interscience, Monograph (2000). https://doi.org/10.1002/0471220418
24. Gómez-Jeria, J.S.: A new set of local reactivity indices within the Hartree-Fock-Roothaan and density functional theory frameworks. J. Can. Chem. Trans. **1**, 25–55 (2013). https://doi.org/10.13179/canchemtrans.2013.01.01.0013
25. Abdallah, H.H., Ishmael, H.M., Abdullah, H.Y., Umar, Y., Ramasami, P.: Theoretical study of the syn- and anti-conformers of 2,2′-bifuran derivatives: rotational barrier and solvent effect. J. Phys. Chem. Liq. **57**, 174–183 (2019). https://doi.org/10.1080/00319104.2018.1437915
26. Wazzan, N.A.: DFT and MP2 study of geometry, IR and UV-visible spectroscopy and first hyperpolarizability of 2-aminopyridine, 3-aminopyridine and 4-aminopyridine in gas phase and in solvents. Asian J. Chem. **27**, 4641–4656 (2015). https://doi.org/10.14233/ajchem.2015.19304
27. Storer, J.W., Giesen, D.J., Cramer, C.J., et al.: Class IV charge models: a new semiempirical approach in quantum chemistry. J. Comp. Aided. Mol. Des. **9**, 87–110 (1995). https://doi.org/10.1007/BF00117280
28. Politzer, P., Murray, J.: The fundamental nature and role of the electrostatic potential in atoms and molecules. J. Theor. Chem. Acc. **108**, 134–142 (2002). https://doi.org/10.1007/s00214-002-0363-9
29. Murray, J., Politzer, P.: The electrostatic potential: an overview. Wiley Interdiscip. Rev. Comput. Mol. Sci. **1**, 153–163 (2011). https://doi.org/10.1002/wcms.19
30. Kulynych, A.D., Shimizu, G.: A pseudo-honeycomb coordination net formed with 5-sulfoisophthalic acid. J. Cryst. Eng. Comm. **4**, 102 (2002). https://doi.org/10.1039/b200503d
31. Leo, A., Hansch, C., Elkins, D.: Partition coefficients and their uses. J. Chem. Rev. **71**, 525–616 (1971). https://doi.org/10.1021/cr60274a001
32. Dennis, W.O., Harold, B.H.,: Effect of the Solid: Liquid Ratio on the Sorption of Sr^{2+} and Cs^{+} on Bentonite. Ads.Me. Geom. Elsevier, pp. 277–289 (1998)
33. Turabdzhanov, S.M., Kedelbaev, BSh, Kushnazarov, P.I., Badritdinova, F.M., Ponamaryova, T.V., Rakhimova, L.S.: New approach to the synthesis of polycondensation ion-exchange polymers for wastewater treatment. J. News. Nat. Acad. Sci. **2**, 206–216 (2019). https://doi.org/10.32014/2019.2518-170X.56
34. Turabdzhanov, S.M., Nazirov, ZSh, Rakhimova, L.S.: Perspective wastewater treatment from Cu^{2+} ions in the mining industry. Symposium. **105**, 02025 (2019). https://doi.org/10.1051/e3sconf/201910502025

Application of Modern Technologies for Planning Improvement and Saving on Costs in the Enterprise of the Industry 4.0

Nazim Tofiq oglu Huseynov(✉)

Azerbaijan State Oil and Industry University, 20 Azadlig Avenue, AZ1010 Baku, Azerbaijan
nazim.huseynov1@gmail.com

Abstract. There are many examples of successful enterprises in the market worldwide. All these successful enterprises have some specific advantages or unique techniques to survive in the market. Sustainable development is the key element to ensure survival of modern enterprises which are based on unique technologies in the Industry 4.0. Regardless of sizes and capabilities of enterprises, the availability of new technologies and correct organization allows these enterprises to respond to changing market and consumer demands immediately. In addition to this there are required well-defined methodologies and experience of employees to ensure qualified and timely feedback to improve conditions. There are suggested some technologies in this study. The objectives are to analyze and show effectiveness of these techniques to improve activities and competitive advantages of the cement factory. Additionally, these technologies resulted in significant saving on costs in the cement factory. Application of these modern methods and technologies resulted in reliable conditions for improvement of competitive advantage, adaptive management, sustainable development, planning as well as optimization of expenses in the cement factory in the Republic of Azerbaijan in the year 2020 as well as in the short and long term.

Keywords: Computerized Maintenance Management Systems (CMMS) · Recurrent Neural Network (RNN) · Condition-Based Maintenance (CBM) · Online Condition Monitoring Systems (CMS) · Sustainable development

1 Introduction

Due to high competition and global recession there are required scientific approaches for efficient management of enterprises in the Industry 4.0 in the market worldwide. Global and local economy challenges prove that the previous model of interaction has to be improved in the economy. The global recession is expected worldwide. Probably demand for crude oil will be less than supply. Actually, profits of some industries decrease along the chain. There are decreasing demand and sustainability of financial chains for small and large enterprises due to the global market shock in the year 2020.

Import and export are certainly limited because of prolongation of consequences due to COVID-19 and recession. There are trends for cost reduction to minimize expenses

© The Author(s), under exclusive license to Springer Nature Switzerland AG 2021
R. A. Aliev et al. (Eds.): WCIS 2020, AISC 1323, pp. 272–279, 2021.
https://doi.org/10.1007/978-3-030-68004-6_35

in business organizations worldwide. There are risks of possible low sales of ready products of the cement plant. There are also additional risks of late and complicated managerial decisions because of uncertainty, fast changing conditions, the huge amount of dynamic data, lack of experience and skills due to COVID-19. These risks result in loss of the time and investment due to outdated existing tools, procedures and control in the processing of technical and commercial data when procurement and delivery of raw materials, equipment and spare parts for the cement factory.

In order to minimize the consequences of recession there is required accurate and qualified management on the example of the cement factory. Therefore, when uncertainty conditions the timely decision making becomes critical for management. Competitive advantage also depends on correct organization, balanced planning and application of state-of-the-art technologies like RNN, CMMS, CBM and online CMS in comparison with preventive maintenance routines on the example of the cement factory. Actually, the prediction technique which is based on the RNN allows to get reliable results on the basis of preceding and current macroeconomic statistics [1]. The RNN prediction technique provides conditions for qualified, effective and balanced planning of raw materials and spare parts as well as optimization of expenses or even direct saving on costs on the example of the cement industry. The concept, principles of operation and training of fuzzy neural networks and technologies of recurrent fuzzy neural networks were given in the works of many researchers, in particular in the research of the internationally recognized scientist Doctor of Technical Sciences Professor Aliev, R.A. [1]. Tested and recognized prediction results and trends for the cement market of the Republic of Azerbaijan were considered in the work [2]. There are specific requirements to apply CMMS and online CMS in view of return on equipment. When purchasing these software and technologies a balanced solution is advised due to the relatively high costs. Application of the above-mentioned technologies provide reliable conditions for improvement of quantitative and qualitative results. These robust results also can create conditions for efficient management, maintenance, planning, saving on costs as well as valid and qualified decision making for the planned production of cement in the Republic of Azerbaijan.

The paper is structured as follows. The Sect. 2 is devoted to general description of technologies in the modern cement industry. The Sect. 2.1 is devoted to description of the RNN structure, functioning principles and application results of modern prediction technologies in the cement industry. The Sect. 2.2 is devoted to description of modern online CMS technologies in the cement industry. The Sect. 2.3 is devoted to description of modern CMMS and planning principles in the cement industry.

2 Application of Modern Technologies in the Cement Industry

Actually, there are some criteria of successful application of modern technologies via adaptive management, first mover advantage, efficiency, feedback, reasonable simplification and rejection of complicated methods on the example of the cement factory.

One of the main conditions of successful implementation is to apply modern marketing analysis and planning via CMMS in the cement factory. This approach is useful for the decision making by management timely. The prediction technique which is based on RNN allows getting reliable results on the basis of historical and current macroeconomic statistics. These computation, analysis and planning can be implemented taking into consideration of macroeconomic statistics on the example of the cement industry. Therefore, this RNN prediction technique can contribute to optimization of expenses or even direct saving on costs on the example of the cement industry. This approach has resulted in the sustainable development and efficiency via timely, sound decision-making on priorities and balanced planning. RNN, CMMS, CBM and online CMS are suitable technologies due to efficient operations when the huge amount of dynamic fast changing technical and commercial data and tasks on the example of the cement factory in the Industry 4.0.

2.1 Modern Prediction Technologies in the Cement Industry

Actually, the accurate qualitative prediction of the cost of cement and the cement market research resulted in confirmation of the increased cost during the years 2017–2018 [2]. The work [2] used fuzzy neural network to do cost forecasting. The results of prediction provided in the work [2] were confirmed to be reliable and accurate in the cement market [2]. The system [2] provided qualitative prediction for years 2017 and 2018. Application of recurrent neural networks is a further step to improve the quality of forecasting due to their inherent dynamic system modeling characteristics. A three-layer recurrent neural network RNN [1, 3] with five inputs, seven hidden neurons and one output was created to predict volumes of cement production for the year 2020.

The architecture (Fig. 1) and working principles of RNN were described in the work [4] as below.

The network uses a no-activation (only-distribution neurons) input (1st) layer, sigmoidal activation function based hidden (2nd) layer, and linear activation function-based output (3rd layer) layer. Three layers of neurons in perceptron networks are proven to be enough for approximation of most of input-output numerical mappings with any desired accuracy [4].

The neurons in the 1st (input) layer only distribute the input signals without modifying their values:

$$\tilde{y}_i^{(1)}(t) = \tilde{x}_i^{(1)}(t) \tag{1}$$

The neurons in 2nd (hidden) layer compute their output signals as follows:

$$\tilde{y}_i^{(2)}(t) = F\left(\tilde{\theta}_i^{(2)} + \sum_j \tilde{x}_j^{(2)}(t)\tilde{w}_{ij}^{(2)} + \sum_j \tilde{y}_j^{(2)}(t-1)\tilde{v}_{ij}^{(2)}\right) \tag{2}$$

where

$$F(s) = \frac{s}{1+|s|}$$

The neurons in the 3rd (output) layer are linear and their outputs are calculated as:

$$y_i^{(3)}(t) = \theta_i^{(3)} + \sum_j x_j^{(3)}(t)w_{ij}^{(3)} + \sum_j y_j^{(3)}(t-1)v_{ij}^{(3)}.$$

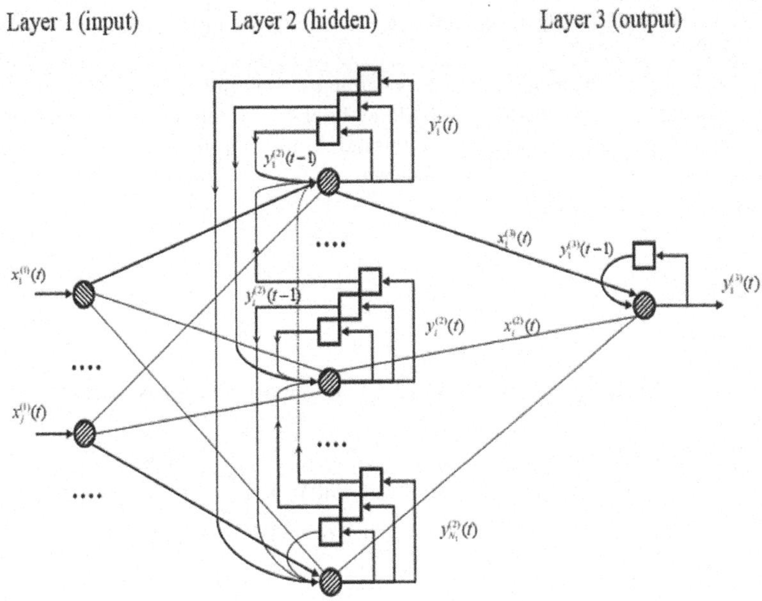

Fig. 1. Architecture of recurrent neural networks [4]

Formally, all signals and parameters (weights and thresholds) of the considered neural network can be arbitrary fuzzy numbers (e.g. triangular fuzzy numbers) or ordinary crisp numbers [4].

Taking into account the macroeconomic statistics, Table 1 and [5–7], the RNN was trained using the data, a segment of which is given below.

Inputs: cement production X1 (t); ready concrete production X2 (t − 2); building bricks production X3 (t − 2); Inflation for consumer prices X4 (t); Oil price average X5 (t). Output: Cement production (sales) X1 (t + 1).

After one thousand training iterations on the basis of the evolutionary (DE) algorithm, RMSE = 1.57E−10 was achieved. The neural network was trained to predict cement production (sales) in the year 2020. The main prediction results were made by RNN and confirmed X1 (2020) = 2746.5 [8].

X1(2018) = 3444.7;
X1(2019) = 3132.9;

Training data were described in the Table 2. The inputs and outputs to the RNN are provided below and shown in the Fig. 2.

X1(2018) = 3444.7;
X1(2019) = 3132.9;
X1(2020) = 2746.5 [8].

The results produced for older years (2017, 2018) are comparable with the work [2]. To check the better performance of the suggested prediction technique, the data were processed with other well-known prediction methods, such as regression analysis and ordinary neural networks (provided in Microsoft Excel XLMINER Platform). RNN was significantly better due to the ability to remember dynamics in historical data.

Table 1. Macroeconomic statistics for RNN inputs [5–8]

Year	Cement production (thousands tons)	Concrete production (thousands tons)	Production of building bricks (thousand cubic meters)	Inflation, consumer prices (annual%)	Average oil price (USD)
2005	1538.0	1916.0	161.0	9.6	48.89
……	……	……	……	……	……
2017	2879.8	–	–	12.9	48.73

Table 2. Actual and predicted data resulted applying RNN [8]

Inputs					Output
X1(t)	X2(t – 2)	X3(t – 2)	X4(t)	X5(t)	X1(t + 1)
1538.0			9.6	48.89	1622.0
…….	…….	…….	…….	…….	…….
2306.9	1692.0	426.1	12.4	38.17	2879.8
2879.8	942.1	280.6	12.9	48.73	

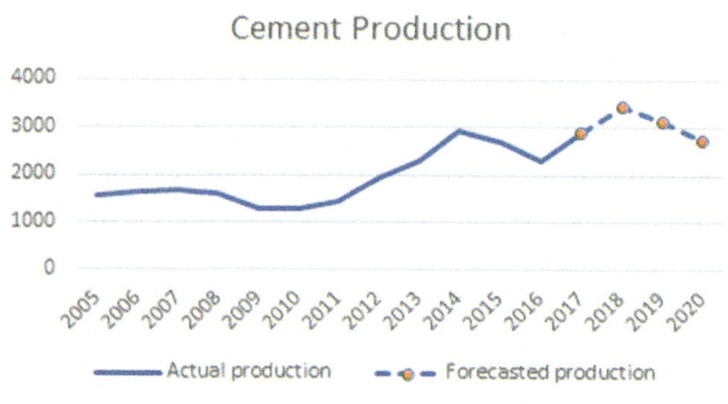

Fig. 2. Actual and predicted results of cement production in the year 2020 [8].

2.2 Modern Online CMS Technologies in the Cement Industry

On the basis of the example of former Garadagh Cement (at present Holcim (Azerbaijan)) there were introduced modern online technologies and condition monitoring program which was encouraged using online CMS of PRÜFTECHNIK GmbH. Prevention of interruptions of production process allowed to ensure successful performance. Therefore, early analysis allowed a planned repair which significantly decreased production and maintenance costs [9]. Therefore Holcim (Azerbaijan) also confirmed application of the online CMS like a modern technology to increase equipment reliability, ensure production availability, decrease maintenance costs. Actually, before installation of the online CMS the maintenance costs were € 51803 in total for the monitored production equipment during five months.

The goal was to decrease the costs to € 14500 that was a decreasing factor of 3.6. Actually, the sum of the maintenance costs reached € 2134 that was a total maintenance costs reduction by a decreasing factor of 24.3 during the next five months. As a result, the maintenance costs were reduced to 4% in comparison with initial status [9].

Additional improvements were made by decreasing of the costs for unplanned stops for the monitored production equipment lines. The frequency and duration were decreased for unplanned stops and unplanned works significantly [9].

2.3 Modern CMMS Technologies and Planning in the Cement Industry

Actually, CMMS ensured conditions for qualified, effective and balanced production and maintenance planning of raw materials, spare parts as well as qualified optimization and scheduling. The planning process required the feedback to ensure the reporting and statistics (Fig. 3). CMMS technology ensured timely decision-making on priorities as well as accurate and balanced planning. Principles and implementing SAP business planning and consolidation were described in [10, 11]. One of the best examples of CMMS is the advanced SAP software platform which was described and applied in many enterprises worldwide [10, 11]. CMMS is the perfect technology due to capability of efficient operations in presence of large data flows and fast changing dynamic process of updating of technical and commercial data and tasks on the example of the cement factory in the Industry 4.0.

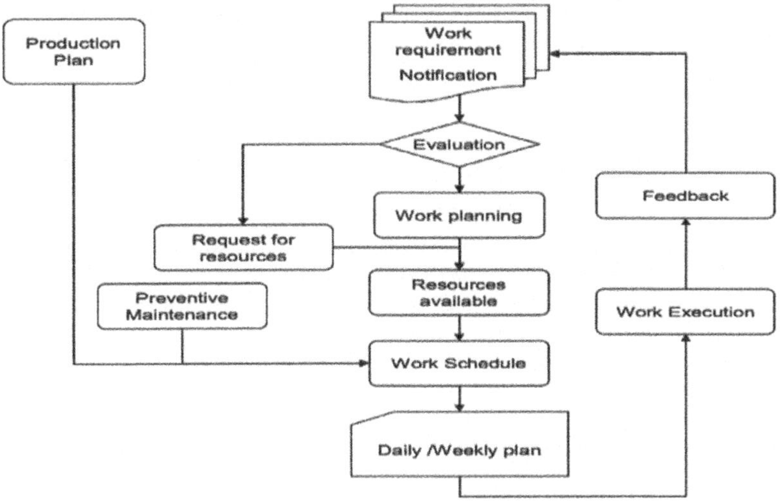

Fig. 3. General planning and scheduling steps.

3 Conclusion

Application of these modern technologies resulted in improvement of the planning and significant costs saving on the maintenance on the example of the cement factory. These modern RNN, CMMS, CMB and online CMS methods and technologies ensured improvement of competitive advantage, adaptive management, sustainable development as well as optimization of expenses in the cement factory in the Republic of Azerbaijan in the year 2020. Application of these advanced technologies confirmed timely adaptation of the cement factory to various changes and its competitive advantage in the market in the Republic of Azerbaijan. On the basis of new challenges of the modern market, results of application helped in decision making for short-term and long-term planning efficiently. Taking into account application of RNN with evolutionary training, these technologies ensured qualified prediction of production volumes in the example of the cement industry X1 (2020), which could be confirmed by the direct links of macroeconomic and digital indicators. In addition to this there were analyzed advantages and practical implementation of CBM and online CMS in comparison with preventive maintenance routines in the cement factory. As a result, the above-mentioned CBM and online CMS implementation ensured prevention of unplanned stops and failures over a long period of time. Efficiency of modern RNN, CMMS, CMB and online CMS technologies can be confirmed for the long-term planning regardless of a type of production and enterprise in the Republic of Azerbaijan and worldwide.

Acknowledgements. I am sincerely thankful to the management and scientists of Azerbaijan State Oil and Industry University. I would like to acknowledge Rector Professor Dr. Babanli M.B., Professor Dr. Aliev R.A. Member of Azerbaijan National Academy of Sciences, MBA Program Director, Professor Dr. Gardashova L.A., Professor Dr. Safarov Q., Professor Dr. Aliyarov R., PhD. Guirimov B.G for their advice and help.

References

1. Aliev, R.A., Guirimov, B.G., Fazlollahi, B., Aliev, R.R.: Evolutionary algorithm-based learning of fuzzy neural networks. Part 2: recurrent fuzzy neural networks. Fuzzy Sets Syt. **160**(17), 2553–2566 (2009). https://doi.org/10.1016/j.fss.2008.12.018
2. Gardashova, L., Huseynov, N.: Cement industry overview and market price forecasting in Azerbaijan. Int. J. Eng. Res. **6**(10), 32–38 (2016)
3. Graves, D., Pedrycz, W.: Fuzzy prediction architecture using recurrent neural networks. J. Neurocomput. Arch. **72**(7–9), 1668–1678 (2009)
4. Aliev, R.A., Fazlollahi, B., Aliev, R.R., Guirimov, B.: Fuzzy time series prediction method based on fuzzy recurrent neural network. In: Lecture Notes in Computer Science, vol. 4233, pp 860–869. Springer, Heidelberg (2006). https://doi.org/10.1007/11893257_95
5. https://www.stat.gov.az/news/macroeconomy
6. https://www.stat.gov.az/source
7. https://www.eia.gov/dnav/pet/hist/
8. Huseynov, N.T.: Application of a neural network for making management decisions in production. In: XXII Republican Scientific Conference of Doctoral Students and Young Researchers, pp. 22–23. Azerbaijan State Pedagogical University (2018)

9. Wirtsch.-Ing., Schmaus, R., Huseynov, N., Federal German Association of the Lime Industry, Federal German Association of the Gypsum Industry: Garadagh cement ensures production uptime using online condition monitoring system (CMS). ZKG Int. **99**, 14–16 (2010)
10. https://www.sap.com/index.html
11. Jones, P., Soper, Ch.: Implementing SAP Business Planning and Consolidation. SAP Press, Bonn (2018)

The Problem of Classifying and Managing Risk Situations in Poorly Formed Processes

Kamilov Mirzayan(✉), Muxamedieva Dilnoz, and Solieva Barno

Scientific and Innovation Center of Information and Communication Technologies (SIC ICT) at the Tashkent University of Information Technologies, 17A Buz 2, 100125 Tashkent, Uzbekistan
kamilov_m@mail.ru, dilnoz134@rambler.ru, barnoxon76@mail.ru

Abstract. In the article a neuro-fuzzy hybrid intellectual model of risk classification and management is developed. Due to the limited capabilities of traditional methods of mathematical modeling, intelligent data analysis technologies are used to solve poorly formed problems, including risk-related problems. They are based on soft computing methods (Soft Calculation, Soft Computing) and Computational Intelligence, which is based on this theoretical and methodological basis. The main components of these directions are fuzzy sets, fuzzy logical inferences, genetic algorithms, artificial neural networks, and neural network-based computational theories. Intellectual computing technologies make it possible to obtain solutions with practically sufficient accuracy through training on the basis of limited, incomplete and qualitatively described initial data.

Keywords: Soft computing · Fuzzy sets · Neural networks · Fuzzy model · Neuro-fuzzy model · Risk · Classification · Decision making

1 Introduction

Depending on the conditions of uncertainty, risk assessment decision-making is divided into two types:

1) decision-making issues in which the initial data are stochastic;
2) decision-making in cases where the initial data are of a non-stochastic nature and the necessary reliable limits of the parameters of the studied processes are unknown or unclear.

Because not only statistical but also linguistic uncertainty is taken into account, the risks are much higher in the second type of issues than in the first type. At the same time, the risk should be considered as an information uncertainty and ambiguity of the "object-subject-environment" system and some of its elements. This uncertainty dimension determines the dangerous extent of possible losses, losses in making a decision, or in the occurrence of events. From this it is possible to distinguish the main properties of risk: risk is only relevant in relation to the future and is closely linked with predictability.

© The Author(s), under exclusive license to Springer Nature Switzerland AG 2021
R. A. Aliev et al. (Eds.): WCIS 2020, AISC 1323, pp. 280–286, 2021.
https://doi.org/10.1007/978-3-030-68004-6_36

Hence, risk assessment is generally associated with decision making. Modeling under uncertainty is an integral tool of risk assessment analysis.

The advantage of fuzzy logic is that it is possible to use expert knowledge about the composition of a given object in the form of "if" inputs, then "outputs". However, the fuzzy logic apparatus does not have a learning mechanism [1, 4]. The combination of fuzzy logic with neural networks provides a new level of quality. The neural network formed as a result of such merging has an intellectual property such as the application of knowledge to natural language. Therefore, the development of algorithmic support for hybrid intelligent systems for the classification and management of risk situations using fuzzy rule conclusions and neural networks, as well as components related to various other areas of intelligent computing, is a pressing issue [7, 8].

Currently, the following scientists are conducting research on modeling based on the theory of fuzzy sets in the United States, England, Japan, Russia, Ukraine, Azerbaijan and other countries: L. Zade, A. Dubois, A. Prada, E. Mamdani, M. Sugeno, T. Takaxi, M. Djamshidiy, N. N. Moiseev, S. A. Orlovskiy, D. A. Pospelov, E. Mushnik, O. I. Larichev, R. A. Aliev, A. O. Nedosekin and others. The evolving fuzzy mathematical modeling and its application is reflected in the research of the following Uzbek scientists: F. B. Abutaliev, T. F. Bekmuratov, A. R. Marahimov, M. A. Rakhmatullaev, N. A. Ignatev, R. N. Usmanov, D. T. Mukhamedieva. In their research, they have applied the theory of fuzzy sets to the study of poorly formed processes and the methodology for formulating decisions based on them.

However, in poorly formed systems or situations, the problem of evaluating risk situations with fuzzy sets and neural network theories, as well as various other intellectual modeling algorithms, has not been sufficiently studied.

While real processes are described by events described as both stochastic and non-stochastic laws, then the study of these complex processes and objects analysis, modeling, decision-making, and management gives a positive impetus to the development of generalized ideas of traditional and new directions.

An analysis of the description of the uncertainties that occur in poorly formed processes allows us to conclude that their appropriate mathematical expression can be carried out within the framework of a non-interval approach.

Currently, decision-making support systems, including advisory and expert systems, are particularly relevant. The use of such systems allows, at the request of the user, to "give" advice on how to act in the conditions that occur in the object under study, and to do this at the level of an experienced specialist. Such systems are not designed to replace the decision-maker (DM). Therefore, decision support systems are usually included in the management system. However, the classical mathematical methods used in database processing do not always allow for an adequate description of complex structural processes and real events. Not all of the managed solutions and conditions that affect this choice can be expressed as quantitative relationships. In such cases, the formed statement of the object of management is considered to be too complex or non-existent; an important part of the information required for the mathematical description of the object will be in the form of requests and assumptions of experts with work experience on the object data. In this regard, the above-mentioned problems can be solved if decision-making systems (DMS) are widely applied in different environments for fuzzy initial information problems that do not have both deterministic and stochastic rigorous formal-mathematical statements [5, 6, 9].

Along with the great achievements in world science, the limitations of opportunities are also obvious. A number of issues related to such limitations were raised by L.Zade at the III International Conference ICAFS, dedicated to the application of fuzzy systems and softcomputing:

- creation of robots that can move with human or animal flexibility;
- automation of driving during high-speed traffic;
- translation of speech and text from one language to another at the level of human interpreter;
- development of programs capable of drawing conclusions from non-trivial messages;
- increase the ability to model poorly formed systems [1, 4].

The problems presented indicate that the application of fuzzy data-based methods in risk assessment of decision-making systems in poorly formed processes is an important issue.

2 Formulation of the Problem

In decision-making systems of poorly formed processes, it is not possible to obtain accurate data or to obtain such information quickly.

In analyzing the state of a poorly formed process, the decision-maker usually acts with fuzzy concepts and categories rather than numbers, and draws a certain conclusion based on professional intuition and experience. Expert participate in the discussion of conclusions, analyze these solutions, and act on appropriate concepts and categories.

Having chosen to explore some aspects of risk assessment in the case of fuzzy data as the object of research, it should be noted that it is a difficult task for researchers to distinguish a problem whose solution can be obtained using mathematical and cybernetic methods. This is why researchers try to limit themselves to verbal methods. Due to the high level of subjectivism inherent in such methods, these methods have not been widespread. The formation of such methods would allow them to be more widely used in science and practice.

By studying these features, it is necessary to generalize the conceptual principles of modeling in conditions of uncertainty:

1. The need for quantitative and qualitative analysis and consideration of risk in assessing the effectiveness of modeling.
2. A fuzzy approach. Process analysis based on fuzzy models is done in the form of a fuzzy approach. As mentioned above, in real processes, strictly unchanging connections are rare.
3. Form. An optimal set of mathematical and verbal estimates of quantitative and qualitative analysis of methods in determining the effectiveness of modeling requires that the processes formed are superior over weakly formed processes, while one serving as the basis for the other.

The need for formulation determines decision-making technologies with all possible options for comparison in the context of risk and uncertainty and the amount of important information that can be processed as a result of their statistical analysis. However, since it is not possible to formulate all of the factors influencing modeling efficiency, the results of the formulated analysis can be viewed as a preliminary information for a poorly formed analysis.

4. Variety of solutions. The fuzzy approach to determining the effectiveness of modeling emphasizes the need to develop a set of solutions using mathematical methods for each situation associated with uncertainty and risk, rather than a single universal solution "for any situation".
5. Optimization. It is necessary to apply optimization methods, as determining the effectiveness of modeling in the context of risk and uncertainty becomes a multi-criteria issue.
6. Prediction. A fuzzy approach to modeling highlights the need to use forecasting methods as design tools, as design means anticipating the future. The use of forecasting methods is also necessary to determine the important parameters of the model under uncertainty and risk conditions.

3 Solution to the Problem

The achievement of fuzzy logic is the linguistic considerations of knowledge about the composition of an object: if there are 'inputs', then it means the possibility of using 'exits'. But the fuzzy logic apparatus does not have teaching mechanisms. Therefore, the results of the fuzzy inference are highly dependent on the appearance of the membership functions, which are interpreted by the fuzzy terms "small", "large", "cold", "hot" and so on.

The main priority of neural networks is their propensity to train. It is done using specially developed algorithms. The most famous of these is the law of back-propagation [7, 8]. No a priori information about the functional connection structure is required to train a neural network. It's necessary for this an instructive sample in the form of an experimental pair of "inputs and outputs". In return, the trained neural network is not subject to structural analysis of the measured arc graph [2, 3].

The object in the following view is considered.

$$y = f(x_1, x_2, \ldots, x_n) \tag{1}$$

It can describe <(x_i) inputs - (y) outputs>.

This matrix corresponds to an fuzzy base of knowledge:

IF $[(x_1 = a_1^{j1})\ AND (x_2 = a_2^{j1}) AND \ldots AND\ (x_n = a_n^{j1})]$ (w_{j1} weight):

OR $[(x_1 = a_1^{jk_j}) AND (x_2 = a_2^{jk_j}) AND \ldots AND (x_n = a_n^{jk_j})]$ (w_{jk_j} weight),

THEN

$$y \in d_j = [y_{j-1}, y_j],\ j = \overline{1, m} \tag{2}$$

The following approximation of the object corresponds to (1) of the fuzzy knowledge base (2):

$$y = (y\mu^{d_1}(y) + y_1\mu^{d_2}(y) + \ldots + y_{m-1}\mu^{d_m}(y)) / \sum_{j=1}^{m} \mu^{d_j}(y) \tag{3}$$

$$\mu^{d_j}(y) = \max_{p=\overline{1,k_j}} \{w_{jp} \min_{i=\overline{1,n}} [\mu^{jp}(x_i)]\} \tag{4}$$

$$\mu^{jp}(x_i) = \exp(-\frac{c_i^{jp}}{2}(x_i - b_i^{jp})^2) \tag{5}$$

where $\mu^{d_j}(y)$ - the membership-function of the d_j class of the y output; $\mu^{jp}(x_i)$ - the membership function of the variable x_i belonging to the term a_i^{jp}; b_i^{jp}, c_i^{jp} - the correction parameters of the membership function.

The content of the training is to select the types of arcs so that the difference between the neuro-fuzzy approximation and the real behavior of the object is small. The following system of recurrent relationships is used for teaching:

$$w_{jp}(t+1) = w_{jp}(t) - \mu \frac{\partial E_t}{\partial w_{jp}(t)} \tag{6}$$

$$c_i^{jp}(t+1) = c_i^{jp}(t) - \eta \frac{\partial E_t}{\partial c_i^{jp}(t)} \tag{7}$$

$$b_i^{jp}(t+1) = b_i^{jp}(t) - \eta \frac{\partial E_t}{\partial b_i^{jp}(t)} \tag{8}$$

$j = \overline{1,m}, i = \overline{1,n}, p = k_j$

They minimize the following criteria, which were used in neural network theory

$$E_t = \frac{1}{2}(\hat{y}_t - y_t)^2 \tag{9}$$

where: $\hat{y}_t$ and y_t - the theoretical and experimental outputs of the object (1.1) in the t-step of the training; w_j^p; c_i^{jp}, b_i^{jp} - parameters of the weights of the laws (w) and (b, c) in the t - step of training.

The specific derivatives (E_t) in the relationship (6)–(8) determine the sensitivity of the error to changes in neuro-fuzzy network parameters and are calculated as follows:

$$\frac{\partial E_t}{\partial w_{jp}} = \varepsilon_1\varepsilon_2\varepsilon_3 \frac{\partial \mu^{dj}(y)}{\partial w_{jp}} \tag{10}$$

$$\frac{\partial E_t}{\partial c_i^{jp}} = \varepsilon_1\varepsilon_2\varepsilon_3\varepsilon_4 \frac{\partial \mu^{jp}(x_i)}{\partial c_i^{jp}} \tag{11}$$

$$\frac{\partial E_t}{\partial b_i^{jp}} = \varepsilon_1\varepsilon_2\varepsilon_3\varepsilon_4 \frac{\partial \mu^{jp}(x_i)}{\partial b_i^{jp}} \tag{12}$$

where

$$\varepsilon_1 = \frac{\partial E_t}{\partial y} = y_t - \hat{y}_t \quad (13)$$

$$\varepsilon_2 = \frac{\partial y}{\partial \mu^{d_j}(y)} = \frac{\overline{d_j} \sum_{j=1}^{m} \mu^{d_j}(y) - \sum_{j=1}^{m} \overline{d}_j \mu^{d_j}(y)}{\left(\sum_{j=1}^{m} \mu^{d_j}(y) \right)^2}, \quad (14)$$

$$\varepsilon_3 = \frac{\partial \mu^{d_j}(y)}{\partial \left(\prod_{i=1}^{n} \mu^{jp}(x_i) \right)} = w_{jp} \quad (15)$$

$$\varepsilon_4 = \frac{\partial \left(\prod_{i=1}^{n} \mu^{jp}(x_i) \right)}{\partial \nu^{jp}(x_i)} = \frac{1}{\mu^{jp}(x_i)} \prod_{i=1}^{n} \mu^{jp}(x_i), \quad (16)$$

$$\frac{\partial \mu^{d_j}(y)}{\partial w_{jp}} = \prod_{i=1}^{n} \mu^{jp}(x_i) \quad (17)$$

$$\frac{\partial \mu^{jp}(x_i)}{\partial c_i^{jp}} = \frac{2c_i^{jp}(x_i - b_i^{jp})^2}{((c_i^{jp})^2 + (x_i - b_i^{jp})^2)^2} \quad (18)$$

$$\frac{\partial \mu^{jp}(x_i)}{\partial b_i^{jp}} = \frac{2(c_i^{jp})^2(x_i - b_i^{jp})}{\left((c_i^{jp})^2 + (x_i - b_i^{jp})^2\right)^2} \quad (19)$$

As a rule, the neuro-fuzzy network training algorithm consists of two stages. In the first step, the model value of the object output (y) corresponding to the given architecture of the network is calculated. In the second step (E_t) the value of the dependence is calculated and the weight of the connections between the neurons according to formulas (10)–(19) is recalculated.

4 Conclusion

Thus, an approach that allows the formation of concepts of professional language has been proposed. The theory of fuzzy sets and the expediency of using neural networks in solving problems with incomplete or linguistic information, as well as in making decisions based on a person's inner feelings, were justified. The application of risk situations to hybrid intelligent assessment systems allows for the formation and adoption of more effective risk management decisions.

References

1. Aliev, R.A., Aliev, R.R.: The Theory of Intellectual Systems and Its Application. Chashyolgy, Baku (2001)
2. Solieva, B., Juraev, Z.: Analysis of models and methods of solving problems of hybrid intelligent decision-making systems. In: Conference 2019, Fergana, pp. 278–280 (2019)
3. Soliyeva, B.T., Sotvoldiev, D.M.: Applied fuzzy systems for predicting the product of a cotton. In: Conference 2019, STEMM, Tashkent, pp. 132–133 (2019)
4. Zade, L.: Foundations of a new approach to the analysis of complex systems and decision-making processes. Math. Today 5–49 (1974)
5. Mukhamedieva, D.T.: Modeling of weakly formalized processes on the basis of processing of fuzzy information. Institute of Informatics of the Academy of Sciences of Uzbekistan, Tashkent (2007). (in Russian)
6. Mukhamedieva, D.T.: Development of fuzzy models of forecasting and optimization problems. Fan va Technology, Tashkent (2012). (in Russian)
7. Rotshtein, A.P.: Intellectual identification technologies: fuzzy logic, genetic algorithms, neural networks. Universum-Vinnitsa, Vinnitsa (1999)
8. Rutkovskaya, D., Pilinsky, M., Rutkovski, L.: Neural networks, genetic algorithms, fuzzy systems. Hot line – Telecom, Moskow (2006)
9. Mukhamedieva, D.T.: Development of fuzzy models of decision making tasks. Palmarium Academic Publishing, AV Academikerverlag GmbH &C.KG Heinrich-Bosking, Saarbrucken (2014)

Z-information Based PROMETHEE Method

K. I. Jabbarova[1(✉)] and Aynur I. Jabbarova[2]

[1] Department of Computer Engineering, Azerbaijan State Oil and Industry University, 20 Azadlig Avenue, AZ1010 Baku, Azerbaijan
konul.jabbarova@mail.ru

[2] Azerbaijan State Economic University, Baku, Azerbaijan
stat_aynur@mail.ru

Abstract. In this paper PROMETHEE method is applied to personnel assessment and selection problem under Z-information. All criteria values are described with Z-numbers. The obtained results show validity and effectiveness the conducted analysis.

Keywords: PROMETHEE method · Z-number

1 Introduction

Decision makers in the manufacturing sector frequently faces the problem of evaluating an extensive range of alternative options and choosing an optimal one based on a set of conflicting criteria [1]. In [1] the PROMETHEE method was used to solve a problem of this kind. The method was improved by integrating with analytic hierarchy process (AHP) and the fuzzy logic. To illustrate the method, four examples are considered. The proposed method is effective for decision making in different real-life situations of the manufacturing environment.

Fuzzy PROMETHEE technique was used to compare suppliers in [2]. Fuzzy sets to get rid of uncertainties in the evaluation process were used. They offer to get the evaluation of decision makers orally, due to easiness [2].

In [3] it is applied the Fuzzy-PROMETHEE method to the textbook selection. The triangular fuzzy number and ranking method are integrated into the PROMETHEE II method. The offered method can manage the uncertainty in the decision making. An example demonstrates the convenience and validity of the proposed method.

In [4] it is represented to application of the fuzzy PROMETHEE method to the investment problem of value investing. For evaluation of the firms growth potential are used five criteria. All these criteria are represented by fuzzy numbers based on the real financial data. The fuzzy PROMETHEE method is used to rank 20 randomly selected stocks in Shanghai Stock Exchange. The obtained results of study shows the effectiveness of the fuzzy PROMETHEE method in the value investing decision making process of in stock markets.

Real-world problems usually are characterized with both fuzziness and partial reliability of information. But this is not taken into account in the works on the MCDM

© The Author(s), under exclusive license to Springer Nature Switzerland AG 2021
R. A. Aliev et al. (Eds.): WCIS 2020, AISC 1323, pp. 287–293, 2021.
https://doi.org/10.1007/978-3-030-68004-6_37

problems. Therefore, Prof. Zadeh offered the Z-number concept. A Z-number consists of a pair of fuzzy numbers Z = (A, B), where A is a soft constraint on a value of a variable of interest, and B is a soft constraint on a value of a probability measure of A, playing a role of reliability of A.

In this article we apply PROMETHEE method to personal selection problem under Z-number- valued information. The paper is organized as follows. In Sect. 2 we give a pre-requisite material used in the sequel. In Sect. 3, we apply Z-number based PROMETHEE method to a considered problem. Conclusion is presented in Sect. 4.

2 Preliminaries

Definition 1. A discrete Z-number [3–6]. A discrete Z-number is an ordered pair $Z = (A, B)$ where A is a discrete fuzzy number playing a role of a fuzzy constraint on values of a random variable X: X *is* A. Is a discrete fuzzy number with a membership function $\mu_B : \{b_1, \ldots, b_n\} \to [0, 1], \{b_1, \ldots, b_n\} \subset [0, 1]$, playing a role of a fuzzy constraint on the probability measure of A: $P(A) = \sum_{i=1}^{n} \mu_A(x_i)p(x_i)$ *is* B.

Definition 2. Operations over Discrete Z-numbers [3–6]. Let X_1 and X_2 be discrete Z-numbers describing information about values of X_1 and X_2. Consider computation of $Z_{12} = Z_1 * Z_2, * \in \{+, -, \cdot, /\}$. The first stage is computation of $A_{12} = A_1 * A_2$.

The second stage involves construction of B_{12}. We realize that in Z-numbers Z_1 and Z_2, the 'true' probability distributions p_1 and p_2 are not exactly known. In contrast, fuzzy restrictions represented in terms of the membership functions are available

$$\mu_{p_1}(p_1) = \mu_{B_1}\left(\sum_{k=1}^{n_1} \mu_{A_1}(x_{1k})p_1(x_{1k})\right), \mu_{p_2}(p_2) = \mu_{B_2}\left(\sum_{k=1}^{n_2} \mu_{A_2}(x_{2k})p_2(x_{2k})\right)$$

Probability distributions $p_{jl}(x_{jk}), k = 1, \ldots, n$ induce probabilistic uncertainty over $X_{12} = X_1 + X_2$. Given any possible pair p_1, p_2, the convolution $p_{12} = p_1 \circ p_2$ is computed as

$$p_{12}(x) = \sum_{x_1 + x_2 = x} p_1(x_1)p_2(x_2), \forall x \in X_{12}; x_1 \in X_1, x_2 \in X_2.$$

Given p_{12s}, the value of probability measure of A_{12} is computed: $P(A_{12}) = \sum_{k=1}^{n} \mu_{A_{12}}(x_{12k})p_{12}(x_{12k})$.

However, p_1 and p_2 are described by fuzzy restrictions which induce fuzzy set of convolutions:

$$\mu_{p_{12}}(p_{12}) = \max_{\{p_1, p_2 : p_{12} = p_1 \circ p_2\}} \min\{\mu_{p_1}(p_1), \mu_{p_2}(p_2)\} \tag{1}$$

Fuzziness of information on p_{12} induces fuzziness of $P(A_{12})$ as a discrete fuzzy number B_{12}. The membership function $\mu_{B_{12}}$ is defined as

$$\mu_{B_{12}}(b_{12}) = \max \mu_{p_{12}}(p_{12})$$

subject to

$$b_{12} = \sum_{i=1}^{n} \mu_{A_{12}}(x_i) p_{12}(x_i)$$

As a result, $Z_{12} = Z_1 * Z_2$ is obtained as $Z_{12} = (A_{12}, B_{12})$.
A scalar multiplication $Z = \lambda Z_1,\ \lambda \in R$ is a determined as $Z = (\lambda A_1, B_1)$.

Definition 3. Fuzzy Pareto optimality (FPO) principle based comparison of Z-numbers [7]. Fuzzy Pareto optimality (FPO) principle allows to determine degrees of Pareto Optimality of multiattribute alternatives. We apply this principle to compare Z-numbers as multiattribute alternatives – one attribute measures value of a variable, the other one measures the associated reliability. According to this approach, by directly comparing Z-numbers $Z_1 = (A_1, B_1)$ and $Z_2 = (A_2, B_2)$ one arrives at total degrees of optimality of Z-numbers: $do(Z_1)$ and $do(Z_2)$. These degrees are determined on the basis of a number of components (the minimum is 0, the maximum is 2) with respect to which one Z-numbers dominates another one. Z_1 is considered higher than Z_2 if $do(Z_1) > do(Z_2)$.

Let us consider a MCDM problem under Z-valued information.

Definition 4 [8]. A distance between Z-numbers. The distance between Z-numbers $Z_1 = (A_1, B_1)$ and $Z_2 = (A_2, B_2)$ is defined as.

$$D(Z_1, Z_2) = \frac{1}{n+1} \sum_{k=1}^{n} \left\{ \left| a^L_{1\alpha_k} - a^L_{2\alpha_k} \right| + \left| a^R_{1\alpha_k} - a^R_{2\alpha_k} \right| \right\} + \frac{1}{m+1} \sum_{k=1}^{m} \left\{ \left| b^L_{1\alpha_k} - b^L_{2\alpha_k} \right| + \left| b^R_{1\alpha_k} - b^R_{2\alpha_k} \right| \right\}$$

where $a^L_\alpha = \min A^\alpha$, $a^R_\alpha = \max A^\alpha$, $b^L_\alpha = \min B^\alpha$, $b^R_\alpha = \max B^\alpha$.

3 Statement of the Problem and a Solution Method

Let us examine MCDM problem with Z-number-valued information. Suppose that a local chemical firm tries to new recruit an online manager [9]. The firm's human resources department offers several corresponding selection tests, as the advantage criteria to be estimated. In this problem has subjective (interviews) and objective (knowledge, skill tests) criteria. Tests of knowledge involve language, professional and safety rule tests. Tests of skill consist of professional and computer skills. Interviews involve panel and 1-on-1 interviews [9]. There are 3 skilled candidates on the list. Objective and subjective criteria are described by Z-numbers $f_{ij} = (A_{ij}, B_{ij})$. The values of some criteria are given in Table 1. All importance weights of objective and subjective criteria are given in Table 2.

Let us solve this problem by using the algorithm of the PROMETHEE method under Z-number-valued information.

At the *first stage*, decision matrix is weighted by multiplying the importance weights of evaluation criteria and the values in the normalized Z-number based decision matrix. The weighted normalized Z-values of some criteria are defined (Table 3):

Table 1. Z-number-valued criteria evaluations

	Language test C_1	Panel interview C_6
f_1	$\{^0/_{0.4}\ ^1/_{0.44}\ ^0/_{0.48}\}\{^0/_{0.8}\ ^1/_{0.9}\ ^0/_1\}$	$\{^0/_{0.42}\ ^1/_{0.47}\ ^0/_{0.52}\}\{^0/_{0.6}\ ^1/_{0.7}\ ^0/_{0.8}\}$
f_2	$\{^0/_{0.42}\ ^1/_{0.47}\ ^0/_{0.52}\}\{^0/_{0.8}\ ^1/_{0.9}\ ^0/_1\}$	$\{^0/_{0.34}\ ^1/_{0.38}\ ^0/_{0.42}\}\{^0/_{0.6}\ ^1/_{0.7}\ ^0/_{0.8}\}$
f_3	$\{^0/_{0.39}\ ^1/_{0.43}\ ^0/_{0.47}\}\{^0/_{0.8}\ ^1/_{0.9}\ ^0/_1\}$	$\{^0/_{0.48}\ ^1/_{0.53}\ ^0/_{0.58}\}\{^0/_{0.6}\ ^1/_{0.7}\ ^0/_{0.8}\}$

Table 2. Importance weights of the criteria

No	Criteria	Weights
C_1	Language test	$\{^0/_{0.059}\ ^1/_{0.066}\ ^0/_{0.073}\}\{^0/_{0.8}\ ^1/_{0.9}\ ^0/_1\}$
C_2	Professional test	$\{^0/_{0.176}\ ^1/_{0.196}\ ^0/_{0.216}\}\{^0/_{0.8}\ ^1/_{0.9}\ ^0/_1\}$
C_3	Safety rule test	$\{^0/_{0.059}\ ^1/_{0.066}\ ^0/_{0.073}\}\{^0/_{0.8}\ ^1/_{0.9}\ ^0/_1\}$
C_4	Professional skills	$\{^0/_{0.117}\ ^1/_{0.130}\ ^0/_{0.143}\}\{^0/_{0.8}\ ^1/_{0.9}\ ^0/_1\}$
C_5	Computer skills	$\{^0/_{0.117}\ ^1/_{0.130}\ ^0/_{0.143}\}\{^0/_{0.8}\ ^1/_{0.9}\ ^0/_1\}$
C_6	Panel interview	$\{^0/_{0.194}\ ^1/_{0.216}\ ^0/_{0.238}\}\{^0/_{0.8}\ ^1/_{0.9}\ ^0/_1\}$
C_7	1-on-1 interview	$\{^0/_{0.176}\ ^1/_{0.196}\ ^0/_{0.216}\}\{^0/_{0.8}\ ^1/_{0.9}\ ^0/_1\}$

Table 3. The weighted Z-number-valued decision matrix

	Language test C_1	Panel interview C_6
f_1	(0.024, 0.029, 0.035)(0.66, 0.82, 0.95)	(0.08, 0.1, 0.13)(0.51, 0.64, 0.78)
f_2	(0.025, 0.031, 0.038)(0.66, 0.82, 0.95)	(0.07, 0.08, 0.1)(0.51, 0.64, 0.78)
f_3	(0.023, 0.028, 0.034)(0.66, 0.82, 0.95)	(0.09, 0.11, 0.14)(0.51, 0.64, 0.78)

At the *second stage,* for comparing two alternatives g and f on each criterion Z- distances $D(Z_{gj}(A,B),(1,1))$, $D(Z_{fj}(A,B),(1,1))$, $D(Z_{gj}(A,B),Z_{fj}(A,B))$ are computed. We consider that $Z_{gj}(A,B) \geq Z_{fj}(A,B)$ iff $D(Z_{gj}(A,B),(1,1)) \leq D(Z_{fj}(A,B),(1,1))$. Then preference function is formalized as:

$$P(Z_{gj}(A,B),Z_{fj}(A,B)) = \begin{cases} 0, Z_{gj}(A,B) \leq Z_{fj}(A,B) \\ D(Z_{gj}(A,B),Z_{fj}(A,B)), Z_{gj}(A,B) > Z_{fj}(A,B). \end{cases}$$

The obtained results are given in Table 4.

Table 4. The preference function $P_j(Z_g(A,B), Z_f(A,B))$

	C_1	C_2	C_3	C_4	C_5	C_6	C_7
$P(1,2)$	0	0.04	0.093	0	0.013	0.12	0
$P(1,3)$	0.013	0	0.12	0	0	0	0
$P(2,1)$	0.04	0	0	0.067	0	0	0.08
$P(2,3)$	0.053	0	0.027	0.037	0	0	0
$P(3,1)$	0	0.16	0	0.03	0.067	0.08	0.08
$P(3,2)$	0	0.2	0	0	0.08	0.2	0

At the *third stage,* Z-number valued preference index is computed to define the value of the outranking relation ($j = 1, 2, \ldots, n$):

$$\pi(g,f) = \sum_{j=1}^{n} \left[w_j P_j(g,f)\right].$$

The obtained results are shown in Table 5.

Table 5. Z-number valued preference index

	f_1	f_2	f_3
f_1	–	(0.037, 0.042, 0.046) (0.48, 0.68, 0.99)	(0.078, 0.088, 0.096) (0.66, 0.81, 1)
f_2	(0.024, 0.027, 0.03), (0.56, 0.74, 1)	–	(0.009, 0.01, 0.011), (0.56, 0.74, 0.99)
f_3	(0.07, 0.077, 0.085), (0.42, 0.63, 0.99)	(0.083, 0.093, 0.1), (0.56, 0.74, 1)	–

At the *fourth stage,* the leaving and entering flows are computed for ranking of alternatives (Tables 6 and 7).

$$\phi^+(g) = \sum_{\substack{f \neq g \\ f=1}}^{m} \pi(g,f),\ \phi^-(g) = \sum_{\substack{f \neq g \\ f=1}}^{m} \pi(f,g) \tag{2}$$

At the *fifth stage,* net flow is computed as follows:

$$\phi(g) = \phi^+(g) - \phi^-(g)$$

Table 6. Z-number valued flows for each alternative

	Leaving flows	Entering flows
f_1	(0.1, 0.12, 0.13), (0.52, 0.67, 0.92)	(0.09, 0.1, 0.12), (0.34, 0.53, 0.98)
f_2	(0.033, 0.037, 0.04), (0.57, 0.7, 0.99)	(0.12, 0.135, 0.146), (0.43, 0.61, 0.94)
f_3	(0.15, 0.17, 0.19), (0.38, 0.55, 0.93)	(0.087, 0.098, 0.107), (0.56, 0.74, 0.99)

Table 7. Z-number valued net flows for the alternatives

	$\phi(g)$
f_1	(0.1, 0.12, 0.15)(0.34, 0.49, 0.75)
f_2	(−0.087, −0.098, −0.106)(0.42, 0.54, 0.75)
f_3	(0.063, 0.072, 0.083)(0.51, 0.61, 0.75)

At *sixth stage*, preference rank of each alternative is evaluated by using Definition 3 (Table 8).

Table 8. Preference rank of each alternatives

Ranking
f_1
f_3
f_2

This implies that the alternative f_1 is the best one.

4 Conclusion

In this paper PROMETHEE is applied to personnel assessment and selection problem under Z-valued-information. For ranking of alternatives, a fuzzy Pareto optimality principle-based comparison of Z-numbers is used.

References

1. Venkata, R.R., Patel, B.K: Decision making in the manufacturing environment using an improved PROMETHEE method. Int. J. Prod. Res. **48**(16) (2010). https://doi.org/10.1080/00207540903049415

2. Baki, R.: For supplier evaluation process through fuzzy PROMETHEE method. Int. J. Mater. Mech. Manuf. **5**(3) (2017). https://doi.org/10.18178/ijmmm.2017.5.3.314
3. Yuen, K.K.F., Ting, T.O.: Textbook selection using fuzzy PROMETHEE II method. Int. J. Future Comput. Commun. **1**(1), 76–78 (2012). https://doi.org/10.7763/IJFCC.2012.V1.20
4. Li, H., Qu, S., Zhu, H.: PROMETHEE: a fuzzy algorithm for decision making in stock market value investing. IREMAN **1**(1), 75–80 (2013)
5. Aliev, R.A., Alizadeh, A.V., Huseynov, O.H.: The arithmetic of discrete Z-numbers. Inf. Sci. **290**, 134–155 (2015)
6. Aliev, R.A., Alizadeh, A.V., Huseynov, O.H., Jabbarova, K.I.: Z-number based linear programming. Int. J. Intell. Syst. **30**(5), 563–589 (2015)
7. Aliev, R.A., Huseynov, O.H.: Decision Theory with Imperfect Information. World Scientific, Singapore (2014)
8. Aliev, R.A., Huseynov, O.H., Aliyev, R.R., Alizadeh, A.V.: The Arithmetic of Z-Numbers: Theory and Applications. World Scientific, Singapore (2015)
9. Aliev, R.A., Huseynov, O.H., Serdaroglu, R.: Ranking of Z-numbers and its application in decision making. Int. J. Inf. Technol. Decis. Making **15**(6), 1503–1519 (2016)
10. Aliev, R.A., Pedrycz, W., Huseynov, O.H., Eyupoglu, S.Z.: Approximate reasoning on a basis of Z-number-valued if–then rules. IEEE Trans. Fuzzy Syst. **25**(6), 1589–1600 (2017)
11. Shih, H.-S., Shyur, H.-J., Lee, S.E.: An extension of TOPSIS for group decision making. Math. Comput. Model. **4**, 801–813 (2007)

Integrated Risk Management System for Sanitary and Epidemiological Control at the Customs Border of the Republic of Uzbekistan

A. A. Saidov[1(✉)], L. T. Pulatova[2], and F. A. Xakimova[3]

[1] Department of Special Sciences, Customs Institute Republic of Uzbekistan, St. Kozirobod Avenue 2nd, 118100071 Tashkent, Republic of Uzbekistan
sobirs59@mail.ru

[2] Department of Customs Expertise and Classification of Goods, Customs Institute of Uzbekistan, St. Kozirobod Avenue 2nd, 118100071 Tashkent, Republic of Uzbekistan
lorena_97@mail.ru

[3] Department of the Management Harmonized System and Customs Examination, State Customs Committee Republic of Uzbekistan, 3- I. Karimov Street, 100003 Tashkent, Republic of Uzbekistan
fazilat.khakimova@mail.ru

Abstract. The paper considers the task of re-engineering the process of organizing customs control of foreign trade goods and the development of the national single-window customs system of the Republic of Uzbekistan. The results of introducing a risk management system in the activities of customs authorities are analyzed and criteria are proposed for identifying possible risks of sanitary and epidemiological control at the border using information modeling methods and an information matrix. The risk management algorithm for control at checkpoints across the customs border on the basis of the Single Window customs information system is presented.

Keywords: Customs control · Certification of foreign trade goods · Customs system "single window · Information modeling · An information matrix of customs risk profiles · An algorithm for managing sanitary and epidemiological risks

1 Introduction

The task of risk management during a control at checkpoints across the customs border is one of the most important and perhaps the most time-consuming stage of the organization of customs control in the process of managing foreign trade. The fact is that in this process, along with customs authorities, a number of ministries and departments for transport, veterinary, phytosanitary, sanitary and quarantine control and their subordinate laboratories and research centers are involved [1].

© The Author(s), under exclusive license to Springer Nature Switzerland AG 2021
R. A. Aliev et al. (Eds.): WCIS 2020, AISC 1323, pp. 294–302, 2021.
https://doi.org/10.1007/978-3-030-68004-6_38

The risk management process in customs is a systematic activity of customs authorities to minimize the likelihood of events related to non-compliance with international treaties and acts in the field of customs regulation and national customs legislation.

In 2019, the risk management system (RMS) and the "Single Window" information system were introduced into the activities of the customs authorities of the Republic of Uzbekistan. The first year of operation showed their certain effectiveness: customs clearance time was reduced, budget revenues from customs payments increased, the opportunity for targeted customs inspection of goods and the identification of violations of customs legislation appeared.

At the same time, the analysis of the results of the application of these systems showed that the risk management system and the Single Window information system of the customs authorities function more independently and are not sufficiently integrated with each other, which affects their effectiveness.

In light of the above, it is important to note that the study of criteria and risk management algorithms at checkpoints across the customs border, including during sanitary-epidemiological control, is quite relevant.

2 Problem Statement

A steady tendency towards an increase in passenger traffic and world foreign trade turnover requires the improvement of methods and means of carrying out control operations, forcing us to look for optimal forms and mechanisms for carrying out their functions.

The appropriateness of total state control is a thing of the past, giving way to selective but much more effective supervisory measures.

The purpose of sanitary and epidemiological control is to prevent the international spread of diseases, control and ensure that the health system responds to them in ways that are commensurate with public health risks, and avoid unnecessary interference in international transport and trade.

The basis for sanitary-quarantine control at the border should be the risks to public health and sanitary-epidemiological well-being.

In those cases when it comes to selective control and risk management, it is necessary to take into account that risks in any system, first of all, must be identified. The main requirement for risking management is the completeness of the list of possible risks. The study of issues related to detailing the completeness of the list of possible risks in the field of sanitary and epidemiological control is the subject of many scientific works, in particular, the following possible risks are given in [2]:

Risk number 1. Arrival of a vehicle from countries with disease-infected areas;
Risk number 2. Failure to provide, in the established manner, preliminary information about the absence on board of vehicles, of people with suspected illness, requiring the implementation of measures for sanitary and epidemiological control;
Risk number 3. The presence on the vehicle of people arriving at an international flight from countries having disease-infected areas or arriving from such countries within the incubation period;

Risk number 4. Detection during the previous sanitary-quarantine control of violations of the law in the field of ensuring the sanitary-epidemiological well-being of the population on a vehicle engaged in international transportation;
Risk number 5. Distressed integrity international mail containing linens, clothes, bedding or other household items, dishes, used toys that come from countries with contaminated areas or from epidemic zones;
Risk number 6.The receipt of information on the presence on the vehicle of people with suspected illness;
Risk number 7. The presence of rodents or traces of their stay in the vehicle;
Risk number 8.The presence of insects in a vehicle that has come from countries with infected areas or from epidemic zones;
Risk number 9. Establishment of the fact of the movement of the vehicle, controlled goods with increased radiation background.

This is not to say that the above list is a complete list in the field of possible risks of sanitary and epidemiological control. However, for the implementation of information modeling of risks of sanitary - epidemiological control, they can be taken as a basis.

The above list indicates that the problem lies in modeling the risks of sanitary and epidemiological control and linking them to countries or regions infected with the epidemic, with goods subject to sanitary and epidemiological control and with international carriers, which requires the development of objective criteria for assessing risks and management algorithms risks in the field of sanitary and epidemiological control.

3 The Concept of Solving the Problem

A conceptual approach to solving the stated problem is the implementation of the Single Window mechanism in the customs authorities and on its basis to introduce a risk management system in the field of sanitary and epidemiological control. According to the UNECE Recommendations No. 33 through the Center for Trade Facilitation and Electronic Business (UN/CEFACT), the Single Window mechanism is a unique tool that allows parties involved in foreign trade operations to submit standardized information and documents using a single transmission channel data (Fig. 1) [3].

Along with this, for the implementation of the Single Window mechanism, important issues are the development of a single package of documents, a single application form for customs clearance of goods and ensuring the completeness of their certification.

As you know, the Cargo Customs Declaration provides the a good deal of complete information about the product, however, certification of goods is carried out before the submission of the Cargo Customs Declaration. In this regard, information on a product presented with an invoice or a shipping document of foreign trade is usually used in practice.

Thus, a single form of a document - an application for customs clearance of goods should provide at least the following information:

- unique number of the application for customs clearance of goods;
- product code;

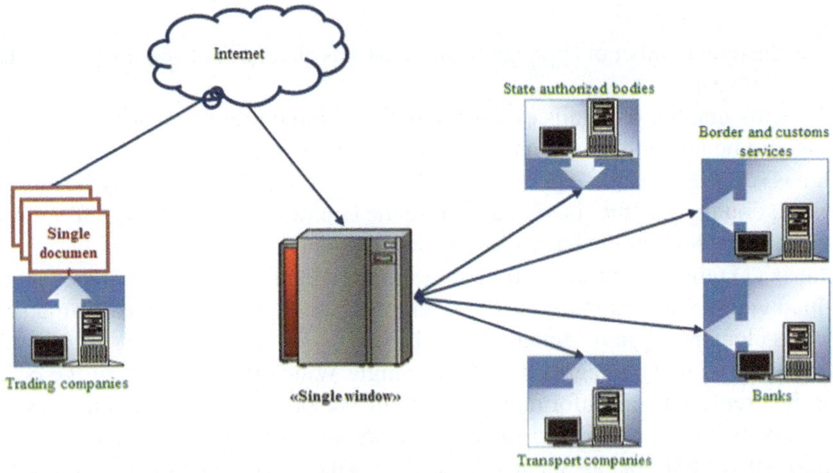

Fig. 1. The mechanism of the "Single Window" of the customs authorities.

- name of product;
- country of origin;
- departure country;
- trading country;
- receiver of product;
- the sender of the goods;
- quantity of goods;
- weight of goods;
- cost of goods;
- carrier;
- country of transit;
- the applicant.

Due to the fact that within one year several million applications for customs clearance of goods are received by the customs authorities of the Republic of Uzbekistan, the formalized data receives the form of a multidimensional matrix, each element which is an information variable z_{nml} and reflects the value of one application parameter.

Z= 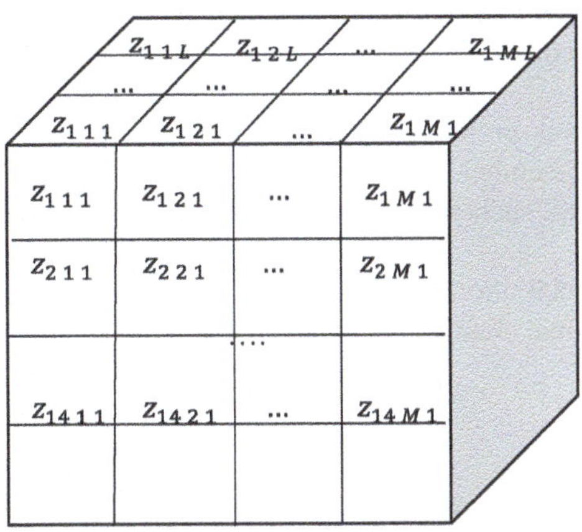 (1)

Here:

L - is the total number of applications for customs clearance of goods received during one calendar year.

M - is the number of details of the details of each request parameter.

It should be noted that each parameter of the application, depending on the level of detail, has several elements. For example, the parameter z_{13ml} - there may be more than one transit country, i.e. the route of arrival in the country of destination of foreign trade cargo may pass through the territories of several countries.

Matrix (1) reflects the information of the application for customs clearance of goods submitted within one year to the customs authorities, and it is an information matrix, i.e. each of its elements is an information variable [4].

The set of data (1), integrated with the Single Window mechanism of the customs authorities, represents an information model of the process of remote customs clearance of goods and is considered to be a conceptual platform for solving the problem of introducing a risk management system for sanitary and epidemiological control at checkpoints across the customs border.

4 Implementation of the Concept

The methodology for the implementation of the above conceptual platform, aimed at solving the problem, is based on the formation of risk profiles of the sanitary and epidemiological well-being of the population.

As the main tool, information risk matrixes of sanitary and epidemiological control are used on the basis of international and national classifiers, approved by the Decree of the State Customs Committee of the Republic of Uzbekistan No. 2773 dated April 6, 2016, registered by the Ministry of Justice Republic of Uzbekistan [5]:

(a) the information matrix of the classifier of states and dependent territories.

It should be noted that ISO 3166 is an international ISO standard that defines the code names of states and dependent territories, as well as the main administrative entities within states. It consists of the following parts:

- short name of states and dependent territories;
- full name of states and dependent territories;
- alpha-2 - two-letter codes of states and dependent territories;
- alpha-3 - three-letter codes of states and dependent territories;
- numeric - digital codes of states and dependent territories.

This classifier was prepared by the ISO/TC 46 secretariat and its latest version was published on December 15, 2011 (final version), the latest geocode changes were identified in 53 states, such as Azerbaijan, Aruba, Afghanistan, Bahamas, Bangladesh and others.

To date, in this classifier, the total number is 250 states and dependent territories. Part of the international classifier of states and dependent territories is given in Table 1.

Table 1. Part of the international classifier of states and dependent territories

Short title	Full name	Letter code		Digital code
		Alpha-2	Alpha-3	
Abkhazia	Republic of Abkhazia	AB	ABH	895
Australia	Australia	AU	AUS	036
Austria	Republic of Austria	AT	AUT	040
Azerbaijan	Republic of Azerbaijan	AZ	AZE	031
Albania	Republic of Albania	AL	ALB	008
………………				

If we classify the classifier of states and dependent territories in the form of an "information matrix", then we obtain the following form:

$$c = \left\{ \begin{matrix} c_{11} & c_{12} & c_{13} & c_{14} & c_{15} \\ c_{21} & c_{22} & c_{23} & c_{24} & c_{25} \\ & & \cdots & & \\ c_{2501} & c_{2502} & c_{2503} & c_{2504} & c_{2505} \end{matrix} \right\} \quad (2)$$

where:

c_{i1} – short name of the *i*-th state;
c_{i2} – full name of the *i*-th state;
c_{i3} – Alpha-2 letter code of the *i*-th state;
c_{i4} – Alpha-3 letter code of the *i*-th state;
c_{i5} – digital code of the i-th state.

Then, the information matrix of the risk profiles of sanitary and epidemiological well-being has the following form:

$$R_c = \left\{ \begin{matrix} rc_{11} & rc_{12} & \cdots & rc_{1x} \\ rc_{21} & rc_{22} & \cdots & rc_{2x} \\ & \cdots & & \\ rc_{2501} & rc_{2502} & \cdots & rc_{250x} \end{matrix} \right\} \quad (3)$$

where:

rc_{i1} – digital code of the i-th state;
rc_{ix} – defined as follows:

$$c_{ix} = \left\{ \begin{matrix} 1, \text{ if in } i - country \\ \text{infection site recorded;} \\ 0, \text{ otherwise} \end{matrix} \right. \quad (4)$$

x – number of risk profiles sanitary - epidemiological well-being of the population, $x = 2,3,\ldots$.

$$i = \overline{1, 250};$$

The information matrix (3) and formula (4) represent a model of risk profiles for the sanitary and epidemiological well-being of the population of countries.

(b) the information matrix of the classifier of international carriers

Similarly, a matrix of risk profiles for sanitary and epidemiological control for carriers is formed:

$$R_p = \begin{Bmatrix} rp_{11} & rp_{12} & \cdots & rp_{1y} \\ rp_{21} & rp_{22} & \cdots & rp_{2y} \\ & \cdots & & \\ rc_{z1} & rc_{z2} & \cdots & rp_{z\,y} \end{Bmatrix} \tag{5}$$

Here:

rp_{il} – unique digital code of the i-th international carrier;

rp_{iy} – defined as follows:

$$rp_{iy} = \begin{cases} 1, \text{fixed} \quad \text{risk} \\ \text{for} \;\; i - \text{carrier}; \\ \\ 0, \text{otherwise} \end{cases} \tag{6}$$

y - the number of risk profiles of sanitary and epidemiological control for international carriers, $y = 2,3,\ldots$.

$$i = \overline{1, z};$$

(c) information matrix of the classifier of goods subject to sanitary - epidemiological control

The next important factor in the field of risk analysis of sanitary and epidemiological control are goods subjects to sanitary and epidemiological control. In accordance with Appendix No. 1 to the Resolution of the Cabinet of Ministers of the Republic of Uzbekistan No. 379 dated December 30, 2012, approved list of goods subjects to sanitary and epidemiological control [6]. To date, 3443 units of the names of such goods are registered in the database of customs authorities of the Republic of Uzbekistan according to the Commodity Nomenclature of Foreign Economic Activity of the Republic of Uzbekistan [7].

From the above analysis of the risk identification algorithm in the field of sanitary and epidemiological control of goods, it clearly follows that the information matrix of goods subject to sanitary and epidemiological control can be presented in the following form:

$$T_{ses} = \left\{ \begin{matrix} t_{11} & t_{12} & t_{13} & t_{14} \\ t_{21} & t_{22} & t_{23} & t_{24} \\ & & & \ldots \\ t_{34431} & t_{34432} & t_{34433} & t_{34434} \end{matrix} \right\} \quad (7)$$

Where:
t_{i1} – product code;
t_{i2} – Name of product;
t_{i3} – basic unit of measurement of goods;
t_{i4} – an additional unit of measure for the goods.

$$i = \overline{1, 3443}.$$

(d) risk management criteria and algorithm in the field of sanitary and epidemiological control at customs checkpoints

The form under consideration indicates that the information models of the process of remote customs clearance of goods (1), as well as the risk profile models of sanitary and epidemiological control (2)–(7), will establish the following risk management criteria for sanitary and epidemiological control at customs checkpoints border:

– risk management criteria related to goods subject to sanitary and epidemiological control:

$$\left.\begin{matrix} Z \cap T_{ses} \neq \emptyset \\ |Z \cap T_{ses}| \geq 1 \end{matrix}\right\} \quad (8)$$

– risk management criteria related to countries or areas infected with the epidemic:

$$\left.\begin{matrix} Z \cap R_s \neq \emptyset \\ \sum_i c_{ix} \geq 1 \end{matrix}\right\} \quad (9)$$

– risk management criteria associated with international carriers that participate in the process of foreign trade operations:

$$\left.\begin{matrix} Z \cap R_p \neq \emptyset \\ \sum_i p_{iy} \geq 1 \end{matrix}\right\} \quad (10)$$

– based on the risk management algorithm of sanitary and epidemiological control at checkpoints across the customs border, the following main indicators can be distinguished, which are a combination of all three types of risk profiles, in particular:

$$R_{total} = (Z \cap T_{ses}) \cup (Z \cap R_s) \cup \left(Z \cap R_p\right) \neq \emptyset \quad (11)$$

It should be noted that the above (8)–(11) criteria give tangible results in those cases when they are be implemented in conjunction with the risk management algorithm of sanitary and epidemiological control at checkpoints across the customs border. For example, in each (8)–(10) criteria, the second condition makes no sense, without taking into account the indicators of the first conditions.

5 Conclusion

From the foregoing, it follows that the problem of system analysis using information technologies in conjunction with risk management has an urgent direction in the development of activities of the customs authorities of the republic with the aim of increasing the effectiveness of actions for transport, veterinary, phytosanitary, sanitary and quarantine control at checkpoints across the state border.

The implementation of the conceptual approach and the algorithm for identifying risks at checkpoints across the customs border described above will integrate the Single Window customs information system and the RMS, and ensure information interaction with other regulatory authorities at the border.

It is important to note that the material presented in this work is the result of a preliminary analysis of the current state of normalization and digital support of customs authorities at the border. To obtain more tangible results, it is necessary to conduct further large-scale and in-depth studies, in close connection with the practical activities of the customs authorities.

References

1. Afonin, P.N., Afonin, D.N., Mutte, G.E., Kondrashova, V.A.: Systemic risk analysis at the checkpoint at the state border of the Russian Federation during the implementation of customs services. Sci. J. Nat. Priorities Russ. **159**(18), 14–18 (2012)
2. Ryzhkov, Y.V., Soloviev, M.Y.: Risk management in quarantine control. Sci. Pract. J. Probl. Espec. Dangerous Infect. (113) 33–37 (2012)
3. Recommendation and Guidelines on establishing a Single Window: UNITED NATIONS New York and Geneva (2005). https://www.unece.org/fileadmin/DAM/cefact/recommendations/rec33/rec33_trd352e.pdf. Accessed 15 Mar 2020
4. Khakimova, F.A., Saidov, A.A.: Methods of information modeling the process of certification of goods in conditions of intensive electronic foreign trade. J. Chem. Technol. Control Manag. - Tashkent (Uzbekistan) **86**(3) 67–73 (2019)
5. Resolution of the State Customs Committee of the Republic of Uzbekistan: on approval of the instruction on the procedure for filling the cargo customs declaration. Registered by the Ministry of Justice of the Republic of Uzbekistan No. 2773 dated April 6, 2016. Collection of legislation of the Republic of Uzbekistan, No. 14, pp. 133–217 (2016)
6. Resolution of the Cabinet of Ministers of the Republic of Uzbekistan: On additional measures to liberalize procedures related to the implementation of foreign trade operations. Behind № 379 dated December 30, (2012), Collection of legislation of the Republic of Uzbekistan. http://lex.uz/ru/docs. Accessed 03 Mar 2020
7. Commodity nomenclature of foreign economic activity of the Republic of Uzbekistan (2017 version), Tashkent, 655 p. (2017)

Estimation of Linguistically Described Weights of Criteria

Shamil A. Ahmadov[1,2](✉)

[1] French-Azerbaijani University, 183 Nizami Street, Baku, Azerbaijan
shamilahmadov@yandex.ru
[2] Azerbaijan State Oil and Industry University, Azadlig 35, Nasimi, Baku, Azerbaijan

Abstract. This study devoted weights of criteria which linguistically described and the problem to find their crisp values are considered. The purpose of this paper is to estimate linguistically described preferences of decision makers and analysis the situation which crisp weights are needed. Usually, in real life multi-criteria problems weights of criteria is describing by words and it characterized by inaccuracy of the information. In such cases if situation depend on human factor then the theory of fuzzy logic is useful. Membership functions are used for representation fuzzy weights, which is able to describe the decisions maker's imprecision linguistic preferences. The basic advantage of used method is including the normalization process into optimization problem and transforming linguistic descriptions into fuzzy numbers. The proposed approach can be referred to different multi-criteria decision making situations, when it is necessary to replace it into a crisp weight.

Keywords: Multi-criteria decision making · Generalized trapezoidal fuzzy number · Linguistically described weights · Optimization problem

1 Introduction

Estimation the weights of the criteria is main problem in multi-attribute decision-making (MCDM). First of all it should be considered that the weights of criteria can critically affect the outcome of the decision-making process. And due to this fact paying attention to factors that clears objectivity for criteria weights are very important. In the past few decades linguistic data has been applied in the fuzzy decision making. Estimate of linguistically described weight of criteria is important problem in decision making models. In multi-attribute decision making models is very difficult to define accuracy weight of criteria and also, this problem exists in Multi-Objective Group Decision Making problems. Usually, in such case preferences of decision maker is describing by word or linguistic term. Data of the real life problems are described by fuzzy data, for instance, in dynamic programming to determine the path in fuzzy environment (in least cost route problem). This case hampers the calculation and to find solution or effective way [1]. Fuzzy set has been applied in the process of decision making characterized by uncertainty. But a problem is that subjectivity of information is not taken into consideration

© The Author(s), under exclusive license to Springer Nature Switzerland AG 2021
R. A. Aliev et al. (Eds.): WCIS 2020, AISC 1323, pp. 303–308, 2021.
https://doi.org/10.1007/978-3-030-68004-6_39

efficiently. A fuzzy set expresses an vague value of the weight, but in the real life problems requires a set the distinct value of weights. In this point of view to find crisp value of the weight via fuzzy data is the necessary problem.

In [2] new study to estimate criteria weights based on a linguistic evaluation is discussed. This method based on linear programming and Bellman's approach. This approach is applying to the construction of crisp criteria weights. Author of paper proposes method mainly related on optimization process, that defines to get the greatest membership degree for all weights.

In [5] discussed method based on Bayesian approach and it evaluates the weights of criteria in utility functions by requesting the DM to rank or choose the perfect alternative in a number of decision alternatives. The proposed approach evaluates criteria weights in different case studies, for example financial investment and university ranking decisions.

Modelling of Decision maker's preferences by using optimization model is discussed in [6]. Authors of the paper simulate real situation using Monte Carlo methods and the best alternatives of mobile phone tariffs have been obtained. They have found the weight vector that expresses of decision maker preferences and alternative obtained by using multi-criteria evaluation is the best for all.

Paper [7] is devoted to weighting Methods for Multi-Criteria Decision Making Technique in crisp environment and given Classification of weighting methods. Here also discussed to the various methods for determining criteria weights in MCDM by considering subjective weighting method, objective weighting method and integrated weights approach. These methods allows the decision maker to identify the difference between the subjective and objective weighting method. All these methods are based on normalization process.

Full Consistency Method (FUCOM) for solving multi-criteria problem is discussed in [8]. The model consists two groups of constraints, which satisfied the optimal values of weight coefficients. The first group is the condition that the relations of the weight coefficients of criteria should be equal to the comparative priorities of the criteria. Mathematical transitivity based conditions are the kernel of the second group.

In literature exist a lot of methods for weight estimation in multi-criteria problem [9–12].

Subjective weight determination is based on expert opinion, the subjective judgments. We are known subjective criteria weight determination is often time consuming especially when there is no agreement between decision makers of the problem under consideration. For instance, Analytical Hierarchy analysis is the subjective weighting method. In objective weight determination method decision maker do not include all factors, when define weight of criteria. Here exist inaccuracy evaluation if weights. In the objective weighting methods, criteria weights are derived from information gathered in each criterion through mathematical models. The integrated weighting approach is a new approach based on the subjective and objective weighting methods. In this work we will discuss subjective weighting approach.

The structure of the paper is as follows. Section 2 gives the definitions of fuzzy set, fuzzy number and linguistic variables. The statement of the problem is given in Sect. 3. In the next section the weight estimation model and numerical example are demonstrated. Last section consists of the conclusion.

2 Basic Definitions

Definition 1 [1]. A fuzzy set is completely characterized by the set of pair: $A = \{(x, \mu_A(x)), x \in X$. X is classical set of objects, which elements are denoted x and $A \in X$.

Definition 2. Fuzzy Number [3]. Fuzzy number is a fuzzy set A on R and its membership function satisfy the following properties:

1) $\tilde{A}$ convex fuzzy set i.e.
2) $\mu_{\tilde{A}}(\alpha x_1 + (1-\alpha)x_2) \geq \min(\mu_{\tilde{A}}(x_1), \mu_{\tilde{A}}(x_2)), \quad \forall x_1, x_2 \in R, \forall \alpha \in [0, 1]$
3) $\tilde{A}$ is normal set. For $\exists x_0 \in R$, $\mu_{\tilde{A}}(x_0) = 1$
4) $\mu_{\tilde{A}}$ is the piecewise continuous.

Definition 3. A fuzzy number $\tilde{A} = (a_1, a_2, a_3)$ is said to be a triangular fuzzy number if its membership function is given by

$$f(x, a_1, a_2, a_3) = \max\left(\min\left(\frac{x - a_1}{a_2 - a_1}, \frac{a_3 - x}{a_3 - a_2}\right), 0\right)$$

Definition 4. A Linguistic Variable [13–15]. This variable is defined using the set (u, T, X, G, M), where name of variable is expressed by u; the term-set denotes T; Syntactic rule is G generating linguistic terms; semantic rule (M) assigns to each linguistic term its meaning, which is a fuzzy set on X.

A meaning $t \in T$ generated using the syntactic rule G is defined a term. Each value of the linguistic term has a corresponding membership function. For example, interpretation of linguistic values like "approximately A", "less than A", "more than A", middle", "much", "high", "low", "near…", "from… to…" and so on; are membership functions. Here the values of variable are triangular fuzzy numbers, which is described as [21]: *neutral:* $(I, I + 2 * Z, I + 3 * Z)$, *more than A:* $(I, A + Z, S)$, *less than A:* (I, A, S), *much:* $(I, S - Z, S)$; *approximately A:* $(A - Z, A, A + Z)$. In these representations I and S are respectively minimum and maximum values of universe, $Z = (S - I) * 0.2$.

3 Statement of the Problem

Considering problem is real-life business problem and it consists of three alternatives and three criteria. A decision maker evaluates values of criteria linguistically. Our aim in this study to find crisp values of the weights on the criteria. There are three alternatives-A, B, C and three states such as No change, Economy declines, Economy expands that we take into consideration for this case. Depend on his own past experience, the investor assigns the following weights to each economic condition:

$\tilde{\omega}$ (economy declines) = (low)
$\tilde{\omega}$ (No change) = (middle)
$\tilde{\omega}$ (Economy expands) = (high)

Decision maker evaluates value of the criteria importance with three terms-high, low, medium. Then the set T is formulated as follows:

$$\mathrm{T} = \{\text{low, medium, high}\}$$

The problem is to determine weights of criteria over the portfolio selection problem.

4 The Weight Estimation Model

The weight estimation model is formulated as follows:

$$\begin{aligned}
&y = \alpha \rightarrow \max \\
&\frac{\omega_i + a_i^l - a_i}{a_i^l} \geq \alpha, \quad \mathrm{i} = 1, 2, \ldots, \mathrm{m} \\
&\frac{a_i^r + a_i - \omega_i}{a_i^r} \geq \alpha, \quad \mathrm{i} = 1, 2, \ldots, \mathrm{m} \\
&\sum_{\mathrm{i}=1}^{\mathrm{m}} \omega_i = 1 \\
&\omega_i \geq 0, \quad \mathrm{i} = 1, 2, \ldots, \mathrm{m} \\
&\alpha \in (0, 1)
\end{aligned}$$

where $\alpha = \min_{\mathrm{i}}\left[\mu_{\tilde{\omega}_i}(\omega_i)\right]$ is the membership degree of the final fuzzy decision, m-number of fuzzy goals $\tilde{\omega}_i = (a_i - a_i^l, a_i, a_i + a_i^r)$.

Where $a_i(\mathrm{i} = 1, 2, \ldots, \mathrm{m})$ is the weight of criterion relevance and a_i^l, a_i^r -tolerance zone for the weight value of the i-th criterion. The membership function of the weight as the fuzzy set $\tilde{\omega}_i$ has the following form:

$$\mu_{\tilde{\omega}_i}(\omega_i) = \begin{cases} 0, & \omega_\mathrm{i} < a_i - a_i^l \\ \frac{\omega_i + a_i^l - a_i}{a_i^l}, & a_i - a_i^l \leq \omega_i \leq a_i \\ \frac{a_i^r + a_i - \omega_i}{a_i^r}, & a_i \leq \omega_i \leq a_i + a_i^r \\ 0, & \omega_\mathrm{i} > a_i + a_i^r \end{cases}$$

4.1 Numerical Example

Linguistically estimated weights of the criteria by decision maker and its fuzzy representation are given in the Table 1.

Table 1. Decision maker's evaluation

Criterion	Relevance	Fuzzy number
Economy declines (c_1)	Low	(0.01, 0.1, 0.3)
No change (c_2)	Medium	(0.1, 0.3, 0.5
Economy expands (c_3)	High	(0.2, 0.6, 0.89)

The structure of the fuzzy number subjectively defined as follow:

$$\mu_L(\omega) = \begin{cases} 0, & \omega < 0.01 \\ \frac{\omega - 0.01}{0.09}, & 0.01 \le \omega \le 0.1 \\ \frac{0.3-\omega}{0.2}, & 0.1 \le \omega \le 0.3 \\ 0, & \omega > 0.3 \end{cases} \quad \mu_M(\omega) = \begin{cases} 0, & \omega < 0.1 \\ \frac{\omega - 0.1}{0.2}, & 0.1 \le \omega \le 0.3 \\ \frac{0.5-\omega}{0.2}, & 0.3 \le \omega \le 0.5 \\ 0, & \omega \le 0.5 \end{cases}$$

$$\mu_H(\omega) = \begin{cases} 0, & \omega < 0.2 \\ \frac{\omega - 0.2}{0.4}, & 0.2 \le \omega \le 0.6 \\ \frac{0.6-\omega}{0.29}, & 0.6 \le \omega \le 0.89 \\ 0, & \omega > 0.89 \end{cases}$$

Triangle type membership functions of the weights of criteria are given in Fig. 1.

By using above given membership function is solved the following linear programming problem:

$$y = \alpha \rightarrow \max$$
$$\frac{\omega_1 + a_1^l - a_1}{a_1^l} \ge \alpha, \quad \frac{\omega_2 + a_2^l - a_2}{a_2^l} \ge \alpha, \quad \frac{\omega_3 + a_3^l - a_3}{a_3^l} \ge \alpha,$$
$$\frac{a_1^r + a_1 - \omega_1}{a_1^r} \ge \alpha, \quad \frac{a_2^r + a_2 - \omega_2}{a_2^r} \ge \alpha, \quad \frac{a_3^r + a_3 - \omega_3}{a_3^r} \ge \alpha,$$
$$\omega_1 + \omega_2 + \omega_3 = 1, \ \omega_1 > 0, \ \omega_2 > 0, \ \omega_3 > 0, \omega_1 < 1, \ \omega_2 < 1, \ \omega_3 < 1,$$
$$\alpha > 0, \ \ \alpha < 1$$

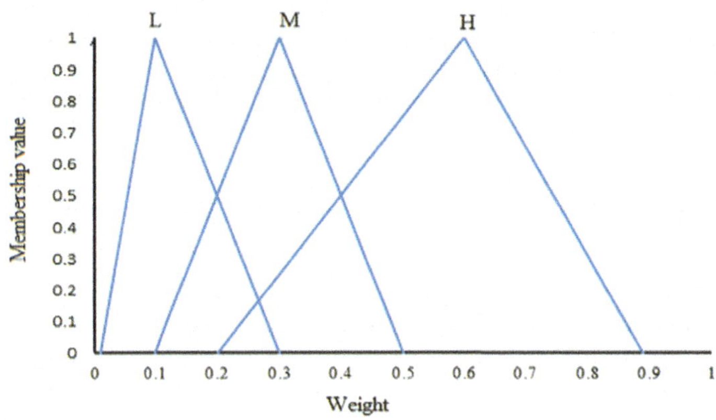

Fig. 1. Representation the weight of criteria by triangle type membership function

So the final result is determined as (value of objective function is 0.25): weight for c_1 is $\omega_1 = 0.25$, for c_2 is $\omega_1 = 0.45$, for c_3 is $\omega_3 = 0.3$

5 Conclusion

This paper introduces a linguistically described weights estimation method. Including normalization in the model is the main preference of the considered approach. This method can be used for defining crisp value of the weights in generalized fuzzy number and applied to a real decision making situation characterized by multiple-criteria.

Here Excel Table processor and Solver are used for all calculations. This model has possibility to define many versions set of weights, using condition of the weight. At last, numerical example is illustrated to process of estimation weight.

References

1. Aliev, R.A., Aliev, R.R.: Soft Computing and its Application. World Scientific, New Jersey (2001)
2. Adam, B.: New approach for estimation of criteria weights based on a linguistic evaluation. Expert Syst. Appl. **125**, 100–111 (2019). https://doi.org/10.1016/j.eswa.2019.01.069
3. Shamil, A.A.: Fuzzy dynamic programming based multi stage transportation problem. In: Advances in Intelligent Systems and Computing, vol. 1306 (2021, in press). https://doi.org/10.1007/978-3-030-64058-3
4. Bellman, R.E., Zadeh, L.A.: Decision-making in a fuzzy environment. Manage. Sci. **17**(4), 141–164 (1970). https://doi.org/10.1287/mnsc.17.4.B141
5. Yet, B., Tuncer, Ş.C.: Estimating criteria weight distributions in multiple criteria decision making: a Bayesian approach. Ann. Oper Res. **293,** 495–519 (2020). https://doi.org/10.1007/s10479-019-03313-z
6. Vladislav, C., Martina, K., Jana, S.: Estimation of weights in multi-criteria decision-making optimization models. https://mme2012.opf.slu.cz/proceedings/pdf/061_Chyna.pdf
7. Odu, G.O.: Weighting methods for multi-criteria decision making technique. J. Appl. Sci. Environ. Manage. **23**(8), 1449–1457(2019). https://doi.org/10.4314/jasem.v23i8.7
8. Dragan, P.; Zeljko, S.; Sinisa, S.: A new model for determining weight coefficient of criteria in MCDM models: full consistency method (FUCOM). Symmetry **10**(9), 1–22 (2018). https://doi.org/10.3390/sym10090393
9. Arvind, J., Janpriy, S.A.: Comprehensive literature review of MCDM techniques ELECTRE, PROMETHEE, VIKOR, and TOPSIS applications in business competitive environment. Int. J. Curr. Res. **10**(2), 65461–65477 (2018)
10. Mirko, S., Edmundas, K.Z., Dragan, P., Zeljko, S., Abbas, M.: Application of MCDM methods in sustainability engineering: a literature review 2008–2018. Symmetry **11**, 350 (2019)
11. Roszkowska, W.: Rank ordering criteria weighting methods-a comparative overview. Optimum Studia Ekonomiczne NR **5**(65), 14–33 (2013)
12. Vinodh, S., Prasanna, M., Praskash, N.H.: Integrated Fuzzy AHP-TOPSIS for selecting the best plastic recycling methods. A case study. Appl. Math. Model. **38**(19, 20), 4662–4672 (2014)
13. Zadeh, L.A.: The concept of a linguistic variable and its application to approximate reasoning–I. Inf. Sci. **8**(3), 199–249 (1975). https://doi.org/10.1016/0020-0255(75)90036-5
14. Zadeh, L.A.: The concept of a linguistic variable and its application to approximate reasoning – II. Inf. Sci. **8**(4), 301–357 (1975). https://doi.org/10.1016/0020-0255(75)90046-8
15. Zadeh, L.A.: The concept of a linguistic variable and its application to approximate reasoning –III. Inf. Sci. **9**(1), 43–80 (1975). https://doi.org/10.1016/0020-0255(75)90017-1

Modeling of the Technological Process of Ammonization of Soda Production Under Conditions of Changing Initial Data

B. T. Kaipbergenov(✉), B. A. Fayzullaev, and M. K. Elmuratov

Nukus Branch of Tashkent University of Information Technology, Nukus, Republic of Uzbekistan
batyrbek.kaipbergenov1970@gmail.com, fayzullaevbayram@gmail.com, m.elmuratov@gmail.com

Abstract. The regularities of the technological process of absorption were studied under conditions of changing initial data. Its features are analyzed in terms of chemical-technological processes of absorption of soda production. A mathematical model of the process based on a fuzzy neural network has been created. On the basis of a fuzzy neural network, graphs of the relative values of computational experiments by the Sub.Clustering algorithms and by the Grid partition algorithm are given.

Keywords: Chemical-technological process · Changing initial data · Neuro-fuzzy networks · Membership function · Absorption column · ANFIS

1 Introduction

In practice, the problem of constructing models for controlling technological processes, as a rule, is complicated by the multidimensionality, uncertainty and nonlinearity of the simulated objects and their inhomogeneous characteristics, the absence (complete or partial) of expert experience and analytical description of dependencies.

The linear control techniques often fail to satisfy the high standards in the control of the contemporary complex plants accounting for the emerging importance of the plant nonlinearity, model uncertainty, variables coupling and disturbances from the industrial environment [1].

The use of modeling and optimization on the technological process allows to increase the efficiency of various systems and mechanisms, contributes to an increase in labor productivity, gives a significant economic effect, and reduces the cost of energy and materials.

The current technical state of chemical-technological production is characterized by several features that are generated by competition problems and the increased requirements for soda production technologies have manifested themselves to such a large extent.

© The Author(s), under exclusive license to Springer Nature Switzerland AG 2021
R. A. Aliev et al. (Eds.): WCIS 2020, AISC 1323, pp. 309–315, 2021.
https://doi.org/10.1007/978-3-030-68004-6_40

In some studies, an adaptive neuro-fuzzy inference system optimized by the genetic algorithm (ANFIS-GA) and an optimized adaptive neuro-fuzzy inference system have been used to predict the critical bending load of steel columns. Particle swarm optimization (ANFIS-PSO) [2].

In [3], an intelligent fuzzy controller is presented, which is used to damp surges and improve the process in a distillation plant, simulated in MATLAB.

The proposed feedback fuzzy control system maintains the asymptotic stability of the system while at the same time increasing the system response time and reducing the effects of voltage surges to achieve better tower concentration control.

In the study [4], both research methods were performed to compare the optimized PID and FGS-PID with a conventional PID in two-wave conditions. The results show the superior performance of both proposed controls over the conventional PID regulator, providing improved power generation in regular and irregular waves.

The paper [5] presents the design of decentralized control systems for the carbonation column of the ammonia-soda production process. In the process, the behavior of the designed control structures is checked in the presence of changes in the setpoint and the action of typical disturbances.

This paper presents a real-time liquid level control implementation for an industrial process. The implementation of the level control is achieved using a fuzzy-based control algorithm [6].

In [7] is given a fuzzy PID self-tuning control method is used to accurately control the speed and steering of the stepping motor, and an Infinite impulse response digital filter is designed to reduce the interference from sea waves.

Some of the existing methods do not fully take into account the peculiarities of the operation of production devices, it requires a relatively large investment of time, a relatively large volume of measurements and a sufficiently high qualification of the service personnel.

The authors consider the problem to be relevant and, within the framework of this work, make an attempt to identify ways to further search for effective control methods and develop appropriate methods and controls.

2 Formulation of the Problem

Real-time automated control system is characterized by a cyclical solution of optimization problems with changing initial data. A dynamic linear programming problem is the problem of finding solutions of the following form:

$$\dot{X}^k = \arg\max\left\{F = {}^{Cx}\!/_{Ax} \leq (B^k)^T,\ x \geq 0\right\},\ k = 1, 2, \ldots; \tag{1}$$

where $C(c_1, \ldots, c_n)$-vector of efficiency factors; $B^k(b_1^k, .., b_m^k)$-vector of constraints of the k-th cycle; $A\left[a_{ij}\right]_{nm}$-cost matrix.

In real-time systems, vector changes from one cycle to another are usually caused by a slow change in the situation requiring a decision, rather large measurement errors. Significant deviations of the decisions of neighboring cycles $\Delta X^k = X^k - X^{k-1}$ do not correspond to the nature of the controlled processes and can lead to various kinds of losses. One of the ways to deal with errors is to replace task (1) with a new task:

$$\dot{X}^k = \arg\max \left\{ \begin{array}{l} F = {}^{Cx}\!/_{Ax} \le (B^k)^T, \\ \left| x - \dot{x}^{k-1} \right| \le E,\ x \ge 0 \end{array} \right\},\ k = 2, 3, ..., \tag{2}$$

where $E\{e_1, ..., e_n\}$ - solution tolerance vector $\dot{X}^k$ and $\dot{X}^{k-1}$. Thus, for k = 1, it is necessary to solve problem (1), and for k = 2, 3, . . ., problem (2), in which the solution of the previous cycle $\dot{X}^{k-1}$ acts as a parameter.

A rather detailed analysis of such processes was carried out in [8]. Thus, the identification problem consists in constructing models that can be represented in a sufficiently general way as follows:

$$\begin{cases} \dot{x}(t) = \dot{A}(u(t-\tau), \mu(t-\tau), \omega(t-\tau)), \\ \dot{q}(t) = \dot{A}(u(t-\tau), \mu(t-\tau), \omega(t-\tau), x(t)), \\ \dot{z}(t) = \dot{A}(u(t-\tau), \mu(t-\tau), \omega(t-\tau), x(t), q(t)) \end{cases} \tag{3}$$

where the various delays in the corresponding channels are denoted by τ for reasons of simplicity. Specific identification tasks will differ in different volumes of a priori information, types of processes, the presence of a delay in the object and communication channels [9].

As an example, consider the H-model of the chemical-technological process of absorption of soda production.

An effective tool for building blocks of technological monitoring and decision-making for control systems are neuro-fuzzy networks (NFN), which combine the advantages of artificial neural networks (the ability to learn and generalize), as well as fuzzy logic (the ability to transform a model into a system of rules like "If-then", convenient for analysis and human perception) [10, 11].

3 Principles of Constructing an NFN for Identifying a Technological Object with Heterogeneous Characteristics

The advantage of NFN is the presence of a trainable network as part of a fuzzy inference system, which independently creates a rule base for a fuzzifier, based on a training sample and specified parameters (type of membership function, number of hidden layers in the network, training epoch, etc.).

ANFIS (Adaptive network-based fuzzy inference system) is designated as follows:

$$R_i : Если\ x_i = A_{ij}\ то\ y_i = b_{i0} + \sum_{j=1}^{n} b_{ij} x_i,\ i = \overline{1, n}; \tag{4}$$

where x and y – fuzzy parameters (for example, flow rate, concentration, temperature, etc.) defined on the sets of real numbers; $b_{io}, b_{i1}, ..., b_{im}$-some real numbers; n is the number of parameters. To obtain differentiable input-output ratios, we will use Gaussian membership functions (MF):

$$\mu_{A_{ij}}(x_i) = \exp\left(-\frac{1}{2}\left(\frac{x_i - c_{ij}}{\sigma_{ij}}\right)^2\right),\ i = \overline{1, n},\ j = \overline{1, m},$$

where $\mu_{A_{ij}}(x_i)$ - MF of a fuzzy set A_{ij}; c_{ij} and σ_{ij} - membership function parameters - maximum coordinate and concentration coefficient. The choice of a Gaussian MF is due to its sufficient flexibility and simplicity - it is specified by only two parameters. This reduces the dimensionality of the optimization problem when tuning a fuzzy knowledge base.

In the NFN:

- the intersection of the rules is determined by the MF in the form of a product;
- Larsen's fuzzy implication is a fuzzy work;
- T - norm - fuzzy product;
- composition is not produced;
- defuzzification method - centroid method.

From (2) for m inference rules, the output variable after defuzzification takes the form

$$y = \frac{\sum_{i=1}^{n}\left(c_{j0} + \sum_{j=1}^{m} c_{ij}x_j\right)\prod_{j=1}^{m}\exp\left[-\left(\frac{x_j^l - a_{ij}}{b_{ij}}\right)^2\right]}{\sum_{i=1}^{n}\prod_{j=1}^{m}\exp\left[-\left(\frac{x_j^l - a_{ij}}{b_{ij}}\right)^2\right]}. \quad (5)$$

NFN using the Takagi-Sugeno algorithm includes five layers [2], and ANFIS contains two parametric layers (first and third). In the process of parametric adjustment of the NFN, it is required to minimize the identification error for all training parameters. When creating this model, the parameters of the absorption process of soda production were used. Input data were obtained on the basis of a real technological process of ammonization (A) from a pressure sensor of gaseous ammonia, at the outlet of the cooler from a PI-423 manometer, MPa; (B) weak ammoniated brine flow rate at the absorber inlet - from the FIC-401 flow meter, m^3/h; (C) gaseous ammonia consumption at the absorber inlet - from the flow meter FIC–402, m^3/h; (D) mother liquor consumption in the distillation section - from the flow meter FIC–315, m^3/h; (E) temperature of weak ammonized brine at the outlet of the cooler - from the temperature sensor TI–407, °C; (F) temperature at the outlet of the ammoniated gas cooler - from the temperature sensor TIC–431, °C. Output data, (Y) temperature of the upper part of the absorber, °C. A huge amount of data was used to train the network and test samples. In this regard, the initial data and the results of information processing are presented in the Table 1.

Table 1. Initial data for training the absorber model

A	B	C	D	E	F	Y
0,1843	58,1500	416.89	58,040	40,600	71,63	70,19
0,1655	58,0588	416,79	58,01	40,62	72,16	70,14
0,1962	58,0603	416,46	57,92	40,65	72,15	70,05
0,1692	59,9102	414,64	58,4	40,78	71,97	70,02
0,0038	60,3941	413,88	58,457	40,97	71,82	70,04
0,037	60,4866	407,06	58,552	41,11	71,72	70,05
0,002	61,9041	427,82	58,522	41,3	71,92	70,01
...	...	...	...	...	...	...
...	...	...	...	...	...	...
...	...	...	...	...	...	...
1,728	54,224	227,38	55,49	41,43	70,97	70,76
A	B	C	D	E	F	Y
1,436	54,679	227,77	55,459	41,41	71,15	70,86
1,401	54,686	227,88	55,357	41,38	71,69	70,69
1,271	54,514	227,9	55,498	41,4	72,21	70,52
1,355	54,425	228,08	55,651	41,4	72,02	70,49
1,519	58,073	228,26	57,794	41,57	71,45	70,72
1,571	58,377	228,35	58,323	41,76	70,94	70,9
...	...	...	...	...	...	...
...	...	...	...	...	...	...
...	...	...	...	...	...	...
3,166	54,42	232,08	55,371	41,15	72,21	67,35
3,29	54,328	225,52	55,504	41,19	71,93	67,07
3,286	54,511	225,86	55,582	41,23	71,27	67,2
2,964	54,3215	225,94	55,571	41,28	70,72	67,56
2,881	54,566	226,13	55,545	41,33	70,67	67,74
2,912	54,528	226,3	55,509	41,35	71,29	67,76
2,874	54,48	226,47	55,452	41,39	71,04	68,19

In the computational experiment (Fig. 1), 1200 samples were used to train the network and 795 samples for the test. The root mean square error is RMSE = 0.081. In the Fig. 2 shows the results in a graphical form of identifying the temperature of the ammoniated brine at the outlet of the absorption column (comparatively based on the developed NFN and the real process).

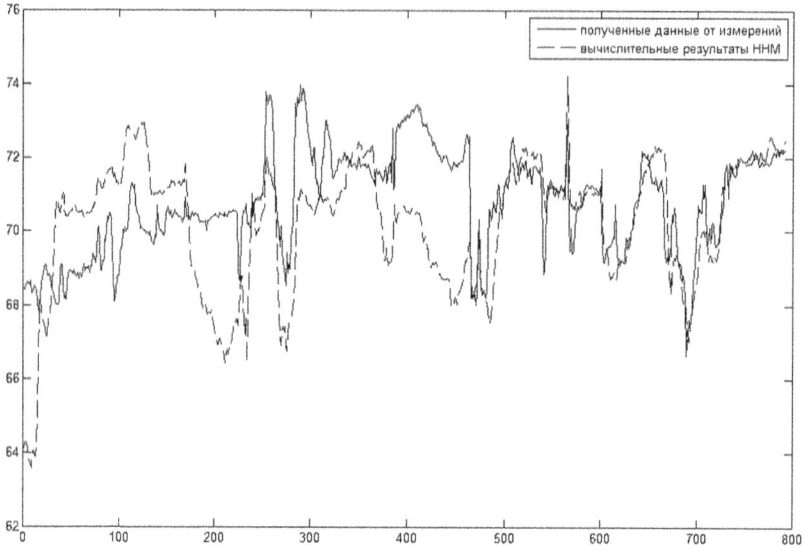

Fig. 1. Identification of the temperature of the upper part of the absorption column (comparatively based on the developed NFN using the Sub.Clustering algorithm and the real process).

In the third computational experiment, 1200 was used to train the network, 795 samples and 3 epochs for the test. The root mean square error was RMSE = 0.072.

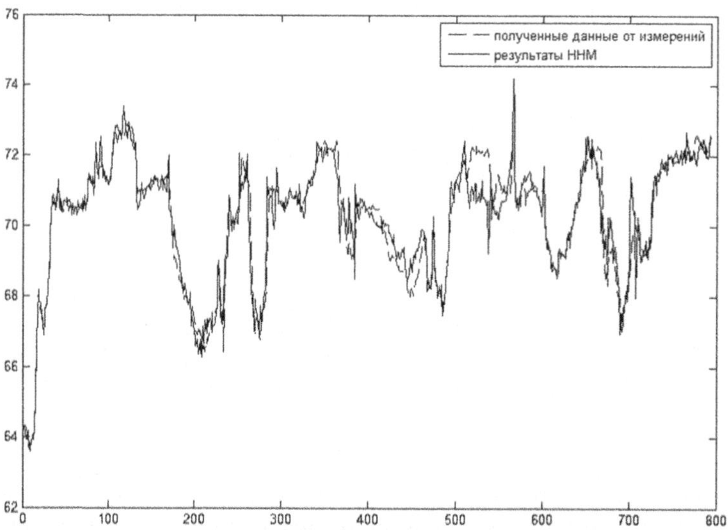

Fig. 2. Identification of the temperature of the ammoniated brine at the outlet of the absorption column (comparatively based on the developed NFN and the real process).

4 Conclusions

The fuzzy neural model ANFIS was chosen as the basis, which adequately proved itself to be simple and flexible in solving such problems.

This model is fundamentally different from the generally accepted ones, first of all, in that they take into account all available variables and the relationships between them in a situation where the discreteness of the control of the latter differ significantly, as well as the levels of a priori information about various channels of the process under study.

References

1. Yordanova, S., Slavov, M., Prolopiev, G.: Disturbance compensation in fuzzy logic control of level in carbonisation column for soda production. WSEAS Trans. Syst. Control **15**, 64–72 (2020). https://doi.org/10.37394/23203.2020.15.8
2. Lu, M.L., Ly, H.-B., Pham, B.T., Le, V.M., Pham, T.A., Nguyen, D.-H., Tran, X.-T., Le, T.-T.: Hybrid artificial intelligence approaches for predicting buckling damage of steel columns under axial compression. Materials **12**(10) (2019). Switzerland. https://doi.org/10.3390/ma12101670
3. Mahmoud, M.: Damping of surge effects and enhancement of dynamic response of distillation plant using MIMO fuzzy control. IFAC Pap. OnLine **51**(30), 15–20 (2018). https://doi.org/10.1016/J.IFACOL.2018.11.237
4. M'zoughi, F., Garrido, I., Bouallegue, S., Ayadi, M, Garrido, A.J.: Intelligent airflow controls for a stalling-free operation of an oscillating water column-based wave power generation plant. Electronics **8**(1), 1–26 (2019). https://doi.org/10.3390/electronics8010070. www.mdpi.com/journal/electronics
5. Cristea, V-M. Control appoarches of the carbonation column for soda manufacturing. Studia UBB Chemia, LXII, **4**, Tom II, 221–229 (2017). https://doi.org/10.24193/subbchem.2017.4.19
6. Aydogmus, Z.: A real-time robust fuzzy-based level control using programmable logic controller. Elektronika ir elektrotechnika **21**(1), 13–17 (2015)
7. Wei, Y., Wu, Y., Zhang, X., Ren, J., An, D.: Fuzzy self-tuning PID-based intelligent control of an anti-wave buoy data acquisition control system. IEEE Access **7** (2019)
8. Ivanov, K.K.: Upravlenie v sistemakh realnogo vremeni. Molodoy ucheniy **20**, 39–141 (2017)
9. Ionov, M.V., Krasnyanskiy, M.N.: Avtotatizirovanniy sistemy tekhnicheskiy технической диагностики химико-технологического oborudovaniya. Voprosy sovremennoy nauki i praktiki **2**(40), 66–73 (2012)
10. Juraev, Z.S., Muhamediyeva, D.T., Sotvoldiev, D.M.: Construction of hybrid intellectual monitoring and decision-making systems. J. Phys. Conf. Ser. **1546**, 012083 (2020)
11. www.mathworks.com/help/fuzzy/building-systems-with-fuzzy-logic-toolbox-software

Analyisis of the Influence of the Shape of the Matrix on the Compaction of Porous Powder Bodies of "Iron-Cast Iron-Glass" Under Imprecise Information

T. G. Jabbarov(✉)

Department of Mechanical and Materials Science Engineering, Azerbaijan State Oil and Industry University, 20 Azadliq Avenue, 1010 Baku, Azerbaijan
tahir196041@gmail.com

Abstract. The article discusses the processes of compaction of porous powder bodies of iron-cast iron-glass in rigid conical and cylindrical matrices using theory, the flow of material.

The processes of pressing and extrusion of a sintered cylindrical porous preform through a conical matrix and two-sided compaction of sintered porous preforms in closed cylindrical matrices are investigated. To solve the problem of the transient process of pressing a sintered cylindrical billet, a conical matrix is used. The zones of large and small deformations are determined by the distributions of porosity. The stress-strain state was assessed during bilateral compaction of sintered porous bodies made of "iron-cast-iron glass".

It has been found that even with excessively high compaction pressures, it is impossible to achieve a non-porous state of the workpiece. This is due to the elastic state of the sintered porous body of the type "iron-cast-iron-glass". In order to account for imprecision of information related to the considered process, we propose to use the fuzzy set theory.

Keywords: Conical matrix · Compaction · Finite element method · Cylindrical billet · Variable cross-section · Extrusion · Strain increments

1 Introduction

In the practice of powder metallurgy of iron-cast-iron-glass blanks, extrusion and multiple pressing with intermediate sintering can be widely used. They make it possible to obtain sintered "iron-cast iron-glass" products from them with an almost low porous structure and high strength properties.

In this regard, of particular interest is the study of the stress-strain state of porous cylindrical billets during extrusion and pressing them in rigid cylindrical matrices. In [1], a methodology for solving problems of pressing porous products, such as bushings, using the dependences of the axial and lateral pressures on the porosity of the sintered billet, is presented. It is shown that in this case there is no need to study the material

© The Author(s), under exclusive license to Springer Nature Switzerland AG 2021
R. A. Aliev et al. (Eds.): WCIS 2020, AISC 1323, pp. 316–321, 2021.
https://doi.org/10.1007/978-3-030-68004-6_41

loading surface. As an example, various schemes for pressing bushings with a large ratio of height and wall thickness are considered. The obtained universal dependences make it possible to calculate the distribution of average porosity over the height of the workpiece for all the above pressing schemes and establish the most optimal ones.

Based on a number of assumptions, including the invariance of porosity along the radius, the stress-strain state of the workpiece is examined and a change in porosity in height is found. However, the determination of porosity both in the radial direction and in the height of the workpiece, as well as the establishment of more accurate ratios of the components of its stress-strain state, is of great importance. The considered process is characterized by imprecision of information. In view of this, we propose to use the fuzzy set theory [3] for modeling.

2 Statement of the Problem

For this purpose, we use the finite element method, which is used [2] to solve important problems in the field of the theory of plasticity of solid and porous bodies. The main dependences of the finite element method for porous preforms are given in.

To analyze the compaction of porous iron-glass-glass blanks in rigid conical and cylindrical matrices, we use the theory of plasticity of porous bodies (flow theory). In order to account for imprecision of information related to the variables of interest, we use the fuzzy set theory. The fuzzy-valued stress column vector $\{\sigma\}$ is related to the strain increment column vector $\{d\varepsilon\}$ by the dependence

$$\{\sigma\} = [D]\{d\varepsilon\}, \tag{1}$$

where the matrix [D] has the form:

$$[D] = \frac{E(1-c)}{(1+c)(1-2c)} \begin{bmatrix} 1 & \frac{c}{1-c} & \frac{c}{1-c} & 0 \\ \frac{c}{1-c} & 1 & \frac{c}{1-c} & 0 \\ \frac{c}{1-c} & \frac{c}{1-c} & 1 & 0 \\ 0 & 0 & 0 & \frac{(1+c)(1-2c)(1+\alpha^m)}{3(1-c)} \end{bmatrix}; \tag{2}$$

$$E = \frac{\beta^{3n}\sigma_{ekv}}{\overline{d}\varepsilon_{ekv}};\ c = \frac{\mu - \alpha^m}{1+\alpha^m};$$

where σ_{ekv} - equivalent voltage; $\overline{d}\varepsilon_{ekv}$ - equivalent increment of plastic deformations; μ is the transverse strain coefficient; *m*, *n* are constants for a given material; α, β are the fuzzy-valued porosity functions (υ) of this material, determined from the expressions:

$$\alpha = \frac{1}{4}\left[\frac{3(1-\upsilon)^{1/3}}{(3-2\upsilon^{1/4})\ln\upsilon}\right]^2; \quad \beta = \left[\frac{3(1-\upsilon)^{1/3}}{(3-2\upsilon^{1/4})}\right]^2. \tag{3}$$

The increase in fuzzy-valued porosity is calculated by the formula:

$$d\upsilon = \frac{q\alpha^m(1-\upsilon)\overline{d}_{ekv}\cdot\sigma_0}{\beta^{3n}\sigma_{ekv}}; \tag{4}$$

where σ_0 is the average normal voltage.

The formed asymmetric body is divided into ring elements with triangular cross sections. Assume that external and internal forces are applied to the nodes of the elements. The remaining equations of the finite element method are classical.

The relationships between the increments of plastic deformations of elements and the displacements of its nodes {dq}, as well as the changes in external forces {dp} and nodal displacements of the finite element mesh {dρ} are expressed by matrix equations:

$$\{d\varepsilon\} = [B]\{dq\}; \tag{5}$$

$$\{\mathrm{dp}\} = [\mathrm{K}]\{\mathrm{dr}\}; \tag{6}$$

where [B] is the fuzzy-valued matrix determined by the approximation of displacements in the volume of the selected finite element; [K] is the general stiffness matrix, an arbitrary element of which is found by summing the corresponding stiffness submatrices of all elements adjacent to neighboring nodes S and 1.

$$|K_{S1}| = \sum [R_{S1}].$$

The fuzzy-valued stiffness matrix of the element has the form:

$$[K] = 2\pi[\overline{B}]^T[D][\overline{B}]\overline{r}\Delta, \tag{7}$$

where r is the radius of the center of gravity of the element relative to which the matrix is defined $[\overline{B}]$; Δ is the cross-sectional area of the finite element; $[\overline{B}]^T$ - transposed matrix.

The diagram of material deformation is approximated by the dependence

$$\sigma_{ekv} = A + B\left(S\overline{d}\varepsilon_{ekv}\right)^b \tag{8}$$

Given a column vector of increments of nodal conditions {dp} and solving a system of nonlinear algebraic Eqs. (6) with respect to increments of nodal displacements of a finite element grid {ρ}, we determine the stress state components {σ by formulas (1), (5) and (4)}, increments of plastic strains { dε} and porosity increment dυ. Then, based on this, we find the equivalent stress and increment of plastic deformations.

The calculation is carried out by successive low loads using the method of variable elastic parameters. At the first step, for all elements, the quantity is considered $\overline{d}\varepsilon_{ekv}$ known and small.

In the preparation of Eqs. (1), (2) and (7), we assume that the entire sample undergoes plastic deformation. Moreover, only the loading process occurs. When taking into account unloading and solving elastic-plastic problems, it is necessary to compile an elastic-plastic stress matrix and perform calculations.

3 Solution of the Problem

As an example, we consider three practically important cases of compaction (loading process) of porous cylindrical billets. Calculations are performed using the «Matlab» program.

Pressing a cylindrical porous preform through a conical matrix is considered as the initial period of an unsteady process. This means that the workpiece has not yet come into contact with the container, that there are points of the lateral cylindrical surface, apart from those lying on the final surface of the matrix and the contact surface with Poisson, they freely moved.

By virtue of symmetry, we consider half the axial section, which was divided into 110 uniform triangular elements with 70 nodes. The movements W, U along the axial Z and radial r directions are free, except for the nodes: a) on the axis of the workpiece AF, $U_{A\Gamma} = 0$; on the surface of the workpiece bounded by the radius AB, $W_{AB} = -a$, $U_{AB} = 0$; b) the node E and other nodes of the outer cylindrical surface, which after deformation of the workpiece will be on the conical surface of the matrix

$$U_{\kappa} = W_{\kappa} tg\gamma; \tag{9}$$

where γ is the angle between the axis of the conical matrix and its generatrix.

Compressing external forces P_{ZAB} are applied to the nodes of the surface bounded by the radius *AB*, and to the nodes of the workpiece located on the conical surface of the matrix

$$P_{rk} = \frac{ftg\gamma - 1}{tg\gamma + 1} \cdot P_{zk}, \tag{10}$$

where *f* is the coefficient of friction.

Taking into account dependences (9) and (10), the systems of algebraic Eqs. (6) were solved. Moreover, we transfer the unknown components of P_{zk} to the left side of the corresponding equations. The components of the external P_{ZAB} forces are modified according to the classical procedure. After solving system (6), the remaining unknown components of displacements U_k and external forces P_{rk} are determined by formulas (9) and (10).

Material: billets-sintered "iron-cast iron-glass" (charge content - iron powder ПЖ2М3-47%, gray cast iron powder – 47%, vacuum glass powder - 6% by weight), ($n = 0{,}3$; $\mu = 0{,}4$; $\upsilon = 0{,}15$; $A = 195$ MPa; $B = 600$ MPa; $b = 0.3$; $f = 0.06$). We assume that $\gamma = 30°$, $h/d_1 = 1{,}0$, where *h* is the height, d_1 is the diameter of the workpiece.

According to the distribution of porosity, zones of large I and small II deformations are determined. In zone I, the average density υ_{medium} was 0.15. The highest degree of compaction was obtained by elements "a" of zone I ($\upsilon_a = 0{,}08$), in zone II, the porosity changed very little. The components of the stress-strain state of the workpiece were appropriately distributed. Their greatest values were achieved in zone I, with the dominant radial σ_r and circumferential σ_0 stresses. The tangential stresses in zone I were tens of times higher than the corresponding stresses of zone II. Thus, zone I was subjected to large local plastic deformations (65%), while the maximum deformations of zone II did not exceed 1percent. This confirms the assumption made above about the free movement of points on the lateral cylindrical surface.

Pressing a porous preform of variable cross section through a conical matrix. In this case, the boundary conditions differed from the above in that there were no nodal forces and nodal displacements on the surfaces bounded by the DEF line were free. We studied a blank of "iron-cast-iron-glass" of variable cross-section $\gamma = 35°$, $d_1/h = 1.15$, $d_1/d_2 = 2.30$, where d_2 is the diameter of the output part of the blank.

The axial section of a deformed billet of variable cross section at $\int d\bar{\varepsilon}_{ekv}^{max} = 7\%$ is shown in Fig. 2. In the elements of zone I, the porosity is most intense. The average porosity in zone I was $\upsilon_{med.I} = 0.155$; the elements of zone *a* I were characterized by the highest degree of compaction ($\upsilon_a = 0.144$).

In zones II and III, $\upsilon_{med.II} = 0.159$ ($\upsilon_b = 0.151$), $\upsilon_{med.III} = 0.166$. In zone IV, the porosity increased ($\upsilon_{med.IV} = 0.172$). The greatest increase in porosity was achieved in elements on the axis of the sample ($\upsilon_c = 0.175$). As calculations show, the axial stress σ_Z along the workpiece height decreased. Surrounded and radial stresses at the output of the conical matrix reached a maximum value.

With an increase in the degree of reduction, the components of the stress-strain state of the workpiece of variable cross section also increased. When $\int d\bar{\varepsilon}_{ekv}^{max} = 15,1\%$ the porosity of the elements and zone I was 0.12, and the porosity of the elements in zone II was 0.13.

The developed technique allows us to study the stress-strain state, both when drawing a porous briquette, and when pressing through a conical matrix of a bimetallic porous preform (for example, placed in a metal shell). For pressing bimetal, as the material of the porous preform (zone V), powdered iron was used, hardened by dispersed inclusions of gray cast iron and vacuum glass ($A = 450$ MPa, $B = 0$, $\upsilon = 0.35$), the shell case was rolled steel grade 3kp ($A = 250$ MPa, $B = 0$, $\upsilon = 0.05$).

Calculations showed that in this case, compared with pressing without a shell, the values of the components of the stress-strain state of the outer layer (shell) increased. The elements of the lower cylindrical part of the shell located at the exit of the matrix were subjected to relatively large plastic deformations. The porosity of the porous preform did not increase.

Bilateral compaction of porous preforms in closed cylindrical matrices by virtue of symmetry considered a quarter of the axial section. In this case, the following differences in the boundary conditions were observed: there were no radial displacements of the nodes of the outer cylindrical surface $U_{BE} = 0$; nodal forces on this surface were related

$$P_{ZBE} = -fP_{ZBE};$$

At nodes of the middle section of the workpiece, limited by the radius EF, $W_{EF} = 0$.

We studied the stress-strain state of the three dimensions of the workpiece.

$h/d_1 = 1.0$; 2.0 and 3.0. Sintered "iron-cast-iron-glass" was used as the material.

Studies have shown that the porosity in the sample was distributed very unevenly. Figure 4 shows the zones of large I and small II porosity ($\upsilon_{med.I} = 0.15$; ($\upsilon_{med.II} = 0.075$) and plots of changes in the components of the stress-strain state of the workpiece at $h/d_1 = 2.0$; $\upsilon_0 = 0.30$ and the maximum plastic deformation of the most loaded element of zone I, equal to 8.9%.

The maximum values they reach on the cylindrical surface of the workpiece. With increasing h/d_1, the non-uniformity of porosity along the z axis increases, along r decreases.

Two-sided compaction of porous bimetallic cylindrical billets in closed cylindrical matrices shows that "plastic" elements of bimetal are subjected to relatively large deformations (and, consequently, compaction). Moreover, a solid metal layer is embedded inside the "plastic", which is confirmed by experiments.

4 Conclusions

A solution to the problem of the unsteady process of pressing a cylindrical porous billet through a conical matrix of the finite element method, as well as pressing a porous and bimetallic billet of variable cross section through a conical matrix of the finite element method, is proposed. The zones of large and small deformations are determined by the distributions of porosity. We proposed to use the fuzzy set theory to account for imprecision of information related to the main variables.

The stress-strain state in bilateral compaction of porous preforms in rigid cylindrical matrices by the finite element method. Cases of compaction of bimetal blanks are also considered.

References

1. German, R.M.: Particle Packing Characteristics. Metal Powder Industries Federation, Princeton (1989)
2. Yamada, Y., Yoshimura, N., Sakurai, T.: Plastic stress-strain matrix and its application for the solution of elastic-plastic problems by the finite element method. Int. J. Mech. Sci. **10**, 343–354 (1968). https://doi.org/10.1016/0020-7403(68)90001-5
3. Zadeh, L.A.: Fuzzy sets. Inf. Control **8**, 338–353 (1965). https://doi.org/10.1016/S0019-9958(65)90241-X

Regular Algorithms for the Parametric Estimation of the Uncertain Object Control

H. Z. Igamberdiev(✉) and U. F. Mamirov

Tashkent State Technical University, Tashkent, Uzbekistan
ihz.tstu@gmail.com, uktammamirov@gmail.com

Abstract. The article presents the regularized algorithms for identification of uncertain dynamic management objects. Various algorithms for identifying dynamic control objects are analyzed in the conditions of approximate assignment of source data. Regularization, moderate spoilage, and imaginary shift methods for extended systems are presented. The results of solving various model problems for different error levels of the initial data are shown. It is shown that the best indicators are provided by the simplified regularization method for an extended system of equations, since this extended system is by definition symmetric and positive definite.

Keywords: Undefined objects · Identification · Extended systems · Regularized algorithms

1 Introduction

In the theory and practice of automation, there are very often problems of managing indeterminate objects [1–7], i.e. objects whose mathematical model at the stage of developing a control system is set up to certain unknown components– constant parameters, variable coefficients, static nonlinearities, etc. To date, the most developed methods are for controlling parametrically indeterminate objects, i.e. objects whose mathematical model uncertainty is reduced to a set of unknown constant parameters [2, 4, 5]. However, under conditions of approximate assignment of the source data, well-known algorithms lose their effectiveness to a certain extent.

When solving problems of synthesis of adaptive and adaptive robust control systems, it becomes necessary to find the desired solutions based on methods for minimizing functionals or matrix equations of the form:

$$x^* = \arg\min Q(x), \quad x \in X,$$

$$Au = f, \quad A : U \to F \tag{1}$$

where $A = (a_{ij}) \in R^{m\times n}, \ f = (f_1, \ldots, f_m)^T \in R^m, \ u = (u_1, \ldots, u_n)^T \in R^n$.

These equations can be solved either by direct variational methods for minimizing functionals [8, 9], or using the ideas of non-variational methods [8–12].

© The Author(s), under exclusive license to Springer Nature Switzerland AG 2021
R. A. Aliev et al. (Eds.): WCIS 2020, AISC 1323, pp. 322–328, 2021.
https://doi.org/10.1007/978-3-030-68004-6_42

Although direct minimization methods are very versatile in their idea, however, when they are used, we have to deal with a number of difficult-to-formalize procedures related primarily to substantiating the model's consistency and quantitative criteria for selecting a solution; identifying and constructively describing a compact containing the desired solution; ensuring convergence of the minimizing sequence, etc. [8–13]. In this respect, non-variational methods of solution are usually more specific and, accordingly, applicable to a narrower range of problems, but due to their specificity and taking into account the deeper properties of the problem, they are usually more convenient to implement [8–11].

The operator approach to the problem of synthesis of indeterminate objects based on the use of concepts of non-variational methods for solving extreme problems determines another requirement for the procedure for constructing models and synthesizing the controller: it should be closely related to the methodology for solving inverse problems in mathematical physics and analysis [8–12]. Therefore, there is a natural tendency to analyze the features of the problems of synthesis of adaptive control systems for indeterminate objects from the standpoint of methods of the theory of incorrectly set problems.

Analysis of the state of work in this area [2, 5–7, 14–17] shows that the greatest success here is primarily associated with the development of algorithms for solving problems of parametric identification and adaptive management of indeterminate objects based on the principle of the reference model and OLS. The problems of regular parametric identification and synthesis of the regulator under conditions of limited a priori information about the desired solution are studied to a lesser extent [6–10]. It is also necessary to note the paucity of available publications on the choice of a class of regularization methods for implementing specific recurrent algorithms for identification and synthesis of control actions. The above brief analysis of the state and trends in the development of the theory of adaptive control systems for indefinite objects and their applications allows us to conclude that despite significant results achieved in the field of methods for the synthesis of adaptive control systems for indefinite objects, regularized algorithms for the synthesis of adaptive control systems for indefinite objects are currently insufficiently developed.

2 Problem Definition

As is known [9, 12], the solution of system (1) from approximate data $d = \{A, f\}$ is an incorrectly set Hadamard problem, since the approximate normal pseudo-solution

$$u_* = A^+ f$$

unstable to infinitesimal perturbations of the original data.

We take the approximation conditions of Eq. (1) as

$$\|A - \bar{A}\| \leq h, \quad \|f - \bar{f}\| \leq \delta, \tag{2}$$

where $\bar{A}$ and $\bar{f}$ – the exact values of the matrix operator and the right side of Eq. (1), respectively.

It is known [10, 11] that the normal pseudo-solution $u_* = A^+ f$ is a normal solution to the normal system of equations

$$A^T Au = A^T f \tag{3}$$

or $A^T r = 0$, where $r = f - Au$.

When solving Eq. (3), the regularization method of A.N. Tikhonov is usually used [10]:

$$(A^T A + \alpha I)^{-1} u = A^T f, \tag{4}$$

where $\alpha > 0$ – regularization parameter, I – is the identity operator.

The use of the regularization method A.N. Tikhonov leads to the system $\alpha u + A^T Au = A^T f$. Here it should be noted that in addition to the well-known advantages of this method, there are serious drawbacks that, in a certain sense, complicate the original ill-posed problem $Au = f$. First, the already poor conditionality of the A matrix is squared during the transition to $A^T A$. As is known [13, 18, 19], the accuracy of the solution when using the method of normal equations is determined by the condition number $\kappa_2(A^T A) = \kappa_2^2(A)$. Thus, in the case under consideration, the error is always limited to a value of type $\kappa_2^2(A)\varepsilon_{mach}$, and not just $\kappa_2(A)\varepsilon_{mach}$. Expect be so that when solving the normal equations may be lost twice true bits than in other methods. Secondly, the characteristic features of changing the data of f are greatly smoothed out as a result of applying the A^T operator. Thirdly, the addition of a small parameter α, although it improves the properties of the operator to be addressed, but at the same time changes the original problem and creates a new problem of choosing the appropriate α.

3 Solution of the Task

It can be shown [13, 18] that the normal system of Eq. (3) is equivalent to the extended system of equations

$$Gz = b, \tag{5}$$

where

$$G = \left[\begin{array}{c|c} I_m & A \\ \hline A^T & 0 \end{array}\right], \quad z = \left[\begin{array}{c} r \\ \hline u \end{array}\right], \quad b = \left[\begin{array}{c} f \\ \hline 0 \end{array}\right],$$

$$z = (r^T, u^T)^T \in R^{m+n}.$$

Extended system (5) for any initial data $d = \{A, f\}$ is compatible and the normal solution to the extended system is defined as $z_* = G^+ b = (r_*^T, u_*^T)^T$, where $r_* = f - Au_*$.

Perturbations in the approximate system $\{G, b\}$ satisfy the inequalities

$$\left\|G - \overline{G}\right\| = \left\|\begin{array}{c|c} 0 & A - \overline{A} \\ \hline A^T - \overline{A}^T & 0 \end{array}\right\| = \left\|A - \overline{A}\right\| \le h,$$

$$\|b-\bar{b}\| = \|f-\bar{f}\| \le \delta.$$

At the same time, an accurate extended system

$$\left[\begin{array}{c:c} I_m & A \\ \hdashline A^T & 0 \end{array}\right]\left[\begin{array}{c} r \\ \hdashline u \end{array}\right] = \left[\begin{array}{c} f \\ \hdashline 0 \end{array}\right] \Leftrightarrow Gz = b$$

always joint [13, 14, 18] and

$$\bar{\mu} = \inf_{z \in R^{m+n}} \|Gz - b\| = \|Gz_* - b\| = 0,$$

where $z_* = G^+ f = \left[\begin{array}{c} f - Au_* \\ \hdashline u_* \end{array}\right]$.

In (5), the matrix G is symmetric, we will use the method of M.M. Lavrentiev [10, 11] to regularize the solution of Eq. (5).

First, we consider the case when the matrix operator G in (5) is symmetric and positive. Then the method of M.M. Lavrentiev leads to an equation of the form [6, 20]:

$$\alpha z + Gz = b, \ \alpha > 0 \tag{6}$$

The solution of Eq. (6) can be expressed in the form

$$z_{(\alpha)} = (\alpha I + G)^{-1} b = \theta_\alpha(G) b, \tag{7}$$

where

$$\theta_\alpha(\lambda) = (\alpha + \lambda)^{-1} \tag{8}$$

– is the generating system of functions, $0 \le \lambda < \infty$ is the spectral parameter.

Based on [13], it can be shown that at $\alpha = h$ the error of deviation is

$$\|z_\alpha - z_*\| = O(h + \delta),$$

where z_α – solution (7).

If G – is a self-adjoint alternating operator, then it is advisable to use the following modification of the method of M.M. Lavrentiev [10, 11, 21], namely, the approximate solution $z_{(\alpha)}$ of Eq. (7) can be determined from the equation:

$$i\alpha z + Gz = b, \quad \alpha > 0, \quad i = \sqrt{-1}. \tag{9}$$

Instead of (9), one can also use an expression of the form

$$z_{(\alpha)} = \mathrm{Re} v_\alpha, \quad v_\alpha = (i\alpha I + G)^{-1} b,$$

or a special computational scheme of A.N. Tikhonov's method [10, 13]:

$$\alpha^2 z_{(\alpha)} + G^2 z_{(\alpha)} = Gb.$$

Taking into account the symmetry of the matrix G, it is possible to reduce the number of conditionality of the considered problems. For this purpose, it is advisable to use the method of imaginary spectrum shift [13, 21–23].

In accordance with this method, instead of Eq. (7), we consider the equation

$$(G + i\sqrt{\alpha}I)z = b, \tag{10}$$

where $i = \sqrt{-1}$ – imaginary unit.

Let $z = x + iy$. Then (10) can be written as an extended real system of linear algebraic equations (SLAE):

$$\left[\begin{array}{c|c} G & -\sqrt{\alpha}I \\ \hline \sqrt{\alpha}I & G \end{array}\right]\left[\begin{array}{c} x \\ \hline y \end{array}\right] = \left[\begin{array}{c} b \\ \hline 0 \end{array}\right]. \tag{11}$$

It follows directly from (11) that $x = \mathrm{Re}z$ is a solution of the equation according to the computational scheme of A.N. Tikhonov

$$\left(G^2 + \alpha I\right)x = Gb, \tag{12}$$

since $G = G^T$ here.

The above means that if z_* – is solution (10), and z_α – is solution (12), then $z_\alpha = \mathrm{Re}z_*$. Thus, the problem of solving SLAE (12) can be replaced by the problem of solving SLAE (10) with an imaginary spectrum shift.

The computational complexity of the solution of SLAE (10) is not much higher than the complexity of the solution of SLAE (12). However, as will be shown, the condition number of SLAE (10) is significantly less than the condition number of SLAE (12).

4 Numerical Examples for Comparing the Given Regular Algorithms

We take the vector

$$z_t = [-1.1201\ 2.5260\ 1.6555\ 0.3075\ \ -1.2571\ \ -0.8655\ \ -0.1765]^T$$

as the exact solution z_t of the system of equations and collapse it according to formula $u = Az_t$ with the matrix operator:

$$A = \begin{bmatrix} 1.0000 & 0.5000 & 0.3333 & 0.2500 & 0.2000 & 0.1667 & 0.1429 \\ 0.5000 & 0.3333 & 0.2500 & 0.2000 & 0.1667 & 0.1429 & 0.1250 \\ 0.3333 & 0.2500 & 0.2000 & 0.1667 & 0.1429 & 0.1250 & 0.1111 \\ 0.2500 & 0.2000 & 0.1667 & 0.1429 & 0.1250 & 0.1111 & 0.1000 \\ 0.2000 & 0.1667 & 0.1429 & 0.1250 & 0.1111 & 0.1000 & 0.0909 \\ 0.1667 & 0.1429 & 0.1250 & 0.1111 & 0.1000 & 0.0909 & 0.0833 \\ 0.1429 & 0.1250 & 0.1111 & 0.1000 & 0.0909 & 0.0833 & 0.0769 \\ 0.4173 & 0.9448 & 0.3377 & 0.1112 & 0.2717 & 0.1320 & 0.5752 \\ 0.0497 & 0.4909 & 0.9001 & 0.7803 & 0.4039 & 0.9421 & 0.0598 \\ 0.9027 & 0.4893 & 0.3692 & 0.3897 & 0.0965 & 0.9561 & 0.2348 \end{bmatrix};$$

As a result, the right-hand side u of the system of equations will take the form:

$$u = [0.3506\ \ 0.4019\ \ 0.3331\ \ 0.2741\ \ 0.2298\ \ 0.1963\ \ 0.1704\ \ 1.9928\ \ 1.5807\ \ -0.3344]^T.$$

Further, simulating the process of occurrence of errors on the right-hand side of the system of equations, we introduce disturbances according to formulas $\tilde{u} = u(1 + 10^{-2} \cdot \nu\theta_i)$ into the values of u, where θ_i are normally distributed random numbers from the interval $(-1,\ 1)$. The relative errors of the values of $\tilde{u}$ compared to u were taken to be 1, 3, and 5%, i.e. $\nu = 1(2)5$.

The resulting systems of equations with a perturbed right-hand side for $\nu = 1(2)5$ were solved by the ordinary method of A.N. Tikhonov regularization (4), as well as by the regularization methods of A.N. Tikhonov (12) and M.M. Lavrentiev (7) for extended systems. The Fig. 1 shows the dependences of $\|\tilde{z} - z\|/\|z\|$ from $|\lg \alpha|$ at $\nu = 1$ and $\nu = 5$.

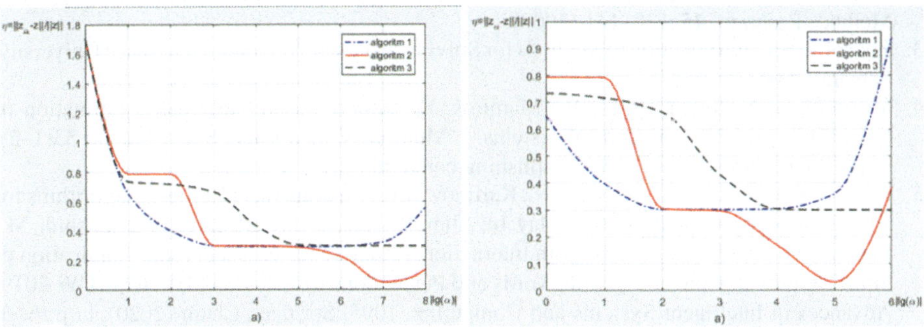

Fig. 1. Graphs of the dependence of $\|\tilde{z} - z\|/\|z\|$ on $|\lg \alpha|$ at $\nu = 1$ (a) and $\nu = 5$ (b): 1) - algorithm (4); 2) - algorithm (7); 3) - algorithm (12)

5 Conclusion

The analysis of the given dependencies shows that the best result is provided by the method of M.M. Lavrentiev for an extended system. This, apparently, is due to the fact that the extended system, by definition, is symmetric and positive definite.

References

1. Egupov, N.D., Pupkov, K.A.: Methods of classical and modern theory of automatic control. Textbook in 5 volumes. Publishing house of MSTU named after N.E. Bauman, Moscow (2004)
2. Afanasyev, V.N.: Control of Indefinite Systems. Peoples' Friendship University of Russia, Moscow (2008)
3. Bobtsov, A.A., Pyrkin, A.A.: Adaptive and Robust Control with Uncertainty Compensation. ITMO National Research University, St. Petersburg (2013)

4. Afanasyev, V.N.: Control of Indefinite Dynamic Objects. Fizmat-ref, Moscow (2008)
5. Nikiforov, V.O., Ushakov, A.V.: Control Under Uncertainty: Sensitivity, Adaptation, Robustness. Gitmo (TU), St. Petersburg (2002)
6. Igamberdiev, H.Z., Yusupbekov, A.N., Zaripov, O.O.: Regular Methods of Estimation and Control of Dynamic Objects Under Uncertainty. TashSTU, Tashkent (2012)
7. Igamberdiyev, K., Mamirov, U.F.: Regularized synthesis algorithms of parametrically invariant control systems of uncertain objects. J. Chem. Technol. Control Manage. **5**, 62–65 (2016)
8. Demyanov, V.F., Tamasyan, G.: On direct Methods for solving variational problems. Proc. Inst. Math. Mech. Ural Branch Russ. Acad. Sci. **16**(5), 36–47 (2010)
9. Vasin, V.V., Ageev, A.L.: Ill-Posed Problems with a Priori Information. Nauka, Yekaterinburg (1993)
10. Tikhonov, A.N., Arsenin, V.Y.: Methods for Solving Ill –Posed Problems. Nauka, Moscow (1986)
11. Tikhonov, A.N., Goncharsky, A.V.: Incorrect Problems of Natural Science. Publishing house of Moscow state University, Moscow (1987)
12. Morozov, V.A.: Algorithmic bases of methods for solving incorrect problems. Comput. Methods Program. **45**, 130–141 (2003)
13. Zhdanov, A.I.: Introduction to Methods for Solving Ill-Posed Problems. Aerospace University, Samara (2006)
14. Yusupbekov, N., Igamberdiev, H., Mamirov, U.: Algorithms of sustainable estimation of unknown input signals in control systems. J. Multiple-Valued Logic Soft Comput. **33**(1–2), 1–10 (2019). https://www.oldcitypublishing.com/pdf/9291
15. Igamberdiev, H.Z., Yusupbekov, A.N., Karimov, D.R., Shukurova, O.P.: Stable algorithms for adaptation of objects with control delay. In: Aliev, R., Kacprzyk, J., Pedrycz, W., Jamshidi, M., Babanli, M., Sadikoglu, F. (eds) 10th International Conference on Theory and Application of Soft Computing, Computing with Words and Perceptions - ICSCCW-2019. ICSCCW 2019. Advances in Intelligent Systems and Computing, 1095. Springer, Cham (2020). http://webvpn.fjmu.edu.cn/10.1007/978-3-030-35249-3_95
16. Sevinov, J.U., Zaripov, O.O., Zaripova, S.O.: The algorithm of adaptive estimation in the synthesis of the dynamic objects control systems. Int. J. Adv. Sci. Technol. **29**(5), 1096–1100 (2020). http://sersc.org/journals/index.php/IJAST/article/view/7887
17. Igamberdiev, H.Z., Yusupov, E.A., Sotvoldiev, H.I., Azamxonov, B.S. Sustainable algorithms for the synthesis of a suboptimal dynamic object management system. In: Aliev, R., Kacprzyk, J., Pedrycz, W., Jamshidi, M., Babanli, M., Sadikoglu, F. (eds) 10th International Conference on Theory and Application of Soft Computing, Computing with Words and Perceptions - ICSCCW-2019. ICSCCW 2019. Advances in Intelligent Systems and Computing, 1095. Springer, Cham (2020). https://doi.org/10.1007/978-3-030-35249-3_120
18. Lawson, C.L., Henson, R.J.: Solving Least Squares Problems. Prentice-Hall, New Jersey, Englewood Cliffs, USA (1986)
19. Godunov, S.K., Antonov, A.G., Kirilyuk, O.P., Kostin, V.I.: Guaranteed Accuracy for Solving Systems of Linear Equations in Euclidean Spaces. Nauka, Novosibirsk (1988)
20. Igamberdiev, H.Z., Boeva, O.H., Sevinov, J.U.: Sustainable algorithms for selecting feedback in dynamic object management systems. J. Adv. Res. Dyn. Control Syst. **12**(7), 2162–2166 (2020). https://doi.org/10.5373/jardcs/v12sp7/20202337
21. Bakushinsky, A.B., Goncharsky, A.V.: Iterative Methods for Solving Ill - Posed Problems. Nauka, Moscow (1989)
22. Nazimov, A.B., Morozov, V.A., Mukhamadiev, E.M., Mullojanov, M.: Shift Regularization Method Theory and Applications. VogTU, Vologda (2012)
23. Igamberdiev, H.Z., Mamirov, U.F.: Regularized algorithms for identification of undefined dynamic control objects. J. Bull. Tashkent State Tech. Univ. **2**, 16–21 (2018)

Decision Making Based on Case Company Target by Using GP Model

Rafig R. Aliyev(✉)

Research Laboratory of Intelligent Control and Decision-Making Systems in Industry and Economics, Azerbaijan State Oil and Industry University, Azadlyg Avenue, 20, AZ1010 Baku, Azerbaijan
rafig.aliyev@hotmail.com

Abstract. In order to find the case company's target, paper requested from CEO to consider the company's current organizational perspective and size of the internationalization project including bottom line of company and local resources such as sufficient international line-manager skills and answer the question of "For successful fulfilment of company's expectation from preferred host location, under which extended level do you want your goals to achieve". In order to ease the job of DM, appropriate score scale of lowest performed and highest performed countries result were presented.

Keywords: Decision making · Goal programming · GP model

1 Introduction

Goal Programming is an extension of Linear Programming (LP) to deal with competing and conflicting objective measure/decision criteria as faced in real-life problems such as international market selection. This indicates that improvement of one criterion may only be possible subject to worsening of some another criterion. That means, due to lack of resources and capabilities in order to achieve some goals, company must sacrifice from other goals [1]. In such cases, one of the possible approaches is to assign desired values for multiple goals - exact ones or threshold conditions and verify which existing alternative is the closest one to these values. In other words, a compromise solution is found as the closest one to the realistic desired alternative [2]. Such type of problems is referred to as goal programming (GP) [3].

2 Goal Programming (GP) Problem

A typical GP problem is formulated as follows [4]:

Determine $X = (X_1, X_2, \ldots, X_n)$.

$$Z = \sum_{i=1}^{n} v_i(d_l^- + d_l^+) \to \min \tag{1}$$

© The Author(s), under exclusive license to Springer Nature Switzerland AG 2021
R. A. Aliev et al. (Eds.): WCIS 2020, AISC 1323, pp. 329–333, 2021.
https://doi.org/10.1007/978-3-030-68004-6_43

Subject to (s.t.)

$$\sum\nolimits_{i=1}^{n} b_{li}A_i + d_l^{-} - d_l^{+} = \alpha_l\,, \quad l = 1, \ldots m \tag{2}$$

$$\sum\nolimits_{i=1}^{n} A_i \leq r_k \quad k = 1, \ldots, K \tag{3}$$

where X represents decision variables and Z represents objective function.A_i is a mean value of variable i. For the considered problem A_i is a binary structural variable, takes on 1 ('locate' decision) or 0 ('do not locate' decision). d_l^{-}, d_l^{+} denotes under and over-achievement amounts for a target value of lth goal, v_l is an importance weight of lth goal. B_{li} is a ith technical coefficient for the lth goal. α_l target value of lth goal, r_k upper limit of the kth resource.

Equation (1) implies minimization of the sum of weighted deviations from target values of goals (Gi). Weights vl are used to account for different priorities of goals. Goals with higher weights vl are achieved prior to others.

Equation (2) represent objectives of a DM, where b_{li} denotes a value of lth indicator for i^{th} country.

Equation (3) expresses limits of resources, in papers' case $\sum_{i=1}^{n} A_i$ as the considered problem is to locate in one country only. The problem (1)–(3) implies determination of values of d_l^{-}, d_l^{+} $l = 1, \ldots, m$ that satisfy constraints (2) and minimize objective function (1), provided (3).

Following steps are for the calculation of GP;

Step 1: Objective function (1) was expressed and given weights w_l computed by using AHP method.
Step 2: Target values α_l, $l = 1, \ldots m$ were assigned for all the m goals and constraints (2) were structured. Additionally, restriction (3) was also designed in this step.
Step 3: (1)–(2) was solved for each alternatives.
Step 4: Ranking of alternatives.

3 Decision Making by Using GP

We apply GP model described in (2) to solve the problem of country selection. Weights for criteria and sub-criteria are given below [5].

w1 = 0.16; w2 = 0.31; w3 = 0.055; … wn = 0.474;
$w11 = 0.25, wj2 = 0.75;$
$w21 = 0.33, w22 = 0.67;$
$w31 = 0.8, w32 = 0.2;$
$w41 = 0.5, w42 = 0.5.$

Paper denote sub-criteria as $G1 = C11$, $G2 = C12$, $G3 = C21$, $G4 = C22$, $G5 = C31$, $G6 = C32$, $G7 = C41$, $G8 = C42$.

The importance weights of $G1,\ldots,G8$ are obtained as aggregated values of the main criteria and subcriteria weights as follows (for convenience of further computations, [0–100] scale is used for aggregated values):

$v1 = w1w11 = 4;$
$v2 = w1w12 = 12;$
$v3 = w2w21 = 10.4;$
$v4 = w2w22 = 20.7;$
$v5 = w3w31 = 1.1;$
$v6 = w3w32 = 4.4;$
$v7 = w4w41 = 23.7;$
$v8 = w4w42 = 23.7.$

Thus, based on the method described in (2) paper has to solve the given problem. At Step 1, we form objective function:

$$4d1 + 12d2 + 10.4d3 + 20.7d4 + 4.4(d5 + d5) + 1.1(d6 + d6) + \\ +23.7d7 + 23.7d8 \rightarrow \min$$

At step 2, we assign targeted values (see Table 1) and formed constraints (1) (see below).

Table 1. Target values for goal

α1	α2	α3	α4	α5	α6	α7	α8
45	70	54,16	63	24	49,4	72	42,5

$$42.87A_1 + 43.66A_2 + 56.92A_3 + 28.80A_4 + 67.55A_5 + 59.98A_6 + 65.32A_7 + 21.82A_8 \\ + 25.85A_9 + d_1 - d_1 = 45$$
$$65.8A_1 + 68.5A_2 + 69.4A_3 + 55.6A_4 + 67.4A_5 + 72.5A_6 + 79.5A_7 + 55.1A_8 + 51.9A_9 \\ + d_1 - d_1 = 70.$$
$$54.3A_1 + 54.3A_2 + 57.2A_3 + 53.3A_4 + 58A_5 + 56.8A_6 + 49.8A_7 + 42.5A_8 + 40.3A_9 \\ + d_1 - d_1 = 54.2$$
$$61.6A_1 + 61.8A_2 + 63.5A_3 + 65.6A_4 + 60.1A_5 + 63.6A_6 + 60.9A_7 + 51.1A_8 + 57A_9 \\ + d_1 - d_1 = 63$$
$$19A_1 + 22.5A_2 + 34.7A_3 + 39.2A_4 + 43.7A_5 + 46.6A_6 + 23.2A_7 + 9.5A_8 + 36.3A_9 \\ + d_1 - d_1 = 24$$
$$41.3A_1 + 45.8A_2 + 17.3A_3 + 23.6A_4 + 17.9A_5 + 18.9A_6 + 21.6A_7 + 26.6A_8 + 16.2A_9 \\ + d_1 - d_1 = 49.4$$
$$74.33A_1 + 77.89A_2 + 72.3A_3 + 77.37A_4 + 71.40A_5 + 71.24A_6 + 83.28A_7 + 55.31A_8 \\ 68.25A_9 + d_1 - d_1 = 72$$

$$46.43333A_1 + 40.73333A_2 + 42.26667A_3 + 48.26667A_4 + 44.4A_5 + 43.93333A_6 \\ + 45.76667A_7 + 26.46667A_8 + 38.3A_9 + d_1 - d_1 = 42.50$$

$$A_1 + A_2 + A_3 + A_4 + A_5 + A_6 + A_7 + A_8 + A_9 = 1$$

For example, coefficient of A_1, $b_{11} = 42.87$ in constraint (1) implies that Governance Performance index of Turkey is 42.87, but the target value of this index is 45.

Therefore, that mean case company's goal is to invest in a country with a governance performance score of 45. In the objective function it illustrates to minimize the amount underachieved from 45.

At step 3, we solve (1)–(3). The problem (1)–(3) is linear programming problem. This problem was solved for each country by using MS Excel. The solutions for Romania, Kazakhstan, Turkey and Georgia are given in Table 2:

Table 2. Solutions for A_1, A_2, A_3, A_7

Weights of		Romania		Kazakhstan		Turkey		Georgia	
Goals		(A_3)		(A_2)		(A_1)		(A_7)	
Goal	weight	–	+	–	+	–	+	–	+
G1	4	0	11.92	1.34	0	2.13	0	0	20.32
G2	12	0.6	0	1.5	0	4.2	0	0	9.5
G3	10.4	0	3.07	0	0.173	0	0,173	4.33	0
G4	20.7	0	0.5	1.2	0	1.4	0	2.1	0
G5	4.4	0	10.7	1.5	0	5	0	0.8	0
G6	1.1	32.1	0	3.6	0	8.1	0	27.8	0
G7	23.7	0	0.3	0	5.891	0	2,33	0	11.28
G8	23.7	0.23	0	1.77	0	0	3,93	0	3.3
$\sum$ Deviation (Z)		Z(3) = 95		Z(2) = 100.6		Z(1) = 119		Z(7) = 123	

At step 4 we finally can rank all countries. The results are given in Table 3.

Table 3. Ranking of countries with respect to *Z(i)*

Romania	*Z(3)* = 95,12
Kazakhstan	*Z(2)*= 100,6
Turkey	*Z(1)* = 118,8
Georgia	*Z(7)* = 123
Bulgaria	*Z(6)* = 151
Croatia	*Z(5)* = 226,8
The Russian Federation	*Z(4)* = 341,5
Ukraine	*Z(9)* = 841
Pakistan	*Z(8)* = 1503,5

4 Conclusion

To find the case company's target for successful fulfilment of their expectation from host location GP approach is used. Decision making problem includes 4 criteria and 8 alternatives countries. Numerical solution as the problem shows validity of used decision making approach.

References

1. Hortacsu, A., Tektas, A.: Modeling the country selection decision in retail internationalization. World Acad. Sci. Eng. Technol. **3**(7), 900–907 (2009)
2. Brans, J.P.: The management of the future. Ethics in operational research. Respect, multicriteria management, happiness. Eur. J. Oper. Res. **153**, 466–467 (2004)
3. Charnes, A., Cooper, W.W., Ferguson, R.O.: Optimal estimation of executive compensation by linear programming. Manag. Sci. **1**(2), 138–151 (1955)
4. Hoffman, J.J.: Evaluating international ethical climates: a goal programming model. J. Bus. Ethics **17**(16), 1861–1869 (1998)
5. Aliyev, R.R.: Country selection decision model enhancement ideal vs. comprehensive solution, MSc Dissertation, The University of Warwick, p. 82 (2019)

Automatic Control of Fuel Efficiency in Gas Burning Furnaces

U. U. Kholmanov(✉) and A. T. Rajabov

Tashkent State Technical University named after Islam Karimov, Tashkent, Uzbekistan
xolmanov0902@mail.ru, azik_2511@mail.ru

Abstract. The paper presents the results of a scientific study of the functioning of gas-burning furnaces as objects for controlling the efficiency of the combustion process using a signal for thermal perception of furnace screens. The results of processing and experimental techniques obtained static and dynamic characteristics of the object on the thermal perception of the furnace screens evidence of high noise resistance of the object under study to intra-current disturbances and the feasibility of using the latter in industrial control systems. It is shown that the proposed optimal control system for the efficiency of the fuel combustion process with a correction for the heat perception index has a higher quality of regulation compared to the standard traditional automatic control system for the efficiency of the combustion process.

Keywords: Gas-burning furnace · Efficiency of combustion · Signal for thermal perception of furnaces · Furnace screen · Regulation of the thermal load of the furnace · Extreme regulation

1 Introduction

One of the most important factors that significantly affect the efficiency of energy-technological complexes and installations is the perfection of the fuel combustion process, which, along with the design and mode of operation of heat-technological equipment, is determined by the operation of automatic control systems for the efficiency of the combustion process. Currently existing traditional methods and typical schemes for controlling the efficiency of the combustion process in fuel-burning furnaces and aggregates using signals for the content of residual oxygen in combustion products have not found wide practical application due to the high inertia of information transmission channels and the lack of reliable and accurate sensors, difficulties in the technical implementation of industrial automation systems.

A comparative analysis of the efficiency indicators of existing automatic heat load control systems and the efficiency of fuel combustion processes in furnaces was carried out, and the limited capabilities of typical systems for optimizing the furnace process were shown. The study of gas-burning furnaces as control objects indicates the feasibility of using a signal on thermal perception $\Delta P_{app.s}$ in systems of extreme control of the efficiency of the combustion process in gas-burning furnaces.

© The Author(s), under exclusive license to Springer Nature Switzerland AG 2021
R. A. Aliev et al. (Eds.): WCIS 2020, AISC 1323, pp. 334–341, 2021.
https://doi.org/10.1007/978-3-030-68004-6_44

2 Study of Dynamic Characteristics of a Gas-Burning Furnace

To assess the accuracy and reliability of the dynamic characteristics of the object of research, experimental dynamic characteristics $h(t)$ of a gas-burning furnace were obtained (Fig. 1) by applying a perturbation on the air flow in the direction of its increase (7%). The measurement $\Delta P_{app.s}$ was performed using differential pressure gauge DM (Fig. 2).

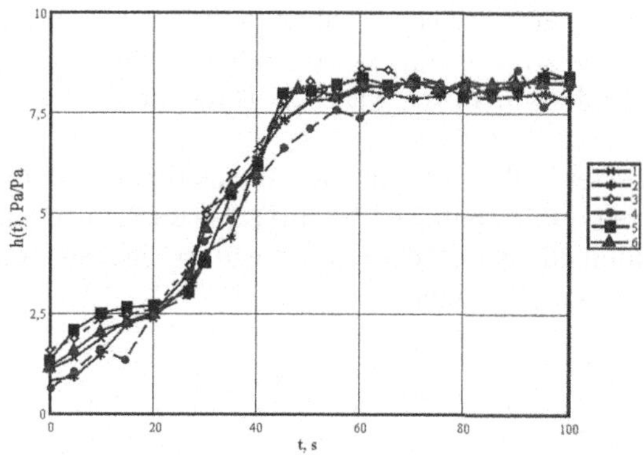

Fig. 1. Transient functions of the $h(t)$ signal $\Delta P_{app.s}$ under perturbation according to the air consumption in the gas-burning furnace.

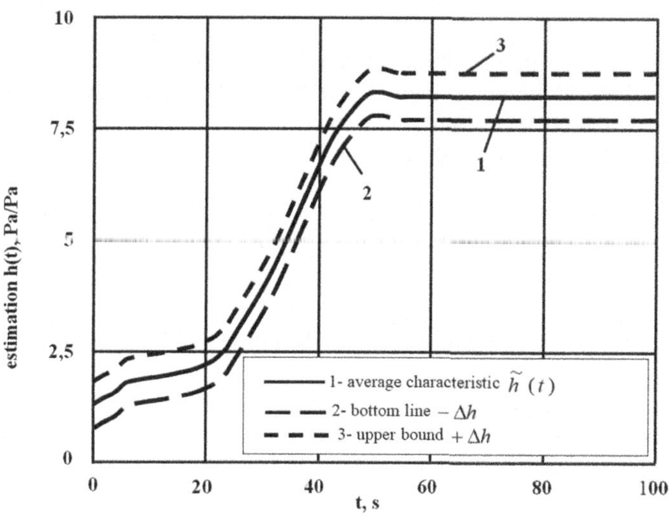

Fig. 2. Estimation $\tilde{h}(t)$ of signal accuracy $\Delta P_{app.s}$: 1 – averaged value of $\tilde{h}(t)$; 2 - lower bound of the confidence corridor ($-\Delta h$); 3 – upper bound of the confidence corridor ($+\Delta h$)

The $h_i^*(t)$ confidence corridor of constant width was used as an estimate of the accuracy of the transition characteristic averaged over a set of implementations $\tilde{h}(t)$

$\pm\Delta h$ [1]. The reliability of the estimate $\tilde{h}(t)$ is understood as the probability p That the confidence corridor $\left(\tilde{h}(t) \pm \Delta h\right)$ will cover the true value of the transition function $\tilde{h}(t)$. The confidence corridor of the function is constructed on the condition that the probabilities of each j-th section $(j = 0,1\ldots m)$ of the function $\tilde{h}(t)$ being inside the confidence corridor are equal and specified:

$$P_1 = P_2 = \ \ldots\ = P_j = \ \ldots\ = P_{adj} \tag{1}$$

The estimate $\tilde{h}(t)$ is calculated at pre-selected points $t_j = j \cdot \Delta t\big(j = 0.1\ldots m;\ \Delta t = t_j - t_{j-1}\big)$. According to the recommendations [1], the sampling step is selected $\Delta t = T_Y/9$, where T_Y is the time when the signal reaches its steady state value.

We denote the discrete values of ordinates of experimental transient characteristics $h_i^*(t)$ by $h_{i,j}^*(t)$, where i is the number of the curve, j is the ordinal number of the section [2]. The estimation of mathematical expectation and variance of a random variable was determined by the following formulas:

$$\tilde{h}_j(1) = \frac{1}{n} \cdot \sum_i^n h_{ij}^* \tag{2}$$

$$\tilde{\sigma}_j^2 = \left(\frac{\sum_i^n (h_{ij}^*)^2}{n} - \tilde{h}_j^2 \right) \cdot \frac{n}{n-1} \tag{3}$$

Where $n = 6$ is the number of experimental dynamic characteristics.

The value of the confidence interval for the mathematical expectation of a random variable distributed according to the normal law is determined by the value of the confidence interval $\pm\Delta h$ in *the* j-th section:

$$\Delta h_j = t_{P_{adj}} \cdot \frac{\tilde{\sigma}}{\sqrt{n}} \tag{4}$$

where: $\tilde{\sigma} = \left(\frac{\sum_o^m \tilde{\sigma}_1^2}{m+1} \right)^{1/2}$ - m the mean-square deviation value averaged over the (m +1) (cross-section) $h_{i,j}^*$; $t_{P_{adj}}$ - quantile of the student's distribution, which depends on the number of degrees of freedom $f = (n - 1)$ and the given probability P_{adj}.

Width of the confidence interval in each j-th section $\tilde{h}_j(t)$

$$I_j = [\tilde{h}_j - \Delta h_j;\ \tilde{h}_j + \Delta h_j] \tag{5}$$

The numerical value I_j depends on the number of n experimental curves $h_i^*(t)$, the given probability P_{adj}, and on the dispersion of interference in the j–th section.

The width of the confidence corridor for the function $\tilde{h}(t)$ as a whole depends both on these factors and on the degree of correlation of estimates $h_{i,j}^*(t)$ in different sections [3].

Assuming that the width of the confidence corridor per function $\pm\Delta h(t)$ is numerically equal to the width of the confidence interval Δh_j, we write the reliability of the transition function $\tilde{h}(t)$ as a probability function [1]

$$P\Big\{h_0, h_1, \ldots, h_j, \ldots, h_m \in \tilde{h}(t) \pm \Delta h(t)\Big\} = P_j^{m+1} \cdot (1 - K_m) + K_m \cdot P_j \quad (6)$$

where: $K_m = \frac{2}{\pi} \cdot \sum_{k=0}^{m-1} \sum_{l=1}^{m} \frac{2 \cdot \arcsin(r_{kl})}{m \cdot (m+1)}$, r_{kl}- coefficients of the normalized correlation matrix $\|r_{kl}\| = \frac{\|K_{kl}\|}{\sigma_k \sigma_l}$.

The coefficients of the correlation matrix $\|K_{kl}\|$ are calculated from experimental data according to the expression:

$$K_{kl} = \left[\frac{\sum_{1}^{n} h_{ik}^{*} \cdot h_{il}^{*}}{n} - \tilde{h}_k \cdot \tilde{h}_l \right] \cdot \frac{n}{n-1} \quad (7)$$

The probability that the built confidence corridor will cover the true value of the function is $P = 0.927$.

Research of the system of regulating the efficiency of the combustion process.

The purpose of the research was to determine the static and dynamic characteristics of the signal $\Delta P_{use.s}$, to verify the effectiveness of the application of this signal in an extreme control system (ESM) efficiency of the combustion process (by comparing the quality of regulation of this system with the design one), acting on the ratio "fuel consumption – air consumption" with a correcting signal for thermal perception $\Delta P_{app.s}$ [4]. Static and dynamic parameters of the control system were calculated. First, we checked the operation of the air regulator at the selected settings. Then, a correction signal for heat perception $\Delta P_{app.s}$ was fed to the input of the regulator, with the help of which the search and maintenance of the optimal excess air ratio in the furnace furnace was carried out. The resulting implementations are shown in Fig. 3, and the numerical values of the dispersion of the studied signals are shown in Table 1.

In parallel, gas analysis was performed at reference points. The resulting implementations are shown in Fig. 3. In the Table 2 presents comparative data of gas analysis obtained as a result of tests, and data taken from the regime maps of the technological regulations.

The efficiency and efficiency of the combustion efficiency control system are derived from a comparison of the obtained results. Obviously, the quality of regulation of the proposed extreme regulatory system (ESM) higher than when using a standard automatic control system (ACR) efficiency of the combustion process with a signal for "heat". The resulting ESM are close to the parameters of the regime map of technological regulations Fig. 4, which confirms its high efficiency [5].

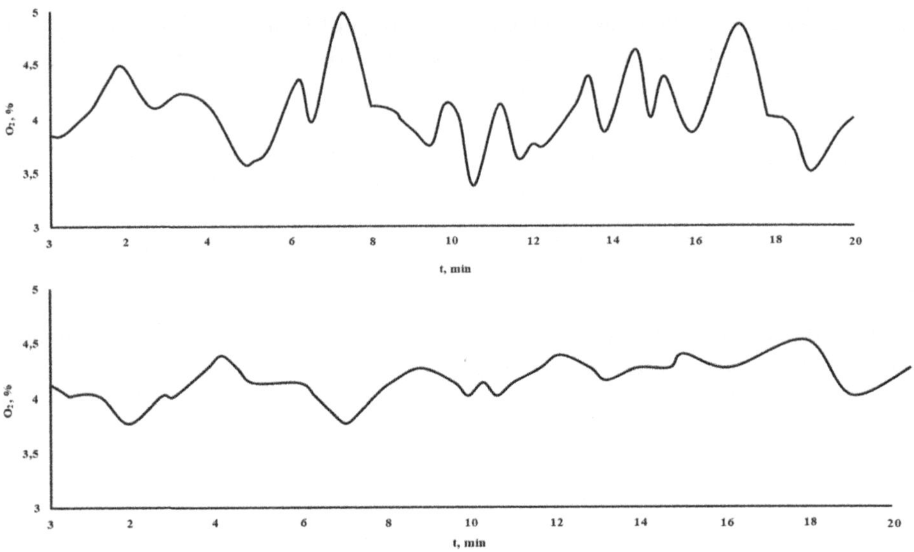

Fig. 3. Implementation of O_2 content in waste gases.

Table 1. Comparative data of the study of the system of regulating the efficiency of the combustion process with signals $\Delta P_{use.s}$ and D_q

Fuel type	Dispersion, $(O_2)^2$	
	$\Delta P_{use.s}$	D_q
Gas	1.25	2.30

Table 2. Results of gas analysis in the study of the combustion efficiency control system

Indicators		Results							
Steam consumption, *t/h*		330	300	260	240	215	230	275	315
Content O_2, %		4,50	4,60	4,90	5,00	6,10	5,80	5,10	4,55
α		1,27	1,28	1,32	1,31	1,42	1,38	1,32	1,26
Technological regulations	About O_2, %	5,1	–	5,4	–	6,5	–	5,5	–
	α	1,32	–	1,34	–	1,47	–	1,36	–
$\Delta P_{app.s}$, *kPa*	5.3	5.0	4.6	3.9	2.8	3.7	4.9	5.0	5.4

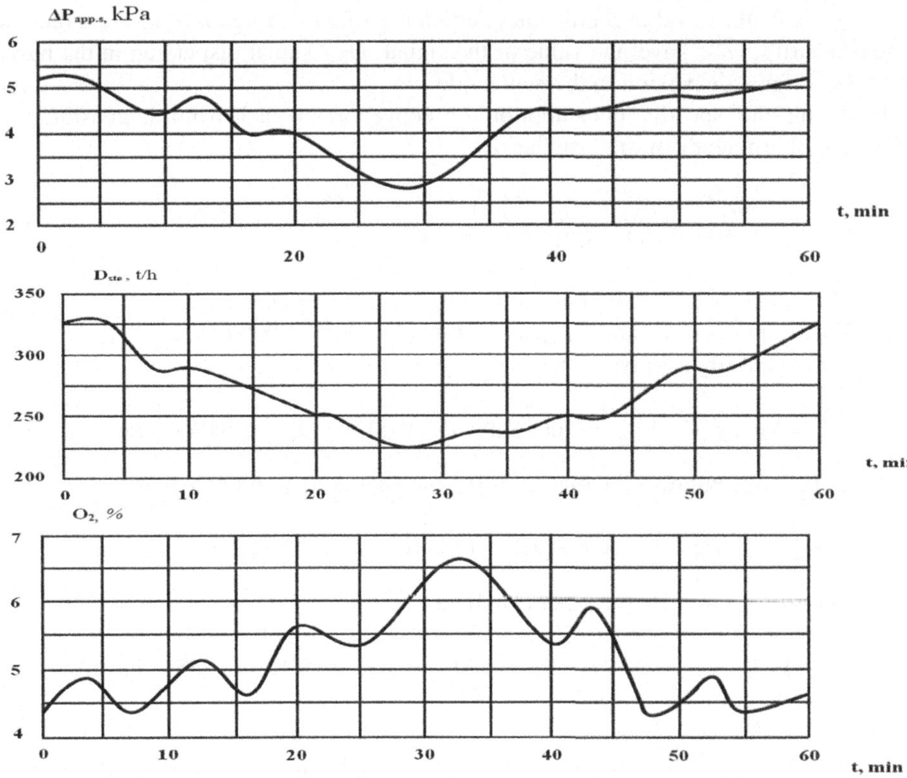

Fig. 4. Realization of the flow rate ($D_{ste.}$), thermal perception $\left(\Delta P_{app.s}\right)$, oxygen content in the outgoing gases (O_2) during the operation of the combustion efficiency control system.

3 Economic Efficiency of the Implementation of a System for Regulating the Efficiency of the Combustion Process

Savings from the operational efficiency of the combustion control system are determined by a higher quality of stabilization of the in-line mode when using the proposed system with a signal $\Delta P_{use.s}$ by reducing the dispersion of heat generation in the furnace, and, consequently, reducing heat losses [6, 7]. The system without correction of the regulator's task is accepted as the starting point, and the extreme control system with a correcting signal is used as the compared $\Delta P_{use.s}$ system. The calculation of the expected economic efficiency is carried out on the basis of the "Methodology for determining the economic efficiency of automatic control systems" [8].The comparison is made based on the calculation tasks for two control systems. The increase in plant coefficient of efficiency is determined by the formula:

$$\Delta\eta_k^{br} = \eta_k^{br} \cdot \left(A_{O_2}/(O_2^{av})^2\right) \cdot (D_b - D_n) = 92 \cdot \left(3,96 \cdot 10^{-4}/0.0036\right) \cdot (0,040 - 0.022) = 0.18\ \% \tag{8}$$

where: η_k^{br} - calculated value coefficient of efficiency of the setting (%); A_{O_2} - coefficient of non-linearity; $O_2^{av.}$ - average value of the signal; D_b - signal dispersion in the basic mode; D_n - signal dispersion in the new mode.

Reducing the specific consumption of conventional fuel from increasing the coefficient of efficiency of the installation

$$\begin{aligned}\Delta b_{EG}^{BR} &= b_{EG}^{BR} - (\eta_k^{br} \cdot b_{EG}^{BR})/(\eta_k^{br} + \Delta\eta_k^{br}) \\ &= 400 - (92 \cdot 400)/(92 + 0,18) = 0,8\end{aligned} \text{ a.c.c.} / \text{KWh}, \quad (9)$$

Where b_{EG}^{BR} is the specific fuel consumption for electricity generation.

Number of hours of installed capacity utilization: hy = 6000 h/year

Annual electricity supply:

$$Э_{Ye}^{SUP} = Ny \cdot hy \cdot (1 - \alpha_{CH}) = 200 \cdot 6000 \cdot (1 - 0.073) = 11{,}1 \cdot 10^5 \text{ MWh/year}. \quad (10)$$

Total fuel consumption calculated during the year at hy = 6000 h/year:

$$\Delta B_{Ye}^{AC} = \Delta b_{EG}^{BR} \cdot Э_{Ye}^{SUP} = 0{,}8 \cdot 10^{-6} \cdot 11{,}1 \cdot 10^5 \cdot 10^3 = 1631{,}7 \text{ t. a. t. / year} \quad (11)$$

Natural fuel savings per year per installation:

$$\Delta B_{Ye}^{N.F} = \Delta B_{Ye}^{AC} \cdot (Q_P^F \text{a.c} / Q_P^F \text{f.s}) = 1631,7 \cdot (29,3/35,8) = 1335,44 \,\text{m}^3 / \text{year}. \quad (12)$$

Annual economic effect (at the price of 1 t.f.e., equal to: S_T = 1200 soums/t.f.e):

$$\Delta S_{Year} = \Delta B_{Ye}^{N.F} \cdot S_T = 1335,44 \cdot 1200 = 1602,528 \,\text{t.soums} / \text{year}. \quad (13)$$

An additional economic effect will be achieved by reducing nitrogen oxide emissions and reducing electricity consumption for the company's own needs.

4 Conclusion

1. The results of processing experimental dynamic characteristics of the signal by thermal perception of furnace screens $\Delta P_{use.s}$ indicate its high noise resistance to in-furnace disturbances and the possibility of using it in industrial control systems.
2. It is shown that the system for regulating the efficiency of the combustion process with a correcting signal for heat perception has a higher quality of regulation compared to the system for automatically regulating the efficiency of the combustion process with a signal for heat.
3. The achieved economic efficiency from the implementation of the proposed system of regulating the efficiency of the combustion process is determined by reducing the calculated and actual value of the specific consumption of conventional fuel in the range of 1 a.c.c/KWh when burning fuel, which in quantitative terms is about 1602,528 t.f.e/year.

References

1. Yusupbekov, N.R., Nurmuhamedov, H.S., Temirov, O.S.H., Turobjonov, S.M., Zokirov, S.F., Tadjihodjaev, Z.A.: Gazlarni qayta ishlash tehnologiyasi, jarayon va qurilmalari. Fan va texnologiya, Toshkent (2016)
2. Yusupbekov, N.R., Gulyamov, Sh.M., Kasymov, S.S., Usmanova, N.B., Mirzaev, D.A.: Software implementation of exchange processes in a distributed network environment of transmission and processing of information. J. Autom. Mob. Robot. Intell. Syst. **12(4),** 64–69 (2018)
3. Toirovich, R.A., Utkiruktamugli, K.: Automatic control and management of technological processes of glassmaking production. J. Adv. Res. Dyn. Control Syst. **12**(6), 764–770 (2020)
4. Dolotovskiy, I.V.: E'nergoe'ffektivnaya gorelka dlya ognetehnicheskogo oborudovaniya neftegazovy'h predpriyatiy. Himicheskoe i neftegazovoe mashinostroenie **1**, 39–40 (2018)
5. Polyakov, A.A.: Optimizaciya processa goreniya v parovom kotle pri sjiganii neskol'kih vidov topliva s primeneniem e'kstremal'nogo regulyatora. Avtoref, Moskva (2005)
6. Huth, A.H.: Fuel flexibility in gas turbine systems: impact on burner design and performance. In: A Volume in Woodhead Publishing Series in Energy, Siemens AG Energy, Germany, pp. 635–684 (2013)
7. Honner, M., Veselý, Z., Svantner. M., Čejka, F.: Optimal heating control system for continuous furnaces. Scand. J. Metall. (2001)
8. Vassilyev, S.N., Novikov, D.A., Bakhtadzc, N.N.: Intelligent control of industrial processes. In: CONFERENCE 2013, pp. 49–57 (2013)

Synthesis of Compound Esters on the Basis of Acid Chloride of Low-Molecular Petroleum Acids of Baku Marine Oil Fields and Chiral Unsaturated Alcohols

A. N. Baghirli[1], S. A. Mamedkhanova[1(✉)], and G. M. Talybov[2]

[1] Azerbaijan State Oil and Industry University, Azadliq Avenue 20, Nasimi, Baku, Azerbaijan
aygunbaghirli91@gmail.com, vahide@rambler.ru

[2] Institute of Petrochemical Processes named after Academician Yusif Mammadaliyev of Azerbaijan National Academy of Sciences, Khojaly Avenue 30, Baku, Azerbaijan
ahmed_adna@rambler.ru

Abstract. Synthesis of Compound Esters based on Acid Chloride of Low-Molecular Petroleum Acids of Baku Marine Oil Fields and Chiral Unsaturated Alcohols. The physicochemical properties of the compound naphthenic acid esters are determined, and the possibilities of their using as plasticizers for high-molecular compounds and antioxidants for fuels are shown. Naphthenic acids extracted from different oils and petroleum products differ widely. The introduction of different radicals into the hydrocarbon molecules of these products gives them special properties: reducing the temperature of cooling, improving the thermal oxidation and hydro-oxidation stability, improving the solvency, temperature-viscous, plasticise, mechanical, noncorrosive, etc. characteristics. The esters of the unsaturated C3-alcohols (propynol, allol) of naphthenic acids were used in multifunctional additives to engine oils, and the additional barium and zinc salts of petroleum naphthenic acids have high inhibitory properties and are used for corrosion of steel in an acidic environment. Halogen anhydrides derived from petroleum acids have a high reactivity, which makes it possible to produce organic compounds with a higher yield. In all cases, the properties of the final products depend to a large extent on the nature and structure of the naphthene radical.

Keywords: Chiral unsaturated alcohols · Petroleum acids · Compound esters

1 Introduction

1.1 Naphthenic Acids

Naphthenic acids extracted from different oils and petroleum products differ widely. Baku oils are the richest naphthenic acids. Petrochemical products derived from petroleum naphthenic acids have special properties. The introduction of different radicals into the hydrocarbon molecules of these products gives them special properties: reducing

© The Author(s), under exclusive license to Springer Nature Switzerland AG 2021
R. A. Aliev et al. (Eds.): WCIS 2020, AISC 1323, pp. 342–347, 2021.
https://doi.org/10.1007/978-3-030-68004-6_45

the temperature of cooling, improving the thermal oxidation and hydro-oxidation stability, improving the solvency, temperature-viscous, plasticise, mechanical, noncorrosive, etc. characteristics [1–3].

Naphthenic acids, which have the same boiling points but are extracted from different oils, differ sharply in the physicochemical constants. Therefore, petroleum naphthenic acids are the most practically valuable compounds and the products extracted from them have useful technical properties.

Unsaturated naphthenic acid esters can be successfully used as plasticisers for polymeric materials, solvents, synthetic lubricants with high operational properties, and as monomers, modifiers for polymeric materials, corrosion inhibitors and as a raw material to produce a number of other substances [4–7].

The esters of the unsaturated C_3-alcohols (propynol, allol) of naphthenic acids were used in multifunctional additives to engine oils, and the additional barium and zinc salts of petroleum naphthenic acids [8, 9] have high inhibitory properties and are used for corrosion of steel in an acidic environment [10].

Halogen anhydrides derived from petroleum acids have a high reactivity, which makes it possible to produce organic compounds with a higher yield.

2 Experimental Work

On the above mentioned petroleum naphthenic acid esters is very promising. Petroleum naphthenic acid esters extracted from halogen-anhydride with chiral unsaturated alcohols have not been fully studied.

The following are some parameters of the compound esters of a number of petroleum acids extracted from the kerosene cuts of the marine oil mixtures:

$$\rho_4^{20}\ 982\ \text{kg/m}^3,\ n_D^{20}\ 1.4772\ \ \nu_{50^0C}\ 27{,}80\ \text{mm}^2/\text{c},\ \nu_{100^0C}\ 57{,}80\ \text{mm}^2/\text{c},\ t^{\text{зас}} =\text{-}52\ {}^0\text{C}$$

According to [11, 12], petroleum acids were chlorinated with thionil chloride after separation from unsaponifiable matters.

For this purpose, an appropriate quantity of petroleum acids was placed in a round bottomed three-neck flask equipped with a power mixer with a mercury cut-off, a thermometer and a capillary vent.

Stirring of the reaction mixture at 40 °C was continued for 30 min until the extraction of SO_2 and HCl according to the reaction shown in Scheme 1:

$$\text{RCOOH} \xrightarrow[\text{SO}_2,\ \text{HCl}]{\text{SOCl}_2} \text{RCOCl}$$

The ROOH: $SOCl_2$ ratio was 1:1. For obtaining narrow fractions, the acid chloride of broad fraction 88–175 °C (0.27–0.4 kPa) were fractionated in vacuum. The physicochemical properties of the obtained acid chlorides are given in Table 1.

Table 1 shows that during the distillation of the wide fraction of acid chlorides of petroleum oils, the main mass is concentrated in a narrow fraction of 130–145 °C (with

Table 1. Physicochemical properties of natural petroleum acids extracted from kerosene fractions of mixtures of Baku Marine Oil Fields.

Components	Cut boiling range, °C (of 0.27-0.4 кPa)	Mass density ρ_4^{20}, кg/m^3	Refractive index n_D^{20}	Acid index mg KOH/g	Fractional yield % (mas.)	Chlorinity, % (mas.)	Acid
Molar mass	88-175	959.7	1.4810	287.5	-	-	195.2
Acid chloride fractions	102-117	1008.7	1.4919		15.3	15.2	229.3
	117-130	1041.5	1.4955		18.3	14.3	
Bottoms				-			
	130-145	1085.7	1.5023		36.4	12.7	233.0
	145-186	1099.0	1.5077		28.1	11.5	
		-	-		1.90	-	

0.27–0.4 kPa). This fraction is selected for the concentration of C_{10}-C_{12} [12] monocyclic naphthenic acids.

The chiral unsaturated alcohols were (1*R*)-[4-(bromomethyl) phenyl]-2-[(prop-2-yn-1-yl) oxy-ethan-1-ol, which we extracted earlier [13].

The experiments were carried out in a three-neck flask equipped with a power mixer, reflux condenser, thermometer and drop funnel.

Usually, the reaction begins simultaneously with the dissolution of the alkoxy-substituted phenols in the acid halides.

The reaction started was determined by the vigorous evolution of hydrogen chloride, and the cessation of the release was the end of the reaction.

The calculated amounts of (1*R*)-[4-(bromomethyl) phenyl]-2-[(prop-2-yn-1-yl) oxy] ethan-1-ol and pyridine were loaded into a reaction flask and diluted with a five-fold volume of benzene. Then, gradually, halogenides were added in drops.

The reaction mixture was heated another 15–20 min at a temperature of 70–75 °C.

The mixture was filtered from the quaternary amine salt distilled in a vacuum, then target products were separated out.

The reaction of interaction of chlorides of petroleum naphthenic acids with chiral unsaturated alcohols proceeds according to Scheme 2:

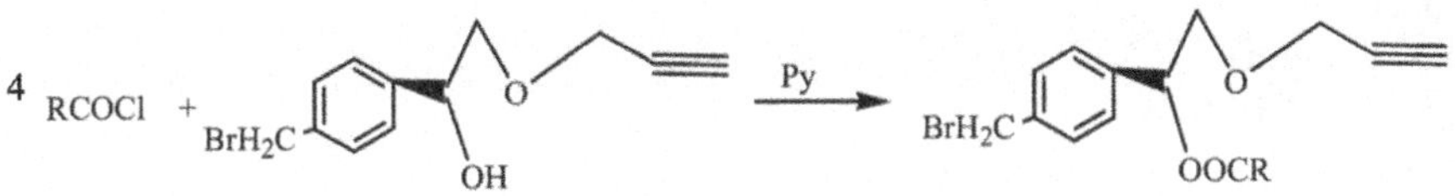

R – Naphtenic Radical

To achieve high yields of target products, the interaction of acid chlorides with (*1*R)-[4-(bromomethyl) phenyl]-2-[(prop-2-yn-1-yl) oxy] ethan-1- olefins was carried out with pyridine which captures the separated hydrogen chloride and thus prevents the of by-products. At the same time, the output of the derived esters is achieved up to 82.3% (mass).

The optimal molar ratio of acid chlorides to *chiral unsaturated alcohols* was 1:1.

Here Neural Network and Matlab environment are used for all calculations.

Fragment of used date for learning (Neural Network) is given below (Table 2) and fragment of computer simulation is the following form:

Table 2. Fragment yields of naphthenic acid esters obtained from unsaturated chiral oxyesters

No	Amount of naphthenic acids (gr)	Amount of unsaturated chiral oxyesters (gr)	Outputs%
5	RCOOH 42,4 gr	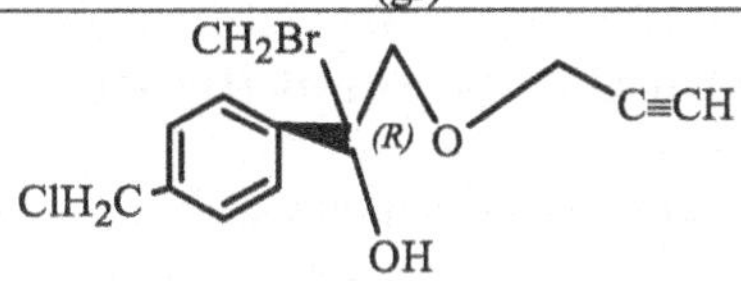 36,2 gr	87
6	RCOOH 42,4 gr	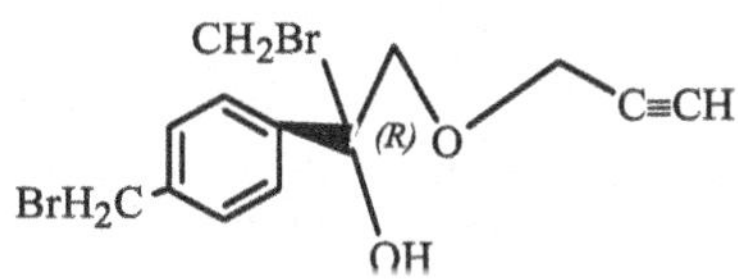 31.7 gr	86
7	RCOOH 42,4 gr	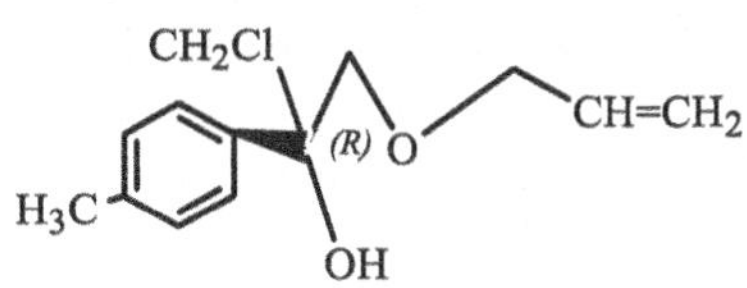 24.1 gr	69
8	RCOOH 42,4 gr	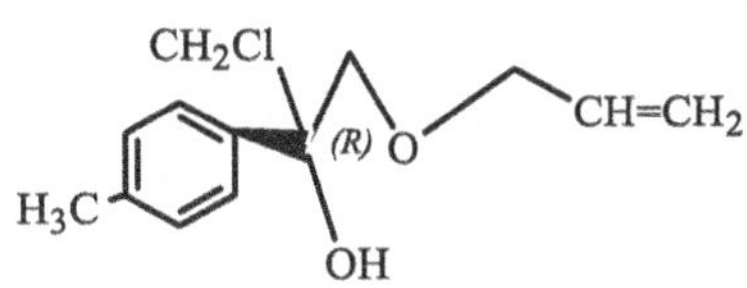 28.1 gr	65

```
[Input1]
Name = 'input1'
Range = [24.1 36.2]
NumMFs = 3
MF1 = 'in1mf1':'trimf', [18.0567335946398 24.947122880534 31.2231819276398]
MF2 = 'in1mf2':'trimf', [23.2972886291887 29.0301040820776 40.340475068587]
MF3 = 'in1mf3':'trimf', [28.100509846698 31.6911023849595 41.2296689466223]

[Output1]
Name = 'output'
Range = [65 87]
NumMFs = 3
MF1 = 'out1mf1':'constant', [70.1244490446538]
MF2 = 'out1mf2':'constant', [61.9560628108825]
MF3 = 'out1mf3':'constant', [104.385943297366]
```

Results of testing: if input1 = 42.4 and input2 = 36.2 then obtained data is 87
if input1 = 42.4 and input2 = 24.1 then obtained data is 69
if input1 = 42.4 and input2 = 33 then obtained data is 86.2
if input1 = 42.4 and input2 = 30.15 then obtained data is 77.5

3 Conclusion and Discussion of the Results

The ultimate composition of the samples was determined on the Karlo Erba 1106 Element Analyzer. The purity of the resulting compounds was controlled by gas-liquid chromatography on the Shimadze GC-9A instrument, GC-2014 [column 2000 × 3 mm, stationary phase-silicone SF-30 (5%), Chromaton N-AW-HMDS carrier, with temperature range from 50 to 270 °C, velocity 8 grad/min and helium carrier gas (47 ml/min)]. For TLC, Sorbfil plates (Russia) and eluent petroleum ester of bp. 40–70 °C were used.

The composition and structures of compounds (1–8) were established on the basis of IR and NMR spectra and elemental analysis data. The IR spectra of the compounds in a thin layer and in KBr were recorded on Specord 75 IR instruments. ^{1}H NMR spectra of substances in $CDCl_3$ were recorded on a Bruker SF-400 instrument (400.134 MHz), ISTD.

The IR spectrum shows absorption bands in the range of 710, 770, 983, 1490, 1720, 3030, 3065, 3085 cm^{-1} which are especially characteristic of monosubstituted benzenes, as well as stretching vibrations at 1119 cm^{-1} and 1762 cm^{-1} which are characteristic of carbonyl ester groups. The absorption bands in the range of 3030, 2110 are characteristic of the final acetylene bond (C≡CH).

The signals of protons of alkyl radicals are usually shown in the range of 0.97-1.45 ppm as multiplet signals. Acetylene protons in the triplet form manifest in the area of 2.51 t (1H, ≡CH, ^{4}J 2.4 Hz). 3.00 d.d (1H, CH_2O, J = 16.5, 7.8 Hz), 3.15 d (1H, J = 16.5, 7.8 Hz). The protons included in the methylenedioxy group of the propargyl fragment manifest in the area of 3.83 д.д (1H, ≡C$\underline{H}_2$O, ^{2}J 16.2 Hz, ^{4}J 2.4 Hz), 4.05 d.d (1H, ≡C$\underline{H}_2$O, ^{2}J 16.2 Hz, ^{4}J 2.4 Hz) respectively.

Protons of methinoxy group **in** 4.13 d.d (1H, CHO, ^2J 5.7, 7.8 Hz), and methyl bromide groups in 4.52 d (1H, Ch_2br, $J = 10.5$ Hz), 4.61 d (1H, Ch_2br, $J = 10.5$ Hz). The methylene protons are diastereotopic and therefore show different signals. Aromatic protons manifest in the 7.03 d.d (2H, $J = 1.5$, 8.7 Hz), 7.38-7.42 (5H, Ph), 7.49 d.d (2H, $J = 1.5$, 8.6 Hz), respectively.

To clarify the suitability of the synthesized esters as plasticizers for polymeric materials and antioxidants for fuels and oils, their physicochemical properties were determined. The synthesized esters of naphthenic acids of chiral unsaturated alcohols can be used as plasticizers for polymeric materials (for example, PVC) and antioxidants for fuels [10].

References

1. Nametkin, N.S., Egorov, V.K., Khamaev, V.K.: Petroleum Acids and Products of their Chemical Processing. Chemistry, Moscow (1982)
2. Abbasov, V.M., Zeynalov, E.B., Mustafaev S.A., et al.: Natural petroleum acids and their derivatives, Elm, Baku (2014)
3. Akhmedov, A.I., Askerova, K.A., Isakov, E.U., Gamidova, D.: Sh: Synthesis of viscous additives for lubricants by copolymerization of butyl methacrylate with allyl naphthenates and C6–C12 a-olefins. Oil Refining Petrochemistry **5**, 31–32 (2009)
4. Kruglov, E.A., Tsypysheva, L.G., Rakhimova, L.A.: Study of the composition of acids derivated in liquid-phase oxidation of naphthenic concentrates. In: The book Oxygen containing compound from crude oil. Chemistry (Works of Neftekhim SRI), vol. 2, 129–133, Moscow (1970)
5. Farzanakh, K.F., Mustafaev, S.A., Mamedov, N.A:. Mixtures of Acids of Baku Marine Oil Field and their acid chlorides. Chem. Probl. **1**, 74–79 (2015)
6. Velieva, M.G., Sadigov, N.A., Mamedova, N.A., Mustafaev, S.A.: Etherification by acetylene alcohols of petroleum naphthenic acids. Petrochemistry **49**, 247–252 (2009)
7. Abbasov, V.M., Mustafaev, S.A., Shakhmamedova, A.G.: Synthesis of unsaturated esters of natural petroleum acids. Oil Refining Petrochemistry **7**, 30–33 (2014)
8. Shahmammadova, A.Q., Mustafaev, S.A., Veliyev, M.: Synthesis and properties of vinyl esters of Cyclic Carboxylic Acid. In: Abstract Book Organic Chemistry Congress, p. 74., Fatiye Mugla, Turkey, 1–6 October 2012
9. Veliev, M.G., Mustafaev, S.A.: Synthesis of unsaturated cetone based on chlorohydrides of cyclic carboxylic acids. Collected Works, Chem. Belarusian Sci. 47–55 (2013)
10. Ragimova, S.N., Mustafaev, S.A., Veliev, M.G., Guseynova, A.G.: Use of esters derived from petroleum naphthenic acids as suppressor additives to diesel fuels. Azerbaijan Oil Ind. **2**, 42–45 (2009)
11. Shikhalizade, P.D., Mustafaev, S.A., Safarov, G.I.: Production of acid chlorides of high molecular petroleum naphthenic acids. Univ. Ed. Oil Gas **4**, 47–50 (1986)
12. Chaudhuri, S.U., Shikhalizade, P.D., Mekhtiev, S.N., Mustafaev, S.A.: Acid chlorides of naphthenic acids of the oil field named after April 28. Univ. Ed. Oil Gas **10**, 22–34 (1986)
13. Talybov, G.M., Baghirli, A.N.: Determination of the absolute configuration of the propargyl oxyether. Russ. J. Org. Chem. **56**(4), 628–630 (2020)

Radiation Resistance of Bituminous Hydroisolation Materials

Islam Mustafayev[1,2], Gulara Imanova[3(✉)], Gulnara Hatamkhanova[1], and Nigar Guliyeva[2]

[1] The Azerbaijan University of Architecture and Construction, Baku, Azerbaijan
[2] Institute of Radiation Problems of Azerbaijan National Academy of Sciences, Baku, Azerbaijan
[3] Azerbaijan State University of Oil and Industry, Baku, Azerbaijan
imanova.gulara@bk.ru

Abstract. The kinetics of gas formation, changes of molecular structure and thermal properties bitum composition in the radiation-chemical transformations in the range of adsorbed dose 0–200 kGy were studied. Composite material was prepared from oil-bitumen, polymer and solvent. Molecular structure was investigated by IR-spectroscopy (Varian-220), gas formation by chromatography (Agilent GC 7890A) and thermal properties by derivatography (Perkin Elmer STA 6000). More than 97% of the gases generated during radiolysis are hydrogen and carbon monoxide and their radiation and chemical yields are G(H2) = 0.12, G(CO) = 0.21 molec/100 eV. Such a low emission of gases during radiolysis indicates the high radiation stability (resistance) of the bituminous material.

Gas molecules are generated in the result of disintegration of hydrogen and oxygen-containing groups from functional groups by radiation exposure. Endurance of bitumen to radiation effect was attested by means of infrared spectroscopy and derivatography. A slight decline in absorption bands corresponding to only oxygen-containing functional groups in the infrared spectrum is observed. In derivatographic studies, the energy of maximum absorbance corresponding to the temperatures of 112–114 °C and 156 °C decreases while the absorption dose increases. The activation energy, heat capacity and enthalpy of the samples are reduced accordingly by decreasing absorbed dose. The mechanism of influence of ionizing radiation on bituminous composition is discussed.

Keywords: Bitumen · Radiation · Gases · IR spectrum · DTA · Chromatography first section

1 Introduction

A study of regularities in regard with the aging of bituminous materials under the effect of radiation has scientific and practical significance. Bitumination of radioactive waste materials are applied in nuclear energetic in more than 10 countries since 60s of the last century and recently the bituminized radioactive waste materials more than 250 thousand tones are available [1–3]. These materials can also be used as construction

© The Author(s), under exclusive license to Springer Nature Switzerland AG 2021
R. A. Aliev et al. (Eds.): WCIS 2020, AISC 1323, pp. 348–357, 2021.
https://doi.org/10.1007/978-3-030-68004-6_46

materials resistant to radiation in the condition of radiation influence in nuclear reactors, accelerators of particles, X-ray facilities, etc. [3] [21]. It should be notified that all petrochemicals and polymers are aged by the effect of radiation and they may lose their initial properties after a certain period of time [4–7]. This may cause severe problems in respect to radiation safety. Learning the radiation resistance of materials and increasing of resistance of materials by modifying with radiation are perspective fields in materials science [8, 9].

Radiation effect is among the best methods to manage bitumen and bitumen-added materials modification processes [10].

For this reason the study of change structure and other physical and chemical properties of bitumen materials (aging) under radiation have a great scientific and practical importance. These researches have independent importance in point of searching new materials for insulation of radioactive waste materials [11]. We should point out that the researches [17–19] on the effect of ionizing radiations are limit though the aging of bituminous materials are studied with the effect of UV.

In this paper results of studies of effects of gamma radiation on the properties of hydroisolation materials based on bitumen produced from Azerbaijani oils [11] are adduced.

2 Experimental Analysis

The bitumen based waterproofing material applied in production has been dried in vacuum condition and sealed in the amount of 10 g. Samples have been irradiated in Co-60 gamma radiation source up to 300 h in P = 0,138 Gy/s dose rate. During irradiation formation kinetics of gases has been studied in Agilent GC 7890A chromatography device. Analyzes have been carried out by using He gas-carrier and 30 m microcolon in FID mode. Infrared spectrums were studied in the Fourier-spectrometer "Varian-220". Bitumen samples have been taken as thin layer in KBr plate, and spectrum have been made within 400–4000 cm^{-1} interval of wave number. Spectrum analysis has been conducted with comparison table [20].

Thermal features of samples has been studied by derivatographic method in Perkin Elmer STA 6000 device, change in enthalpy, activation energy, heat-mass exchange up to 900 °C temperature has been observed depending on absorbed radiation dose.

3 Results and Discussion

Infrared Spectra of Bitumen. The IR-spectroscopy method revealed a change in its structural group composition during the irradiation of the bitumen mastic with γ-rays. Irradiation under the γ-rays was done at intervals of D = 2, 5–100, 0 kGy (irradiation time 5–200 h). The spectra of 3 samples as a examples shown in the Figs. 1, 2 and 3.

The following characteristic absorption bands are observed in the IR spectrum (Fig. 1) of the initial non-irradiated sample of bitumen:

- Valent oscillations of CH groups in 5 and 6-membered saturated rings at 2925, 2865 cm^{-1} -CH_2 and CH_3 groups;
- Deformation vibrations of CH_2 and CH_3 groups in branching structures of paraffin at 1455, 1375 cm^{-1};
- Oxygenous components (C—O—C), (C—OH) at 1100 cm^{-1};
- Valent oscillations of aromatic ring compounds, naphthenes and condensed aromatic compounds at 3018, 1032, 799 cm^{-1}

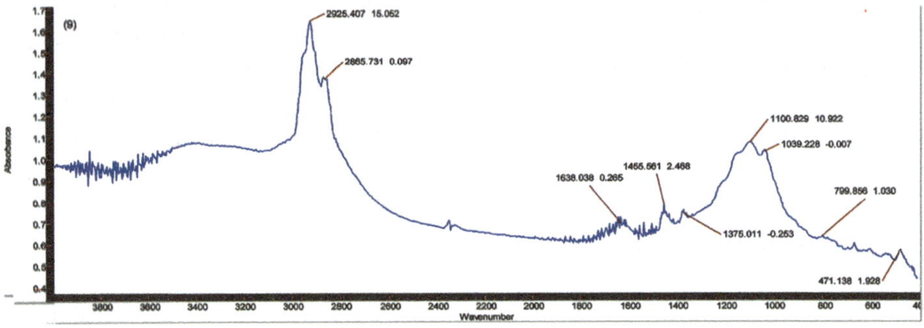

Fig. 1. İR spectrum initial (non-irradiated) samples of bitumen composite, D = 0

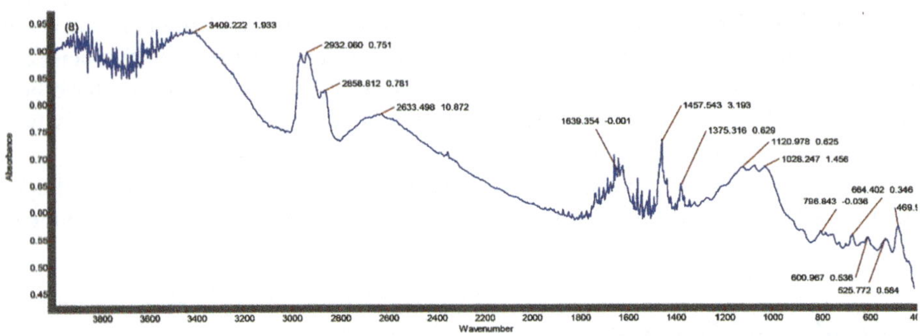

Fig. 2. IR spectrum irradiated samples of bitumen composite, D = 100 kGy (200 h)

Changes in the absorption bands at 3018, 1030, 1100 cm^{-1} occur with increasing of radiation dose, which is due to the reduction of oxygenous and aromatic groups. Additionally, changes in CH_2 and CH_3 groups occur.

Changes in the IQ-spectrum absorption bands are observed at 100 kGy (Fig. 6) - 900–730 cm^{-1} and 1639–1650 cm^{-1} intervals - the intensity of the absorption bands corresponding to deformation oscillations of multicolor aromatic compounds are increasing. It is assumed that the condensed aromatic compounds related to polymerization are

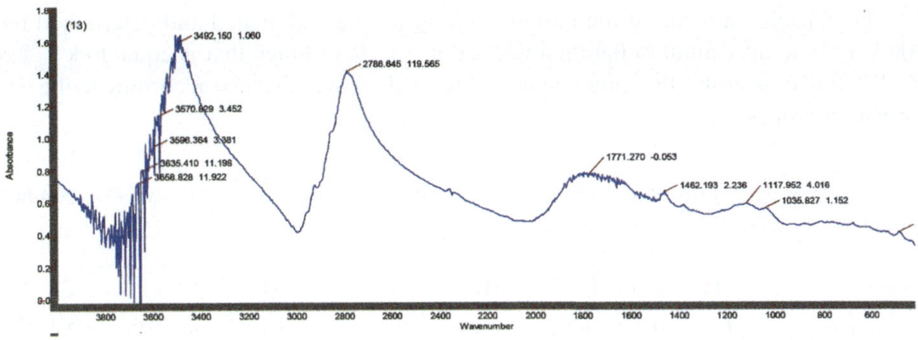

Fig. 3. IR spectrum irradiated samples of bitumen composite, D = 150.0 kGy (300 h)

increasing. Based on IR spectra, it can be said that the bitumen mastic is resistant to the radiation.

Gas Formation. The radiation and chemical decomposition of the bitumen were investigated in the D = 0–100 kGy range of doses of hydrogen, carbon monoxide and C_1-C_6 hydrocarbons. Gas formation kinetics is given in Fig. 4. It is clear from the comparison of kinetic curves that the yield from carbon monoxide and hydrogen is at least a thousand times bigger than gases C_2-C_6. These yields are typical for compounds rich in polycyclic aromatic compounds. The relatively high amount of hydrogen and carbon monoxide yields is due weakly bound oxygenous functional groups in the bitumen [10, 18] [21].

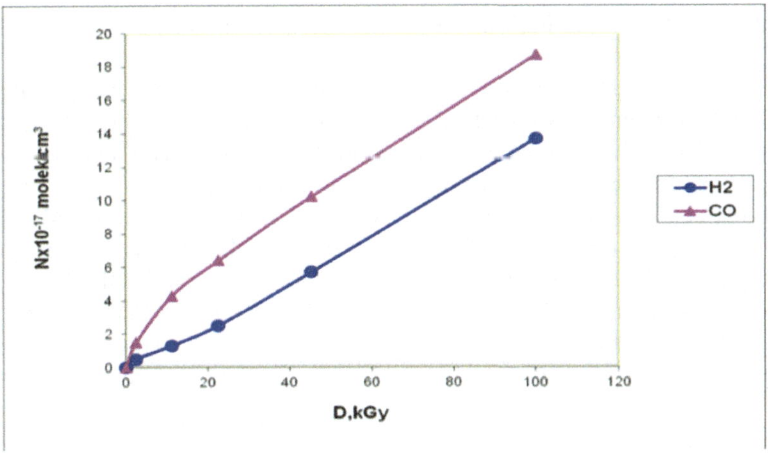

Fig. 4. Dependence of hydrogen and carbon monoxide formation on the absorbed dose

The average value of radiation- chemical yield (produced molecules per 100 eV absorbed radiation energy) of gases from these curves are given in Table 1.

The conversion value of the bitumen into gases was calculated and determined for 100 kGy - the maximum radiation dose, and it was determined that is equal to k = 2.3 $10^{-3}\%$. It demonstrates that bitumen has a high radiation resistance according to the gas formation process.

Table 1. Average grades of radiation-chemical yields of gases at the radiolysis of bitumen composite

Gases	H2	CO	CH_4	C_2H_6	C_2H_4	C_3H_8	ΣC_5	ΣC_6
G (average) molecule/100 eV	0.12	0.21	0.008	2.3 10^{-5}	4.3 10^{-5}	2.3 10^{-5}	3.9 10^{-5}	2.2 10^{-5}

High radiation resistance of bitumen is related with being a great quantity of polycyclic aromatic compounds in its composition and occurring disintegration only in peripheral functional groups as a result of radiolysis:

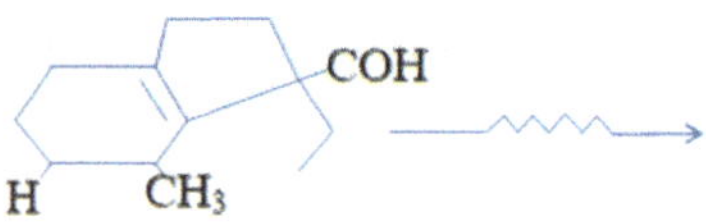

H, CH_3, CO (1)

$$H + H \rightarrow H_2 \tag{2}$$

$$CH_3 + H \rightarrow CH_4 \tag{3}$$

H and CH3 radicals in regard with majority of unsaturated chemical bonds are attained with these bonds and the emission of molecular products become less. These unsaturated bonds provide high radiation resistance of bitumen by playing a play of "antirad".

Radiation-initiated destruction and polycondensation reactions by participation of polycyclic macroradicals ultimately determine the radiation stability of the bitumen composition. The main parameters that determine the competitiveness of these reactions are the intensity of ionizing radiation (P) and temperature (T).

Yuxarıda verilmiş sadə kinetik sxemə politsiklik makroradikalların (M) aşağıdakı kimyəvi reaksiyaları da daxil etsək destruksiya və polikondensasiya reaksiyalarının sürətlərinin P və T parametrlərindən asılılığnı matematik ifadə edə bilərik:

If we include the following chemical reactions of polycyclic macroradicals (M*) in the above simple kinetic scheme (1–3), we can mathematically express the dependence of the rates of destruction and polycondensation reactions on the parameters P and T:

$$M^* + M \rightarrow M_1^* + RH\,(gas) \quad (Destruction) \tag{4}$$

$$M^* + M^* \rightarrow M - M \quad (Polycondencation) \tag{5}$$

Condensation processes increase the radiation resistance of the material. The predominance of reaction (5) in the competition of reactions (4) and (5) ensures the radiation resistance of bitumen.

The change in the concentration of polycyclic macroradicals from a simple kinetic model is described by the formula:

$$\frac{dM^{\cdot}}{dt} = G_{M*} P10^{-2} - K_4[M][M^*] - K_5[M^*]^2 \quad (6)$$

where Gm* - radiation-chemical yield of active macroradicals, 1/100 eV, P-radiation dose rate, eV/g s, Ki- the rate constant of the i reaction, [M] - the density of the initial conditional molecules, [M *] - concentration of macromolecules activated by radiation, Ei - activation energy of i reaction, In the quasi-stationary case, this formula is as follows.

$$K_5[M^*]^2 = G_{M*}P10^{-2} + K_4[M][M^*] \quad (7)$$

Condition for the polycondensation process will prevail under destruction is

$$K_5[M^*] \gg K_4[M] \quad (8)$$

If we replace the expression M*

$$[M^*] = \sqrt{\frac{G_{M*}P10^{-2}}{K_s}} \quad (9)$$

Finally, for the lower limit of the radiation dose rate, we obtain the following expression

$$\boldsymbol{P} > \frac{K_{04}{}^2[M]^2 e^{-2E4/RT}}{K_5\, G_m \cdot 10^{-2}} \quad (10)$$

Polycondensation processes prevail in the values of dose rate determined by this expression and the radiation resistance of the material increases.

The effect of temperature is in different directions. Acceleration of diffusion in the temperature range 150–160 °C increase the reactions rates of macroradicals. Increasing the rates of detachment reactions at higher temperatures (160–400 °C) accelerates destruction processes (10).

Pure thermostimulated destruction reactions occur at temperatures above 400 °C. The ratio of the rates of radiation-thermal (rt) and thermal (t) reactions (Wrt/Wt) under the simultaneous action of radiation and thermal effects is determined by the following formula:

$$W_{rt}/W_t = \frac{W_t + W_r}{W_t} = 1 + \frac{G_0 P10^{-2}}{A\exp(-E_0/RT)[M]} \quad (11)$$

where A is the steric factor of the thermal reaction rate, Eo is the activation energy of that reaction.

Thus, it is possible to purposefully manage the radiation resistance of bitumen by pre-selecting the dose rate and temperature.

4 Derivatographic Measurement

Through derivatographic measurements, we can observe the investigation of heat exchange capacity with environment and disintegration of organic mass under the heat. From this perspective, the changing regularity of these properties of the bitumen under the radiation effect is studied and as examples initial and irradiated sample indicated in Figs. 5 and 6.

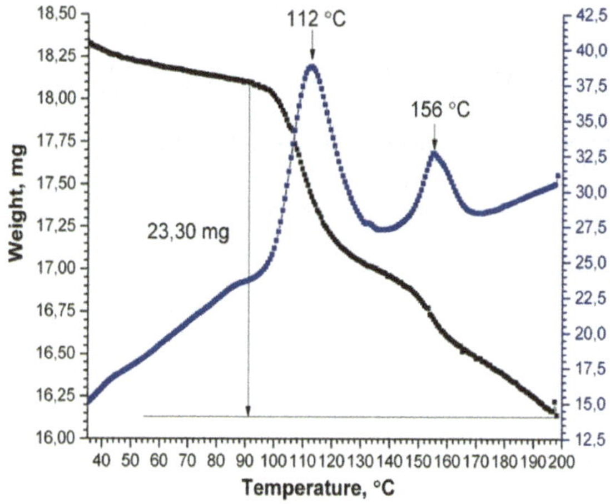

Fig. 5. Bitumen D = 0

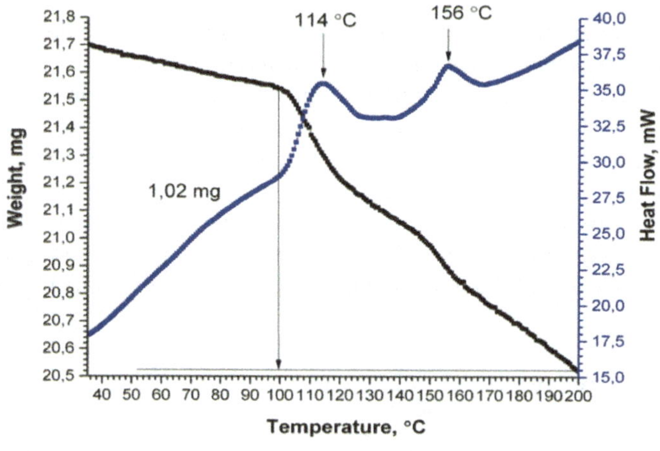

Fig. 6. D = 100 kGy (200 h)

Figure 5 shows the thermogravimetric analysis of the non-irradiated sample. The black line shows the kinetics of mass change and the blue ones shows the change of heat flux in the graphic description. Practically no change occurs in the mass up to T

= 97 °C. The absence of a specific change in the mass is observed in the heat flux as the specimen absorbs energy up to T ≤ 970 °C (linear part of the blue line). However, starting from T ≥ 97 °C there is a noticeable change both in the mass and in the energy line (the amount of heat absorbed).

The kinetic description of rapid disintegration in the T ≥ 97 °C–139 °C interval in the mass spectrum consists of two parts. The deep disintegration with 112 °C central peak is noted with endo-effect observed in the heat flux that the activation energy-one of the heat values of this process is 0.0619 kC/mol. The second effect - T = 126 °C, is marked with a special bent in mass kinetics.

The third change of mass, value of activation energy of the second endo-effect with 156 °C central peak is 0,0154 kC/mol.

The thermogravimetric description specimen irradiated (D = 12 kGy) for 5 h. Unlike the irradiated specimen, the change of both mass and heat amount consist of two parts. The same amount of heat is absorbed by the example up to T = 97 °C. The central peak of endo-effect is 108 °C whist starts at T = 97 °C and ends at T = 136 °C. The activation energy of this process is equal to 0.0465 kC/mol. There was a mass loss of 2.41% during the mass kinetics of this process. The activation energy of the second endo-effect in the heat flux spectrum is 0.146 kC/mol. It is recorded with 0.940% loss in the mass kinetics.

Analyzes were made for specimens irradiated in 4 more higher dozes (12, 24, 48,100 kGy) (Example for D = 100 kGy, see Fig. 6) and it was found that the thermal effects (endo-thermic) observed at two temperatures (112 °C and 156 °C) are diminished as the dose increases. Accumulation of radiation energy in the specimen and its segregation in the temperature rise reduce the endothermic effect. The complexity of the structure and composition of the specimen do not allow for a clear description of the mechanism of processes occurring.

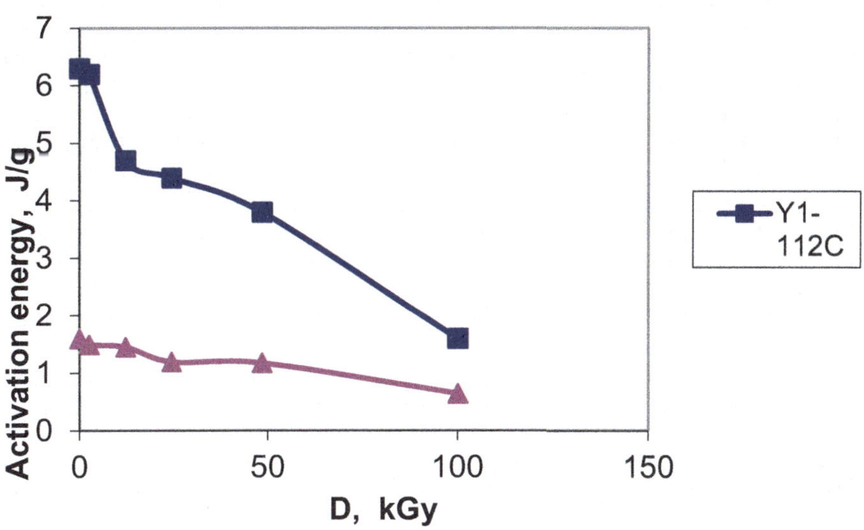

Fig. 7. Change of the activation energy of the specimen depending on dose

The numerical analysis of the derivatograms of the initial and irradiated samples was carried out in accordance with the instructions of the device and the dependence of these parameters (heat capacity, activation energy, enthalpy and field energy) on the absorbed dose were established. The dependence of activation energy on absorbed dose indicated in the Fig. 7 as an example. All these 4 parameters related to each other diminish depending on the specimen absorption dose for both temperatures (112 °C və 156 °C).

Thus, the heat absorption capacity of the specimen decreases in a result of the accumulation of radiation energy during radiation of bitumen specimen before of effect of heat and this effect is even stronger as the dose increases.

5 Conclusion

In the oil bitumen that forms the basis of hydroisolation materials, polycondensation processes dominate under radiation, small gas separation and a decrease in heat exchange parameters with the environment up to 200 °C occur during these processes.

Studies conducted with every 3 methods show that as the radiation dose increases, the heat absorption capacity of the samples decreases. 0,0023% of the Conversion rate of the specimen into gases and the absence of significant change in the radiation exposure in the IR spectrum indicate the high radiation resistance of the bitumen.

References

1. Ojovan, M.I., Lee, W.E.: An Introduction to Nuclear Waste Immobilisation, 2nd edn., Elsevier, Amsterdam, 362 p. (2014)
2. Sercombe, J., Gwinner, B., Tiffreau, C., Simondi-Teisseire, B., Adenot, F.: Modelling of bituminized radioactive waste leaching. Part I: Constitutive equations. J. Nucl. Mater. **349** 96–106 (2006)
3. Mobasher, N., Bernal, S.A., Kinoshita, H., Sharrad, C.A., Provis. J.L.: Gamma irradiation resistance of an early age slag-blended cement matrix for nuclear waste encapsulation. J. Mater. Res. **30**, 1563–1571 (2015)
4. Menapace, I., Yiming, W., Masad, E.: Chemical analysis of surface and bulk of asphalt binders aged with accelerated weathering tester and standard aging methods. Fuel **202**, 366–379 (2017)
5. Lins, V.F.C., Araújo, M.F.A.S., Yoshida, M.I., Ferraz, V.P., Andrada, D.M., Lameiras, F.S.: Photodegradation of hot-mix asphalt. Fuel **87**, 3254–3261 (2008)
6. Durrieu, F., Farcas, F., Mouillet, V.: The influence of UV aging of a Styrene/Butadiene/Styrene modified bitumen: comparison between laboratory and on site aging. Fuel **86**, 1446–1451 (2007)
7. Feng, Z., Yu, J., Zhang, H., Kuang, D., Xue, L.: Effect of ultraviolet aging on rheology, chemistry and morphology of ultraviolet absorber modified bitumen. Mater. Struct. **46**, 1123–1132 (2013)
8. Xu, S., Yu, J., Wu, W., Xue, L., Sun, Y.: Synthesis and characterization of layered double hydroxides intercalated by UV absorbents and their application in improving UV aging resistance of bitumen. Appl. Clay Sci. **114**, 112–119 (2015)
9. Yakubov, K.M.: Obtaining a radiation-resistant bitumen composition. Oil Refining Petrochemicals **8**, 22–24 (2004)
10. Samedova, F.I., Allahverdiyev. A.A.: Technology for bitumen. Baku. Elm. 155 p. (2007)

11. Mouillet, V., Farcas, F.: Study of UV rays effects on the evolution of bituminous mix behavior. In: E&E Congress 2016, Prague, Czech Republic (2016)
12. Zeng, W., Wu, S., Wen, J., Chen, Z.: The temperature effects in aging index of asphalt during UV aging process. Constr. Build. Mater. **93**, 1125–1131 (2015)
13. Liu, X., Wu, S., Liu, G., Li, L.J.M.: Effect of ultraviolet aging on rheology and chemistry of LDH-modified bitumen. Materials **8**, 5238–5249 (2015)
14. Feng, Z., Bian, H., Li, X., Yu, J.: FTIR analysis of UV aging on bitumen and its fractions. Mater. Struct. **49**, 1381–1389 (2016)
15. Mouillet, V., Farcas, F., Besson, S.: Ageing by UV radiation of an elastomer modified bitumen. Fuel **87**, 2408–2419 (2008)
16. Mello, M.S., Braz, D , Motta, L.M.G.: Investigation on the effects of gamma-irradiation on bitumen. In: International Nuclear Atlantic Conference - INAC 2011 Belo Horizonte,MG, Brazil, 24–28 October 2011
17. Kovaleva, O.V., Yushkin, N.P.: Effect of radiation on the molecular structure of natural solid bitumen. Rep. RAS **410**(4), 516–518 (2006)
18. Michael, I., Ojovan, M.I.: Cement and Bitumen: Radiolysis Effects/Nuclear Engineer. Department of Nuclear Energy, IAEA, Vienna (2013)
19. Tarasevich, B.N.: IR spectrums of basic classes of organic compounds. Handbook. Moscow, 55 p. (2012)
20. Pikayev, A.K.: Sovremennaya radiasionnaya ximiya. Moskva, "Nauka", 448 p. (1987)

Analysis of Influence of Imagination, Fantasy, Exaggeration, and Hoax on a Level of Lie Under Linguistic Information

Farida Huseynova(✉)

Azerbaijan State Oil and Industry University, Azadlig Avenue, 20, 1010 Baku, Azerbaijan
farida_hus@hotmail.com

Abstract. In the article we try to attempt discerning a difference between imagination, fantasy, exaggeration, and hoax. Being able to use them the person will make it much easier to manipulate the others by their behaviors. Our analysis is aiming to reveal it's good and bad sides, and explain what motivates individuals to resort to such questionable communication methods to influence on human affairs. Information related to the considered four factors is often described linguistically. We propose to model influence of the considered four factors to a level of lie by using fuzzy rules.

Keywords: Imagination · Fantasy · Exaggeration · Deception · Lie · Relationship · Trust

1 Introduction

And what's the difference between the four? In unfamiliar, pervasive environments it is hard to overcome initial suspicion and allow secure collaboration to take place without differentiation of them.

All of them refer to whether to spare someone's feelings or to dupe someone in high-stakes negotiation that, occurs every day in life and in business. People with a good imagination, fantasy, exaggeration, and hoax are better in terms of reproductive success compared to those who were always honest. Being able to use imagination, fantasy exaggeration and hoax, the person will make it much easier to manipulate the others by their communication and behaviors, as there are no strictly controlled environments. People will have to be capable of dealing with unforeseen circumstances from unexpected interactions with other people.

People are constantly looking to see who they can trust and who they can't. People are actually much more perceptive and aware of who tells the truth and who doesn't. However, it is not always possible to see what is right or what is wrong. Then, how can you trust and respect people more and more. Over time, honesty shows itself as a trait that is beautiful and deeply respected.

Our analyses is aiming to reveal what motivates individuals to resort to such questionable communication methods, influencing on human affairs. It can be the security

© The Author(s), under exclusive license to Springer Nature Switzerland AG 2021
R. A. Aliev et al. (Eds.): WCIS 2020, AISC 1323, pp. 358–363, 2021.
https://doi.org/10.1007/978-3-030-68004-6_47

policy that encompasses billions of potential collaborators. Information related to the considered four factors is often described linguistically. We propose to model influence of the considered four factors to a level of lie by using fuzzy sets [3].

2 Objectives

Imagination, fantasy, exaggeration, and hoax- all these four factors are products of lies, which have good and bad sides. The deeper we have gone into the subject, the more we see it's affects to everyone involved. We are always taught to be a person of integrity, it means your walk have to match your talk. We've learned to an open interpretation of that what we see, hear or read. Since, we are usually open-minded and in various circumstances we take a chance to differentiate them according to our judgments.

Actually, all of us have been affected by all these four factors, whether our own or someone else's. Sometimes we are encouraged by them as they put a lot more relief in our life. At times, they can increase people's capacity to build lasting relationships. They are mainly used to avoid punishment or to protect resources, to maintain relationships or bonds, to preserve "face" or the self-image [1].

All people like movies, cartoons, fictions, tales, drawings, music, dances, tricks, juggles and etc. which are based on these four factors. According to Walt Disney all cartoons, fables must be exaggerations. It's very nature of fantasy, fable and caricatures. But still we do not find this difference and examine the delicate lies in the sequence of events in a play, novel, movie, cartoon or similar works. Additionally, findings suggest that deception is not always harmful to subjects. Contrariwise, we enjoy them with pleasure, as we are entertained by them very much. In some cases, using deception is the only way to obtain certain kinds of information, and that prohibiting all deception in research would "have the egregious consequence of preventing researchers from carrying out a wide range of important studies" [2].

Around age two or three, children start to tell lies to hide wrongdoings. At this stage, children are not able to craft lies specific to the listener, meaning they might tell an adult that an elephant came into their apartment and ate all the dinner.

Being adults, they have much more advanced thinking than young children. More realistic and believable than primary lies, children around age four tell stories. The difference at this stage is that children are able to tell lies tailored to the understanding of the listener. So they begin to describe events, narrate tales and stories according to their own imagination, fantasy, exaggeration and hoaxes which improve their thinking abilities. These lies are more thought out, harder to pick out and include supporting statements to convince the listener.

Deception is a social norm. Politicians. Businessmen lie and become successful. Though, everybody knows that it is unethical behavior, in a lot of situations, it is considered acceptable. In social discourse, we lie daily so as not to offend other people. We say you look beautiful, instead of you look ugly. This is something we teach our children to do whether or not it is intentional. We guide them to avoid saying true things that might hurt others, or teach them to lie in order to spare their feelings (i.e., "Soup is delicious!").

In our opinion all these four factors- imagination, fantasy, exaggeration, and hoax that benefits from lying, can be manipulative in some situations. People who are more accurate in judging, can easily reveal it. All of them are the lies, that people appear deceptive when motivated to be believed. They can cause a loss for the partners. People judge others' deceptions more harshly than their own. Lying has always posed a moral problem. Aristotle wrote that "falsehood is in itself culpable;" St. Augustine believed that every lie is a sin; and Kant regarded truthfulness as an "unconditional duty which holds in all circumstances". Having been a moral issue for millennia, deception came also to be viewed as a legal challenge.

Let's analyze separately what are imagination, fantasy, exaggeration, and deception and see their differences.

2.1 The Imagination

It is one of the most unique and important aspects of being a human being. It is the synthesis of mental images into new ideas. It is the power of the mind to form mental representations of a thought, concept, dream, and symbol. The imagination has the power to create or re-create any sensation perceived or possibly perceived by the mind. Since sensation is reliant upon the body and the brain, the imagination cannot be entirely separated from either.

Imagination can be the opposite of reality. Instead, it is not always a means of adapting to reality. The degree of mental life could not exist without the imagination and it is quite different in different minds. Certainly creativity would be impossible without the imagination.

2.2 Fantasy

It is a product of the imagination. It is an imagined sequence of events or mental images (for example, a daydream). It is a form of a story from the self with one common factor: The subject imagining the fantasy appears as one of the actors in the story. A fantasy may originate from conflicts, desires, frustrations, or wishes when the imagination interacts with reality. A fantasy may substitute for action or pave the way for later action.

In Freudian terms, the fantasy may itself afford gratification for impulses (our darkest and most hidden drives), may serve the ego as a defense (our recognized self), or may take over superego functions by providing the imagery on which concepts are based (our self we project to others).

A fantasy can be a conscious or unconscious construction. When it is unconscious, it is sometimes spelled "phantasy". Again, mental life could not exist without fantasy. And fantasy is also a unique and important aspect of being a human being.

2.3 Exaggeration

It represents something as better as or worse than it really is. It means the describing of something and making it more than it really is. Exaggeration is truth that has lost its temper. According to American humorist Josh Billings "There are some people so addicted to exaggeration that they can't tell the truth without lying".

Experts say that 'yet in milder doses, exaggeration can be nearly impossible to detect". Exaggeration is very different psychologically from other forms of truth twisting.

The findings imply that some kinds of deception are aimed more at the deceiver than at the audience, and they may help in distinguishing braggarts and posers from those who are expressing personal aspirations, however clumsily.

Psychologists have studied exaggeration from all sides and have found that it usually puts a psychological or physical strain on the person.

An example of exaggeration would be: "I was walking along when suddenly this enormous dog walked along. It was as big as an elephant". The dog may have been big, but it was certainly not as big as that. Another example of exaggeration would be: "I caught a fish as big as my house."

Overstatement is another word that means almost the same thing. The opposite of overstatement is understatement.

People exaggerate things because they have strong feelings about something. People may exaggerate to make people listen to what they say. They may do it to emphasize something. They may also exaggerate just to sound funny: "I've heard that a million times".

2.4 Hoax

A hoax is a falsehood deliberately fabricated to masquerade as the truth. It is distinguishable from errors in observation or judgment. It is often intended as a practical joke or to cause embarrassment, or to provoke social or political change by raising people's awareness of something. It can also emerge from a marketing or advertising purpose.

A hoax is an act or statement which misleads, hides the truth, or promotes a belief, concept, or idea that is not true. It is often done for personal gain or advantage.

It includes several types of communications or omissions that serve to distort or omit the whole truth. Examples of hoax range from false statements to misleading claims in which relevant information is omitted, leading the receiver to infer false conclusions. For example, a claim that "coffee is beneficial to brain health more than other beverages, it is a stimulant for health, with continuing research on whether long-term consumption reduces the risk of some diseases, and they must be used 7–9 times a day" may be misleading, as it leads the receiver to believe it. In fact, so much consumption of coffee can lead to high heart pressure as it is not particularly good for brain health. This claim leads the receiver to infer false information.

Sometimes a person can disguise by an appearance to create the impression of being somebody or something else; for a well-known person this is also called incognito. Passing involves more than mere dress and can include hiding one's real manner of speech.

3 Methods

Lying has gotten a bad rap. When you deal with people all should take into consideration the value of honesty. Theodor Schaarschmidt tells "in fact, lie is among the most sophisticated accomplishments of the human mind. But how can one tell if a person is fibbing? Today many philosophers take a more nuanced view. German philosopher Bettina Stangneth argues that lying should be an exception to the rule because, in the final analysis, people rely on being told the truth in most aspects of life. Among the reasons they lie, she notes in her 2017 book "Deciphering Lies", is that it can enable them to conceal themselves, hiding and withdrawing from people who intrude on their comfort zone. It is also unwise.

We think that, people should be able to have a moral understanding to differentiate mistakes and "bad" or antisocial lies. Misunderstanding of good lies and bad lies can bring to bad results. Of course, not everyone agrees that some lying is necessary, it is the product of our brain in the form of imagination, fantasy, exaggeration or hoax. Some of thinkers have lined up against this perspective. Yet how much do researchers know about lying in our daily existence? How ubiquitous is it? When do children usually start engaging in it?

In this article, we consider trustful judgments to insight the double standard for evaluating imagination, fantasy, exaggeration and hoax. We got different answers, as evaluation was inappropriately affected by variations in judgmental bias. In our own kind, small children love to make up stories, but they generally tell their first purposeful lies at about age four or five. Before starting their careers as con artists, writers, producers and etc. children must first acquire important cognitive skills, which can result their creativity. But, at the same time they can become great liars as we don't prevent their mastering lies. One is deontic reasoning: the ability to recognize and understand social rules becomes very harsh.

We have surveyed 20 people of different ages, asking whether it is good or bad to lie? Whether they lie a lot? In which circumstances do they do it? If other people always trust them, detecting their lie? Have children to learn to lie? Are you good at detecting lies? Is discovering lying in young children a sign that they have mastered some important cognitive skills? It's good if you are having exaggerated imagination and thinking.

Current research shows that 12 people of elder generations under the age 40–65, typically tell the truth more easily than they tell a lie and that lying requires far more cognitive resources. They say, what they think. Sometimes, they can tell lies, by not noticing it, because of facts and details. They are against fantasy and exaggeration, and hoax and children must be taught not to lie. As such there are no paths to overcome it but you must control it, you must be good at perceptions and judgments to notice lie as it can inherently be linked to higher risks.

They told that, deception violates relational rules and is considered to be a negative violation of expectations. Most people expect friends, relational partners, and even strangers to be truthful most of the time. If people expected most conversations to be untruthful, talking and communicating with others would require distraction and misdirection to acquire reliable information. Moreover, when liars' communication and behaviors are so artificial, they can hardly have anything in common with other people. Hoax often leads to feelings of betrayal and distrust between relational partners. Using exaggeration, fantasy as one of their tools people can be manipulated or led astray.

6 people under age 17–30 answered that, imagination, fantasy may lead to make people more creative. Who doesn't want creativity? Young people are good in exaggerations. Even in their communication they prove it, they persuade others very easily. In most cases, at different situations or hard times people themselves need to be deceived.

4 people under age 30–40 answered that, exaggeration, fantasy are very useful and young generation must be motivated to use their creativity as it is necessary for all them. We have to overcome the stereotypes that they do harms to people.

But perhaps just as important, survey shows that lying is a major component of the human behavioral repertoire; without it, we would have a hard time coping. When people speak unvarnished truth all the time, they tend to be judged tactless and hurtful. In everyday life, we tell little white lies all the time, if only out of politeness: Your homemade pie is awesome (it's awful). No, Grandma, you're not interrupting anything (she is). A little bit of pretense seems to smooth out human relationships without doing lasting harm.

Due to imprecision of information related to imagination, fantasy, exaggeration, and hoax, we propose to use fuzzy If-Then rules to describe influence of this factors to a level of lie. The fuzzy rule may have the following form:

Rule *i*: If imagination is A_{i1} and fantasy is A_{i2} and exaggeration is A_{i3} and hoax is A_{i4} Then lie is B_i.

In this rule, $A_{i1},\ldots, A_{i3}$ are linguistic evaluations of levels of the considered factors, and B_i is the corresponding level of lie. The linguistic evaluations are described by fuzzy sets.

4 Conclusion

Our research considers that people regard lying as unexceptional skill. Evaluating four factors -imagination, fantasy, exaggeration and hoax as truths or lies, the answers were quite different, because they were affected by variations of judgmental biases.

According to our queries of surveys we learned that some people;

1. are against of any lie.
2. lies are inevitable, it is part of our life.
3. accept them as important cognitive skills which must be acquired from early childhood.

To model influence of the considered four factors to a level of lie under linguistic information, we assume to use fuzzy rules.

References

1. Buller, D.B., Burgoon, J.K.: Interpersonal deception theory. Commun. Theory **6**(3), 203–242 (1996). https://doi.org/10.1111/j.1468-2885.1996.tb00127.x
2. Kimmel, A.J.: In defense of deception. Am. Psychol. **53**(7), 803–805 (1998). https://doi.org/10.1037/0003-066X.53.7.803
3. Zadeh, L.A.: Fuzzy sets. Inf. Control **8**, 338–353 (1965). https://doi.org/10.1016/S0019-9958(65)90241-X

Governance Model the Stochastic Process of Rectification of Multicomponent Mixtures Based on Fuzzy Logic

Avazov Yusuf Shodiyevich(✉)

Department of Automation of Production Processes, Tashkent State Technical University named after Islam Karimov, Tashkent, Uzbekistan
yusufbek_avazov@mail.ru

Abstract. The problem of controlling a stochastic process of separating multicomponent mixtures based on fuzzy logic is considered. Taking into account that the process of separation of multicomponent mixtures is one of the most difficult in order to confirm the multi-dimensional and multi-connectedness of the rectification column, considered as a control object, the parameters of the selected control object are shown. The distillate concentration is taken as the main parameter and a dependence is revealed to determine its distribution over the column plates. To simplify obtaining a mathematical model of the process, the influence of the steam flow supplied to the column was not taken into account. The terms of the membership function are formulated through the linguistic variable "distillate concentration" and its graphs are obtained. With the help of three fuzzy logic-based pulsed values inherent in the regulation of distillate concentration: "very low", "low" and "normal", terms-sets are formed for each state. Transition matrices are constructed describing changes in accordance with the concentration of the distillate. Graphs were obtained on the basis of the values of the accessory function "increase the concentration of the initial mixture". Controlling the concentration of alcohol in the distillate by changing the temperature difference between the top of the column and the feed tray is considered. At the same time, on the basis of production experiments, it was revealed that the value of the optimal temperature difference is (10–11) °C. A description of the regulation of the distillate strength by changing the amount of reflux using linguistic variables has been made. On the basis of a fuzzy decision-making model, a state control method was introduced and a state function was obtained at the ith step of the process control. According to the number of marks of the terms "Concentration of the initial mixture", "Temperature difference" and "Amount of reflux", a list of states is formed.

Keywords: Linguistic variables · Rectification of multicomponent mixtures · Membership function · Concentration of distillate · Amount of reflux · Temperature drop · Situational control

1 Introduction

Based on the analysis of the process of rectification as a control object, it is established that the control object is multidimensional and multi-connected (Fig. 1). This complicates

© The Author(s), under exclusive license to Springer Nature Switzerland AG 2021
R. A. Aliev et al. (Eds.): WCIS 2020, AISC 1323, pp. 364–376, 2021.
https://doi.org/10.1007/978-3-030-68004-6_48

the control and management systems of the entire technological process. Therefore, there is a need to reduce its multi-dimensionality and multi-connectivity. One option is to reduce the process in columns to the separation of a pseudo-binary ethanol-water system (main components).

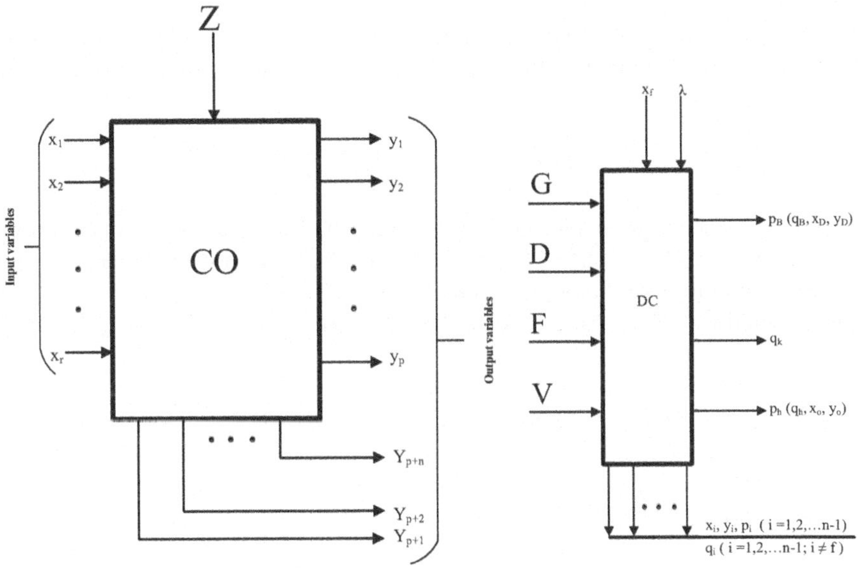

Fig. 1. The rectification column as a multidimensional object.

Then the distribution of concentrations in the column is uniquely determined by the distribution of only one component on the plates – ethanol, and the regulated values are determined by the dependencies:

$$x_i = f(P_i, t_i), \quad (1)$$

$$t_i = \phi(x_i), \quad (2)$$

where x_i, t_i, P_i – the concentration of alcohol (ethanol) in the liquid phase, the temperature and pressure on the i-th plate, respectively.

A significant reduction in the impact of multi-dimensionality and multi-connectivity can be achieved through the technological decomposition of the distillation unit (DU) into simpler blocks and considering the processes occurring in each column as an independent control object. The possibility DU decomposition is related to its technological features. From the point of view of technological topology, the indirect-action DU is a weakly closed chemical-technological system [1].

2 Main Part

In particular, at the production site under study (JSC "Biochim" in the city of Yangiyul) the technological process and, consequently, its mathematical model contain only one reverse technological relationship between the rectification column and the epuration column for the consumption of non-standard alcohol (the head fraction of the epuration column). It is important to note that the process flow between the columns is in a liquid state and, therefore, there is no connection between the columns along the steam flow [2]. This significantly reduces the influence of subsequent columns on the previous one's operating mode.

Using the example of the linguistic variable "Distillate concentration", we will form the terms of the membership function. The linguistic variable concentration has the form [3]:

<"Distillate concentration-C", C', X>, (3)

where $C' = \{$"concentration very low", "concentration low", "normal concentration"$\}$;

X = {0; 0,1; 0,2; 0,3; 0,4; 0,5; 0,6; 0,7; 0,8; 0,9; 1,0} (deviation of percent by volume from the concentration of the output distillate 96.6% by volume).

These functions of linguistic variable have the form:

1. <"Concentration very low", [0, 1], $C_{v.low}$>
   $C_{v.low}$ = {<1/0>, <0,609/0,1>, <0,381/0,2>, <0,254/0, 3>, <0,179/0,4>, <0,132/0,5>, <0,101/0, 6>, <0,08/0,7>, <0,065/0,8>, <0,054/0, 9>, <0,055/1>}.
2. <"Concentration low", [0, 1], $C_{low.}$>
   C_{low} = {<0,582/0>, <1/0,1>, <0,582/0,2>, <0,347/0, 3>, <0,23/0,4>, <0,164/0,5>, <0,123/0, 6>, <0,096/0,7>, <0,077/0,8>, <0,063/0, 9>, <0,066/1>}.
3. <"Normal concentration", [0,9], C_{normal}>
   C_{normal} = {<0.069/0>, <0,199/0,1>, <0,535/0,2>, <1/0,3>, <0,572/0, 4>, <0,232/0,5>, <0,144/0,6>, <Is 0.102/0, 7>, <0,075/0,8>, <0,059/0, 9>, <0,519/1>}.

Let's present the membership functions in graphical form (Fig. 2).

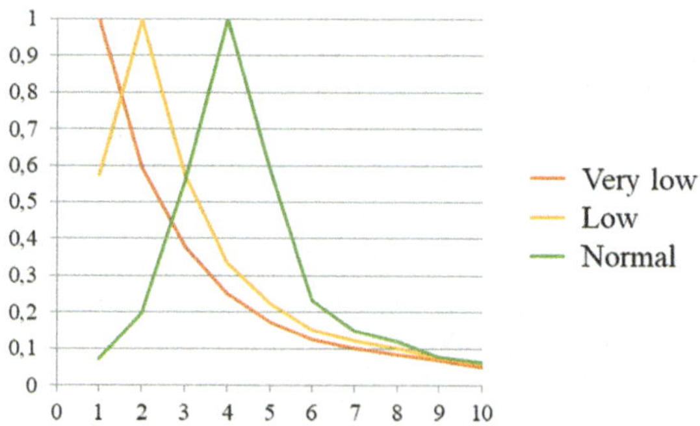

Fig. 2. The membership function of the "Concentration of the distillate".

3 Description of Possible Control Methods Based on Fuzzy Logic

The control variables are represented as:

1. C, <C_1>, X), the state "very low" for the "Distillate Concentration" parameter is the worst option, but with the help of the following two transitions, you can go from it to normal:

   "very low" → "low" → "normal".
   Term set for the "very low" state:

$$<C_1> = \{\text{"Significantly increase", "Increase"}\}. \tag{4}$$

2. (C, <C_2>, X), the "normal" state for the "Distillate Concentration" parameter is an ideal option and does not require changing the values of the controlled parameters.

   Term set for the "normal" state:

$$<C_2> = \{\text{"Do not modify"}\}. \tag{5}$$

Let's represent the state of the system, which is subject to various influences, in the form of a transition matrix. From the two control variables listed above, 3 transition matrices are formed as a result [4].

Denote the term set C = {"very low", "low", "normal"} with the following set:

$$C = \{C_1, C_2, C_3\}. \tag{6}$$

Transition matrix and possible transitions under the influence:

a) Significantly increase the

$$M_{C_{11}} = \left\| \begin{array}{cccccc} & C_1 & C_2 & C_3 & C_4 & C_5 \\ C_1 & 0 & 0,1 & 1 & 0,3 & 0 \\ C_2 & 0 & 0 & 0,1 & 1 & 0,3 \\ C_3 & 0 & 0 & 0 & 0,1 & 1 \\ C_4 & 0 & 0 & 0 & 0 & 1 \\ C_5 & 0 & 0 & 0 & 0 & 1 \end{array} \right\| \tag{7}$$

b) Increase

$$M_{C_{12}} = \left\| \begin{array}{cccccc} & C_1 & C_2 & C_3 & C_4 & C_5 \\ C_1 & 0,1 & 1 & 0,2 & 0 & 0 \\ C_2 & 0 & 0,1 & 1 & 0,2 & 0 \\ C_3 & 0 & 0 & 0,1 & 1 & 0,2 \\ C_4 & 0 & 0 & 0 & 0,1 & 1 \\ C_5 & 0 & 0 & 0 & 0 & 1 \end{array} \right\| \tag{8}$$

c) Do not change

$$M_{C_{22}} = \left\| \begin{matrix} & C_1 & C_2 & C_3 & C_4 & C_5 \\ C_1 & 1 & 0 & 0 & 0 & 0 \\ C_2 & 0 & 1 & 0 & 0 & 0 \\ C_3 & 0 & 0 & 1 & 0 & 0 \\ C_4 & 0 & 0 & 0 & 1 & 0 \\ C_5 & 0 & 0 & 0 & 0 & 1 \end{matrix} \right\| \quad (9)$$

4 Adjust the Distillate Concentration in Terms of Fuzzy Logic

Concentrations of distillate components during rectification of multicomponent mixtures are regulated by changing the following technological parameters (Table 1).

Table 1. Change of technological parameters when adjusting the distillate concentration

№	Process parameter	Dimension
1	Concentration of components in the initial mixture	% vol. fraction
2	Temperature drop (point of feed and the top of the column)	°C
3	The amount of phlegm	M^3/h

Representation of regulation by changing the concentration of components of the initial mixture (1-parameter in Table 1). Changing the concentration of the components of the head fraction (ether-aldehyde fraction–EAF) in the initial mixture allows you to qualitatively change the process of rectification. When changing the concentrations of EAF components in the initial mixture, control actions should be applied to the entire process of rectification. Changing the feed quantity of the initial mixture does not affect the concentration of EAF in the distillate, since the concentration of the initial mixture is related to the indicators of the brew. The fermentation process can last a maximum of 72 h (the imaginary concentration of alcohol in the brew is (10–12)% volume fraction).

The concentration of the initial composition is measured in the fermentation compartment and the brew is sent to the heat exchanger before the rectification process. The readings of the control devices differ from the actual data.

This means that the measured concentration value has some error and significant inertia in its dynamics. In practice, the reading error is $\pm$ (0.5 $\div$ 1)% of the volume fraction.

Let's denote the interval of maintaining the concentration of the initial mixture [Cmin, Cmax] within certain limits of which it is necessary to maintain the value of the concentration of alcohol in the feedstock.

The absolute value is not used, because other parameters of the production process can affect this value regardless of the initial conditions and current state of the rectification process.

Linguistic variables [5] are used to describe the effect of concentration on the initial mixture and to estimate the relative change in concentration in the distillate:

1) <"Increase C_{mix}", $C_{mix\ inc.}$, $X_{mix\ inc.}$>, where $X_{mix\ inc.} = \{$"Increase", "Significantly increase"$\}$, $X_{mix\ inc.} = \{1, 2\}$;
2) <"Do not change C_{mix}", $C_{mix\ is\ not\ change}$, $X_{mix\ is\ not\ change}$>, $C_{mix\ is\ not\ change} = \{$"Do not change"$\}$, $X_{mix\ is\ not\ change} = \{0\}$.

The function "Increase C_{mix}" is shown in Fig. 3.

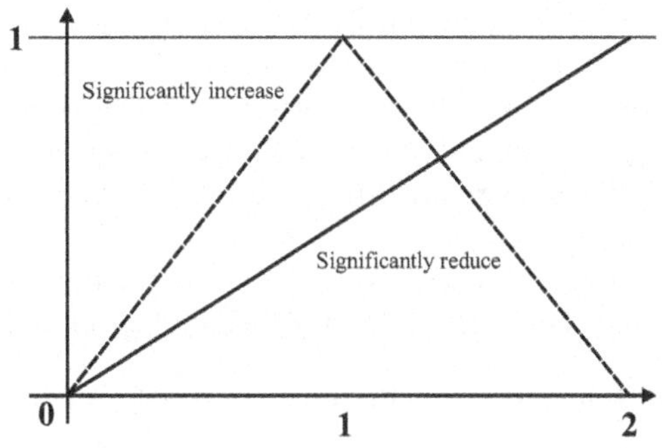

Fig. 3. Membership function "Increase the concentration of the initial mixture".

For temperature States under the control "Increase C_{mix}", we will form a matrix of transitions "Increase" (Table 2).

Table 2. Transition matrix of "Increase"

	"very low"	"low"	"normal"
"very low"	0,2	1	0,1
"low"	0	0,2	1
"normal"	0	0	0,2

For temperature States under the control <"Increase Cmix">, we will form a matrix of transitions "Significantly increase" (Table 3).

Table 3. Transition matrix of "Significantly increase"

	"very low"	"low"	"normal"
"very low"	0,1	0,3	1
"low"	0	0,1	0,3
"normal"	0	0	0,1

It should be noted that the regulation of the alcohol concentration in the initial mixture is possible due to changes in the aging time of the mixture in the breading vats (from 48 to 72 h).

If it is necessary to significantly change the concentration of alcohol in the initial mixture, it is added a certain amount of yeast and adjust the temperature regime of the fermentation process.

Regulation of alcohol concentration in distillate by changing the temperature difference between the feed plate and the top of the column (2-parameter in Table 1). In the process of rectification, the temperature difference between the food plate and the top of the column plays an important role [6]. According to the regulations, it is required to maintain the optimal value of the temperature difference in the processing areas of (15–18) °C. The results of experiment carried out in real production processes showed that the value of this temperature difference was 11 °C. This difference creates optimal conditions for the accumulation and concentration of alcohol in the upper part of the rectification column. If the temperature difference values are deviated, the concentration of alcohol in the distillate decreases.

Type of linguistic variables for regulating alcohol concentration in distillate by using the temperature difference between the food plate and the top of the column:

$$\langle \text{Increase the temperature difference} - \Delta T_{inc}, \Delta T_{inc}, X_{\Delta Tinc} \rangle,$$

where $\Delta T_{inc} = \{$"Increase", "Significantly increase"$\}$, $X_{\Delta Tinc} = \{5, 7, 9, 10\}$;

$$\langle \text{Reduce temperature difference} - \Delta T_{red}, \Delta T_{red}, X_{\Delta Tred} \rangle,$$

where $\Delta T_{red} = \{$"Reduce", "Significantly reduce"$\}$, $X_{\Delta Tred} = \{-10, -9, -7, -5\}$;

$$\langle \text{"Do not change } \Delta T\text{"}, \Delta T_{not\ change}, X_{C\ not\ change} \rangle,$$

where $\Delta T_{not\ change} = \{$"Do not change"$\}$, $X_{not\ change} = \{0\}$.

Due to the fact that the temperature difference affects the performance of the entire fracking process, the value of this parameter must be maintained within (10–11) °C.

For the effect of this parameter on the concentration of alcohol in the distillate, we present the relationship matrices. For temperature difference States under the control "Increase temperature difference – ΔT_{inc}", we will form a matrix of transitions "Increase" (Table 4).

Table 4. Transition matrix of "Increase"

	"very low"	"low"	"normal"	"high"	"very high"
"very low"	0,1	0,3	1	0	0
"low"	0	0,1	0,3	0	0
"normal"	0	0	0,1	1	0
"high"	0	0	0	0,1	1
"very high"	0	0	0	0	1

For the temperature difference States under the control "Increase temperature difference – ΔT_{inc}", we will form a matrix of transitions "Significantly increase" (Table 5).

Table 5. Transition matrix of "Significantly increase"

	"very low"	"low"	"normal"	"high"	"very high"
"very low"	0	0,3	1	0,1	0
"low"	0	0	0,3	1	0,1
"normal"	0	0	0	0,3	1
"high"	0	0	0	0	1
"very high"	0	0	0	0	1

For temperature difference States under the control "Reduce temperature difference – ΔT_{red}", we construct a matrix of transitions "Reduce" (Table 6).

Table 6. Transition matrix of "Reduce"

	"very low"	"low"	"normal"	"high"	"very high"
"very low"	1	0	0	0	0
"low"	1	0,1	0	0	0
"normal"	0	0	0,1	0	0
"high"	0	1	1	0,1	0
"very high"	0	0	0	1	0,1

For the temperature difference States under the control "Reduce temperature difference – ΔT_{red}", we construct a matrix of transitions "Significantly reduce" (Table 7).

Table 7. Transition matrix of "Significantly reduce"

	"very low"	"low"	"normal"	"high"	"very high"
"very low"	1	0	0	0	0
"low"	1	0	0	0	0
"normal"	1	0,3	0	0	0
"high"	0,1	1	0,3	0	0
"very high"	0	0,1	1	0,3	0

For temperature difference States under the "Do not change temperature difference – $\Delta T_{not\ change}$" control, we will construct a matrix of "Do not change" transitions (Table 8).

Regulation of distillate strength by changing the amount of phlegm (3-parameter in Table 1). In the process of rectification, the amount of reflux plays an important role in the concentration of distillate. Theoretically change the value of the amount of reflux in the interval $[0, \infty]$.

The optimal reflux quantity or optimal reflux number (R) is determined experimentally or by calculation based on a mathematical model. When the reflux increases, the

Table 8. Transition matrix of "Do not change"

	"very low"	"low"	"normal"	"high"	"very high"
"very low"	1	0	0	0	0
"low"	0	1	0	0	0
"normal"	0	0	1	0	0
"high"	0	0	0	1	0
"very high"	0	0	0	0	1

separation capacity of the distillation column increases, due to this, the purity of the components in the distillate increases, but the amount of steam supplied to the column cube increases and the yield of distillate decreases [7].

The type of linguistic variables regulating the concentration of alcohols distillate by changing the amount of reflux can be written as follows:

$$<\text{Increase the amount of reflux} - R_{inc}, R_{inc}, XRinc>, \quad (10)$$

where R_{inc} = {"Increase", "Significantly increase"}, $X_{Rinc} = \{0, 1, 0{,}2\}$;

$$<\text{Reduce reflux} - R_{red}, R_{red}, X_{Rred}>, \quad (11)$$

where R_{red} = {"Reduce", "Significantly reduce"}, $X_{Rred} = \{-10, -9, -7, -5\}$;

$$<\text{"Do not change the amount of reflux} - R_{not\ change}\text{"}, R_{not\ change}, X_{Rnot\ change}>, \quad (12)$$

where Rnot change = {"Do not change"}, XRnot change = {0}.

To describe the effect of this parameter on the distillate concentration, we present the relationship matrices. Changes in the quality of distillate under the control "Increase the amount of reflux – R_{inc}" are formed by transition matrices "Increase" (Table 9).

Table 9. Transition matrix of "Increase"

	"very low"	"low"	"normal"	"high"	"very high"
"very low"	0,1	0,3	1	0	0
"low"	0	0,1	0,3	0	0
"normal"	0	0	0,1	1	0
"high"	0	0	0	0,1	1
"very high"	0	0	0	0	1

To change the quality of distillate under the control "Increase the amount of reflux – Rinc", we will form a matrix of transitions "Significantly increase" (Table 10).

Table 10. Transition matrix of "Significantly increase"

	"very low"	"low"	"normal"	"high"	"very high"
"very low"	0	0,3	1	0,1	0
"low"	0	0	0,3	1	0,1
"normal"	0	0	0	0,3	1
"high"	0	0	0	0	1
"very high"	0	0	0	0	1

For the States of distillate concentration under the control "Reduce the amount of reflux – R_{red}", we construct a matrix of transitions "Reduce" (Table 11).

Table 11. Transition matrix of "Reduce"

	"very low"	"low"	"normal"	"high"	"very high"
"very low"	1	0	0	0	0
"low"	1	0,1	0	0	0
"normal"	0	0	0,1	0	0
"high"	0	1	1	0,1	0
"very high"	0	0	0	1	0,1

To change the quality of distillate under the control "Reduce the amount of reflux – R_{red}", we construct a matrix of transitions "Significantly reduce" (Table 12).

Table 12. Transition matrix of "Significantly reduce"

	"very low"	"low"	"normal"	"high"	"very high"
"very low"	1	0	0	0	0
"low"	1	0	0	0	0
"normal"	1	0,3	0	0	0
"high"	0,1	1	0,3	0	0
"very high"	0	0,1	1	0,3	0

To change the quality of distillate under the control "Do not change the amount of reflux – $R_{not\ change}$", we will construct a matrix of transitions "Do not change" (Table 13).

Table 13. Transition matrix of “Do not change”

	“very low”	“low”	“normal”	“high”	“very high”
“very low”	1	0	0	0	0
“low”	0	1	0	0	0
“normal”	0	0	1	0	0
“high”	0	0	0	1	0
“very high”	0	0	0	0	1

5 The Method of Situational Control Based on a Fuzzy Model of Decision-Making

Here is a fuzzy situation of the control process that may occur at the *i*-th step:

$\overline{S}_i$ = {< μ_1^1 /“very low”>, < μ_2^1 /“low”>, < μ_3^1 /“normal”> / “*the concentration of the initial mixture*”>,

<< μ_1^2 /“very low”>, < μ_2^2 /“low”>, < μ_3^2 /“normal”>, < μ_4^2 /“high”>, < μ_5^2 /“very high”> / “*temperature difference*”>,

<< μ_1^3 /“very low”>, < μ_2^3 /“low”>, < μ_3^3 /“normal”>, < μ_4^3 /“high”>, < μ_5^3 /“very (13)

We denote μ_i^j the value of the degree of membership of *j*-th characteristic, (e.g. $j = 1$, 1st sign (“concentration of initial mixture”) in the above situation) the *i*-th term (e.g., $i = 2$, 2nd term (“low”) of the attribute “concentration of the initial mixture”).

Let’s create a list of reference features, which will be limited by the number of terms. Denote y_j by a sign, where *j* is its number.

Let “Concentration of the initial mixture” – y_1, “Temperature difference” – y_2, “Amount of reflux” – y_3.

Then for the attribute “Concentration of the initial mixture” we have:

1) ≪0,09/“very low”>, <0,1/“little”>, <0,12/“normal”>/y_1>;
2) ≪0,06/“very low”>, <0,8/“little”>, <0,1/“normal”>/y_1>;
3) ≪0,5/“very low”>, <0,7/“little”>, <0,95/“normal”>/y_1>.

For the “Temperature difference”:

1) ≪0,02/“very low”>, <0,3/“low”>, <0,9/“normal”>, <0,91/“high”>, <0,93/“very high”>/y_2>;
2) ≪0,3/“very low”>, <0,35/“low”>, <0,85/“normal”>, <0,86/“high”>, <0,92/“very high”>/y_2>;
3) ≪0,4/“very low”>, < 0,51/“low”>, <0,89/“normal”>, <0,95/“high”>, <0,97/“very high”>/y_2>;

4) ≪0,43/“very low”>, <0,48/“low”>, <0,88/“normal”>, <0,91/“high”>, <0,95/“very high”>/y_2>;
5) ≪0,32/“very low”>, <0,45/“low”>, <0,86/“normal”>, <0,89/“high”>, <0,94/“very high”>/y_2>.

For the “Amount of reflux”:

1) ≪0,14/“very low”>, <0,16/“low”>, <0,22/“normal”>, <0,35/“high”>, <0,48/“very high”>/y_3;
2) ≪0,11/“very low”>, <0,17/“low”>, <0,23/“normal”>, <0,45/“high”>, <0,50/“very high”>/y_3;
3) <0,17/“very low”>, <0,20/“low”>, <0,24/“normal”>, <0,36/“high”>, <0,49/“very high”>/y_3;
4) ≪0,19/“very low”>, <0,21/“low”>, <0,25/“normal”>, <0,38/“high”>, <0,47/“very high”>/y_3;
5) ≪0,20/“very low”>, <0,23/“low”>, <0,26/“normal”>, <0,37/“high”>, <0,43/“very high”>/y_3.

From the obtained features of reference States, we will form a set of reference situations.

The terms of the “Concentration of the initial mixture” attribute are denoted by: C_1^1 – “very low”, C_2^1 – “low”, C_3^1 – “normal”.

The terms of the “Temperature Difference” feature are denoted by: T_1^2 – “very low”, T_2^2 – “low”, T_3^2 – “normal”, T_4^2 – “high”, T_5^2 – “very high”.

The terms of the “Amount of reflux” attribute are denoted by: R_1^3 – “very low”, R_2^3 – “low”, R_3^3 – “normal”, R_4^3 – “high”, R_5^3 – “very high”.

$$\tilde{S} = \{\ll C_1^1>, <C_2^1>, <C_3^1>/y_1>, \ll T_1^2>, <T_2^2>, <T_3^2>, <T_4^2>, <T_5^2>/y_2>, \ll R_1^3>, <R_2^3>, <R_3^3>, <R_4^3>, <R_5^3>/y_3>\}. \quad (14)$$

The rest of the reference situations are formed by any possible combination of sets of state features listed above. In total, 75 reference situations are formed 3 × 5 × 5 = 75 [8].

6 Conclusions

1. For the development of a control action in the object control system, in addition to the main measured parameters, it is also necessary to take into account the uncertainties of the values of the measured quantities.
2. On the basis of the method of comparison in pairs, the membership functions of the controlled parameters of the rectification process are revealed, the value of these functions allows you to enter the appropriate settings for the parameters of fuzzy controllers of the control process.

3. A fuzzy model with a linguistic variable for fuzzy situations in the control of rectification processes has been obtained. Using the resulting model, you can form a list of fuzzy situations and make optimal decisions.

References

1. Ahmetov, S.A., Ishmiyarov, M.H., Verevkin, A.P., Dokuchaev, E.S., Maly'shev, Yu.M.: Tehnologiya, e'konomika i avtomatizaciya processov pererabotki nefti i gaza. -Moskva: Himiya,. -736s. (2005)
2. Tehnologicheskiy reglament polucheniya e'tilovogo spirta iz zerno. –Yangiyul', 41s. (2015)
3. Yusupbekov, N.R., Aliev, R.A., Aliev, R.R., Yusupbekov, A.N.: Intellektual'ny'e sistemy' upravleniya i prinyatiya resheniy. -Tashkent: Gosudarstvennoe nauchnoe izdatel'stvo «U'zbekiston milliy e'nciklopediyasi»,.-490 s. (2014)
4. Avazov, Yu.Sh.: Mathematical simulation and development of the system of control of processes of rectification of compounds in production of ethyl alcohol. Int. Sci. Tech. J. Chem. Technol. Control Manage. **3**(81), 57–64 (2018)
5. Zadeh, L.A.: The Concept of a linguistic variable and its application to approximate reasoning, Part 1, 2 and 3. Inf. Sci. **8**, 199–249 (1975)
6. Yusupbekov, N.R., Gulyamov, Sh.M., Muhitdinov, D.P., Avazov, Yu.Sh.: Matematicheskoe modelirovanie processov rektifikacii mnogokomponentny'h smesey. Monografiya. -Tashkent: TashGTU, 159s. (2014)
7. Yusupbekov, N.R., Nurmuhammedov, H.S., Zakirov, S.G.: Kimyoviy texnologiya asosiy jarayon va qurilmalari. -Tashkent: Fan va texnologiyalar, 848b. (2015)
8. Avazov, Yu.SH., Usmonov, U.T.: Forimirovanie neche'tkih e'talonny'h situaciy s pomosch''yu lingvisticheskih peremenny'h v upravlenii processami rektifikacii. Agentstvo po intellektual'noy sobstvennosti RUz. Svidetel'stvo № DGU 04970 ot 18.01.2018 g.

Mental-Computing Based Investigation of Kinetic and Mechanisms of Homogeneous Gas-Phase Oxidation of Methane by Hydrogen Peroxide

L. M. Gasanova(✉)

Institute of Catalysis and Inorganic Chemistry named after Academician M. Nagiyev, Azerbaijan State Oil and Industry University, Baku, Azerbaijan
gasanova.48@mail.ru

Abstract. Experimental and kinetic studies of incomplete direct oxidation of formaldehyde and methanol, deep oxidation to $H_2 + CO_2$ gas without the use of a catalyst under homogeneous conditions with "green oxidizer" hydrogen peroxide, which has inductive properties of methane. Each process was carried out under determined conditions and high selectivity of formaldehyde (t = 520 °C, CH_2O-39%, sel. 94%) and methanol (t = 400 °C, P = 7 atm, sel. 99%) was obtained under optimal conditions. Complete conversion of methane to $H_2 + CO_2$ gas was obtained at t = 880 °C. The coherent-synchronized (conjugated) nature of methane oxidation reactions with hydrogen peroxide was determined by the determinant equation and the free-radical chain mechanisms of methane direct oxidation reactions to formaldehyde, methanol and $H_2 + CO_2$ gas were described within the conjugated reaction principle. Kinetic analysis of reactions was performed on the basis of the given mechanisms and using the determinant equation.

Keywords: Oxidation · Kinetic analysis · Homogeneous · Hydrogen peroxide · Conjugate · Coherent-synchonization

1 Introduction

According to recent statistics, the Earth has 350–500 • 10^{12} trillion m^3 of gas fields, of which 2 trillion m^3 enter the atmosphere each year. This figure corresponds to the world's annual gas formation [1]. Of this, 24% falls on Russia, 16% on Iran, 12% on Qatar, 8% on Turkmenistan, 5% on the United States and 1.4% on Azerbaijan [2]. The amount of hydrocarbons released from the depths of the Earth to the Earth's surface is a thousand times greater than global oil fields, and the location of the gases released from the Earth's surface is restored by gases in the depths of the Earth and new gas resources are created. Namely, the presence of large natural gas fields in the Azerbaijan Republic creates opportunities for the widespread use of methane in the chemical industry, which makes up 94–96% of its composition. One of the main goals of chemistry is to develop methods for the effective conversion of methane into chemical compounds.

© The Author(s), under exclusive license to Springer Nature Switzerland AG 2021
R. A. Aliev et al. (Eds.): WCIS 2020, AISC 1323, pp. 377–384, 2021.
https://doi.org/10.1007/978-3-030-68004-6_49

For this reason, more attention has been paid to the development of methods for oxidizing methane to more efficient direct-purpose products. Despite a lot of work in this direction, they have not found application on an industrial scale due to low conversion and selectivity.

The high-strength C-H bond in the methane molecule, the main component of natural gas (bond breaking energy 435.76 kJ/mol) makes it difficult to use in technological processes. High temperature (900–1000 °C) and high pressure (up to 100 atm) [3] are required to activate this bond, or catalysts consisting of precious metals or metal oxides are used [4].

The chances of obtaining high selectivity with oxidants such as O_2, air or air + O_2 mixture are very $H_2O_2 \rightarrow H_2O + O^{\cdot}$ low. Based on the research conducted in [5, 6], a new technology of hydrocarbon oxidation processes was developed, in which using hydrogen peroxide as an oxidant, it was possible to achieve in addition to increasing the conversion also increased the selectivity for the target products.

The effective action of hydrogen peroxide on oxidation processes is due to its specific physicochemical properties, mainly inductive effect. Also, hydrogen peroxide, called "green oxidizer" in the international concept of "Green Chemistry", ensures that the process is both environmentally friendly and safe.

As a result of the inductive effect of H_2O_2, the oxidation processes that take place in its presence occur by a mechanism different from the mechanism of oxidation reactions with molecular oxygen.

2 The Mechanism of Oxidation Reactions

The mechanism of oxidation reactions in the presence of H_2O_2 is primarily based on the mechanism of its decomposition. This decomposition reaction proceeding with the formation of water and molecular oxygen $2H_2O_2 \rightarrow 2H_2O + O_2$ is not a simple reaction, as it seems, but a complex reaction with a free-radical mechanism:

$$
\begin{array}{rr}
H_2O_2 + M \rightarrow 2\dot{O}H + M & (1) \\
H_2O_2 + \dot{O}H \rightarrow H_2O + H\dot{O}_2 & (2) \\
H\dot{O}_2 + \dot{O}H \rightarrow H_2O + O_2 & (3) \\
H\dot{O}_2 + \text{wall} \rightarrow H\dot{O}_2(\text{ads.}) & (4) \\
\dot{O}H + \text{wall} \rightarrow \dot{O}H(\text{ads.}) & (5) \\
\hline
2H_2O_2 = 2H_2O + O_2 &
\end{array}
$$

The determination of intermediate free $\dot{O}H$ and $HO_2^{\cdot}$ radicals by EPR research method shows that the formation of molecular oxygen (O_2) during the decomposition of H_2O_2 is $H_2O_2 \rightarrow H_2O + \dot{O}$ not due to intermediate atomic oxygen rather, it is the result of the formation and interaction of $\dot{O}H$ and $HO_2^{\cdot}$ radicals (reactions (1), (2), (3)).

$$2H_2O_2 \rightarrow \left[\dot{O}H, HO_2^{\cdot}\right] \rightarrow 2H_2O + O_2 \quad (6)$$

Here, a system of reactions is formed in which one of the H_2O_2 molecules is oxidizing and the other is reducing, interacting with another compound (coherent-synchronized), in which case one of the H_2O_2 molecules acts as an actor and the other as an inductor [7].

Chemical induction in oxidation processes with H_2O_2 results in a system of chemically conjugated reactions in which the decomposition of H_2O_2 generates active $\dot{O}H$ and $HO_2^{\cdot}$ radicals in the system as a primary reaction, and these radicals are consumed to accelerate oxidizing any compound (Acc), proceeding the secondary reaction [8, 9]. Academician T.M. Nagiyev gave the determinant equation characterizing the quantitative and qualitative chemical interference (induction) [7]:

$$D = \nu\left(\frac{r_1}{r_{acc}} + \frac{r_2}{r_{acc}}\right)^{-1} \tag{7}$$

Where r_1 and r_2 are the amount of actor consumed on the primary (1) and secondary (1) reactions, respectively, and the values of $D = (0 \div \nu)$ represent a system of conjugated or coherent-synchronized reactions.

It is precisely this inductive effect of H_2O_2 that ensures that the oxidation processes in its presence are carried out in the direction of the selective formation of target products.

The process of gas-phase methane oxidation under the conditions of chemical conjugation with H_2O_2 is carried out in an open system in a flow quartz reactor with a special design. This design, which allows H_2O_2 to enter the reaction zone directly, breaks it down into active free radicals, causing it to interact directly with the methane in the reaction zone and to obtain the target product in the conjugated reaction system. A refrigeration system installed adjacent to the reaction zone ensures rapid refrigerating of the reaction products, thus preventing additional reactions.

The conjugated reaction of methane with hydrogen peroxide was carried out in a wide range of parameters: $t = 450 \div 900$ °C, $CH_4:H_2O_2 = (1:1) \div (1:4)$, $C_{H2O2} = 15 \div 35\%$.At the initial stage, the process was carried out at low temperatures (460–560 °C) in the direction of incomplete oxidation of methane – to obtain formaldehyde or methanol. Very effective oxidation of methane to formaldehyde occurs at a temperature of 500–560 °C, in the ratio $CH_4.H_2O_2$ – 1:1 and with 25% H_2O_2. Methane activation begins at a temperature of 500 °C (formaldehyde yield 10%, select. 100%) and at a contact time of 520 °C, τ = 1.2 s, formaldehyde is obtained with a yield of 25% H_2O_2 and 94% selectivity with a yield of 39.0% [13]. At temperatures above 520 °C, methanol is obtained, and as the temperature increases, formaldehyde decreases, and methanol increases, and at temperatures above 560 °C, deep oxidation products – CO and CO_2 – are obtained.

As we have already mentioned, in the primary reaction of the conjugated reaction system, the CH_3 radical and the formaldehyde in the other stage are obtained by the very rapid activation of the C-H bond of methane under the influence of the active free radicals introduced into the system by H_2O_2.

Under optimal conditions for the maximum yield of formaldehyde (t = 520 °C, $CH_4:H_2O_2 = 1:1$, $C_{H2O2} = 25\%$, $\tau = 1.2$ s), the mechanism of its formation is described as follows:

	lgA,sm³/mol.s	E.kJ/mol	
$H_2O_2 \rightarrow 2\dot{O}H$	17.08	190.4	(1)
$H_2O_2 + \dot{O}H \rightarrow H\dot{O}_2 + H_2O$	13.00	7.5	(2)
$CH_4 + \dot{O}H \rightarrow \dot{C}H_3 + H_2O$	12.5	8.4	(8)
$\dot{C}H_3 + H\dot{O}_2 \rightarrow CH_2O + H_2O$	13.5	0	(9)
$H\dot{O}_2 + wall \rightarrow \dot{O}H(ads.)$	0.3	29.3	(4)
$\dot{O}H + wall \rightarrow H\dot{O}_2(ads.)$	1.0	34.3	(5)
$\dot{C}H_3 + wall(ads.) \rightarrow \dot{C}H_3(ads.)$	0.68	55.5	(10)
$CH_4 + 2H_2O_2 = CH_2O + 3H_2O$			

The decrease in formaldehyde yield with increasing temperature is accompanied by its gradual decomposition ($CH_2O \rightarrow CO + H_2$), and the process of increasing methyl alcohol in the system in connection with the change in the direction of increasing the concentration of O˙H radicals of $HO_2^{\cdot}/O^{\cdot}H$ ratio is associated with the acceleration of the following elementary stages.

	lgA, sm³/mol·s	E, kJ/mol	
$\dot{C}H_3 + H\dot{O}_2 \rightarrow CH_3\dot{O} + \dot{O}H$	11.2	0	(11)
$\dot{C}H_3 + \dot{O}H \rightarrow CH_3OH$	13.16	0	(12)
$CH_3\dot{O} + CH_4 \rightarrow CH_3OH + \dot{C}H_3$	12.1	46.55	(13)

Thus, based on the results of experimental studies and the mechanism we have described, we can say that the reactions for the formation of formaldehyde and methanol in this reaction system proceeded independently of each other. Each of these reactions occurs under the coherent-synchronized influence of radicals, conjugated with the decomposition of H_2O_2 under conditions of chemical interference. This is also confirmed by the value of D = 0.2 calculated by the determinant Eq. (7).

On the basis of the free-radical mechanism of the reaction to obtain formaldehyde, the kinetic equation, which adequately expresses the experimental parameters of the reaction, was developed by applying the method of stationary concentrations:

$$\frac{d[CH_4]}{d\tau} = \frac{2k_1k_8[CH_4][H_2O_2]}{k_1[H_2O_2] + k_1[CH_4] + k_4} \qquad (14)$$

The experimental results show that despite the formation of a certain amount (about 10%) of methanol at temperatures above 520 °C, there is no increase in methanol emissions at temperatures above 560 °C. Therefore, in order to obtain high-yield methanol directly from methane, the gas-phase reaction of methane with hydrogen peroxide is carried out under a certain pressure. This is due to the fact that, according to a number of well-known studies, pressure has a positive effect on the mechanism of methanol formation [11].

The process was carried out in a flow micropilot installation in an ideal displacement mode at a temperature of 350–475 °C and a pressure range of 4–12 atm. At t = 400 °C, CH_4:H_2O_2 = 1:1.4 and 30% H_2O_2 and at a pressure of 7 atm, the selectivity of methanol reached 18% [12]. This pressure (7 atm) is much lower than the pressures used in the formation of methanol directly from known methanol.

According to the determined values of the concentrations of active radicals using the method of stationary concentrations by means of differential kinetic equations compiled on the mechanism of predominance of the elementary stages (11), (12), (13) of the methanol formation reaction:

$$[\dot{O}H] = 1.5 \cdot 10^{-11} \frac{mol}{sm^3}, \qquad \frac{[\dot{O}H]}{[H\dot{O}_2]} = 0.25 \cdot 10^3, \qquad \frac{[\dot{O}H]}{[\dot{C}H_3]} = 0.9 \cdot 10^2$$

It can be said that the $\dot{O}H$ radicals play a leading role in the formation of CH_3OH.

The kinetic analysis of reaction can also be performed using the determinant Eq. (7), which characterizes the conjugated reactions without the application of differential kinetic equations. The value of D = 0.18 was determined for the conditions of methanol formation.

By making some changes to the Eq. (7) the rate of the methanol formation reaction can be expressed as follows:

$$r_{Acc} = r_{CH_4} = \frac{D}{v}(r_1 + r_2) = \frac{D}{v}(k_2[H_2O_2] + k_3[CH_4]) \cdot [\dot{O}H] \tag{15}$$

$$[\dot{O}H] = \frac{r_{CH_4} \cdot v}{D(k_2[H_2O_2] + k_3[CH_4])} \tag{16}$$

$r_{CH_4} = 0.17 \cdot 10^{-5} mol/s \cdot sm^3$, $[\dot{O}H] = 0.179 \cdot 10^{-10} mol/sm^3$

The value of $[\dot{O}H] = 0.17 \cdot 10^{-10}$ mol/sm^3 calculated using the determinant equation almost coincides with the value of $[\dot{O}H] = 1.5 \cdot 10^{-10}$ mol/sm^3 calculated using the kinetic equations based on the above mechanism.

Thus, the determinant equation allows to analyze the kinematics of complex reactions in a very simple way, to determine the reaction rate, as well as to characterize the chemical induction quantitatively and qualitatively.

The study of the coherent-synchronized oxidation reaction of methane with hydrogen peroxide was continued at high temperatures (800–900 °C) at atmospheric pressure. $H_2 + CO_2$ gas mixture was obtained by complete conversion of methane with 15% hydrogen peroxide under homogeneous conditions at 880 °C. In this case, the amount of CO gas was at least 0.4%, and the emission of H_2 gas was 74% (according to the stoichiometric number). It is known that the complete conversion of methane to a synthesis gas (H_2 + CO) in the presence of O_2 or O_2 + air takes place at temperatures above 1000 °C. A gas mixture of H_2 and CO_2 is the most suitable ingredient for the ammonia formation. Thus, an additional device is required for the conversion of CO gas into synthesis gas (H_2 + CO) for the formation of ammonia.

The effective action of the inductive properties of hydrogen peroxide is manifested even at high temperatures. The calculated value of the determinant D = 0.5 for this condition is a quantitative indicator of the formation of coherent-synchronized reactions system (chemical induction).

Based on the above mechanisms and the experimental results, a free-radical chain mechanism of methane complete oxidation to $H_2 + CO_2$ gas at high temperatures (880–900 °C) is given:

$$H_2O_2 \rightarrow 2\dot{O}H$$
$$H_2O_2 + \dot{O}H \rightarrow H\dot{O}_2 + H_2O$$
$$CH_4 + \dot{O}H \rightarrow \dot{C}H_3 + H_2O$$
$$CH\dot{O} \rightarrow CO + \dot{H}$$
$$CH_4 + \dot{H} \rightarrow \dot{C}H_3 + H_2$$
$$\dot{C}H_3 + H\dot{O}_2 \rightarrow CH_2O + H_2O$$
$$\dot{C}H_3 + \dot{O}H \rightarrow CH_2O + H_2$$
$$CH_2O + \dot{O}H \rightarrow CH\dot{O} + H_2O$$
$$CO + H\dot{O}_2 \rightarrow CO_2 + \dot{O}H$$
$$CO + H_2O \rightarrow CO_2 + H_2$$

The given mechanism also involves elementary reactions of active radicals breaking down on the reactor wall. The mechanism shows that the deep oxidation of methane to H_2, CO_2 and small amounts of CO gases goes through the process of obtaining the intermediate CH_2O, and the active radicals $\dot{O}H$ and $HO_2^{\cdot}$ that play a leading role in the formation and consumption of formaldehyde. The possibility that the formation of H_2, CO and CO_2 gases will pass through the formation of CH_2O has been confirmed by experimental investigations.

Experiments were carried out on the thermal decomposition of formaldehyde both by the injection of formaldehyde into the reaction system and by the injection of pure formaldehyde and its aqueous solution to the reactor together with H_2O_2, provided that the reaction conditions remained the same. Experimental results show that the injection of formaldehyde into the $CH_4 + H_2O_2$ system dramatically increases the emission of H_2 gas and, accordingly, increases CO gas. Formaldehyde alone decomposes completely into CO and H_2 gases at the reaction temperature, and H_2 and O_2 are obtained as the final product together with an aqueous solution of H_2O_2. These results fully confirm the mechanism by which CO and H_2 gases are associated with the formation of intermediate product CH_2O, and the resulting CO gas is completely converted to CO_2 gas by the action of H_2O_2 and water vapor in the system. The simple oxidation of formaldehyde to formyl ($CH\dot{O}$) radicals and the easy dissociation of formyl radicals under high temperature conditions are reactions that are practically proven and thermodynamically possible.

3 Conclusion

In this paper, Fuzzy neural network based methods is used to learning experimental data (Fuzzy neural network characteristics are: Type = 'sugeno', Version = 2.0, NumInputs = 1, NumOutputs = 1, NumRules = 3, AndMethod = 'prod' ,OrMethod = 'probor', ImpMethod = 'min', AggMethod = 'max', DefuzzMethod = 'wtaver'). Fragment of initial data is given in Table 1.

Table 1. Fragment of initial data

r_{exp}	r_{theor} (1) –	r_{theor} (2) –
0,11	0,104	0,0986
0,12	0,112	0,1005
0,15	0,14	0,12
0,1687	0,1611	0,145
0,175	0,1668	0,1536
0,18	0,175	0,1592

where r_{exp} – experimental speed, r_{theor} (1) – speed calculated with using the determinant equation, r_{theor} (2) – speed calculated from the kinetic equation.

Computer simulation results based on Fuzzy neural network is given in Table 2.

Table 2. Computer simulation results

t	r exp	rther [1}	r ther (2)
1,203	0,175	0,156128	0,144092
1,218	0,173096	0,157854	0,145566
1,233	0,174797	0,15958	0,147039
1,248	0,176499	0,161306	0,148512
1,263	0,178201	0,163032	0,149985
...	...	....	...
1,323	0,185008	0,169937	0,155878
1,338	0,18671	0,171663	0,157352
1,368	0,188412	0,173389	0,158825

Experimental results compared to the theoretical results. All calculations are made in MATLAB environment and MS Excel.

Comparison results is given below. Root Mean Square Error (RMSE) on r_{theor} (1) using the determinant equation – 0,051439, r_{theor} (2) – using the kinetic equation – 0,095475, learning RMSE by using fuzzy neural network –0,001052275.

Results of calculations of determinant (1) obtained using equation and equations obtained on the basis of law of acting masses (2) are combined in Fig. 1. It can be seen from the figure, in comparison with the experimental curve, curve (1) more adequately repeats the nature of the curve, and also gives a smaller absolute error.

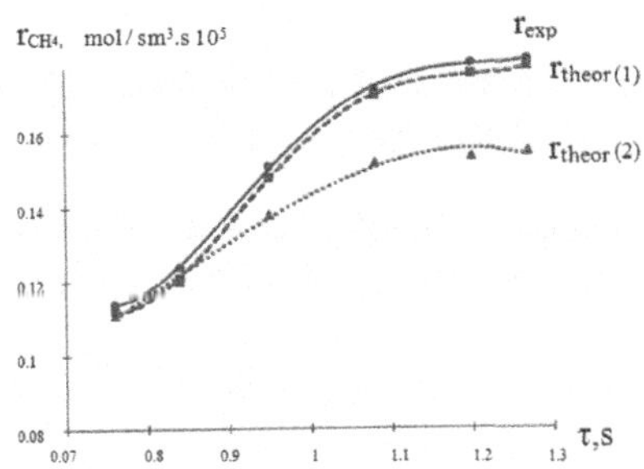

Fig. 1. Dependence of experimental and design consumption rates methane from contact time.

Oxidation of methane by hydrogen peroxide under homogeneous conditions without a catalyst at relatively low temperatures (400–520 °C) with high selectivity (94 and 99%) and yield (39 and 18%), respectively, as well as complete conversion to $H_2 + CO_2$ gas at 880 °C was possible. Based on the experimental results, free-radical chain mechanisms of formaldehyde and methanol formation reactions under chemical induction, as well as complete conversion to $H_2 + CO_2$ gas are given. It has been determined that it is possible to carry out a kinetic analysis of the rate of any complex reaction without the use of differential kinetic equations using the determinant equation characterizing chemical induction.

References

1. Statistical Review of World Energy (2015). http://www.bp.com/en/qlobal/corporatc/energy-economics/statistical-review-of-worldenergy.html
2. Skorobogatov, V.A., Staroselsky, V.I., Yakushev, V.S.: World reserves and resources of natural gas. Gas Industry **7**, 17–20 (2020)
3. Vedeneev, V.I., Goldenberg, M.Y., Gorban, N.I., Teitelboim, M.A.: The mechanism of the initial stage of methane oxidation at high pressures. Khim. Phys. **6**, 626 (1987)
4. Claridge, J.B., Green, M.L.H., Tsang, S.C.: Methane conversion to synthesis gas by partial oxidation of methane and by reforming over rhenium catalysts. Catal. Today **21**, 455–460 (1994)
5. Nagiev, T.M.: Chemical Conjugation. Science, Moscow (1981)
6. Aghamammadova, S., Nagieva, I., Gasanova, L., Nagiev, T.: Kinetics and mechanism of the reaction of coherently synchronized oxidation and dehydrogenation of cyclohexane by hydrogen peroxide. Int. J. Eng. Res. Appl. **7**(2), 36–40 (2017)
7. Nagiev, T.M.: Coherent Synchronized Oxidation Reactions by Hydrogen Peroxide. Elsevier, Amsterdam (2007)
8. Nagiev, T.M.: The Interaction of Synchronous Reactions in Chemistry and Biology. Elm, Baku (2001)
9. Nagiyev, T.M.: Conjugate reactions in the context of contemporary ideas. Baku (2020)
10. Nagiyev, T.M., Hasanova, L.M.: Conjugate oxidation of methane to formaldehyde with hydrogen peroxide. Chem. Phys. **10**, 1461–1467 (1984)
11. Vedeneev, V.I., Goldenberg, M.Y., Gorban, N.I., Teitelboim, M.A.: Quantitative model of methane oxidation at high pressures: I. Description of the model. Kinet. Catal. **29**(1), 7–13 (1988)
12. Nagiev, T.M., Faradzhev, E.G., Gasanova, L.M., Mamedov, E.M.: Kinetic modeling of direct oxidation of methane to methanol by hydrogen peroxide under pressure. Azerb. Chem. Mag. **1**, 101–108 (1988)
13. Gasanova, L.M., Ramazanova, ZYu., Nagiev, T.M.: Free-radical conjugated oxidation of natural methane by hydrogen peroxide. J. Chem. Chem. Eng. **5**, 159–164 (2011)

Fuzzy Assessment of Corrosion Risks

H. G. Ismayilova, F. B. Ismayilova, E. Kh. Iskandarov(✉), and S. G. Hajizade

Azerbaijan State Oil and Industry University, Azadlıq 20, Baku AZ1010, Azerbaijan
ismayilova.hecer@bk.ru, fidan.ismayilova.2014@mail.ru,
e.iskenderov62@mail.ru

Abstract. The probability of accidents in oil and gas pipeline systems for one reason or another and oil spills into the environment increases during operation. This is mainly due to the corrosion of the pipe metal and technological and construction defects. Recent studies show that internal corrosion factors play a greater role in the failure of pipelines.

In the article internal and external corrosion factors were investigated, and corrosion risks were evaluated. Based on the examination of internal and external corrosion factors and fuzzy assessment of corrosion risks, it was determined that the internal corrosion risks are higher. The authors noted that this is mainly due to the multiphase flow inside the pipe

Keywords: Oil pipeline · Corrosion · Environmental risks · Fuzzy · Multiphase flow · Linguistic values · Erosion

1 Introduction

Corrosion is one of the important factors that reduce the reliability and operational efficiency of oil and gas pipelines. Corrosion risk factors and their classification are given in the table. As can be seen from the table, these risks are related to internal and external corrosion factors. It should be noted that pure oils and gases that do not contain harmful compounds have very low corrosive wear ability. The analysis shows that the factors affecting the internal corrosion of pipelines are followings [1–4]:

Local corrosion of the lower parts of the pipelines and accidental spills of oil pipelines occur mainly due to the water cutting of oils and the instability of oil emulsions and the separation of water in a separate phase

High mineral content of produced water;

Presence of harmful compounds such as carbon and hydrogen sulfide in produced water and associated gas

External corrosion is associated with the environmental conditions, the lack of insulation coatings or incongruousness, and stress corrosion.

One of the interesting results of the analysis is that in most cases, local corrosion (corrosion-erosion) occurs in the lower parts of the pipelines, despite the fact that the transported product is not active in terms of corrosion. The analysis shows that large-diameter oil and gas collecting pipelines are more prone to corrosion cracks in the

© The Author(s), under exclusive license to Springer Nature Switzerland AG 2021
R. A. Aliev et al. (Eds.): WCIS 2020, AISC 1323, pp. 385–388, 2021.
https://doi.org/10.1007/978-3-030-68004-6_50

lower parts of the pipes. Initially, corrosion destructions occur in the form of separate "wounds", but later they deepen and merge into cracks 20–60 mm wide and 5–20 m long. At present, the causes of such corrosion-erosion processes are explained differently by different researchers. A group of researchers relates the occurrence of such corrosion to the acid-carbon mechanism and explains it as follows: Although CaCO3 deposits accumulate on the inner surface of the pipeline, in some places the protective salt layer formed for various reasons (mechanical, chemical, etc.) breaks off from the surface, forming macro pair- the metal and the other surface of the pipe covered with sediment (metal as anode, pipe surface as cathode). Thus, an intensive corrosion process takes place and its speed can reach 5–8 mm/year. Despite the above-mentioned mechanical and chemical effects, they are unpredictable and, most importantly, do not allow to explain the local corrosion destructions in the lower part of the pipe (Table 1).

Table 1. Linguistic estimates of corrosion in multiphase pipelines

	Corrosion risk factors and their classification	Existing opinion					Authors' opinion				
		Very weak	Weak	Medium	High	Very High	Very weak	Weak	Medium	High	Very High
Internal corrosion	**Internal erosion** – f_1: Mechanical wear due to the effects of suspended particles in liquids and gases or high-speed flow		●								●
	Internal total corrosion – f_2: Metal damage in large areas due to the impact of transported products	●								●	
	Different types of internal local corrosion – f_3: Small - local metal dissolution under the influence of transported products		●								●
External corrosion	**Lack or weak insulation (ineffective electrochemical protection)**– f_4: Increased probability of external corrosion due to soil			●				●			
	Electric current – f_5: Electrochemical dissolution of metal at the outlets of external electric current from the soul to the pipeline				●			●			
	External soul corrosion – f_6: Electrochemical dissolution of metal under the influence of the soil				●				●		

2 Solution and Discussion of the Problem

Some researchers, including the authors, believe that the phenomenon of localization of corrosion destructions in the lower parts of the pipelines is due to the hydrodynamic properties of multiphase flows. Thus, in many cases, as the flow velocity is quite low (0.1 ÷ 0.8 m/s) stratified flow forms exist in multiphase systems. *As water sometimes moves with mechanical particles to the bottom of the pipe in a separate phase, and oil emulsion and gas move in layers over the water, waves are created due to the different viscosities at the boundaries of the liquid phases.* As these waves replace along the flow, water droplets and mechanical particles (iron carbonates and sulfides, sand and clay) move along the lower surface of the pipeline, causing a permanent hydroerosive effect on the surface. As a result, the bottom of the pipe gets exposed to corrosion-erosion (Fig. 1).

Taking into consideration that, the mentioned corrosion mechanism is not conventional, it is not possible to prevent such local corrosion by inhibitors. This is because the protective layer of the inhibitor is periodically cleaned from the metal surface by a multiphase flow. Therefore, in order to prevent local corrosion processes in the bottom

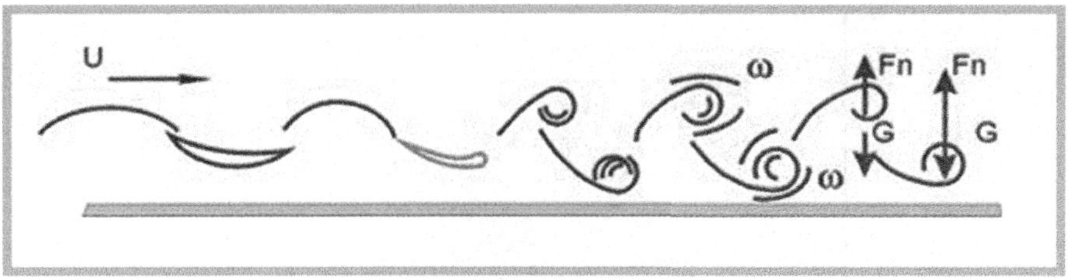

Fig. 1. Scheme of forming vortices on the wavy surface of oil and water phases

parts of the pipelines, it is very important to take into account the hydraulic properties of multiphase flows. First of all, during the design (or operation) of oilfields, such diameters of oil and gas pipelines must be selected which allow keep the velocity of multiphase flows at an optimal level, in other words, to prevent the separation of water from the system. If this is not possible (if the water encroachment of extracted oil is exceeded), it is important consider the discharge of water that has accumulated in parts of the pipeline along its route during the high encroachment period. In terms of corrosion activity, plugged and layered (divided into separate phases) structural regimes are considered dangerous. Thus, the movement of the gas "plug" generates strong vibrations in the pipeline. The interval between the passages of these plugs in the pipeline can vary from 1–2 min to 15–25 minutes per hour. As a result, the oil storage collector is exposed to cyclic (variable) loads, which leads to intensive local mechanical and chemical corrosion and corrosion-fatigue cracks in the pipeline. Although, there is a classification of risk factors for pipeline corrosion, there is very little accurate research on the relative impact of individual factors, i.e. the share. Analysis of statistical data shows that in many cases, corrosion-erosion of pipelines can not be explained only by external corrosion factors. Such erosion-corrosion conditions, which are typical of multi-phase flows, lead to short-term failure of oil and gas pipelines. Taking into account the abovementioned, the assessment of the probability of damage to the pipelines from internal and external corrosion factors was considered. Linguistic estimates (very weak, medium, high and very high) related to internal and external corrosion factors in multi-phase pipelines are given in the table. In addition to the existed opinion, the linguistic assessments proposed by the authors are also reflected in the table. According to the authors, the most characteristic factors for multiphase flows are factors associated with internal corrosion.

Based on the values in the table, both opinions on internal and external corrosion factors were aggregated. The linguistic values shown in the table have become fuzzy numbers on the scale (Fig. 2). Then, according to the method [5], it was determined that the sum vectors obtained for internal and external corrosion factors differ from the "very high" risk category. These differences showed a higher risk of internal and external corrosion (70%). The results obtained for internal and external corrosion factors were [0.41,0.58,0.80] and [0.23,0.41,0.65], respectively. As can be seen, these values differ from each other and can be assessed as "high" for internal corrosion factor and "medium" for external corrosion on the scale [0.1] (Fig. 2).

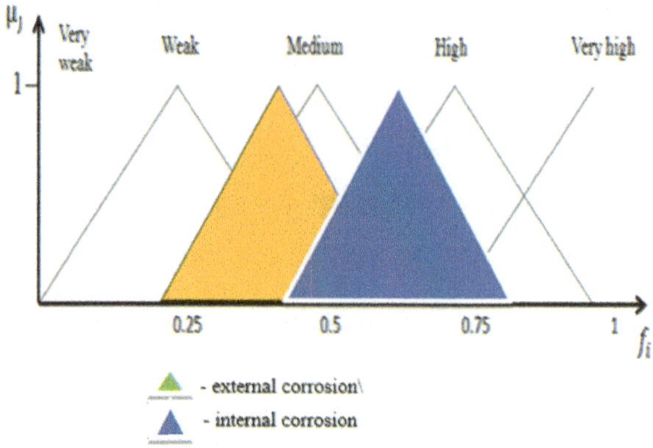

Fig. 2. Evaluation on the scale of fuzzy numbers

3 Conclusion

1. Based on the fuzzy assessment of internal and external corrosion factors in oil and gas pipelines, it was determined that the risk of internal corrosion is higher than external corrosion. Thus, this risk is "high" and about 70%. However, this risk for external corrosion is less than 60%;
2. The preventing internal corrosion in the pipeline is impossible without taking into account the hydrodynamic properties of multiphase flows. It is more expedient to regulate the corrosion process in multiphase flows by hydrodynamic methods.

References

1. Khodjaeva, G.K.: Emergency assessment of oil pipeline systems in terms of geodynamic processes, 132 p. Publishing House of Nizhnevartovsk State University, Nizhnevartovsk (2016)
2. Vorobiev, U.L., Akimov, V.A, Sokolov, U.I.: Warning and liquidation of accidental spills of oil and oil products. Institute of Risk and Safety, 368p (2007)
3. Ketilman, G.S.: Environmental Protection During Exploration, Transportation and Storage of Oil and Gas, 256p. Nedra, Moscow (1981)
4. Telegin, L.G., Kim, B.I., Zonenko, V.I.: Environmental Protection During the Construction and Operation of Oil and Gas Pipeline, 188p. Nedra, Moscow (1988)
5. Babanli, M.B.: Synthesis of new materials by using fuzzy big date concepts. Proc. Comput. Sci. **120**, 104–111 (2017)

Intellectual Decision-Making Support System

Yusupbekov Azizbek Nodirbekovich and Turaev Khurshid Salim Ugli(✉)

Tashkent State Technical University, Tashkent, Republic of Uzbekistan
abek71@mail.ru, xturaev@mail.ru

Abstract. The article examines the issues of developing intelligent systems located in an environment capable of perceiving this environment and acting autonomously to achieve the assigned tasks. To create technological intelligent process control system, it is necessary to monitor the mode ensuring high quality of finished products with minimal energy consumption at a given productivity of the entire complex. It is important to have a stream of input data to provide the system with operational information to support decision making. The proposed intelligent control system gives the possibility of analyzing the results of production activities for making decisions of an organizational and technical nature.

Keywords: Intelligent system · Decision making support · Technological process control · Intelligent control · Multi-agent systems · Life cycle

1 Introduction

At present, interest in the use of intelligent systems has increased in the chemical, petrochemical and other industries. With the advent of distributed software applications, traditional approaches to programming interactions between subsystems have turned out to be laborious and insufficiently reliable. Developed software products for distributed systems must interact with the external environment. Interaction with the environment is especially problematic in fields where a lot of external influencing factors that need to be taken into account when monitoring and controlling, maintaining or modeling production processes [1–5].

In the past, the implementation of such systems required explicit programmatic interaction with each external influencing factor. If the considered influencing factors have complex interactions between the measured parameters, then it is difficult to implement an application that supports a process control system. Moreover, the application may not have the required flexibility or precision with respect to the external environment. Intelligent agent-based approaches have proven to be well suited where complex interactions with a constantly changing external environment are required. Perhaps the most important attribute of multi-agent systems is that each agent is an autonomous computational object that, combined with the ability to perceive the environment, supports the decision-making system in the event of structural and parametric deviations of the control object [6, 7].

An intelligent system refers to a system located in an environment which is capable of perceiving a given environment and acting autonomously to achieve its objectives.

© The Author(s), under exclusive license to Springer Nature Switzerland AG 2021
R. A. Aliev et al. (Eds.): WCIS 2020, AISC 1323, pp. 389–393, 2021.
https://doi.org/10.1007/978-3-030-68004-6_51

Agents receive information about the state of their environment through sensors of the lower level of the process control system ASUPT, and they can also receive information from cloud storage or the Internet of things. The main properties of an agent are autonomy, reactivity, cooperation, learning, and mobility [8].

The proposed multi-agent system can work integrated with the information-control system of the production cluster of the Ustyurt gas chemical complex, and can also independently perform its functions in the environment for which it was directly intended. Given method of building distributed systems is very effective, since each agent in the system is responsible for performing its tasks, responding to events and communicating with other agents in the system. The main advantage of the developed multi-agent system is that there is no need to explicitly program each interaction of influencing factors of the entire system [11].

The proposed multi-agent system consists of several autonomous and dissimilar agents. Each agent in the system has limited set of capabilities to solve an immediate task. This approach to creating a multi-agent system implies that each agent can work in a decentralized manner, as well as in conjunction with a global management system. The need for interaction between autonomous agents of the system is necessary for a successful search for solution to the problem. An information database of agent-based interactions from simple information exchange to decision support are fundamental because agents have different and limited capabilities.

None of the existing software and hardware sets (SHS) fully meets the above-mentioned principles of construction of intelligent integrated control systems (ICS) and does not ensure the implementation of all stages of the product life cycle (PLC). And in this regard, the priority direction is the development of such a software-technical and software-methodological set, which would include the following functions:

1. accelerated creation of an information management system for an individual order;
2. necessary software tools;
3. orientating on the active participation of the user in the creation of a software product to equip his workplace;
4. organization of production activities of the enterprise using the enterprise's IUS;
5. software-methodological support for creating a unified information environment for an industrial enterprise.

Of course, the development of such complex requires the involvement of a large number of engineering resources of various specializations, but in the end, this technology will provide the information technology industry with another real solution to increase the economic efficiency of industrial enterprises and ensure their safety.

The multi-agent system of corporate design is based on a multi-level model which provides parallel design and providing the development of standard systems that solve the problems of automating a number of engineering stages during the product life cycle (PLC).

The functional structure of the multi-agent corporate design system is shown in Fig. 1.

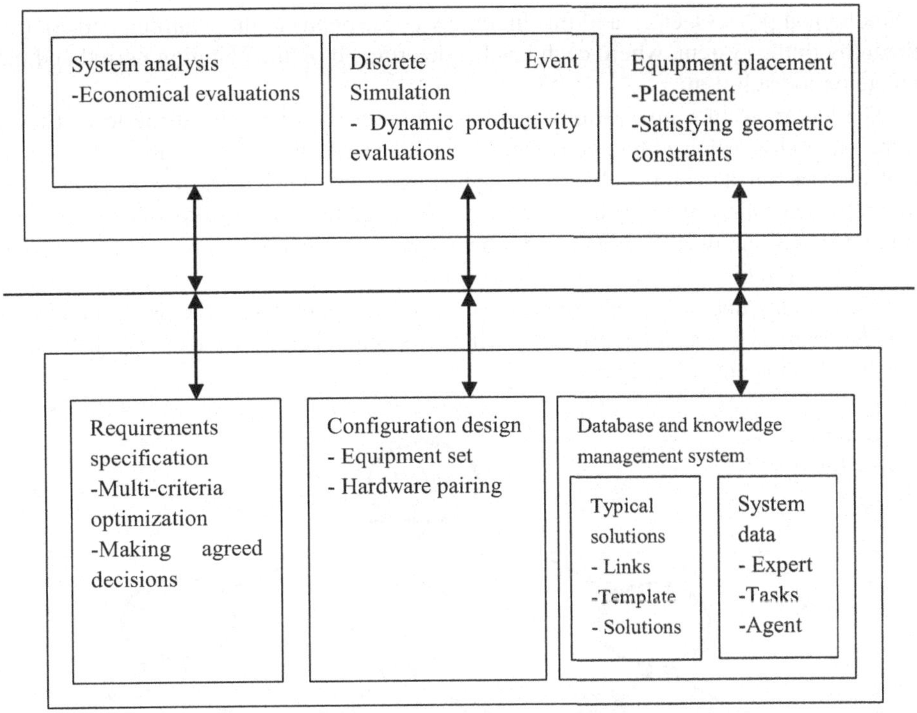

Fig. 1. The system's functional structure.

The given structure implies the development within the framework of a specific project of a necessary standard set of architectural elements, including the division into subsystems according to purpose, a description of interfaces and protocols of interaction among subsystems.

Such a system is a kind of testing ground where technical solutions for the implementation of the concept of multi-agent systems are being worked out.

Typically, each individual SAPR tool is developed to solve a specific narrow specialized design problem and has its own specific structure and mechanism of operation. The multi-agent enterprise design system with parallel design provides a real-time multitasking solution. Thus, this technology can be used as the basis for the system integration of a multitude of interacting autonomous agents linked into a single multi-agent system.

The considered approach is focused not on a completely decentralized coordination structure, when each agent participates in the coordination process, but on a partially centralized structure. In this case, one of the agents acts as a person who makes decisions, and the rest, acting as subcoordinators, solve subproblems to achieve a common global goal. The distributed intelligent system is implemented as a feedback system and has a fan-shaped structure, in the nodes of which there are coordinating intelligent agents connected with local intelligent agents.

In the architecture of a distributed intelligent system, the control levels are determined by the aggregation of resources (enterprise, production, workshop, installation, technological process, etc.), and this hierarchy corresponds to the planning and control hierarchy in the system, where each level is determined by the planning period and the update period solution.

The hierarchy of a given multi-agent system consists of agents at the lowest level of the hierarchy, followed by agents who control a group of agents at a lower level, and finally, one agent at the top at the top of the hierarchy. The agent at the lowest level of the hierarchy determines the optimal time, which is based on information collected at each of the production areas. Agents at a higher level of the hierarchy change the decision of lower hierarchical agents. From time to time, there might be a conflict of interest or general delay due to the selected action. To solve this problem, the developed MAS includes a supervisor agent application, which has general control and evenly distributes tasks between agents (Fig. 2).

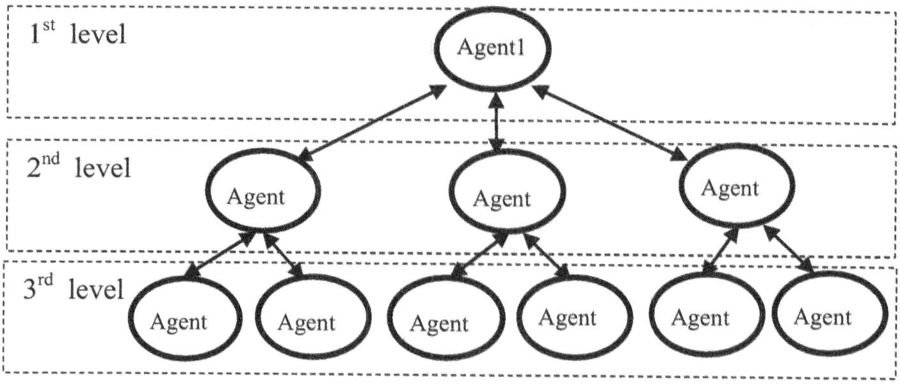

Fig. 2. MAC levels' hierarchy.

2 Conclusion

The constructed conceptual model of a multi-agent system (MAS) for decision making support on life cycles managing of industrial enterprises makes it possible to substantiate general approaches to solving the problems of choosing the topological structure of MAS. The limitations imposed by the technical characteristics of the software and hardware modules, as well as the multi-agent system standard, lead to the conclusion of using a fan topological structure to support decision-making for industrial enterprises with a continuous nature of production processes throughout the entire cycle of automation objects.

References

1. Stark J.: Production Lifecycle Management: 21st Century Paradigm for Production Realization, 2nd edn. Springer, London, 581 p. (2011)

2. Jones, K., Romatier, C.: Leveraging the Cloud to Drive Complex-wide Profitability. Honeywell UOP. 25 East Algonquin Rd. Des Plaines, IL 60016 (2017)
3. Yusupbekov, N.R., Marahimov, A.R., Gulyamov, Sh.M., Igamberdiev, H.Z.: APC fuzzy model of estimation of cost of switches at designing and modernizations of data-computing networks. In: 4th International Conference on Application of Information and Communication Technologies, AICT 2010, 5612015. https://doi.org/10.1109/ICAICT.2010.5612015
4. Yager, R.R, Zadeh, L.A. (eds): Fuzzy Sets, Neural Networks and Soft Computing. VeN Nextremal Renold, NewYork, p. 440 (1994)
5. Broy, M.: Dynamic Distributed Systems - Towards a Mathematical Model, Extended Abstract (2000). 5
6. Yusupbekov, N.R., Abdurasulov, F.R., Adilov, F.T., Ivanyan, A.I.: Application of cloud technologies for optimization of complex processes of industrial enterprises. Adv. Intell. Syst. Comput. **896**, 852–858. https://doi.org/10.1007/978-3-030-04164-9_112
7. Liang, T.-Y., Shieh, C.-K., Zhu, W.: Task mapping on distributed shared memory systems using Hopfield neural network. In: Proceedings of Communication Networks and Distributed Systems Modeling and Simulation Conference (1997)
8. Yusupbekov, N., Igamberdiev, H., Mamirov, U.: Algorithms of sustainable estimation of unknown input signals in control systems. J. Autom. Mob. Robot. Intell. Syst. **12**(4), 83–86. https://doi.org/10.14313/JAMRIS_4-2018/29
9. Tarasov, V.B.: Kontseptsiya METAKIP: ot komp'yuterno - integrirovannogo proizvodstva k INTERN ET/INTRANET - setyam predpriyatiy. Programmnyye produktyy i sistemy. **3**, 19–22 (1998)
10. Yudov A.I., Trofimov V.M.: Intellektual'naya sistema avtomatizatsii deyatel'nosti predpriyatiya. V trudax konferentsii. Intelligent Systems and Information Technologies in Control (IFAC) SPb/Pskov, 19–23 iyunya 2000 g., Izd-vo. SPbGTU, pp. 380–384 (2000)
11. Yusupbekov, N.R., Gulyamov, S.M., Yusupbekov, A.N., Kabulov, N.A.: Simulation of chemical-technological complexes. In: Advances in Intelligent Systems and Computing, AISC, vol. 1095, pp. 588–595 (2020). https://doi.org/10.1007/978-3-030-35249-3_75

A Strategy of Prosthesis Control Using Artificial Intelligence and Execution on Robotic Fingers

Elbrus Imanov[1(✉)] and Ehsan Faramarzi Asli[2]

[1] Department of Computer Engineering, Near East University, North Cyprus Via Mersin 10, Mersin, Turkey
elbrus.imanov@neu.edu.tr

[2] Research Assistant Area of Biomedical Engineering, 19166 Tabriz, Iran
ehsanfaramarziasli32@gmail.com

Abstract. Nowadays, pervasiveness of advanced technology provides new methods in various fields of trade, industry, medicine, etc. One of those important technologies is biomedical engineering, which uses engineering principles to reduce gap between engineering and medicine, including diagnosis, monitoring and treatment, has become a strong assistant for doctors. Orthopedic Prosthesis and Orthotics is an important field that many researchers are working in, and many hardware and software equipment are developed accordingly. Prosthetics has developed in the recent years to the extent that give their users an ability to replicate the natural human motion on certain limb disorders. These 'smart' limbs require the use of multiple technologies and knowledge from a broad field of areas including biomechanics, biomedical, electronics, mechanics, mechatronics and software engineering. For bionic limb applications, multitude of sensors are required to measure and predict the user intent and then a set of mechatronic actuators are utilized to conduct the resulting motion or the gesture. However, for severely injured or disabled patients that can only move their finger or one hand, the use of such control techniques is very limited due to the restricted muscle movement. In this study, a new command and control approach for bionic limbs based on Android application MIT App Inventor 2, which is one of the most ideal platforms to be used for developing handsets and controlling the hand is analyzed. For implementation and demonstration, inertial sensors of mobile phones are used to detect the gesture of the user to control prototype robotic arm prosthesis.

Keywords: Artificial Intelligence · Prosthetics · Control Systems · Mobile Platforms · Bionics · Robotic Arm

1 Introduction

Natural cause, disease, nutrition, environmental cause or accident-based and even hereditary limb or organ loss are very common today. Scientists want to copy human organs and limbs from the past to the present and to increase the quality of life of the person. Many projects, experiments and researches have been done about artificial organs and artificial limbs. Replication of human reasoning processes and behavior with the

© The Author(s), under exclusive license to Springer Nature Switzerland AG 2021
R. A. Aliev et al. (Eds.): WCIS 2020, AISC 1323, pp. 394–401, 2021.
https://doi.org/10.1007/978-3-030-68004-6_52

help of computers and other artificial technologies' is the main issue of artificial intelligence [1]. Many world-famous medical device companies have artificially developed the internal organs and limbs of the human body. These devices can never be used for definitive treatment. Intelligent system today is in this way discovered promoting medication prescribing in clinical laboratory and educational settings [2]. These artificial organs are only applied to the patient for the purpose of performing the functions of other organs of the body until the organ transplantation is performed for a certain period. The examination of human brain goes back thousands of years. It was inevitable to try to take advantage of this thinking process with the introduction of modern electronics [3]. Today, artificial limbs are implanted in many amputated patients. Most important of all is that these artificial limbs are biocompatible with the human body. Robotics science is a branch of the AI. Robotics is the discipline and high-tech robots, counting their plan, construct, and implementation. In manufacturing, robotic manipulators are commonly used [1]. Robots are a mixture of one or all of the above capabilities with the ability to move and handle objects on the ground [4]. One of the reasons why these prostheses are not widely used is the high price. Researchers have attempted many articles to solve this problem and have examined implantable prostheses in terms of materials, mechanisms and structures. For many years, human beings have been able to solve the amputee barrier, but as technology advances, human comfort and health have a different importance. The high production costs and the lack of powerful control platforms have led to the slow realization of these limbs. Recently the powered prosthetic limbs industry has been flourished. The availability of low-cost 3D printing technology enables the production of special prosthetic limbs for different structures and purposes helped greatly in this direction [5]. Also, powerful and cheap microcontroller boards are introduced with user-friendly environments to program and control. The microcontrollers are very popular in electronics and computer-controlled systems overall, in the world [6]. There are many such projects introduced various methods to control powered limbs. It's worked on the Electromyography signal generated from neural system were used for the first time to activate the arms [7]. This method requires special electronics to perform, which increases the cost of the device. The image processing has been used to estimate the movement of skin thus introducing signal which can be used for controlling the limbs [8]. The upper limbs are more challenging to control unlike the lower limbs. Since the lower limps has less complex movement comparing to upper limbs which perform multi-joints accurate movements [9]. Mobile smart phones are becoming an essential device in our daily life. It's a normal development that these devices would take more tasks related to mental and physical performance [10]. Capability of controlling the hand by this method was impractical. The use of touch screen of the mobile phone to takes more time to handle. Sensors such as flex sensors are used to control the prosthetic hand [11]. Besides, motion sensors have been introduced as new controlling method of prosthetic limbs in the touch [12]. These sensors are embedded in almost all the currently used mobile phones. In order to minimize the cost of manufacturing a new control joystick or glove, the smart phone can be used. By using simple programming app which is named as "MIT appinventor2" can be designed to analyze the sensors data equipped within the phone. Creating smart mobile phone application has been simplified and introduced to public especially in the android operating system. The "MIT appinventor 2" which is a visual programming environment has been used to create productive applications with a simple knowledge of software engineering [13]. The pause of the study is structured as accompany. Section 2 involves materials and methodology used for the research and is followed by Sect. 3, which gives the presentation of the expend system. Finally, in Sect. 4 conclusion points are given.

2 Materials and Methodology

It is very difficult for the rehabilitation patients to receive medical care for a long-time interval because of its high cost. As of late, the plausibility of combining mechanical and rehabilitative advances has been communicated with extraordinary rehabilitative applications. Based on this innovation, analysts have begun testing on the EMG control of multi-fingered fist, tale-operation of complicated robotic prosthetic fist, usage of drive tangible criticism on my electrical direct lower arm prostheses and plan of multifunctional robot prosthetic hand mechanisms. Currently, the effective controlled prosthesis may have one or two controlled degrees of freedom driven via moo control motors. The currently used limit prostheses are either an end assembly (such as hand or snare drum) controlled by the movements of a bearish stem transmitted by a cable. Within the last-mentioned control strategy, terminals implanted within the socket of a prosthesis distinguish EMG signals produced by withdrawal of the leftover muscles of a removed limb. These signals give a trigger for battery-powered DC motors that enable the prosthesis to move the hand, or wrist. Flow chart of the process is shown in Fig. 1.

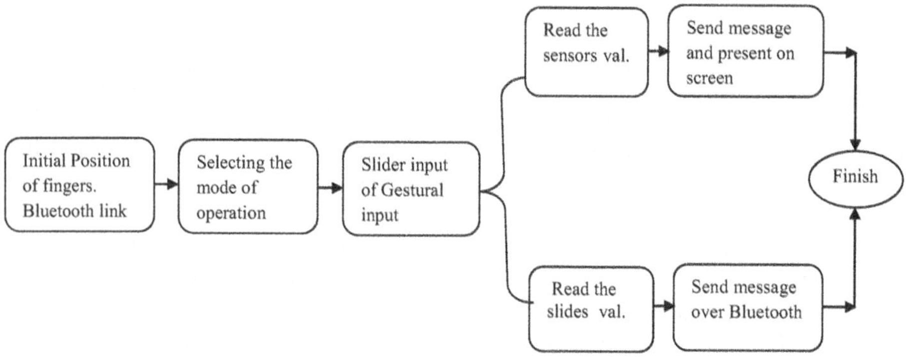

Fig. 1. Flow chart of the process.

However, these types of prostheses support restricted capability to operate a mini piece. The control of the holding force is not strong as there is no sensory feedback. In addition, the support structures and gear systems that provide the weight of the motors enable the prosthesis to increase its weight. It also eliminates the significant aesthetic qualities of a device with excessive system noise. Refine control of several point of liberty and sensory reaction will greatly refine the action of these prostheses. In the case of technical studies, it is very important and guided by limb designers to choose these features from all normal hand characteristics. It is adaptable as a means for many different types of activities such as rubber bands, hook prostheses and the like, there has been a long yeoman ship in the health service of thousands of arms amputees, without significant limitation in shape and appearance [14]. The hand of human could be a complicated mechanical construction contains tendons, bones freely interfacing bones, muscles providing as pressure engines, ligaments performing as cables joining muscles to bone, and a covering of defensive delicate skin and tissue. Muscles can make joint movements or torque via weight, and for each muscle there are one or more muscles which provide opposite via counter-torque or restricting movement. First mode the fingers of the prosthetic hand are controlled straightforwardly by 5 slide bars on the android mobile phone application. Movements of these fingers are produced by servo motors inside the arm. Second mode Gestural input mode was outlined to change over gestural movement of the mobile phone into complex movements of prosthetic hand. In this mode, inertial sensors counting accelerometer and whirligig implanted within the mobile phone are utilized to distinguish the hand gestures of the client to form more complex and foreordained hand and finger movements. Arduino has different items like boards, Lily pads and shields. The point of this paper is to investigate the world of Arduino innovation in terms of Boards; Lily pads and Shields covering in depth regarding Technical Details, features and real-world applications. Arduino innovation has empowered different manufactures and investigate devotees to come out with their customized boards and shields as per their inquire about requirements and area of executions. Arduino Open Source community is additionally giving stage for analysts to come up with inventive inquire about applications and showcase prepared items in terms of domestic automation, robotics, wireless connectivity, rambles and numerous others. Arduino, an open source equipment panel is found on Atmega328 microcontroller found on 8, 16, 32-bit AVR innovation. Substantially, Atmega328 is an AVR 8-bit RISC controller based on double inline bundle innovation and has 20 MHz clock oscillator, 32 kB streak, 1 kb stream, 23 programmable pins, 6 channels 10-bit ADC and 6 PWM outputs [15]. Arduino Integrated Development Environment (IDE) is an independent base for Arduino equipment and can operate in multiple operating frameworks. It is a multi-tier application based mainly on Java Innovation and the Processing Programming dialect with the established Cabling Initiatives. The Arduino IDE can be a solid stage for all analysts, software engineers and other industries to expand their progress experts to build initiatives on Arduino Controllers and other sensors. A robotic arm is proposed for servomotor controller design. Any mechanical movement around is achieved by an electric motor which is capable of converting energy by taking electrical energy and producing mechanical energy. Electric motors provide power for hundreds of devices used in daily life. Other small motors find applications in micro-machines for separating red blood cells according to its size [16]. For the sufferers with serious motor function loss, care assistants can aid them for their daily activities [17]. A robotic arm is proposed for servomotor controller design. This servomotor is used to control an arm, which has servomotors MG995, to accomplish the movement of gripping some goods [18]. Bluetooth communication was used to interact drivers and android. The drivers can interface to the Bluetooth share despite UART rules. Robot movement can be controlled by accepting commands from Android. The reliability of a mechanical frame in combination with quality and repeatability is unmatched. Select and Place robots can be reprogrammed and can be tooled to supply different applications. Android application, MIT App Inventor 2 is one of the most ideal platforms used to develop handsets and to control the hand. App Inventor 2 is a block editor style: instead of typing lines of code in the traditional way, drags and drops blocks to represent designer functions and variables. You can create applications by reviewing the following steps: App Inventor Designer for which you choose components for your application. App Inventor Blocks Editor, where you integrate plan lump that specify how elements should act. Bring programs together visually; assemble pieces like pieces of a puzzle. When adding tracks, your app will appear step by step on the phone, so you can test your work while creating it. Once finished, you can package and upload your application to create a stand-alone application. Processing of MIT App Inventor is shown by a schematic description, in Fig. 2.

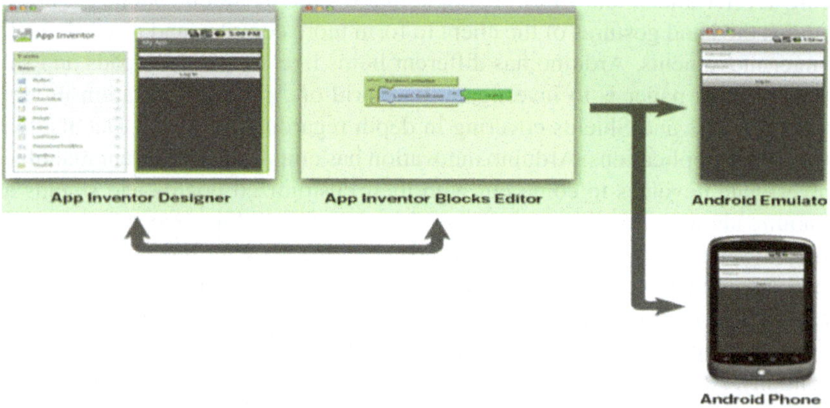

Fig. 2. MIT App Inventor's schematic description.

3 Presentation of the Expend System

Furthermore, the prototypes and methods of this disclosure show that intelligent human muscles can be used in place of the actuator in artificial human limbs. However, both medical and non-medical applications and designs can be realized by the instruction of the present invention. For this research the results were taken by monitoring the response of the bionic hand to the commands issued from the mobile application and transmitted with Bluetooth communication protocol. The accuracy of system was calculated as 100% when each command was issued 50 times. The baud rate of communication was 9600 bit/sec. The motion of the hand was generated by 5 servo motors. Each of the motors commanded by signal from Microcontroller board. The control and driving approach of the motors is based on Pulse Width Modulation (PWM) technique. The duty cycle of the command signal PWM will determine the position of the motor shaft. Processing of the position of the finger is shown Fig. 3.

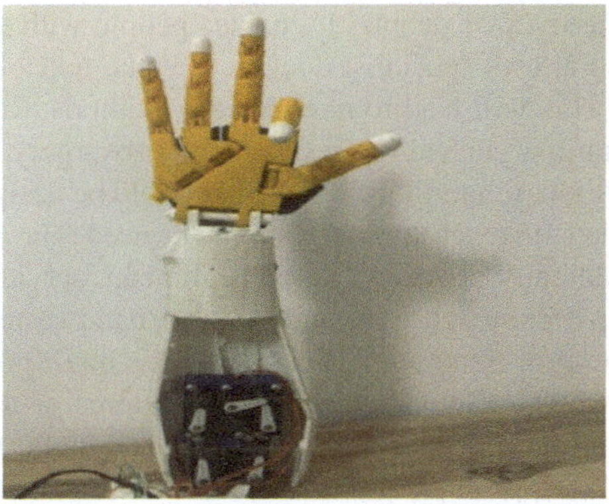

Fig. 3. Position of servomotors in flexion motion

The pulse width of the PWM driving signal is related to the position of the finger. For example, if the pulse width is 0.5 ms the finger is in final flexion position. When the width of PWM is 2.5 ms the finger is in total extension position. Also Fig. 4 shows the PWM driving signals for two fingers positions.

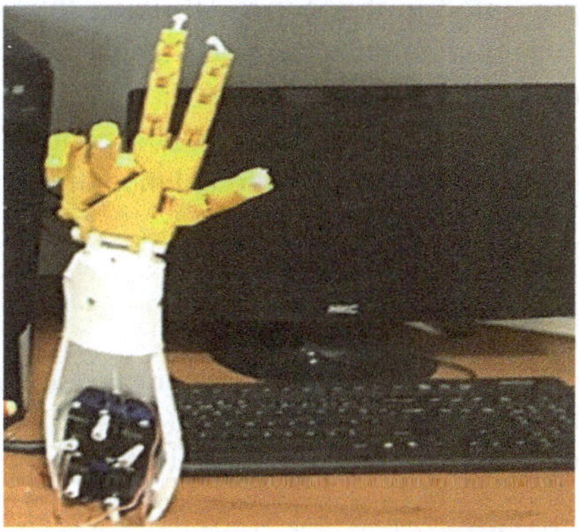

Fig. 4. PWM driving signals for two fingers positions.

The most important advantage is that it is a remote-controlled prosthetic arm that is easy for the user. This is because more cables are eliminated. Furthermore, this study will provide the user with a good understanding of the object and ease the hand movements for the patient's daily life. By using the smart phone's gyroscope feature, it can transfer the biological hand movements to the prosthetic arm.

4 Conclusion

As a result of the literature researches, it was realized that the studies and inventions on biomedical, rehabilitation, biomechanical, prosthetic and orthotic engineering have an important impact on the lives of people. It enables people with disabilities to perform delicate procedures and allows specialist doctors to make quick decisions and provide successful treatments. This will lead to new devices methods. Robotic systems can be applied to achieve operations such as brain surgery, neurosurgery, and knee arthroplasty and eye surgery. In addition, rehabilitation systems should be designed and controlled as assistive tools for people with disabilities and should be used for robotic and mechatronic technologies by automatically providing home health care services. The main point of this paper is to develop a compact, light and portable hand structure for people in the field of rehabilitation and for treatment. This structure is made possible by the addition of two cable-modular cable systems to traditional systems of the past, so that it can execute the correct configuration and movement of the robotic fingers like the finger of an adult person. MIT App Inventor, which is one of the most optimal platforms is used to develop handsets and to control the hand. The achievements of this research are listed below: Firstly, two proposed coupling systems work as desirable so that they can eliminate the imperfections associated with the incompatibility of human natural finger and these artificial hand movements in rehabilitation applications. Secondly, the whole structure is compact and light weight, so that it can easily be covered with gloves, and even in the skillful movements there is no disturbance with the coating gloves, which is complicated issue to achieve by robotic fingers with bridge mechanisms. In the final step, based on the achievements of this study, an example of this finger was constructed using a 3D printer and presented in this paper. Utilizing the mobile phone sensors to command a bionic hand was implemented with a low-cost material. This approach will have a positive impact on the life quality of disabled patients, especially with upper limbs disability since the method is simple and besides, controlling lower limbs is more complex and needs investigating the weight and balance of human body.

References

1. Aliev, R.A., Fazlollahi, B., Aliev, R.R.: Soft Computing and its Application in Business and Economics. Springer, Heidelberg, pp. 1–2, 381–382 (2004)
2. Ahmet, Y.: Application of soft computing to medical problems. In: International CONFERENCE 2009 on Intelligent System Design and Applications, pp. 614–619 (2009)
3. Sonali, B.M., Priyanka, W.: Research paper on basic of artificial neural network. Int. J. Recent Innov. Trends Comput. Commun. **2**(1), 96–100 (2014)
4. Avneet, P.: Artificial intelligence and its application in different areas. Int. J. Eng. Innov. Technol. **4**(10), 79–83 (2015)
5. Pîrjan, A., Petroşanu, D.M.: The impact of 3D printing technology on the society and economy. Romanian Econ. Bus. Rev. **7**(2), 360–370 (2013)
6. Imanov, I.: Line following robot control using IF-THEN rule. In: International Symposium on Advances in Applied Mechanics and Modern Information Technology, Baku, pp. 45–88 (2011)
7. Kato, I.: Electro-pneumatically controlled hand prosthesis using pattern recognition of myo-electric signals. In: 7th ICMBE, p. 367 (1967)
8. Martin, H., Donaw, R., Kelly, Jung, Y., Kim, J.: A novel approach of prosthetic arm control using computer vision, biosignals, and motion capture. In: IEEE Symposium on Computational Intelligence in Robotic Rehabilitation and Assistive Technologies, pp. 26–30 (2014)

9. Aprile, I., Rabuffetti, M., Padua, L., Di Sipio, E., Simbolotti, C., Ferrarin, M.: Kinematic analysis of the upper limb motor strategies in stroke patients as a tool towards advanced neurorehabilitation strategies: a preliminary study. Biomed Res. Int. **2014**, 636123 (2014)
10. Tezel, C., Gunay, O., Kayisli, K.: Implementation of a robot hand controlled with android software. Int. J. Eng. Sci. Appl. **1**(4), 137–141 (2017)
11. Özkan, S.S., Karayel, D., Atali, G., Gokbayrak, İ: Robotic hand controlling based on flexible sensor. Acad. Plat. J. Eng. Sci. **5**(3), 35–40 (2017)
12. Kyberd, P.J., Poulton, A.: Use of accelerometers in the control of practical prosthetic arms. IEEE Trans. Neural Syst. Rehabil. Eng. **25**(10), 1884–1891 (2017)
13. Xie, B., Abelson, H.: Skill progression in MIT app inventor. In: IEEE Symposium on Visual Languages and Human-Centric ComputingVL/HCC, pp. 213–217 (2016)
14. Mavroidis, C., Pfeiffer, C., DeLaurentis, K.J., Mosley, M.J.: U.S. Patent No. 6,379,393. U.S. Patent and Trademark Office, Washington, DC (2002)
15. Nayyar, A., Puri, V.: A review of Arduino boards, Lilypad's & Arduino shields. In: 3rd International Conference on Computing for Sustainable Global Development INDIA Com, pp. 1485–1492 (2016)
16. Pahuja, R., Kumar, N.: Android mobile phonecontrolled bluetooth robot using 8051 microcontroller. Int. J. Sci. Eng. Res. **2**(7), 14–17 (2014)
17. Kottaimalai, R.,Pallikonda, R.M.,Selvam, V.,Kannapiran, B.: EEG signal classification using principal component analysis with neural network in brain computer interface applications. In: International CONFERENCE 2013 on Emerging Trends in Computing, Communication and Nanotechnology, India, pp. 227–231 (2013)
18. Li, S.A., Weng, C.W., Chen, Y.H., Lo, C.H., Yang, M.H., Lin, Y.C., Wong, C.C.: Servo motor controller design for robotic manipulator. In: International Symposium on Intelligent Signal Processing and Communications Systems, pp. 254–257 (2012)

Analysis of the Main Trends and Directions of Automation in the Formation of the "Smart Archives" System

Ahmedov Barat[1](✉) and A. Abdullayev[2]

[1] Director of the National Archive of Scientific, Technical and Medical Documentation of Uzbekistan, Tashkent, Uzbekistan
bm_axmedov@umail.uz

[2] Deputy Chairman of the State Committee for Land Resources, Geodesy, Cartography and State Cadastre of the Republic of Uzbekistan, Tashkent, Uzbekistan
rustam.metrolog@mail.ru

Abstract. The article examines the current trend in the development of information technology in the Republic of Uzbekistan. Several approaches to the creation of an automation system and the intellectualization of the activities of organizations and enterprises are analyzed. On the example of the archival activities of the Republic, it is proposed to create and implement SMART technologies – "Smart Archive" covering the full range of automation of the issues of packaging, storage, digitization and use of archive documents. The system and subsystems Smart Archive, its functional diagram, main elements are proposed and the effectiveness of using such systems in modular application is analyzed.

The rapid development of modern information and communication technologies and integrated automation systems in all areas of human activity, has served as a great impetus for the creation of innovative intellectualized, so-called "Smart systems" in many areas of the economy.

Keywords: Smart system · Smart sit · Smart farm · Smart school · Smart archive

1 Introduction

"Smart city" - a city, or part of an urban network where the control of processes of one direction or another is carried out without the participation or with partial participation of a person (like a complex for managing urban traffic, urban transport traffic, city security, market zones, large retail chains, etc.);

"Smart farm" - a farm where all or part of the processes of growing crops are completely carried out by automatic systems (automated control of a drip system, a system for maintaining a thermo-humidity regime in a greenhouse, power plants, etc.);

"Smart school" is a general education school (college) with the total use of automated complexes in the management of educational processes, information about the progress of teachers, the organization of distance courses, etc.

© The Author(s), under exclusive license to Springer Nature Switzerland AG 2021
R. A. Aliev et al. (Eds.): WCIS 2020, AISC 1323, pp. 402–408, 2021.
https://doi.org/10.1007/978-3-030-68004-6_53

SMART is an acronym for a five criteria system. It structures and tracks the "route", rejects uncertainty, creates verifiable trajectories to a designated point with clear milestones, an estimate of reachability. Each task, from an intermediate step to a global concept, can be solved with the help of SMART / 1 /.

The term was first used in the November 1981 issue of George T. Doran's Management Review magazine, in the following capital letters:

Specific (specific, simple, reasonable, lean);

Measurable (measurable, motivated);

Achievable or Attainable (consistent, achievable, weighted);

Relevant (reasonable, realistic, resourced, results oriented);

Time (timely, limited, time-dependent). Sometimes two more conditions are added to the standard decryption:

In recent years, articles with the use of the terminology "Smart archive" have become increasingly common.

Let's take a closer look at the use of this term in the archival field.

The archival sphere in Uzbekistan, as in other countries, is characterized by the following 4 directions/1/:

- completing archives with archival documents;
- records of documents accepted for archival storage;
- ensuring the safety of archival documents;
- organization of the use of archival documents.

The main task of the archives is to meet the needs of the state, organizations and citizens in retrospective information. The most important factor in the quality organization of this task is the level of automation of the above areas of archiving, the effectiveness of the introduction of modern automated technologies in various business processes of archives.

The 21st century is often called the "information age", because wide reforming of the social order is accompanied by a high growth in the need for "information", while simultaneously causing the value of archival information. The main tool in the qualitative satisfaction of the information needs of society for "archival information" is the informatization of business processes in archives, and the effective implementation of automatic and automated control systems.

Currently, systems for automated information retrieval, establishing relationships between applicants and archival institutions, reference information retrieval systems that provide automation of accounting, analysis and technologies for performing the main functions of archives, as well as automated information systems and databases, including search systems are relevant. And presentation of archival information, as well as a system for automated control of the storage conditions of archival documents.

Information support and automation of business processes in archives is carried out by creating compatible software in the form of a complex of interconnected databases, its implementation and practical use. If a few years ago these information systems and automated technologies were used to solve specific problems and automate individual archival processes, the rapid growth of information demand for archival information over the past quarter century tenfold requires the development of more promising technologies and systems that ensure interaction in the operation and control. The quality of all or several areas of the field of archiving. Thus, the complex application of Smart archive (SA) technologies is exactly the answer to the requirements.

Optimally, the SA databases should function in the local computer network of the archival institution as a complex of local or distributed databases. The core of this complex is a database containing accounting information at the fund level and inventory on the preservation of archival materials, on the composition and content of documents. This can be characterized by the following expression:

$$Sa = \int Si$$

Wherein:

Sa - Aggregate system SA
Si - Components of SA

The SA components are expressed as the sum of all automated elements in all 4 archive directions:

$$Si = \sum \quad Sa = S\text{к} + S\text{у} + S\text{х} + Si$$

where: $S\text{к}$, $S\text{у}$, $S\text{х}$, $S\text{и}$– separate automated complexes in the segment of acquisition, accounting, storage and use of archival documents;

Analysis of the developments and engineering solutions used by SA at the present time shows that the smart archive system is like a software package for solving some problems regarding archival documents.

One of the striking examples of SA technologies is the Smart Archive software package of the Company, which is an IT solution designed specifically for a wide range of tasks for managing archived media materials in various environments, combining materials accounting and rights management tools/3/.

Designed specifically for managing media archives in a variety of environments. Material accounting and rights management tools. Based on open solutions Web-client, cross-platform.

PC “Smart Archive” is a set of closely integrated systems, developed on a modular basis, allowing you to manage the archive as a whole and materials, media, their storage and accounting, logistics, customer relations, workflow, reporting and other areas in particular. At the same time, the modules of the software complex are configured to interact with the client’s familiar software tools from third-party developers, supplement them or replace them. The structural diagram of the Smart Archive PC is shown in Fig. 1.

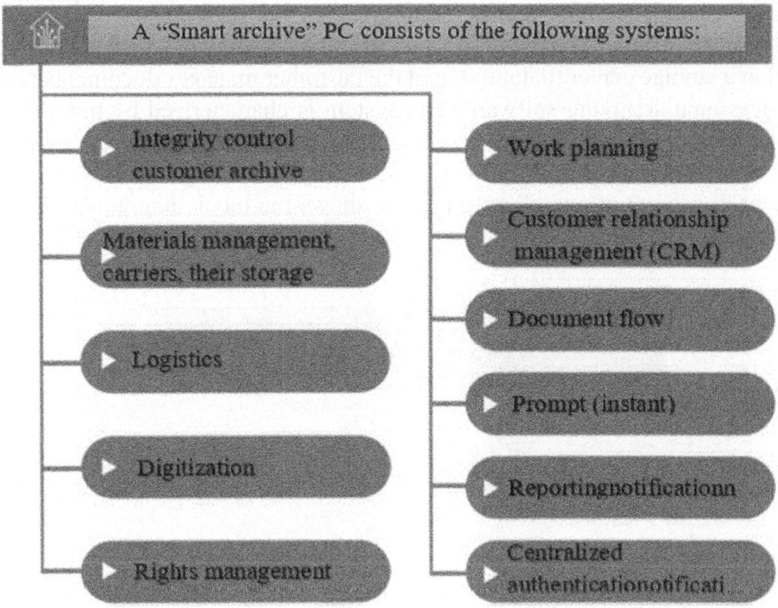

Fig. 1. Block diagram of the smart archive PC.

The next solution for smart archiving is the complex automated system Smart ArchiveHelper/4/.

Smart ArchiveHelper is a set of technological solutions that allows you not only to create an electronic archive with automatic classification of documents and check completeness, but also to process existing image archives, significantly reducing the size (up to 90%) and increasing the information content of the contained elements. This system includes embedded Smart ReSquare technology, which enables developers to integrate functionality for searching and highlighting document boundaries in an image into their information systems.

Smart ReSquare technology allows you to automatically find the borders of a document in a photo and restore the image of the document while maintaining its original geometric and color properties. This technology allows you to turn a modern camera or a mobile phone equipped with a camera into a full-fledged scanner of business documents, which ensures further text recognition from a picture using OCR technology.

The features of the algorithms implemented in the technology make it possible to eliminate most of the geometric distortions and make it possible to apply the technology in a wide range of mutual positions of the document and the camera. Special algorithms for working with lighting and color allow the system to suppress color distortions introduced by photographing processes and compensate for the effects of shadowing in areas of the document.

This technology can be used in conjunction with Smart PDF/A technology, allowing not only to highlight documents in photos, but also convert them into a compact form, which is extremely important for mobile platforms.

The next example is Smart-archive databox.

Smart-archive databox is a hybrid document storage system in which paper originals are stored in a storage center (databox), and the customer manages documents remotely using professional archiving software. The system is characterized by monitoring the integrity of the customer's archive, accurate accounting of documents, establishing a sequence of interaction, controlling costs, maintaining several legal entities (subordinate independent divisions) of one client. Figure 2 shows the block diagram of the Smart-archive databox.

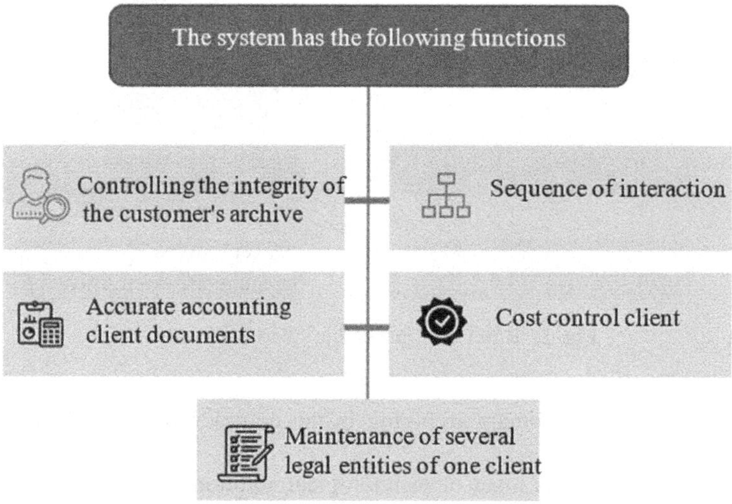

Fig. 2. An approximate structural diagram of smart-archive databox in the activities of the agency "Uzarchive".

Another of the most common systems is the "Archive - SMART" software package, which was developed for organizing an electronic archive of documents (Archive-SMART).

The main purpose of the system is aimed at ensuring the automation of archival records management, which makes it possible to store all electronic documents of organizations formed and received in the process of working activities, with the consistent integration of all information systems of the organization and providing a structured hierarchical scheme for storing documents in a single centralized storage. The main functionality of this software package is maintaining an electronic archive of documents, the ability to flexibly configure the structure of storing documents, creating a card - a form for each document in automatic mode with the ability to specify filling rules, working with an electronic archive from the Keysystems software navigator, transferring documents from interface of software products in the form of a fileless exchange, signing documents with an electronic signature, searching for documents in file storage and the ability to view saved documents for an unlimited period of time. That is, the system is aimed at automating the activities of archives by 90–95% (Fig. 3).

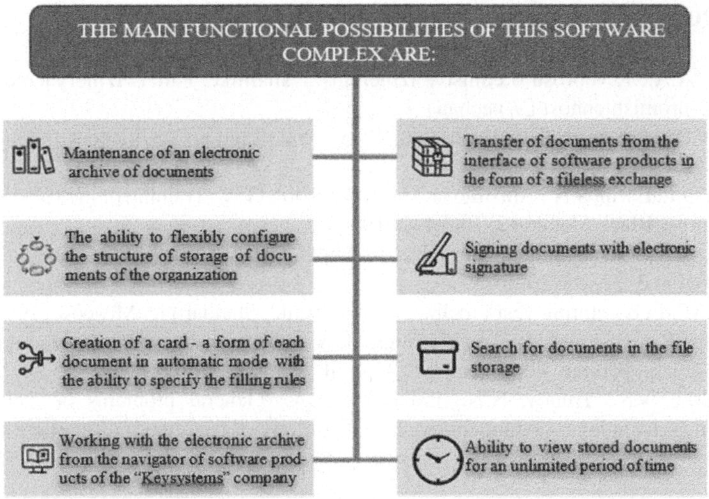

Fig. 3. Capabilities of the "Archive – SMART" software package

In addition to the above systems, there are also a large number of similar software complexes aimed at the complex automation of archives.

Their analysis shows the inherent aspects of each, which are inherent in solving a number of tasks related to the storage and use of archival documents.

As a result of the analysis of these features, advantages and disadvantages of these systems, as well as taking into account the modern requirements of international rankings for the assessment of doing business (Doing Business), a general structural diagram of the use of smart technologies for archival institutions has been developed.

In drawing up this structural diagram, the main principles of archival institutions were also taken into account in the complex of the information environment of the Electronic Government, as well as in the implementation of interactive state services by archival institutions.

This scheme was also developed from the point of view that "smart archive" means an archival institution, whose activities are based on an integrated system of comprehensive automation of all available areas of activity of this archive, taking into account various services to users at all stages of their relationship with the archive (preliminary order of archival documents, notification - an invitation to the archive, provision of a desktop in the reading room at the time the user visits the archive, early preparation of ordered archival documents, etc.).

This scheme was also developed from the point of view that "smart archive" means an archival institution, whose activities are based on an integrated system of comprehensive automation of all available areas of activity of this archive, taking into account various services to users at all stages of their relationship with the archive (preliminary order of archival documents, notification - an invitation to the archive, provision of a desktop in the reading room at the time the user visits the archive, early preparation of ordered archival documents, etc.).

References

1. Begunov, A.A.: K voprosu o edinstve izmereniy v analitike. Part 1. Izmeryaemoe svoystvo. Pishevaya promishlennost (V pechati)
2. Vasilev, V.P., Morozova, R.P., Kochergina, L.A.: Praktikum po analiticheskoy ximii. Ximiya, Moscow (2000)
3. Boriskin, O.I., Larin, S.N., Anisimova, M.A., Nuzhdin, G.A.: Training of metrology specialists in metallurgy. Black Metals **12**, 70–74 (2019)
4. Begunov, A.A.: Pretsizionnie izmeritelnie yacheyki. CH. 1. Ustroystvo i metodika izmereniya j. Metrologiya **3**, 22–26 (2016)
5. Rudo, N.M.: Laboratornie vesi i tochnoe vzveshivanie. Standartgiz, Moscow (1963)
6. Korenman, Y.A.I.: Praktikum po analiticheskoy ximii. Analiz pishevix produktov: V 4-x knigax. 2-e izd., pererab. i dop. Kniga 1, 239p: il. Kolos S, Moscow (2005)
7. Aleksandrov, V.S., Trunov, N.N., Lobashev, A.A.: General problems of metrology and measurement technique: system approach to metrology of quantum multiparticle systems

Modelling of Environmental and Economic Factors in the Azerbaijani Oil Industry in the Context of Sustainable Development

Ata Babayev(✉)

Azerbaijan State Oil and Industry University, Azadlig Avenue, 20, AZ1010 Baku, Azerbaijan
babayevata1964@mail.ru

Abstract. In the article, sustainable development issues and environmental and economic problems in the context of global problems have been researched. The essence and effectiveness of the sustainable development processes that began in the 1960s have been elucidated. The strategic importance of the environmental factor in strengthening sustainable development at the global level has been substantiated. The article highlights the growing urgency of these problems, especially in recent years due to climate change. The importance of defining optimal solutions for environmental problems and more effective coordination at the international level has been emphasized. The issues of special importance of the environmental factor in the oil sector and oil companies have been researched. The article also contains the author's approach to the problems of improving and developing the ecological and economic system in the oil industry enterprises in accordance with the requirements of modern times. The need for conceptual approaches and effective mechanisms for environmental protection and solution of environmental problems in the oil industry of Azerbaijan has been stated. The country's investment in environmental protection and more efficient use of natural resources has been researched. Indicators of air pollutants and hazardous industrial waste have been analyzed. The losses in the Azerbaijani crude oil balance have been noted. The amount of pollutants released into the atmosphere in the structural units of SOCAR has been analyzed. In line with global challenges, proposals have been made to address environmental and economic problems in Azerbaijan's oil industry.

Keywords: Sustainable development · Ecology · Ecological and economic factors · Oil industry · Ecological problems in oil industry · Global ecological problems · Efficiency of ecological and economic system

1 Introduction

Global problems and trends are reflected significantly in the oil industry as in all areas. In this respect, ecology problems and optimization of the ecological and economic system are of strategic importance. Furthermore, the solution of environmental problems in the context of sustainable development and ensuring sustainable development at the global

© The Author(s), under exclusive license to Springer Nature Switzerland AG 2021
R. A. Aliev et al. (Eds.): WCIS 2020, AISC 1323, pp. 409–417, 2021.
https://doi.org/10.1007/978-3-030-68004-6_54

level is considered to be one of the most difficult. Thus, the intensifying climate change in recent years is a clear example of this. Moreover, these concerns have been exacerbated by interstate coordination and international tensions, particularly in the United States, the world's largest economy, and its indifference to climate change. As environmental problems increase, they require greater resources and well-thought-out strategic approaches. With regard to oil production, environmental pollution and the escalation of environmental problems are not new. No matter how advanced the technology is, crude oil is released into the environment during oil production. Such cases also occur during the transportation and refining of oil [1]. In addition, since 1960–70s, world countries have been interested in developing long-term strategies to ensure sustainable development and solving global problems, including environmental problems, provided that states make international commitments. At the international conference held in Rio de Janeiro, Brazil on 3–4 June 1992, a Declaration was adopted with the participation of UN experts and representatives of the world to solve these issues [2]. At that conference, 27 principles on the world's sustainable development problems were developed and adopted. In addition, at the Summit, held on September 25, 2015, the growing concern about environmental issues, which are quite problematic in today's global context, have been highlighted. At that Summit, the world countries have made certain commitments for sustainable development and reflected on key issues in solving current problems [3].

One of the factors that negatively affect the environmental situation in the world is related to oil production. As oil is the main fuel, the demand for it is constantly increasing and therefore 100 billion dollars are invested in the development and exploitation of new reserves. As the world's first industrial oil producer, Azerbaijan is also active in developing its oil reserves and exporting them to world markets. In this respect, resource extraction processes from the earth's crust are among the factors that can have a serious impact on the environment and the balance of the ecological and economic system. Each country should develop and implement a strategy to evaluate this factor in strengthening its ecological and economic system and ensuring sustainable development. Some of the world's oil-producing countries have not been able to achieve the desired level of efficiency and optimum management in oil production, export and transportation to world markets. All this prevents the systematic solution and international coordination of the problems in the oil sector and the unification of the world's efforts. The issue of effective management of oil production, transportation, storage, refining and delivery to end producers, such as the most complex and specific problems in the oil industry, is especially important. In this case, the important issue is to create the maximum correct approach to the requirements of the ecological and economic system. In other words, the strategic issues of the ecological and economic system and the optimal management and regulation of problems in the process of its operation should be considered and resolved at the strategic level [4]. In the enterprises engaged in oil production and refining, the issues of complex and systematic regulation of economic factors and aspects, management mechanisms should be more highlighted in order to ensure environmental safety [5].

2 Method

In the article, the systematic and comparative analysis of the studied problems, the most effective methods and techniques for determining the direction of improving the management mechanisms of environmental and economic problems in the oil sector in Azerbaijan and the works of various researchers on these problems have been used. During the research, different approaches have been made by the author. In forming the views, the realities of the current environmental problems at the global level have been considered.

3 Results and Discussion

World countries and international organizations should take sustainable measures to protect the environment, strengthen environmental safety and increase the efficiency of the ecological and economic system, and should be open to international cooperation, strive for close cooperation with countries in these areas, actively participate in projects of international and regional importance, develop and implement national environmental safety programs. Although leading world powers and international organizations have openly expressed concern about these problems, transnational companies continue to implement new large-scale projects to increase the use of oil resources and production in various parts of the world, and in the structure of investments in these issues, there is a significant disproportion between the funds allocated for production and environmental measures. Necessary measures should be taken at all stages of oil production and transportation during organization and strengthening of management mechanisms of environmental and economic problems in oil-producing enterprises. Thus, the continuous release of gases, other pollutants, including solid, gaseous and liquid bodies, and other gaseous substances into the environment and the atmosphere in the oil sector should be considered as factors complicating the environmental situation. In these processes, it is important to strive to increase the effectiveness of the ecological and economic system by increasing the sense of responsibility and strengthening the management system. However, for this, it is necessary to ensure that the ecological-economic system itself consists of working mechanisms [6]. In the constituent components of the ecological-economic system, increasing the sense of responsibility of oil producers to protect environmental requirements, increasing motivation, developing incentives, as well as the availability of other regulatory tools, such as tax regulations, the formation and development of reserve funds, are important conditions. In particular, the issues of developing environmental and economic mechanisms, financing of measures and waste management at each oil production and refining enterprise should be considered and ensured in a systematic and consistent manner [7]. Awareness-raising activities should be carried out on the regulation of ecological and economic relations, the preference for maximum waste-free technologies and the formation of conceptual approaches to issues related to environmental pollution, and the effective implementation of economic-organizational and economic-legal mechanisms should be adressed. In the formation of the ecological and economic system in the oil sector, the protection of nature should be prioritized and specific approaches should be made to related issues [8]. The issues of operating low-efficiency oil wells should be approached strategically and environmental factors should be taken into account in the implementation of the necessary technological measures to increase production. In this regard, the operational processes in the oil sector should be regularly regulated and improved.

3.1 Issues of Ensuring Economic and Ecological Balance in Accordance with the Principles of Sustainable Development

In modern times, the issues of ensuring economic and environmental balance and solving related problems should be very important in the economic policy of each state, and especially in energy policy, including oil resources [9]. Therefore, there should be mechanisms to systematically and effectively address the problems that arise during oil production. During oil production, crude oil is released into the environment, as well as sludge from oil production and drilling operations, contaminated water, soil and water basins. Moreover, the volume of gases emitted into the atmosphere and released without burning is quite large. Continuous burning flames can be seen even from a distance in the oil producing areas. Losses and environmental pollution can also be seen during the treatment of oil and gas wastes remaining in the system [10]. The systematic solution of environmental problems in the petroleum sector and the solution of strategic approaches to strengthen the ecological and economic system in general is one of the important conditions. In addition, the recycling of waste generated in the production and refining processes in the oil sector should be provided in accordance with the requirements of modern technology and necessary measures should be taken [11]. The level of measures that will regulate ecological and economic relations on the basis of objective facts, ensure the health of the population and the protection of the environment should be met by society and the impact of these measures should be reflected in real life [12]. At the same time, the state should pay special attention to the improvement and functioning of environmental and economic policy and apply flexible practical tools for their implementation [13].

The oil industry in Azerbaijan is the driving force and leading locomotive of the economy, so the effective use of the potential of this sector is of strategic importance. Thus, the fuel and energy complex has an important role in improving the welfare of the population from Azerbaijan's natural and national resources. Currently, 71 oil and gas fields have been discovered in the country and most of them are being exploited. Major projects in the oil and gas industry are being implemented. Our country has a very large electricity and heat production and consumer delivery network. The main task of the fuel and energy complex, which combines all this, is to meet the continuous and uninterrupted demand of the national economy and the country's population for energy resources. In addition, taking measures to use of underground resources in our country and to extract hydrocarbon resources from the deep layers of the earth in compliance with safety regulations are kept in the center of special attention. Thus, the pollution caused by radioactive elements released from the deep layers of the earth by oily water has a negative impact on the environment [14]. These should be prevented and sustainable measures should be taken. Investments should be made and financial resources should be allocated in research to neutralize associated gases emitted into the atmosphere at oil and oil refineries, in the development of refining technologies [15].

The implementation of the oil strategy in our country and large-scale investments in this area require great responsibility. Sustainable solution of environmental problems in the efficient use of existing oil resources will allow the organization of activities in this direction in the framework of national interests and national economic interests.

After the restoration of independence, as a result of the reconstruction of the material and technical infrastructure of our country's oil industry and the commissioning of facilities based on modern technologies, it has been possible to increase oil production significantly. The volume of oil production in 2010 was 4 times higher than in 1990 and a record 50.8 million tons of oil was produced that year. During 2011–2018 years, oil production decreased relatively. Thus, oil production in 2012 was 43.4 million tons, in 2015 - 41.6 million tons, in 2016 - 41 million tons, and in 2018 - 38.8 million tons [16]. During environmental monitoring, more emphasis is placed on ensuring effective quantitative and qualitative control of harmful gases, liquid and solid wastes released into the environment as a result of natural and anthropogenic impacts in oil production, refining and transportation processes. Elimination of emerging problems and shortcomings, restoration of ecological balance, evaluation of the factors that negatively affect the ecological and economic system and taking appropriate measures are of strategic importance [17]. Waste management issues also has a special role in maintaining the ecological balance. For effective waste management, there are mechanisms and rules for the collection, sorting, transportation, placement and burial of waste generated in oil companies. This is all reflected in "Waste management Plan" (WMP) of SOCAR [18].

3.2 Analysis and Evaluation of Environmental and Economic Factors in the Azerbaijani Oil Industry

Innovative approaches should be used more to strengthen environmental safety and to increase the effectiveness of environmental protection measures. Our country's oil industry is carrying out innovation processes at a higher level than in other sectors of the economy. The technologies used in this area are characterized by modernity and productivity, as well as high standards, but in all cases the protection of environmental requirements requires additional and sustainable work, and new mechanisms need to be developed and implemented. Issues of ensuring the balance of the ecological and economic system in the process of modeling the evaluation system of new oil projects are of special importance. In this regard, the development of criteria and principles for determining environmental and economic damage and the formation of a system of indicators are the important conditions [19]. One of the main issues is to ensure a strategic approach to issues related to oil refining processes, and secondary approaches to these issues are unacceptable. Thus, the violation of ecological and economic balance and the increase in environmental risks may hinder the development expectations of the sector and become questionable. The dynamics of the funds directed to fixed capital for the protection of the environment and the efficient use of natural resources in Azerbaijan in 2005–2018 are shown in Fig. 1. As can be seen, in 2018, these funds amounted to about 248 million manats, which is 86.5% more than in 2017.

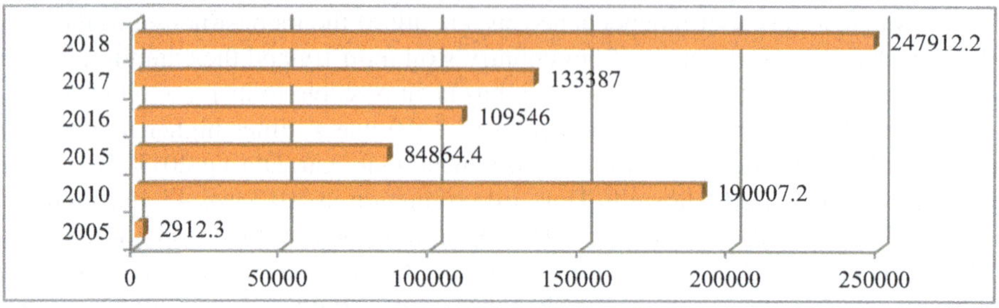

Fig. 1. Funds directed to fixed capital for environmental protection and efficient use of natural resources in Azerbaijan, thousand manats, 2005–2018 years (Designed by author in accordance with the data of the SSC of Azerbaijan).

In Fig. 2, the total volume of pollutants emitted into the atmosphere in Azerbaijan in 2005–2018 has been analyzed, and no increase has been observed in these substances for the last 4 years.

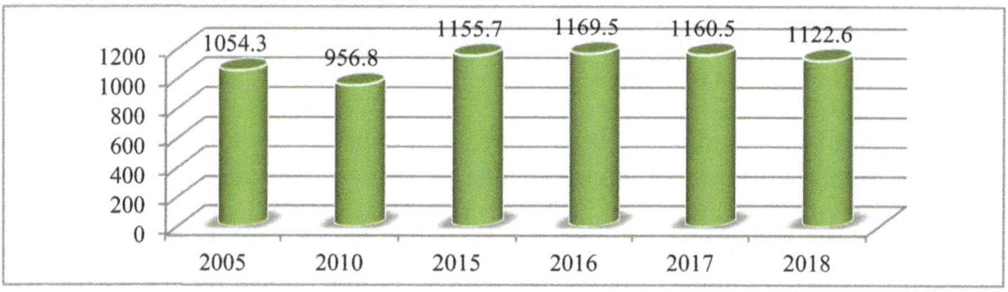

Fig. 2. Pollutants emitted into the atmosphere in Azerbaijan, thousand tons, 2010–2018 years (Designed by author in accordance with the data of SSC of Azerbaijan).

In Table 1, indicators of hazardous industrial wastes and pollutants emitted into the atmosphere from stationary sources in Azerbaijan in 2010–2018 has been indicated. It should be noted that the volume of hazardous waste generated during this period has not decreased at all, the volume of hazardous waste used has remained at the same level, the volume of solids released into the atmosphere has not changed much, and the volume of gaseous and liquid substances has decreased slightly.

In general, it should be noted that the level of hazardous waste disposal in our country is many times lower, for instance, while this indicator was 2.4 times in 2010, it was about 5.3 times in 2018, and the average annual volume of waste generated during that period increased more than 2.4 times. During 2017, a total of 32 types, more than 80,800 tons of various types of waste were collected and handed over accordingly. During the year, 6.9 thousand tons of drilling cuttings were recycled [20]. In Fig. 3, the dynamics of losses in the crude oil balance of Azerbaijan in 2010–2018 has been indicated. The analyze shows that during this period, the level of losses has not decreased, on the contrary, in 2018 it has increased by 89.2% compared to 2017 and amounted to 76.8 thousand tons.

Table 1. Indicators of hazardous industrial wastes and pollutants emitted into the atmosphere from stationary sources in Azerbaijan in 2010–2018, thousand tons

s/s	Indicators	2010	2015	2016	2017	2018
01	Generated hazardous waste	140,0	262,6	632,6	266,0	338,7
02	Used hazardous waste	5,5	5,2	47,8	5,5	5,7
03	Disposed of hazardous waste	58,4	210,9	25,9	35,8	64,2
04	Solids released into the atmosphere	19,3	6,6	6,1	6,2	6,5
05	Gaseous and liquid substances emitted into the atmosphere, including:	195,5	171,4	181,5	177,9	165,9
	Sulfur dioxide	2,2	3,8	8,6	4,9	0,7
	Carbon monoxide	27,2	27,9	24,3	25,5	16,7
	Nitric oxides	19,8	18,9	18,2	22,1	20,8

Note: Designed by author in accordance with the data of SSC of Azerbaijan

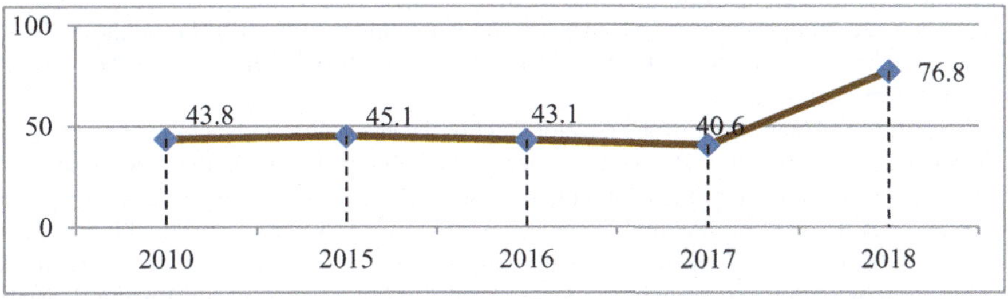

Fig. 3. The dynamics of losses in the crude oil balance of Azerbaijan, thousand tons, 2010–2018 years (Designed by author in accordance with the data of SSC of Azerbaijan).

4 Conclusion

In the current situation, the oil sector in Azerbaijan is strategically important as a key sector of our economy and in this respect, balancing the environmental problem and strengthening the ecological-economic system, which is a global problem, is important for achieving better results in the oil sector. A group of issues related to these should especially be considered:

- Measures should be taken to establish high-tech processing enterprises by concentrating the cycle of collection, storage and processing of waste generated during oil production in a single complex, and the involvement of private sector entities, foreign investors and firms and companies with experience in this field in the world should be ensured in these processes;
- Strategic approaches for the most efficient organization of production processes in the oil sector and enterprises of the country, the development and implementation of optimal management mechanisms should be maximized and harmonized with international practices;

- Given the global challenges depending on the current situation in the oil industry in the process of improving the management mechanisms of environmental and economic problems, the development and implementation of maximum more efficient mechanisms has emerged as an objective requirement and so on.

References

1. Burakov, A.A., Tyurina, S.G.: On the environmental problem of oil pollution. https://scienceforum.ru/2017/article/2017040361. (original in Russian)
2. Armenia carries out environmental terror against Azerbaijan. https://www.un.org/en/development/desa/population/migration/generalassembly/docs/globalcompact/A_CONF.151_26_Vol.I_Declaration.pdf. (in Azeri)
3. UN Summit on Sustainable Development. 25–27 September 2015. UN Headquarters, New York, United States of America. https://www.who.int.ru. (original in Russian)
4. Nikitina, Y.A.: Economic aspects of environmental safety in the oil and gas industry (international experience and Russian practice). In: Synopsis of Dissertation, Candidate of Economic Sciences, Moscow, p. 32 (2012). (original in Russian)
5. Reznik, G.A., Malyshev, A.A.: Mechanisms for managing the sustainability of the ecological and economic system. News High. Educat. Inst. Volga Region. Soc. Sci. **4**(24), 138–146 (2012). (original in Russian)
6. Elkina, L.G.: Management of the development of the ecological and economic system of an industrial enterprise. Diss. Dr. econ. sciences. Ufa, 2009, p. 331. (original in Russian)
7. Redina, M.M.: Ecological and economic analysis of the activities of enterprises in the oil and gas industry. Diss. Cand. Econom. Sciences, Moscow, p. 164 (2009). (original in Russian)
8. Paitaeva, K.T.: Ecological and economic aspects of the effectiveness of environmental protection in the oil and gas industry. Russ. Acad. J. **4**, 37–39 (2013). (original in Russian)
9. Nurtdinov, A.R.: Institutional support for sustainable development in conditions of ecological and economic imbalance. Diss. Cand. Econom. Sciences. Kazan, 195 (2011). (original in Russian)
10. Khomutko, V.: Oil and related environmental problems. https://neftok.ru/raznoe/neft-i-ekologiya.html. (original in Russian)
11. Govorushko, S.M.: Environmental problems of extraction, transportation and processing of fossil fuels. Dalnauka, Vladivostok, p. 208 (2014). (original in Russian)
12. Akhmetshina, A.R.: Regulation of economic and environmental relations in the framework of sustainable development of the economic system. Diss. Dr. Econ. Sci. Kazan, p. 415 (2011). (original in Russian)
13. Sheveleva, A.V.: Development of institutions and tools to improve the environmental and economic policy of oil and gas enterprises. Synopsis of Dr. Econ. Sciences, Rostov-on-Don, p. 60 (2018). (original in Russian)
14. Ecological policy. State Oil Company of the Republic of Azerbaijan. Baku, p. 40 (2008). (original in Azerbaijani)
15. The environment of Azerbaijan. Statistical Data Collection. Baku, p. 140 (2018). (original in Azerbaijani)
16. Aliyev Sh.T. Azerbaijan's economy. SSU, p. 384–158 (2018). (original in Azerbaijani)
17. Environmental monitoring. https://www.socar.az/socar/en/environment-and-safety/menu/environmental-monitoring. (original in Azerbaijani)
18. Waste management. https://www.socar.az/socar/en/environment-and-safety/menu/waste-management. (original in Azerbaijani)

19. Pinaev, V.E.: Modeling of systems of assessment of projects of oil and gas industry on the example of the Yamalo-Nenets Autonomous Okrug. E-J. Sci. **8**(4) (2016). https://naukovedenie.ru. (original in Russian)
20. SOCAR, 2017. Annual Report. https://www.socar.az. (original in Azerbaijani)

Method of Managing Configurations of Engineering Systems

Mikhail Ovsiannikov(✉), Sergey A. Bukhanov, and Sergey A. Podkopaev

Bauman Moscow State Technical University, Building 1, 5, 2nd Baumanskaya Street, 105005 Moscow, Russia
mvo50@mail.ru, bukhanov@yandex.ru, sergey0511@mail.ru

Abstract. Management of configurations on different stages of the product life cycle (PLC) includes the problem of identification, configuration synthesis, checking compliance with requirements and constraints. Generally, these problems are solved differently. This paper proposes the approach that allows solving all configuration management problems on a unified entity-relationship model and a class of algorithm.

The approach is based on the development of PLC ontology, constructing ontologies of the application area describing the entities and relationships between them, as well as the algorithm for synthesizing configurations based on concepts of applied ontologies.

The universal algorithm of solving the problem by a constraint satisfaction method allows the user to adjust the problem statement during the solution of the problem.

The proposed solutions were verified for the problem of synthesis of product configuration, creation of production and power system, for planning maintenance procedures and so on.

The proposed methods allow solving important engineering problem of determining the structure of objects in the product being designed or processes being developed, which can be applied to different stages of the PLC.

Keywords: CIM · Configuration management · Ontology model · Constraint satisfaction problems · CSP

1 Introduction

Configuration management is one of the main technologies of PLC aimed at establishing and maintaining the conformity of the functional, physical and operational properties of the product to requirement during its design and transformation over the whole lifecycle.

Configuration management refers to a complex of measures taken during the lifecycle of a product. The measures are aimed at establishing and maintaining compliance between the requirements, the characteristics of the developed product, and the configuration documentation.

© The Author(s), under exclusive license to Springer Nature Switzerland AG 2021
R. A. Aliev et al. (Eds.): WCIS 2020, AISC 1323, pp. 418–426, 2021.
https://doi.org/10.1007/978-3-030-68004-6_55

In particular, such problems arise in developing a product that meets the requirement of a specific customer, creation of computer and power systems, as well as constructing maintenance procedures and so on.

The regulatory base in the direction of «configuration management» emerged back in the 1950s. The Quality Management - Guidelines for Configuration Management (ISO 10007-95) [5] can be considered a foundational regulation.

ISO 10007-95 describes the following forms of representing configurations:

Functional configuration (FC) is the structure of a product family where a Configuration Item (CI) is a function of the final product identified for the goal of configuration management and having a certain set of attributes (properties, characteristics).

Project configuration (PC) is an FC-based structure of a product model (version) where CI is any hardware or software performing a finite task (or a certain function of the final products) identified for the goals of configuration management and having a certain set of attributes (properties, characteristics).

Physical configuration is the structure of a real product manufactured according to PC, where CI are specific instances.

The base elements of the technology of configuration management in a machine-building plant focus on solving the problems of managing configuration, storing, and providing access to it over the whole product LC [14, 17].

Generally, based on the literature review, it can be concluded that there is no unified approach efficiently solving all problems of configuration management [3].

The proposed approach will allow you to solve configuration management problems using a single method that includes 3 stages:

- building a formal applied ontology of the problem domain,
- configuring and adjusting restrictions and configuration requirements;
- application of an algorithm for solving a combinatorial problem by the constraint satisfaction method using the ontology of the problem domain.

2 Product Life Cycle Ontology

FC can be defined as the structure having the items of functional configuration, their properties, relationships of properties and items, as well as statements based on the properties of items and variables.

Formally speaking, the information model of FC is an ontology of the "Product configuration management" application area; this ontology allows performing the procedures of synthesizing and recording configuration status based on the full data set accumulated over all stages of the product life cycle. This data representation format allows taking into account the operational changes of the product characteristics in the following processes of synthesizing similar configurations.

In the foundation of the proposed PLC ontology, the ontology of relationships presents the properties of entities as binary relationships. This allows creating a metaontology as a set of classes of objects and relationships and the top-level ontology as a classifier of objects and their properties.

This approach allows applying universal algorithms based on constraint satisfaction method programming methods for a wide range of problems of structural synthesis and configuration management.

The common feature of the approaches using binary models is the representation of elements by implementing binary associations, i.e., statement with only two terms. The main concepts are singlets so that constraint can be explicitly formulated. The changes in the application area are reflected in adding or removing types of binary relationships, entities, and entity names, as well in changing constraints.

This approach is based on the works of Ovsyannikov and Bukhanov on the creation of algorithms and construction of ontologies of different PLC problems [1, 7, 13].

The developed metaontology of the life cycle is based on schemes and description language RDF and OWL, as well as the main entities and provisions in the data model of the standard «Industrial automation systems and integration.

Integration of lifecycle data for process plants including oil and gas production facilities» (ISO 15926). The main requirement to the developed model is to enable the use of universal algorithms based on the constraint satisfaction method or other AI algorithms.

In this paper, a method interpretation description is used; the method is developed to utilize the Constraint Satisfaction Algorithm. For solving the problem of configuration synthesis, Bukhanov S.A. developed a specialized language for the description of constraints [1].

The applied ontology of the configuration synthesis problem based on the proposed metaontology includes classifiers of objects, properties, and the ontology of the problem (synthesis or determination of the degree of configuration compliance).

The method of constructing an applied ontology has been tested on the problems of synthesizing the design configuration of a car, the local electrical network, the maintenance process, and drawing up an individual curriculum.

Using the proposed ontology will solve the problems using the constraint satisfaction method and enable interaction with external data sources to fill the classifiers and relationships according to ISO 15926.

3 Solving the Configuration Management Problem Based on an Ontological Model

The main problem of the existing approaching the automating the solutions of this class of problem is the need to time-consuming code implementation of the search algorithm and maintaining the developed software taking into account all changes in the calculation method.

On the values of properties of components defined in the component dictionary, expressions of constraints defined in the functional configuration and requirements to the indicator values are applied. Resolving constraints is done iteratively by searching through the space of axioms with the calculation of intermediate values of parameters until all parameters listed in the requirements are found and tested, or the fact of incompleteness or infeasibility of meeting the constraints for the configuration tested is established.

In the process of solutions, the scores of compliances of the actual parameters of the configuration to the required values are formed.

The obtained scores allow evaluating the quality of the configuration tested (separately or by comparing it with other similar configurations) and choose the plan for further work on changing the configuration.

To solve these problems, a method of configuration management based on the Constraint Satisfaction Problem (CSP) was developed. The method allows an engineer to describe and solve applied problems without having to code complicated algorithms [6, 8].

The utilized paradigm called constraint programming requires only problem description and does not require a developer to define the algorithm of solving the problem [12]. In this case, the problem and the model are represented as an unstructured set of relationships corresponding to connections existing between problem variables. These relationships commonly referred to as "constraints" can have the form of equations, inequalities, logical expression, etc. [4].

Definition 1. The configuration management problem in CSP terms can be defined by the following structure:

$$(X, D, S, C, F), \tag{1}$$

where: the set of "positions" ("variables" in CSP terminology) $X = \{X_1,...,X_n\}$ correspond to terminal objects of the functional configuration (OFC).

Set of component domains that implement functions of position $D = \{D_1,..., D_n\}$, где D - where D is a function(mechanism) that connects position X_i to its domain $D(X_i)$. The domain of a variable is a finite set of directory objects;

Each domain contains a set of objects $D_i = \{d_1,..., d_m\}$;

Set of position and component properties $S = \{S_1,..., S_r\}$;

The set of constraints $C = \{C_1\{S(X)\},..., C_o\{S(X)\},\}$. Where the constraint C_j is defined on a subset of the position properties $S(X)$.

Objective function (preference function) $F = f\{S(X)\}$.

Definition 2. A configuration for (1) (in ISO terms, a project configuration) is a set of P assignments of *dm*j instances from the *DJ* component domains for positions xi

$$P = \{p_1 = d_{mi} \in D_i, \ldots p_I = d_{mI} \in D_I\} \tag{2}$$

The configuration (2) is consistent if the assignments in P are consistent with the constraints in C. In addition, the configuration is complete if specific instances of d_{mi} components are assigned to all positions in X. Finally, a configuration is valid if it is both consistent and complete.

For configuration management there are three class of problems for different stages configuration management process (CMP) (see Fig. 1):

- synthesis software based on objects and constraints of functional configuration (FC) and the requirements for a particular project,

- assessment of the extent to which the values of the product characteristics to customer requirements based on the facilities and limitations of functional configuration (FC) for all stages of life cycle,
- determination of characteristic values of the software, when running functional configuration (FC) limitations and requirements for the product.

Definition 3. The solution to the configuration synthesis problem is to find a tuple of component assignments that satisfies all the constraints.

Then the synthesis problem can be formulated:

Find such an assignment of components *P*, from domains $D = \{D_1, \ldots, D_n\}$, for positions.

$X = \{X_1, \ldots, X_n\}$ where all the constraints $C = \{C_1, \ldots, C_o\}$, defined with the value of the properties of the assigned d_{mj} components are valid, and the preference function has the best value.

$$F = f\{S(P)\} \rightarrow extr. \tag{3}$$

$$\ddot{Y}\{C = \{C_1\{S(P_1)\}, \ldots, C_o\{S(P_o)\} = true\},$$

Where: *S(P)* – values of the properties of the components assigned to the corresponding positions.

For CSP problems, there is a special class of modified methods of a different kind having an increased efficiency due to specifics of working with constraints. For solving the problems with constraint satisfaction methods, specifics of the problems are often used to determine a set of heuristics.

The existing method of finding solutions for the CSP class can be subdivided into four main classes:

- Systematic search methods: the Backtracking method, upon which many depth-first methods are based. Modifications of the Backtracking methods differ by the heuristics that increase the efficiency of the main algorithm and are used in the search with the Constraints Propagation method [9];
- Stochastic and heuristic search methods (random search, local search, ant colony method, G-algorithm, simulation annealing, variable neighborhood search, genetic algorithm, tabu-search, and so on) [2].
- Methods of limiting the search space, also called Consistency Techniques: Node Consistency, Arc Consistency, Path Consistency, Directional Arc Consistency, Directional Path Consistency, Restricted Path Consistency and so on. [15, 16]
- Methods of resolving underdetermined models [4].

The developed hybrid metaheuristic algorithm for (3.3) is comprised of a sequence of the main stages inherited from the unified local search method: initial solution selection, the formation of the neighborhood, checking feasibility and evaluation of the utility of each solution in the neighborhood, rating the solution by their utility, checking the search termination criterion, and choosing one or several solutions as the initial solutions for the next iteration [1, 10].

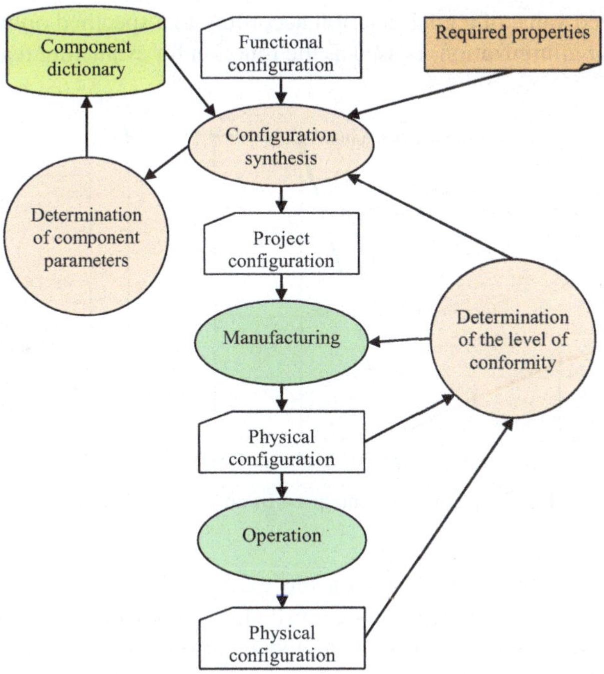

Fig. 1. Configuration management process

The algorithm is extended with methods of satisfying constraints and reinforced with a set of heuristics considering specifics of the considered problem and elevating the efficiency of the search of feasible solutions close to the optimal solution by using a goal function.

Formulation of the problem in terms of CSP yields additional advantage, whereby the information of the number of violated constraints (MinConflicts heuristic) can be used as a utility function for comparing solutions of the search neighborhood. Thus, the goal function does not have to be calculated in the initial area of unfeasible solutions.

The main feature of the implemented stochastic search algorithm is the strategy of search in two subspaces: a subspace of unfeasible solutions and a subspace of feasible solutions, allowing the search to be initiated with any random solution, including an unfeasible solution, Fig. 2.

For unfeasible solutions, the value of the goal function is determined by a complex indicator called the degree of deviation, which evaluates the deviation of the current configuration from a feasible one. The degree of deviation is formed as a weighted convolution of four criteria for the considered unfeasible solution.

For calculating the degree of deviation, the following indicators of an unfeasible solution are considered as criteria:

- Number of violated constraints;
- Number of items in the configurations that violate constraints;
- Number of parameters of the configuration items constituting the violated constraints;
- Maximum deviation (error) of the values of parameters in the violated constraints.

The calculated value of a goal function according to a specified optimality criterion (maximization or minimization) is used as the function for evaluating feasible solutions.

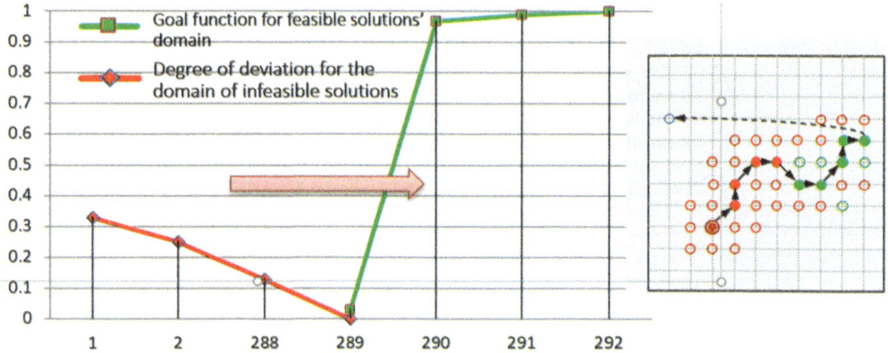

Fig. 2. Exiting the subspace of unfeasible solutions

The influence of these indicators on the total value of the complex indicator is determined by a set of specified coefficients. This approach allows transforming the landscape of the evaluation function from a stepped function to a smoother one, thus reducing the algorithm of search of feasible solutions to an optimization algorithm.

In the algorithm, the search strategy can be fine-tuned by activating individual heuristics solving individual subproblems of the method. The following subproblems can be highlighted: method of selecting the initial set of solutions, selection of the form of the neighborhood of each landscape, selection of the character of the local phase of the search between local search and global scanning, selection of between the trajectory-based or population-based exploration of the landscape, selection of the local search termination criterion, parameters of controlling the global search phase.

The proposed algorithms were implemented as software for solving configuration synthesis problems CMSolver [7].

The solved is developed to solve a relatively specific class of problems of structural synthesis, in particular, the problem of product configuration synthesis.

The solves allows conduction solution on all stages of the problem, from problem statement to the analysis of results.

A system of constraints is an important part of the problem statement. Constraints link the parameters of the configuration objects. For the statement of the problem of configuration synthesis as a constraint satisfaction problem, the structure of relations of the parameters of the configuration objects can be represented as a graph called the constraint graph [16].

During preparation for the search, the hypergraph is analyzed. The processing of hypergraph reveals and implements the method of calculating aggregate properties and testing the feasibility of solutions. Also, unresolved arguments, expression redundancy, and graph acyclicity are checked.

Evaluation and checking of constraints for each neighborhood configuration are conducted by applying the map of configuration parameters to each solver one by one, and the solvers form the implementation of the constraint hypergraph of functional configuration.

4 Analysis of the Results of Performance Evaluation of Algorithms

The proposed method and the algorithms were tested on a simple example of forming the configuration of an automobile (characteristics of the VW cars were used in the example). The process is based on forming configuration by exhaustive search of variants with prior constraining of the search space.

In the performance evaluation of the developed algorithms, a series of runs of the algorithms on different problem modification and different dictionary volumes was conducted.

5 Conclusion

Ontological approaches to data modeling and semantic technologies for processing and storing them were created for the operations with quasi-structured data conducted in the problems of analyzing texts in natural languages or social network data. The development of semantic technologies led to the expansion of their application areas, including the use of them on structured life cycle data of complex engineering objects. These approaches allow effective communication of a large number of computers and people in the framework of increasingly complex engineering projects.

The proposed approach will allow you to solve configuration management problems using a single method that includes 3 stages:

- building a formal applied ontology of the problem area,
- configuring and adjusting restrictions and configuration requirements;
- application of an algorithm for solving a combinatorial problem by the constraint satisfaction method using the ontology of the problem domain.

The approach proposed in this paper is based on expanding the procedure for identifying the functional configuration (FC) by constructing an ontology of the subject area with the definition of regularities and restrictions for forming the product structure: selecting possible components from among alternative ones and determining requirements for designing missing components.

The developed methods allow the following:

- Maximum utilization of existing design solutions;
- Simplify components selection for forming a new product configuration;
- Select and use existing components whose parameters meet requirements in a new configuration even if the components were created for the use in other products;
- Define design requirements for the missing components;

- Avoid component selection mistakes by constant comprehensive configuration checking.

The proposed method allows solving crucial engineering problems of determining the structure of the objects of products designed or processes formed, which can be applied at different stages of the product life cycle.

References

1. Bukhanov, S.A., Ovsyannikov, M.V.: Configuration management using constraints programming. In: Efficient Methods of Automation of Production Engineering and Planning: Collection of Research Paper. "Spektr" Publishing House, Moscow, pp. 150–156 (2011)
2. Karpenko, A.P.: Modern algorithms of search optimization. In: Karpenko, A.P. (ed.) Algorithms Inspired by Nature: A University Textbook. Bauman Moscow State Technical University Press, Moscow, p. 446 (2014)
3. Levin, A.I., Sudov, E.V.: Methods and technologies of management of configuration of complex products – Instrument engineering technologies, № 4 (2003)
4. Nariniani, A.S.: Introduction in under definiteness. In: Appendix to the Journal Information Technologies, no. 4, p. 32. New Technologies Press (2007)
5. Norenkov, I.P.: International standards of information support of the product life cycle stages. Inf. Technol. **4**, 49–51 (1999)
6. Ovchinnikov V.A.: Algorithmization of Combinatorial Optimization Problems in Design of Computers and Computer Systems. Bauman Moscow State Technical University Press, Moscow (2001)
7. Ovsyannikov, M.V., Bukhanov, S.A.: Configuration management via programming in constraints. BMSTU J. Mech. Eng. **14**, 70–75 (2012)
8. Papandimitriu, K., Steiglitz, K.: Combinatorial Optimization. Algorithms and Complexity, p. 512. Mir Press, Moscow (1985)
9. Rassel, S., Norvig, P.: Artificial Intelligence: A Modern Approach, p. 1408. "Willams" publishing house, Kyev (2006)
10. Bochkarev, S.V., Petrochenkov, A.B., Bukhanov, S.A., Ovsyannikov, M.V.: Structural synthesis of complex electrical equipment based on the method of satisfaction of constraints. Electr. Eng. **6**, 63–69 (2015)
11. Scherbina, O.A.: Satisfaction of constraints and constraint programming. Intell. Syst. **15**, 53–170 (2011)
12. Apt, K.R.: Principles of Constraint Programming. Cambridge University Press (2003)
13. Petrochenkov, A.B., Bochkarev, S.V., Ovsyannikov, M.V., Bukhanov, S.A.: Construction of an ontological model of the life cycle of electrotechnical equipment. Russ. Electr. Eng. **86**(6), 320–325 (2015). https://doi.org/10.3103/S1068371215060115
14. Hass, A.M.: Configuration Management Principles and Practice, 1st edn, p. 432. Addison-Wesley Pub Co (2002)
15. Muller, T.: Constraint-based Timetabling. Ph.D. Thesis. Charles University in Prague, p. 120 (2005)
16. Rossi, F., van Beek, P., Walsh, T. (eds.) Handbook of Constraint Programming, p. 978. Elsevier (2006)
17. Watts, F.B.: Engineering Documentation Control Handbook: Configuration Management for Industry, 2nd edn, p. 265. Noyes Publications (2000)
18. Wielinga, B., Schreiber, G.: Configuration design problem solving. IEEE Intell. Syst. **12**, 49–56 (1997)

Iodinated Multi-walled Carbon Nanotubes and Their Antioxidant Profile by Using ANFIS

E. B. Zeynalov[1], M. M. Agahuseynova[2](✉), A. B. Huseynov[1], and N. I. Salmanova[2]

[1] Nagiyev Institute of Catalysis and Inorganic Chemistry, Azerbaijan National Academy of Science, H. Javid Avenue 113, AZ1143 Baku, Azerbaijan
elzey@mail.ru

[2] Azerbaijan State University of Oil and Industry, Ministry of Education, 20, Azadlig Avenue, Baku, AZ 1010, Azerbaijan
minira_baku@yahoo.com, nazilya.salmanova@gmail.com

Abstract. Iodine grafted multi-walled carbon nanotubes (I-MWCNTs) were sybthesized and tested with respect to their capacity to retard the aerobic oxidation of hydrocarbons. The model cumene initiated (2,2′-azobisisobutyronitril) oxidation was employed to estimate the oxidation rates. This approach was based on the known intrinsic anti-oxidtive ability of MWCNTs which should be synergetically combine with capability of covalently anchored iodine groups to quench free alkyl radicals. The kinetic experiments have shown that the presumption was substantive and expected effect has been achieved. The iodination allows to shift the electron density along with surface of the nanotubes urging intensive scavenging free radicals and thereby provokes the heightened anti-oxidative activity of I-MWCNTs. The iodinated MWCNTs demonstrate feasibility to be used as an effective constituent part of a stabilizer's package for polymers since even tiny iodine covalently linked groups afford strong additional anti-oxidative effect. Many points in this field are so far open but one is undoubted - the results create a definite ground for further fruitful investigations and discussions.

Keywords: Multi-walled carbon nanotubes · Iodine doped carbon nanotubes · Antioxidative activity

1 Introduction

Polymeric materials of the present day are multicomponent systems doped and mixed with various types of additives, each of them to strictly oriented functioning. At that, the additives content in polymer composites can vary within a very wide range [1].

Stabilization of the original properties of polymers against thermo-oxidative onrush is usually doing by means of an addition of low-molecular substances- primary and secondary antioxidants, thermo-stabilizers, anti-ozonants, deactivators of metals, multi-functional stabilizers and others, is vital problem which does not need to have additional explanation. We would like to note only that the researcher's community came up in this

© The Author(s), under exclusive license to Springer Nature Switzerland AG 2021
R. A. Aliev et al. (Eds.): WCIS 2020, AISC 1323, pp. 427–434, 2021.
https://doi.org/10.1007/978-3-030-68004-6_56

way to conclusion that stabilization cannot be accomplished for all polymers by any single method. Hence, the particular problem that forms the basis for polymer breakdown must be defined in each case [2].

Among these existing approaches, nanomaterials represent one of the most promising frontiers in the research for improved antioxidants. Some nanomaterials exhibit intrinsic redox activity that is often associated with radical trapping and/or with superoxide activities [3].

This paper presents results of a study of iodine doped multiwalled carbon nanotubes (MWCNTs) to ability to reveal anti-oxidative activity.

2 Background

Our previous investigations as well as data of other authors unambiguously proved the antioxidative activity of pristine patterns of MWCNTs comprising in hydrocarbons or polymeric oxidation zones [4–12].

However, there are serious obstacles which alleviate and even suppress the intrinsic anti-oxidative capability of MWCNTs. First and main, it is remains of transitions metals which almost always present in the inner channels of as-prepared MWCNTs produced by CVD method. During this pyrolytic synthesis, the organic part of precursor-catalyst (usually metal-organic compounds) completely crashed under high temperatures and further synthetic growth of CNTs occurs around of metal particles. Resulting CNTs contain inner metals which are hard to reach and unamenable by any treatment.

We have already shown that the presence of transition metals in MWCNTs structure provokes decay of hydroperoxides and suppresses route of scavenging free radicals [13, 14]. Thus, instead of anti-oxidative power, finally, we observe catalytic oxidative activity of MWCNTs This phenomenon has been analyzed in details in our previous work [15].

Second problem connects with unsolubility of CNTs in organic liquids and bad compatibility upon preparation of polymer composites.

One of the best ways to solve these above mentioned problems is purposeful chemical modification of MWCNTs. The grafted functional groups impart to CNTs any desirable property, in particular, heightened antioxidative capacity [16–18].

In this study we undertake the effort to display potential antioxidative activity of the iodine- grafted MWCNTS. Moleculat iodine and the iodine compounds are related to the antioxidants operating as alkyl radical acceptors [19, 20]. These types of oxidation inhibitors are usually effective when they are used in combination with basic peroxy radicals acceptors, unless have higher rate constants of alkyl radical quenching. In order to realize this postulate we conceived to add to the inherent property of CNTs the additional antioxidative power stipulated by the grafted iodine moieties.

3 Synthesis of I-MWCNTs

MW carbon nanotubes were synthesized by the aerosol - assisted chemical vapour deposition.

Method/The synthesis and purification procedures describe in details in [21, 22].

The required amount of a sample of synthesized MWCNTs and commercial crystalline iodine was weighed to the accuracy 0.1 mg and placed in a thick-walled quartz reactor, the open end of which was sealed. The reactor, containing the sample was placed in a tubular muffle furnace and keeps there at 350 °C for 3 h. As the temperature was raised, the crystalline iodine passed to the vapor state and reacted with MWCNTs to form the C-I bond. At the end of the reaction time, the reactor was cooled, the quartz tube was unsealed and the obtained target product was recovered from the reactor and carefully washed by ethanol [22].

The MWCNTS and iodinated patterns structure were analyzed and proved by scanning electron microscopy (SEM) (see Fig. 1) and energy-dispersive X-ray spectroscopy (see Fig. 2) [22]. The presence of a chemical bond C-J in the synthesized samples is proved by the Fourier transform infrared (FTIR) spectroscopy method [22].

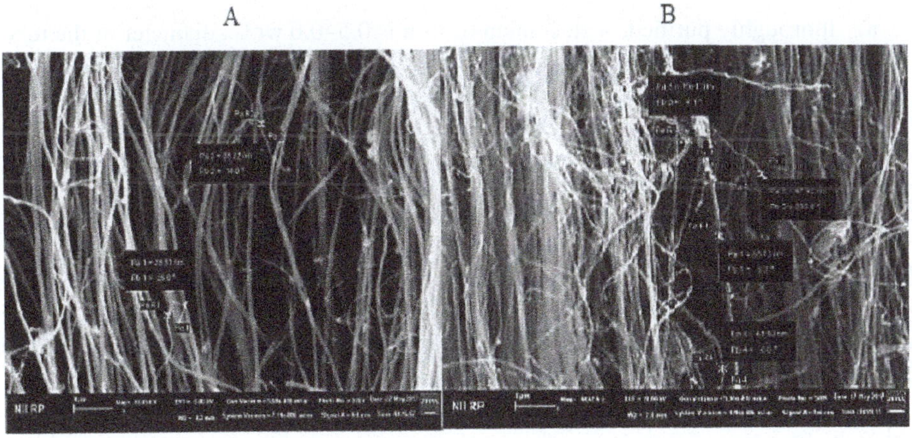

Fig. 1. SEM snapshots of the MWCNTs: **A**- purified nanotubes, **B** – iodinated nanotubes

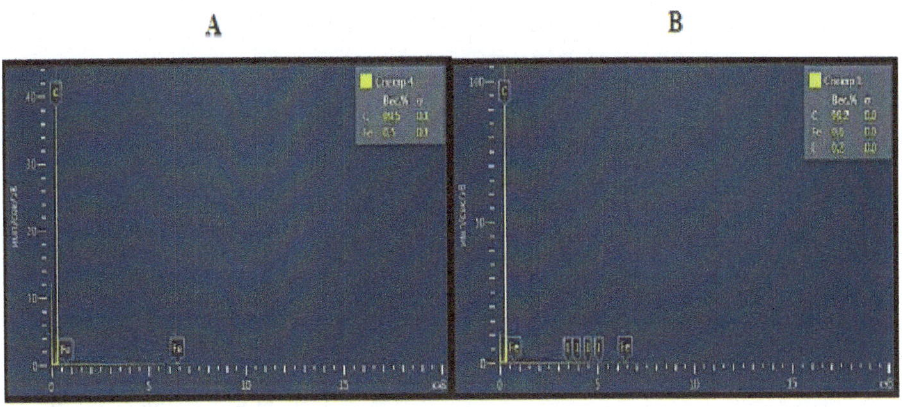

Fig. 2. EDX spectra of the MWCNTs: **A**- purified nanotubes, **B** – iodinated nanotubes

4 Kinetic Measurements

For the kinetic study the following practical facilities and compounds were used:

1. Hydrocarbon - cumene (Alfa Aesar), purity of the reagent was 99%

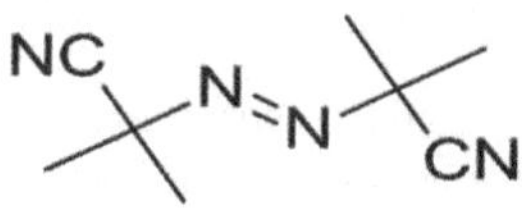

2. Radical initiator - azobisisobutyronitrile (AIBN)
3. Multi-walled carbon nanotubes (MWCNTs) samples [24]:

   a) thoroughly purified, with content of iron is 0.5–0.6 wt.%, diameter of the tubes within 35–40 nm
   b) iodinated pattern, content of chemically bonded iodine is 0.2 wt.% ($1.6 \bullet 10^{-2}$ mol/kg), iron is 0.5–0.6 wt.%, diameter of the tubes within 35–40 nm.

5 Oxygen Adsorption Technique (Cumene)

The reaction of cumene aerobic oxidation was carried out in a reaction vessel with a volume of 30–40 ml. The vessel had the form of a closed cylinder with an exit through a thin section into the gasometric installation system, as shown in Fig. 3. The reaction rate was measured by the amount of absorbed oxygen in the air.

Procedure and details of the oxygen uptake measurement are described in Ref. [23–25].

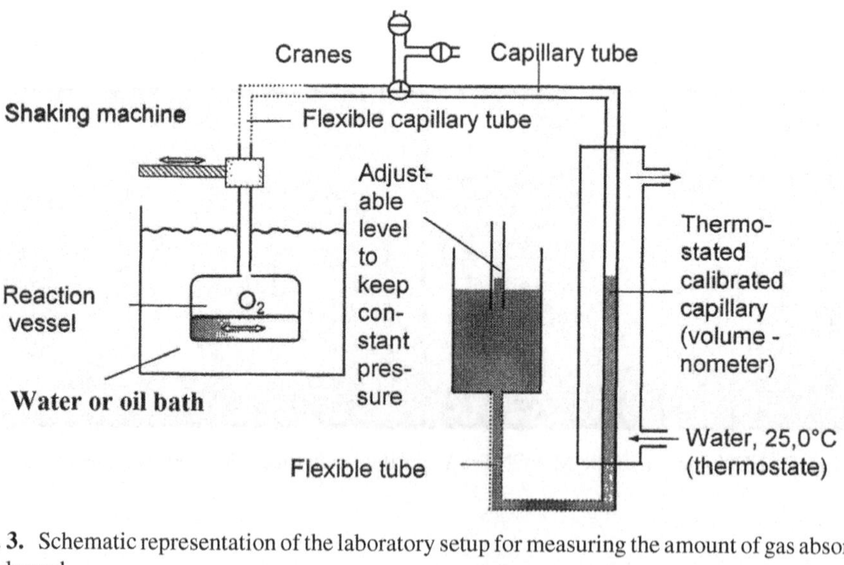

Fig. 3. Schematic representation of the laboratory setup for measuring the amount of gas absorbed or released.

6 Results and Discussions

Figure 4 represents results of kinetic measurements of oxidative activity of the additives.

The kinetic outcomes show that oxidation rate of the cumene initiated oxidation conducted without additives is equal 102 mm^3/min or 6.9 • 10^{-6} molO_2/l • s i.e. it is in accordance with theoretically calculated oxidation rate for assigned conditions of the reaction [23–25]. Additions of the purified MWCNTs reduce oxidation rate to the magnitude 79 mm^3/min (5.4/ • 10^{-6} molO_2/l • s), the iodinated sample - to 46 mm^3/min (3.1 • 10^{-6} molO_2/l • s).

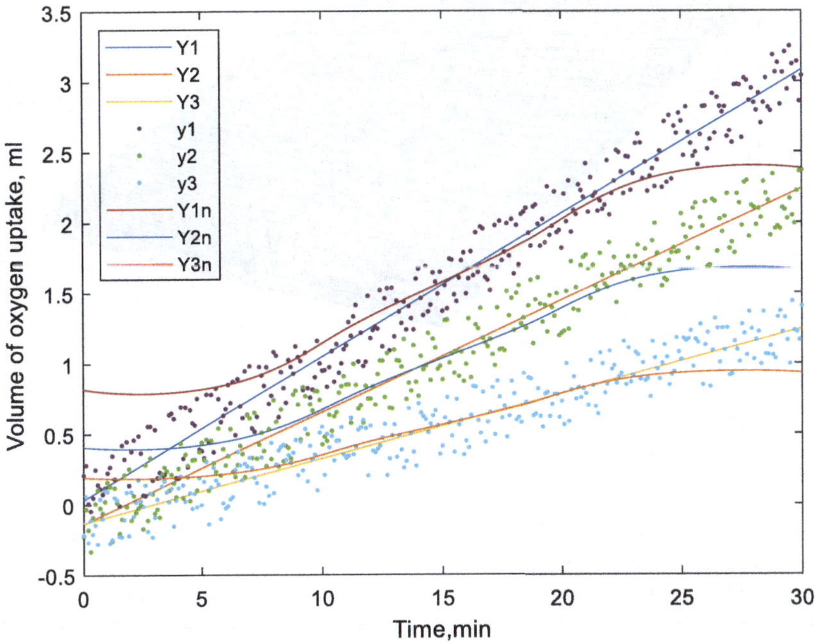

Fig. 4. Oxygen absorption kinetics in the initiated oxidation of cumene in the absence of (y1) and the presence of as synthesized-purified (y2) and iodinated (y3) MWCNTs. Temperature 60 °C, oxygen pressure 20 kPa (air). The initiator is azobisisobutyronitrile (AIBN), the initiation rate $W_i = 6.8 \bullet 10^{-8}$ Ms^{-1}. Nanotube concentrations: - 0; 1 g/l (MWCNTs); 1 g/l (I-MWCNTs). Were $Y1 = 0.1016x + 0.049$, $Y2 = 0.0791x - 0.1329$, $Y3 = 0.0458x - 0.129$ are regression model outputs, and $Y1n$, $Y2n$, $Y3n$ are ANFIS outputs respectively.

Due to complexity of the considered relationship, we intent to construct a model in the form of fuzzy If-Then rules. The rules are derived by means of Fuzzy C-means method-based clustering of the data in Fig. 4. The derived rules are shown below:

1. If (in1 is in1cluster1) then (out1 is out1cluster1)(out2 is out2cluster1)(out3 is out3cluster1) (1)′
2. If (in1 is in1cluster2) then (out1 is out1cluster2)(out2 is out2cluster2)(out3 is out3cluster2) (1)′

3. If (in1 is in1cluster3) then (out1 is out1cluster3)(out2 is out2cluster3)(out3 is out3cluster3) (1)′

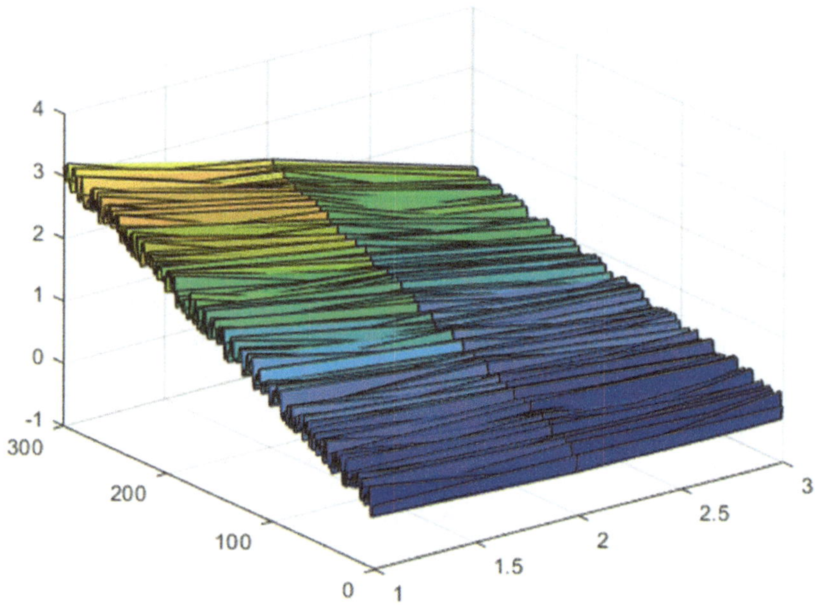

Fig. 5. A graph of relationship between X and Y_1, Y_2.

Constructed by using the If-Then rules, the surface of relationship between input variable X and output variables Y_1, Y_2 is given in Fig. 5.

Thus, the foregoing presumption is confirmed and consequently one may assert that even the tiny concentrations of the grafted I-groups (in our case it is $1.6 \bullet 10^{-5}$M) are able to retard the oxidation.

Moreover, it should be noted that the kinetic curves of inhibited oxidation have the same profile that implies identity of mechanism of inhibition. In both cases a small induction period is observed that can be ascribed to scavenging peroxy radicals by core of the MWCNTs.

7 Conclusions

1. The remains of transitions metallic catalyst which are usually present in channels of as-CVD-prepared MWCNTs samples are a sturdy barrier to uncover the inherent anti-oxidative activity of the nanotubes. The problem can be solved by stepwise processing of the nanotubes by reagents which can be rapidly react with carbides and metallic residues [15]. By such a manner an initial metallic inclusions content can be maximally decreased.

2. The iodinated samples of MWCNTs have been obtained and characterized by SEM Fourier transform–IR and EDX-ray spectroscopy
3. The heightened ant-oxidative capacity of I-MWCNTs has been concisely and convincingly shown by the kinetic experiments. The effect was affected with both the intrinsic electron affinity of the MW nanotubes configuration and grafted iodine groups.
4. The iodinated MWCNTs "work" even in negligible concentrations that therefore may be considered as an effective constituent part of a stabilizer's package for polymers. At that, the objects are also able to bear functions of compatibilizers and fillers. Anyway, it gives substantial impetus and basis for the further elaborations.

References

1. Zeynalov E.B.: Antioxidants for polymeric materials, p. 342. Elm, Baku (2009). (in Azerbaijani)
2. Neiman, M.B.: Aging and Stabilization of Polymers, p. 364. Springer Science & Business Media (2012)
3. Valgimigli, L., Baschieri, A., Amorati, R.: Antioxidant activity of nanomaterials. J. Mater. Chem. B **6**(14), 2036–2051 (2018)
4. Guadagno, L., Naddeo, C., Raimondo, M., Gorrasi, G., Vittoria, V.: Effect of carbon nanotubes on the photo-oxidative durability of syndiotactic polypropylene. Polym. Degrad. Stab. **95**(9), 1614–1626 (2010)
5. Martínez-Morlanes, M.J., Castell, P., Alonso, P.J., Martinez, M.T., Puértolas, J.A.: Multiwalled carbon nanotubes acting as free radical scavengers in gamma-irradiated ultrahigh molecular weight polyethylene composites. Carbon **50**(7), 2442–2452 (2012)
6. Shi, X., Jiang, B., Wang, J., Yang, Y.: Influence of wall number and surface functionalization of carbon nanotubes on their antioxidant behaviour in high density polyethylene. Carbon **50**(3), 1005–1013 (2012)
7. Sreekanth, P.R., Kumar, N.N., Kanagaraj, S.: Improving post irradiation stability of high density polyethylene by multi walled carbon nanotubes. Compos. Sci. Technol. **72**(3), 390–396 (2012)
8. Watts, P.C.P., Fearon, P.K., Hsu, W.K., Billingham, N.C., Kroto, H.W., Walton, D.R.M.: Carbon nanotubes as polymer antioxidants. J. Mater. Chem. **13**(3), 491–495 (2003)
9. Zeynalov, E.B., Friedrich, J.F.: Anti-radical activity of fullerenes and carbon nanotubes in reactions of radical polymerization and polymer thermal/thermo-oxidative degradation: a review. Mater. Test. **49**(5), 265–270 (2007)
10. Zeynalov, E.B., Friedrich, J.F.: Antioxidative activity of carbon nanotube and nanofiber. Open Mater. Sci. J. **2**, 28–34 (2008)
11. Zeynalov, E.B., Wagner, M., Friedrich, J., Magerramova, M.Y., Salmanova, N.I.: Antioxidant traits of some carbon moieties integrated in polymer materials (A review). Chem. Chem. Technol. **10**(4s), 581–586 (2016)
12. Zeynalov, E., Nagiyev, T., Friedrich, J., Magerramova, M.: Carbonaceous nanostructures in hydrocarbons and polymeric aerobic oxidation mediums. In: Grumezescu, A.M. (ed.) Fullerenes, Graphenes and Nanotubes: A Pharmaceutical Approach. Edited. Elsevier – William Andrew Publishing House, 631–681 (2018). chapter 16
13. Zeynalov, E.B., Friedrich, J.F.: Antioxidant properties of multiwall carbon nanotubes: first measurements using a model oxidative reaction. Polym. Polym. Compos. **14**(8), 779–785 (2006)

14. Zeynalov, E.B., Aliyeva, A.Z., Friedrich, J.F.: Factors affecting the intrinsic anti-oxidative activity of carbon nanofibers: metallic impurities materials testing (materials and components, technology and application) section nanotechnology and polymer materials **51**(4), 210–214 (2009)
15. Zeynalov, E.B., Allen, N.S., Salmanova, N.I., Vishnyakov, V.M.: Carbon nanotubes catalysis in liquid-phase aerobic oxidation of hydrocarbons: Influence of nanotube impurities. J. Phys. Chem. Solids **127**(4), 245–251 (2019)
16. Zeynalov, E.B., Magerramova, M.Y., Salmanova, N.I.: Radical scavenging properties of piperidine derivatives of fullerene C_{60}/C_{70} and multi-walled carbon nanotubes. Mol. Cryst. Liq. Cryst. **640**(1), 152–157 (2016)
17. Shieh, Y.T., Wang, W.W.: Radical scavenging efficiencies of modified and microwave-treated multiwalled carbon nanotubes. Carbon **79**, 354–362 (2014)
18. Zeynalov, E.B., Wagner, M., Friedrich, J., Magerramova, M.Y., Salmanova, N.I., Gundula Hidde, G., Meyer-Plath, A.: The peculiar behavior of functionalized carbon nanotubes in hydrocarbons and polymeric oxidation environments. J. Adhes. Sci. Technol. **31**(9), 988–1006 (2017)
19. Smirnova, A.I., Osovskaya, I.I.: Functional materials in plastics production, p. 31. St. Petersburg State Technological University of Plant Polymers, St. Petersburg (2015). (in Russian)
20. Fomin, V.M.: Radical-chain oxidation of organic compounds. In: Electronic Textbook. Nizhny Novgorod State University, Nizhny Novgorod, p. 12 (2010). (in Russian)
21. Abdullayeva, S.H., Musayeva, N.N., Frigeri, C., Huseynov, A.B., Jabbarov, R.B., Abdullayev, R.B., Sultanov, C.A., Hasanov, R.F.: Characterization of high quality carbon nanotubes synthesized via Aerosol-CVD. J. Adv. Phys. **11**(4), 3229–3237 (2015)
22. Abdullayeva, S.H., Huseynov, A.B., Nazarov, I.G., Israfilov, A.O., Madatly, N.M.: Fabrication and investigation of conductive polymeric nanocarbon composites based on iodine-functionalized multi-walled carbon nanotubes. Azerb. J. Phys. **22**(3), 53–61 (2017)
23. Emanuel, N.M., Gladyshev, G.P., Tsepalov, V.F., Piotrovskiy, K.B.: Testing of chemical compounds as stabilizers for polymer materials. Chernogolovka, Institute of Chemical Physics of the USSR Academy of Sciences, p. 52 (1973, preprint). (in Russ.)
24. Zeynalov, E.B., Vasnetsova, O.A.: Kinetic screening of inhibitors of radical reactions, p. 228. Elm, Baku (1993)
25. Zeynalov, E.B., Allen, N.S.: Simultaneous determination of the content and activity of sterically hindered phenolic and amine stabilizers by means of an oxidative model reaction. Polym. Degrad. Stab. **85**(2), 847–853 (2004)

Algorithm for the Synthesis of a Self-tuning Neural Network Control System for Multicomponent Dynamic Objects

Siddikov Isamiddin Hakimovah(✉) and Iskandarov Zohid Ergashbaevich

Tashkent State Technical University, Tashkent, Republic of Uzbekistan
isamiddin54@gmail.com, zohid_1981@mail.ru

Abstract. The issue of creating a self-adjusting control system for a multi-connected dynamic object based on neural network technology is considered. The presence of mutual influence of control loops significantly complicates the problem of choosing the structure and setting up the parameters of regulators, andт certain difficulties arise in the technical implementation of compensating devices designed to ensure the autonomy of transmission channels. To solve these problems, we propose the use of a neural network controller that can adjust its parameters depending on changes in the properties of the object itself, as well as external disturbing influences. The structure of the proposed neural network is represented in the form of a multilayer perceptron. For training and correction of weight coefficients, a method of interactive adaptation is proposed, the essence of which is to train the neural network in an implicit way. Self-tuning of parameters of the neuro-controller is carried out by taking into account the speed of change of any parameter of the control object. It presents descriptions of the dynamic system and the main calculation relations. A simulation experiment was performed to test the operability of self-adjusting control systems of a multi-linked dynamic object, which showed the effectiveness of the proposed approach, due to the short time of self-tuning of the system. A self-tuning control system can be implemented on any modern programmable logic controller.

Keywords: Neural network · Adaptation · Neural network regulation · Neuro-controller · Frechet derivative · Weight coefficient

1 Introduction

Many real multicomponent control objects are characterized by significant non-linearity and stochastic behavior, a wide range of parameter changes in the conditions of its functioning. To achieve high quality management indicators for such objects, the management system must have adaptation elements, which will allow the system to adjust parameters when changing the object's properties.

One of the possible solutions for building a control system is to use the theory of neural networks.

© The Author(s), under exclusive license to Springer Nature Switzerland AG 2021
R. A. Aliev et al. (Eds.): WCIS 2020, AISC 1323, pp. 435–441, 2021.
https://doi.org/10.1007/978-3-030-68004-6_57

This article discusses the problem of constructing a self-tuning neural network control system for multicomponent dynamic objects. In the work, the object of regulation is a multicomponent statically controlled series of objects, for which the number of regulated parameters is equal to the number of regulatory actions.

The proposed self-tuning neural network control system is constructed in such a way that the error vector after the comparison elements enters the input of the neural network. Then, depending on the error signal, at each discrete time t, the weighting coefficients are corrected. The vector of control signals from the output of the neural network is fed to the input of the control object.

2 Method of Solution

A neural network for managing multicomponent objects is a multi-layer neural network with a single intermediate layer, which contains N_0 neurons in the input layer and N_2 neurons in the output layer, with $N_2 = N_0 = n$, where n is the numbers of controlled variables. The network is characterized by the number of neurons N_1 in the inner layer. The input layer consists of nodes-receivers of the error signal, and the output layer consists of neurons-sources of the control signal.

At the first stage, the superposition of input signals of the neuron is calculated:

$$z_i^l = \sum_{j=1}^{N_{l-1}} w_{ij}^l o_j^{l-1} \tag{1}$$

where w_{ij}^l is the weight coefficient, which is a configurable parameter and characterizes the connection of the j-th neuron $(l-1)$ of the layer with the i-th neuron of the l–th layer; o_j^{l-1} - the output scalar.

Then the$z$ value is converted to the output value of the neuron

$$o_i^l = f(z_i^l). \tag{2}$$

The nonlinear transformation (2) is given by the activation function, which is often defined by the sigmoid function, i.e.

$$f(z) = \frac{1}{1 - \exp(-z)}. \tag{3}$$

An important property of this activation function is its ease of definition the derivative of this function

$$f'(z_i^l) = f(z)(1 - f(z)). \tag{4}$$

With the accepted notation the mathematical description of the neural network network is written in the form of a system of equations

$$\left.\begin{array}{c} u_j(t) = o_j^2,\ j = \overline{1,n}; \\ o_i^l = f(z_i^l),\ i = \overline{1,N_l},\ l = \overline{1,2}; \\ z_i^l = \sum\limits_{j=0}^{N_{l-1}} w_{ij}^l o_j^{l-1},\ i = \overline{1,N_l},\ l = \overline{1,2}; \\ o_j^0 = e_j(t),\ j = \overline{1,n}; \\ o_j^0 = o_0^1 = 1 \end{array}\right\} \tag{5}$$

When a self-adjusting neural network control system is operating, the system settings – the neural network weights-are changed in such a way that the value $E = \|e\| \to 0$ of the control system changes. The value *of* E is determined by the formula:

$$E = \frac{1}{2}\sum_{i=1}^{n} w_{ij}e_i^2,\ j = \overline{1,n} \tag{6}$$

where α_i are the coefficients that determine the weight of each control channel in the total error E.

Correction w_{ij}^l of neural network weights (neural network training) is performed using the method of interactive adaptation [2]. The essence of which is that the error that is required for training is calculated implicitly.

When using the method of interactive adaptation, the system is divided into N-subsystems, each of which has an integrated output signal y_n and an integrated input signal x_n, the relationship between them is represented as a functional relationship

$$F_n : X_n \to Y_n,\ n = \overline{1,n} \tag{7}$$

The relation of the i-th element of the system has the form:

$$y_i(t) = F_i[x_n(t)],\ i = \overline{1,n} \tag{8}$$

Let the interaction between the elements and the external signal $u_i(t)$ be linear and described by the equation:

$$x_i(t) = u_i(t) + \sum_{j\in J_i} w_{ij}y_i(t),\ i \in n, \tag{9}$$

$J_i = \{K : y_i = i\}$ - sets of connected inputs of the i-th element; w_{ij} - weights of connections.

In this case, the ratio of input and output of the i-th element is described by the following equation:

$$y_i(t) = F_i\left[u_i(t) + \sum_{j\in J_i} w_{ij}y(t)\right],\ i \in n \tag{10}$$

The purpose of training is to adjust the weights of links w_{ij} in such a way as to minimize the loss function $E(y_1, \ldots, y_n, u_1, \ldots, u_n)$, which is the error function of the system.

Training of neural networks consists in minimizing the error of the control system. This is done by adjusting the weights w_{ij} of neural network connections.

In this case, the link weights w_{ij} are configured using the following rule:

$$\dot{w}_{ij} = F_{ij}'^{\text{вх}}[x_{ij}^{\text{вх}}] \cdot \left(\frac{y_{ij}^{\text{вых}}}{y_{ij}^{\text{вх}}}\right) \sum w_{ij} u_{ij} - \gamma \cdot F_{ij}'^{\text{вх}}[x_{ij}^{\text{вх}}] \cdot y_{ij}^{\text{вых}} \cdot \frac{\partial E}{\partial y_{ij}^{\text{вх}}}, \tag{11}$$

where $\gamma > 0$ is the coefficient that determines the learning speed; $F_{ij}'^{\text{вх}}[x_{ij}^{\text{вх}}]$ is the Frechet derivative [3]; E is the loss function (error).

Provided that Eq. (11) has a unique solution for w_{ij}, where the loss function $E(y_1, \ldots, y_n, u_1, \ldots, u_n)$ decreases monotonically over time and the following equation is satisfied:

$$\dot{w}_{ij} = -\gamma \cdot \frac{\partial E}{\partial y_{ij}^{\text{вх}}}, \tag{12}$$

With this approach, the neural network can be decomposed into composite elements, represented as an elementary neural network (Fig. 1).

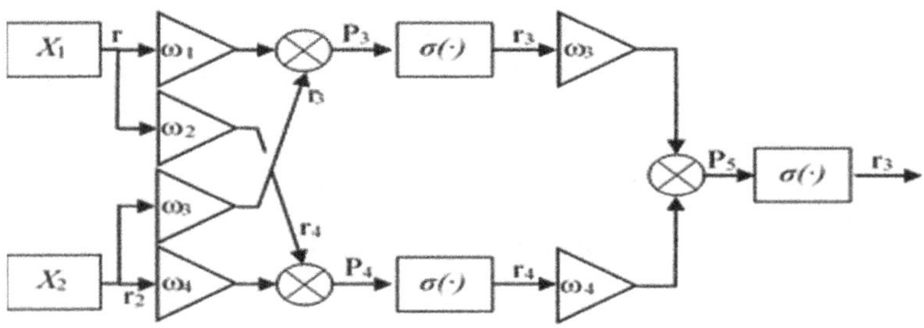

Fig. 1. Block diagram of an elementary neural network

Mathematically, the neural network learning algorithm can be represented as:

$$Pn = \sum_{s \in Dn} \omega_s \cdot r_{pres} \tag{13}$$

$$r_n = \sigma(pn),$$

where: n – the index of the neuron; Dn – the set of input synapses of neuron, n; *pres* and *post* – presynaptic and postsynaptic neuron corresponding to synapse, s; ω_s is the weight of synapse, s; pn is the membrane potential of neuron, n; r_n is the frequency of excitation of the neuron n; σ is the activation function of sigmoid type, which is a:

$$\sigma(x) = \frac{1}{1 + e^{-x}} \tag{14}$$

In this case, the weight of synapses is determined by the formula:

$$\dot{\omega}_s = r_{pres}(\phi_{posts}\sigma(-P_{posts}) + \gamma \cdot f_{posts}), \tag{15}$$

$$\phi_n = \sum_{s \in A_n} \omega_s \cdot \dot{\omega}_s.$$

where: γ – the direct feedback coefficient for all neurons, ϕ_n- the signal of the direct feedback error.

It should be noted that this training method is equivalent to a reverse propagation error algorithm, but it does not require the use of a reverse propagation neural network to transmit an error from the network's output to its input.

3 Sample Solution

Let's consider the possibilities of this algorithm for the design of an adaptive neurocontroller (Fig. 2). Let the neurocontroller have two inputs e_1-an error signal, e_2-a delayed signal e_1. Maintaining the signal e_2 consists in the fact that the neurocontroller is a device without memory. The output signal of the control object depends on both the current control signal and previous control signals.

Let the controller's output signal be generated in two ways:

1. On the way out – tangent sigmoidal function;
2. At the output of the amplifier link.

With this representation of the neural network, the mathematical relations of the controller have the following form:

$$\begin{aligned} & r_1 = x_1, \quad r_2 = x_2; \\ & P_3 = \omega_1 \cdot r_1 + \omega_2 \cdot r_2; \quad P_4 = \omega_3 \cdot r_1 + \omega_4 \cdot r_2 \\ & r_3 = \delta(P_3), \quad r_4 = \delta(P_4) \\ & P_5 = \omega_5 \cdot r_3 + \omega_6 \cdot r_4 \end{aligned} \tag{16}$$

Then the loss function (error signal) is written as follows:

$$E = x_1^2 = (r - y)^2 \tag{17}$$

then $\frac{dE}{dy} = -2r + 2y = -2(r - y) = -2x_1$.
In this case, the neural network weights are set up as follows:

$$\begin{aligned} & \dot{\omega}_1 = x_1 \cdot \phi_3 \delta(-p_1) \\ & \dot{\omega}_2 = x_2 \cdot \phi_3 \delta(-p_1) \\ & \dot{\omega}_3 = x_1 \cdot \phi_4 \delta(-p_2) \\ & \dot{\omega}_4 = x_2 \cdot \phi_4 \delta(-p_2) \end{aligned} \tag{18}$$

where $\phi_3 = \omega_5 \dot{\omega}_5, \ \phi_4 = \omega_6 \dot{\omega}_6$.

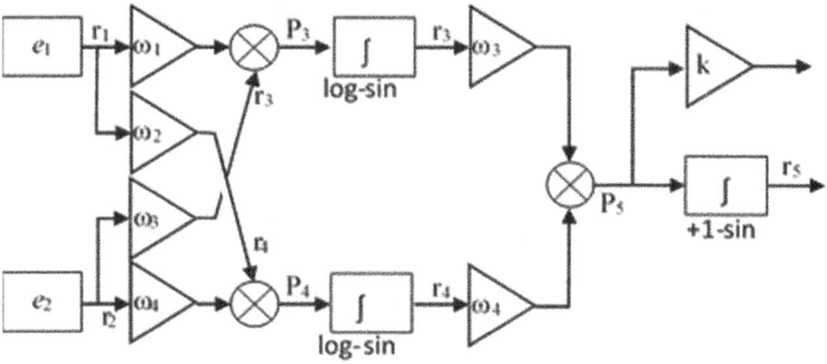

Fig. 2. The structure of the adaptive neurocontroller.

The function ϕ_n is defined from the output synapses of neuron n as follows:

$$\phi_n = \frac{1}{2}\frac{d}{dt} \cdot \sum_{s \in An} \omega_s^2 = \sum_{s \in An} \omega_s \cdot \dot{\omega}_s, \tag{19}$$

where An is the set of output signals of the neuron n. To calculate $\dot{\omega}_5$ and $\dot{\omega}_6$ use the Frechet derivative, we obtain:

$$\dot{\omega}_5 = -\gamma \cdot F'_{posts}[u] \cdot r_3 \cdot (-2x_1), \tag{20}$$

where γ – the learning speed; $F'_{posts}[u]$- the Frechet derivative; u – the external signal.

Assuming that if the Frechet derivative is approximated by a constant, then including it in the learning rate γ we obtain:

$$\begin{aligned} \dot{\omega}_5 &= \gamma \cdot r_3 \cdot x_1 \\ \dot{\omega}_6 &= \gamma \cdot r_4 \cdot x_2 \end{aligned} \tag{21}$$

To reduce the time of adjustment and re-adjustment of the system, it is necessary to change the initial weights of the system, taking their values equal to the steady ones.

The presence of a priori information about the control object, its dynamic and static characteristics will significantly reduce the deviations of controlled values in the process of self-tuning at the technological control object. Related to selfой-tuning of the system based on an approximate model of the object.

For simplicity and clarity of demonstrating the consistency of the selected control system, the object control channels are described by an aperiodic link of the first order with a delay, while the model parameters are defined approximately. The error in the parameter estimation of $\pm 100\%$ has little effect on the final result of self-tuning, but the highest sensitivity is observed for the gain factor.

Using the proposed approach, a series of simulation experiments were conducted, which showed that the system under consideration adapts well to changes in the time and lag constants – changing them even by two or three times does not significantly affect the change in the quality of regulation in the system.

It should also be noted, with adaptation of a neural system to change object properties control more important, not the amount of change of any parameter of the control object, and the rate of change of this parameter.

4 Conclusion

A neural network control system can be successfully applied to manage both stationary and non-stationary multicomponent objects. The use of this management system contributes to:

- yreducing system self-tuning time;
- вability тои адаптивноadapt подстройки to changes in the dynamic properties of the control object;
- the relative ease of using the algorithm and narrowing a number of tuning parameters.

References

1. Siddikov, I.X., Iskandarov, Z.E.: Synthesis of neuro fuzzy PID-regulator for controlling the pressure pipelines. Chem. Technol. Control Manage. **3–4**, 127–130 (2015)
2. Aliev, R.A., Aliev, R.R.: The Theory of Intelligent Systems. Chashyoglu, Baku (2001)
3. Yusupbekov, N.R., Aliev, R.A., Aliev, R.R., Yusupbekov, A.N.: Intelligent Control Systems and Decision Making. Uzbekiston Milliy Encyclopedia, Tashkent (2014)
4. Rubanov, V.G., Titov, V.S., Bobyr, M.V.: Adaptive Systems for Making Fuzzy-Logical Decisions. Monograph, BSTU, Belgrade (2015)
5. Frolov, S.V., Elizarov, I.A., Soludanov, M.N.: Self-adjusting neural network control system. Bull. TSTU **4**, 848–854 (2005)
6. Suhail, M.: Using an adaptive neuro-fuzzy inferense system (AnFis) algorithm for automatic diagnosis of skin cancer. J. Commun. Comput. **8**, 751–755 (2011)

Modeling the Complex Formation of Rhenium with Organic Nitrogen-Containing Ligands by Using Fuzzy Inference

F. D. Gudratova and M. M. Agaguseynova(✉)

Azerbaijan State Oil and Industry University, Azadlıq Avenue 34, Baku Az1010, Azerbaijan
qudretova85@mail.ru, minira_baku@yahoo.com

Abstract. The paper presents the results of research on the development of effective methods for the synthesis of complex rhenium compounds with a number of organic nitrogen-containing ligands. The composition, structure, and physical and chemical properties of new complexes of this transition-valence metal are studied and patterns of changes in physical and chemical properties depending on the nature of the ligand are revealed. The properties of the synthesized compounds are established on the basis of a set of data from elemental analysis, IR spectroscopy, ESP and thermogravimetry. Based on the obtained results type of coordination of ligands: bidentate coordination of o-phenanthroline and monodentate benzimidazole coordination through the pyridine nitrogen atom have been established. Simulate the process of formation of new rhenium complex compounds show the dependents of the rate of complex formation and the influence of the concentration of HCl and the number of coordinated ligand molecules on the thermodynamic characteristics of the process.

Keywords: Complexes of rhenium · Nitrogen-containing organic ligands · O-phenanthrolin · Benzimidazol

1 Introduction

Currently, a huge amount of experimental and theoretical material has been accumulated in the field of coordination compound chemistry, which allows us to predict the activity of complexes in relation to living organisms, polymer-composite materials, and other objects [1–8]. Today it is possible to perform complex syntheses of complexes with previously defined properties. Moreover, these complexes may have different structures and levels of complexity. It is well established that a number of complex compounds of rhenium exhibit antitumor properties. The coordination chemistry of rhenium has developed significantly in recent years, driven largely by interest in the development of radiopharmaceuticals. It was found that rhenium complex compounds can be successfully used for the diagnosis and treatment of cancers of various human organs. In this case, the radiotherapy drug can be delivered directly to the patient's diseased organ using specific molecular carriers, i.e. in the form of coordination compounds with a certain type of ligands and a certain structure [9–12].

© The Author(s), under exclusive license to Springer Nature Switzerland AG 2021
R. A. Aliev et al. (Eds.): WCIS 2020, AISC 1323, pp. 442–451, 2021.
https://doi.org/10.1007/978-3-030-68004-6_58

However, much remains unexplored in the chemistry of these metals, and no regularities have yet been established in changing the physical and chemical properties of their compounds depending on the nature of the ligands.

Against the background of research in the field of complex compounds of rhenium and ruthenium with S- and P-containing ligands, there is a certain gap in the study of compounds with N-containing ligands [11]. This fact played a significant role in the formulation of this study.

All this makes it necessary to find new, convenient and purposeful ways to synthesize rhenium (V) complexes with nitrogen-containing ligands and to study the resulting compounds in more depth.

The aim of this work is to develop effective methods and optimal conditions for synthesis, to study the composition, structure and physical and chemical properties of new rhenium complexes with a number of organic ligands containing N-donor atoms. Determining the features of their structure, finding out the regularities of changes in physical and chemical properties depending on the nature of the ligands, modeling of complex formation and study dependents of some factors on thermodynamic characteristics.

2 Experimental Part

The choice of ligands is not accidental, but is justified on the one hand by their wide use in medical chemistry, on the other hand by the fact that they contain donor nitrogen atoms in their composition.

It is known that coordination compounds of transition metals usually enhance the physiological activity of their constituent organic molecules. However, some metal ions have toxic effects and other harmful side effects on the body.

In this respect, the selected metal is an exception, since rhenium is non-toxic element. According to the literature data, some complex compounds of these enzymes show a protective effect in ulcerative diseases, have anti-carcinogenic activity, and are growth-regulating drugs of a number of agricultural crops. In this regard, it is of great interest to obtain and study Re complexes with the above ligands.

$H_2[ReOCl_5]$, where (x = Cl, Br), was used as the starting substances, which we obtained using the methods given in. Rhenium was determined as perrenate of Nitron, carbon, hydrogen and nitrogen-on an automatic analyzer of the company “Karlo Erba”. The IR spectra of the obtained complexes were taken using the Specord M-80 and Specord 75-IR spectrometers, and the electronic absorption spectra were taken using the Specord UV VIS device. Thermal analysis was performed on the q-1500 d derivatograph of the “Paulik-Paulik-Erdei” system. Molar electrical conductivity of solutions of complexes was determined by the conductometric method. EPR spectra were taken using Radiopan SE/x2542 radio spectrometers. The mass spectra were obtained on the “Kratos” MS 890 mass spectrometer.

Synthesis of [PhenH$_2$][ReOCl$_5$] (I). A Solution of 1.10- phenanthroline in hydrochloric acid was added to a solution of $H_2[ReOCl_5]$ in HCl. The reaction takes place at room temperature and stirring for 1.0 h. the fine-crystalline precipitate of light brown color

Was filtered out after a day, treated with a weak HCl solution, washed with ether and dried at 105 °C to constant mass.

Found, %: Re-25.01; C-9.10; H-1.72; N-4.03

For $[PhenH_2][ReOCl_5]$; Calculated,%: Re-24.44; C-18.90; H-1.58; N-3.68

Synthesis of $[PhenH_2][ReOBr_5]$ (II). An equivalent amount of 1.10-phenanthroline solution in HBr was added to the oxobromorenic acid solution in concentrated HBr. When stirring for 1.5 h, the precipitation is completed in the form of small reddish crystals. The precipitate was filtered out a day later, washed with 5 mol/l HBr, benzene, ether, and dried in a thermostat at 105 °C to constant mass.

Found, %: Re-23.68; C-17.64; H-1.81; N-3.22

For $[PhenH_2][ReOBr_5]$; Calculated,%: Re-23.22; C-17.96; H-1.50; N-3.49

Synthesis of $[ReOCl_3Phen]$ (III). 0.5 g of $PhenH_2[ReOCl_5]$ was placed in a test tube with two taps. For 1.5 h, dry argon was passed through the test tube and heated in the argon current at a temperature of 195 °C. Then the test tube was cooled, without stopping the argon supply, to room temperature. After thermal decomposition (I), the color of the substance changes to yellow-green.

Found, %: Re-38.24; Cl-21.40; N-5.74

For $[ReOCl_3Phen]$; Calculated,%: Re-38.10; C-21.76; N-5.73

Synthesis of [ReOBr3Phen] (IV). In the argon current, the $PhenH_2[ReOBr_5]$ suspension (0.5 g) in a test tube with two tubes was maintained at 190°C for 3 h, then cooled in an inert atmosphere. After dehydrohalogenation and rearrangement (II), substance IV has a light green color.

Found, %: Re-29.36; C-22.85; H-1.07; N-5.10

For $[ReOBr_3Phen]$; Calculated,%: Re-29.94; C-23.15; H-1.29; N-4.50

Synthesis of $[HL]_2[ReOCl_5]$ (V). 0.46 mmol of benzimidazole was added to 40 ml of a solution of 0.15 mol $H_2[ReOCl_5]$ in HCl (8 mol/l) with strong stirring. The mixing was continued for 24 h, the yellow precipitate was filtered out, successively washed with HCl (30 ml) and hexane (150 ml), and dried to a constant mass over KOH in a vacuum desiccator.

Found, %: Re-29.97; Cl-28.31; N-9.24; C-27.35; H-2.38

For $[HL]_2[ReOCl_5]$; Calculated,%: Re-30.14; Cl-28.73; N-9.07; C-27.20; H-2.27

Synthesis of $[HL]_2[ReOBr_5]$ (VI). 0.91 mmol of benzimidazole was added to 40 ml of a solution of 0.29 mol $H_2[ReOBr_5]$ in HBr (7 mol/l) with intensive stirring. After 16 h of mixing, a reddish-yellow precipitate fell out. The precipitate was filtered out and washed with HBr (60 ml) hexane (120 ml). Dried over dry AIR in a vacuum desiccator to a constant mass.

Found, %: Re-22.02; Br-46.89; N-6.56; C-20.13; H-1.47

For $[HL]_2[ReOBr_5]$; Calculated,%: Re-22.16; Br-47.61; N-6.67; C-19.99; H-1.67

Synthesis of $[ReOL_2Cl_3]\cdot 2H_2O$ (VII). 0.45 mmol of benzimidazole was added to 40 ml of a solution of 0.15 mol $H_2[ReOCl_5]$ in HCl (6 mol/l) while stirring. After 24 h of mixing, the light marsh-colored precipitate was separated by filtration. After cleaning, 40 ml of HCl and 30 ml of hexane were dried over solid KOH in a vacuum desiccator to a constant mass.

Found, %: Re-31.80; Cl-18.24; N-9.47; C-29.09; H-2.81

For $[ReOL_2Cl_3]\cdot 2H_2O$; Calculated,%: Re-32.06; Cl-18.34; N-9.64; C-28.93; H-2.75

Synthesis of $[ReOL_2Br_3]\cdot 2H_2O$ (VIII). 0.80 mmol of benzimidazole was added to 40 ml of the mixture of 0.30 mmol $H_2[ReOBr_5]$ in HBr (5 mol/l) with strong stirring. After 24 h of mixing, the green precipitate was separated from the reaction mixture by filtration. Washed with 5 mol/l HBr (70 ml) and hexane (120 ml), dried over dry KOH in a vacuum desiccator to a constant mass.

Found, %: Re-25.89; Br-33.65; N-7.76; C-23.60; H-2.20

For $[ReOL_2Br_3]\cdot 2H_2O$; Calculated,%: Re-26.07; Br-33.60; N-7.84; C-23.52; H-2.24

3 Results and Discussion

The interaction of oxochlororhenic acid solution in hydrochloric acid with an equivalent amount of hydrochloric acid solution of o-phenanthroline isolated the compound $[PhenH_2][ReOCl_5]$ - (I). The complex $[PhenH_2][ReOBr_5]$ - (II) was obtained by adding a solution of o-phenanthroline in HBr to a solution of oxobromorhenic acid in hydrobromic acid. Oxochloro - and oxobromorenates of o-phenanthroline were deposited as small crystals with light brown (I) and reddish (II) color:

$$\text{FoPhen} + H_2[ReOX_5] \rightarrow [PhenH_2][ReOX_5]$$

These formulas were written on the basis of data from elemental analysis and IR spectroscopy. Thus, the IR spectra of compounds I and II have wide absorption bands in the region of 3300–3340 cm^{-1}, which are typical for protonated nitrogen atoms in a heterocycle and confirm the formation of an outer-sphere cation $[PhenH_2]^{2+}$.

On the heating curves of compounds I and II, decomposition is accompanied by two endothermic effects at 84 and 186 °C for $[PhenH_2][ReOCl_5]$ and 87, 188 °C for

$[PhenH_2][ReOBr_5]$. The first endothermic effect corresponds to the removal of crystallization water, the second is associated with dehydrohalogenation. In order to obtain substances formed after removal of HX (X = Cl, Br), compounds I and II were subjected to thermal action in an inert atmosphere. By heating at slightly higher temperatures (190–195 °C) the second endothermic effects were obtained by compounds of the composition $[ReOCl_3Phen]$ - (III) and $[ReOBr_3Phen]$ - (IV). When heated, dehydrohalogenation occurs and the ligand passes from the outer sphere to the inner sphere of the complex, and the color of the substances changes to yellow-green (III) and light green (IV). The reaction proceeds according to the scheme:

$$[PhenH_2][ReOX_5] \rightarrow [ReOX_3Phen] + 2HX\ (X = Cl,\ Br).$$

In support of this scheme of thermal decomposition with the release of two hydrogen halide molecules and the entry of the o-phenanthroline molecule into the inner sphere of the complexing atom, data from elemental analysis and a comparative study of the IR spectra of the initial and obtained compounds speak. In the IR spectra of compounds III and IV, absorption bands in the region of 3300–3340 cm^{-1} disappear due to the presence of protonated nitrogen atoms of the heterocycle in the outer-sphere cation $[PhenH_2]^{2+}$ of compounds I and II. The intense band in the region of 848 cm^{-1} caused by fluctuations of free o-phenanthroline in the IR spectra of compounds III and IV splits into two bands: less intense at 848 cm^{-1} and low-intensity at 843 cm^{-1} (III) and 840 cm^{-1} (IV), which confirms the entry of the phenanthroline molecule into the internal coordination sphere of the rhenium atom (V). Changes in intensities and shifts in absorption bands on the electronic spectra of compounds III-IV relative to I-II in the visible and near UV regions that characterize the phenanthroline chromophore system indicate the transition of the phenanthroline molecule from the outer sphere (I-II) to the inner sphere of the complexing atom (III and IV). The band at 19050 cm^{-1} splits into two bands with a significant hyperchromic shift at 23500 and 22500 cm^{-1}, and for the long-wave absorption band in the region of 12020 cm^{-1}, a hyperchromic shift is observed (Δ550 cm^{-1}). Based on the above and taking into account the fact that the complex anions $[MeOX_5]^{2-}$ (x = Cl, Br) have the configuration of a flattened octahedron, the structure of compounds (III and IV) is established.

Interaction of initial rhenium compounds with benzimidazole in the corresponding solutions of HX (x = Cl, Br), the type of complexes obtained depends on the concentration of HX. Thus, when adding benzimidazole to solutions of $H_2[ReOX_5]$ in 8 mol/l HCl and 7 mol/l HBr, respectively, compounds of the composition $[HL]_2[ReOCl_5]$ - (V) and $[HL]_2[ReOBr_5]$ - (VI) (L-benzimidazole) were obtained. These are fine-crystalline substances of yellow (V) and reddish-yellow (VI) colors, slightly soluble in polar and insoluble in non-polar organic solvents.

Using the rhenium compounds trans $[ReOCl_3(Ph_3P_2)]$ and $H_2[ReOBr_5]$, benzimidazole complexes ($C_7H_6N_2L$) were synthesized. The structure and composition of the obtained coordination compounds were established on the basis of IR and UV spectroscopic data and the results of elemental analysis. In the synthesized complexes, on the basis of a set of physicochemical studies, the monodentate coordination of benzimidazole was determined.

The IR spectrum of complex (V) contains characteristic absorption bands of individual functional groups and bonds in the regions indicated below.

IR spectrum (ν, $cм^{-1}$) 3412, 3104, 2924, 1624, 1500
1480, 1424, 1248, 930, 748, 620, 420, 320, 306.

The localization of the places of coordination in the complex was determined based on the results of the analysis of IR spectroscopy data.

The absence of the absorption bands of stretching vibrations, characteristic of the conjugated C = C and C = N bonds in the region of 1700–1800 $cм^{-1}$, indicates the delocalization of the electron density inside the imidazole ring as a result of the coordination of the imidazole ligand and the formation of complex V. The shift to the high-frequency region (3000–4400 $cм^{-1}$) and an increase in the intensity of the absorption bands corresponding to the stretching vibrations of the imidazole ring (1480–1580 $cм^{-1}$) indicate the coordination of rhenium with benzimidazole due to the nitrogen atom of pyridine benzimidazole. This fact is confirmed by the appearance in the spectrum of the complex compound of intense absorption bands in the regions of 748, 620, 420 $cм^{-1}$. These lines can be attributed to vibrations of the Re – N bond. The absorption bands of stretching vibrations of the NH group (3400–3100 $cм^{-1}$) remain unchanged. The absorption bands located in the range 320–306 $cм^{-1}$ are related to the vibrations of the Re – Cl bond. The absorption bands at 930 $cм^{-1}$ indicate the retention of the Re = O bond. The absence of νP–C-aril absorption bands (510–530 $cм^{-1}$) in the complex compound (V) indicates that all triphenylphosphine ligands are replaced by benzimidazole.

Based on the data of infrared spectroscopy, it can be argued that the coordination of the rhenium atom occurs with the help of the nitrogen atom of pyridinebenzimidazole, and the ligand is monodentate.

Analysis of the electronic absorption spectra of solutions of benzimidazole and complex (V) in water and methanol showed that the absorption bands in the regions 35760, 36600, 37100, 38700 (α-band) and 39800, 40700 $cм^{-1}$ (p-band) change intensity, and are also characterized by a hypsochromic shift (1000–1500 $cм^{-1}$).

The shift of the π-band, in comparison with the-band, is much larger. This is the result of the coordination of benzimidazole at the expense of the tertiary nitrogen atom.

Based on the data of IR spectra, electronic absorption spectra and elemental analysis, the structural structure of the$[ReOL_3Cl_2]Cl$ complex was established.

Another complex VI of rhenium with benzimidazole was obtained using rhenium oxohalide $H_2[ReOBr_5]$ as a starting reagent. As a result, an orange-colored complex (VI) $[LH]_2[ReOBr_5]$ was obtained, here LH^+ is a benzimidazolium ion. The IR spectrum of the obtained compound (VI) contains the following absorption bands, ν, $cм^{-1}$: 3380, 1573, 1240, 1113, 980, 953, 920, 880, 726, 713, 580.

The presence of a broad absorption band in the region of 3380 $cм^{-1}$ in the IR spectrum of complex II indicates protonization of nitrogen atoms of the heterocycle.

A clear peak at 953 $cм^{-1}$ indicates retention of the Re = O bond.

The totality of the results obtained makes it possible to establish the structure and composition of the synthesized complexes of rhenium with benzimidazole and determine the monodentant coordination of the latter. Data from elemental analysis and measurements of molar electrical conductivity of compounds V and VI indicate the formation of onium complexes. Along with the characteristic bands of benzimidazole, bands appear in the IR spectra of these compounds at 1000–1010 cm^{-1} due to valence fluctuations of the onium complex V (Re = O). The bands characteristic of protonated benzimidazole δ(NH+) appear in the region of 1600–1610 cm^{-1}. The expanded singlet in the PMR spectra of compounds V and VI at 7.2 m. d. corresponds to two NH groups.

A more detailed study of the complex formation process at various temperatures and concentrations of HCl was studied using the example of phenanthroline complexes of rhenium with the aim of modeling this process.

The dependence of the values of the thermodynamic functions of the process of complexation of rhenium(V) with phenanthroline on the concentration of HC1 are shown in Figs. 1, 2 and 3. From the data in Fig. 1 it can be seen that the value of ΔG for the monosubstituted complex increases with increasing concentration of HC1. However, this increase on going from 5 mol/l to 6 mol/l HC1 becomes more noticeable. The dependences $\Delta G_i = f(C_{HCl})$ for two, three, four, and five-substituted complex forms are extreme. However, to the least extent, this character is manifested for the disubstituted complex.

When the dependence $\Delta G_i = f(C_{HCl})$ is expressed through a fuzzy inference system, the knowledge base can be defined as follows:

1. If (C-HCl is mf1-low) then (dG is mf1-low) (1)
2. If (C-HCl is mf2-medium) then (dG is mf2-medium) (1)
3. If (C-HCl is mf3-high) then (dG is mf1-low) (1)

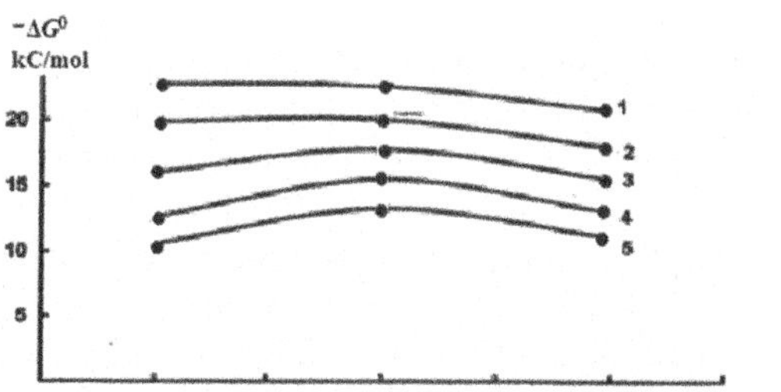

Fig. 1. Dependence of ΔG^0 on the HCl concentration for oxochloro-phenanthroline complexes of rhenium (V) 1-pK_1 2-pK_2; 3-pK_3; 4-pK_4;5-pK_5

For the dependences $\Delta H_i = f(C_{HCl})$ (Fig. 2) for the first complex forms formed in the solution, an increase in the value of ΔH is observed with an increase in the concentration of HC1. This experimental fact can presumably be associated with the fact that with an increase in the HC1 concentration, the competition between chloride ions and phenanthroline molecules for a place in the internal coordination sphere increases, as a result of which the heat generation process in the system intensifies. For three, four and five-substituted phenanthroline complexes of rhenium (V), the dependence $\Delta H_i = f(C_{HCl})$ has an extreme character and the maxima on these curves appear at an HCl concentration of 5 mol/l. This means that at this concentration the reaction proceeds with the greatest heat release.

When the dependences $\Delta H_i = f(C_{HCl})$ is expressed through a fuzzy inference system, the knowledge base can be defined as follows:

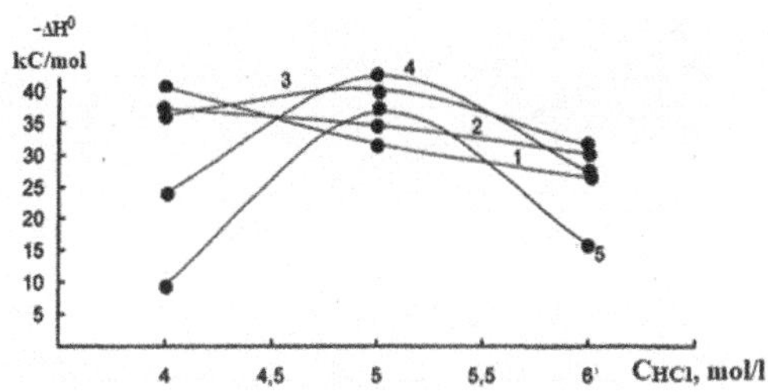

Fig. 2. Dependence of ΔH^0 on HCI concentration for oxochloro-phenanthroline complexes of rhenium (V) 1-pK_1 2-pK_2; 3-pK_3; 4-pK_4;5-pK_5

1. If (C-HCl is mf1-low) then (dH is mf3-high) (1)
2. If (C-HCl is mf2-medium) then (dH is mf2-medium) (1)
3. If (C-HCl is mf3-high) then (dH is mf3-high) (1)

Similar dependences are observed for the value of ΔS_i (Fig. 3).

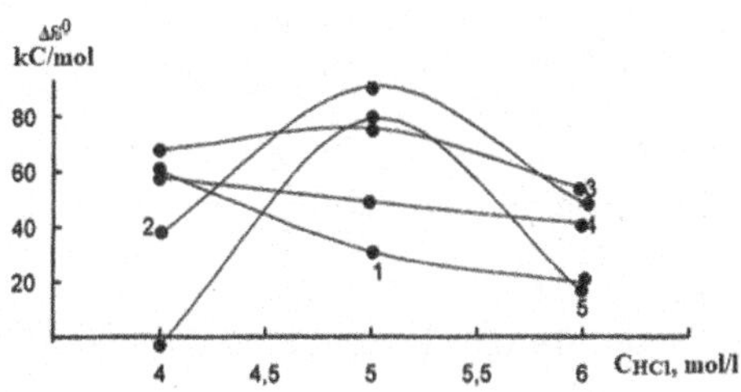

Fig. 3. Dependence of ΔS^0 on the HCI concentration for oxochloro-phenanthroline complexes of rhenium (V) 1-pK_1 2-pK_2; 3-pK_3; 4-pK_4;5-pK_5

When the dependences $\Delta S_i = f(C_{HCl})$ is expressed through a fuzzy inference system, the knowledge base can be defined as follows:

1. If (C-HCl is mf1-low) then (dS is mf1-low) (1)
2. If (C-HCl is mf2-medium) then (dS is mf3-high) (1)
3. If (C-HCl is mf3-high) then (dS is mf1-low) (1)

Figure 4–5 shows the dependence of the thermodynamic functions on the number of attached phenanthroline molecules. Figure 4 shows that the dependence ΔGi = f (nEtthio) in media with an HCl concentration of 6 mol/l and 5 mol/l has a straight-line character, that is, with an increase in the number of coordinated phenanthroline molecules, the value of ΔG increases. In a medium of 4 mol/l HCl, the value of ΔG increases with increasing n, but not according to a linear law, which is associated with the influence of hydrolytic processes. The fact that the value of ΔG increases with increasing n is associated with a decrease in the probability of the entry of subsequent phenanthroline molecules into the inner sphere of the complexes.

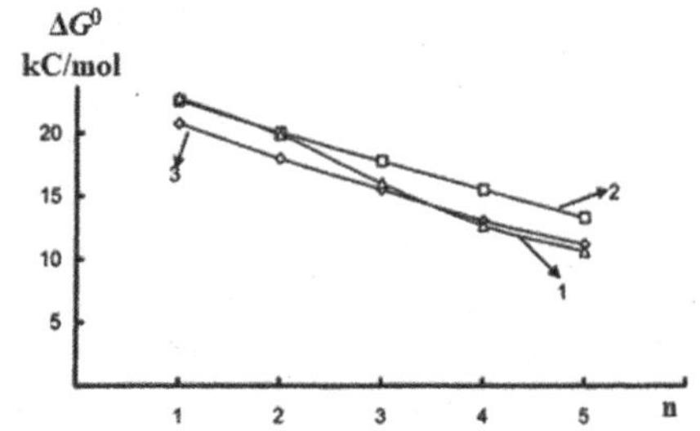

Fig. 4. Dependence of ΔG^0 on the number of attached ligand molecules in media of 4 mol/l (1), 5 mol/l (2), 6 mol/l (3) HCl for phenanthroline complexes of rhenium

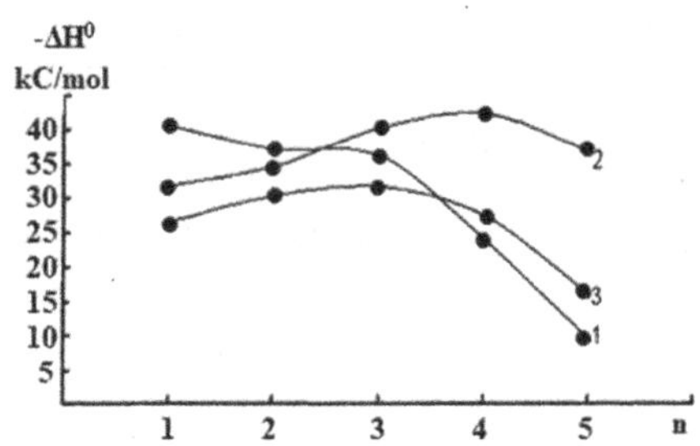

Fig. 5. Dependence of ΔH^0 on the number of attached ligand molecules in media of 4 mol/l (1), 5 mol/l (2), 6 mol/l (3) HCl for phenanthroline rhenium complexes

The dependence $\Delta Gi = f$ (nEtthio) is expressed through a fuzzy inference system, the knowledge base can be defined as follows:

Rules of inference:

1. If (n is mf1-low) then (dG is mf2-medium) (1)
2. If (n is mf2-medium) then (dG is mf3-high) (1)
3. If (n is mf3-high) then (dG is mf1-low) (1)

In a medium of 4 mol/l HCl, the value of ΔHi increases with increasing n (Fig. 5), and in a medium of 5 mol/l HCl, it decreases upon going from monosubstituted to four-substituted, and then increases again on going from four to five-substituted. In a medium of 6 mol/l HCl, the value of ΔH decreases when going from mono- to two- and three-substituted complexes, then increases again when going to four and five-substituted complexes.

The dependence $\Delta Hi = f$ (nEtthio) is expressed through a fuzzy inference system, the knowledge base can be defined as follows:

Rules:

1. If (n is mf1-low) then (dH is mf2-medium) (1)
2. If (n is mf2-medium) then (dH is mf2-medium) (1)
3. If (n is mf3-high) then (dH is mf1-low) (1)

4 Conclusions

1. Methods of Synthesis have developed and Re(V) complex compounds with three types of N-donor containing ligands (phenanthroline, benzimidazole) have been obtained.
2. Based on the cumulative results of the conducted studies it was established: -during the synthesis of phenanthroline complexes, a rearrangement occurs due to dehydrohalogenation and the transition of the outer-sphere ligand into the inner sphere of the complex.
3. The composition of benzimidazole complexes depends on the concentration of HX (x = Cl, Br) in the medium. At high concentrations, onium - type complexes are formed; at low concentrations, complexes with monodentate-coordinated ligands are formed through an atom of unsubstituted nitrogen of the imidazole ring
4. In order to simulate the process of formation of new rhenium complex compounds the dependence of the rate of complexformation and the influence of the concentration of HCl and the number of coordinated ligand molecules on the thermodynamic characteristics of the process was situated, which made it possible to calculate the coefficients of the equations of linear functions, enthalpy, entropy and Gibbs energy of this process.

References

1. Espenson, I.H.: Inorg. Chem. **40**, 67/7 (2001)
2. Henderson, N., Mc.Coffrey, L.I., Nicholson, B.K.: I. Chem. Soc. Dalton Trole, **2753** (2000)
3. Dublin, M., Villamil, SH.F., Stopani, AO.M.: Agent. Medicina **61**, 345 (2002)
4. Banerjee, S., Bhattacharyya, S., Dirghandi, B.K., Menon, M., Chakiavorty, A.: Inorg. Chem. **39**, 6 (2000)
5. Paula, A., Domingos, A., Gargia, R., Santos, Z.: Inorg. chem. **39**, 5669 (2000). https://doi.org/10.1021/ic000481n
6. Alessia, E., Zangrando, E., Iengo, E., Macchi, M., Marzi, P.A., Marzilli, L.G.: Inorg. Chem. **39**, 294 (2000)
7. Agaguseynova, M.M., Jabbarova, N.E.: Coordination Compounds of Transition Metals in Catalysis, p. 244. Elm Publishing House, Baku (2006)
8. Aminjanov, A.A., Verbıtskaya, N.A., Kadlimova, O.A., Slivinskiy, V.I.: Reqularities of structure and material properties of Re (V) complex compounds. Compos. Mater. **5**(1), 57–61 (2011)
9. Yoshimira, T., Sno, C., Tsuge, K., İshiroka, S., Nozaki, K., Kitamera, N., Shinohara, A.: Excited-state properties of octahedral hexarhelium complexes with redox-active heteroaromatic ligands. J. Inorg. Chem. **49**, 531–540 (2010)
10. Baker, E.B., Haase, A.A., Reich, R.M., Crans, D.C., Kühn, F.E.: Orgametallic and coordinatium rhenium compaunds and their polential is cancer therapy. Coord. Chem. Rev. **393**, 79–11 (2019)
11. Adams, R.D., Captain, B., Smith, M.D., Beddie, C., Hall, M.B.: Unssaturated Platinum-Rhenium cluster complexes. https://doi.org/10.1021/ja070773o
12. Wilson, J.J., Lippard, S.J.: Synthetic methods for the preparation of platinum anticancer complexes. Chem. Rev. **114**(8), 4470–4495 (2014). https://doi.org/10.1021/cr4004314

Traffic Sign Recognition Using Histogram of Oriented Gradients and Convolutional Neural Networks

Sanan Abizada(✉)

Department of Electrical and Electronic Engineering, Near East University, Northern Cyprus North Nicosia, Turkey
abizadasanan@gmail.com

Abstract. This paper presents the design of the hybrid system that uses the HOG (Histogram of oriented gradients) with convolutional neural networks (CNN) for traffic sign recognition. The system implements two basic operations- feature extraction and classification. Using HOG and convolutional layers the features of images are extracted. The extracted images are used for classification. This is implemented using fully-connected neural networks. GTSRB (German Traffic Sign Recognition) is used to design the recognition system. During designing stochastic gradient descent algorithm is applied for the learning of CNN parameters. The accuracy of the CNN based traffic recognition system was obtained 99.87% for training data, 99.74% for validation data and 98.2% for test data. The obtained results demonstrate the applicability of the designed system in traffic sign recognition.

Keywords: Traffic sign recognition · Deep learning · CNN

1 Introduction

Traffic signs are vitally important for drivers which provide both reliable and safety guiding information for them in motorway traffics and urban environments. Sometimes, oscitation, lack of attention of drivers may cause serious traffic accidents. Therefore, different driver-assistant systems are designed for assisting drivers in their long way journey. These systems are used for informing traffic signs, warning the unusual situations in the driver cabin. These systems allow to inform, warn and guide drivers all days and night in traffic. Traffic sign recognition acquires importance not only in driver assistance and safety, and also in traffic surveillance, autonomous driving of cars. Nowadays online, fast and accurate recognition of traffic signs is an significant problem in traffic.

Several studies have been developed for recognition of traffic signs. Traffic signs are very essential assets in the safety of driving. Drivers need to observe the traffics' signs precisely regardless of the time and conditions. Deficiency in the observations can cause safety faults. Distraction in the view of the driver, change in illumination, weather conditions, sometimes risky point of view can cause drivers to observe the signs wrong. Therefore, traffic sign recognition systems are made to assist the drivers for safety and

© The Author(s), under exclusive license to Springer Nature Switzerland AG 2021
R. A. Aliev et al. (Eds.): WCIS 2020, AISC 1323, pp. 452–459, 2021.
https://doi.org/10.1007/978-3-030-68004-6_59

reducing the risks. They have a supportive role not only for drivers also for pedestrians and residents. This system guides the drivers in traffic for their safety.

Traffic signs have colour attributes (white, blue and red) which can be easily extracted from the background. Many benchmarks have promoted the traffic-sign recognition and detection task [1–6]. [4] uses HOG features for identification of traffic signs and test their model on the GTSRB. [7] used HOG features, Gabour features combined local binary patterns and ELM classification for traffic sign recognition. In the design of the traffic sign detection systems mostly hand-crafted features have been applied. The HOG [8, 9], scale-invariant feature transform (SIFT) [10] are used for extraction of features. Many machine learning methods have been promoted using support vector machine (SVM) [9, 7], random forests [11], artificial neural networks as an extreme learning machine (ELM) [6, 7].

Recently a number of studies have been done using deep learning for recognition of traffic signs [1, 2, 6, 12–14]. Some research papers used CNN and other feature extraction techniques. The paper [1] uses CNN and other traditional techniques, such as SVM, HOG that extracts features of traffic signs from hand-crafted images for traffic sign identification. The paper [2] used simplified Gabor feature maps to get image features and convolution neural networks to classify the traffic signs. [12] used classifier based on CNN in order to classify traffic signs and published the preformances of the state-of-the-art methods. The paper [14] used combination of colour segmentation procedure, CNN and HOG (Histogram of oriented gradients) to recognize and classify traffic signs. The paper achieved good classification accuracy. [15] used CNN for identification of German and Belgium traffic sign datasets using different learning algorithms. [16] used Mask R-CNN for recognition of street traffic signs. The paper [17] firstly used HOG for feature extraction of pedestrians. For better recognition and classification of traffic sign by using deep learning methods, to satisfy the demands of real-time application, we need further research and study the deep learning structure for improvement the performance of classification system.

The existing studies in traffic sign recognition have shown that still there are problems related to colour distortions, change of viewpoints or bad viewpoints of signs, blurring, degradation of contrast, occlusion of traffic signs that complicate the accurate recognition of signs. Also sometimes the presence of objects of the same colour around traffic signs affects the accuracy of detection. In this paper, deep learning based on CNN is promoted for traffic sign identification. The structure and design algorithms of the constructed recognition system are presented.

The paper is shown as follows. Section 2 presents the architecture of CNN used for traffic sign recognition. Section 3 describes the modeling of the traffic sign recognition system. Section 4 presents the conclusions of the paper.

2 CNN Based Vehicle Recognition System

In this study deep learning using CNN is applied identification of traffic signs. The designed system is implemented for detection and classification of traffic signs. CNN contains several layers for the extraction of features and classification purposes. The convolutional, rectified linear units, pooling and flatten layers are used for feature extraction

[18–21]. For classification purpose often fully-connected networks are used. In the paper, we used double convolutional layers for extracting the features [18]. The structure of the CNN used for traffic sign recognition is given in Fig. 1. Convolutional layers using two-dimensional filters perform convolutional on two-dimensional images. The selection number of layers and also the structure of convolutional layers affect the performance of feature extraction. The convolutional layer consists of neurons that are characterized by weight coefficients. The features are extracted during feed forward processing in the convolutional layers. Each filter is convolved with the image map. If the input of CNN is denoted with x matrix, the weights of CNN are denoted as w matrix, and outputs of the convolutional layer with y then output lth layer is determined as

$$y^{l}_{i,j}=\sum_{m}\sum_{n} w^{l}_{m,n} y^{l-1}_{i+m,j+n} + b^{j} \tag{1}$$

The ReLU activation is applied to the output of the convolutional layers, that is y = ReLU(y). output. These signals are input for pooling layer which is determined as

$$u^{l}_{i,j} = pool\left(f\left(\sum_{m}\sum_{n} w^{l}_{m,n} y^{l-1}_{i+m,j+n} + b^{j}\right)\right) \tag{2}$$

In CNN flatten layer is used for arranging of obtained features in the one-dimensional matrix. Its output is determined as $u^{l}_{flatten} = flatten\left(u^{l}_{ij}\right)$. This output vector is input for the fully-connected network where the classification feature map is performed. This will be determined as o $F\left(u^{l}_{flatten}\right)$.

Training of CNN is performed using network output signals and training image patterns. For training of the network, the loss function is applied.

$$L = \frac{1}{N}\sum_{i=1}^{N} E\left(\theta; o^{(i)}, p^{(i)}\right) \tag{3}$$

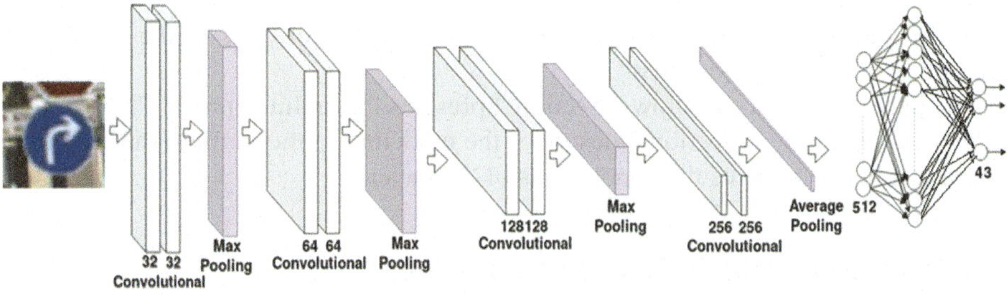

Fig. 1. The architecture of CNN used for traffic sign recognition

Using loss function stochastic gradient algorithm is applied for training network parameters. As a result of training, the parameters of the network is determined.

3 System Modelling

The constructed CNN based traffic recognition system has two main stages which are extraction of feature and classification. In feature extraction stage the basic features of the traffic signs are extracted for recognition purpose. For this purpose Histograms of Oriented Gradient (HOG) technique and CNN are used. The output of the feature extraction stage is sent to the input of the classification block. In the paper, we used a fully connected network for classification purpose.

The input images entering the system are scaled as 48 × 48, then the extraction of the features is performed. HOG is proposed for pedestrian detection [17]. Here the gradient of colour is used for detection of features. Using these gradients of colours different normalized and weighted histograms are determined. These are performed for small overlapping and larger overlapping blocks of multiple pixels. To extract the features of the images three HOG descriptors are applied. Figure 2 demonstrates the image with extracted HOG feature visualization. The HOG features provide a good representation of the traffic sign. These signals are entered to the input of CNN. By applying filters CNN extract deeper features of the traffic sign images. These features after flatten operation are entered fully connected layers for classification purpose.

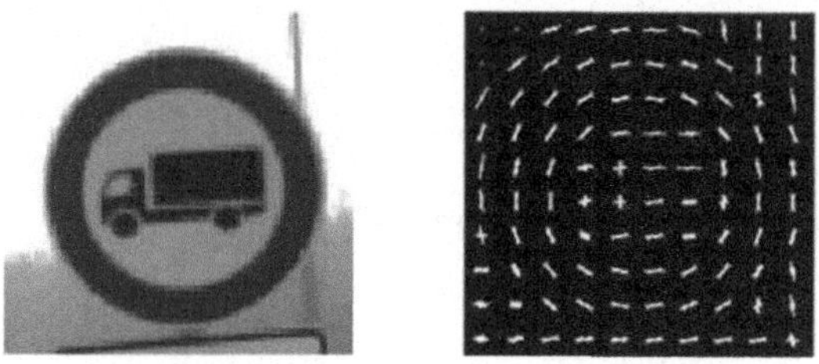

Fig. 2. Image and HOG feature visualization

During modelling GTSRB (German Traffic Sign Recognition Benchmark) dataset is used. The dataset includes 51838 images of the 43 classes. The images used are shown in Fig. 3. The GTSRB data sets are divided into three subsets. The division is implemented randomly. These are training, validation and testing images. Training sets include 39209 images, testing sets 12629 images. Training sets includes the images used for training and images used for validation. Therefore, the training sets are divided into two sets- training and validation. We have used 50% for training images, 25% for validation images and 25% for testing images. In the paper cross-validation approach was employed for implementation of training.

In modelling, the images are rescaled 48 × 48 and entered to the system. Besides HOG features, the convolution neural network (CNN) extracts the features of images for classification. CNN is a multi-layer neural network. It gets the features from the pattern by using combined convolution, pooling, and activation layers. CNN is a technique most widely used in the pattern recognition field. Table 1 shows the CNN network architecture used in this paper. The size of the traffic signs images in GTSRB dataset changes from 15 × 15 to 250 × 250. Therefore, images are uniformly cropped and resized to 48 × 48.

Fig. 3. The 43 traffic signs used in GTSRB [4]

Table 1. CNN architecture and parameters.

Layer (type)	Output shape	Parameters
conv2d_1 (Conv2D)	(None, 32, 48, 48)	896
conv2d_2 (Conv2D)	(None, 32, 46, 46)	9248
max_pooling2d_1 (MaxPooling2)	(None, 32, 23, 23)	0
dropout_1 (Dropout)	(None, 32, 23, 23)	0
conv2d_3 (Conv2D)	(None, 64, 23, 23)	18496
conv2d_4 (Conv2D)	(None, 64, 21, 21)	36928
max_pooling2d_2 (MaxPooling2)	(None, 64, 10, 10)	0
dropout_2 (Dropout)	(None, 64, 10, 10)	0
conv2d_5 (Conv2D)	(None, 128, 10, 10)	73856
conv2d_6 (Conv2D)	(None, 128, 8, 8)	147584
max_pooling2d_3 (MaxPooling2)	(None, 128, 4, 4)	0
dropout_3 (Dropout)	(None, 128, 4, 4)	0
flatten_1 (Flatten)	(None, 2048)	0
dense_1 (Dense)	(None, 512)	1049088
dropout_4 (Dropout)	(None, 512)	0
dense_2 (Dense)	(None, 43)	22059
Total params: 1,358,155		

CNN is trained using stochastic gradient descent algorithm. The proposed CNN architecture use combination of several layers such as convolutional, max-pooling layers, spatial transformer, normalization and non-linearity. This architechture extract the feature which is classified by two fully connected layers. Spatial transformer layers performs geometric operations on input images and feature maps. Therefore the system focuses on object and can learn the object by removing background and geometric noises. We optimize the all parameters in all layers by reducing the classification error of the training set. An algorithm called Stochastic gradient descent is applied for learning CNN parameters. The learning has been done using 30 epochs. As a result of the simulation, the training accuracy was obtained as 99.98% and validation accuracy-99.74%. The plot of accuracies values obtained during learning for training data was depicted in Fig. 4(a). Figure 4(b) presented the plot of the values of loss function obtained during learning for training data. Here we plotted the values of accuracies and loss functions obtained at the end of each epoch. As shown from the figures in each epoch the values of accuracies increasing towards 100%, the values of the loss function were decreasing towards 0%. Figure 5(a) depicted the plot of values of accuracies obtained for validation data. Figure 5(b) depicted the plot of values of loss function obtained for validation data.

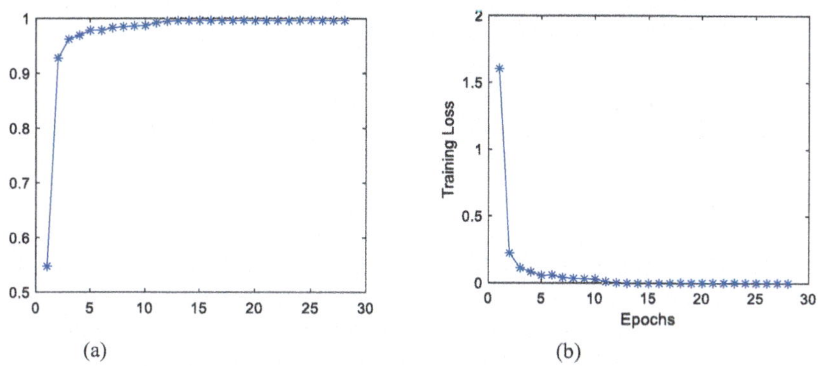

Fig. 4. Plots of accuracy (a) and loss function (b) obtained for training data

The efficiency of the designed model is tested through comparison of other models. The test result of the proposed model for the GTSRB dataset has achieved an accuracy of 98.50%. Comparative results are given in Table 2. The accuracy of reference [15] is 99% used other methods, Spatial Transformer Networks with their CNN networks. Our aim is to improve our network and the methods for accurate identification of traffic signs in future.

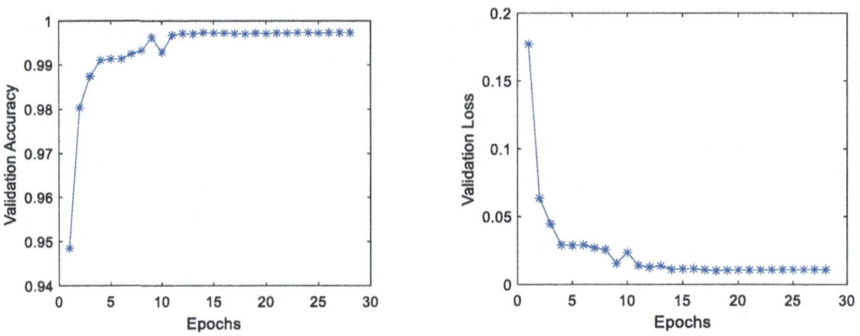

(a) (b)

Fig. 5. Plots of accuracy (a) and loss function (b) obtained for validation data

Table 2. Recognition-rate accuracy of different methods.

Paper	Method	Accuracy (%)
Persson [22]	VGG16	74.50%
Álvaro Arcos-García [15]	Single CNN with 3 STNs	99.71%
Lu [23]	Multi-Scale CNNs	98.31%
Current paper	CNN with HOG	98.50%

4 Conclusion

In this paper, the combination of CNN based deep learning architecture and HOG is proposed for traffic sign recognition. The system implements feature extraction and classification stages. The feature extraction is implemented using HOG methods and convolutional layers of CNN. Obtained features are classified using fully-connected neural networks. The training has been done using stochastic gradient descent algorithm. At the results of training, the values of accuracies for training and validation data were obtained as 99.88% and 99.74% correspondingly. The value of accuracy for test data was 98.5%. The obtained results show that the presented CNN architecture could achieve competitive results with much simpler architecture.

References

1. Yao, C., Wu, F., Chen, H., Hao, X., Shen, Y.: Traffic sign recognition using HOG-SVM and grid search. In: 12th International Conference on Signal Processing, Hangzhou, pp. 962–965 (2014)
2. Shao, F., Wang, X., Meng, F., Rui, T., Wang, D., Tang, J.: Real-time traffic sign detection and recognition method based on simplified Gabor wavelets and CNNs. Sensors **18**(3) (2018). https://doi.org/10.3390/s18103192
3. Stallkamp, J., Schlipsing, M., Salmen, J., Igel, C.: Man vs. computer: benchmarking machine learning algorithms for traffic sign recognition. Neural Netw. **32**, 323–332 (2012)

4. Houben, S., Stallkamp, J., Salmen, J., Schlipsing, M., Igel, C.: Detection of traffic signs in real-world images: the German traffic sign detection benchmark. In: IJCNN, pp. 1–8. IEEE Press (2013)
5. Mogelmose, A., Trivedi, M., Moeslund, T.B.: Vision-based traffic sign detection and analysis for intelligent driver assistance systems: perspectives and survey. Trans. Intell. Transp. Syst. **13**(4), 1484–1497 (2012)
6. Zaklouta, F., Stanciulescu, B.: Real-time traffic sign recognition in three stages. Robot. Auton. Syst. **62**(1), 16–24 (2014)
7. Azizi, S., Mohamed, E.A., Youssef, F.: Traffic sign recognition based on multi-feature fusion and ELM classifier. Procedia Comput. Sci. **127**, 146–153 (2018)
8. Zaklouta, F., Stanciulescu, B.: Real-time traffic-sign recognition using tree classifiers. IEEE Trans. Intell. Transp. Syst. **13**(4), 1507–1514 (2012)
9. Greenhalgh, J., Mirmehdi, M.: Real-time detection and recognition of road traffic signs. Trans. Intell. Transp. Syst. **13**(4), 1498–1506 (2012)
10. Kus, M.C., Gokmen,, M., Etaner-Uyar, S.: Traffic sign recognition using scale invariant feature transform and color classification. In: 23rd International Symposium on Computer and Information Sciences, Istanbul, pp. 1–6 (2008)
11. Ellahyani, A., Aansari, M.E., Jaafari, I.E.: Traffic sign detection and recognition using features combination and random forests. IJACSA **7**(1), 6861–6931 (2016)
12. Qian, R., Yue, Y., Coenen, F., Zhan, B.: Traffic sign recognition with convolutional neural network based on max pooling positions. In: 12th International Conference on Natural Computation; Fuzzy Systems and Knowledge Discovery, Changsha, China, 13–15 August 2016
13. Zhu, Z., Liang, D., Zhang, S., Huang, X., Li, B., Hu, S.: Traffic-sign detection and classification in the wild in CVPR, pp. 2110–2118 (2016)
14. Youssef, A., Albani,, D., Nardi, D., Bloisi, D.D.: Fast traffic sign recognition using color segmentation and deep convolutional networks. In: ACIVS 2016: Advanced Concepts for Intelligent Vision Systems, Lecce, Italy (2016)
15. Arcos-García, Á., Soilán, M., Álvarez-García, J.A., Soria-Morillo, L.M.: Deep neural network for traffic sign recognition systems: an analysis of spatial transformers and stochastic optimisation methods. Neural Netw. **99**, 158–165 (2018)
16. Tabernik, D., Skočaj, D.: Deep learning for large-scale traffic-sign detection and recognition. arXiv:1904.00649v1 (2019)
17. Dalal, N., Triggs, B.: Histograms of oriented gradients for human detection. In: IEEE Comp. Society Conference on Computer Vision and Pattern Recognition, pp. 886–893 (2005)
18. Abiyev, R., Arslan, M., Idoko, J.B., Sekeroglu, B., Ilhan, A.: Identification of epileptic EEG signals using convolutional neural networks. Appl. Sci. **10**(12) (2020). https://doi.org/10.3390/app10124089
19. Abiyev, R.H., Arslan, M.: Head mouse control system for people with disabilities. Expert Syst. **37** (2020). https://doi.org/10.1111/exsy.12398
20. Abiyev, R.H., Idoko, J.B., Arslan, M.: Sign language translation using deep convolutional neural networks. KSII Trans. Internet Inf. Syst. **14**(2) 2020. https://doi.org/10.3837/tiis.2020.02.009
21. Abiyev, R.H., Idoko, J.B., Arslan, M.: Impact of machine learning techniques on hand gesture recognition. J. Intell. Fuzzy Syst. **37**(3), 4241–4252 (2019). https://doi.org/10.3233/JIFS-190353
22. Persson, S.: Application of the German Traffic Sign Recognition Benchmark on the VGG16 network using transfer learning and bottleneck features in Keras (Dissertation) (2018). http://urn.kb.se/resolve?urn=urn:nbn:se:uu:diva-344672
23. Lu, K., Ding, Z., Ge, S.: Sparse-representation-based graph embedding for traffic sign recognition. IEEE Trans. Intell. Transp. Syst. **13**(4), 1515–1524 (2012)

Algorithms for the Synthesis of a Neural Network Regulator for Control of Dynamic Objects

H. Z. Igamberdiev[1(✉)] and T. V. Botirov[2]

[1] Tashkent State Technical University, Tashkent, Uzbekistan
ihz.tstu@gmail.com
[2] Navoi State Mining Institute, Navoi, Uzbekistan
btv1979@mail.ru

Abstract. The article deals with the synthesis of adaptive systems with a reference model based on the method of Lyapunov functions. An algorithm for the synthesis of an asymptotically stable adaptive system with a reference model based on the direct Lyapunov method and recurrent target inequalities is presented. The proposed algorithm allows synthesizing an asymptotically stable self-tuning algorithm and obtaining a sufficient criterion for its stability.

Keywords: Adaptive system · Reference model · Lyapunov method · Asymptotically stable algorithm · Recurrent target inequalities

1 Introduction

Within the framework of the theory of adaptive control, the typical problem is the synthesis of a controller, which should ensure the stabilization of the control object in conditions when some of the parameters of the object are not known.

To solve this problem, combinations of identification and optimization methods are widely used [1–3]. The use of identification methods is complicated if the object is affected by uncontrollable disturbances (noises), especially if the latter do not possess useful statistical properties (stationarity, uncorrelatedness, Gaussianity, etc.).

The use of Lyapunov functions makes it possible to determine the structure of the adaptive controller, and in some cases, the method of adjusting the controller parameters, which ensures the achievement of control objectives for any object from a given class. Such ideas are widely used in search-free self-tuning methods using sensitivity functions, in gradient self-tuning schemes with a reference model and in the synthesis of controls according to the criterion of generalized work in methods based on the hyperstability of control objects [4–11].

In adaptive systems with a reference model, the stability region and the quality of control by tunable parameters largely depend on the type and intensity of control actions [12, 14].

© The Author(s), under exclusive license to Springer Nature Switzerland AG 2021
R. A. Aliev et al. (Eds.): WCIS 2020, AISC 1323, pp. 460–465, 2021.
https://doi.org/10.1007/978-3-030-68004-6_60

A change in the intensity and type of control actions in a fairly wide range leads to a change in the duration and nature of transient processes in the adaptation loop. The indicated drawback limits the field of application of adaptive systems with reference models [15].

2 Problem Definition

This article develops an approach to the synthesis of algorithms for constructing adaptive control systems with a reference model. In adaptive systems, with significant tracking errors, continuous self-tuning is performed, which ensures that the special quadratic Lyapunov function decays in time. The Lyapunov function is constructed in the form of a function that admits an infinitely small upper limit [16–18], so that its design takes into account the dependence on control actions to a lesser extent. With such a special choice of the Lyapunov function, the quality of the parameter setting increases significantly. The duration of transient processes in terms of adjustable parameters is less dependent on the type and intensity of control actions. This combination of these methods allows the use of Lyapunov functions of a rather general form, in contrast to the well-known self-tuning methods, in which the matrix of a quadratic form must depend on the tuned parameters in a very special way. The proposed method for constructing the Lyapunov function allows one to synthesize an asymptotically stable self-tuning algorithm and obtain a sufficient criterion for its stability.

Suppose that the controlled object is described by the equation

$$\frac{dx(t)}{dt} = \left[A_M + \Delta A + \Delta\alpha(t, x, x^*)\right]x(t) + (B_M + B)f(t) + v(t), \tag{1}$$

where the vector $x(t)$ describes the state of the object at the moment of time-control action.

The matrices A_M, B_M of the corresponding dimensions are assumed to be known and define the reference model:

$$dx_M(t)/dt = A_M x_M(t) + B_M u(t). \tag{2}$$

the coordinate x_M of which is the desired state of the OS under the initial conditions

$$x(t_0) = x_0\,,\;\; x_M(t_0) = x_0^*, \tag{3}$$

in the area of $t \geq t_0$, $\|x\| \leq H_1$, $\|\alpha\| \leq H_2$,

$$H_1,\; H_2 = const > 0. \tag{4}$$

$$\|x\|^2 = (x, x),\;\; \|\alpha\|^2 = Sp(\alpha', \alpha). \tag{5}$$

$$\alpha = \Delta A + \Delta\alpha(t, x, x^*), \tag{6}$$

$x,\; x^* \in E_n$, where E_n-n-dimensional phase space; $\alpha \in \Gamma_{n,n} - n^2$-dimensional parameter space with elements of the $\alpha_1', \ldots, \;\alpha_n'$; $\alpha_i' - i$ -th row of the matrix $\alpha(t, x, x^*)$.

In (1) and (2), the following designations are adopted: vectors of phase x, x^* –coordinates of the system and reference model of dimension $[n \times 1]$; $f(t)$- vector-function of control actions of dimension $[m \times 1]$; A_M-constant matrix, whose eigenvalues are different and have negative real parts; ΔA – $[n \times n]$ - matrix of object parameters; $\Delta\alpha(t, x, x^*)$ – $[n \times n]$-matrix of regulatory influences of adaptation loops; B_M - a given constant $[n \times m]$-matrix, is $v(t)$ a vector of disturbing influences.

3 Formulation of the Problem

It is assumed that the elements of the matrix vary within a certain area H_2 of the parameter space $\Gamma_{n,n}$ and can take any, a priori unknown constant values from this region

$$\Delta\alpha_{i,i} \in H_2, \quad \Delta\alpha_{i,j} = const. \tag{7}$$

We assume that the coordinates of the vector of control $f(t)$ actions are such that

$$R \geq x_i(t) \geq r > 0 \,\text{and}\, R \leq x_i(t) \leq -r < 0, \tag{8}$$

where r and R are some given positive constants. The functions $f_i(t)$ are defined in domain (4), where they are such that the conditions for uniqueness of the solution are provided for Eqs. (1), (2). We assume that the system described by Eq. (2) is completely controllable and completely observable [12].

In the terminology of A.M. Lyapunov, the solutions $x(t)$ end $x^*(t)$ of Eqs. (1) and (2) will be called, respectively, perturbed and unperturbed trajectories [17–19].

We denote the vector of deviation from the unperturbed trajectory by

$$\varepsilon = x - x^*. \tag{9}$$

As a control goal, we set the condition

$$\lim_{t\to\infty} \varepsilon(t) = 0, \tag{10}$$

if there is no interference ($v(t) \equiv 0$), and

$$\overline{\lim_{t\to\infty}} \|\varepsilon(t)\| \leq c_\varepsilon, \tag{11}$$

with non-zero interference. Here the constant c_ε determines the admissible value of the mismatch error, denotes $\|\varepsilon\|$ the Euclidean norm of the vector ε.

$$\frac{d\varepsilon}{dt} = A_0\varepsilon + \Delta\alpha(t, x, x^*)x. \tag{12}$$

Let the laws of change of the output coordinates of the self-tuning contours have the form:

$$\frac{d\Delta\alpha(t, x, x^*)}{dt} = -\Psi(t, x, x^*), \tag{13}$$

$$\psi_{i,j}(t,0) = 0 \text{ by } t \geq t_0, \quad (i,j = 1,2\ldots,n), \tag{14}$$

where the matrix Ψ elements $\psi_{i,j}$ are still unknown functions of the vectors x and x^*.

We define a new space as a direct product of spaces E_n and $\Gamma_{n,n}$, which has the following form:

$$E_n \times \Gamma_{n,n} = \left\{\left(\varepsilon; \alpha_1^{'}, \ldots \alpha_n^{'}\right) : \varepsilon \in E_n, \alpha_i^{'} \in \Gamma_{n,n}\right\}(i = 1,2\ldots,n) \tag{15}$$
with the norm $(\|\varepsilon\|, \|\alpha\|)$.

If there is no relation $\alpha x(t) = 0$ that is identical with respect to t on the interval $[t_0, \infty)$, then the system of equations:

$$\begin{aligned} \frac{d\varepsilon}{dt} &= A_0\varepsilon + \Delta\alpha(t,x,x^*)x, \\ \frac{d\alpha}{dt} &= -\Psi(t,x,x^*), \end{aligned} \tag{16}$$

in the area

$$(\|\varepsilon\|, \|\alpha\|) \leq H, t \in [t_0, \infty), \tag{17}$$

space has an equilibrium position at the origin

$$\varepsilon = 0, \quad \alpha_1^{'} = \alpha_2^{'} = \ldots.\alpha_n^{'} = 0. \tag{18}$$

The control task is to choose a method for the formation of control actions $f(t)$ and the law of formation of adjustable parameters, which ensure the control goal (10) or (11). In this case, the value of the parameter T is not assumed to be known, but only the set T of its possible values is known.

It is required to find an algorithm $\Psi(t,x,x^*)$ for an adaptive system for which the zero solution to (18) in domain (17) is asymptotically stable in the sense of A.M. Lyapunov [15, 16].

4 Solution of the Task

Let us consider the problems of adaptive control when adaptation (self-tuning) algorithms can be obtained using a Lyapunov function of a special form.

Suppose the equation has the form

$$\dot{\varepsilon} = A_M\varepsilon + \psi\big[t, \ \hat{\tau}_2(t)\big]\big[\hat{\tau}_2(t) - \tau_2\big], \tag{19}$$

with Hurwitz matrix A_M. Define a positive matrix from the equation

$$HA_M + A_M * H = -G, \tag{20}$$

where is a G-positive definite matrix, and we introduce the Lyapunov function

$$V(t) = \varepsilon^*(t)H\varepsilon(t) + \big[\hat{\tau}_2(t) - \tau_2\big]^*\Gamma^{-1}\big[\hat{\tau}_2(t) - \tau_2\big], \tag{21}$$

where G is a positive matrix.

Calculating the derivative of this function by virtue of Eq. (19), we find

$$\dot{V}(t) = -\varepsilon^*(t)G\varepsilon(t) + 2[\tau_2(t) - \tau_2]^* \Gamma^{-1}[\tau_2(t) - \tau_2]^* \psi\left[t, \hat{\tau}_2(t)H\varepsilon(t) + \Gamma^{-1}\frac{d\tau_2}{dt}\right]. \tag{22}$$

Choosing the parameter $\hat{\tau}_2$ self-tuning equation (adaptation algorithm) in the form

$$\hat{\tau}_2 = -\Gamma\psi * \left[t, \hat{\tau}_2\right] H\varepsilon(t),$$

we obtain $\dot{V}(t) = -\varepsilon(t)G\varepsilon(t)$, that, due to the positiveness of the matrix G, it allows us to establish relation (3).

5 Conclusion

Thus, regardless of the value of the parameter $\tau_2 \in T$ in Eq. (19), adaptation algorithm (23) ensures that the goal of the equation is achieved. The proposed algorithm for constructing the Lyapunov function makes it possible to synthesize an asymptotically stable self-tuning algorithm and obtain a sufficient criterion for its stability.

References

1. Chernorutsky, I.G.: Optimization methods in control theory. St. Petersburg, Peter (2004). ISBN 5-94723-514-5
2. Yusupbekov, N.R., Gulyamov, S.M., Kasimov, S.S., Usmanova, N.B.: Knowledge-based planning for industrial automation systems: the way to support decision making. In: Advances in Intelligent Systems and Computing, vol. 896, pp. 873–879. https://doi.org/10.1007/978-3-030-04164-9_115
3. Yusupbekov, N.R., Igamberdiev, H.Z., Mamirov, U.F.: Algorithms of sustainable estimation of unknown input signals in control systems. J. Autom. Mobile Robot. Intell. Syst. **12**(4), 83–86 (2018). https://doi.org/10.14313/JAMRIS_4-2018/29
4. Yusupov, R., Rozenwasser, E.: Sensitivity of Automatic Control Systems. CRS Press, London (1999)
5. Krasovsky, A.A.: Handbook on The Theory of Automatic Control. Science. Ch. ed. physical-mat. lit., Moscow (1987)
6. Botirov, T.V.: Procedure formalization for constructing an adaptive control system with a reference model based on the Lyapunov function. Int. Sci. Tech. J. Chem. Tech. Control Manage. **3**(93), 45–48(2020). ISSN 1815–4840
7. Trigeassou, J.C., Maamri, N., Sabatier, J., Oustaloup, A.: A Lyapunov approach to the stability of fractional differential equations. Signal Process. **91**(3), 437–445 (2011)
8. Botirov, T.V., Buranov, B.M., Latipov, S.: About one synthesis method for adaptive control systems with reference models. J. Phys: Conf. Ser. **1515**, 022078 (2020). https://doi.org/10.1088/1742-6596/1515/2/022078
9. Proskurnikov, A.V., Yakubovich, V.A.: Synthesis of stabilizing regulators in the tracking problem. Doklady Math. **5**(1), 819–823 (2005)

10. Yusupbekov, N.R., Abdurasulov, F.R., Adilov, F.T., Ivanyan, A.I.: Application of cloud technologies for optimization of complex processes of industrial enterprises. In: Advances in Intelligent Systems and Computing, vol. 896, pp. 852–858. https://doi.org/10.1007/978-3-030-04164-9_112
11. Igamberdiev, H.Z., Boeva, O.H., Sevinov, J.U.: Sustainable algorithms for selecting feedback in dynamic object management systems. J. Adv. Res. Dyn. Control Syst. **12**(7), 2162–2166 (2020)
12. Botirov, T.V., Buranov, B.M, Latipov, Sh.B.: Barakayev methods for synthesizing adaptive control with reference models using adaptive observers. In: IOP Conference Series: Materials Science and Engineering, vol. 862 (2020). https://doi.org/10.1088/1757-899X/862/5/052012
13. Åström, K.J., Wittenmark, B., Adaptive Control. Courier Dover (2008)
14. Fomin, V.N.: Some general principles of constructing adaptive control systems. Soros Educ. J. **12**, 102–108 (1996)
15. Rutkovsky, V., Glumov, V.M., Sukhanov, V.M.: Physically realizable adaptive control algorithm with a reference model. Avtomat. i Telemekh. **8**, 96–108 (2011)
16. Tsykunov, A.M.: Adaptive and Robust Output Control of Dynamic Objects. FIZMATLIT (2009). ISBN 978–5–92–211094–5
17. Krasovskii, N.N., Kotelnikova, A.N.: One problem of stable tracking of motion. Proc. Steklov Inst. Math. **12**, 142–156 (2006)
18. Aksenov, G.S., Fomin, V.N.: Synthesis of adaptive control using a degenerate Lyapunov function. Bull. Leningrad Univ. **19**, 5–12 (1979)
19. Krasovskii, N.N., Kotelnikova, A.N.: On the method of Lyapunov functions for stability problems in systems with after effect. Sci. World J. **3**(37), 73–74 (2006)

A Generic Model-Based on Probabilistic Reasoning for Path Loss of Wireless Indoor Channel

Bülent Bilgehan(✉) and Fahreddin Sadikoglu

Near East University, Nicosia, Cyprus
bulent.bilgehan@neu.edu.edu, fahreddin.sadikoglu@neu.edu.tr

Abstract. The probabilistic and statistical methods help to understand the dependencies of the parameters representing the data. This research aims to encourage engineers of a new universal probability distribution function which in its mathematical form is a powerful instrument for the solution of practical problems. The universal probability distribution function offers a great advantage of representation without prior knowledge of data type. The introduced model is in its most general form to apply the most related engineering applications without the need to know the behavior of the data. The model disregards preliminary work in engineering problems that reduces complexity and execution time. The proposed model uses an exponential based function. The introduced model produces the best representation to randomly varying data such as the wireless indoor channel with many varying parameters. The success of the introduced model is due to the parametric representation of the exponential function. The method applied to wireless indoor channel representation with great success.

Keywords: Wireless channel · Channel model · Channel characteristics · Probability density function · Data representation

1 Introduction

An accurate representation of wireless indoor channel is very significant in the design procedure of the communication systems [1]. The real signal strength at the receiver can only be determined by the accurate channel representation. Some receivers operate based on the threshold signal strength value [2]. Therefore, it is important to determine accurately the channel characteristics in transmissions. The classical models are simple in the calculation but do not represents all the varying parameters accurately [3]. This produces less sensitive channel characteristics for different indoor applications[4]. The classical models are tested in paper [5].

The wireless channel model in [5] produced a more accurate channel representation for the recorded data. However, the introduced model is data-driven and the data include all variations during recording. Therefore, the model has the advantage of a better representation of the channel variations.

© The Author(s), under exclusive license to Springer Nature Switzerland AG 2021
R. A. Aliev et al. (Eds.): WCIS 2020, AISC 1323, pp. 466–471, 2021.
https://doi.org/10.1007/978-3-030-68004-6_61

Several different methods are applied in applications. The paper [6] applied wavelet artificial intelligence for wireless indoor signal detection. The research paper [7] contributed towards channel path loss prediction in an office environment. The path loss prediction becomes more difficult to model at the wireless sensors [8]. The ANN application resulted in an accurate wave propagation model in [9]. The accurate wireless indoor channel model is closely linked with the localization of the transmitting source [10]. The model-driven deep learning method is applied as in [11].

This research paper uses the parametric approach and produces a new mathematical model to represent wireless indoor channel characteristics. The newly introduced parametric model produces a very accurate probability distribution function that represents random variations during the transmission of a signal.

The performance of the newly introduced model is compared with the ideal exponential method. The measure of the performances was based on the p-test and threshold value.

2 Method

This section is divided into two main parts. The first part explains the recorded data and the second section describes the newly introduced universal probability distribution function.

2.1 Signal Measurements

An experimental wireless indoor channel characteristic has been analyzed. The newly.

introduced probabilistic method is applied to the measured data and compared with the ideal (exponential) model. The power delay of the transmitted clusters measured at different points on the same floor. The measurement was performed in an indoor 2nd floor of the Electrical and Electronic Engineering building, Near East University, Cyprus. The construction material of the building is made up of concrete and brick walls. The aim is to apply the newly introduced method to statistical channel characterization of the 1600–2000 MHz bands in an indoor environment. The measurement uses a vector network analyzer (VNA) measurement technique to scan the channel in the frequency domain with the receiving antenna moved to different locations in the horizontal direction to capture the channel characteristics. The time resolution is 2.5 ns and space resolution is 0.75 m. The block diagram representation is given in Fig. 1.

The type and specification of the equipement is shown in Table 1.

Table 1. Types and specifications of equipments

Equipment	Description	Specification
AMP1	Amplifier Miteq	0.1–4 GHz, 30 dB
AMP2, AMP3	Amplifier	17 dB, 1–2 GHz
CAB1, CAB2, CAB3	Cable	RG213/U, 0.5 dB/m
ANT1, ANT2	Omnidirectional Antenna	900–1800 MHz, 3 dB
VNA(HP8720A)	Vector Network Analyzer	130 MHz–20 GHz

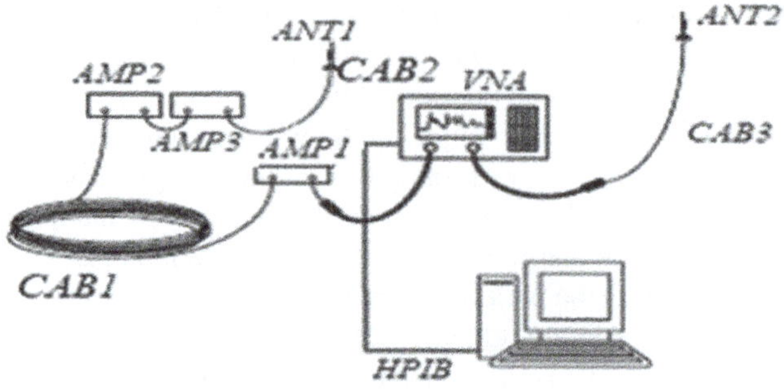

Fig. 1. The measurement system

The transmitter was fixed at a location and the receiver moved to different locations where at least 20 measured data were recorded at each location. The test was repeated at different locations and recorded 1000 measured values. The measured power delay cluster demonstrates a random variation at the receiving point. The different objects in the test room cause reflection and refraction path to the receiver location with random time variation for the clusters at the receiving antenna. The interarrival time of clusters can be characterized by the probability density function. The classical process requires the histogram plot of the data. This identifies the possible distribution to fit. The histogram of the measured random data was plotted in Fig. 2 for statistical evaluation.

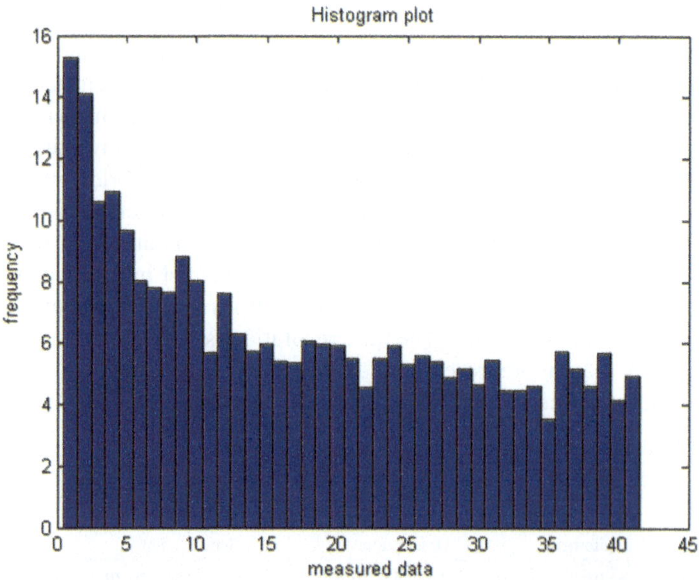

Fig. 2. Histogram plot to visualize the variation of the measured wireless indoor channel data

The histogram in Fig. 2 indicates the exponential type probability distribution function to be a suitable fitting model. Therefore, we will use the exponential type probability distribution function and compare it with the universal probability distribution function using the fitness measures described in the statistics theory.

2.2 Universal Probability Distribution Function

Initially, in paper [12] a general exponential function introduced to represent exponential data variations as:

$$f(\boldsymbol{b}, x) = x^{b_1} e^{b_2 x^2 + b_3 x + b_4} \tag{1}$$

The major advantage of Eq. (1) is that the varying parameters b_1, b_2, b_3, and b_4 allow representing the Rayleigh, Gaussian, and exponential functions. Let us now consider Eq. (1) and define a universal model as:

$$g(x) = (2ax + b)(ax^2 + bx)^q e^{(-(ax^2 + bx))}, a, b \geq 0, x > 0, Re(q) > -1 \tag{2}$$

The exponential probability function in Eq. (2) can now be defined as:

$$f(x; a, b) = \frac{1}{\Gamma(q + 1)} g(x), 0 < x < \infty \tag{3}$$

The probability distribution function can be derived through the expected function by assigning the variables in Eq. (3) to mean and standard deviation to follow:

$$E\left(x^2\right) = \sigma^2 + \mu^2 \tag{4}$$

Where E represents the expected function, σ is the standard deviation, and μ is the mean value of the data. Let us now generate $F_x(x)$ as a cumulative distribution function so that random variable x is based on the model defined in Eq. (1). The expected function by definition follows as:

$$E(x) = \frac{1}{\Gamma(q + 1)} \int_0^{\infty} (2ax + b)^2 (ax^2 + bx)^q e^{-(ax^2 + bx)} dx \tag{5}$$

The cumulative distribution function of a random variable x for the newly introduced universal function using Eq. (5) can be formulated as:

$$F_x = \frac{1}{\Gamma(q + 1)} \int_0^{x} (2ax + b)^2 (ax^2 + bx)^q e^{-(ax^2 + bx)} dx \tag{6}$$

The substitution of the variables into Eq. (6) and evaluation process yields the universal probability distribution function as:

$$F_x(x) = \frac{\Gamma(q + 1, 0, ax^2 + bx)}{\Gamma(q + 1)} \tag{7}$$

The parameter values of a, b, and q derived from the recorded data and substituted into Eq. (7) for all data points x. The calculated parameter values are:

$a = 0.00019$, $b = 0.005$, and $q = -0.32$.

The plots of the universal and the exponential probability density functions are displayed in Fig. 3.

The goodness of fit test for the universal probability distribution function executed using Minitab software. It provides a comparison of two different functions at once. Distribution tests that have high p-values are suitable candidates for data distribution. The testing criteria is that the p-value in this experimental data should be 0.5 or above. The values below 0.005 indicate that the function is not suitable to represent the data. The goodness fit test produced a p-value of 0.345 for the exponential distribution whereas it is 0.567 for the universal function. The second test follows the threshold value on the probability density plot Fig. 3. The Minitab software produced a threshold point estimation as 0.03. The test result identifies if the selected probability distribution function is biased or unbiased. If the estimator equals the initial cluster delay parameter, the estimator is unbiased. Otherwise, the probability distribution function is biased and it is not suitable to apply in such an application. The plots in Fig. 3 indicates that the universal distribution function stops abruptly at the determined threshold value. However, the exponential distribution does not converge. Therefore, it is classified as biased for such data. The two goodness fit test results are sufficient to conclude that the universal probability distribution function is a better candidate than the ideal exponential function.

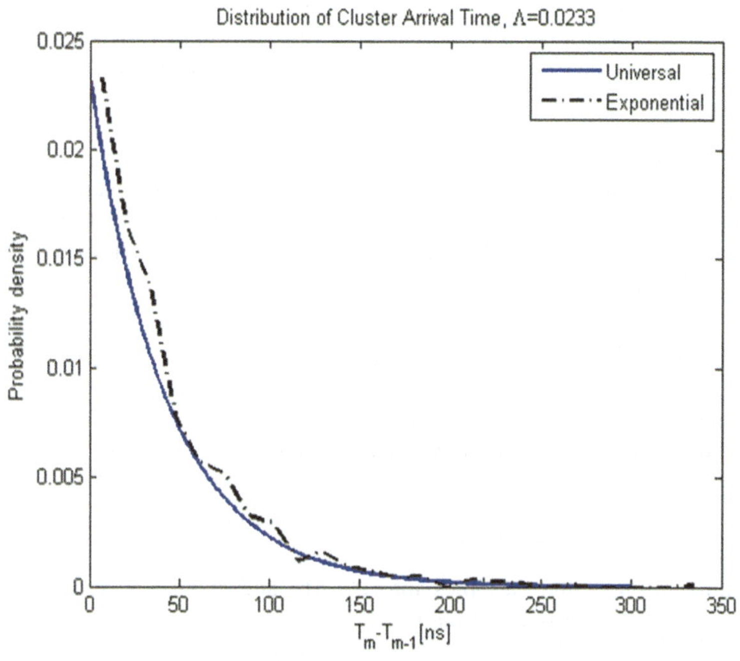

Fig. 3. Probability distribution plots to visualize and compare the universal (new) with the exponential distribution

3 Conclusion

This work is based on the wireless indoor channel model. The universal probability distribution function tested, compared with the ideal exponential model and the goodness of fit test applied based on the standard statistical analysis methods. The newly introduced model gives the highest p-test value of 0.567 which indicates to be the candidate for such an application. The new model also has another advantage that does not require prior knowledge of the randomly varying data and does not require analysis such as plotting a histogram for the data. The universal model covers successfully the most important distribution (Gaussian, Rayleigh, and exponential) with improved accuracy. The promising results will encourage the engineers to apply the universal probability distribution function in many fields of engineering applications.

References

1. Feng, L., Yang, H., Hu, R.Q., Wang, J.: MmWave and VLC-based indoor channel models in 5G wireless networks. IEEE Wirel. Commun. **25**(5), 70–77 (2018)
2. Ghendir, S., Sbaa, S., Al-Sherbaz, A., Ajgou, R., Chemsa, A.: Towards 5G wireless systems: a modified Rake receiver for UWB indoor multipath channels. Phys. Commun. **35**, 100715 (2019)
3. Greenstein, L.J., Ghassemzadeh, S.S., Hong, S., Tarokh, V.: Comparison study of UWB indoor channel models. IEEE Trans. Wireless Commun. **6**(1), 128–135 (2007). https://doi.org/10.1109/TWC.2007.04691
4. Sun, S., et al.: Investigation of prediction accuracy, sensitivity, and parameter stability of large-scale propagation path loss models for 5G wireless communications. IEEE Trans. Veh. Technol. **65**(5), 2843–2860 (2016). https://doi.org/10.1109/TVT.2016.2543139
5. Bilgehan, B., Ojo, S.: Multiplicative based path loss model. Int. J. Commun. Syst. **31**(17), e3794 (2018)
6. Dickenson, R.J., Ghassemlooy, Z.: Wavelet-AI equalization and detection for indoor diffuse infrared wireless systems. Int. J. Commun. Syst. **18**(3), 247–266 (2005)
7. Rubio, L., Torres, R.P., Rodrigo Peñarrocha, V.M., Pérez, J.R., Fernández, H., Molina-Garcia-Pardo, J.-M., Reig, J.: Contribution to the channel path loss and time-dispersion characterization in an office environment at 26 GHz. Electronics **8**(11), 1261 (2019). https://doi.org/10.3390/electronics8111261
8. Li, Z., Hong, T., Ning, W., Hong, Y., Wen, T., Li, J.: Path-loss prediction for radio frequency signal of a wireless sensor network in field based on artificial neural network. Trans. Chin. Soc. Agric. Eng. **26**(12), 178–181 (2010)
9. Benmus, T.A., Abboud, R., Shatter, M.K.: Neural network approach to model the propagation path loss for great Tripoli area at 900, 1800, and 2100 MHz bands. In: 2015 16th International Conference on Sciences and Techniques of Automatic Control and Computer Engineering (STA), pp. 793–798. IEEE (2015)
10. Bilgehan, B., Abdulbari, A.: Fast detection and DOA estimation of the unknown wideband signal sources. Int. J. Commun. Syst. **32**(11), e3968 (2019)
11. He, H., Jin, S., Wen, C.K., Gao, F., Li, G.Y., Xu, Z.: Model-driven deep learning for physical layer communications. IEEE Wirel. Commun. **26**(5), 77–83 (2019)
12. Bilgehan, B., Özyapici, A., Sensoy, Z.B.: An experimentally approved generalized model for circuit applications. Int. J. Circ. Theory Appl. **46**(3), 599–611 (2018)

Ecological Cybernetics of Oil Processing and Petrochemical Processes

A. I. Babayev(✉)

Azerbaijan State Oil and Industry University, Baku, Azerbaijan
abulfaz47@mail.ru

Abstract. Ecological cybernetics of technological processes, both as a scientific and educational subject, along with the improvement of theoretical knowledge in the application of technological processes improved by nanotechnologies, where the design technologies are intensively updated with new hardware and software of design technologies, will allow providing professional approach to the project development, ecological expertise as an environmental engineer. It is known that the performance of technological processes optimally reduces its impact on the environment. However, this does not answer the question of how much. In my view, studying ecological cybernetics of technological processes will provide the evaluation of ecological level in technological processes performed under complex technological, constructive design conditions. In other words, each optimum oil processing and petrochemical processes must be investigated as a research object, and technological and constructive parameters of the equipment, efficiency of the catalysts in the process, the composition of the raw material, processing depth must be studied, improved and brought to more advanced stages. It is known that re-refining of oil and petrochemical processes are performed in the combination of more complex conversions and more complex structural design of equipment and one of the useful ways of improving ecological condition is to determine a new level by increasing cybernetic bases of the process.

Keywords: Ecology · Industrial symbiosis · Cybernetics · Processes · Wastes · By-products · Pyrolysis · Synthesis · Acetic acid · Methanol · Mathematical modeling · Management systems

1 Introduction

At the current development of computing technics and technologies, the importance of managing complex environmental and technological processes during designing production units in the design of a complementary system to ensure the global sustainability of the environment should be emphasized. Industrial symbiosis, environmental modeling, research, study, design of ecosystems, scientific researches in this area, all components studied separately to ensure at least the stability of CO_2 concentration in the atmosphere, which is a clear indicator of the current environmental situation, should be consolidated in a single system (Copenhagen forum could not solve this problem). Ecological cybernetics aims at developing theoretical and practical bases for the creation of

© The Author(s), under exclusive license to Springer Nature Switzerland AG 2021
R. A. Aliev et al. (Eds.): WCIS 2020, AISC 1323, pp. 472–478, 2021.
https://doi.org/10.1007/978-3-030-68004-6_62

single management systems that can create conditions for the reduction and elimination of atmospheric emissions, technical, chemical, thermodynamic, aggregative changes in the target facility within the intended physical parameters. The main purpose of the system is to collect, save, process and provide the data on the parameters that reflect the ecological characteristics of all devices and units that are part of the production process and lead to deviations, to provide environmentally optimal management with operational instructions to make the necessary corrections. One of the main stages of ecological cybernetics is the identification of the parameters that can compare, interact in the process, the transformation of these parameters into information, the possibility of their transmission, the creation of working conditions within the system. The concept of information, its formation, collection, transmission are of great importance as a key factor in ecological cybernetics as in all areas of cybernetics. The significant difference is harsh variables that arise from the processes which are observed with complex, sometimes unintended transformations in oil refining and petrochemical processes. At the same time, in the case of systematization of complex technological processes, in many cases, data frame interval is used, rather than single-digit data. Unlike abstract cybernetics, ecological cybernetics focuses on a known, existing subject. Therefore, this begins with the collection of data, their processing by mathematical methods, and the analysis of the environmental parameters of the technological process under consideration. If any parameter needs to be determined during studies, additional testing may be required. The main goal is to create mechanisms that allow implementing them in optimal management systems by directing the logical interactions of the individual elements of the parameters involved in the process to achieve the goal. The analysis of the ecological cybernetics of designed processes is carried out in contrast to the existing processes.

2 Applied Researches, Analysis of Results

As it was mentioned during the creation of ecological cybernetics systems of oil processing and petrochemical process the technological parameters collected in database are given in the form of interval figures. This is related both to the process and equipment indicators. The mixed processes are also observed. The rectification process going with azeotropic drying in the same column, the rectification of diene hydrocarbons which goes with the delay of polymerization by adding special inhibitors into bottom of the column can be shown as a simple example. Both cases can be mathematically modelled as an ecological object and analyzed as a research object. But this will be one component of an industrial symbiosis. It will be an ecological component of the current technological process. In other words it is a part of an ecological system. Let's consider another example typical of the system. It is known that during methanol synthesis many additional reactions occur in parallel with the main reactions that results in the formation of other large amount of organic compounds together with a target product.

a) Dimethyl ether $2CH_3OH \leftrightarrow (CCH_3)_2O+H_2O$

b) Organic acids $2CO+2H_2 \leftrightarrow CH_3COOH$

$CH_3OH+CO \leftrightarrow CH_3COOH$

$CO+H_2O \leftrightarrow HCOOH$

c) Alcohols $_{n}CO+2_{n}H_2 \leftrightarrow C_nH_{2n+1}OH+(n-1)\ H_2O$

d) Complex ethers $CH_3OH+CH_3COOH \leftrightarrow CH_3COOCH_3+H_2O$

e) Aldehydes $CH_3COOH+H_2 \leftrightarrow CH_3CHO+H_2O$

$HCOOH+H_2 \leftrightarrow HCHO+H_2O$

f) ketons $2CH_3COOH \leftrightarrow CH_3COCH_3+CO_2+H_2O$

g) amines (can be obtained from ammonia, nitrogen and hydrogen)

$CH_3OH+NH_3 \leftrightarrow CH_3NH_2+H_2O$

$CH_3OH+CH_3NH_2 \leftrightarrow (CH_3)_2NH+H_2O$

$CH_3OH+(CH_3)_2NH \leftrightarrow (CH_3)_3N+H_2O$

h) saturated paraffin hydrocarbons

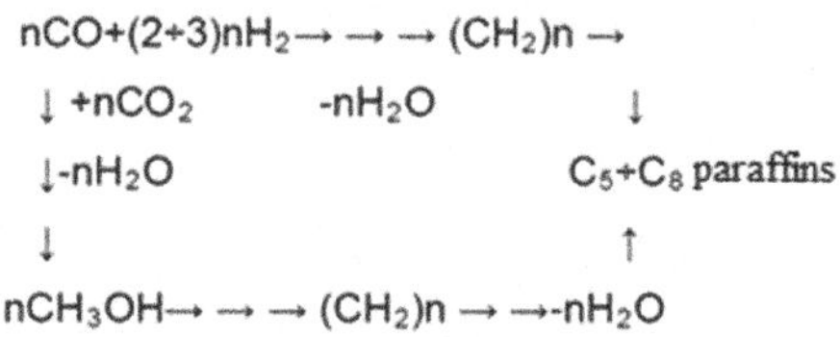

Based on the research of technological factors which impact on the obtaining of by-products: temperature, pressure, catalyst activity, reaction time, flow rate of reaction mass, construction of a reactor, communication design, automatic control system perfection, composition of a raw material, composition of recirculation components and functional ratio, equilibrium condition of the reactions, number of cycles of recycle gas, composition of a synthesis gas, efficiency of heat-exchange- cooling systems, composition of sewage and etc. advanced control algorithm can be developed by creating general mathematical model of the technological process, and according to it the process can be managed under optimum conditions in terms of high efficiency and eliminating the factors which impact on the environment. The process efficiency is characterized by its selectivity under general approach and can be calculated by the following equation:

$$I = \frac{S_1}{S_1 + S_2 + S_3}$$

Here, S_1-amount of a target product produced in the process; S_2-total amount of by-products for raw material produced during the process; S_3-amount of wastes released into atmosphere. As the technological process is improved theoretically, S_2 or S_3 approach zero and the selectivity increases and reaches up to 100%, the amount of wastes will be zero. In real processes in most cases it is impossible to get the amount of by-products or wastes down to zero. In most cases they become the elements of the process. We can only talk about optimizing their limits in the process. This means that there will always be a need for the improvement of the processes that we encounter on a large scale. The parameters affecting S_1 and S_2 which are the expression of variations in oil processing and petrochemical processes consist of numerous variations and have abovementioned wide ranges. These parameters can be identified by various experiments in the operating system or in its real physical model.

Let's review two-stage catalytic process of producing acetic acid and ethylacetate as a specific example [1]. At the first stage of the process a target product – acetic acid is formed from ethanol, acetaldehyde and carbon dioxide are formed. Acetaldehyde is recycled to the system, but carbon dioxide is released into atmosphere. In this case the selectivity is determined by the following way:

$$I = \frac{S_1}{S_1 + S_3}$$

At the second stage a target product – ethylacetate is produced from ethanol and acetic acid. Mathematical model for the first stage is given by:

$$\frac{dA_1}{dl} = S \cdot \frac{k_3 b_1 b_3 P_1 P_4}{\left(1 + b_1 P_1 + b_2 \sqrt{P_2} + b_3 P_4 + b_4 P_5\right)^2}$$

$$\frac{dA_2}{dl} = S\left(\frac{k_2 b_4 b_2 P_5 \sqrt{P_2}}{\left(1 + b_1 P_1 + b_2 \sqrt{P_2} + b_3 P_4 + b_4 P_5\right)^2} - \frac{k_3 b_1 b_3 P_1 P_4}{\left(1 + b_1 P_1 + b_2 \sqrt{P_2} + b_3 P_4 + b_4 P_5\right)^2}\right)$$

$$\frac{dA_3}{dl} - S\left(\frac{k_1 b_1 b_2 P_1 \sqrt{P_2} - k_2 b_2 b_4 P_5 \sqrt{P_2}}{\left(1 + b_1 P_1 + b_2 \sqrt{P_2} + b_3 P_4 + b_4 P_5\right)^2} \quad \frac{k_4 b_4 P_5 \sqrt{P_2}}{\left(1 + b_1 P_1 + b_2 \sqrt{P_2} + b_3 P_4 + b_4 P_5\right)^2}\right)$$

$$\frac{dA_4}{dl} = \frac{2 k_4 b_4 P_5 P_2}{1 + b_1 P_1 + b_2 \sqrt{P_2} + b_3 P_4 + b_4 P}$$

$$\frac{dT}{dl} = -\frac{\sum_{i=1}^{4} r_i \Delta H_i}{\sum_{i=1}^{8} n_i C_{pi}} - \frac{\alpha (T - T_x)}{\sum_{i=1}^{8} n_i C_{pi}}$$

Table 1. Results of theoretical optimization on the mathematical model are presented in the

α, %	5.96	6.7	7.6	7.8	9.35	14.1
Gy.к., kg/hour	2612	2649	2785	2785	3255	4914
Xк, %	20	21	22	22	27	40
CO_2, kg/hour	675	788	874	896	1074	1619
Acetaldehyde (recycle)	1177	2292	3296	3654	3872	3423

The Table 1 shows that as optimum values of ethanol conversion to oxygen at different ratios of ethanol increase the yield of by-product CO_2 which is released into atmosphere, increases. Due to the use of a new catalyst at the second stage by-products and emissions are not formed, i.e. the catalyst works with 100% of selectivity. In other words in the current case the solution of an ecological problem in the process which excludes emissions is solved through the improvement of a catalyst quality, i.e. the awareness of determining a proper direction of scientific study by using the results and developing the implementation of a technological process by stage. This is an example of the logical interaction of individual elements of the system to obtain a specific result, which is one of the elements of the logical basis of ecological cybernetics.

Let us review pyrolysis process of ethane with feedback [2–4]. Pyrolysis is a rather complex process, in which a large number of different components are involved, and, accordingly, its chemistry and kinetics are expressed by several hundreds of reactions and equations, that significantly increases the role of mathematical modeling of the optimum process modes. However, the methods of mathematical modeling are not considered in this work, but the possibility of using the results as a component of ecological cybernetics is investigated. Pyrolysis of ethane is accompanied by side conversions in parallel resulting in reduced yield of a target product of ethylene. Definite conversion level will conform to the maximum yield of ethylene during single processing of a primary raw material. Further deepening of the process in a single pass may lead to the increase in the yield of by-products. Therefore, the highly relevant for the most favorable process is to find that conversion level and share of recycled material in which the highest yield of a target product and the lower emissions into atmosphere can be achieved. To study these problems first of all it is necessary to provide a mathematical description of the process which reflects the features of using recycling.

The kinetic model of the process developed in accordance with proposed stoichiometric diagram is presented by the following equations.

$$\frac{dn_1}{dl} = -u^{-1}\left[k_1 n_1 + k_7 n_1 \cdot \frac{n_2 P}{\sum n_i RT} - \frac{k_1}{k_{p1}} \cdot n_2 \frac{n_3 P}{\sum n_i RT}\right]$$

$$\frac{dn_2}{dl} = -u^{-1} n_2 \left[\begin{array}{l} k_1 \frac{n_1}{n_2} - \frac{k_1}{k_{p1}} \cdot \frac{n_3 P}{\sum n_i RT} - k_2 \left(\frac{n_3 P}{\sum n_i RT}\right)^2 + \frac{k_2}{k_{p2}} \cdot \frac{n^2 P}{n_1 \sum n_i RT} - k_3 - k_4 \\ -k_5 - k_6 - k_7 \frac{n_1 P}{\sum n_i RT} \end{array}\right]$$

$$\frac{dn_3}{dl} = u^{-1} \left[\begin{array}{l} k_1 n_1 + \frac{k_2 n_4^2 P}{k_{p2} RT \sum n_i} + n_2 \left(\begin{array}{l} -\frac{k_1 n_3 P}{k_{p1} RT \sum n_i} - 2k_2 \left(\frac{n_3 P}{RT \sum n_i}\right)^2 + \frac{1}{8} k_3 + k_4 + \\ +k_5 + 2k_6 + \frac{93}{150} k_7 \frac{n_i P}{RT \sum n_i} \end{array}\right) + \\ \frac{n_{15} P}{RT \sum n_i} (k_8 n_{10} + k_9 n_{13}) \end{array}\right]$$

$$\frac{dn_4}{dl} = 2k_2 u^{-1} \left[n_2 \left(\frac{n_3 P}{\sum n_i RT}\right)^2 - \frac{1}{k_{p2}} \left(\frac{n_4 P}{RT \sum n_i}\right)^2\right], \quad \frac{dn_5}{dl} = u^{-1} k_3 \frac{n_2}{4}, \quad \frac{dn_6}{dl} = u^{-1} k_3 \frac{n_2}{8}$$

$$\frac{dn_7}{dl} = u^{-1} k_3 \frac{n_2}{8}; \quad \frac{dn_8}{dl} = u^{-1} k_4 \frac{n_2}{3}; \quad \frac{dn_9}{dl} = u^{-1} k_5 n_2; \quad \frac{dn_{10}}{dl} = u^{-1}\left[2k_6 n_2 - k_8 n_{10} \frac{n_{15} P}{RT \sum n_i}\right]$$

$$\frac{dn_{11}}{dl} = \frac{143}{150} u^{-1} k_7 n_2 \frac{n_1 P}{RT \sum n_i}; \quad \frac{dn_{12}}{dl} = \frac{57}{150} u^{-1} k_7 n_2 \frac{n_1 P}{RT \sum n_i};$$

$$\frac{dn_{13}}{dl} = u^{-1} k_8 n_{10} \frac{n_{15} P}{RT \sum n_i} - k_9 n_{13}; \quad \frac{dn_{14}}{dl} = u^{-1} k_9 n_{13} \frac{n_{15} P}{RT \sum n_i}.$$

where k_i – reaction rate constant; k_{p1}, k_{p2} – equilibrium constants; n_i – number of moles of ethane, ethylene, hydrogen, methane, divinyl, butylene, butane, benzene, acetylene, carbon, propylene, propane, carbon monoxide, carbon dioxide and water; u – linear rate of a raw material; l – reactor length; R – gas constant.

The kinetic parameters of the proposed model have determined on the basis of experimental data of the unit EP-300 of Sumgait plant «Ethylene- Propylene».

In the present process the target products are ethylene + propylene; by-products are hydrogen, methane, divinyl, butylene, butane, benzene, acetylene, emissions are carbon, carbon monoxide and carbon dioxide. Optimization studies were carried out under the conditions of a non-recirculating ethane pyrolysis process in the following intervals of changing the factory input parameters: $g_{ethane} = 2 \div 5$ t/h; $t_O = 775 \div 850\ ^{\circ}$C; the ratio of water vapor: raw material - 1:1; 1.5:1 and 2:1; $P_O = 3 \div 4.2$ atm - depending on the water steam content. As a result of the optimization it was determined that at the beginning of the process primary reactions prevail, therefore, at the first section of the pipe from the beginning of the reactor, ethylene accumulation rate will be maximum, then as ethane is consumed it decreases gradually reaching minus values, in which consumption rate of ethylene in secondary and tertiary reactions begins to increase the rate of its accumulation. Therefore, up to this point (equilibrium point of accumulation rates and consumption of ethylene) the amount of C_2H_4 along the reactor length increases up to the maximum value of 826.4 kg/h. Further increase in consumption rate of ethylene to secondary and tertiary reactions reduces the amount of ethylene up to 692,1 kg/h and increases emissions into atmosphere. However, due to the reversibility of the reaction, through which ethylene is partially formed from methane, the ethylene consumption rate decreases, but not so much as to result in an increase in the absolute yield of ethylene. Therefore, at the outlet of the reactor, the amount of ethylene is lower than in the previous sections, with the maximum ethylene values with an increase in charge for all temperatures reach near the end of the reactor. Therefore, starting from the loading of $g_{ethane} = 3.25$ t/h, the yields of target products are maximized at the end of the reactor. Thus, with increasing loading for each temperature, the selectivity of the process rises and with increasing temperature - decreases. The selectivity of pyrolysis of hydrocarbon raw material is determined as a ratio ethylene + propylene to the amount of converted raw material, ethylene and propylene are formed as a result of primary reactions, increase of selectivity will contribute the suppression of secondary decomposition reactions of target products and thereby reduce, in separate cases, eliminate emissions into the atmosphere. This requires decreasing the residence time of feedstock in reaction zone and reducing partial pressure of hydrocarbons by selecting optimum level of dilution rate of raw material with water steam for each load, and transferring this command to the center of optimum control of technological processes in the context of ecological cybernetics. Data source for the command is the change in the output of by-products and the amount of carbon monoxide in it, obtained from flow chromatographs that provide the information to the technological process. In this mathematical model since the formation of carbon monoxide formed from the waste is dynamic, and the formation of deposited carbon layers on inner surface of coils in pyrolysis furnace where conversion processes occur, is static, the variation criteria in the amount of carbon monoxides in the flow are taken into account to give a quick ecological improvement command to the technological process. This means that ecological cybernetics of technological processes allows studying the flexibility and dynamic activity of the impact in management systems in depth, thereby reducing emissions by increasing the sensitivity of the management system as a whole. Each data delivered to intelligent management centers of technological processes must contain the information which allows managing the process to carry out environmental optimization. The comprehensiveness and completeness of carrier database is determined by local (at plant level), regional (territorial units where many plants operate), global environmental analysis of technological processes. When reviewing the global approach to environmental problems as an example of control over carbon monoxide, indicators that reflect the performance of all sectors, including transport and other utilities, should be included in the database. The research that we are reviewing will be an integral part of the global control and management system. The same approach applies to gases, liquids, solid wastes and sewage. This means that ecological cybernetics is a new direction that summarizes the results of scientific researches, studies, serves to improve intelligent management systems which aimed at solving environmental problems at different levels. This enables high-skilled engineers, mathematicians, programmers to work together creatively, to meet modern regulatory requirements in the field of environmental safety, to create more flexible management systems, to solve environmental problems through the improvement of technological processes in a new approach and scale.

References

1. Safarov, A.R.: Modeling and optimization of the processes of obtaining acetic acid and ethylacetate by combined technology. Abstract of Ph.D. in technical sciences, Baku p. 26 (2006)
2. Aliyev, A.M., Babayev, A.I., Tahirov, A.Z., Huseynova, A.M., Ismayilov, N.R.: Mathematical modeling and optimization of industrial process of pyrolysis of ethane with butane-butylene fraction in view of feedback. Azerbaijan Chem. J. **2**, 16–24 (2010)
3. Babayev, A.I., Aliyev, A.M., Tahirov, A.Z., Huseynova, A.M.: Modeling and optimization of industrial processes of pyrolysis of paraffine hydrocarbons and their mixtures with feedback. In: II International Conference of D.I. Mendeleyev Russian Chemical Society «Innovative Technologies and Biotechnology of Materials and Products», Theses, Moscow, pp. 11–13 (2010)
4. Babayev, A.I., Aliyev, A.M., Tahirov, A.Z., Ismayilov, N.R., Huseynova, A.M.: Modeling and optimization of pyrolysis of ethane with feedback. Azerbaijan Chem. J. **3**, 16 (2008)

Industry 4.0 Technologies: Impact on Increasing the Productivity of Industrial Enterprises

O. O. Zaripov[(✉)] and U. Sh. Khamrakulov

Tashkent State Technical University, Tashkent St University 2, Tashkent, Uzbekistan
o.zaripov@edu.uz, umid198924@gmail.com

Abstract. The article discusses one of the most important problems, the ways of integration into modern internet technologies and the main characters of cyber physical systems of Cyber-Physical system builds the technologies of Industry 4.0. Besides that, it is given the concept of Internet of Things of creating the production by the help of cloud technologies.

Keywords: Industry 4.0 technologies · Control · Internet of Things · Big data · Cyber-physical systems · Cloud · Integration · Concept

1 Introduction

Increasing the efficiency of the economy and forming an innovative way of its development is currently impossible without its digitalization and restructuring based on various tools and mechanisms. One of these tools is the implementation of Industry 4.0 technologies [1]. Industry 4.0 corresponds to new digital technologies – cloud services, mobile devices, augmented reality (portable gadgets), the Internet of Things, geolocation (location determination), improved interfaces for interaction between an individual and a computer, authentication and others, 3D printing, technologies within artificial intelligence, Big data, advanced algorithms, personification by customer profile, etc., allowing you to create an effective business model of an enterprise.

At the same time, with the introduction of digital technologies and Cyber-Physical Systems into production processes, a significant part of production will take place without human participation [2, 3].

2 Economic Transformations Associated with the Implementation of Projects for the Development and Implementation of Industry 4.0 Technologies

Economic transformations associated with the implementation of projects for the development and implementation of Industry 4.0 technologies. A key driver of Industry 4.0 technologies is the increased integration of "Cyber-Physical Systems," or CPS, into manufacturing processes. Possible economic transformations associated with the implementation of projects for the development and implementation of Industry 4.0 technologies

© The Author(s), under exclusive license to Springer Nature Switzerland AG 2021
R. A. Aliev et al. (Eds.): WCIS 2020, AISC 1323, pp. 479–486, 2021.
https://doi.org/10.1007/978-3-030-68004-6_63

include the following possible changes in production [4]. 1. Increase in labor productivity by reducing the number of personnel, the emergence of new technologies that can increase the efficiency of services provided and products. 2. The emergence of new types of industrial technologies and new industrial production, caused by: - development of mobile Internet and cloud technologies; the development of 3D printing; the emergence of new materials, genetic and biotechnologies, which will give a new impetus to the development of medicine, agriculture, pharmaceuticals, plastics and polymers, biofuels. 3. Widespread use of innovative solutions, a new level of automation and robotization of production processes, an increase in the computing power of computers and the use of the capabilities of large data arrays, artificial intelligence and machine learning, which make it possible to automate operations that were previously perceived as impossible for machines to perform. 4. Development of remote access and control over production equipment with the development of the Internet of Things, i.e., the ability to control physical objects using the Internet. 5. The emergence of new sources of energy and energy technologies, increasing the energy efficiency of production. 6. Blurring the boundaries between industries and activities.

At the same time, changes are also forecasted in the sphere of consumption, [5] including: 1. Development of the sharing economy; 2. Increased interest in ecological products; 3. The growing importance for consumers of ethical aspects of consumption and problems of personal data protection; 4. Development of new specific products, services and business models due to an aging population and increasing life expectancy.

2.1 Industry 4.0 Today and Tomorrow

As the Internet of Things rolls out, global networks will emerge that will connect machines, management and storage systems, and the manufacturing facilities of smart factories. A conditional model of the future intelligent enterprise is shown in Fig. 1.

All objects included in a common network can exchange data, initiate certain actions and control each other. This allows you to simplify the modernization of manufacturing processes, design, supply chain, materials use and life cycle management. Also, Industry 4.0 makes it possible to improve the quality, flexibility and reliability of the entire operation of the enterprise [6]. At the heart of the fourth industrial revolution are smart factories. Such enterprises can implement production processes of any complexity, while minimizing the risk of disruption and ensuring the effective creation of "smart" products. One of the important components of such industries are wireless networks that cover all processes, machines, resources and employees, and also allow the exchange of data between companies.

Within the framework of the new production, "smart" products can be accurately identified, their current state can be found out, which production processes they have already passed, and which ones are just coming - in all details. Depending on the information received, "smart" factories build a product route and equipment operation. This approach allows for mobility and improved logistics.

With global networks, smart products can be monitored in real time throughout the entire production cycle. In some cases, they will even be able to almost autonomously control their production process, ensuring the optimization of these stages in terms of logistics, service and integration with the rest of the enterprise processes. Also, in the

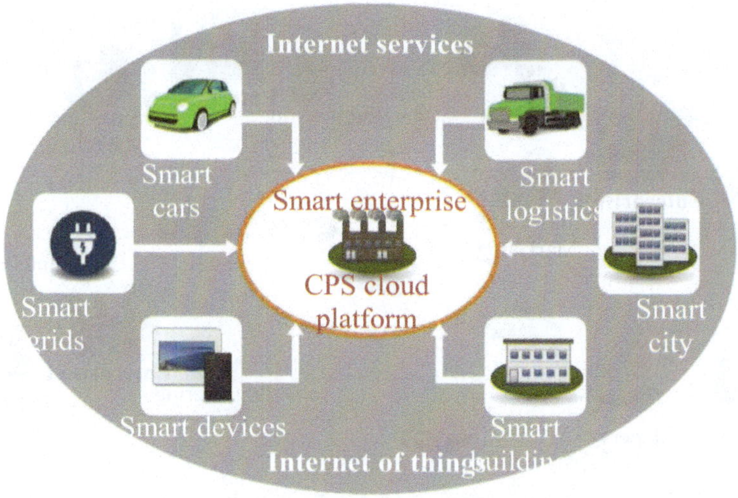

Fig. 1. Model of a network node of a "smart" enterprise

future, at any time, it will be possible to integrate some service functions into smart products and set specific design parameters, order formation, production planning, operation and disposal, which is especially important when producing small batches of goods. Smart factories will be able to take into account the individual requirements of customers, change the operating mode of production at any moment and quickly respond to disruptions in the work of suppliers. Full transparency of production processes allows you to make optimal decisions and create new business models. It is important to note that in smart factories, workers will be freed from the need to perform routine tasks due to the ability to monitor and manage production processes through networks. Professionals will be able to pay more attention to professional development and more creative assignments. At the same time, older employees will be in demand for a longer time due to the lack of qualified workers familiar with the production. The flexible organization of production will also contribute to a more efficient combination of work and quality of life of employees.

Also, for the implementation of "Industry 4.0" it is necessary to develop the network infrastructure, increase the bandwidth for resource-intensive applications and improve the quality of service of the network, especially in cases where the task execution time is critical. A possible option for building global networks in a "smart" enterprise, indicating the links between smart objects and services, is shown in Fig. 2. All of the above technologies can be implemented on the basis of a CPS. Let's consider its advantages in more detail.

2.2 Cyber-Physical Systems

For the first time the term "Cyber-Physical Systems" was introduced in 2006 by the director of the National Science Foundation for Embedded and Hybrid Systems of the United States Helel Jill to define complexes containing natural objects, artificial subsystems

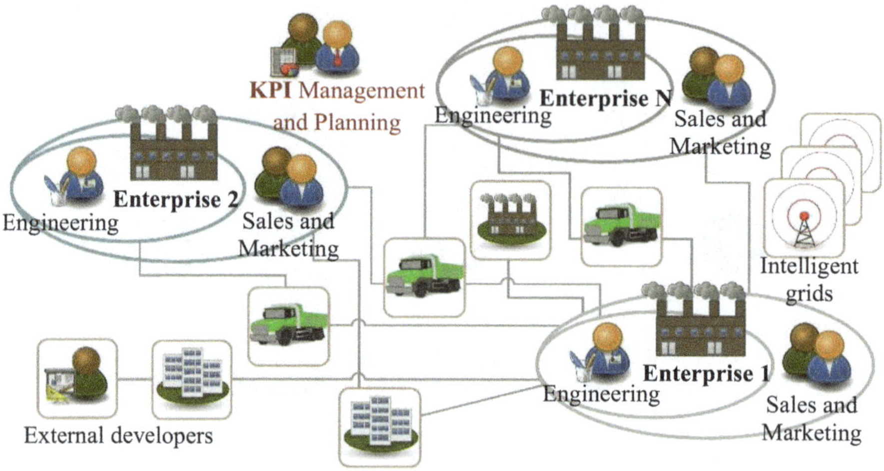

Fig. 2. Architecture of the network of "smart" enterprises and their management.

and controllers [7]. These systems were supposed to solve the problem of modernizing production processes by reforming the management organization. CPS an information technology concept that implies the integration of computing resources into physical entities of any kind, including biological and man-made objects. In CPS, the computing component is distributed throughout the entire physical system, which is its carrier, and is synergistically linked with its constituent elements. The set of technologies that ensure interaction between the virtual and physical world is called CPS (Fig. 3). The overall CPS architecture is broken down into five fundamental layers: physical layer (the physical layer lays the foundation for the CPS architecture); network layer (routing of packets based on the conversion of a unique identifier assigned to each active equipment in the network); transport layer (packets are split into small fragments); intermediate level (terminal management, protocol conversion); application layer (stores, analyzes and updates information).

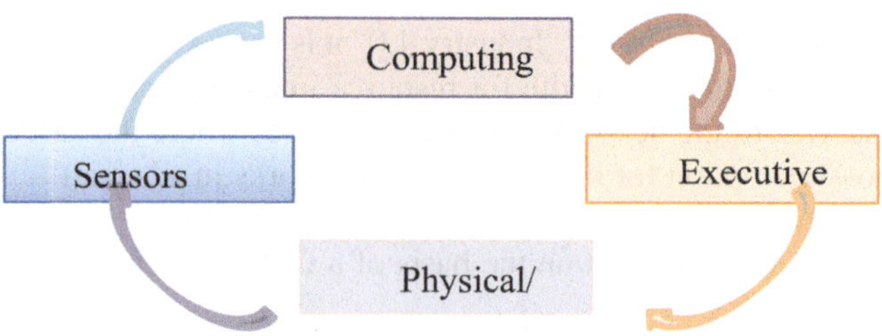

Fig. 3. CPS architecture.

The generalized structure of a CPS contains a network of intelligent measuring and computing nodes, united by a switching environment and supported by high-performance computing facilities and information security facilities connected to information collection and processing centers.

2.3 Cyber-Physical Systems as Technologies of Information Subsidiary Management

Increased investment in new technologies. Industry 4.0 assumes significant growth in investment in fixed assets and intangible assets. This can be seen if you highlight the investment in the described Industry 4.0 technologies. At the current level of development, information technologies are widely used in the management of economic systems. The information management strategy uses the principle of subsidiarity, according to which management tasks should be solved at the lowest, smallest or remotest level from the center, at which their solution can bring a positive effect [8]. Modern subsidiary management is based on the use of information management technologies, for the implementation of which information structures, information models and a specific unit of information are used. Information constructions reflect the essence of the information approach in modeling, are a conceptual description of management technologies. Information models are used to describe the processes of situations and objects, for the quality of separation of control components in the form of scenes of situations and management processes. The information construct performs the functions of a model, but at a higher level of abstraction. Information construction is used to identify implicit knowledge and its transition into explicit knowledge [9]. From a managerial point of view, the CPS is a distributed subsidiary system [10]. As mentioned above, this system has many levels, but at each level of scale and complexity, new properties appear that are not reducible to those of simple levels. The concept of a CPS is closely related to the concepts of robotics and sensor networks with intelligent mechanisms of computational intelligence. Despite the high relevance of scientific research in the field of the digital economy and theoretical research on the development of CPS, many problems of creating the latter remain unresolved. So, the theoretical substantiation of the principles of building CPS is insufficient. The issues of functional completeness and synergetic effect from combining various components into a system and other problems remain unexplored. The approaches to the construction of CPS are based on the analysis of the peculiarities of the interaction of physical processes of the surrounding world with cybernetic means (measuring and computing, communication, control and executive). These features include: the speed of the flow of physical processes (in comparison with the computing and communication capabilities of cybernetic means); type of physical processes (linear, nonlinear, synergistic); the ability to determine the state of the physical process (completeness of information, accuracy); the ability to influence the change in the states of a physical process. A universal theoretical description and the corresponding theoretical solutions that would make it possible to move from the analysis of these features to the synthesis of CPS are not yet available [11]. In Fig. 4 shows the dynamics of growth in the automation market. Virtually all sectors creating Industry 4.0 technologies have shown significant growth. A particular example is the dynamics of sales of industrial robots (Fig. 5). These and other trends are shaping the future of the industry to a large extent; they should be taken into account by both the government and business. In addition, these trends indicate the growth of several markets, which creates new opportunities for companies aimed at finding their niches [12].

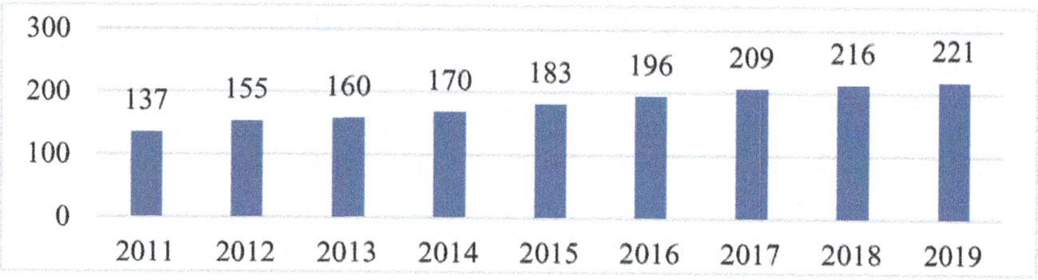

Fig. 4. Global dynamics of the industrial automation market volume, USD.

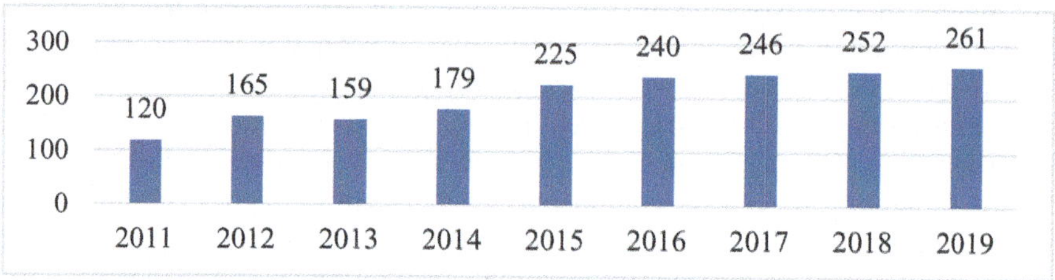

Fig. 5. Dynamics of sales of industrial robots around the world, thousand units.

In the scientific literature, there are no studies of CPS in the field of research and development: new ways of combining and organizing the interaction of components of CPS to ensure high efficiency of planning and implementation of complex tasks for the research and management of target objects; research into the principles of building intelligent sensor systems and sensor networks; new mathematical models and algorithms for transforming informative, multidimensional, non-periodic signals. This is only a small part of the problems facing the creation, development and implementation of CPS. New principles of planning and distribution of tasks that will be solved by the CPS make it unique in terms of the physical environment.

2.4 Synergy of Cyber-Physical System and Intelligent Control

A CPS is an organizational and technical concept for managing information flows, the integration of computing resources into physical production processes for continuous real-time control. In such a system, sensors, controllers and information systems are united into a single network throughout the entire product life cycle. The control and synchronization of a huge variety of electronic devices must be performed in order to achieve optimal performance. The communication system (wireless sensor networks, wired networks and mixed data transmission systems) demands the need for two-way communication to transmit impact signals to physical objects. The nodes of this network contain universal intelligent control modules with the possibility of self-learning and self-organization. Thus, a CPS is a non-equilibrium dynamic system based on discrete intelligent modules. A CPS is a complex distributed system controlled and controlled by computer algorithms and tightly integrated with the Internet and its users. The CPS is characterized by the possibility of parallel information flows, making asynchronous decisions, and intelligent information processing. An important circumstance for a CPS is that networked intelligent nodes for processing information rebuild the system from a communication to an intelligent one. The management methods that are used in cybernetic systems, hierarchical, matrix, situational and network-centric management do not ensure the effectiveness of management in the face of the complexity of management tasks when using distributed systems. This problem is exacerbated by the problem of big data due to its large volumes and large number of connections. There is an intranet management problem for complex distributed systems. For complex distributed systems, only intelligent control is effective, which is based on neurotechnologies, evolutionary technologies [13]. The synergy of computational and physical components of CPS ensures the joint operation of elements of cyber and physical spaces, integrating computing resources via the Internet and cloud technologies. CPS combine information from intelligent sensors distributed in the physical environment to make management decisions. CPS carry out computational procedures in their distributed structure, based on information from smart sensors, they make it possible to reconfigure information flows in the network depending on conditions. Thus, CPS are distributed systems with the ability to intelligently process and reconfigure flows through intelligent control [14].

3 Conclusion

The formed problems of building CPS and the proposed principles of building a hardware-software platform for creating applied CPS require the need for their further development. Now it is extremely important to accelerate the development of fundamental research in the field of the general theory of intelligent information technologies, which will make it possible to create classes of new promising technologies that actively interact with the external environment and its objects. Such technologies should be based on: speech technologies for natural environments; technologies of mass use for processing text information, recognition and use of knowledge; highly dynamic technologies for meaningful analysis of information in computer networks and their filtering; technologies for setting and solving problems according to previously specified algorithms, intelligent technologies in solving problems "smart enterprise", "smart home", "smart city", "smart transport", "smart health care", "smart school" etc.

The development of work on the problems of intellectualization requires that promising means of information processing, along with a computing platform, have a thinking platform.

References

1. Arseniev, D.G., Overmeyer, L., Kälviäinen, H., Katalinić, B.: Cyber-Physical Systems and Control. Lecture Notes in Networks and Systems, 1st edn. Springer, Cham (2020)
2. https://tass.com/sp/948066

3. Sevinov, J.U., Zaripov, O.O., Zaripova, S.O.: The algorithm of adaptive estimation in the synthesis of the dynamic objects control systems. Int. J. Adv. Sci. Technol. **29**(5), 1096–1100 (2020)
4. Horvath, I., Gerritsen, B.: Cyber-physical systems: concepts, technologies and implementation principles. In: Conference 2012, TMCE, pp. 19–36 (2012)
5. Zaripov, O.O., Akhmedov, D.A., Mamirov, U.F.: Adaptive estimation algorithms for the state of nonlinear dynamic systems. Int. J. Psychosoc. Rehabil. **24**(3), 247–253 (2020)
6. Jamaludin, J., Rohani, J.: Cyber-physical system (CPS): state of the art. In: Conference 2018, ICE Cube (2018). https://doi.org/10.1109/icecube.2018.8610996
7. Zaripov, O.O., Shukurova, O.P., Sevinov, J.U.: Algorithms for identification of linear dynamic control objects based on the pseudo-concept concept. Int. J. Psychosoc. Rehabil. **24**(3), 261–267 (2020)
8. Fradkov, A.L., Miroshnik, I.V., Nikiforov, V.O.: Nonlinear And Adaptive Control Of Complex Systems. Kluwer, Dordrecht (2018)
9. Alguliev, R., Imamverdiev, Y., Sukhostat, L.: Ensuring information security of cyber-physical systems. In: Conference, pp. 40–45 (2017). https://doi.org/10.25045/NCSoftEng.2017.07
10. Igamberdiyev, H.Z., Yusupbekov, A.N., Zaripov, O.O., Sevinov, J.U.: Algorithms of adaptive identification of uncertain operated objects in dynamical models. Procedia Comput. Sci. **120**, 854–861 (2017)
11. Rawat, P., Jayant, P.: A review of challenges in implementation of industry 4.0 in indian manufacturing industry. In: Conference 2019, ICRTIEST, pp. 289–297 (2019)
12. Zaripov, O.O., Khamrakulov, U.Sh.: The models of changing and giving information in integrated informational-analytical systems. In: Conference 2020, Tashkent, Uzbekistan, pp. 63–66 (2020)
13. De Ville, B.: Microsoft Data Mining: Integrated Business Intelligence for e-Commerce and Knowledge Management. Digital Press, Boston (2011)
14. Tarasov, I.V.: Industry 4.0: technologies and their impact on productivity of industrial companies. SDRM **2**, 62–69 (2018). https://doi.org/10.17747/2078-8886-2018-2-62-69

Fuzzy C-Means Based Synthesis of Ion-Exchange Materials

F. A. Amirov, G. A. Aliyeva(✉), F. R. Rahimova, and G. F. Rustamova

Azerbaijan State Oil and Industry University, Azadlıq Avenue 34, Baku Az1010, Azerbaijan
fariz_emirov@mail.ru, gulnaraaliyeva@mail.ru,
firengizrehimova@gmail.com, rustamova-guler@mail.ru

Abstract. The presented article is devoted to the synthesis of sulfocationites with simplified technology and with improved properties. For this purpose, polymer wastes taken in different proportions were mechanochemically modified with polyvinyl chloride. Samples of polystyrene taken from the interior walls of the household "ORSK" refrigerator were used as polymer waste. The extracted polymer samples were first cleaned, washed, crushed and then some of their parameters (molecular weight, density, solubility, swelling rate, chemical resistance) were determined. The process of mechanochemical modification of polymers was carried out at a temperature of 150–160 °C in a wide range of shear strain. The rheological properties of the polymer alloy were studied to confirm the homogeneity of the obtained polymer mixtures. Ion exchange resins were synthesized by sulfonation on the basis of a polymer mixture obtained with mechanochemical method. Using anhydrous $AlCl_3$ as a catalyst, the sulfonation process was carried out with the concentrated sulfuric acid at different temperatures and times. Optimal conditions for this process have been determined. In order to confirm the course of mechanochemical modification and sulfonation process, IR spectrum of both the primary components and the modified composition, as well as sulfocationites were taken. Satisfactory results were obtained by studying the main properties of the obtained sulfocationites such as swelling coefficient, static and dynamic exchange capacity, mechanical resistance, and the use of ion exchange resins in desalination of water was proposed.

Keywords: Sulfocationite · Waste polystyrene · Polyvinylchloride · Mechanochemical modification · Effective viscosity · Shear strain · Shear rate · Sulfonation · Static and dynamic exchange capacity · Mechanical resistance

1 Introduction

Synthesis of sulfocationites was carried out by sulfonation of mechanochemically modified polymer mixtures of decommissioned polystyrene with polyvinyl chloride. Waste polystyrene and polyvinyl chloride were used as polymer carcasses in the synthesis of sulfocationites. The interior walls of the 1979 "OPCK" refrigerator were used as waste polystyrene (WPS). In order to increase the impact strength and mechanical durability

© The Author(s), under exclusive license to Springer Nature Switzerland AG 2021
R. A. Aliev et al. (Eds.): WCIS 2020, AISC 1323, pp. 487–495, 2021.
https://doi.org/10.1007/978-3-030-68004-6_64

of waste polystyrene and sulfocationites obtained on its basis, waste polystyrene was modified mechanochemically with polyvinyl chloride (PVC).

Waste polystyrene, i.e. a sample taken from the interior walls of the "OPCK" refrigerator, was selected, cleaned, washed and crushed. Some properties of polystyrene waste were determined – solubility, swelling coefficient, molecular weight, as well as IR spectrum of samples. It was determined that the density of the polystyrene sample is 1,069–1,125 g/cm^3, the melting point is 240 °C, it is soluble in dichloroethane, acetone, toluene, chloroform, CCl_4 and esters. Polyvinyl chloride obtained in suspension with a density of 1.35–1.45 g/cm^3, melting point 150 °C, crystallization rate 10–15% was used as a modifier in the mechanochemical method of obtaining the polymer composition.

This polymer is soluble in esters, chlorinated hydrocarbons and acetone.

It is known that the ion-exchange capacity of ion exchangers of any structure does not depend on the size of the grains, because ionite can participate in the ion-exchange reaction throughout its volume. However, in most cases, porous ionites have advantages over smooth ionites. The porosity of ion exchangers is an important factor and accelerates the ion exchange process. For this reason, volatile solvent is added to the reaction mixture before the sulfation process begins, and after some time, after solvent leaving the system, pores form in the modified mixtures, as a result, the sulfonation process is enhanced and porous ionites are formed. We used dichloroethane and CCl_4 as solvents.

Taking into account the differences in the chemical nature of solid-phase polymers, at different values of shear strain and shear rate in the capillary viscometer WPS:PVC compositions were obtained in the following ratios, % (mass): 99:1; 98:2; 97:3; 96:4; 95:5. Polymer mixtures based on polyvinyl chloride modified polystyrene waste were granulated after cooling in water, 2–3 mm grains were dried at a temperature of 30–40 °C for 5–10 min. The modified samples were then subjected to a sulfonation process to obtain sulfocationites.

Ion-exchange materials based on styrene modified with functional group monomers have complex properties and they are considered promising polymer compositions. However, the production of modified polymers in the industry using the chemical method of graft copolymerization is not economically viable. This is due to the low yield of the modified polymer (3–5% by weight), the deterioration of the initial properties of the polymer matrix and the high consumption of electricity. It is shown that the effectiveness of chemical modification can be obtained during the combined processing of functional group polymers with the above-mentioned copolymers under certain temperature and stress conditions.

However, multistage technology has been used in the synthesis of sulfocationites based on the copolymer of styrene obtained in suspension, and this process is not suitable for high power consumption.

It is known that the method of adding small amounts of other polymers and oligomers to large tonnage industrial polymers is widely used to obtain polymer materials with improved properties.

The method of modification of polymers is based on the effect of the molecular structure, as well as the conditions of the relaxation process on the properties of polymers. In this case, the complex effect of additives on the structure and properties of the polymer is accompanied.

When a small amount of modifier is added, the physico-mechanical properties of the obtained polymer improve, and the service life of the products increases.

The effect of the modifier on the polymer properties exists not only in the solid state, but also in the form of an alloy. This is explained by the fact that the molecular structure of polymers is maintained at temperatures higher than their melting points. The rheological properties of polymer mixtures depend on the morphology and compatibility of the system, as well as the chemical nature and amount of additives.

2 Experimental Part

Considering all this, we propose to synthesize ion-exchange materials with a more simplified technology, i.e. on the basis of a ready-made polymer carcass. A polymer mixture based on polystyrene household waste, was used as the polymer carcass. Optimal parameters of thermomechanical action on different polymer mixtures were determined and the mechanism of mechanochemical modification of mixtures was proposed.In order to improve the complex properties of waste polymer, functionalize, strengthen the polymer carcass and further sulfonate, the waste polystyrene sample was modified mechanically and chemically in a laboratory capillary viscometer (IIRT-1) with polyvinyl chloride. Polyvinyl chloride is a polymer with the active functional group – *Cl* [1–3]. Thermomechanical action on the WPS:PVC mixture was carried out in order to obtain graft and block copolymers. The addition of a polymer containing a functional group aims to increase the impact viscosity of the obtained copolymers. As a result of the increase in the impact viscosity of polystyrene and the weak structuring with the addition of the functional group, it was expected to increase the mechanical resistance of sulfocationites obtained on the basis of this polymer mixture and improve ion exchange rates.

In the first stage of the research, some properties of polystyrene waste were determined – solubility, swelling coefficient, molecular weight, as well as IR spectrum of the samples.In addition, the rheological properties of the polymer waste and the flow rate of the polymer alloy were studied for comparison.

Depending on the chemical nature of the raw material used in the capillary viscometer, WPS:PVC polymer compositions have been developed under different conditions (i.e. shear strain and shear rate). Mechanical mixtures of the above-mentioned polymers were prepared, and then mechanochemical modification of this mixture was carried out. Mixtures based on waste polystyrene: polyvinyl chloride in different proportions were prepared (%, mass): 99:1, 98:2, 97:3, 96:4, 95:5. Mechanochemical modification was carried out by extrusion under the following conditions:

- temperature – 150–160 °C;
- time – 5 min.

Under these conditions, the polystyrene sample and all modified mixtures have a stable flow. The samples are stored in a desiccator for 24 h after crushing to a size of 2–3 mm, then which they are dried at 75 °C for 2–3 h.

3 Results and Discussion

One of the important factors during sulfonation is the homogeneity of the polymer body. It is known that modification of polymers leads to the formation of a heterogeneous structure, which complicates the process of sulfation and the process does not go smoothly. For this reason, in order to confirm the homogeneity of the system obtained by mechanochemical modification, the rheological properties of the modified polymer composition WPS:PVC were studied. The effective viscosity was calculated under constant displacement stress ($\tau = 5{,}073 \cdot 10^5$ Pa). The results of the obtained effective viscosity (η_e) are given in Table 1 and shown in Fig. 1.

Table 1. Rheological parameters of WPS:PVC polymer composition alloy (t = 150 °C)

Composition of composite material, % mass		Q, cm^3/sec	lgτ, Pa	lgγ, se^{-1}	lgη_e, Pa · se	AFR, g/10 min
WPS	PVC					
99	1	$4{,}06 \cdot 10^{-10}$	5,073	$2{,}75 \cdot 10^2$	2,323	0,830
98	2	$9{,}60 \cdot 10^{-10}$	5,073	$3{,}12 \cdot 10^2$	1,952	1,311
97	3	$9{,}76 \cdot 10^{-10}$	5,073	$3{,}13 \cdot 10^2$	1.943	1,910
96	4	$14{,}11 \cdot 10^{-10}$	5,073	$3{,}29 \cdot 10^2$	1,783	2,438
95	5	$21{,}36 \cdot 10^{-10}$	5,073	$3{,}47 \cdot 10^2$	1,406	2,825

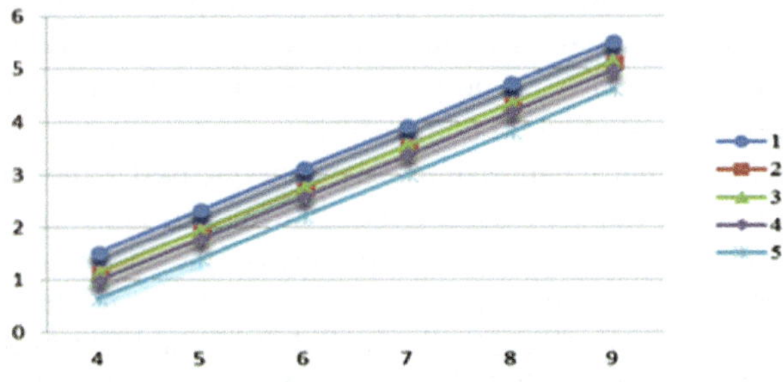

Fig. 1. Dependence of the displacement rate of the WPS:PVC polymer composition on the displacement stress, (T = 150 °C): The amount of PVC in the composition of 1, 2, 3, 4, 5% by weight of 1, 2, 3, 4, 5 curves, respectively

The data in the table show that the amount of modifier in the polymer mixture increases from 1% to 5%, the effective viscosity of the mixture decreases and the consumption of exudate and the rate of displacement of the polymer alloy increase. Therefore, the effective viscosity of the alloy at the same displacement stress varies depending on the amount of modifier added to the polymer mixture i.e. polyvinyl chloride.

The homogeneity of polymer compositions obtained in different proportions is confirmed by the non-Newtonian nature of the flow of polymer alloys (Fig. 1). The linear nature of the displacement velocity dependence on the displacement voltage and the monotonous change in the effective viscosity depending on the amount of modifier added to the modified composition indicate that all samples are homogeneous.

Thus, the obtained results show that when up to 5% mass of PVC is added to the polystyrene waste, the components blend well with each other and a homogeneous system is obtained in all mixtures.

Modified systems for the production of sulfocationites were sulfonated with concentrated sulfuric acid.

To make the sulfation process more efficient, a modified WPS:PVC polymer mixture is first placed in a conical flask and solvent in ratio 1:13 is added. Swelling of the copolymer in the solvent occurs within 20–30 min.

The excess solvent is then emptied and concentrated sulfuric acid (93,6%) is added to the swollen copolymer granules. Copolymer:sulfuric acid ratio = 1:8. Anhydrous $AlCl_3$ used as a catalyst in the sulfation reaction is taken as 1% (mass) for the copolymer [4].

To determine the optimal conditions of the sulfation process, this process was carried out over a wide temperature and time range:

- temperature: $30 \div 80$ °C;
- time: $2 \div 8$ h

Due to the high value of indicators of static and dynamic exchange capacity of sulfocationites obtained at a temperature of 40° IRSC and a sulfonation time of 4 h, we can take these parameters as optimal conditions.

At the end of the sulfonation reaction, the sulfonated sample is transferred to a glass filter and washed with a solution of sulfuric acid, the concentration of which gradually decreases from 75% to 5%. Finally, the sulfocationite granules washed with different concentrations of acids are washed with distilled water until neutral reaction and kept in water for 24 h. The sulfocationites washed in this way are dried at room temperature and then in a vacuum oven at 40 °C until constant weight.

In order to confirm the occurrence of the sulfation reaction, IR spectrum of sulfocationites obtained on the basis of WPS:PVC mechanochemically modified polymer composition were taken.

In the IR spectrum of sulfocationite obtained on the basis of WPS:PVC polymer composition, specific to sulfogroups $\nu = 460\text{–}470\,\text{cm}^{-1}$ (arylsulfides), $\nu = 1045\text{–}1060\,\text{cm}^{-1}$ (alkylsulfooxy), $\nu = 1180\text{–}1210\,\text{cm}^{-1}$ (sulfogroups SO_4^-), $\nu = 1350\text{–}1400\,\text{cm}^{-1}$ (alkylsulfones) have been identified and this proves the formation of a new polymer, i.e. sulfocationite.

One of the main properties of ion-exchange materials is ion exchange capacity, which is characterized by static and dynamic exchange capacity (SEC and DEC). This value is determined by the absorption of the ionite mass in the gram-equivalent amount of ions. Static exchange capacity is the capacity of ion exchange under static conditions, and the capacity of dynamic exchange is the exchange capacity when ionites “slipp”

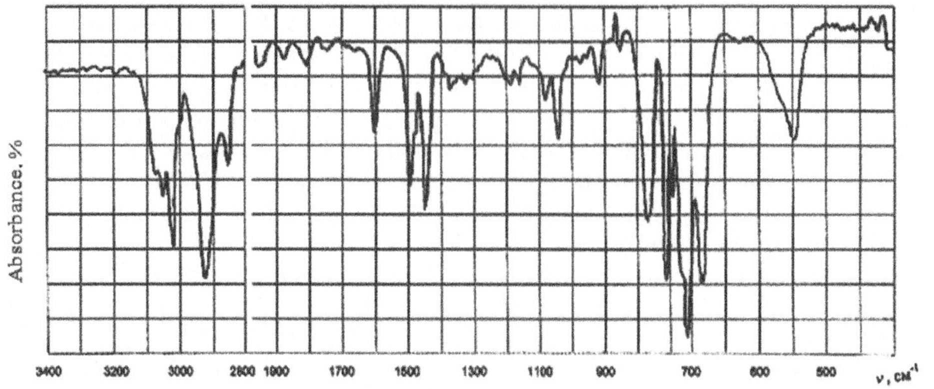

Fig. 2. IR spectrum of sulfocationite obtained on the basis of WPS:PVC mixture

in the filtrate under filter conditions. Therefore, high ion exchange capacity is a basic requirement for both cationites and anion exchange resins.

Ion exchange materials must also have high kinetic properties (high ion exchange rate) and resistance to various acids and alkalis. Ion exchange resins are water-insoluble and have low swelling properties.

Taking into account the above-mentioned, the main parameters of ion-exchange materials have been determined – static and dynamic exchange capacity, swelling rate, mechanical resistance and resistance to various aggressive environments. The named properties of the ion-exchange materials were tested. Table 2 shows the main indicators of synthesized ion exchangers [5, 6].

Table 2. The main parameters of sulfocationites obtained on the basis of WPS:PVC polymer composition.

№	Sulfocationite based on polymer body, (%) mass		Swelling coefficient in water	SEC based on NaOH, mg-equi/g	DEC based on $CaCl_2$, mg-equi/g	Mechanical durability after 10 h of shaking,%
	WPS	PVC				
1	99	1	1,04	5,9	0,59	90
2	98	2	0,91	6,2	0,69	90
3	97	3	0,75	5,6	0,62	90
4	96	4	0,57	5,7	0,61	90
5	95	5	0,49	6,7	0,79	90

We consider a problem of modeling of relationship between input and output variables of 15 process. Input variables are: the amount of polystrene waste (WPS) mass fraction (X_1); the amount of (PVC) mass fraction (X_2); sulfation time. Hour (X_3); sulfation temperature ^{0}C (X_4). Output variables are NaOH according SDT. mg-equi/g (Y_1);

$CaCl_2$ according DDT. mg-equi/g (Y_2). The experimental data on the values of the variables are given in Table 3.

Table 3. Dependence of ion exchange indicators of sulfocationites on the amount of primary components, sulfonation time and temperature

№	X_1	X_2	X_3	X_4	Y_1	Y_2
1	99	1	2	20	5.3	0.52
2	99	1	4	40	5.9	0.59
3	99	1	6	60	5.8	0.53
4	98	2	2	20	5.4	0.51
5	98	2	4	40	6.2	0.69
6	98	2	6	60	5.9	0.58
7	97	3	2	20	5.1	0.49
8	97	3	4	40	5.6	0.62
9	97	3	6	60	4.9	0.46
10	96	4	2	20	5.1	0.48
11	96	4	4	40	5.7	0.61
12	96	4	6	60	4.9	0.47
13	95	5	2	20	5.7	0.57
14	95	5	4	40	6.7	0.79
15	95	5	6	60	5.6	0.57

Due to complexity of the considered relationship, we intent to contruct a model in the form of fuzzy If-Then rules. The rules are derived by means of Fuzzy C-means method-based clustering of the data in Table 3. The derived rules are shown below:

IF X_1 is *medium* and X_2 is *medium* and X_3 is *high* and X_4 is *high* THEN Y_1 is *low* and Y_2 is *low*.

IF X_1 is *low* and X_2 is *low* and X_3 is *low* and X_4 is *low* THEN and Y_1 is *low* Y_2 is *low*.

Constructed by using the If-Then rules, the surface of relationship between input variables X_1, X_2 and output variable Y_1 is given in Fig. 3.

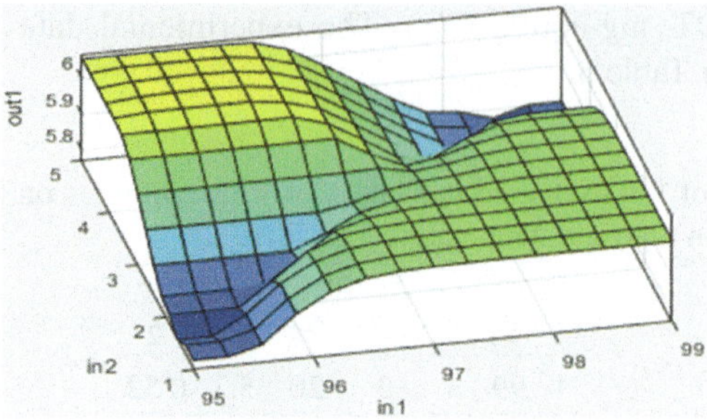

Fig. 3. A graph of relationship between X_1, X_2 and Y_1

4 Conclusions

1. Polystyrene waste was mechanochemically modified with polyvinyl chloride. Optimal conditions of the modification process (temperature, pressure, time) were determined. The initial components for the preparation of the compositions were taken in several proportions.
2. The rheological properties of WPS:PVC polymer compositions prepared in different conditions were studied. The results confirmed the homogeneity of the modified systems. The melting index of polymer alloys was determined.
3. Ion exchange materials were obtained by sulfonation on the basis of mechanochemically modified polymer compositions. Optimal conditions of sulfonation process (time and temperature) have been determined.
4. IR spectrum of primary components, modified systems and sulfocationites were extracted and the course of mechanochemical modification, as well as sulfonation reaction was confirmed.
5. The main parameters of the synthesized sulfocationites, such as static and dynamic exchange capacity, mechanical resistance, swelling coefficient were studied. It was noted that the above-mentioned indicators of sulfocationites obtained on the basis of WPS:PVC mechanochemically modified systems are satisfactory.
6. We offer the use of synthesized ion exchange resins in industrial water desalination.

References

1. Aliyeva, G., Safarov, J., Rahimova, F., Khalilov, V.: Mechano-chemical modification of wastes of Polystyrene with Polyvinyl Chloride. In: THERMAM 2019 Conference, Turkey, Izmir, Cheshme, p. 61 (2019)
2. Rahimova, F.R.: Recycling of waste polystyrene. In: Conference 2019, pp. 33–35. Baku Higher Oil School, Baku (2019)
3. Aliyeva, G.A., Rahimova F.R.: Research of the rheological properties of waste polystyrene. Actual Problems of Modern nature and Economic Sciences, pp. 203–205 (2019)
4. Amirov, F.A., Aliyeva, G.A., Abdullayeva, I.G.: Synthesis and properties of high molecular weight compounds. Training manual. Baku, 157p. (2019)

5. Aliyeva, G.A., Rahimova, F.R.: Synthesis of sulfocationites based on polymer waste. In: Conference 2019ç Reagent, Ufa, pp. 99–100 (2019)
6. Aliyeva, G.A., Rahimova, F.R.: Synthesis of sulfocationites based on waste polystyrene. Educ. Sci. Russ. Abroad **6**(70), 24–29 (2020)

The Impacts of Genetic Studying on the Cost Analysis Based on Labour Market

Hüseyin Özdeşer(✉) and Günay Kuşaf

Near East University, Nicosia, North Cyprus
huseyin.ozdeser@neu.edu.tr, kusafgunay02@gmail.com

Abstract. In the study the genetical studying methods are mentioned about preventing cancer diseases at the early stages. Genetical studyings play an important role to lower the death rates of the people at early ages because of cancer as the cancer illness can be prevented to move to the advanced stages. This advantage is very important for labour market; where life – expectancy of labours can be increased. An increased life – expectancy for labours will improve the labour market in the economy which has positive contribution in economic growth. In economics, labour as a factor of production has potential to effect positively the other factors of production in the economy (capital, and natural resources). Improvements in labour market is also very important for technological developments, where technological developments are essential for the economy to reach its potential level of output (GDP). All the countries should take genetical studyings into consideration seriously and arrange the required regulations in their countries urgently. Genetical studyings can be expensive at the beginning; but in long term; it will create a lot of cost benefits for economies.

Keywords: Genetic · Genetical studying · Labour · Labour market · Efficiency · Potential output

1 Introduction

The genetic examinations can be classified as a kind of medical analysis which are playing a significant role in observation of any type of differentiations that taken place in chromosomes, genes, proteins or their functions [1]. There are different types of genetic testing that can be applied in various periods of lifetime such as; newborn screening, diagnostic testing, carrier testing, prenatal testing, preimplantation testing, predictive and presymptomatic testing, and forensic testing. The newborn screening are applied just after the born is completed for the prevention of some type of disorders which can be treatable such as phenylketonuria (PKU). For the identification of any genetic or chromosomal disorder which are related with the physical sign and manifestations, diagnostic testing is implemented and can be applied in any time of life. The carrier testing is performed when the recessive disorder is carried, in other Word, one gene which is mutated is carried and if two copies of that gene is carried will lead to genetic disorder. This test should be applied for the people if they have family history about this health problem or if one of the couples carry the recessive gene mutation [2].

© The Author(s), under exclusive license to Springer Nature Switzerland AG 2021
R. A. Aliev et al. (Eds.): WCIS 2020, AISC 1323, pp. 496–503, 2021.
https://doi.org/10.1007/978-3-030-68004-6_65

The genetic studyings in a country will play an important role for increasing the life expectancy of the people in other words the people who contribute to GDP in the country which is very important to increase the contribution of labour into the economy. The higher life expectancy means longer working life of the employees in an economy, where this is essential for achieving the economic growth.

2 Cost-Effectiveness Analysis and Hereditary Non-polyposis Colorectal Cancer

The Lynch syndrome (LS) also called as hereditary non-polyposis colorectal cancer (HNPCC). The inheritance of HNPCC is accepted as an autosomal dominant, and the reason of this disease is the mutations which takes place in genes of DNA mismatch repair (MMR) system. The mutL homolog 1 (MLH1), mutS homolog 2 (MSH2), mutS homolog 6 (MSH6), and PMS1 homolog 2 (PMS2) are the mismatch repair genes which are playing role in the reparation of errors during DNA replication [3]. The risk of cancer to be existing in other places such as; stomach, small intestine, liver, gallblader ducts, urinary tract, brain, and skin is increased by the people who have HNPCC. Furthermore, the risk of being ovarian and endometrial cancer is observed as in a level of rising in the people who are female and suffering from HNPCC [4]. According to the HNPCC patients who are living in Europe and North America, their 2.2% are identified as colorectal cancer (CRC); and this lead to determination of colorectal cancer as hereditary CRC predisposing syndrome. Different type of genetic tests can be preferred for the diagnosis of HNPCC, such as DNA sequencing for MMR genes, tumor testing by applying microsatellite instability (MSI), and immunohistochemical analysis (IHC). The genetic testing for HNPCC is committed proposal for applying the patients who are recently analyzed as CRC by The Evaluation of Genomic Application in Practice and Prevention (EGAPP) Working Group and it is supported by Centers for Disease Control and Prevention in the United States [3].

2.1 Four Different Methods on Hereditary Non-polyposis Colorectal Cancer

By the EGAPP and EWG were commissioned demonstration commentaries of screening methods of HNPCC and suggestion for offering to screening the all CRC patients who recently identified irrespective of age or family history in January 2009. Applying of the mutation analysis for identification of mutation that is carried by which relatives were established as a question of primary search of EWG suggestion, and it has been justified in an economic way by national health care system. The second question of search has been related with analysing methods for identification of mutations which associated with HNPCC, and these methods could be considered as provided plausible rate for cost [5].

The four different methods for identification of HNPCC have been used for the recently diagnosed CRC patients who were less than fifty years old by both using no testing and targeted testing. The first two methods were started by applying IHC analysis to all patients who recently identified with CRC utilization antibodies to the MMR proteins procreated by four MMR genes (MLH1, MSH2, MSH6, and PMS2) which are related with HNPCC. If the stain is lacked for the protein, it is demonstrated as likely mutation for genes that are related. In method 1, people with absent MLH1 protein dying has been analyzed for the BRAF V600E mutation. If the test was resulted as positive, then it is evaluated as person has not HNPCC. When the individuals' MSH2, MSH6, or PMS2 protein staining is found absent, and if the MLH1 protein staining is absent also BRAF mutation is not presented, sequencing or rearrangement analysis of the MMR gene related with the absent protein is concluded. The same procedures in method 1 is applied to method 2, but BRAF mutation analysis is not tested in method 2. In method 3, microsatellite instability (MSI) is used as anterior test. The MSI analysis is different from IHC because which MMR gene is not functioned can not be predicted. If MSI results were concluded as high, sequencing or deletion analysis is implemented. The whole four MMR genes that are associated with HNPCC are recommended for performing as sequencing or rearrangement analysis to all patients in method 4 [5].

The principal quantity of effectiveness has been applied for discounted life years (LYs) to be protected. The amount of LYs lost to CRC is measured as the total of the products of the amount of deaths coming out with every ten year age intervals and the amount of discounted LYs for each interval resulted from United States life table for the median age in the interval and measured by the use of 3% discount rate. The count of deaths appeared within every period is the possibilities of being CRC within a dedicated ten year age period, the possibility of dispensation of phases at diagnosis, and the one subtraction the cognate survive possibilities by phase at diagnosis. In general the conclusion of quality-adjusted life years (QALYs) is committed for cost-effectiveness analysis. Utilization of QALYs for the cost-effectiveness analysis is noticed as an analysis of sensitiveness. The benefit of patients who continue to survive with CRC that has not been in the stage of metastasis, is decreased by 0.1 for two years. Any negative effects have not been observed yet to the quality of life by the analysis of mutations that are related with MMR gene has been applied between relatives. According to the suggestion of United States Multi–Society Task Force on Colorectal Cancer, the kin of the probands who with Lynch Syndrome would be applied colonoscopy every two years between the age 20 and 79 [5].

3 Economics and Labour

Labour as onę of the factors of production has a very important contribution to economics. In economics, there are mainly three factors of production, which are labour, capital and natural resources, where among these factors of production labour is the most important one. Labour as a factor of production has impacts on the efficiency of the other factors of production. For example capital means; the buildings, factories, universities, machines, and etc.... where all can be constracted and put into action by the labours. To build a factory; there is a need of a civil engineer, a mechanical engineer or an architect. At the same time all the machines which are used for the production of the capital goods are produced by human power and all the natural resources which are also put into the service for economic production is the result of the human power. The labours work in the undergrounds to bring out many natural resources such as gold, iron, steel, coal, petrol and many others. As can be seen; labour as a production factor also plays an important role for the other factors of production to be in the service of economy. Technology is also produced and developed by the brain power of the labours. The labours who work in Electric and Electronic Departments or Mechanical Engineering Departments in the universities are the ones who produce and develop technology. As it's known technological developments in a country is one of the most important elements which can improve the production capacity of a country, and pushes the economies to reach their potential level of output; where all economies aim to produce at the potential level of output to minimize the unemployment in the economy. Today; all the countries which are in the category of developed countries; they are the ones which apply the latest technology and also the countries which have the capability of transferring (exporting) technology can move much easier to economic and political leadership in the World.

The countries for achieving the economic development and growth also needs high quality educated economists, businessmen, accountants, sociologist and all the other professionals in social sciences where all of them are the labours to be employed in the economies. At the same, time there is a need of high quality professors employed as teaching labours in the universities to have a chance of having high quality educated social scientists. The labour in the economy of production can be liked to the energy used for any machine; as long as the energy used for a machine is in high quality and as long as there is the continuity of high quality energy, then the machine is going to have a possibility working long in a high efficiency.

All the sectors to be employed with high quality educated employees is essential for an economy. The high quality of work can only be given out by high quality educated employees. In a country, it's not possible to realize high quality education in the universities as long as the students study in the universities do not have a high quality educational background built in the primary and high schools. This means; it is an important need for the high quality teachers to be employed in the schools where students study before they go to universities. The primary schools and high schools in a country are the roots of high quality education, and also roots of developed and technology producing countries; which has a very significant role for the welfare of a country. As can be understood; it is still human (labour) which is the guarantee of a developed and welfared country.

4 Health of Human

If labour is the most important factor of production in the economy, then the point is; how it is possible to benefit from high quality labours longer and more efficiently. The answer is simple, to keep them healthy as much as possible, and as long as possible. For realizing this, there is no any doubt that good doctors (labours) and hospital (good equipped) are needed. But; the more important thing here is, high quality genetical studyings are required. High quality educated genetic specialists will be heavily required to be able to keep the workers (labours) healthy and benefit from them longer. By genetic specialists; it is going to be possible to find cures for many kind of illnesses which have a genetical background. If many kind of preventable illnesses can be prevented at early stages by genetical studyings, then it is also possible to increase the continuity of high quality labours in the economy much longer. For example; it's a big waste for a good engineer or a professor to be lost at early ages due to a preventable genetical illnesses. In all the countries; this issue should be taken into consideration very importantly; and special hospitals should be constracted for genetical studyings and researches.

Labour cost is one of the highest shares in the cost of many kind of productions. If the amount of employees can be kept at high rates in an economy, then this means a higher supply of employees in the economy; which is the guarantee of lower wages in the economy. Lower wages in the economy means the lower cost of production, which is essential to keep the economy moving to producing at its potential level of output. Reaching potential output is one of the main target for all the economies to improve the welfare of a country. Because; if the economies can not reach to their potential level of outputs; then, this means in the economy; there are some sources of production which have been wasted; because potential output in an economy means; the maximum level of output that an economy can produce when all the sources of production in the economy is fully utilized with the available technology.

5 Employment Market Analysis

Supply and demand for labour will have effects in the operations of economies at local, national and international levels [7]. As in all the markets what determines the price (wage) of the employee is the supply and demand conditions. As an economic rule; higher demand for labours increase the wages and lower demand lower the wages of the labours. On the other hand, when the supply of the labours increase; then wage rates fall, and when the supply of the employees fall, the wage rates rise. The process in the labour market works as in the case of good and service market if labour as a factor of production is accepted as a product.

In labour market, the aim of all the countries should be increasing the supply of labour more than demand. Normally; all the countries should have an economic growth higher than population growth, which means the demand rise for labour in economics will be in continuity. According to the estimations of the World Bank within next ten years more than 600 million jobs will be needed to be created for preventing the the unemployment rate to rise [6]. Saving a lot of expenses that government could spend for its people for many preventable illnesses based on genetical reasons which may cause the employees to be out of working force and cause the governments to be face to face with unemployment payments is another important advantage of illness preventing genetical studyings in a country. If it is calculated that the health expenses and unemployment payments are two important budget shares of the governments; then, the importance of genetical studyings in a country can be figured out easier.

Knowledge economics is a rising trend for all the in the world. Knowledge economics has important impacts on the economic development and on the economic growth. Beside this, knowledge economics has important reflections on the labour market as well. These reflections can be given as, entrepreneurship and firm, working people, employment, new working kind, the rising trends in skills, education for labour market, the importance of policies such as brain migration and labour mobility, syndicalism, the possibilities of cooperation between countries, and the roles of the public and private sectors. No doubt, an economic structure which is based on knowledge will have impacts on the labour markets. According to the studyings of OECD, its mentioned that by the knowledge economics the importance of human capital is rising. This means, the demand for high qualified employees keeps rising. As long as there are developments in the use of technologies; then, demand for high qualified employees keep rising where demand for the low qualified employees fall. In some cases; because of the technological changes; it may cause for some of the high qualified employees to loose their jobs (Ahmet Ali Kelleci).

If it's accepted that the technological improvements in the world will never end, that means the need for improvement in knowledge as well will never stop. So, there will always be a need for training of the employees to keep themselves to be challenging with changing and improvements in technologies. At the same time, it's very important of the employees to be trained by the high qualified trainers.

By the knowledge economics, learning has a critical issue and also it's very important for increasing the competiveness of the firms (Ahmet Ali Kelleci). In the perfectly competitive markets; the competition will force the firms to be always dynamic in satisfying the needs and wants of the consumers. Competition in the economy requires perfection and perfection means technological improvements. Competitive markets will encourage the firms always keeping their employees to be in a continuous training process. Also, the firms should spend a certain amount of money for research and development activities to improve the skills of their staff. The requirements and needs of the consumers never stay the same. The reason for this is, the technological developments cause the new demands to be created. New demands in the competitive structure means; new markets. Research and development activities of the firms no doubtedly will shape the demand of the consumers continuously.

Entrepreneurship in a competitive economy is a very significant factor to achieve the economic growth. An effective entrepreneur is the person who puts the factors of production into the service of the economy. The effective labour market as mentioned before plays a very developer role about the efficiency of the other factors of production which means labour market has significantly contribution to the success of the entrepreneurs.

No doubt, advanced economies are in need of high skilled labours, and high skilled labours are needed for potential use of natural resources and capital, where the countries can easily reach to optimum level of output or natural rate of unemployment rate. In any advanced economy, the entrepreneurs can not be successful in attempting any working area if the working area they are attempting to work do not include an efficient labour market which is created by the knowledge economics.

Creativity is the main issue in the competitive markets to boom the economic production. Creativity can affect the production level in an economy as long as it is supported by technological improvements which is realized by humans, in other words labours in the economy. High skilled employees in an economy is the guarantee of dynamic and efficient working economy. Due to this reality; the education, and development of the labour market should be indispensable for all the countries.

6 Conclusion

Knowledge economics is a rising trend for all the in the world. Knowledge economics has important impacts on the economic development and on the economic growth. Beside this, knowledge economics has important reflections on the labour market as well. These reflections can be given as, entrepreneurship and firm, working people, employment, new working kind, the rising trends in skills, education for labour market, the importance of policies such as brain migration and labour mobility, syndicalism, the possibilities of cooperation between countries, and the roles of the public and private sectors. No doubt, an economic structure which is based on knowledge will have impacts on the labour markets. According to the studyings of OECD, its mentioned that by the knowledge economics the importance of human capital is rising. This means, the demand for high qualified employees keeps rising. As long as there are developments in the use of technologies; then, demand for high qualified employees keep rising where demand for the low qualified employees fall. In some cases; because of the technological changes; it may cause for some of the high qualified employees to loose their jobs [8].

If it's accepted that the technological improvements in the world will never end, that means the need for improvement in knowledge as well will never stop. So, there will always be a need for training of the employees to keep themselves to be challenging with changing and improvements in technologies. At the same time, it's very important of the employees to be trained by the high qualified trainers.

By the knowledge economics, learning has a critical issue and also it's very important for increasing the competiveness of the firms [8]. In the perfectly competitive markets; the competition will force the firms to be always dynamic in satisfying the needs and wants of the consumers. Competition in the economy requires perfection and perfection means technological improvements. Competitive markets will encourage the firms always keeping their employees to be in a continuous training process. Also, the firms should spend a certain amount of money for research and development activities to improve the skills of their staff. The requirements and needs of the consumers never stay the same. The reason for this is, the technological developments cause the new demands to be created. New demands in the competitive structure means; new markets. Research and development activities of the firms no doubtly will shape the demand of the consumers continuously.

Entrepreneurship in a competitive economy is a very significant factor to achieve the economic growth. An effective entrepreneur is the person who puts the factors of production into the service of the economy. The effective labour market as mentioned before plays a very developer role about the efficiency of the other factors of production which means labour market has significantly contribution to the success of the entrepreneurs.

No doubt, advanced economies are in need of high skilled labours, and high skilled labours are needed for potential use of natural resources and capital, where the countries can easily reach to optimum level of output or natural rate of unemployment rate. In any advanced economy, the entrepreneurs can not be successful in attempting any working area if the working area they are attempting to work do not include an efficient labour market which is created by the knowledge economics.

Creativity is the main issue in the competitive markets to boom the economic production. Creativity can affect the production level in an economy as long as it is supported by technological improvements which is realized by humans, in other words labours in the economy. High skilled employees in an economy is the guarantee of dynamic and efficient working economy. Due to this reality; the education, and development of the labour market should be indispensable for all the countries.

There are mainly three factors of production in the economies; but the most important factor of production is labour as it also affects the other two factors of production which are capital and natural resources. It's the physical and brain power that builds the buildings, factories, machines, and all the other capital goods; also the natural resources such as petrol, iron, steal, and etc. are presented to the service of production as well by the physical and brain power of the labour. Improvements in labour market has direct impacts on the improvements of the other factors of production as well; then the production efficiency in the economy can be increased as a whole. Improvements in the market of factors of production in a country is essential for the economy to be carried to its potential level of output, where the unemployment rate could be minimized. An increased life-expectancy in the labour market due to the genetical studyings will increase the contribution of labour market to the economic growth.

The genetical studyings in the countries are so important for improving the healthy life-expectancy. The countries should establish institutions which will be responsible for genetical researches. The genetical studyings and researches will provide the countries to save a lot of money which would be paid for the health expenses for preventable illnesses and unemployment payments due to the early retirements related to health reasons.

References

1. Genetics Home Reference: What is genetic testing? Genetics Home Reference (2018). https://ghr.nlm.nih.gov/primer/testing/genetictesting
2. Genetics Home Reference: What are the types of genetic tests? Genetics Home Reference (2019). https://ghr.nlm.nih.gov/primer/testing/uses
3. Chen, Y.-E., Kao, S.-S., Chung, R.-H.: Cost-effectiveness analysis of different genetic testing strategies for lynch syndrome in Taiwan. PLoS ONE, **11**(8), 160599 (2016). https://doi.org/10.1371/journal.pone.0160599
4. Genetics Home Reference: Lynch syndrome. Genetics Home Reference (2014). https://ghr.nlm.nih.gov/condition/lynch-syndrome#inheritance
5. Mvundura, M., Grosse, S.D., Hampel, H., Palomaki, G.E.: The cost-effectiveness of genetic testing strategies for Lynch syndrome among newly diagnosed patients with colorectal cancer. (2010). https://doi.org/10.1097/GIM.0b013e3181cd666c
6. Labour Markets Overview. https://www.worldbank.org/en/topic/labormarkets
7. Labour markets and Employment. https://www.lse.ac.uk/business-and-consultancy/consulting/research/labour-markets
8. Kelleci, M.A.: Bilgi Ekonomisi, İşgücü Piyasasının Temel Aktörleri ve Eşitsizlik, Eğilimler, Roller, Fırsatlar ve Riskler.T.C. Başbakanlık, Devlet Planlama Teşkilatı (2003)

Drive Control Algorithm in Conditions of Insufficient Information

F. O. Qosimov[1(✉)], E. H. Nematov[2], A. H. Ahmedov[2], and M. A. Ismailov[3]

[1] Tashkent Chemical Technological Institute, 32, Navoi Street, 100011 Tashkent, Uzbekistan
fqosimov@gmail.com

[2] Tashkent State Technical University named after Islam Karimov, 2, University Street, 100095 Tashkent, Uzbekistan

[3] Scientific Innovation Center for Information and Communication Technologies (SIC ICT), Tashkent University of Information Technologies, 17A Boz-2, 100125 Tashkent, Republic of Uzbekistan

Abstract. This work is devoted to the construction of a screw drive control system in the absence of information. A hybrid control structure is developed as an element of an intelligent control system and a mathematical model of the actuator is formed as a complement to this structure. The screw actuator is considered as a control object. The issue of forming a more accurate mathematical model is considered. The structure and algorithm of an intelligent control system based on a mathematical model of an actuator are proposed.

Keywords: Screw drive · Insufficient information · Intelligent control system · Hybrid control algorithm · Simulink model · Torsion torque · Extinguishing power · The force spreads along the axis

1 Introduction

The problem of process control in conditions of lack of information is found in many practical issues. Control devices do not always have the ability to configure feedback. The reason for this is the impossibility of measuring the parameters of the object of interest to us, the impossibility of installing sensors under certain conditions, the presence of unpredictable stochastic effects. In addition, built-in sensors sometimes provide erroneous information depending on the situation. For example, the installation of a sensor to assess the state of the mechanism in screw drives of robotics causes an increase in the size of the mechanism. In many cases, such mechanisms are governed by the principle of open control.

In processes regulated by the principle of open control, it is natural that information about the state of the object will be insufficient [1–3]. However, to do the management properly we need complete information. In such situations, a hybrid control structure is generally used [4–6]. Hybrid control systems are one of the elements of intelligent control systems (Fig. 1).

© The Author(s), under exclusive license to Springer Nature Switzerland AG 2021
R. A. Aliev et al. (Eds.): WCIS 2020, AISC 1323, pp. 504–511, 2021.
https://doi.org/10.1007/978-3-030-68004-6_66

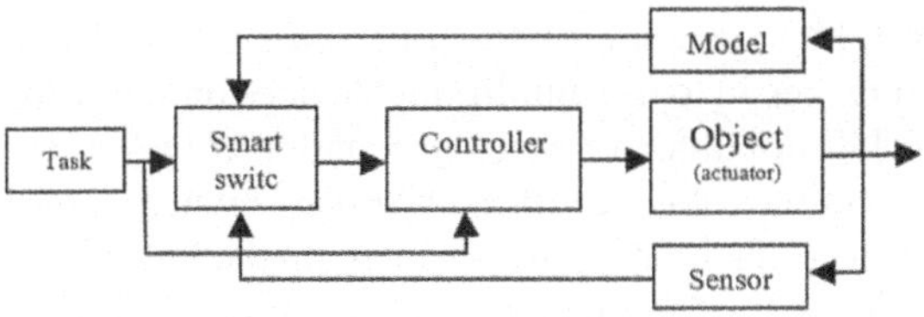

Fig. 1. Block diagram of a hybrid control system.

In this case, if information about the state of the object is insufficient, the control system relies on the model of the object and fills in the empty information with the results of the model.

This requires that the adequacy of the object model should be maximized. To this end, this article discusses the design of a more detailed screw drive model.

2 Formulation of the Problem

Although screw electric drives used in robotics and manipulators are widely used in practical work, their dynamic characteristics usually are not studied in detail. In many cases, models that represent linear characteristics are used. Ignoring some forces acting on the elements allows us to simplify the model. However, the use of simplified models leads to structural errors during operation. In fact, the dynamic characteristics of helical gears are not always linear. The presence of an electric motor in the design of such a drive also indicates that the system is actually non-linear.

A more accurate assessment of the state of actuators requires a more in-depth study of their properties. In this case, it is necessary to consider the actuator as a system and analyze each of its elements as a subsystem.

3 Solution to the Problem

Screw electric drives include all screw electric drives controlled by electric drives. As electric drives, AC and AC motors are mainly used. Changing the state of the drive is controlled by such parameters as the number of engine revolutions, speed, torque. When analyzing these types of actuators as a system, they can be divided into two main components: electrical and mechanical (Fig. 2).

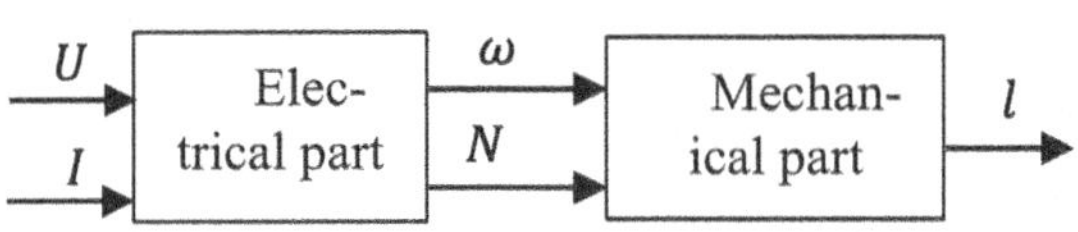

Fig. 2. Block diagram of the electric actuator.

The dynamics of the electrical part is written based on the laws of electrical engineering [7, 8].

Recently, to control the speed of rotation of electric motors by changing the voltage applied to the armature, began to use amplifying devices on semiconductor triodes or on silicon elements [9]. It is possible to derive the equation of the DC motor of independent excitation. If we neglect the influence of the transverse reaction of the armature (in the reversible electric motors, the brushes are installed on the line of geometric neutral, and the longitudinal component of the reaction of the armature is absent), take the value of the inductance of the armature winding as constant, neglect the influence of eddy currents and hysteresis, i.e. if we consider the magnetic flux $\Phi = \text{const}$, then the initial expressions for deriving the equations of engine dynamics can be written as:

$$\begin{cases} U = E + ir + L\frac{di}{dt}; \\ J\frac{d\omega}{dt} = M_{д} - M_c'; \\ E = c\Phi\omega; \\ M_d = c\Phi i; \end{cases} \tag{1}$$

where U is the voltage of the network to which the motor is connected; L and r are the inductance and resistance of the anchor circuit; I is the armature current; c is a constant coefficient; M_d is the moment developed by the engine; M_c' is the static moment of resistance on the motor shaft; J is the total moment of inertia; ω- rotation speed of the anchor.

Equations (1) can be written as:

$$\begin{cases} U = c\Phi\omega + ir + L\frac{di}{dt}; \\ J\frac{d\omega}{dt} = c\Phi i - M_c'; \end{cases} \tag{2}$$

From the second Eq. (2) we define

$$i = \frac{J}{c\Phi} * \frac{d\omega}{dt} + \frac{M_c'}{c\Phi} = \frac{J}{c\Phi} * \frac{d\omega}{dt} + i_c \tag{3}$$

where I_c is the armature current corresponding to the moment of resistance M_c'.

Assuming $M_c' = const$ ($i_c = const$), and differentiate with respect to t Eq. (3):

$$\frac{di}{dt} = \frac{J}{c\Phi} * \frac{d^2\omega}{dt^2} \tag{4}$$

Substituting expressions (3) and (4) in the first Eq. (2), we get

$$U = c\Phi\omega + \frac{Jr}{c\Phi} * \frac{d\omega}{dt} + i_c r + \frac{JL}{c\Phi} * \frac{d^2\omega}{dt^2} \tag{5}$$

Having divided all the equations into $c\Phi$, we can write further

$$\frac{JL}{(c\Phi)^2} * \frac{d^2\omega}{dt^2} + \frac{Jr}{(c\Phi)^2} * \frac{d\omega}{dt} + \omega = \frac{U}{c\Phi} - \frac{i_c r}{c\Phi} \tag{6}$$

where $\frac{Jr}{(c\Phi)^2} = T_M$ is the electromechanical constant,

$$\frac{JL}{(c\Phi)^2} * \frac{r}{r} + \frac{Jr}{(c\Phi)^2} * \frac{L}{r} = T_M T_e \tag{7}$$

where $\frac{L}{r} = T_e$ - electromagnetic constant; $\frac{U}{c\Phi} = \omega_0$ - ideal idle speed; $\frac{i_c r}{c\Phi} = \frac{M_c' r}{(c\Phi)^2} = \frac{M_c'}{\beta}$; $\beta = \frac{(c\Phi)^2}{r}$ is the rigidity of the mechanical characteristic.

There are three equivalent options for recording the right side Eq. (6), reflecting the influence of the load moment M_c'.

1. Considering that $\frac{i_c r}{c\Phi} = \Delta\omega_c$ is a static drop in speed, in accordance with Eq. (6) we obtain $\omega_0 - \Delta\omega_0 = \omega_c$ - speed, corresponding in mechanical characteristic to the static load moment M_c'. Then Eq. (6) is reduced to the form:

$$T_e T_{\text{M}} \frac{d^2\omega}{dt^2} + T_{\text{M}} \frac{d\omega}{dt} + \omega = \omega_0 - \Delta\omega_0 = \omega_c \tag{8}$$

2. Denoting $\frac{i_c r}{c\Phi} = \frac{M_c' r}{(c\Phi)^2} = \frac{M_c'}{\beta}$, we rewrite Eq. (6) in the form:

$$T_e T_{\text{M}} \frac{d^2\omega}{dt^2} + T_{\text{M}} \frac{d\omega}{dt} + \omega = \frac{1}{c\Phi} U - \frac{M_c'}{\beta} \tag{9}$$

3. Introducing the notation $\frac{M_c' r}{(c\Phi)^2} = \frac{i_c r}{c\Phi} = \frac{\Delta U_c}{c\Phi}$,

$$T_e T_{\text{M}} \frac{d^2\omega}{dt^2} + T_{\text{M}} \frac{d\omega}{dt} + \omega = \frac{1}{c\Phi} U - \frac{\Delta U_c}{c\Phi} \tag{10}$$

In Eqs. (9) and (10), the value $\frac{1}{c\Phi} = k$ is called the gain, counting the speed as the output value, and the main voltage U and the voltage ΔU_c proportional to the moment M_0' as the input value. Physically, the voltage ΔU_c is (in steady-state) the difference between the voltage of the network and the magnitude of the counter-electro stress force E:

$$\Delta U_c = U - E = c\Phi\omega_0 - c\Phi\omega_c \tag{11}$$

With an ideal idle, E = U. The lower the value of the steady-state speed ω_c, the higher the value of ΔU_c. If the output coordinate of the electric motor is the angle of rotation of the motor shaft α_d, the transfer function is written as

$$\text{W(s)} = \frac{\alpha_d(s)}{U(s)} = \frac{k}{\left(T_e T_{\text{M}} s^2 + T_{\text{M}} s + 1\right)s} \tag{12}$$

By differentiating the expression (12) over time, it can be written that the angular velocity of the motor depends on the input voltage.

$$\text{W(s)} = \frac{\omega_c(s)}{U(s)} = \frac{k}{T_e T_{\text{M}} s^2 + T_{\text{M}} s + 1} \tag{13}$$

The dynamic properties of an electric motor are determined by its gain coefficient k, as well as constant time coefficients T_e, T_{M}, $\omega(\tau)$ is the angular velocity of the shaft, k is the motor coefficient, $U(\tau)$ is the voltage applied to the motor. From expression (1), one can determine the rotation angle by integrating the angular velocity:

$$\varphi(\tau) = \int \omega(\tau) \quad (14)$$

We use the balance of forces in modeling the mechanical part of the system. According to a large-scale literature study in this area, the driving force is used to overcome the force propagating along the axis, the quenching force, and the inertia forces. The mathematical expression for this conclusion is as follows [10]:

$$F_D = F_{ov} + F_q + F_{in} \quad (15)$$

where F_D is the driving force; F_{ov} is the force propagating along the axis; F_q is a force that adversely affects system performance. This force is a function of the shape parameters and the friction coefficients in the system. F_{in} is the inertia force of the linear moving part.

The force propagating along the axis falls on the shaft of the screw. This means that the driving force is distributed along the length of the rod:

$$F_D = (C_s + C_t)l \quad (16)$$

where C_s is the stiffness coefficient of the threaded rod, C_t is the transmission coefficient of the screw system, l is the working length of the rod.

The extinguishing force in the system depends on the speed of the linear moving part.

$$F_q = \psi V \quad (17)$$

where ψ is the coefficient of friction, V is the linear velocity of the moving part.

Inertial force is the combined force of a linear moving part and the load attached to it. According to Newton's second law, this force can be written as follows.

$$F_{in} = a \sum m \quad (18)$$

where $\sum m$ is the sum of the masses of the moving part and the elements attached to it. a is the modulus of acceleration of the moving part.

The driving force expended to overcome all of the above forces is proportional to the angle of rotation supplied by the engine or reducer.

$$F_D = \varphi K C_s \quad (19)$$

where φ is the angle of rotation of the drive shaft, K is the transmission coefficient of the spherical screw drive.

Substituting the resulting expressions (16)–(19) into Eq. (15), we obtain the following general equation:

$$\varphi K C_s = (C_s + C_t)l + \psi V + a \sum m \quad (20)$$

Determining the constant coefficients in the resulting expression (20), Eq. (13) is written if the working length of the rod is written in relation to l:

$$\varphi KC_s = Cl + \psi \frac{dl}{dt} + M \frac{d^2 l}{dt^2} \tag{21}$$

where $C = C_s + C_t$; $M = \sum m$.

An integral mathematical model representing the dynamic characteristics of the actuator is represented by the following system of equations:

$$\begin{gathered} T_e T_{\mathrm{M}} \frac{d^2\omega(\tau)}{dt^2} + T_{\mathrm{M}} \frac{d\omega(\tau)}{dt} + \omega(\tau) = kU(t), \\ \varphi(\tau) = \int \omega(\tau), \\ Cl + \psi \frac{dl}{dt} + M \frac{d^2 l}{dt^2} = \varphi KC_s. \end{gathered} \tag{22}$$

From the generated differential equations of motion, a computer model is created in modern programming tools, in particular, in the Simulink (Fig. 3).

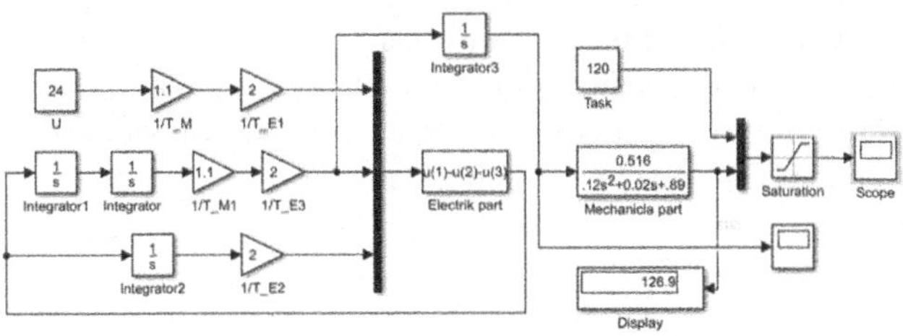

Fig. 3. Computer model of screw execution mechanism.

In the constructed computer model $U = 24\,\mathrm{V}, k = 1, T_e = 2, T_m = 1.1$ the following result was obtained when the values were given (Fig. 4).

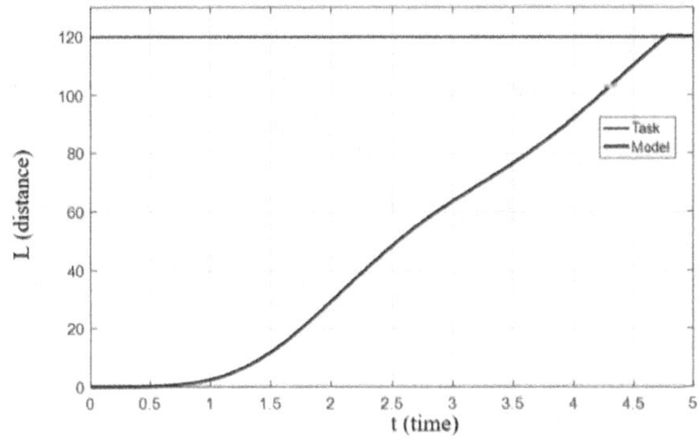

Fig. 4. Dynamics of screw execution mechanism

The results of this model show the difference between the two models compared to the ideal model constructed in the traditional way (Fig. 5). Due to this difference, hysteresis occurs in ideal hybrids.

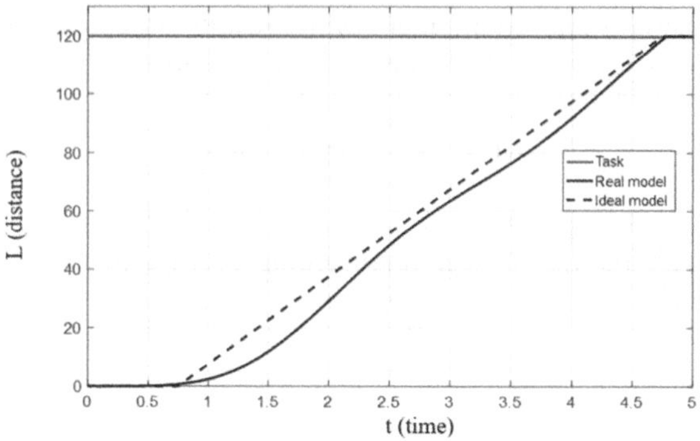

Fig. 5. Ideal and real model results

Based on the above results, the following universal algorithm is proposed to control the execution mechanism in the case of insufficient information (Fig. 6). The algorithm works in the following order: If the system has a sensor element that provides complete information about the state of the object, the adjuster evaluates the state of the object based on this information. If the system does not have a sensor, then the intelligent block receives information about the object from the model.

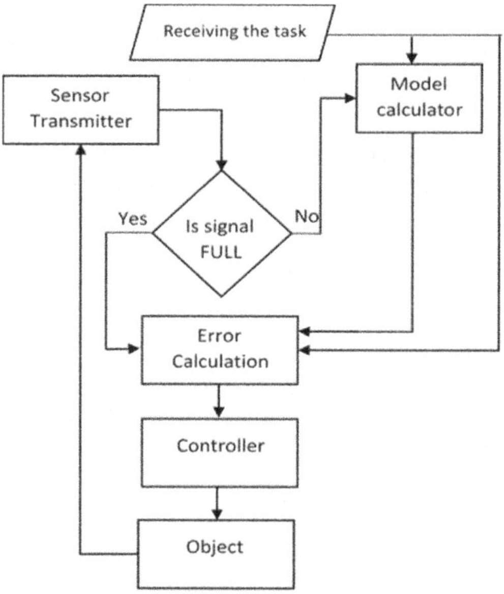

Fig. 6. An intelligent universal algorithm for controlling the execution mechanism

4 Conclusion

The developed control algorithm allows to control the screw execution mechanism in the absence of information. If the information is complete, then the proposed algorithm allows you to control the object according to the feedback effect. If the information is insufficient, the algorithm relies on the model of the execution device.

The proposed model more accurately reflects the dynamic characteristics of the screw actuator. In most cases, the use of a simplified ideal model leads to the accumulation of static errors over time, leading to significant errors in the system. The use of the proposed model gives good results in many repetitive processes and their imitation.

References

1. Aliev, R.A., Aliev, R.R.: The Theory of Intelligent Systems. Chashyolgy, Baku (2001)
2. Yusupbekov, N.R., Aliyev, R.A., Aliyev, R.R., Yusupbekov, A.N.: Boshqarishning intellektual tizimlari va qaror qabul qilish. O'zbekiston milliy ensiklopediyasi, Uzbekistan (2015)
3. Narendra, K.S., Balakrishnan, J.: Improving transient response of adaptive control systems using multiple models and switching. IEEE Trans. Autom. Control **39**(9), 1861–1866 (1994). https://doi.org/10.1109/9.317113
4. Chen, F.C., Tsao, W.C.: Adaptive control of linearizable discrete-time systems. In: Proceedings of American Control Conference, vol. 1, no. 11, pp. 880–881 (1994). https://doi.org/10.1109/acc.1994.751869
5. Lutsenko, E.V., Laptev, V.: System analysis in adaptive management: intelligent scalable open interactive online environment for teaching and researching on the basis of ASC-analysis and "EIDOS" system (2019)
6. Tsygankova, M.V.: The study of the dynamic properties of the feed drive model of a small-sized 3-coordinate engraving machine. Bull. Tomsk Polit. Univ. **2**, 14–17 (2018)
7. De Silva, C.W.: Sensors and Actuators Engineering System Instrumentation, 2nd edn. Taylor & Francis Group, London (2015)
8. Ida, N.: Sensors, Actuators, and Their Interfaces: A Multidisciplinary Introduction. SciTech Publishing, London (2014)
9. Chinayev, P.I., Chumakov, N.M.: Raschet ispolnitelnix, korrektiruyushix preobrazovatelnix elementov avtomaticheskix system. Texnika (1971). (in Russian)
10. Krasovsky, A.B.: Research of the DC Motor of Independent Excitation: Guidelines for Laboratory Work in the Discipline "Electrical Engineering and Electronics." Publishing House MSTU. N.E. Bauman, Moscow (2014)

Fuzzy Clustering of Influence of Oxygen on Oxidation of Isobutylene in Pulse Mode on Tin-Antimony Catalyst

A. N. Guliyeva(✉) and K. A. Gazvini

Azerbaijan State University of Oil and Industry, 20 Azadlig Avenue, AZ1010 Baku, Azerbaijan
afag.guliyeva1@gmail.com, gazvini@bk.ru

Abstract. The research of impact of oxygen on oxidation of isobutylene on tin-antimony catalyst was carried out by using impulsive microcatalytic methods. The effect of oxygen in the gas phase on the state of the catalyst surface, which consists in a change in its chemical composition, leading to a change in the specific activity of the catalyst, has been experimentally proved. The dependence of the activity of a catalyst that has undergone preliminary exposure in an oxygen-containing gas stream on the concentration of oxygen in a pulse is found. The obtained data are comparable with the results of a similar study carried out under conditions excluding changes in the state of the catalyst surface under the influence of oxygen in the gas phase. The observed differences in the nature of the dependence of the rates of deep and soft oxidation of isobutylene reflect the effect of oxygen on the state of the working surface layer of the catalyst.

Keywords: Isobutylene · Oxidation · Tin-antimony catalyst · Pulse mode catalyst · Catalytic surface

1 Introduction

The aim of this work was to study a less studied aspect of the mechanism of the catalytic oxidation of isobutylene - the mechanism of oxygen participation in the formation of reaction products. In kinetic experiments performed earlier, the effect of oxygen on the formation of oxidation products of isobutylene-methylacrolein and carbon oxides under stationary conditions was established. It is known that the activity and selectivity of a catalyst is determined by the conditions in which it operates. In the initial period of operation of the tin-antimony catalyst under the influence of the reaction mixture, the state of its surface changes (partial recovery of product deposits, compaction), which affects its activity and selectivity. Therefore, the study of the kinetics and mechanism of non-stationary stages is a necessary addition to the study of the process under stationary conditions of catalyst operation.

This work reflects the results of studies of the effect of oxygen on the oxidation of isobutylene in a pulsed mode on a tin-antimony catalyst.

© The Author(s), under exclusive license to Springer Nature Switzerland AG 2021
R. A. Aliev et al. (Eds.): WCIS 2020, AISC 1323, pp. 512–519, 2021.
https://doi.org/10.1007/978-3-030-68004-6_67

2 Experimental Part

The oxidation of isobutylene was studied on a tin-antimony catalyst with an atomic ratio $S_n/S_b = 1$. The specific surface area of the catalyst, determined chromatographically by thermal desorption of nitrogen, is 16.5 m^2/g. The catalyst was prepared from metallic tin and antimony with pure metallic tin (granulated), ТУ-6-09-2704-73; antimony metal (in ingots), СУ-00 and chemically pure nitrogen being used. The calculated amount of powder was slowly added with constant stirring to a boiling solution of nitric acid (specific weight 1.16). In a similar manner, the required amount of powdered antimony was added to a hot solution (80–90%) of nitric acid HNO_3 (specific weight 1.34). The formed suspensions were boiled with stirring until the excretion of nitrogen oxides ceased, and then they were poured together and boiled for another 15–20 min. The residue was filtered off, washed with distilled water (hot, then cold) and dried at 100–120 °C for 4–5 h. The dried precipitate was calcined in air at 820–840 °C. The duration of the high-temperature treatment was 8 h. The rate of temperature rise was 100 degrees per hour. The calcined mass was thoroughly ground and tableted under a pressure of 140 kg/sm^2, after which, to relieve stress, it was additionally held at 802 °C for 2 h. The required fraction (0.4–0.6) mm was prepared from the tablets (Fig. 1).

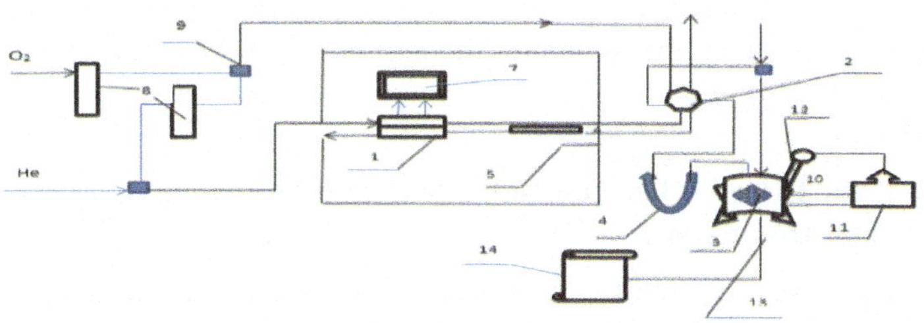

Fig. 1. Technological scheme of the impulse installation. 1-thermal conductivity detector; 2-six-way crane; 3-microreactor; 4-trap; 5-chromatographic columns; 6-sampler; 7-chromatograph potentiometer; 8-capillary rheometers; 9-gas mixer; 10-electric furnace; 11-thermostat; 12, 13-thermocouple; 14-recorder potentiometer;

2.1 Installation Scheme and Experimental Technique

The studies were carried out on a pulsed installation in a micro-catalytic reactor, included in the circuit of the chromatograph ЛХМ-8МД. Carrier gas - helium (frequency 99.9%) was divided into 2 streams. The first stream sequentially passed through the comparative cell of the thermal conductivity detector (1), by means of the valve (2), through the reactor (3), then a trap cooled with liquid nitrogen (4), chromatographic columns (5) and the working cell of the detector. A certain volume of isobutylene - oxygen mixture was pulsed into the sampler (6) and was fed into the reactor by the flow of helium. The products of reaction and unreacted starting materials after the reactor enter the chromatographic columns and then into the detector. The detector signal was recorded by a potentiometer (7) in the form of a chromatographic peak. The second helium stream after passing through the rheometer (8) was mixed in general stream (9) with a given amount of oxygen. The formed mixture, if necessary, was fed using a six-way valve (2) into the reactor, the latter being disconnected from the chromatograph, in the system of which pure helium passed (first stream). With the cessation of the supply of the helium-oxygen mixture to the reactor, the valve (2) was again included in the chromatograph circuit. The microcatalytic reactor is made of Pyrex glass and has a sliding thermocouple pocket. The volume of reactor is 2 sm^3, internal diameter was 8 mm. The thermostating of the reactor was carried out in a quartz electric furnace (10) with a photoelectric regulator (2) in combination with a Chromel-Copel thermocouple (12). The temperature in the reaction zone was recorded using a Chromel-Copel thermocouple (I3) with a recording potentiometer ПС-I (14). Under the conditions of a microcatalytic reactor, due to the small thermal effect of the reaction in a pulse, the temperature gradient across the catalyst bed is practically absent; moreover, to remove the temperature gradient, the catalyst was diluted with quartz grains of the same size as the catalyst. The reactor was loaded with 2 g of catalyst; the free volume of the reactor was filled with broken quartz to exclude the possibility of homogeneous oxidation. The studies were carried out in a pulsed non-chromatographic mode in the kinetic region of the process. For this, taking into account also the need for a clear separation of the mixture components on a chromatographic column, the velocity of the transporting gas (helium) was chosen 2 liters per hour, pulse volume was 2 ml, and catalysis grain size was (0.4–0.6) mm. These conditions ensured implementation in a non-chromatographic mode and exclude external and intra-diffusion complications.

2.2 Study of the Effect of Oxygen on the Oxidation of Isobutylene (Experimental Results and Discussion)

The role of oxygen in the oxidation of isobutylene is one of the main issues in the mechanism of oxidative conversion of olefin. Oxygen is one of the reaction components, and its partial pressure in the absorption layer influences the oxidation rate in a certain way. Meanwhile, a change in the partial pressure of oxygen can lead to a change in the chemical composition of the working surface layer of the catalyst (oxygen content in it, valence of cations), which results in a change in the specific activity of the catalyst. These changes occur relatively slow in the flow and therefore have an effect when the catalyst is operating under stationary conditions. Tin-antimony catalysts are referred to the contacts, the initial period of which is characterized by a change in the structure of their surface and activity. Therefore, it can be expected that the nature of the oxygen dependence of the rate of isobutylene oxidation on a tin – antimony catalyst observed in the flow is the result of its double action. The results of our study of the effect of oxygen on the oxidation of isobutylene in the pulsed mode made it possible to clarify this issue. Three series of experiments were carried out. In the first series, the effect of an oxygen pulse on the rate of isobutylene oxidation on a surface of constant composition was studied. After its compulsory preparation, the catalyst was pulsed with a standard mixture containing isobutylene, oxygen, and helium in a molar ratio of 1:1:3 until a steady state characterized by some constant activity (total isobutylene conversion 5 mol.%, selectivity for methyl acrolein-61 mol.%) was reached. One pulse of the working isobutylene-oxygen-helium mixture with a given oxygen content was applied to the sample prepared in this way, and the yield of oxidation products was determined. Then the sample was treated again with pulses of the standard mixture until reaching the initial activity, and a pulse of the working mixture with a different oxygen content was used. The catalyst was treated with a standard mixture before each new pulse of the working mixture was applied. The oxygen content in the working mixture was varied from 0 to 70 mol.%, the concentration of isobutylene remaining constant and equal to 20 mol.%. The results of the experiments are presented in Fig. 2.

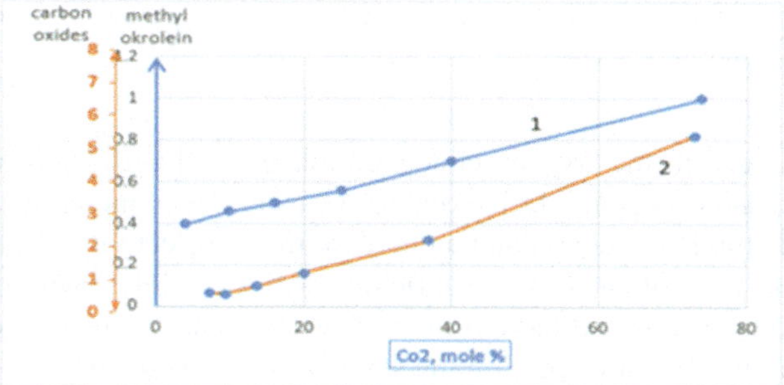

Fig. 2. Dependence of the rate of formation of isobutylene oxidation products on the oxygen content in the pulse.

As can be seen, the rates of formation of methylacrolein and carbon dioxide (expressed, respectively, through the number of moles of methylacrolein and carbon dioxide formed in one pulse) continuously increase with an increase in the oxygen content in the pulsed working mixture, while the order of the reaction of mild olefin oxidation by oxygen is fractional, while deep oxidation reactions - close to one. Since the change in the state of the catalyst surface under the influence of the pulse oxygen is practically excluded during the pulse, the data obtained show a purely kinetic effect of oxygen on the oxidation of isobutylene. The task of the next series of experiments was to reveal the possible effect of oxygen in the gas phase on the change of the state of the working surface of the catalyst. For this purpose, each time after the usual mandatory preparation the catalyst sample was kept for a certain constant time in a helium flow containing different amounts of oxygen varying from 0 to 25 mol.%, and was in contact with isobutylene pulses of an oxygen mixture of constant composition (i-C_2H_4: O_2:He = 1:1:3) until a steady state of the surface was reached, characterized by a certain catalyst activity in the oxidation of isobutylene. The results of experiments showed that the state of the contact surface and its stationary activity achieved by holding the catalyst in an oxygen-containing helium flow depend on the oxygen content in the helium flow (Fig. 3).

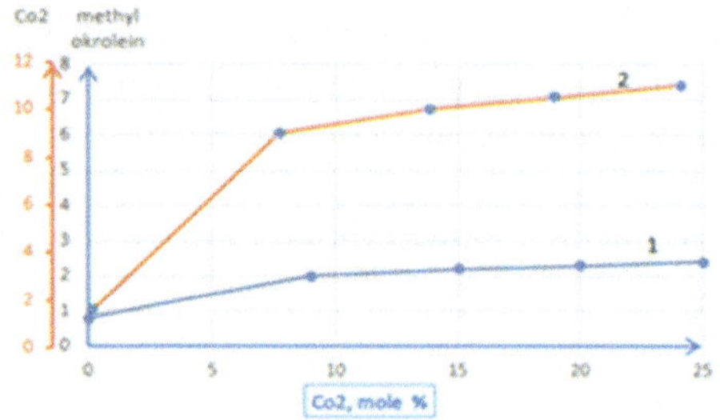

Fig. 3. Dependence of the yield of soft (1) and deep (2 oxidation) products on the oxygen content in the carrier gas flow.

With an increase in the oxygen concentration up to 15–20 mol. %, the yield of products of mild and especially deep oxidation of isobutylene increases, which leads to an increase in its total conversion and a decrease in the selectivity of partial oxidation (Fig. 4). This can be explained by the enrichment of the catalyst with oxygen, mainly weakly bound to the surface. If there is more than 15–20 mol.% of oxygen in helium, it stabilizes the nature of its influence on the state of the catalyst surface.

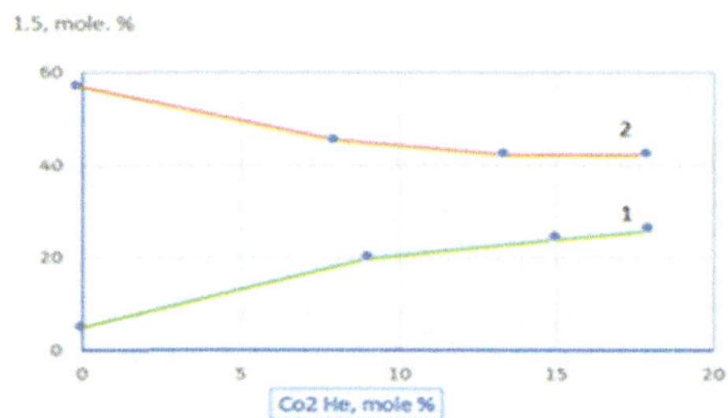

Fig. 4. Total activity (1) and selectivity of catalyst samples kept in a flow of helium with different oxygen content.

The data obtained allow us to assume that under flowing conditions of the catalyst operation, with a change in the oxygen content in the reaction gas mixture, the state of the contact surface changes, which results in a change in the specific activity of the catalyst. In the third series of experiments, as in the first, the dependence of the catalyst activity on the oxygen content in the pulse of the isobutylene-oxygen mixture was studied. In contrast to the first series, in the last experiments the sample, after the usual obligatory preparation, was kept in a flow of helium with 25 mol.% oxygen. After bringing the catalyst to a steady state by pulses of a standard mixture (i-C_4H_8: O_2:He = 1:1:3), one pulse of the working mixture with different oxygen content in it was fed to it. Comparison of the results of the experiments of the first and third series shows that at any oxygen content in the pulse of the working mixture, the total activity of the most oxidized catalyst (the third series) is significantly higher, and its selectivity for mild oxidation products is noticeably lower than in the corresponding experiments of the first series (Fig. 5).

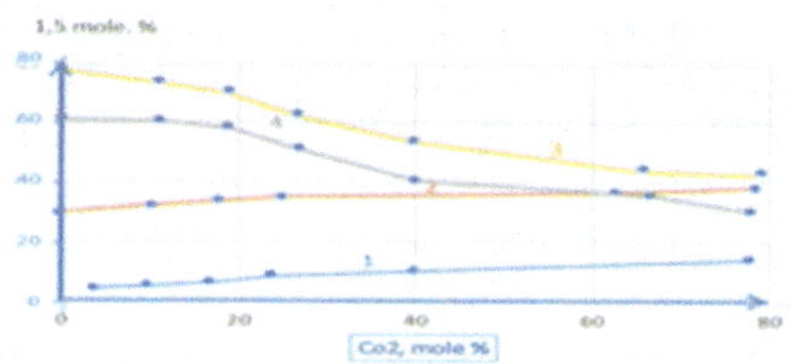

Fig. 5. Dependence of the total activity and selectivity of catalysts treated with oxygen (cr. 2.4, respectively) and without it (cr. 1.3, respectively) on the oxygen content in the pulsed working mixture.

As can be seen from Fig. 6, the rate of deep oxidation of isobutylene on a catalyst in a helium flow with a maximum oxygen content (25 mol.%) increases with an increase in the oxygen concentration in the pulse, but not as strongly as in the first series. Methylacrolein yield decreases as the amount of oxygen in the pulse increases.

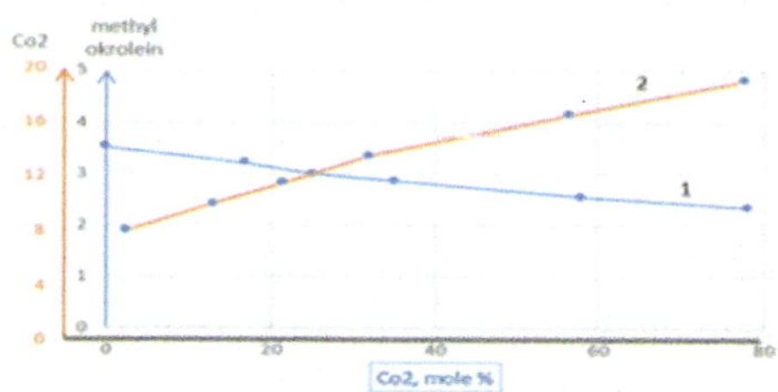

Fig. 6. Dependence of the yield of methylocrelein (1), and Co_2 (2) on the oxygen concentration in the pulse for a carrier gas containing 25 mol.% O_2 .

The observed character of the dependence of the rates of deep and soft oxidation of isobutylene on the oxygen content in the pulsed gas mixture differs from that for the first series of experiments and reflects the effect of oxygen on the state of the catalyst surface. Thus, in the process of isobutylene oxidation on a tin-antimony oxide catalyst, a change in the oxygen content in the gas reaction mixture affects the rate of oxidative conversion of olefin not only directly, purely kinetic, but also through a change in the state of the contact surface (its chemical composition), leading to a change its specific activity. This must be taken into account when using its kinetic data obtained at a steady state of the catalyst to clarify the mechanism of the process.

3 Solution of Problem

We deal with a problem of modeling of relationship between input and output variables of research of influence of oxygen on oxidation of isobutylene in pulse mode on tin-antimony catalyst process. Input variables is: time (X). Output variables is rate of mild and deep oxidation of isobutylene. The experimental data on the values of the variables are given in Table 1.

Table 1. Experimental data

X	Y
0	76
10	72
17	69
24	62
36	54
60	45
72	44

Due to complexity of the considered relationship, we purpose to build a model in the form of fuzzy If-Then rules. The rules which obtained by Fuzzy C-means method-based clustering are shown in Table 1. The obtained rules are shown below:

IF X is low THEN Y is high.

.

.

IF X is high THEN Y is low.

Constructed relationship between input variables X1, X2 and output variable Y1 is given by using the If-Then rules of in Fig. 7

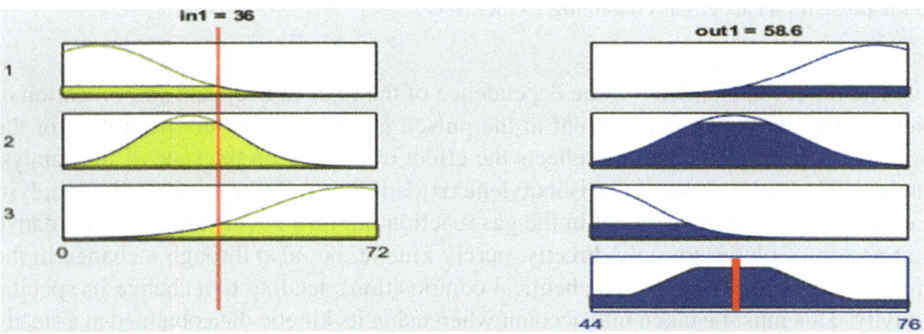

Fig. 7. IF-THEN rules

This conclusion can be extended, apparently, to a whole series of oxide catalysts for partial oxidation, which, like tin-antimony contacts, have a low mobility of the bulk oxygen of the crystal lattice.

4 Conclusions

1. The effect of oxygen on the oxidation of isobutylene on a tin-antimony oxide catalyst (atomic ratio $S_n/S_b = 1$) in a pulsed mode was studied. 2. The dependence of the catalyst

activity on the oxygen content in the pulsed working isobutylene-oxygen mixture was revealed. It is shown that the rate of mild and deep oxidation of isobutylene on a catalytic surface of constant composition increases with an increase in the oxygen content in the pulse; at the same time, the order of the reaction of mild oxidation with respect to oxygen is fractional, while for the reaction of deep oxidation is close to unity. The total conversion of isobutylene under these conditions also continuously increases, and the selectivity for methyl-acrolein decreases. 3. It has been experimentally proved that the effect of oxygen in the gas phase on the state of the catalyst surface, is in changing its chemical composition (oxygen content, valence of cations) leading to a change in the specific activity of the catalyst. The data obtained allow us to assume that these changes occur in the catalyst under stream conditions of its operation. 4. The dependence of the catalyst activity, preliminary exposured in an oxygen-containing gas, on the oxygen concentration in the pulse of the working isobutylene-oxygen mixture has been established. The observed character of the dependence of the rates of deep and soft oxidation of isobutylene on the oxygen content in the pulsed gas mixture differs from that for a catalyst that has not undergone preliminary exposure to oxygen and, therefore, reflects the effect of oxygen on the state of the catalyst surface. 5. On the basis of the obtained experimental data, it was concluded that in the process of isobutylene oxidation on a tin antimony catalyst, gas-phase oxygen affects the rate of oxidative transformations of olefin not only directly (purely kinetic), but also through a change in the state of the catalyst surface, leading to a change in its specific activity. 6. This conclusion can be extended to a number of oxide catalysts for partial oxidation, characterized, like tin-antimony contact, by low mobility of bulk oxygen in the crystal lattice, and must be taken into account when using kinetic data obtained at a steady state of the catalyst to clarify the mechanism of the catalytic process.

References

1. Boriskov, G.K.: Kinet. Catal. **14**(7) (1973)
2. Kazansky, V.B.: All-Union Conference on the Mechanism of Heterogeneous Catalytic Reactions, Moscow (1974)
3. Sokolovskiy, V.D.: Abstract of doctoral dissertation. Institute of Physical Chemistry. L.V. Pisarzhevsky, K. (1975)
4. Boreskov, G.K., Venyaminov, S.A., Sazanova, N.N., Pankratyev, Yu.D., Pitaeva, A.N.: All-Union Conference on the Mechanism of Heterogeneous Catalytic Reactions, Moscow (1974)
5. Anshim, A.G., Sokolovsky, V.D., Boreskov, G.K., Davydov, A.A.: Kinet. Catal. (1978)
6. Ranveer, S.M., Rouleau, D.J.: Catalytic oxidation of isobutylene to methacrolein over copper oxide catalysts. Adv. Chem. **51**, 41–51 (1965). https://doi.org/10.1021/ba-1965-0051.ch004
7. Song, N., Rhodes, C., Bartley, J.K., Taylor, S.H., Chadwick, D., Hutchings, G.J.: Oxidation of isobutene to methacrolein using bismuth molybdate catalysts: comparison of operation in periodic and continuous feed mode. J. Catal. **236(2)**, 282–291 (2005). https://doi.org/10.1016/j.jcat.2005.10.008

Toward Analysis of a Process of Obtaining High-Dispersive Stable Oil/water Emulsions

G. S. Hasanov(✉), R. G. Hasanov, N. F. Aliyeva, and V. K. Valiyev

Azerbaijan State Oil and Industry University, Azadlig Avenue, 20, Baku, Azerbaijan
gaman51@mail.ru, gasanovr935@gmail.com, nuranafalieva@gmail.com, Vaqifveliyev1953@gmail.com

Abstract. The conditions of obtaining suspension-emulsion system, high-dispersive oil/water emulsions in the composition of wastes stabilized by a solid emulsifier with natural bentonite were studied in the work for development of emulsion method of neutralizing wastes. It is shown that a sufficiently effective emulsification in an alkaline aqueous medium is achieved by using only bentonite suspension with a concentration of higher than 5.0% mass. To enhance the emulsifying ability of bentonite an activator $FeCl_3$ was added into system by reducing its hydrophilic property. The addition of electrolyte activator $FeCl_3$ to the aqueous phase (0.03% mass) during the oil emulsification process causes a sharp increase in emulsion stability with a bentonite content of 5.5% mass in aqueous suspension. The lifetime of an oil drop interfacial oil-0.5% suspension of Na-bentonite activated with 0.03% mass of $FeCl_3$ significantly depends on pH of an aqueous phase, the optimum value of which is 8.0 to form a stable oil-water emulsion. Due to imperfect experimental data, we propose to use fuzzy If-Then rules for modeling of the considered dependence.

Keywords: Oil sludge · Oil-contaminated soil · Waste liming · Decontamination · High-dispersive emulsions · Solid emulsifier · Activator · Surfactants

1 Introduction

Known chemical decontamination technologies of drilling and oil production wastes, oil-contaminated soils (oil sludges), without considering the features, are based on liming methods by mixing them with burnt and ground lime and water [1–5]. These technologies are attractive for their simplicity and rapid technological fulfillment. However, their general disadvantage is the dependence of the degree of neutralization of toxicants on the efficiency of mixing waste mixtures with reagents and non-absolute resistance of capsules with immobilized toxicants to atmospheric influence and microbial impact, which can lead to secondary pollution of environment. In environmental projects related to man-made contact with nature the prerequisite must be performed for the environmental safety of developed technologies accepted for use.

In this regard, to develop an efficient emulsion technology for decontamination of drilling and oil production wastes, which includes emulsifying oil toxicants in wastes,

© The Author(s), under exclusive license to Springer Nature Switzerland AG 2021
R. A. Aliev et al. (Eds.): WCIS 2020, AISC 1323, pp. 520–528, 2021.
https://doi.org/10.1007/978-3-030-68004-6_68

the conditions for obtaining high-dispersive oil/water emulsions were studied. Due to imperfect experimental data, we proposed to use fuzzy If-Then rules for modeling of the considered dependence.

2 Experimental Part

Research object is one type of system - oil-in-water (O/W) emulsions. 20% of toluene solution of oil recovered from contaminated soil in Balakhany district was used as a model system for obtaining emulsion. To stabilize oil/water emulsions, a solid emulsifier Na-form of natural bentonite was used, in some experiments the latter was activated with electrolyte $FeCl_3$. Dispersion was carried out resulting in an emulsion in 250 ml glass cylinder with continuous stirring using a propeller type four-bladed electric mixer with a constant rotation number of ~2000 rpm. Toluene oil solution was poured with a continuous flow at a specified rate (0.8 cm^3/s.) to distilled water in the cylinder (based on the final volume ratio of both phases 1:1) at temperature of $20 \pm 1\ ^\circ C$.

The resulting emulsions were tested for stability depending on the content of solid emulsifier in the aqueous suspension, the concentration of electrolyte in the aqueous phase and its pH. Stability of the emulsions was estimated by the time (τ, min) of their spontaneous destruction when left to stand with the release of "creams".

3 Results and Discussion

When the oil emulsions are destructed only the amount of released aqueous layer can be clearly noted. It is almost impossible to consider the release of oil, so we considered the entire upper layer to be "cream". 50% of "cream" corresponds to the complete separation of emulsion into aqueous and oily layer. Along with emulsions comprising large drops of oil (with an average diameter of ~100 μm), during the experiments emulsions containing small drops of oil were formed (with an average diameter of ~1 μm.

Both relatively fast destroying emulsions and practically stable emulsions (τ > 2–3 days) were formed. As seen from the dependences emulsion stability τ – emulsifier concentration of C_{TB}, high-dispersive bentonite particles form emulsions of different stability in an alkaline aqueous medium at different concentrations of solid phase.

Sufficiently effective emulsification in an alkaline aqueous medium is achieved only using bentonite suspension with a concentration of higher than 5.0% mass. At lower concentrations of emulsifier under the same conditions, the stability of emulsion drops sharply. These differences in the stability of emulsions are associated both with the natural inhomogeneity (mosaic) of the surface of bentonite particles referring to selective wetting of its individual areas by both emulsion phases, and with the possibility of a hydrophobic coagulation structure in the emulsion system. Amount of acidic substances in residual oil (oil acidity) is sufficient to form hydrophobic coatings on the surface of the clay particles - the adsorption layer of soap necessary for its subsequent structure formation. Only under this condition, and at a relatively high concentration of bentonite in C_{TB} suspension (not less than 5.0% mass), M/B type stable emulsion is formed in alkali medium.

Thus, emulsifying ability of bentonite is enhanced by acidic substances in emulsified oil. Confirmation of this is [6] an example of using SiO_2 as a solid emulsifier, which, due to its inertia to acidic compounds in oil composition, does not form emulsions at all, regardless of its quantitative content in oil-water system. Based on the above, a solid high-dispersive emulsifier in the form of natural bentonite clay was used in developed emulsion technology of waste disposal at its first stage of emulsification of oil-toxicants.

The effect of such emulsifier is related to the fact that its particles are wetted simultaneously by both phases of the emulsion - water and oil, as a result of which these particles move to the interface of the resulting oil drops while dispersing oil toxicants, armoring them with solid coatings. With optimal balance in the distribution of clay particles of hydro- and oleophilic fields, M/B type stable oil-in-water emulsions are formed with a protective coating of a clay film attached to the interfacial surface of oil drops with its oleophilic fields. At the same time, the other part of the surface of a clay film on oil drops remains hydrophilic, it is selectively better wetted with the external phase - water. Such feature of a protective coating of oil drop of O/W emulsion is a determining factor in subsequent interaction of armored oil drops with hydrophilic quicklime in the second stage of the technology, ensuring their adsorption on the surface of lime particles and their further encapsulation and neutralization.

The adsorption capacity of armored oil drops with respect to the surface of hydrophilic lime particles can be enhanced to increase the efficiency of waste disposal. To this end, it is necessary to increase the number of hydrophilic fields that are adsorption centers of lime adsorbate on the surface of bentonite clay particles, which serves as a protective coating of oil drops. Oil phase, which is a continuous phase of inverse emulsion in oil sludge, at the stage of preparing suspension-emulsion system while mixing it with bentonite clay and surfactant OP-10 is continuously dispersed, while changing both the physical-chemical nature of the surface of armored oil drops and the ratio between them and the dispersion medium - water. To enhance the adsorptivity of armored oil drops with respect to the surface of hydrophilic lime particles, it is necessary to increase the hydrophilicity of their armor coating and increase the number of hydrophilic fields on the surface of its particles that have firmly inherent water. To increase the amount of strongly inherent water on the surface of bentonite clay particles, it is necessary to disperse (peptize) clay particles.

Structural damage of clay mud is performed by mechanical effect on it by mixing with dispersing devices, dispersants at a mixing rate of more than 500 rpm. To increase the amount of strongly inherent water on the surface of bentonite clay particles and thereby increase its hydrophilicity, the disperse system and the greatest mutual mobility of its components should be mixed intensely to disaggregate the structure of primary particles at the first stage of the technology while producing a suspension-emulsion system. Therefore, while using intensive mechanical effects by mixing the dispersion system, when the clay aggregates break down into the primary particles and their initial structural bonds are broken, the surface of the primary particles will be watered not by capillary suction, but by diffusion processes. Thus, the amount of strongly inherent water can be increased on the surface of the primary particles of bentonite clay, and their hydrophilicity. As a result, in preparation process the suspension-emulsion system, using intensive mixing of the mixture consisting of an aqueous bentonite suspension, dispersion from its particles will be ensured efficiently, followed by armoring the resulting oil drops in waste composition with more hydrophilic coating of clay particles.

To improve the economic performance of waste disposal using emulsion technology, the studies were carried out to reduce the amount of natural bentonite used (lower than 5.0% mass) as a solid emulsifier and to achieve an optimal balance in the distribution on the surface of their particles of hydro- and oleophilic fields. With the purpose of forming high-dispersive stable emulsions, it was necessary to increase sharply the natural low grade mosaic surface of the particles, i.e., to activate them, to increase the emulsifying ability of bentonite by reducing its hydrophilicity. Such activation of clay particles was achieved by using $FeCl_3$ as an electrolyte activator in the aqueous phase. The addition of $FeCl_3$ to the aqueous phase during the oil emulsification process causes a sharp increase in emulsion stability. At the same time stable emulsion is formed in aqueous suspension with bentonite content less than (0.5% mass) in absence of electrolyte (5.0% mass). A significant reduction in the solid emulsifier using the electrolyte $FeCl_3$, while obtaining a stable emulsion, is due to optimally hydrophobization of the surface of its particles and formation a protective structured layer on the surface of armored oil drops.

Both the first and the second are related to the optimum concentration of hydrogen ions in the aqueous phase. pH value of natural Na-bentonite suspension is in the region of 7.5–9.5, therefore, acidic substances of acidic that are part of oil, as they are neutralized, at the time of emulsion formation, transform from an oil phase to an aqueous one in the form of sodium soaps, and the electrolyte $FeCl_3$ forms the main iron salt in the equilibrium reactions of its hydrolysis. Accordingly, the dispersion of oil upon its emulsification occurs in an aqueous solution of sodium soaps which are effective surfactants, and the resulting stable emulsion is stabilized by bentonite particles (with enhanced emulsifying ability) covering the surface of the dispersed oil drops with an armored structured film. As hydrophobization increases and, as a result, the adhesion strength of bentonite particles to the surface of oil drops increases, the stability of the emulsion increases. Enhancement of emulsifying ability of bentonite occurs by reducing hydrophilicity of its particles, hydrophobization of their surface by basic iron salts. Thus, as regulators of hydrophilic-hydrophobic balance of bentonite particles and dispersion degree of emulsion, along with the electrolyte, there are also acidic substances that are part of the emulsifying oil and the pH of the aqueous phase, that is impossible to obtain a stable oil emulsion with high-dispersive fragmentation without controlling and giving them optimal values.

The optimum value of the regulators of hydrophilic-hydrophobic surface balance of the bentonite particles used to enhance their emulsifying ability is determined by the optimal balance on the surface distribution of clay particles of hydro- and oleophilic fields. It is not possible to predict their optimal value for the formation of stable emulsions, since it depends on many factors, such as the composition and properties of oil and water phases, the size of the dispersed oil drops, the interaction of a drop surface with a regulator, etc. Based on this, the effect of the concentration of electrolyte $FeCl_3$ and pH of the medium on the hydrophilic-hydrophobic balance of the surface of bentonite particles was determined by studying the stability of emulsions by the lifetime of an oil drop in oil interface - 0.5% of bentonite suspension [7, 8]. For each of regulators its optimal values and amount necessary to stabilize oil emulsions thickened with 0.5% suspensions of Na-bentonite were found. The efficiency of activating bentonite suspensions with electrolyte $FeCl_3$ depends on its concentration in the dispersion medium. Therefore, to obtain maximum stable oil emulsion the maximum lifetime of an oil drop τ_m on boundary of phases in a system oil - water $FeCl_3$ solution was determined. In oil system, 0.5% of Na-bentonite suspension activated with a different amount of $FeCl_3$, τ_m oil drops takes different stability, reaches several hours, while τ_m oil drops in non-activated Na-bentonite suspension for only 20 s. As Table 1 shows the addition of small amounts of $FeCl_3$ significantly increases the stability of emulsions. The effect of adding electrolyte to bentonite suspension is that hydrophobic fields which are coagulation centers of the particles occur on the surface of bentonite clay particles.

Table 1. Dependence of lifetime of an oil drop on the boundary of oil - 0.5% of Na-bentonite suspension on the concentration of $FeCl_3$ in an aqueous phase (pH of an aqueous phase is 8.0).

Concentration of $FeCl_3$, % mass	Maximum lifetime of an oil drop, τ_m, sec
–	20.0
0.01	128.0
0.02	403.0
0.03	7000.0
0.04	7010.0

When shaking suspension with oil phase (oil), a contact with the resulting oil phase drops occurs along these areas, and the particles move to the drop surface - the medium, attaching to it more firmly than in the absence of an electrolyte, providing a certain drop of oil before its coalescence (merging). In this regard, the optimum concentration in the aqueous phase of the electrolyte $FeCl_3$ which is sufficient to activate bentonite and form a stable oil-water emulsion can be considered to be 0.03% mass. As mentioned above the concentration of hydrogen ions also affect on the dispersity and stability of oil-water emulsion obtained using bentonite asan emulsifier activated with $FeCl_3$ electrolyte. Dependence of lifetime of an oil drop in bentonite suspension activated by $FeCl_3$ electrolyte on pH of aqueous phase is presented in Table 2. Data from the table show that the lifetime of an oil drop on the boundary of oil - 0.5% of Na-bentonite suspension activated by 0.03% mass of $FeCl_3$ significantly depends on the pH of an aqueous phase. Optimum value pH medium for the formation of stable oil-water emulsion equals to 8.0–9.0.

Table 2. Dependence of lifetime of an oil drop on the pH change of an aqueous phase in oil - 0.5% of Na-bentonite suspension activated by 0.03% mass of $FeCl_3$.

pH of an aqueous phase	9.0	8.0	7.0	6.0
Lifetime of an oil drop, τ_m, sec	6000.0	7000.0	700.0	26.0

The results of experiments on the stability of oil emulsion are confirmed. Thus, when emulsifying oil in a bentonite suspension and electrolyte $FeCl_3$ in an aqueous phase with pH of the latter 8.0, a stable emulsion is formed at a concentration of bentonite in 0.5% mass suspension. Also, under the same conditions, but at a neutral medium (pH 7.0), a stable emulsion is not formed. In these experiments, the dispersity of emulsions, the size distribution of oil drops depending on pH change of the aqueous phase are of interest. Experimental results of emulsions (micrographs) showed the average size of oil drops of dispersed phase of these emulsions depending on pH change of an aqueous phase (Table 3). As table shows the reaction of an aqueous phase medium is of great importance in emulsification process of oil under the conditions studied. So, with a relatively small content of bentonite (0.5% mass), activated with 0.03% mass of $FeCl_3$ at pH of an aqueous phase of 8.0 the average size of oil drops is equal to 0.96 μm.

Table 3. Dependence of an average size of oil drops on pH change of an aqueous phase of oil-water emulsions.

Emulsification condition	Concentration of $FeCl_3$, % mass	0.03	–	0.03
	Concentration of bentonite, % mass	0.5	5.0	0.5
	pH of an aqueous phase	7.0	8.0	8.0
Average size of an oil drop, μm		126.0	1.3	0.96
Emulsion stability, min		3.0	∞	∞

In the absence of an activator $FeCl_3$ in the emulsion system, but with a relatively high content of bentonite (5.0% mass) and pH 8.0, stable emulsions with a drop size of 1.3 μm are obtained. Under the same emulsification conditions, but in a neutral aqueous phase medium, the average oil drop size increases sharply up to 126 μm. In compared experiments, the stability of the emulsions at their non-comparable values is very different, respectively, ∞ and 3.0 min.

An important factor in the formation of an emulsion system using an emulsifier is the emulsifying ability of the latter, that is, the maximum amount of an oil phase that can be dispersed in a given volume of the emulsifier solution (suspension) under certain conditions. Emulsifying capacity for a number of concentrations of Na-bentonite suspensions was determined by the emulsion preparation method described above.

As it is seen from the dependence (ratio V_{oil}/V_{water} – Na-bentonite concentration), the emulsifying ability, expressed in ratio units, grows disproportionally to the Na-bentonite concentration in the suspension, but is much slower, tending towards some limit at high concentrations. At concentration of Na-bentonite suspension in water which is equal to 5.0% mass an emulsion containing 40.0% mass of oil and 60.0% mass of water is formed. Flow rate of Na-bentonite is 8.0% mass for oil.

To neutralize oil sludge, inherent water in their composition in the emulsified state of reverse emulsion must be released to interact with quicklime in liming process of the suspension-emulsion system of the second stage of the technology [9–11]. The exceptional high stability of oil emulsions of "water-in-oil" type is determined by the stabilization by their structural surface layers of complex natural emulsifiers - salts of naphthenic acids, resins, asphaltenes and particles of finely dispersed phases. The most common physical and chemical way of destroying such emulsions is to affect them with surfactant demulsifiers, which adsorbing on drop-medium interface, displace the protective coating by adsorption, thereby disrupting the aggregative stability of the emulsion release inherent water. In [12], a method was developed to estimate the demulsifying effect of substances while selecting effective demulsifiers of inverse emulsions.

Using the developed method, we defined change in the aggregative stability of oil emulsions of reverse type during their interaction with demulsifiers was defined and highly estimated the demulsifying effect of non-ionic surfactant OP-10 which characterizes the release time from the W/M emulsion in the free form of half amount of emulsified water in 60 min. To release inherent water as part of oil sludges and increase the efficiency of process of their neutralization at the first stage of technology we used the demulsifier OP-10 surfactant with concentration of 5.5% mass in a water phase.

To achieve the efficiency of the process of decontamination of oil sludge in the mixing reactor, the maximum homogeneity of the charge must be achieved, provided by the design parameters of the mixing reactor and the mixing device. The efficiency of decontamination of oil sludge was calculated from the indicators of the concentration of oil (oil products) in the water extract from the products of decontamination and the original oil sludge. As can be seen from the results of the experiments (Table 4), at a rotation frequency of the shaft of the mixing device 800 ÷ 900 min-1, the efficiency of oil sludge neutralization reaches 92.2%.

Table 4. Efficiency of oil sludge neutralization depending on the rotation frequency of the stirrer shaft of the reactor-mixer.

Shaft rotation frequency, min-1	Concentration of oil (petroleum products) in water extract, mg/l	The efficiency of oil sludge neutralization,%
–	3.611	–
600	0.410	88.6
700	0.351	90.3
800	0.280	92.2
900	0.280	92.2

The data on dependence of efficiency of oil sludge neutralization on the influential factors are characterized by imprecision. In this case, fuzzy sets may be used for modeling [14]. We described this dependence by using fuzzy If-Then rules derived by using fuzzy C-means clustering algorithm. The surface of dependence produced by the obtained rules is shown in Fig. 1.

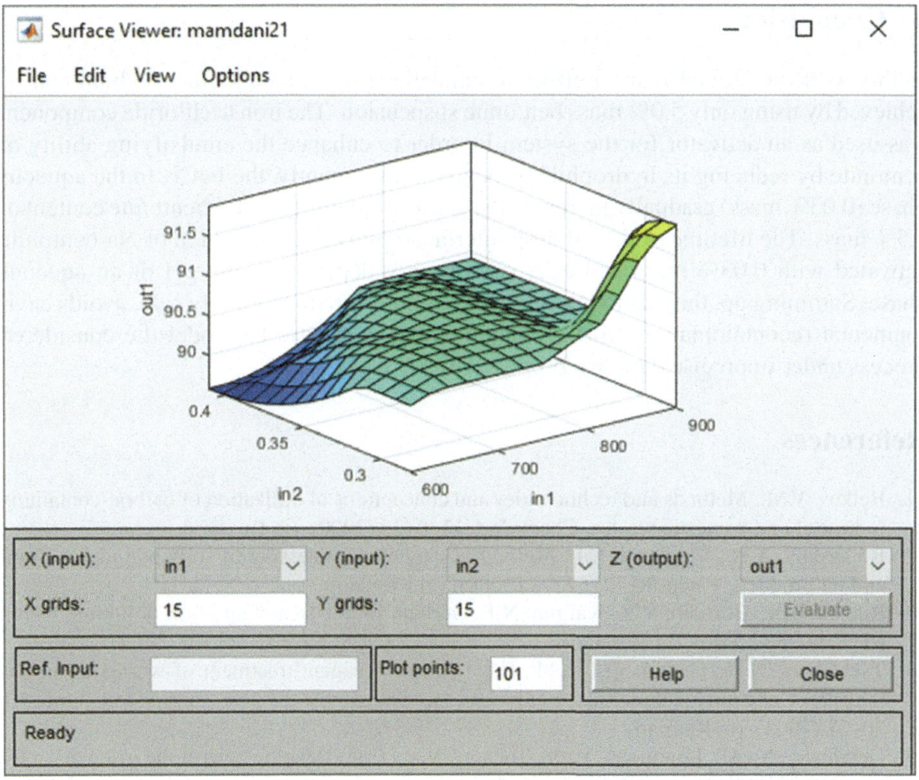

Fig. 1. Dependence of efficiency of oil sludge neutralization on the influential factors produced by fuzzy If-Then rules.

The calculation of the power consumed by the blade mixer drive was carried out according to the method to determine the resistance coefficients of the mixed media in the mixers with blade and frame mixers [13] according to the formula;

$$N = \int_{r}^{R} dP \cdot v,$$

where dP is the force acting on an elementary site, N; v is a rotation speed of an elementary area, m/s; r is the inner radius of the stirrer blade, m; R is the outer radius of the agitator blade, m.

Results of researches on obtaining high-dispersive stable emulsions oil/water were used for the development of technology to neutralize waste of drilling and oil production, to eliminate mud pits and to improve the environment on their territories. Disposal effect of wastes was estimated by contamination parameters of water extract from neutralization products. It was found that under optimum conditions of waste neutralization, water extract parameters comply with sanitary standards: pH 8.5; absence of compounds of heavy metals and oil (petroleum products).

4 Conclusion

In this work, we found that an effective emulsification in an alkaline medium can be achieved by using only 5.0% mass bentonite suspension. The iron trichloride component was used as an activator for the system in order to enhance the emulsifying ability of bentonite by reducing its hydrophilic property. Subsequently, the $FeCl_3$ in the aqueous phase (0.03% mass) gradually increases the emulsion stability with a bentonite content of 5.5% mass. The lifetime of the oil drop interfacial oil-0.5% suspension of Na-bentonite activated with 0.03% mass of $FeCl_3$ significantly depends on the pH of an aqueous phase. Summing up, the method developed is simple, environmentally safe, avoids environmental recontamination with decontamination products. To model the considered process under imprecise data, we proposed to use fuzzy sets.

References

1. Belkov, V.M.: Methods and technologies and conceptions of utilization of carbon-containing industrial and domestic wastes. Chem. Ind. **11**, 8–25 (2000). (in Russian)
2. Bolshakov, A.A., Vyalkova, E.I.: Method of chemical processing of soils polluted with oil and oil products. Patent RU 2152274 (2000). (in Russian)
3. Rudnik, M.I., Borodin, V.V., Kalinin, N.F.: Sorbent for the treatment of oil pollution. Patent RU 2160758 (2000). (in Russian)
4. Tskhadaya, N.D., Gerzhberg, Yu.M., Popov, A.N.: Reagent treatment of wastes of oil-gas industry. Construction materials, equipment, technologies of the 21st century, vol. 3, no. 50, 30–31 (2003). (in Russian)
5. Gerzhberg, Yu.M., Logunova, Yu.V., Tokarev, V.V. et. al.: Development of the installation for treatment of hydrocarbon containing industrial wastes by reagent capsulation method. Omsk Sci. Bull. **2(68),** 76–81 (2008). (in Russian)
6. Koretskiy A.F., Taubman A.B.: On emulsifying activity of fine-grained ыщдшвы. Proc. USSR Acad. Sci. **124,** 358–361 (1959). (in Russian)
7. Babalyan, G.A., Kravchenko, I.I., Markhasin, I.A., Rusakov, G.V.: Physicochemical fundamentals of surfactant application in oil reservoirs development. Gostoptexizdat, M., Russia, 152 (1962). (in Russian)
8. Rebinder, P.A., Venstem, E.K.: To the physics of foam and emulsion. Phys. Chem. J. **2**(5–6), 754–763 (1931). (in Russian)
9. Hasanov G.S., Abdullayev F.Z.: Sorbent for neutralization of the oil-contaminated soils. Patent EAPO 018089 (2013). (in Russian)
10. Hasanov G.S., Abdullayev F.Z.: Sorbent for neutralization of the oil-contaminated soils. Patent EAPO 024776 (2016). (in Russian)
11. Hasanov, R.G., Abdullayev, F.Z., Hasanov, G.S., et al.: Sanation of soils contaminated with drilling and oil production wastes. In; 3rd International Turkic World Conference on Chemical Sciences and Technologies 103 (2017)
12. Filatova, M.A., Nikitina, S.A., Taubman, A.V.: Demulsifying action of surfactants and structural-mechanical properties of their adsorption layers. Proc.USSR Acad.Sci. **140(4),** 874–876 (1961). 10.31857/S0869-5652. (in Russian)
13. Kuryavtsev, V.A., Levina, N.S., Svetlov, S.V.: Study of mixing process of the components of mixing contents in arm mixer. Polzunovsky Bulletin **3**, 141–144 (2009). https://doi.org/10.25712/ASTU.2072-8921. (in Russian)
14. Zadeh, L.A.: Fuzzy sets. Inf. Control **8**, 338–353 (1965). https://doi.org/10.1016/S0019-9958(65)90241-X

Multidimensional Classifier of Risk Analysis Methods

Chereshkin Dmitriy[1], Royzenson Gregory[1,2,3(✉)], and Britkov Vladimir[1,2]

[1] Institute for System Analysis (FRCCSC of RAS, ISA), Federal Research Center "Computer Science and Control" of the Russian Academy of Sciences, Vavilova Street 44/2, 119333 Moscow, Russian Federation
dchereshkin@yandex.ru, rgv@isa.ru, britkov@mail.ru

[2] Moscow Institute of Physics and Technology (State University) (MIPT), Institutskiy per. 9, 141701 Dolgoprudny, Moscow Region, Russian Federation

[3] National Research University "Moscow Power Engineering Institute" (MPEI), Krasnokazarmennaya 14, 111250 Moscow, Russian Federation

Abstract. A scheme for constructing a multidimensional classifier of risk analysis methods is proposed. The proposed multidimensional classifier of intelligent risk analysis methods is a convenient analytical tool that enables identification of "gaps" in the development of these methods. The approach proposed makes it possible to define for specific tasks a list of already available methods or the methods, the development of which is a promising task, taking into account properties of socio-economic systems (uniqueness, concurrent presence of more than one weakly formalized goal, significant complexity of determining optimality, dynamic nature, availability of many criteria, human errors, and interpretability of the result); it also suggests the approaches to measure risk (engineering, model, expert, and sociological).

Keywords: Socio-economic systems · Strategic decisions · Crisis situations · Crisis management · Risk management · Artificial intelligence

1 Introduction

Several attempts have been made over the past 100 years to develop a general theory of risk [7,14]. First of all, the pioneering work of A.A. Bogdanov on tektology [7] should be noted that laid the foundations of the theory of systems and cybernetics. A.A. Bogdanov's theory of tektology is valued because it was the first attempt to combine the most diverse areas of human knowledge (associated with the nature and human society) within a unified theory. In accordance with the definition of A.A. Bogdanov's crisis is a change in the organizational form of the complex. A.A. Bogdanov divides crises into two main categories (C and D).

The reported study was funded by RFBR, project numbers 19-07-00522 and 20-57-82004

© The Author(s), under exclusive license to Springer Nature Switzerland AG 2021
R. A. Aliev et al. (Eds.): WCIS 2020, AISC 1323, pp. 529–536, 2021.
https://doi.org/10.1007/978-3-030-68004-6_69

The first category (C) of crises arises from the violation of complete disingressions, i.e. rupture of tectological boundaries, which leads to the formation of new links within the system (complex). The second category, on the contrary, arises from the formation of complete disingressions, the creation of new boundaries where they did not exist, that is, from the breaking of ties. The first category is therefore designated as "crises C", that is, conjugation, connecting; second category as "crises D", I.e.—"disjunctive"—dividing. The developed classification allowed A.A. Bogdanov to formulate another definition of the crisis. A crisis is an imbalance and, at the same time, a process of transition to some new equilibrium. Many studies note [22,34] the importance of A.A. Bogdanov's work as a basis for creating a theory of risk.

New opportunities for developing a risk theory are offered to date by the emergence of artificial intelligence (AI) methods [19], which is an important development of Bogdanov's interdisciplinary tektological approach.

We present first the basic definitions for the terms used.

The measurement of risk is understood in general as the determination of the hazard that emerges from a particular source (type of activity) for an individual or a group [18,30].

The situation is somewhat more complicated in what regards definition to be used for AI research. Scientific community has not developed as yet a definition that would be well-established and unanimously accepted. An immense number of various definitions of AI have been proposed, the comparison and analysis of which is apparently beyond the scope of this work. We define AI as a group of methods and approaches focused on solving weakly structured problems.

Similar to the definition of the AI research area, it is rather difficult to formulate an exact definition of a socio-economic system (SES). The main reason is the immense number of definitions of the concept of "system". Therefore, a constructive way to overcome this problem is to describe SES by means of a set of properties or features [1,24]. Thus, if the object of research fully satisfies certain requirements pertaining to the indicated characteristics, it may be considered as a full-fledged SES. If satisfaction is not complete, the object under study is referred to a quasi SES (or partially SES). Consequently, if unsatisfactory estimates are obtained for most of the features, the object under study cannot be considered as a full-fledged SES.

We now consider in more detail the features that can be used to describe a SES.

The first feature is the *uniqueness* of the SES. The most clear-cut examples of this feature are unsuccessful attempts to apply the health care system developed for one region in another region, various environmental problems, and transport infrastructure.

The second important feature of SES is the *presence of more than one weakly formalized concurrent goal.* The city is a good example in this context. The main goal of the city's development is to provide comfortable conditions for work, living, and leisure for all residents. To fulfill this main task, several subgoals should be set, which can be considered as weakly formalized (for example, transport accessibility, standards of medical services, etc.).

The third feature that characterizes SES is difficulty in determining *optimality* (or even its absence). It is simply impossible for many SES to concurrently "play" several different scenarios of its functioning. An example is various large man-made disasters (the accident at the Chernobyl nuclear power plant, Fukushima power plant, etc.). Main resources (organizational, financial, personnel, etc.) are usually spent after such accidents on eliminating their impacts, and the options to analyze scenarios that could prevent a catastrophic development of the situation are in a number of cases very limited (the possibility of implementing options is an even more challenging task).

The fourth inherent feature of SES is its *dynamism*. Main problems are associated from this perspective with the difficulties of constructing an accurate mathematical model of the SES, which could fully take into account the changes in its state over time.

The fifth important feature that characterizes SES is the *presence of multiple criteria*. A typical example is the multicriteria assessment of credit risk [16] or maintenance of safety standards at an enterprise [30]. It is especially important in case of multicriteria problems to focus on the tasks of strategic choice.

The sixth feature that should be taken into account in describing SES is the *presence of free will* (or *taking into account the human factor*). It is appropriate here to consider various cases of transport accidents (errors of pilots, air traffic controllers, train drivers, etc.).

And, finally, the seventh important feature that characterizes SES is the *interpretability of the result*. In solving problems of risk analysis in SES, recognition of crisis situations, etc. it is of utmost importance not only to fix an abnormal state of the system, but also to develop measures (scenarios) that will prevent catastrophic consequences. Thus, it may be discussed what SES properties/characteristics may be changed and in what way to avoid undesirable impacts or to mitigate them. Therefore, the development of various tools that enable a meaningful risk analysis in SES and various interpretable scenarios is highly needed. It is also necessary to stress that the methods for selecting strategic decisions in SES enable in some cases taking into account various kinds of uncertainties [30].

2 Classification of Methods

We now consider the classification of various methods (including AI) from the perspective of risk measurement [15,18].

The first direction is well known as *engineering* method [15,18]. The main efforts are focused in this direction on collecting statistical data on failures and accidents that involve leakage of harmful substances into environment. The engineering direction concentrates on quantitative calculation of the probability of failures, malfunctions, and other undesirable events. We mention here various methods of pattern recognition, support vector machines, methods of randomized machine learning [21] and neural network approach [11]. The listed methods are intended primarily for solving the classification problem. As already noted, an important characteristic of SES is its uniqueness. In this context, the use of probabilistic approaches can involve certain difficulties (in fact, a lack of precedent-setting information) [28]. An aspect of importance for many SES is the interpretability of the result; therefore, using, for example, a neural network approach not always allows obtainment of an acceptable explanation of the decisions made.

The second direction is usually called the *model* method. This direction involves modeling processes that can provoke various undesirable consequences (accidents, etc.) A problem with this direction is that the construction of a SES model can take considerable time. By the time the model is completed, the SES itself can experience significant transformations, so the modeling process actually needs to be started anew. Thus, the model direction to measuring risk is closely related to such a property of SES as dynamism. In what pertains to the model direction, works on the mathematical theory of risk, the multi-model approach [9,10], and a number of studies should be noted. AI methods (in particular, multi-agent systems and collective behavior of automata) are actively used now in modeling, for example, in design, for developing digital twins, in robotics, for creating smart enterprises and smart cities, etc. The combined use of AI and simulation methods is also closely related to such a property of SES as optimality (for example, a trade-off between the cost of product development, development time, functionality, and reliability).

The third direction to measuring risk is known as an *expert* method [15–18,25]. As already noted, situations often arise in applying the engineering and model directions, when statistical data are insufficient (or there are doubts regarding their reliability). Moreover, in constructing models, it is difficult in some cases to identify various dependencies (the so-called weakly structured problems [17,25]).

Experts are in such a situation virtually the only source of information. Problem of this direction are associated with the subjectivity of expert judgments and the mechanisms for processing both quantitative and qualitative expert estimates [5,25]. As already noted, an important property of SES is multicriteria. Consequently, the use of various multicriteria approaches for expert estimates of SES is an important area of application of AI methods. Multicriteria expert methods can also be subdivided in terms of measurements [16,25]. Another important aspect of using expert methods for risk analysis in SES is the account of important properties of SES, namely, the human factor, obtainment of optimal (preferable) decisions, and the interpretability of the result (decisions).

The fourth direction to measuring risk is known as *sociological* method [6, 15,18]. This direction assumes measuring the perception by the population and its individual groups of a specific risk [12,29]. Among AI methods that have been successfully applied in the sociological direction to measure risk, first of all, the JSM method [8] should be noted (a system for managing flight safety risk factors, analysis and prevention of strikes at enterprises, etc.). The JSM method is designed to establish an unknown property of an object based on the analysis of a set of structures of the object-property type (inductive method

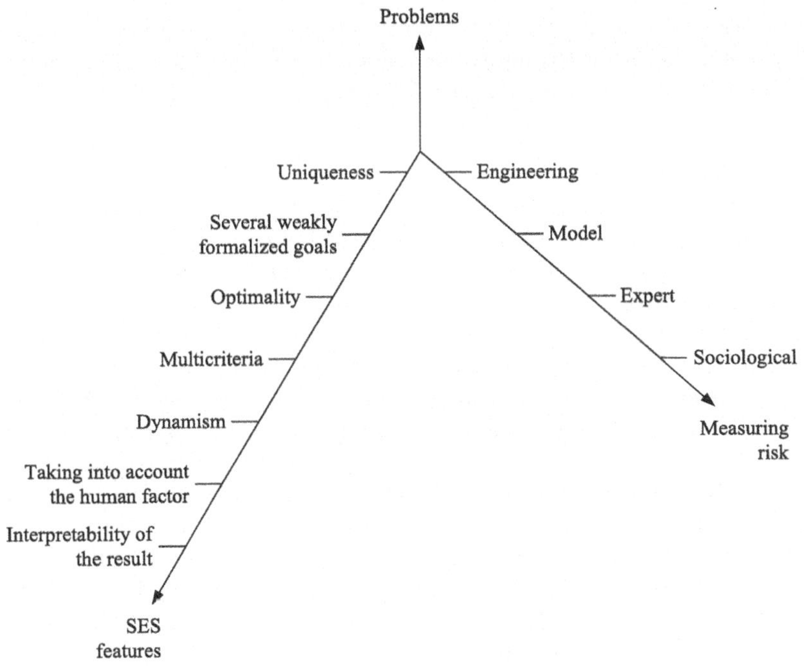

Fig. 1. Multidimensional classifier of risk analysis.

of generating hypotheses). The sociological direction to measuring risk makes it possible to analyze such properties of SES as account of the human factor, concurrent presence of several weakly formalized goals, and multicriteria.

Thus, the main properties of SES, the four ways to measure risk (engineering, model, expert, and sociological), and a list of specific tasks enable development of a multidimensional classifier of risk analysis methods (Fig. 1).

3 Discussion

Nowadays, intensive penetration of AI technologies into everyday life (self-driving cars, face recognition technologies, smart enterprises, etc.) is witnessed. The implementation of such technologies is intrinsically linked with the need to assess new risks. This circumstance, in turn, has prompted many international agencies to initiate activities in the area of AI ethics [13]. These initiatives involve development of a series of standards that will regulate both the development of AI technologies and their application in practice. It is worth noting, in particular, the initiatives of the Institute of Electrical and Electronics Engineers (IEEE) (series of draft standards P7000 - P7014) [2,13,26]; the International Organization for Standardization (ISO) [4], the United Nations Educational, Scientific and Cultural Organization (UNESCO) [3], etc. Success in the development and implementation of these standards apparently is directly related to the creation of a theoretical base required for the development of new methods of risk analysis. Therefore, the proposed multidimensional classifier of risk analysis methods may be used as a convenient analytical tool for solving such problems.

4 Conclusion

Some preliminary conclusions are summarized below.

Obtainment of new basic results in the *engineering* direction to risk analysis is in the near future most likely problematic. This is primarily related to the theory of probability being currently very well developed owing to the efforts of Russian scientific community (A.N. Kolmogorov and others). Nevertheless, should new results be obtained that will make the operation of neural networks more transparent and understandable, some perspective may also emerge in this area.

New results may be associated in the *model* direction with the universal introduction of new digital technologies (the fourth industrial revolution [27,31]), which will strongly expand the usage of a multi-agent approach (collective behavior of robots, smart enterprises, roads, cities, unmanned vehicles, etc.) and various smart sensors (soft measurements and calculations), to which independent decision-making will be delegated.

Emphasis should be made in the *expert* direction in the risk analysis tasks on the development of the main methods to obtain and present knowledge (ontologies (semantic networks), frames, products, and inference methods) [20]. The most promising are the approaches focused on handling non-quantitative (qualitative) information (VDA methods [15–18,25,26], fuzzy logic [5,32,33] and soft computing [31], interaction with computer in human language [19,23], etc.).

However, the most significant new results may be expected in the *sociological* direction to risk analysis. Here, the studies of Nobel prize winners in economics Daniel Kahneman (2002) [12] and Richard Thaler (2017) [29] in the area of behavioral economics should be noted.

The use of several different directions for measuring risk in solving a specific problem and comparing the results, which are actually obtained on the basis of completely different mathematical constructions, makes it possible to more reasonably verify the results obtained and choose one or several (set) methods for solving the problem at once [25].

References

1. Working group on unification of software interfaces intelligent modules (2002). http://www.raai.org/resurs/resurs.shtml?arcim. (in Russian)
2. IEEE: Ethically Aligned Design (2016). https://ethicsinaction.ieee.org
3. Elaboration of a recommendation on the ethics of artificial intelligence (2019). https://en.unesco.org/artificial-intelligence/ethics
4. ISO/IEC CD 23894: Information Technology. Artificial Intelligence. Risk Management (2020). https://www.iso.org/en/standard/77304.html?browse=tc

5. Aliev, R.A.: Fundamentals of the Fuzzy Logic-Based Generalized Theory of Decisions. Springer, Heidelberg (2013)
6. Beck, U.: Risk Society: Towards a New Modernity. SAGE Publications, London (1992)
7. Bogdanov, A.: Essays in Tektology: the General Science of Organization, 2nd edn. Intersystems Publications, Seaside (1984)
8. Finn, V.K.: Epistemological foundations of the JSM method for automatic hypothesis generation. Autom. Document. Math. Linguist. **48**(2), 96–148 (2014)
9. Gelovani, V.A., Britkov, V.B., Dubovskiy, S.V.: USSR and Russia in the global system (1985–2030): Results of global modeling. URSS, Moscow (2018). (in Russian)
10. Gelovani, V.A., Britkov, V.B., Yurchenko, V.V.: An interactive modeling system for analysis of alternative decisions. In: IIASA Proceedings Series, pp. 149–151. Pergamon Press, Oxford (1980)
11. Haykin, S.: Neural Networks and Learning Machines, 3rd edn. Pearson Education, Singapore (2016)
12. Kahneman, D., Slovic, P., Tversky, A. (eds.): Judgment Under Uncertainty: Heuristics and Biases. Cambridge University Press, Cambridge (1982)
13. Karpov, V.E., Gotovtsev, P.M., Roizenzon, G.V.: To the issue of ethics and artificial intelligence systems. Philos. Soc. **2**(87), 84–105 (2018). (in Russian)
14. Knight, F.H.: Risk, Uncertainty and Profit. Houghton Mifflin Co, Boston (1921)
15. Larichev, O.: Comparison of decision alternatives with regard to risk and safety considerations: methodological problems, pp. 189–196. No. WP-88-34, International Institute for Applied Systems Analysis (1988)
16. Larichev, O., Borisenkov, P., Asanov, A., Naryzhny, Y., Royzenzon, G.: Development and implementation of automated system for evaluation of commercial bank credit quality. Audit Financ. Anal. **1**, 176–180 (2001)
17. Larichev, O., Moshkovich, H.: Verbal Decision Analysis for Unstructured Problems. Kluwer Acedemic Publishers, Boston (1997)
18. Larichev, O.I.: Factor-based decision making problems risk and safety. Vestnik AN SSSR **57**(11), 38–45 (1987). (in Russian)
19. Osipov, G.S.: Methods of Artificial Intelligence. Fizmatlit, Moscow (2011). (in Russian)
20. Pedrycz, W.: Knowledge-Based Clustering: from Data to Information Granules. Wiley, New York (2005)
21. Popkov, Y.S.: Soft randomized machine learning procedure for modeling dynamic interaction of regional systems. Entropy **21**(4), 424 (2019)
22. Popkov, Y.S., Tishchenko, V.I. (eds.): Systems Research. Methodological Problems. Yearbook 2013-2014. Thirty seventh issue. URSS, Moscow (2014). (in Russian)
23. Popov, E.V.: Talking with Computers in Natural Language, 2nd edn. Springer, Heidelberg (2011)
24. Pospelov, D.A.: Situation-Driven Control: Theory and Practice. Nauka, Moscow (1986). (in Russian)
25. Royzenson, G.V.: Synergistic effect in decision making, pp. 248–272. No. 36. 2011–2012, URSS, Moscow (2012). (in Russian)
26. Royzenson, G.V.: Problems of formalization of the concept of ethics in the artificial intelligence. In: Sixteenth Russian Conference on Artificial Intelligence (RCAI-2018). Conference Proceedings, vol. 2, pp. 245–252. RKP, Moscow (2018). (in Russian)

27. Schwab, K.: The Fourth Industrial Revolution. Crown Publishing Group, New York (2017)
28. Taleb, N.N.: Black swans and the domains of statistics. Am. Stat. **61**(3), 198–200 (2007)
29. Thaler, R.H.: Misbehaving: the making of behavioral economics. W.W. Norton and Company, New York (2015)
30. Tsygichko, V.N., Chereshkin, D.S., Smolyan, G.L.: Safety of critical infrastructures. URSS, Moscow (2019). (in Russian)
31. Yusupbekov, N.R., Tarassov, V.B.: Integrated intelligent systems for industrial automation: the challenges of Industry 4.0, information granulation and understanding agents. Chem. Technol. Control Manage. **6**(84), 62–75 (2018)
32. Zadeh, L.A.: Fuzzy sets. Inf. Control **8**(3), 338–353 (1965)
33. Zadeh, L.A., Kacprzyk, J.: Fuzzy Logic for the Management of Uncertainty. Wiley, New York (1992)
34. Zelený, M.: Tectology. Int. J. General Syst. **14**(4), 331–342 (1988)

Modelling of the Droplet Velocity of the Dispersed Phase Components in Liquid Extraction

A. A. Hasanov(✉)

Azerbaijan State Oil and Industry University, Azadliq Avenue 20, Nasimi, Baku, Azerbaijan
alakbar48-48@mail.ru

Abstract. Existing methods (physicochemical, chemical, biological, etc.) of wastewater treatment from organic solvents have a number of disadvantages. To eliminate these disadvantages, liquid extraction in a spray extractor has been applied, in which one liquid can be dispersed into another in the form of small drops, resulting in formation of a large contact surface between two phases. Mass transfer from single droplets rising in the second liquid was researched. It has been determined that amount of the substance extracted during the contact between the phases can be divided into three parts during the periods: formation, rise, and coalescence of drops. The study of extraction during the period of droplet formation showed that up to 20% of the maximum possible amount of the distributed component corresponding to the equilibrium can pass from one phase to another. It was found that the rate of mass transfer increases with an increase in the motion velocity of the liquid flowing around the drop, since mixing in the drop promotes a high speed of movement. In the spray extractor, droplets are formed in the holes of the distributor, through which the liquid is fed into the column. The scheme of the extractor, the composition of waste water, calculations of the hydrodynamic parameters of the process are presented in the article. Mathematical model of the process has been obtained on the basis of the experimental data taking into account the hydrodynamic and mass transfer parameters, such as the true and average concentration of components in the continuous and dispersed phases, the volume of the extractor, the coefficients of molecular diffusion, mass transfer, mass efficiency, contact surface area, the component distribution coefficient, the number of stages, the volumetric velocities of a droplet of components (butyl glycol, phenylisopropanol, acetic acid, acetylacetone) of the dispersed phase, which play an important role in the design of extractors have been calculated.

Keywords: Extraction · Waste water · Droplets · Volumetric velocity · Diffusion · Component · Design

1 Introduction

The mechanism of mass transfer with known surface and phase contact time has been investigated by single drops. A device designed to carry out mass transfer between two liquids can be made quite compact if one liquid is dispersed to the other in the form of

© The Author(s), under exclusive license to Springer Nature Switzerland AG 2021
R. A. Aliev et al. (Eds.): WCIS 2020, AISC 1323, pp. 537–544, 2021.
https://doi.org/10.1007/978-3-030-68004-6_70

small drops. The advantage of this device is that a very large contact surface is created between the two phases. In this case, one of the liquids can be introduced into the other by feeding it through nozzles, as a result of which drops are formed, which rise or fall in the second (continuous) phase [1]. The data of the results of works related to both droplet mechanics and mass transfer are contained in [2].

In a typical experimental study, the continuous phase is in a vertical transparent tube. The second liquid is introduced through a small capillary from the bottom, resulting in a sequence of droplets that rise upward. At the end, they coalesce and are removed as a continuous phase. Thus, the total amount of the substance extracted during the contact time between the phases can be divided into three periods: formation, rise, and coalescence of drops.

The study of extraction during the period of droplet formation showed that up to 20% of the maximum possible amount of the distributed component corresponding to the equilibrium one can pass from one phase to another [3, 4]. During the period of droplet formation, mass transfer can be very significant. After the formation and beginning of movement, single drops soon reach their maximum speed [5].

2 Experimental Part

In spray columns, the contact time can be significantly longer, since back mixing leads to retention of the dispersed phase. In the case of somewhat large droplets, internal circulation occurs due to the presence of frictional resistance.

As the droplet size grows, its limiting velocity reaches a maximum and then slowly decreases. In addition, as the droplet size increases to the maximum limiting velocity, the droplet shape is distorted. Drops of even larger sizes tend to disintegrate with the formation of two or several smaller droplets. Therefore, we tried to carry out experiments so that the droplet sizes remained constant.

In our case [6], the working area, in which the extraction takes place, does not contain any devices and the droplets of the dispersed liquid rise freely through the continuous phase (Fig. 1).

In a spray extractor, droplets are formed in the holes of the distributor through which the liquid is fed into the column. The speed of movement of the dispersed liquid droplets relative to the walls of the column depends on the viscosity, the difference in density, as well as the linear (volumetric) velocity of the dispersed phase. Therefore, the determination of the velocity in the dispersed phase is of great interest.

When studying the process of liquid-phase extraction, samples of wastewater containing butyl glycol, phenyl-isopropanol, acetic acid, acetylacetone were used (Table 1).

Isopropyl ether was used as an extractant.

3 Results and Discussion

For the first time a formula for determining the velocity of components in the dispersed phase based on the equations of mathematical modeling has been proposed, taking into account the true and average concentrations of the components in the continuous and

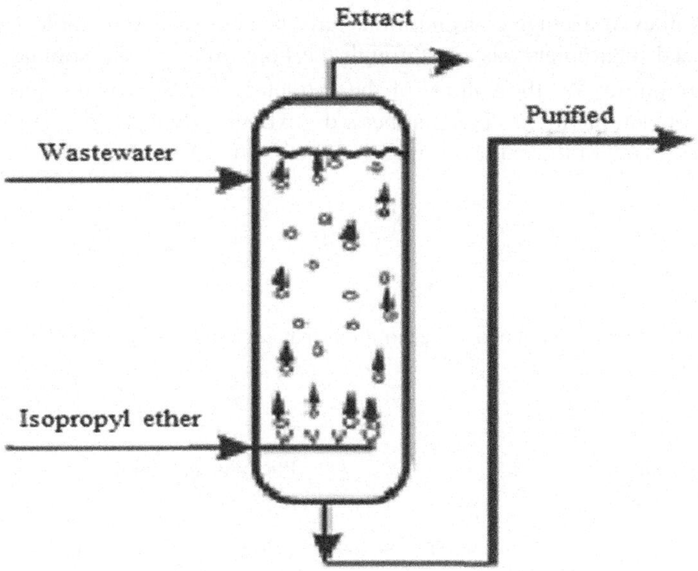

Fig. 1. Spray column

dispersed phases, the volume of the extractor, the coefficients of molecular diffusion, mass transfer, mass efficiency, contact surface area, the component distribution coefficient (phase constant equilibrium), the number of steps, the volume fraction of the dispersed phase.

Table 1. Composition of waste water

Composition	Consumption, mol/h	Concentration, mol/dm^3
Butyl glycol	0,339	0,113
Phenylisopropanol	0,143	0,048
Acetylacetone	0,036	0,012
Water	162,433	54,144

The equation is written as follows

$$W_i = \frac{-k_i \cdot FV\left[(C_i)_i^{av} - (\alpha_i c_i)_i\right]}{(1-\varepsilon)\left[(\alpha_i c_i)_{i-1} - (\alpha_i c_i)_i\right]n}$$

where C_i^{av} – average concentration of a component in a cell [7].

$$(C_i)_i^{av} = (\alpha_i C_i)_i - \left[(\alpha_i C_i)_i - (C_i)_{i-1}\right]\frac{\varepsilon U_d^0}{K \cdot Fh_i}\left(1 - e^{-\frac{K_i Fh_i}{\varepsilon U_d^0}}\right)$$

The found values of average concentrations have been presented in Table 2 and 3.

U_d^0—speed of droplets passing through the hole, m/s; ε - the volume fraction of the dispersed phase; V - the volume of the extractor, m^3; F - specific surface area of interphase contact, m^2/m^3; i - cell numbers; d - extractor diameter d = 0.12 m; S - the cross-sectional area of the extractor, m^2; $S = 0.785 \cdot (0.12)^2 = 0.0113\ m^2$.

$$S = \frac{\pi d^2}{4}$$

Table 2. Distribution of component concentrations in phases (Butyl glycol and Phenylispropanol consentrations)

№	Concentration, mol/dm^3					
	Butyl glycol			Phenylisopropanol		
	C_1	$\alpha_1 C_1$	C_1	$\alpha_1 C_1$	C_1	$\alpha_1 C_1$
1	0,0397	0,00056	0,0397	0,00056	0,0397	0,00056
2	0,0493	0,00069	0,0493	0,00069	0,0493	0,00069
3	0,0548	0,00072	0,0548	0,00072	0,0548	0,00072
4	0,0677	0,00095	0,0677	0,00095	0,0677	0,00095
5	0,0753	0,00106	0,0753	0,00106	0,0753	0,00106
6	0,0873	0,00118	0,0873	0,00118	0,0873	0,00118
7	0,0930	0,00131	0,0930	0,00131	0,0930	0,00131
8	0,1149	0,00162	0,1149	0,00162	0,1149	0,00162
9	0,1419	0,00200	0,1419	0,00200	0,1419	0,00200
10	0,1577	0,00222	0,1577	0,00222	0,1577	0,00222
11	0,1753	0,00247	0,1753	0,00247	0,1753	0,00247
12	0,1948	0,00274	1,3319	0,2508	0,00343	0,6619
13	0,2406	0,00339	1,4113	0,2753	0,00377	0,7017
14	0,2972	0,00419	1,4954	0,3024	0,00414	0,7337
15	0,3671	0,00517	1,5845	0,3261	0,00446	0,7883
16	0,4079	0,00574	1,6790	0,3503	0,00479	0,8357
17	0,4533	0,00639	1,7791	0,3750	0,00513	0,8859
18	0,5037	0,00710	1,8852	0,3750	0,00513	0,9390
19	0,5597	0,00789	1,9977	0,3750	0,00513	0,9954
20	0,6219	0,00885	2,1169	0,3750	0,00513	1,0552
21	0,6547	0,00923	2,2431	0,3750	0,00513	1,1183
22	0,7275	0,01026	2,3766	0,3750	0,00513	1,1495

H-height of the extractor, determined using equations.

H = n · h.
where n is the number of theoretical steps, n = 22;
h - is the distance between theoretical steps, h = 0.05 m;
Then H = 22 · 0.05 = 1.1 m.
k_i - mass transfer coefficient of components, is determined by the formula [6]:

$$\frac{1}{k_i} = \frac{1}{k_i^c} + \frac{\alpha_i}{k_i^D} \tag{1}$$

where α_i is the distribution coefficient of components in phases and is determined from the equilibrium equation. k_i^c и k_i^D. Are the coefficients of mass transfer of the i-th component in the continuous and dispersed phases, respectively [8, 9]:

Table 3. Distribution of component concentrations in phases (Acetic acid and Acetylacetone consentrations)

№	Concentration, mol/dm^3					
	Acetic acid			Acetylacetone		
	C_3	$\alpha_3 C_3$	$C_3^{av.}$	C_4	$\alpha_4 C_4$	$C_4^{av.}$
1	0,0503	0,00070	0,0066	0,0018	0,024	0,0004
2	0,0665	0,00092	1,0827	0,0024	0,032	0,0851
3	0,0739	0,00102	1,1423	0,0034	0,046	0,0902
4	0,0822	0,00114	1,2052	0,0043	0,058	0,0955
5	0,0914	0,00127	1,2716	0,0054	0,073	0,1012
6	0,1016	0,00141	1,3417	0,0067	0,091	0,1071
7	0,1129	0,00156	1,4156	0,0075	0,102	0,1134
8	0,1255	0,00174	1,4935	0,0094	0,128	0,1201
9	0,1395	0,00193	1,5758	0,0105	0,142	0,1273
10	0,1550	0,00215	1,6627	0,0131	0,178	0,1348
11	0,1723	0,00239	1,7543	0,0146	0,198	0,1487
12	0,1915	0,00266	1,8509	0,0163	0,221	0,1576
13	0,2128	0,00295	1,9528	0,0182	0,247	0,1670
14	0,2365	0,00328	2,0603	0,0226	0,307	0,1770
15	0,2628	0,00365	2,1739	0,0252	0,347	0,1875
16	0,2921	0,00406	2,2938	0,0280	0,380	0,1987
17	0,3607	0,00501	2,4202	0,0347	0,471	0,2106
18	0,4454	0,00619	2,5536	0,0429	0,583	0,2232
19	0,5499	0,00764	2,6444	0,0477	0,648	0,2365
20	0,7817	0,01086	2,8429	0,0531	0,722	0,2506
21	0,8686	0,01207	2,9995	0,0591	0,803	0,2656
22	1,0725	0,01490	2,9995	0,0657	0,897	0,2815

$$k_i^c = \frac{2\pi^2 D_i^c}{3d_d};\ k_i^D = \frac{2\pi^2 D_i^D}{3d_d}$$

Here, D_i^C and D_i^D are the coefficients of molecular diffusion of the components in the continuous and dispersed phases, respectively [10].

d_d - drop diameter d = $4 \cdot 10^{-3}$ m.

All process parameters are involved in determining the velocity of the components in the dispersed phase. Therefore, this formula is assumed to be universal for determining the velocity of components in the dispersed phase in all cells of the extractor.

For all four components (butyl glycol, phenylisopropanol, acetic acid, acetylacetone), Eq. (1) takes the following form

$$(W_1)_i = \frac{-k_1 F \cdot V\left[(C_1)_i^{av} - (\alpha_1 c_1)_i\right]}{(1-\varepsilon)\left[(\alpha_1 c_1)_{i-1} - (\alpha_1 c_1)_i\right]n} \quad (2)$$

$$(W_2)_i = \frac{-k_2 FV\left[(C_2)_i^{av} - (\alpha_2 c_2)_i\right]}{(1-\varepsilon)\left[(\alpha_2 c_2)_{i-1} - (\alpha_2 c_2)_i\right]n} \quad (3)$$

$$(W_3)_i = \frac{-k_3 FV\left[(C_3)_i^{av} - (\alpha_3 c_3)_i\right]}{(1-\varepsilon)\left[(\alpha_3 c_3)_{i-1} - (\alpha_3 c_3)_i\right]n} \quad (4)$$

$$(W_4)_i = \frac{-k_4 FV\left[(C_4)_i^{av} - (\alpha_4 c_4)_i\right]}{(1-\varepsilon)\left[(\alpha_4 c_4)_{i-1} - (\alpha_4 c_4)_i\right]n} \quad (5)$$

where indices 1, 2, 3, 4 refer to butyl glycol, phenylisopropanol, acetic acid, acetylacetone, respectively.

The volume fraction of the dispersed phase in the continuous phase is determined by formula [6]

$$\varepsilon = \frac{U_d}{W_d}$$

where $U_d = \frac{V_e}{S_{fl.rate}}$ (V_e - isopropyl ether volumetric flow rate).

$V_e = 80$ cm^2/h = 0,022 $\cdot$ 10^{-6} m^3/s.

W_d - true linear velocity of the dispersed phase droplet, which is determined by the Stokes equation [6, 9, 10]

$$W_d = \frac{d_d^2(\rho_c - \rho_D) \cdot g}{18\mu_c} \quad (6)$$

Here d$_d$ - diameter of the dispersed phase droplet in the column, $d_d = 4 \cdot 10^{-3}$ m;

ρ_c - solid phase density;

$\rho_c = 1 \bullet 10^3$ кg/m^3;

ρ_D - dispersed phase density,

$\rho_D = 728$ кg/m^3;

μ_c - continuous phase viscosity,

$\mu_c = 1 \bullet 10^{-3}$ Пз.

Substituting these values into (6), $W_d = 0{,}242$ m/s will be obtained Then ε ≈ 0,02.

Specific surface of interphase contact F = 30 m^2/m^3.

The results of calculating the coefficients of molecular diffusion, distribution of components, mass efficiency and mass transfer are shown in Table 4.

Table 4. Coefficients of molecular diffusion, distribution of components, mass transfer and mass efficiency.

Component	$D_i^{c.v.} \cdot 10^9$	$D_i^d \cdot 10^9$	$K_i^{c.v.} \cdot 10^5$	$K_i^d \cdot 10^5$ m²/sec	$K_i \cdot 10^5$	α_i
Butyl glycol	0,894	4,18	0,147	0,687	0,141	0,2139
Phenylisopropanol	0,658	3,07	0,108	0,504	0,103	0,2142
Acetic acid	1,280	5,98	0,210	0,983	0,201	0,2136
Acetylacetone	0,831	3,88	0,136	0,673	0,130	0,2135

Table 5. Velocity distribution of dispersed phase components in extractor cells

№	Component velocities in continuous phase, hour^{-1}			
	W_1	W_2	W_3	W_4
	Butyl glycol	Phenylisopropanol	Acetic acid	Acetylacetone
1	0,0306	0,0284	0,0475	0,0322
2	0,031	0,0284	0,0461	0,0322
3	0,0295	0,0261	0,468	0,0360
4	0,0304	0,0261	0,0467	0,0360
5	0,0309	0,0261	0,0457	0,0403
6	0,0304	0,0261	0,0457	0,0403
7	0,0299	0,0268	0,0461	0,0302
8	0,0306	0,0268	0,0461	0,0302
9	0,0306	0,0258	0,0461	0,0169
10	0,0304	0,0258	0,0461	0,0169
11	0,0304	0,0254	0,0468	0,0291
12	0,0300	0,0254	0,0468	0,0291
13	0,0307	0,0250	0,0457	0,0266
14	0,0296	0,0250	0,0468	0,0266
15	0,0307	0,0257	0,0461	0,0248
16	0,0298	0,0257	0,0468	0,0248
17	0,0304	0,0262	0,0461	0,0338
18	0,0305	0,0262	0,0461	0,0338
19	0,0302	0,259	0,0461	0,0270
20	0,0328	0,0259	0,0461	0,0270
21	0,0305	0,0261	0,0461	0,0353
22	0,0305	0,0261	0,0461	0,0353

Substituting the found values of the above parameters characterizing the hydrodynamics and mass transfer in the spray column into the mathematical modeling Eq. (2), (3), (4), (5), the values of the droplet velocity of the components in the dispersed phase were obtained (Table 5).

With an increase of number of cells, the volume rate of butyl glycol henylisopropanol and acetic acid decrease, while the volume rate of acetylacetone increase. All calculation were made in MS Excel. The model presented in the article can be used in the analysis of chemical processes. This indicate that the solvent is more effective against butyl glycol, phenylisopropanol and acetic acid.

4 Conclusion

Mass transfer from single droplets rising in the second liquid was subjected to research. It has been determined that the total amount of the substance extracted during the contact between the phases can be divided into three parts during the periods: formation, rise, and coalescence of drops. The study of extraction during the period of droplet formation showed that up to 20% of the maximum possible amount of the distributed component corresponding to the equilibrium one can pass from one phase to another. It was found that the rate of mass transfer increases with an increase in the motion rate of the liquid flowing around the drop, since mixing in the drop promotes a high speed of movement. According to the experimental data obtained, using the equations of a mathematical model, which take into account the true and average concentrations of the components in the continuous and dispersed phases, the volume of the extractor, the coefficients of molecular diffusion, mass transfer, mass efficiency, contact surface area, the component distribution coefficient, the number of steps, the volume velocities of the component droplet have been calculated (butyl glycol, phenylisopropanol, acetic acid, acetylacetone) of the dispersed phase, which play an important role in the design of extractors.

References

1. Sherwood, T., Pigford, R., Wilkie, C.: Mass Transfer. Chemistry, Moscow (1982)
2. Braunstein, B.I., Shchegolev, V.V.: Hydrodynamics, Mass and Heat Exchange in Column Apparatuses. Chemistry, Leningrad (1988)
3. Treybal, R.: Liquid Extraction. Publishing house Chemistry (1996)
4. Zulkovsky, Z.: Liquid Extraction in Chemical Industry. State chem. Publication (1963)
5. Kelbaliyev, G.I., Suleimanov, G.Z., Phariborz, Zorofi, A., Gasanov, A.A., Rustamova, A.I.: Extraction separation and cleaning of sewage waters by organic solvents with recirculation. Russ. J. Appl. Chem. **83**(6), 1114–1119 (2011)
6. Gasanov, A.A. : The developing of liquid extraction of paint industry wastewater in spray column. J. Water Chem. Technol. **38**(4), 407(2016)
7. Gasanov, A.A., Suleimanov, G.Z., Kelbaliev, G.I.: Analyses of liquid extraction cleaning process of lacquer dye production waters in the extractor with mixer. J. Water Chem. Technol. **38**(6), 607–617 (2016)
8. Gasanov, A.A.: Mathematical model of extraction process of multi-component wastewaters. J. Water Chem. Technol. **3**(40), 300–308 (2018)
9. Romankov, P.G., Frolov, V.F.: Mass Exchange Processes of Chemical Technology (Systems with a Dispersed Solid Phase). Chemistry, Leningrad (1990)
10. Dytnersky, Y.: Processes and Devices of Chemical Technology. Chemistry, Moscow (1995)

Implementation in Analytical Cloud the Algorithm of Optimization of MDEA Consumption

Farukh Adilov[1(✉)], Fakhritdin Abdurasulov[2], Andrey Korotaev[1], Fedor Davidov[1], and Sitora Isakova[1]

[1] LLC "XIMAVTOMATIKA", Tashkent, Republic of Uzbekistan
{Farukh.Adilov,Andrey.Korotaev,Fedor.Davidov, Sitora.Isakova}@himavtomatika.uz
[2] Oltin Yo'l GTL" Company, Tashkent, Republic of Uzbekistan
info@oltinyolgtl.com

Abstract. This article describes the implementation in a virtual cloud IT-infrastructure of an algorithm developed by a group of scientists to optimize the consumption of methyldiethanolamine solution (MDEA), which is widely used in the technological processes of natural gas purification at gas processing plants. The algorithm assumes implementing a mathematical model of a process unit deployed in the cloud and adjusting the operation mode of the unit based on the results of interaction between the digital twin and the production facility itself in real time mode.

Keywords: Methyldiethanolamine (MDEA) · Algorithm · Analytical cloud · UniSim digital twin · Cloud infrastructure · Gas processing plant · Pressure · Temperature · Concentration

1 Introduction

The so-called upstream part of the oil and gas sector is not so complicated in terms of control algorithms for technological processes, compared to the processes of deep processing of hydrocarbons, but on this sector of processes group of scientists-co-authors of this article has developed and tested the functionality of cloud solutions for optimization of control processes. In this case, we mean the technological process of amine purification of natural gas using a solution of methyldiethanolamine (MDEA) as a chemosorbent.

A large number of articles [1–4] are devoted to solutions for creating cloud infrastructure in general and cloud solutions for creating specific digital twins in this cloud environment, but they are mostly of a conceptual scientific nature and rarely describe specific technological processes.

In this article, we would like to open a series of articles devoted to the application of specific optimization algorithms in the cloud infrastructure to improve the efficiency of the selected technological process or process unit [5, 6].

© The Author(s), under exclusive license to Springer Nature Switzerland AG 2021
R. A. Aliev et al. (Eds.): WCIS 2020, AISC 1323, pp. 545–552, 2021.
https://doi.org/10.1007/978-3-030-68004-6_71

2 Parameters Selection for Process Optimization

The amine treatment unit is part of a process unit for removing acidic components of natural gas, designed to remove acid gases, H_2S and CO_2 contained in raw gas. This technological unit is an integral part of any modern gas processing plant, and optimization of its operation is one of the most significant factors in assessing the technical and economic efficiency of the entire enterprise [7–10].

The classic technological scheme for the removal of acidic components of natural gas can be presented in the following view (see Fig. 1).

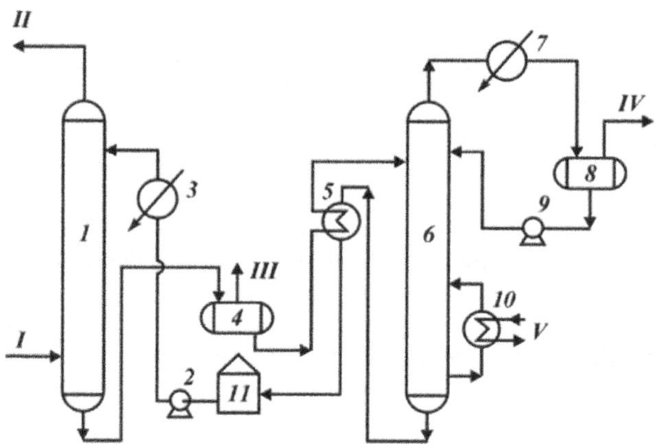

Fig. 1. Scheme for removing acid gases with ethanolamine solutions: I – gas to purification; II – cleaning gas; III – expansion gas; IV – acid gas; V – steam; 1 – absorber; 2,9 – pumps; 3,7 – cold storages; 4 – expander; 5 – heater; 6 – desorber; 8 – separator; 10 – boiler; 11 – regenerated amine drum.

In the study, which is the subject of this article, as a chemical absorbent was considered the solution of methyldiethanolamine (MDEA). MDEA reacts slowly with CO_2, significantly less than with H_2S. In this case, the reaction occurs mainly:

$$H_2S + R_2R'N \leftrightarrow HS^- + R_2R'NH^+ \tag{1}$$

Therefore, solutions of tertiary amines selectively absorb H_2S in the presence of CO_2.

The frequency of absorbent circulation is one of the most important indicators of gas treatment plants, since it accounts for approximately 60% of the energy costs of the cleaning process. The consumption of the absorbent must be such as to ensure the required degree of gas purification from acidic components with minimal process costs. A higher consumption of the absorber compared to the required one will lead to an increase in the cost of cleaning, and a lower one will not provide the specified degree of gas cleaning. In turn, the required consumption of the absorbent is determined by the working degree of saturation of the solution with acidic components, which is taken on the basis of data on the maximum equilibrium degree of saturation of the amine [11, 12].

The total capacity is equal to the sum of the chemical and physical capacity:

$$\alpha_i = \alpha_{ix} + \alpha_{i\phi} \tag{2}$$

The maximum chemical capacity in accordance with the above equations for the reactions of H_2S and CO_2 with amines is equal to:

$$1 - \alpha_{ax} - n \cdot \alpha_{Bx} = 0 \tag{3}$$

where α_{ax}, α_{Bx} - chemical capacity by H_2S and CO_2, mol/mol.

In the temperature range of 10–100 °C and partial pressures up to 2 MPa the physical capacity of the amine can be determined from the data on the solubility of H_2S and CO_2 in water:

$$\alpha_{a\phi} = y_a * К/m_a * C_{Ж0} \tag{4}$$

$$\alpha_{в\phi} = y_в * К/m_в * C_{Ж0} \tag{5}$$

where y - molar fraction of the component in the gas;

$C_{Ж0}$ - amine concentration in solution, mol/l;

m_a, $m_в$ - phase equilibrium constant of H_2S and CO_2 in water, (kmol/m^3)/(kmol/m^3):

$$m_{H2S} = 0,257 + 0,00716 * t \tag{5}$$

$$m_{CO2} = 0,6 + 0,028 * t \tag{6}$$

К - conversion rate to normal conditions:

$$К = 10 * p * 273/22{,}4 * (273 + t) \tag{7}$$

where p - the total absorption pressure, MPa; t – solution temperature, °C.

The combined solution of these equations allows us to find an expression for determining the minimum specific consumption of the absorber:

$$q_p = \frac{0{,}446 \cdot [y_{H_2S} + (1-\varphi) \cdot n \cdot y_{H_2S}]}{C \cdot [1 + {}^{K}⊠_{C_Ж}(y_{H_2S}/m_{H_2S} + [(1-\varphi) \cdot n \cdot y_{CO_2}]/m_{CO_2}) - (\alpha_{H_2S_p} - n \cdot \alpha_{CO_2})_p]} \tag{8}$$

where φ – selectivity of H_2S absorption from its mixtures with CO_2. $\varphi = (C_{yx\ CO2} / C_{BX\ CO2})$;

For DEA - n = 2, for MDEA - n = 1.

The above equation is valid only for simultaneous absorption of H_2S and CO_2 by aqueous solutions of amines at a high gas pressure (more than 2 MPa) and a high content of acidic components.

3 Conclusions on the Influence of Process Optimization Parameters

3.1 Influence of Pressure

Increasing the pressure at a constant temperature and amine concentration increases the degree of gas purification from acidic components. Usually, gas purification with amine solutions is carried out at a pressure of 2 to 7 MPa (see Fig. 2).

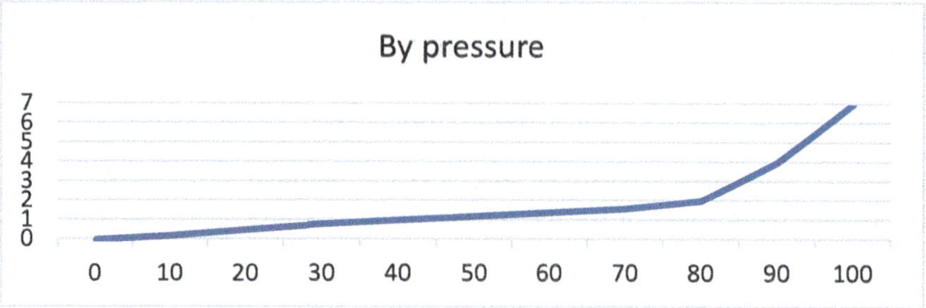

Fig. 2. Graph of pressure dependence on the additional coefficient K.

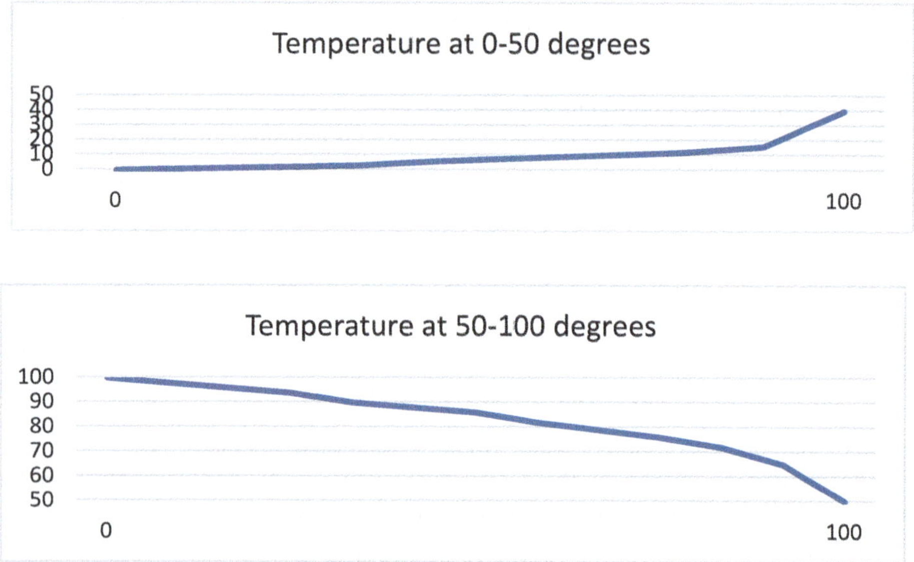

Fig. 3. Graph of temperature dependence on the additional coefficient K.

3.2 Influence of Absorption Temperature

A decrease in the absorption temperature leads to an increase in the degree of extraction of the target components, but does not reduce the selectivity of the process due to an increase in the solubility of hydrocarbons in amine solutions (see Fig. 3).

An increase in temperature increases the selectivity of the process with respect to acidic components, but may lead to an increase in the residual content of acidic components in the purified gas, and to an increase in the moisture content of the purified gas.

The interaction of hydrogen sulfide with amine occurs in the range from 30 to 50 °C.

3.3 Influence of Concentration of Amine Solution

The use of high-concentration amine solutions makes it possible to reduce the volume of the circulating solution and, consequently, reduce the cost of pumping the solution, but at the same time:

- the temperature of the amine solution increases excessively.
- increases steam consumption for regeneration.
- the loss of the amine solution increases due to evaporation (see Fig. 4).

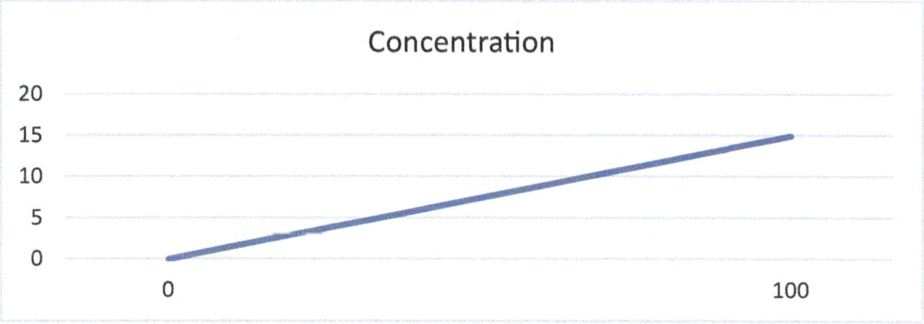

Fig. 4. Graph of concentration dependence on the additional coefficient K.

4 Algorithm of Optimization of MDEA Consumption

Based on these parameters of influence on the process, an algorithm can be constructed for calculating the optimal value of the theoretical additional coefficient depending on several values (temperature, pressure, amine concentration, acid gas concentration in the incoming stream), for each of which the optimal additional coefficient is calculated. This algorithm helps to significantly reduce the energy consumption of the process and the loss of MDEA solution due to entrainment. The structure and flow diagram of the algorithm for calculating the optimal additional coefficient with subsequent adjustment of data from the gas analyzer measurements and flow parameters is presented below (see Fig. 5).

As can be seen from the flowchart, K_{1T}, K_{2P}, $K_{3\,\text{amine conc.}}$, $K_{4\,\text{acid gases conc.}}$ – additional coefficients, each of which is defined as follows:

$$K = \frac{(100 - K_{current\,value\%})}{100} \tag{9}$$

To define the value of K_{total} let us use following formula:

$$K_{total} = \sum\nolimits_{i=1}^{n} K_i/n \tag{10}$$

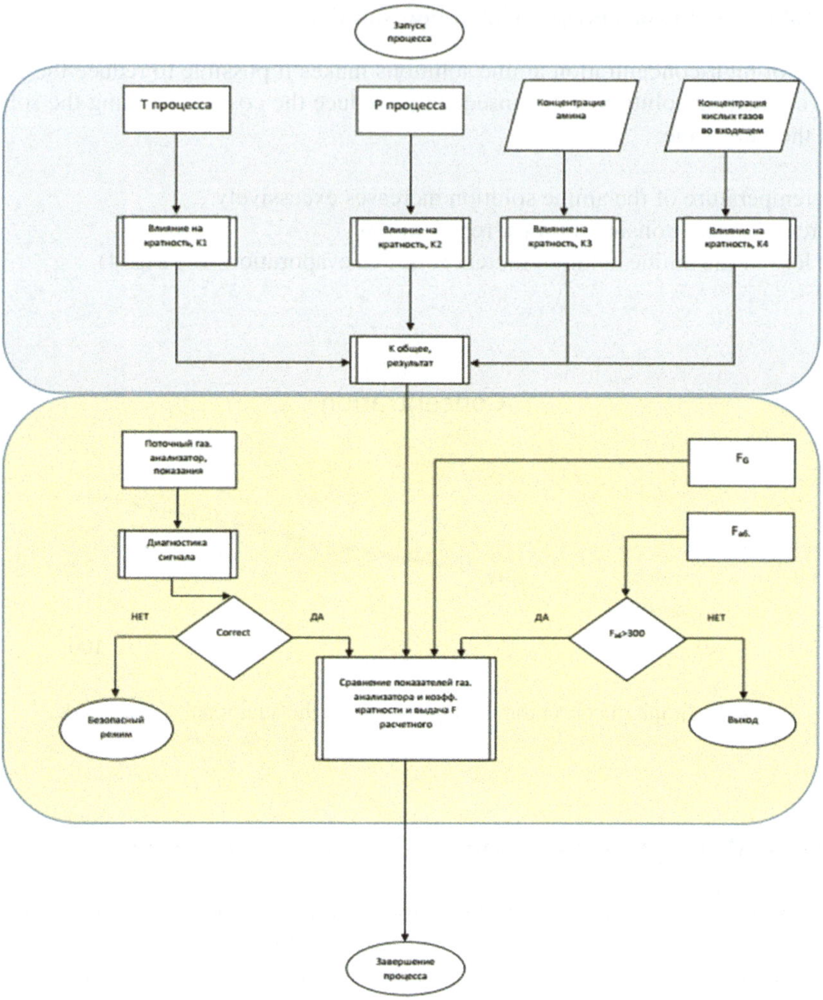

Fig. 5. 4Algorithm of optimization of MDEA consumption.

where

n = 4 (4 additional coefficients);

K_{1K} - the dependence on the concentration is linear, which is expressed in Fig. 4 as a straight line.

K_{1T}, K_{2P} – the dependence on the parameters, shown by experimental measurements on a mathematical model in the UniSim Design software environment, used as a software shell deployed in the cloud infrastructure. This dependence tends to level functions, as can be seen in Figs. 2 and 3.

Simplified architecture of the “cloud” solution for creating a “digital twin” based on the Unisim technology is shown in Fig. 6 below [3].

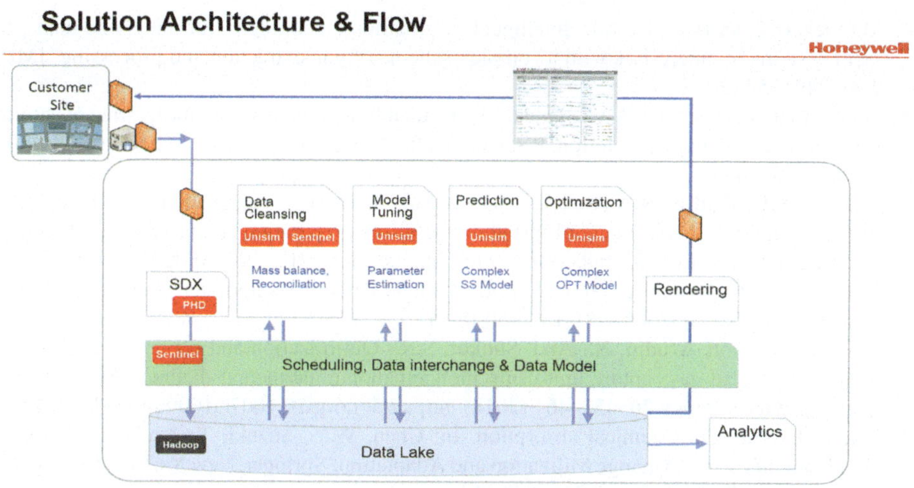

Fig. 6. Simplified architecture of the "cloud" solution for creating a "digital twin" based on the Unisim technology.

5 Conclusion

The solution described for implementation in a cloud infrastructure mathematical models of process unit for removal of acid gases with built-in optimization algorithm of MDEA consumption allows you to increase the capacity by sour gas, to provide the desired composition of the clean gas and to reduce the consumption of MDEA by more precise operational control of the technological process and prevent losses of MDEA due to the entrainment.

References

1. Yusupbekov, N., Sattarov, Sh., Doudin, I., Adilov, F., Ivanyan, A.: Application of solutions of connection of production cluster to analytical cloud in chemical industry. In: Conference 2017, vol. I, pp. 246–252 (2017)
2. Yusupbekov, N., Somakumaran, S., Narwadkar, A., Jurayev, T., Sattarov, S., Adilov, F., Ivanyan, A.: Using of Sentience platform for integration of intelligent systems and devices into cloud. Chem. Tech. Control Manage. **N4–5**, 67–71 (2018)
3. Yusupbekov, N., Abdurasulov, F., Adilov, F., Ivanyan, A.: Application of cloud technologies for optimization of complex processes of industrial enterprises. In: Conference 2018, ICAFS, pp.1–7 (2019)
4. Yusupbekov, N., Somakumaran, S., Narwadkar, A., Abdurasulov, F., Adilov, F., Ivanyan, A.: Advantages of analytic solutions of sentience cloud platform. Chem. Tech. Control Manage. **4–5**, 9–12 (2018)
5. Igamberdiev, H.Z., Boeva, O.H., Sevinov, J.U.: Sustainable algorithms for selecting feedback in dynamic object management systems. JARDCS **12**(7), 2162–2166 (2020)
6. Igamberdiev, H.Z., Yusupov, E.A., Sotvoldiev, H.I., Azamxonov, B.S.: Sustainable algorithms for the synthesis of a suboptimal dynamic object management system. In: Advances in Intelligent Systems and Computing, pp. 902–907 (2020)

7. Matvienko, E.V., Ivanyan, A.I.: Intelligent information technologies for integrated management systems of enterprises with a complex scheme of gas-extraction and processing. JMIS **2**(3), 249–254 (2015)
8. Golubeva, I.A., Dashkina, A.V., Shulga, I.V.: Demanding problems of amine treating of natural gas: analysis and ways of solution. Pet. Chem. **60**, 45–50 (2020). https://doi.org/10.1134/S0965544120010089
9. Sviridov, V.P., Vasil'ev, R.A., Sokolova, O.N., et al.: Analysis of the effectiveness of operation of the technological equipment of installations for the preparation and processing of natural gas. Chem. Petrol. Eng. **32**, 596–598 (1996). https://doi.org/10.1007/BF02416688
10. Dai C., Zhao F.: Natural gas treatment. In: Oilfield Chemistry. Springer, Singapore (2018). https://doi.org/10.1007/978-981-13-2950-0_14
11. Law, L.C., Yusoff Azudin, N., Abd. Shukor, S.R.: Optimization and economic analysis of amine-based acid gas capture unit using monoethanolamine/methyl diethanolamine. Clean Techn. Environ. Policy **20**, 451–461 (2018). https://doi.org/10.1007/s10098-017-1430-1.
12. Fang, M., Zhu, D.: Chemical absorption. In: Chen, W.Y., Suzuki, T., Lackner, M. (eds.) Handbook of Climate Change Mitigation and Adaptation. Springer, New York (2015). https://doi.org/10.1007/978-1-4614-6431-0_38-2.

The Concept of Connectivity Model Measurements of Geographic and Information Space of the Region

Vladimir V. Khramov(✉)

Private Educational Institution of Higher Education «Southern University (IMBL)», Rostov-on-Don, Russia
vxpamov@inbox.ru

Abstract. The article considers the basis of the concept of a single geographic and information (geo-informational) space as an objective reality of the world around us. Variants of its soft model are proposed in relation to the socio-economic problems of the region. Approaches to forming the architecture of such a model and its implementation in the framework of a fuzzy knowledge base are discussed

Keyword: Information space · Interoperability · Multi-dimensionality · Dimensions · Functional connectivity · Integrity

1 Introduction

One of the most important concepts used in the process of transforming socio-economic systems within the framework of the project "Digital Earth" is the Unified geographic information Space (UGIS). Taking into account the fact that there is currently no unambiguous interpretation of some terms, we will fix the variant of their use in this work.

The concept of space itself is interpreted "as an objective reality, a form of existence of matter characterized by length and volume" [1]. Among the many known spaces, we would like to concentrate on the information and geographical spaces. Moreover, the latter occupies a central place in the Earth sciences. It is understood as "the form of existence of geographical *objects* and *phenomena* within the geographical shell; a set of relations (including information – *auth.*) between geographical objects located on a specific territory and developing over time" [2]. It is assumed [3] that space, like any complex system (more precisely, System of Systems), includes three components – matter, energy and information [3, 4], the description of which reflects its own characteristics of the entire system as a whole [5].

In turn, the *territory* is a terrestrial (land, two-dimensional, surface) [6], space with certain borders.

Based on the goals of this study, we will consider in more detail the *pragmatic* properties of the geo-information space *model*, which characterize the degree of usefulness of the information available in it for the practice of solving socio-economic problems.

© The Author(s), under exclusive license to Springer Nature Switzerland AG 2021
R. A. Aliev et al. (Eds.): WCIS 2020, AISC 1323, pp. 553–558, 2021.
https://doi.org/10.1007/978-3-030-68004-6_72

2 Properties of a Soft Model of the Geo-information Space

The properties of the UGIS model that prove its usefulness, first of all, include "the presence of *meaning* (*semantics*), relevance (and *novelty*). These properties encourage "to develop, self-organize, compare one information with another" [6], generate new knowledge, using, among other things, "interoperability - the ability of two or more information systems or components to exchange information and to use information obtained as a result of exchange" [7].

In recent years, quite a lot of systems have appeared that implement certain elements of the geo-information space and claim to be "local" [8] UGIS within certain regions of Russia. There were studies of the stratification of geo-information space [1, 6], the formation of "semantic layers", however, in relation to one, "own" territory. The issues of interrelations and mutual influences of socio-economic objects were also studied, mainly within the same territory.

For further description, we note the requirements formulated in official documents [5, 7, 9] for the model (according to the above terminology) of a single information space, in relation to a specific territory.

The model currently referred to as the "Unified information space" consists of the following main components:

- information resources containing data, information and knowledge recorded on the appropriate media;
- organizational structures that ensure the functioning and development of a single information space, in particular, the collection, processing, storage, distribution, search and transfer of information;
- means of information interaction between citizens and organizations that provide them with access to information resources based on appropriate information technologies, including software and hardware and organizational and regulatory documents [9].

When the model works smoothly, all its components must have interoperability, that is, "the ability of two or more information systems or components to exchange information and use the information obtained as a result of exchange" [7]. Models UGIS should be interoperable, although they "work according to independent algorithms, do not have a single control point, and the *overall* control is determined only by a *single* set of standards - the interoperability profile" [7], that is why we have to deal with System of Systems [10]. Accordingly, there is a problem of determining the requirements for these standards, criteria for their adequacy to solving problems, on the one hand, and the adequacy of the functional relationships of the real UGIS and its model at least at the level of semantic layers.

3 Features of the Nonlinear Model of Objects Interaction

The limited distribution of the interaction of objects can also display a quadratic relationship: many researchers believe that this influence is inversely proportional to the square of the distance from the object's border – the so-called "gravitational" model [8, 9].

The spatial influence model (SIM), [4] is usually set by the membership function of the form:

$$S_{i,j} = S_i(1 - \frac{k(j)}{r^n(i,j)})$$

where n - the order of the influence function.

Research has shown that for identification of physical processes of distribution (mutual)influence of objects, a soft model of propagation of this influence is also convenient,which can [6, 9] have the form of:

$$S_{i,j} = S_j \exp(-a(j)r(i,j) \ln b(j)),$$

where S(i,j), S(j) - are the degrees of influence.

Obviously, by varying the parameters b(j)) and a(j), it is possible to adapt such a function of belonging to experimental data, thereby performing parametric identification of the rate of change of this influence in the necessary ranges.

If the connectivity functions reflect the very fact of mutual influence of socio-economic properties of the geo-information space, then the parameters of models are determined, as a rule, by methods of expert assessments, or experimentally, using fuzzy neural systems [6, 8]. "The use of expert assessments, which are unclear, is a certain compromise. To determine these functions, as well as to determine the coefficients of significance or weight of the influence of objects, we use the method of transition from non-strict, approximate expert estimates of the impact of objects on the surrounding territory to an analytical form of presenting these estimates based on the use of continuous typical membership functions" [8, 9].

In a number of works [6, 10, 11], we have shown that soft models adequately describe the properties of functional interconnection of layers of geo-information space and can be applied to implement pragmatic goals in the management of socio-economic systems in the region.

To build a management decision support system (MDSS), a mechanism is needed for an adequate assessment of the current state of the required properties of the management object. In relation to the region, it is advisable to use a methodology for assessing its socio-economic status as a cumulative (and multidimensional) property (potential) to ensure the harmonious implementation of production, economic and social activities [1, 12]. This potential of the territory is defined by the relevant properties of the territory and the physical and socio-economic impact on the territory of a uniform space formed by the entire set of objects located in this territory, and beyond it in its media projections [2, 11, 13]. An example of this is the so-called "virtual enterprises", the number of which is constantly increasing in the world [14, 15].

To develop a detailed methodology for taking into account the mutual information influence of natural and artificial objects in different regions when assessing the socio-economic properties of the territory, it is necessary to form a developed knowledge base. The authors considered these issues when developing the principles of interconnection of individual semantic layers on the example of the so-called digital plan schemes [11].

4 A Method for Forming a Cognitive Two-Dimensional Space - a Digital Plan-Scheme of the Territory by Algorithms of the Main Procedures for Its Implementation

1. The system performs data mining (DM) on the presented images and forms the initial semantic layer of the DSS.
2. Intelligent forms a virtual image of the territory model that corresponds to the setting of a specific socio-economic task.
3. A set of scenarios for solving problems is formed, containing solution algorithms that are ranked according to the selected fuzzy proximity measure.
4. Consistently selects and implements algorithms corresponding to their ranked descriptions.
5. Analyses obtained after running each algorithm, the situation of the task descriptions. A decision is made to continue the DM.
6. The automated formation of the DSS is carried out.
7. Development of ways to account for geo-information in socio-economic systems.

At the "early stages of the theory of spatial planning and assessment of the success of placement of socio-economic objects, models were used in which the spatial factor was taken into account mainly through additional transport costs. Models of spatial placement (Weber, Tunen, Lesh, Kristaller, etc.) were more demonstrative and informative than practical. Attempts at later stages of research to cover and take into account the most significant natural, infrastructural and socio-economic properties have transferred the problems of spatial planning to macroeconomics" [4]. In recent decades, there have been studies related to the transfer of solutions to such problems in the information plane.

In a number of studies, "gravitational" models of mutual influence propagation were proposed to solve certain local problems of managing socio-economic development of the territory [4, 6, 12]. "The corresponding economic and mathematical modeling provides new fundamental opportunities in solving problems of placement and subsequent logistics, which have a spatial nature. However, first of all, these models are general at this stage, and second, they take into account the influence of important, from the information point of view, objects only in this territory. Finally, in each individual case, it is necessary to develop a specific methodology for their application in a given territory" [13], as well as parameterization of the corresponding models.

5 Non-orthogonal Models of Interaction of Socio-Economic Objects

Even at the beginning of the formation of spatial research of socio-economic orientation (as shown above, Kristaller, Lesh, etc.), many scientists paid attention to the prospects of using non-orthogonal (triangular and hexagonal) lattices in the spatial description of the properties of territories and objects located on them. In this study, when forming the concept of a Unified geo-information space, non-orthogonal lattices, including irregular ones (Delaunay triangulations), found their important place.

Description of useful properties of non-orthogonal lattices in the description of territories, individual objects on them, morphological characteristics of semantic relief structures are described in many (more than three hundreds) works of the authors. We note in this paper one original approach to the use of such lattices for the identification of extended socio-economic objects.

In addition, with the help of non-orthogonal rasters, the integrity and dynamics of the development of a single geo-information space as a complex System of Systems and the new properties inherent in this integrity, in particular, emergence, can be sufficiently taken into account [14]. At the moment, we have not studied the impact of a holistic UGIS on the functioning of a separate information object in a given territory.

To solve the pragmatic problems of spatial planning and "management, it is necessary to perform two types of tasks: analysis, when the existing socio-economic state of territories is studied" [15], and synthesis, when the problem of territorial placement of various socio-economic objects is solved [14, 15].

6 Conclusions

The analysis of existing approaches to spatial modeling of information objects of socio-economic systems in order to manage their state and development is carried out. Trends of research and modeling of information connectivity of objects of these systems are defined, problems of increasing the efficiency of economic and social activities are shown, the solution of which will allow to harmonize the design and management of territorial complexes.

The main directions of research in the field of development of manageability of territories based on the use of the properties of a single geo-information space are shown.

References

1. Yurasov, A.V.: The basics of e-commerce. Textbook for high schools. Hot line Telecom, Moscow (2008). (in Russian)
2. Nefedov, S.P., Khlebnikova, T.A.: Technology for Creating and Updating a Unified Geo-Information Space. Sibgeoinform, Novosibirsk (2017). (in Russian)
3. Prigozhin I., Stengers, I.: Time, chaos, quantum: to solve the paradox of time. LIBROKOM (2014)
4. GOST P57193—2016: System and software engineering. System lifecycle processes (ISO/IEC/IEEE 15288: 2015, NEQ) [Electronic resource]
5. Gnatyuk, A.B.: Assessment of the territory based on methods of spatial analysis and modeling. Works of KSTU. Kostroma, pp. 52–54 (2005). (in Russian)
6. GOST R 55062–2012: National standard of the Russian Federation. Information technology. Industrial automation systems and their integration. Interoperability. Basic provisions [Electronic resource]
7. Akperov, G.I., Khramov, V.V.: A fuzzy semantic data triangulation method used in the formation of economic clusters in Southern Russia. In: Advances in Intelligent Systems and Computing, vol. 1095, pp. 340–344 (2020). https://doi.org/10.1007/978-3-030-35249-3_43
8. Jamshidi, M.: System of Systems Engineering: Innovations for the Twenty-First Century. Wiley, Hoboken (2008)

9. Akperov I. G., et al. Method for forming a digital plan-scheme of agricultural objects and a system for its implementation. Patent for invention RUS 2612326 24.02.2015
10. Rosenberg, I.N., Starostina, T.A.: Solution of Placement Problems with Fuzzy Data Using Geo-Information Systems. Scientific World, Moscow (2006).(in Russian)
11. Akperov, I.G., Khramov, V.V.: Development of instruments of fuzzy identification of extended objects based on the results of satellite monitoring. In: Advances in Intelligent Systems and Computing, vol. 896, pp. 325–332 (2019). https://doi.org/10.1007/978-3-030-04164-9_44
12. Kramarov, S.O., Khramov, V.V.: Methodology of formation of unite geo-informational space in the region communications in computer and information science. In: CCIS, vol. 1201, pp. 309–316 (2020). https://doi.org/10.1007/978-3-030-46895-8_24
13. Khramov, V.V.: Method of aggregation of several sources of fuzzy information. Izvestiya of TSURE **3**(21), 52–53 (2001)
14. Lukyanov, G.V., Nikishin, D.A.: Applied aspects of cognitive modeling in the design of complex information systems. Syst. Comput. Sci. Tools **27**(1), 134–154 (2017)
15. Akperov, I.G., Khramov, V.V., Lukashevitsh, V.I., Mityasova, O.Yu.: Fuzzy methods and algorithms in data mining and formation of digital plan-schemes in earth remote sensing. Procedia Comput. Sci. 120–125 (2017). https://doi.org/10.1016/j.procs.2017.11.218

Development of Criteria for Synthesis of the Optimal Structure of Monitoring and Control Systems

E. Uljayev[1], U. M. Ubaydullayev[1], G. T. Tadjitdinov[2], and Sh.N. Narzullayev[1](✉)

[1] Tashkent State Technical University named after Islam Karimov, 2, University Street, 100095 Tashkent, Uzbekistan
narzullayevsh1993@mail.ru

[2] Military Technical Institute, National Guard of the Republic of Uzbekistan, Chigatay-Aktepa, Zangiata District, 100017 Tashkent, Uzbekistan

Abstract. In this paper, the substance of structural and parameter optimization is considered. The attempt of development of the main criteria, synthesis of optimum structure of a control and management system for parameters of technological processes or other objects is made. At the same time, optimization by one, two or three criteria may not meet another, less important, to indicators of a control system and diagnostics. Parameters that are decisive for a given device or production are selected during the synthesis of the optimal structure. General criteria (requirements) and examples of technical specifications for the design of continuous and discrete objects, as well as an example of technical specifications for the designed control and diagnostics system are given.

Keywords: Parametrical optimization · Structural optimization · Criteria · Specification · Discrete objects · Continuous objects · Parameters · Specification forms · Remote control · Diagnostics

1 Introduction

At synthesis of the structural diagram of a system according to [1–4] the issues connected by minimization of number of blocks, structures of connection between them are resolved and also the questions connected by minimization of structure of creation of internal schemes of blocks are considered. In the literature, the main criteria of optimization of the structural diagram of control and controlling systems are not completely disclosed. In this regard, an attempt is made to more detailed develop the main criteria that should be paid attention to when synthesizing the optimal structure of the system for monitoring and controlling the parameters of technological processes or other objects in this work. At the same time, optimization by one, two or three criteria may not meet another, less important, to indicators of a control system and diagnostics. Therefore, during the synthesis of the optimal structure, it is necessary to select such parameters that are

© The Author(s), under exclusive license to Springer Nature Switzerland AG 2021
R. A. Aliev et al. (Eds.): WCIS 2020, AISC 1323, pp. 559–563, 2021.
https://doi.org/10.1007/978-3-030-68004-6_73

decisive for a given device or production (there can be three or four), these include: the reliability of the system, the device; the minimum number of blocks; minimum number of connections; uncomplicated manufacturing technology; minimum cost; performance; high load capacity; minimum power dissipation; powered from a single source; linearity of characteristics; ease of use; great information content; simplicity of device operation; the possibility of expanding memory; flexibility of programming; interchangeability of blocks (modules); high speed of information exchange and processing; use of standard modules for building a system, device; simplicity, convenience of rewriting the program; the ability to replace expensive modules, elements; the ability to conduct test checks; high noise immunity and etc.

According to a design requirement of any control and monitoring system, it is necessary to give a task (to develop), at least the approximate specification. Let's give structure of the specification and examples of parameters on projectable (some) control and manipulating systems. Projectable systems, devices can be continuous (analog), discrete or combined. An automation system or device for monitoring and control consists of series, parallel or mixed-connected functional blocks that perform different functions. The internal structures of building functional blocks, depending on the function performed, may be different.

The typical content of a technical specification (TS) for a certain block is as follows: specific numerical requirements for the output parameters; specific digital data characterizing the ranges of variation of external parameters; a verbal (qualitative) description of restrictions, requirements and conditions that cannot be directly quantified.

Give the TS form on development of a continuous object. Example. Consider TS on development of a schematic scheme of the electronic amplifier [5, 6, 8].

The amplification coefficient K_o must be at least 104 at medium frequencies, the input resistance $K_{вх}$is at least 1 MΩ at medium frequencies, the output resistance $K_{вых}$is not more than 200 Ω. The upper boundary frequency $f_в$is less than 100 kHz, the temperature drift of zero $U_{др}$is not more than 50 μV/deg; the amplifier should function normally in the temperature range from −50 to +60° C. The voltage of the power supplies is +5 and −5 V, the maximum voltage deviations should be no more than ±0.5%, the amplifier is operated in a stationary installation. Installation: it necessary to do on a 60 × 40 mm printed circuit board. This is a typical TS for the development of a continuous object. In this case, the output parameters are gain, input and output resistances, boundary frequency, temperature drift, i.e.

$$Y = \left(K_o, K_{вх}, K_{вых}, f_в, U_{др}\right)$$

.

Ambient temperature and tension of power sources are belong to external parameters. Internal parameters are not mentioned in TS, their list and sense becomes known after synthesis of structure of the scheme. Parameters of resistors, condensers, transistors, chips (parameters of circuit elements) belong to internal parameters.

From the given example, it is visible that the main part of TS is made by requirements to output parameters of the scheme. The values characterizing these requirements are called technical requirements TT_j. Let's designate a vector of technical requirements through a TT, i.e.

$$TT = (104,\ 1\ МОм,\ 200\ Ом,\ 100\ кГц,\ 50\ мкВ/град)$$

The desired relationship between output and technical specification is called serviceability. In the example that is given above, the operability conditions have the form of the following inequalities: $K_o > 104$; $R_{вх} > R_{вых} < 0,2$ $кОм$; input resistance $K_{вх}$at medium frequencies is not less than 1 MΩ, output resistance $K_{вых}$is not more than 200 Ω, upper boundary frequency $f_в > 100$ $кГц$; $U_{др} > 50$ $мкВ / град$. In cases where the output parameters have a two-sided restriction according to the TS, the operability condition is written in the form of two inequalities or in the form of equality:

$$Y_j = TT_j \pm y_j$$

Where Y_j – output parameter, y_j – the admissible deviation of this parameter from the value specified in the TS TT_j. Consequently, all conditions of operability for convenience of consideration can be reduced to a single form: $y_j < TT_j$.

Conditions of working capacity are important at design as the problem of design is formulated as follows: develop the block such in which all conditions of working capacity in all range of change of external parameters are in the best way satisfied and at execution of all qualitative requirements of TS. Often in TS restrictions for many internal parameters directly enter or are meant. In the example above, natural restrictions are imposed on the resistances of the resistors and the capacitances of the capacitors, which do not allow negative numerical values; with an integral version of the same amplifier, restrictions on the same parameters would also be imposed from above, since the integral technology would not allow obtaining large resistances and capacitances.

Here are the forms of technical specifications for the development of discrete objects.

TS usually have a slightly different form for discrete objects. Consider a technical specification for the development of a block of some logical device of an automation system: develop a device scheme that has four inputs and two outputs. The signals at the inputs and outputs are expressed by logical variables, i.e. can take only two possible values, one of which is designated 1, the other 0; the functioning of the block must correspond to the truth table.

The delay of the output signals in relation to the input should not exceed 0.3 μs.

a_1	a_2	a_3	a_4	b_1	b_2
0	0	0	0	0	0
0	0	0	1	1	0
0	0	1	0	0	1
1	1	1	1	1	0

In the validity table: a_1 – a_4- input, b_1 and b_2 - output variables of the block, each line specifies the required values of the signals at the outputs for the next combination of values of the input signals.

The main task at execution of these TS at the level of design of logic circuits: wiring design, providing functioning in compliance with the validity diagram and being optimal in some cases, for example, in the sense of the minimum quantity of the used logic gates. The technical requirement on such output parameter of the block as signal delay, will practically not be used at this level, and will automatically pass into TS to development of elements, at the same time delays will be already specified separate elements instead of a total delay.

A typical technical specification for the design of a discrete object contains a description of the operability conditions and some additional conditions and restrictions; although its presence is always necessary, an indication of the optimality criterion may not be explicitly contained. The peculiarity of such TS is in the form of working conditions: the latter do not have to be a set of equalities or inequalities, they can be in the form of logical equations, textual or validity tables [8].

Let us provide simpler specification for development of the system of remote control and diagnostics of technological parameters of objects:

Number of monitored parameters, pcs.	6
Diagnostics type	remote
Types of monitored parameters	3-analog and 3-digital signals

Operating conditions (external parameters):

ambient temperature, °C	from -30 to +60
humidity,%	$10 \div 80$
Operating conditions	ground
vibration level	not less $25 \cdot g$
supply voltage, battery or power supply current, V	5B±5%
Information display type	digital or graphic
Type of monitor used	microprocessor with extended memory or (Devkit 7000LCD)
Sensor types	contactless
Method of selecting the monitored parameter	manual or automatic

Types of controlled parameters: emergence of a flame $(Д_{пл})$; emergence of smoke $(Д_{дым})$; pressure change out of limits $(Д_{дав})$; temperature changes $(Д_T)$; changes of humidity $(Д_{вл})$; emergence of a gas leak $(Д_{уг})$, etc.

2 Conclusion

The provided technical requirements or criteria can be useful to undergraduates, doctoral candidates and design engineers of control and management systems of technological parameters of different objects.

References

1. Kalyavin, V.P., Ribakov, L.M.: Reliability and diagnostics of electrical installation elements, Sank-Peterburg, Elmor, p. 336 (2009)

2. Uljaev, E., Igamberdiev, X.Z., Uljaev, Z.E.: Reading automatic schemes. Cho'lpon nashr, p. 212 (2012)
3. Yusupbekov, N.R., Igamberdiev, Kh.Z., Marakhimov, A.R.: Strukturno-parametricheskij sintez intellektualnykh sistem upravleniya. «Universitet », p. 395 (2018)
4. Shaxnova, V.A.: Design and technological design of electronic equipment, Izd-vo MGTU, p. 568 (2002)
5. Norenkov, I.P.: Basics of Computer Aided Design. Ucheb. dlya vuzov. Izd-vo MGTU im. Baumana, p. 360 (2000)
6. Klyuev, A.S., Glazov, B.V., Dubrovskiy, A.X., Klyuev, A.A.: Proektirovanie sistem avtomatizatsii texnologicheskix protsessov [Design of automation systems for technological processes]. Energoatomizdat, p. 464 (1999)
7. Tregub, G., Ladanyuk, A.P., Plujanov, L.N.: Design, installation and operation of automation systems in the food industry, Vo. Agropromizdat (1991)
8. Aripjonov, M., Uljaev, E.: Lecture notes for the course: Basics of designing automation systems, TGTU, Tashkent (2001)

On the Use of Fuzzy Logic in Criminal Justice

Ed. A. Vasil'ev(✉) and A. I. Fomenko

Private Educational Institution of Higher Education, Southern University (IMBL), Rostov-on-Don, Russia
akadem_prava@iubip.ru, fomenko300078@mail.ru

Abstract. The article attempted to raise questions about the possibility of using certain methods of fuzzy logic in the field of criminal justice, in the formation of modern criminal policy, as well as the criminal legal reaction to the results of the use of artificial intelligence in the disclosure and investigation of criminal activities in modern society.

Keywords: Fuzzy logic · Neural networks · Methods · Criminal liability · Linguistic variables · Neural models · Crime · Criminal policy

1 Introduction

Scientific and technological progress is one of the most significant socio-economic processes of the modern world. What seemed completely unrealistic fifteen to twenty years ago has been embodied and continues to be translated into concrete results available to an increasing number of ordinary people. Indeed, in the early 2000s, few could imagine what role a mobile phone would play in the life of a resident. However, it is generally recognized that scientific and technological progress has, like any social phenomenon, its flip side, expressed in the increasing number of its negative consequences (in particular, in the form of man-made accidents and disasters), as well as in the increasing possibility of using its results to the detriment of individuals, States and the entire world community.

The legal doctrine is dominated by the judgment that in the modern era of the development of society, issues of legal evaluation of the achievements of scientific and technological progress take on particular importance. It should be noted that scientific progress as a phenomenon accompanies the development of law (including criminal) over a long period of time and, as correctly noted by many domestic and foreign authors, cannot but affect the content of criminal law. However, criminal (and not only criminal) legislation is constantly lagging in its, if you can put it that way, reactions to undeniably grandiose achievements and side effects of scientific and technological progress [1–4].

At the same time, the results of scientific and technological progress in the field of criminal justice usually continue to be considered in terms of their use in the commission of "traditional" crimes. For example, the possibility of using the results of scientific and technological progress to carry out terrorist actions is widely discussed in both domestic [2, 8–10; 3, 283–288] and foreign doctrine [4].

To date, it is obvious that the current security system should be organized in a new way, the need for the use of fuzzy logic methods is ripe.

© The Author(s), under exclusive license to Springer Nature Switzerland AG 2021
R. A. Aliev et al. (Eds.): WCIS 2020, AISC 1323, pp. 564–570, 2021.
https://doi.org/10.1007/978-3-030-68004-6_74

In this regard, the importance of research into the methods of fuzzy logic in the field of criminal justice, since it contributes to the realization of such key duties of the State as ensuring security, the creation of optimal conditions for combating crime, and the development of a scientifically sound theory of the use of fuzzy logic methods in the formation of an offensive strategy in combating manifestations of modern crime.

In this article, we will try, first of all, to raise questions about the possible directions for using methods of fuzzy logic and criminal legal reaction to the result of scientific and technological progress that has not yet been received in a formalized form, but the receipt that is expected in the foreseeable future. According to the authors' purely subjective opinion, the doctrine should already (at least hypothetically) prepare for the possibility of using fuzzy logic methods in the formation of criminal law doctrine, its strategy and criminal law assessment of its effectiveness in ensuring criminological security of the interests of society, the individual and the state through the lens of the results of scientific and technological progress.

The state of scientific development of the problem. The interdisciplinary problem of using fuzzy logic methods in the humanities, in jurisprudence is the subject of research by KozaevN.Sh., Tislenko D.I., Kibalnik A.G., Kravtsov D.A., Zadeh L.A., Shestak V.

2 Research Materials and Methods

In addressing the criminological problems of ensuring the security of public relations, the question arises of the effectiveness of the analysis methods commonly used in jurisprudence, in which judgments and sometimes human biases have a significant role. It should be noted that legal phenomena and processes related to the exercise of criminal responsibility cannot be well described as mechanical and economic systems that allow for numerical description and measurement.

The complexity of phenomena and the blurring of concepts in the legal field require an approach different from the generally accepted quantitative methods of analysis. The study of legal and criminological problems, including those based on the sampling methods of the general population, can confirm the principle of incompatibility known in mathematics, which argues that high accuracy is incompatible with the great complexity of the system. More specifically, it can be argued that the complexity of the system and the accuracy at which it can be analyzed are in inverse proportionality. Here there is an analogy with the known fundamental principle of uncertainty of Werner Heisenberg, according to which it is possible to accurately establish either the speed or the position of an elementary particle, but not both [5].

In this regard, some scholars propose not to strive for the use of quantitative methods, but instead to use the methods of fuzzy logic and linguistic variables, which allow as values of variables not only numbers, but also words and sentences of natural language [6, 43–80]. Such an approach may be more effective in assessing the complexity and inaccuracy of the legal field system than conventional methods of analysis.

The formal definition of linguistic variables, their properties and operations with them in 1965 were proposed by Lofty Zade. The term "soft computing" was introduced by Lofty Zade in 1994. This concept combines such areas as: fuzzy logic, neural networks, probabilistic reasoning, trust networks and evolutionary algorithms; which complement each other and are used in different combinations or independently to create hybrid intelligent systems [7, 23–27].

Therefore, the creation of systems operating with uncertainty should be understood as an integral part of "soft" calculations.

We believe that such an approach in the implementation of the tasks of criminal justice is well associated with the intuitive perception of a person who operates with well-defined concepts, but at the same time does not impose strict conditions on them. It is convenient and correct to describe many quantities in criminology and criminal law in this way: the concepts of "criminal," "especially dangerous relapse" "high latency," "low motives," "self-serving motives," "significant damage," etc. The analysis of the general and selective totality of crime, its quantitative indicators taking into account many dependent factors, is always a range of values, which, taking into account certain rules of fuzzy logic, also reflects reliability in them.

We will outline a number of directions for using the rules of fuzzy logic and the possibilities for their application in solving criminological and criminal legal problems.

The first and, it seems to us, the main possibility is to use the rules of fuzzy logic in analyzing the dynamics of crime and its structure. By highlighting the states of a number of homogeneous groups of individuals relevant to crime, and considering the links between them, which are determined by the set of circumstances, it is possible to draw a relationship between these values. Such models establish a statistical relationship between the number of individuals who will be in these states at one time or another. There are ratios linking the number of criminologically active population in a single region, country, with its constituent groups: Law-abiding; under investigation or trial; latent group; recidivists.

The use of fuzzy logic methods involving the use of linguistic concepts is also possible when constructing a model of the genesis of criminal behavior. At the same time, the use of neural networks in modeling the genesis of criminal behavior, in which the values of the elements of the genesis of crime were not quantitative, but qualitative variables, is considered very promising.

The genesis of the crime can be imagined as a chain of the following links of the mechanism of criminal behavior: the needs and interests of the offender – the possibilities available to him to satisfy these needs – a system of value orientation of the offender, including his legal consciousness: – deciding to act – an act (crime or abandoning it). Unfortunately, none of these elements are quantifiable to date. Therefore, you have to turn to fuzzy logic. It is possible to construct the following neural model, with characteristic elements and their qualitative characteristics:

1. Requirements – normal (perverted);
2. Opportunities – legal (illegal);
3. Value orientations – positive (negative);
4. Motivation – low-lying (devoid of low-lying content)
5. Solutions are adequate (inadequate).
6. Mental state – sanity (not sanity, borderline state).

The output neuron (i.e., act) is also measured on the same scale ~ criminal (lawful).

The training sample is formed by an expert, training is carried out by providing examples consisting of sets of input data and matching with corresponding results at the network output. Based on the results of the computational experiment, it is concluded that at the initial stage of the formation of the training sample, the fuzziness of the concepts used in the "qualitative model" appeared. Apparently, the characteristic blocks should be linguistic variables and represented by fuzzy sets, and the corresponding neural network algorithms are used to conduct training on such samples.

The experience of a group of scientists from Shanghai University who studied the ability of people to distinguish criminals from other people by simply viewing their photographs seems very informative. Xiaolin Wu and Xi Jan used various machine vision algorithms to study the faces of criminals and law-abiding citizens, and then checked whether the machine could detect the difference.

Offering its automated system for predicting criminal inclinations based on face shots, Chinese scientists noted that unlike people, the computer algorithm for classifying images is not at all burdened by the baggage of subjectivity, does not have emotions and inaccuracies related to past experience, racial, religious or political preferences, with an assessment of a person's personality by sex, age and so on, it does not tire or suffer from the effects of lack of sleep or food.

Four approaches to automatic classification of objects were used by scientists:

- method of reference vectors;
- method k (method of nearest neighbors);
- logistic regression;
- use of convolutional neural network.

They offered these algorithms a set of 1856 photo portraits of men, aged 18 to 55 years, without a beard and mustache, without scars and tattoos, with a neutral facial expression, of which 730 came under suspicion of the police or had criminal experience (including 235 of them in connection with serious crimes). The authors separately note that the faces of the criminals were shown in ordinary photographs, and not pictures taken for police archives.

After training, all four algorithms demonstrated a certain ability to identify the faces of criminals among law-abiding citizens. The convolutional neural network proved to be better than others, the accuracy of which reached almost 90%. Moreover, the authors point to specific features allegedly characteristic of a criminal personality, including a more pronounced bend of the upper lip, a smaller angle between the corners of the mouth and the tip of the nose, an increased distance between the inner corners of the eyes.

However, it is necessary to remember the experience of past years of studying crime. As you know, one of the founders of the anthropological theory of crime was the Italian criminologist Cesare Lombroso and his formula: to measure, weigh and hang. He believed that criminals can be identified by special features: a beveled forehead, the specific structure of the auricles, various facial asymmetries and long arms. To prove his point of view, he took many measurements. Since Lombroso was an educated doctor, he made hundreds of comparative brain autopsies - deceased criminals and ordinary people.

The possible connection between a person's appearance and complex personality traits, such as honesty or cruelty, is not without a certain relationship. However, modern science denies such unambiguous and rude relationships. Thus, a correlation between the propensity for aggressive behavior and testosterone levels is known; it is also known that testosterone affects facial features. However, the connection between face and aggression is almost disappearing.

All this does not bother Chinese scientists at all, they also note that although the computer algorithm is really "not burdened by the baggage of subjectivity," the selection of persons on which he studied and checked can suffer from it. The authors could, without realizing it themselves, select persons convenient for such an interpretation. Judges can make stricter decisions against people with harsh and cruel facial features. Unlike an expert or judge, the computer vision algorithm does not have emotions, biases about experience, race, religion, political addictions, it does not get tired, it does not need sleep or food.

The experience of the Russian company Tselina Data LAB, which uses deep machine learning algorithms called Fraudoscope to create software, which determines lies on emotions on a person's face, is also very promising. Trained with CUDA and TITAN X graphics processors and a false recognition application uses a high-resolution camera to capture interrogations and decode results [8, 108–110]. The camera is aimed at the person being questioned, and the software notes on the video a change in pixels that correspond to breathing, pulse, pupil expansion, facial tics. Already, an incomplete application shows accuracy of more than 75%.

As with polygraph tests, when using Fraudoscope for calibration, the interviewer needs to ask questions that are well known. Using the example of how a person thinks about fictional answers, the system learns to recognize the lies of a particular test subject.

A group of Perm scientists, together with law enforcement agencies of the region, attempted to create an intelligent program for calculating polygrams using a fundamentally new approach based on the use of modern neural network technologies and social genetic algorithms, allowing to reduce the influence of the human factor and, thus, significantly increase the accuracy of the polygraph.

The fundamental differences of the above approaches are that the developers have applied modern methods of fuzzy logic and artificial intelligence. They as much as possible refused to use the known laws and rules traditionally laid down in reading programs. These rules are implicitly generated by the computer program itself during the exchange of information between the device and the subject. The computer program, being a system of artificial intelligence [9, 547–554], itself extracts and formalizes in the form of rules of the regularity of the body of the person being interrogated, automatically adjusts to its individual physiological features, screening out possible artifacts. In the end, the volume and time of the polygraphologist's work is reduced, the influence of the human factor decreases, and the computer reliability of the calculations increases.

It can be stated that in existing theoretical and practical studies defining the formal rules for the use of fuzzy logic methods in mathematical calculations, there is already a certain slimness and completeness. But now these rules are not scientific methods; rather, they draw a perspective for the future [10, 805–808].

The apparatus of fuzzy set theory, demonstrating a number of possibilities for its application, turned out to be difficult to implement in the field of jurisprudence. Given the scientific complexity of concepts and rules in fuzzy logic, we ask ourselves the question: are there tools that are clear and easy to use, which can make the calculations necessary for criminologists and legal scholars, fast enough and with acceptable accuracy? We believe that yes! Fuzzy logic suggests solving various problems according to the following scheme, you describe any qualitative concepts (crime, crime detection, victimhood, damage, etc.), some distribution functions similar to probabilistic functions, and then use them as supposedly accurate, not caring more about their "fuzzy" nature. The theory of fuzzy logic allows you to perform all logical operations on such quantities: union, intersection, negation, etc. It has been proved that any mathematical system can be approximated by a system based on fuzzy logic and in criminological research will eliminate unnecessary efforts to collect information, increase the reliability and scientific completeness of the results. And for this, it is necessary for lawyers to acquire mathematical and logical knowledge, to master the Internet and computer technology.

However, to become a really massive tool for criminologists and legal scholars, fuzzy logic must take another step, find a simple shell, familiar to an ordinary lawyer. To do this, it is necessary to combine the mathematical apparatus of fuzzy logic with the linguistic concepts of jurisprudence, criminology and sociology, which allows you to make quick calculations in various fields with an acceptable degree of accuracy.

The development and implementation of fuzzy logic techniques in the area of criminal justice will contribute to the more effective implementation of such key State responsibilities. as ensuring the security of the interests of the individual, society and the state, the development of a scientifically sound theory of the use of methods of fuzzy logic for analyzing the manifestations and dynamics of modern crime, creating optimal conditions for the fight against crime and the transition from a defensive strategy to an offensive one when developing the directions of modern criminal policy of the state.

References

1. Kozaev, N.Sh.: Contemporary problems of criminal law due to scientific and technical progress. Author's abstract … doctor of legal sciences, Krasnodar, pp. 2–4 (2016)
2. Tislenko, D.I.: Environmental terrorism: criminal-law and criminological problems: author's abstract … candidate of legal sciences, Saratov, pp. 8–10 (2012)
3. Fomenko, A.I.: Socially dangerous acts related to artificial intelligence: criminal legal aspects. Topical problems of the application of criminal legislation. In: Sorokun, N.S. (ed.) Collection of Materials of the International Scientific and Practical Conference, pp. 283–288 (2020)
4. Kibalnik, A.G.: International criminal law problems in international thesis researches. Criminol. J. Baikal National Univ. Econ. Law **2**, 171 (2014). EID: 2-s2.0-84918843482
5. Heisenberg's uncertainty principle. https://elementy.ru/trefil/21096/Printsip_neopredelennosti_Geyzenberga
6. Zadeh, L.A.: The concept of a linguistic variable and its application to approximate reasoning, Part 1, 2 and 3. Inf. Sci. **8**, 199–249 (1975). 301–357 and Inf. Sci. **9**, 43–80 (1976)
7. Zadeh, L.A.: The calculus of fuzzy if-then rules. AI Expert **7**, 23–27 (1992)
8. Kravtsov, D.A.: Artificial intelligence: crime prevention and prediction. Bull. Moscow Univ. Russ. MIA **3**, 108–110 (2018)

9. Shestak, V.A., Volevodz, A.G., Alizade, V.A.: On the possibility of doctrinal perception of artificial intelligence as the subject of crime in the system of common law: using the example of the U.S. criminal legislation. Vserossiiskii kriminologicheskii zhurnal = Russ. J. Criminol. **13**(4), 547–554 (2019)
10. Kuznetsova, I.S.: Operations research methods for rapid assessment of small business. Econ. Entrepreneurship **10-3**(75), 805–808 (2016)

Development of an Integrated Security Monitoring System Based on the Analysis of the Goals of Actions of Subjects of Computer Systems

M. M. Karimov[1(✉)] and S. M. Khamdamova[2]

[1] State Test Center Under the Cabinet of Ministers of the Republic of Uzbekistan, Bogishamol 12, Tashkent 100202, Uzbekistan
dr.mmkarimov@rambler.ru

[2] Tashkent State Technical University in the name of Islam Karimov, 2, Universitetskaya Street, Tashkent 100095, Uzbekistan
informtgtu@mail.ru

Abstract. Security monitoring systems for computer systems and networks (CSN) are classified by areas of information infrastructure where security policy violations are detected. At the same time, there are security monitoring systems (SMS) focused on: workstation; applications; database management system (DMS); computer network. SMS of a workstation are executed on the protected node and monitor various security events, collect and analyze information reflecting the activity taking place in the operating system of a separate computer. Application-level SMSs and DMSs collect and analyze information from specific applications such as Web servers, firewalls, or database management systems. Network layer SMSs collect information from network traffic. They can be performed on ordinary or specialized computers, and also can be integrated into routers or switches. The article discusses the features of the practical implementation of SMS in CSN. Based on the analysis of the problems arising during the functioning of security monitoring tools in modern CSN, the main research area is highlighted - the development of methods and means of security monitoring that ensure an increase in the effectiveness of SMS against the actions of malicious intent by: predicting possible actions of intruders; dynamic analysis of risks of implementation of threats to the security of information resources; recommendations to the adaptive security management system for reconfiguring/modifying protection when the threat level changes. The developed complex SMS based on the analysis of the goals of the actions of the subjects of computer systems is described.

Keywords: Security monitoring system · Computer systems and networks · Vulnerability · Attack detection tools · Means of security

1 Introduction

Security monitoring is a set of measures and measures (organizational, technical and legal) aimed at the implementation of observation, analysis and forecasting of the safety

© The Author(s), under exclusive license to Springer Nature Switzerland AG 2021
R. A. Aliev et al. (Eds.): WCIS 2020, AISC 1323, pp. 571–579, 2021.
https://doi.org/10.1007/978-3-030-68004-6_75

states of CSN. The need to use SMS in CSN is determined by the fact that the use of conventional means and mechanisms for protecting the processed information is insufficient, since they support only the basic functions of ensuring the security of the system and do not allow monitoring the security of the functioning of information systems [1–4]. The main criteria for choosing and using a specific security monitoring tool are the goals and objectives of ensuring the protection of information processed by computer systems and networks. At the same time, first of all, it is determined: from what threats it is supposed to build protection, what means can be used to prevent these threats, and what measures have already been used for protection purposes [5]. In general, CSN security monitoring and analysis systems allow collecting and analyzing information from various security system components (firewalls, intrusion detection systems, routers, VPN servers, operating systems, DMS, antivirus tools, etc.). Then, based on the correlation analysis of events, threats are detected and the process of resolving emerging incidents is monitored [6, 7]. In addition, some manufacturers prefer to release not complex solutions, but monitoring and analysis tools aimed at identifying and analyzing emergency situations in specific components of the CSN. Existing security monitoring tools developed by manufacturers generally remain "closed" for users and cannot be modified to meet the requirements of specific institutions, corporations and possible special conditions for their use, which results in the problem of developing new protection tools and analysis with their capabilities. Targeted adaptation. This problem requires the creation of models and tools for modeling and verification of the developed and applied protection means. When developing mechanisms and means of protection, it is necessary to take into account the existing security threats in the CSN [4]. In its classical form, SMS is understood as means of detecting attacks (intrusions). But, taking into account the disadvantages of this approach shown below, it is more expedient to raise the CSN security monitoring system to the same level as the safety controls. At the same time, the SMS receives information from protection means, attack detection tools and systems for monitoring the functioning of the CSN. In addition, SMSs are able to analyze data from deceptive systems, security analysis tools and analyzers of processes occurring in the CSN. Thus, SMSs allow building a holistic picture of the actions of intruders and making correct decisions on further countermeasures against them.

Based on the comparison of the main modern means of safety monitoring, the following general disadvantages are highlighted:

- high level of errors of the first and second kind, because the formation of an assessment of the probability of intrusion of intruders into the CSN does not take into account the prediction of their behavior;
- the priorities of reaction to security events are set statically, which does not allow adaptive control of the CSN security system.

Thus, it is urgent to create new methods, as well as modified security monitoring tools that provide a high probability of detecting and timely prevention of malicious attacks.

2 Integrated Security Monitoring System CSN

An analysis of existing SMSs and the identified shortcomings of their implementations made it possible to propose a comprehensive CSN security monitoring system (Fig. 1) that provides an increased efficiency of intrusion detection by attackers by predicting the development of attacks, as well as emulating vulnerabilities and system information. As a means of gathering information and checking the status CSN protrude existing intrusion detection systems, protection means (antivirus software, firewalls, hardware protection and access control), as well as means of security analysis and analyzers processes work-stations and corporate network [8]. Intrusion detection systems are configured on the false positive principle, i.e. if it detects any suspicious activity that cannot be reliably identified, the system classifies it as a potential intrusion.

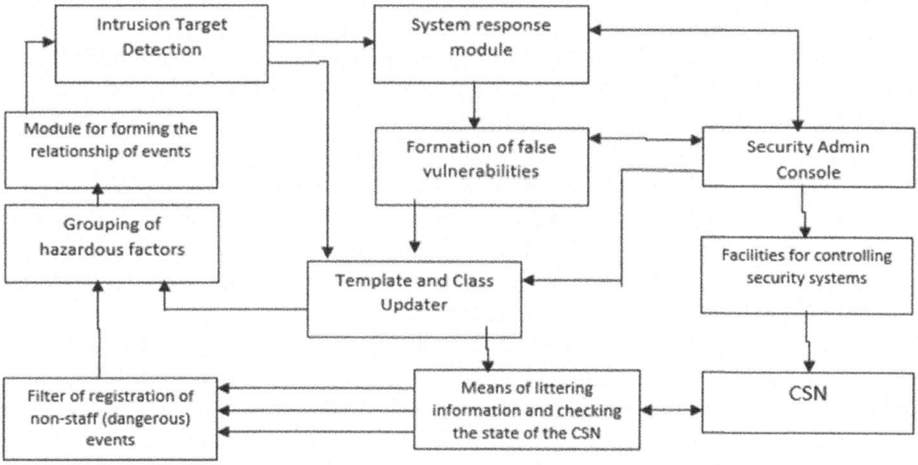

Fig. 1. The structure of the software of the integrated security monitoring system CSN

Security analysis systems conduct comprehensive studies of controlled resources in order to detect vulnerabilities that can lead to a violation of the CSN security policy. [9]. The SMS evaluates the levels of intrusion threats based on an appropriate method that allows predicting the development of attacks. Proposed and developed method for analyzing the goals of actions of the subjects of CSN on the basis of determining the correlations of hazard factors. The security event registration filter receives information from the means of collecting information and checking the state of the system, acting as application agents, using a special protocol for receiving data. In accordance with this protocol, the application agents' event logs are converted into the internal data format - the vector X, which is used in the proposed SMS. The undoubted advantage of the proposed approach is the possibility of a fairly simple connection of new agents for collecting information.

Each record of a suspicious activity, following the rules of formalization, is converted into a specific format. An example of such a format is a data tuple:

$$(\{s_1, \ldots s_n\}, \{a_1 \ldots a_m\}, \{g_1 \ldots g_k\})$$

where s_i is the subject-initiator of the event, a_i - parameters of the subject's actions to implement the invasion, g_i - possible goals.

In turn, the intrusion parameters are also represented as a tuple of data:

$$(\{t_1...t_c\}, \{l_1 \ldots l_d\}, \{r_1...r_e\}, \{\gamma_1...\gamma_f\})$$

where t_i is the time of the event, l_i is the location of the event, r_i is the funds involved, γ_i is the success rate of the invasion.

Using templates of matrixes of weighting coefficients xa_{ij}, the grouping of hazard factors and the transformation of the extended session vector X into the vector A of the aggregate of dangerous actions a_j by intruders are performed in accordance with

$$A = F_{xa}(X),$$

where F_{xa} is the transformation function of vector X into vector A.

To determine the relationship of events in the proposed method, the module for generating the relationship of events builds the probability distribution matrices of pairwise exploitation of certain vulnerabilities - the targets of intrusions, and forms a template of the matrix of correlation coefficients for the purposes of the actions of intruders.

The intrusion target detection module uses this information to create a vector G and a graph of possible priorities for implementing vulnerabilities. To correct the graph of potential goals of the subjects, a feedback with the module for forming the relationship of events is proposed. As a result, we obtain more reliable values of the transition probabilities in the sequence of using the targets - vulnerabilities.

At the same time, the intrusion detection module analyzes the existing causal relationships indicating the relationship between intrusions differentiated by time, place, method of attack, and the means involved. Also, a relationship is formed between the events, the initiator of which was not identified, and the events being analyzed. The implication of the parameters of various intrusions into a single assessment is made by concatenating the goals (to obtain the final result, the offender first needs to achieve intermediate goals g_1, g_2, and only then the final g_m) or on the basis of collecting information about the actions of a certain subject.

Based on the developed method for assessing the level of intrusion threats, it is proposed to implement the following stages of collecting and preparing data to support decision-making on security management:

1) Analysis of the configuration of a fragment of the corporate network, carried out to identify the main information resources that can be targeted by intruders.
2) The analysis of CSN vulnerabilities is performed based on the network topology, statistical data on events in the computer system obtained using the specialized software Microsoft Baseline Security Analyzer, as well as information on known intrusions. At this stage, the vector of goals G is determined and a graph model of the vulnerabilities of this fragment of the CSN is built (Fig. 2). The vertices of the graph represent vulnerabilities, the exploitation of which makes it possible to exert a certain effect on a computer system, and the arcs show the degree of the possibility of transition from one vulnerability to another.

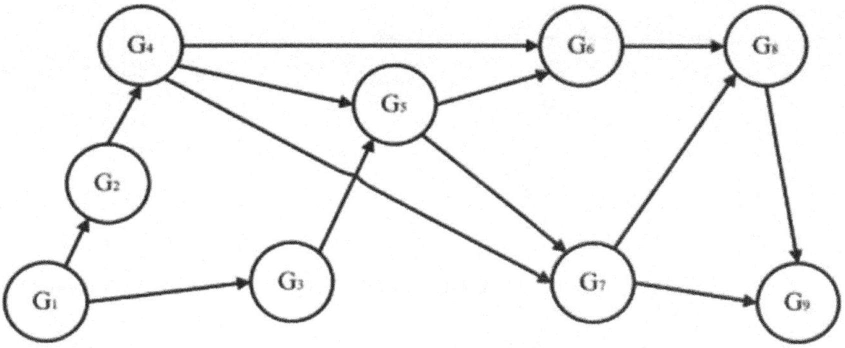

Fig. 2. Possible priority of vulnerabilities implementation (graph of potential targets of attackers)

When constructing this graph, in addition to the above information, expert assessments are used, which introduces some element of subjectivity. However, this circumstance is compensated for by automatic updating of the correlation coefficients of safety factors during the operation of the CSN.

Vertices of graph G_i represent the following vulnerabilities: G_1 - open communication channel; G_2 - IS2004 proxy server configuration vulnerabilities; G_3 - unprotected network access point (the ability to run software and access local resources) - vulnerability in "both directions"; G_4 - WinXP vulnerabilities (getting the local Administrator password when deleting a hash file/brute-force password); G_5 - DNS vulnerabilities (getting the DNS table; remote access using corrupted RPC packets); G_6 - vulnerabilities in Active directory WinServer 2003 (reading the active directory when installingAdminPack.msi); G_7 - vulnerabilities in the TCP/IP protocol stack (transmission of passwords in an unprotected form); G_8 - vulnerabilities in the configuration of the Acronis server (non-allocation of the protected data area); G_9 - MS SQL2005 vulnerabilities (possibility of not domain, but sql-authentication); and others.

Determining the potential targets of attackers allows one to determine the appropriate actions by which these vulnerabilities can be implemented.

3) Analysis of the stages and traces of attacks is carried out to track the current activity of the attackers and form the sequence of actions to be performed. To exploit the above vulnerabilities, an attacker can perform a specific combination of a_i actions: a_1 - use of fragmented packets; (G_1); a_2 - Dos - attacks; (G_1); a_3 - connection to a modem; (G_3); a_4 - local analysis of logs; (G_4); a_5 - local password guessing; (G_4); a_6 - launches Trojan horse programs; (G_4); a_7 - getting ip addresses and DNS names of servers; (G_5); a_8 - change/capture the ip address; (G_5); a_9 - obtaining complete information about domain users; (G_6); a_{10} - get user passwords using TCP packet analysis; (G_7); a_{11} - copy/modify backups of MS SQL databases; (G_8); a_{12} - access to the working MS SQL database (G_9) (Fig. 3).

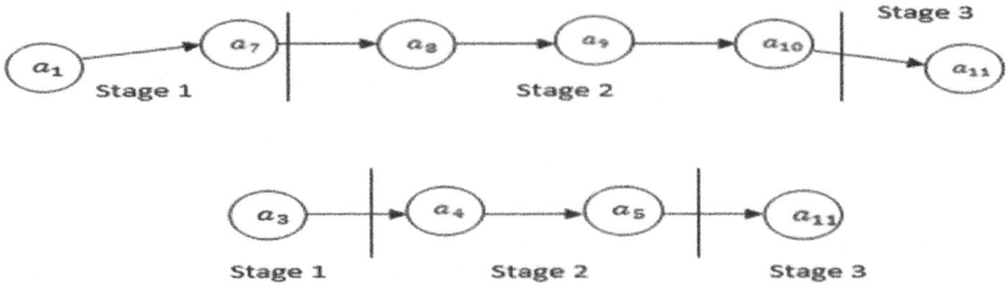

Fig. 3. Stages and routes of attacks. Examples of the development of 2 specific attacks

To monitor user actions, it is necessary to adjust security tools, such as intrusion detection systems, protection tools (antiviruses, firewalls, hardware protection and access control) and CSN process analyzers, to control security factors.

4) Consideration of safety factors xi that need additional control: traffic control of IPv6 packets and I CMP protocol; (a_1); control of port scanning (responses without requests); (a_2); number/intervals of connections; (a_2); initialization of connection to the modem; (a_3); attempts to access the files SecEvent.Evt, AppEvent.Evt, AppEvent.Evt, application logs; (a_4); control of unsuccessful attempts to authenticate SecEvent.Evt; (a_5); attempts to connect to the remote registry service; (a_6); installation of new services, putting the software into autorun; (a_6); installation of AdminPack.msi, access to the security folder of the domain objects. (a_9); and others….

For example: control factor (a_4) - attempts to access SecEvent.Evt, AppEvent.Evt, AppEvent.Evt files, application logs are performed using the File System Viewer (HP Open View) module, which scans the file system and generates age reports files, their size, type and frequency of user accesses to them, on the basis of which you can make decisions on user behavior. The Chargeback, Report Designer, and Global Reporter utilities are designed for storage billing, reporting, and aggregation of reports from multiple copies of Storage Essentials, respectively.
Journal fragment TCPdump:

07:11:38.123565 200.0.0.200 > 200.0.0.34: icmp: echo request
07:11:51.456342 200.0.0.200 > 200.0.0.47: icmp: echo request
07:11:04.678432 200.0.0.200 > 200.0.0.3: icmp: echorequest
07:12:18.985667 200.0.0.200 > 200.0.0.12: icmp: echo request
07:12:31.024657 200.0.0.200 > 200.0.0.11: icmp: echo request
07:12:44.044567 200.0.0.200 > 200.0.0.9: icmp: echorequest

File System Viewer generates the following report based on TCPdump log data:

Resource criticality	Event	Number of events	Status
High	NMap-Scan	139	Impact unknown

Thus, after the 4th stage, information resources, vulnerabilities, possible actions of intruders and security factors that allow identifying these actions have been identified. Further, using the means of protection and checking the state of the system, acting as application agents, information is collected about the events occurring in the CSN.

5) Monitoring of user actions is aimed at identifying abnormal events in the CSN and assessing the level of intrusion threats. At this stage, the factors xi of individual, including low-level, actions of the intruder are grouped into higher-level events of a security violation, and a vector of actions A is formed. In this case, the correlation coefficient is used that characterizes the degree of dependence of the factors x_a and x_b. A positive correlation between factors means that as the values of one of them increase, the other tends to increase. Further, the actions of the attackers are reduced to their corresponding goals and the focus on specific information resources is determined. The graph of vulnerabilities allows you to build a chain of possible actions of an intruder, based on his current position in the space of targets (vulnerabilities) and current activity, as well as predict his next steps and determine potential and possible ultimate goals, and, accordingly, possible actions. This allows you to control possible intrusions at their initial stage. Based on the constructed prediction of the behavior of attackers using a reverse traversal through the graph of the attack model, the security threat factors that need additional control and strengthening of protection means are determined (they are used to reconfigure/modify the protection system). The stages of attack development make it possible to control security factors dynamically, i.e. if the attack has passed the first stage of its development, then it is necessary to control the factor 1.4. If the attack has passed the first stage of development and is in the second, then factors 5, 6, 9 must be taken into account, and additional control of factors 1, 2, 3 can be disabled.
6) Carrying out a risk analysis of the implementation of threats to the security of information resources allows you to rank resources according to the degree of risks for CSN.

The level of security risk of a computer system is assessed on the basis of a modification of the econometric model of credit and financial risks in relation to the tasks of protecting information that has a certain value. To analyze the risks of the implementation of threats to the security of information resources, it is necessary to compare the goals of the attacker and the specific resources being attacked (Table 1).

Table 1. Risk levels for information resources

Resources threats	R1	R4	R5
G6	R16	R46	–
G8	–	R48	R58

7) Recommendations for adaptation of the protection vector

At the last stage, recommendations are made to the security system for modifying the protection. For example, an attacker targets a host and a DNS server. Considering the value of information, as well as the degree of criticality of resources, the decision support system offers the security administrator the following options for action:

- prohibit modem connections on hosts, create a modem pool;
- increase the security of the DNS server with specialized protection tools.

Thus, the system of adaptive security management of the CSN, which is distinguished by taking into account the risks of the implementation of threats to the security of information resources, makes it possible to provide a proactive response of protection means to the actions of intruders.

The graph shown in Fig. 2 is a simplified example (without defining transition probabilities) of the graph of potential targets of subjects and shows the sequence of actions of violators when moving between vertices. The vertices of the graph represent the goals, the achievement of which allows you to have a certain impact on the computer system, and the arcs are actions that contribute to the transition from one goal to another. This graph allows you to build a chain of possible actions of an intruder based on his current position in the target space, as well as predict his next steps and determine the potential and possible ultimate goals of the intruder. The system reaction module forms the vector R of the level of threats to the available information resources. Further, an automatic ranking is carried out according to the danger levels of threats to the actions of the intruder, based on the analysis of the possible damage to the CSN. If the impact on the system is critical, the system reaction module attempts to automatically neutralize the attacker's actions and notifies the Administrator about it. If the degree of threat does not exceed the threshold level, then the system offers the administrator several ways for the further development of the situation, ranging from complete tolerance towards the offender to identify his motives and further plans to conduct attacks on CSN, and ending with a "masquerade" (replacing their real critical resources imitation), which is implemented using the block for generating false vulnerabilities.

The block for generating false vulnerabilities can emulate vulnerabilities and system information ("masquerade"), relying on the forecasting data of the intrusion target determination module. Moreover, depending on the nature of the actions and goals of the attacker, both concealment and disinformation can be used, as well as a complete substitution of critical data. Based on information about the intrusions and the actions of intruders, the template and class update module ensures timely correction of the interconnection matrices that are used when grouping factors and determining goals. It is also possible to modify the rules for collecting information by agents regarding the tracking of certain security factors.

3 Conclusion

The developed complex SMS based on the method of analyzing the goals of the actions of the subjects of the CSN provides the ability to easily connect new agents for collecting information and increases the likelihood of correct detection of intrusions and control possible intrusions at their initial stage. The administrator performs the functions of managing the entire monitoring system as a whole, and is also responsible for the communication and interaction of the SMS with the CSN and the means of collecting information and checking the state of the system. He also makes the final decisions regarding the formation of false vulnerabilities, configuration of the CSN protection system and further actions in relation to violators.

References

1. Schubert, M., Bennett, D., Gines, J., Hay, A., Strand, J.: Nagios.: Enterprise Network Monitoring: Including Plug-Ins and Hardware Devices. Publishing Syngress (2008)
2. Mark, R.-O.: Information Security, 2nd edn. Graw Hill (2016)
3. Stamp, M.: Information Security: Principles and Practice, 2nd edn. USA (2011)
4. Khamdamova, S.M.: Methods and algorithms for the implementation of security monitoring in computer networks of automated process control systems. Int. J. Adv. Sci. Technol. **29**(11s), 1464–1469 (2020)
5. William, S.: Network security essentials: Applications and Standards, 4th edn. Prentice Hall, Upper Saddle River (2011)
6. Bejtlich, R.: The Tao of Network Security Monitoring: Beyond Intrusion Detection. Pearson, London (2004)
7. Ou, X, Govindavajhala, S., Appel, A.W.: MulVAL: a logic-based network security analyzer. In: Proceedings of the 14th USENIX Security Symposium, pp. 113–128. Princeton University (2005)
8. Northcutt, S., Zeltser, L., Winters, S., Kent, K., Ritchey, R.W.: Inside Network Perimeter Security, 2nd Edition. Sams Publishing, Indianapolis (2005)
9. Ammann, P., Wijesekera, D., Kaushik, S.: Scalable, graph-based network vulnerability analysis. In: Conference 2002, pp. 217–224, ACM Press (2002)

Application of Fuzzy Models in Building a Digital State in the Russian Federation

N. V. Mamitova(✉) and M. A. Makhotenko

Private Educational Institution of Higher Education "Southern University (IMBL)", Rostov-on-Don, Russia
nvmamitova@mail.ru, mahotencko@iubip.ru

Abstract. The article is devoted to the topical issue of introduction and use of digital technologies of public administration in the process of digitalization of the state and society. The article describes the features of building a digital state and its relationship with automation based on "soft models". It is necessary to modernize the existing social infrastructure. The problems of uncertainty or underdetermination in the management of the economy, social sphere, legal regulation, etc. are primarily faced with the challenges in the field of digitalization facing the country. The development of legal mechanisms to combat corruption is becoming a significant and urgent problem for the development of modern Russian society and the state. For the implementation of fuzzy control systems, on the one hand, there is an uncertainty associated with both the lack of information and the complexity of the system, and the impracticability or inexpediency of describing it in traditional ways, and, on the other hand, the presence of the necessary control actions, disturbances, etc. about the object.

Keywords: Digitalization · Digital economy · Digital public administration · Digital transformation · Big data · Data mining · Mobile applications · Analytical tools · Fuzzy models · System of Systems (SoS)

1 Introduction

The modern period of development of the Russian state is characterized by a systemic reform of the political, economic and social foundations of society. In the Address of the President of the Russian Federation to the Federal Assembly of the Russian Federation dated March 1, 2018, the thesis was put forward that "Digitalization of the public administration system, increasing its transparency is a powerful factor in combating corruption." In this regard, the task was set to ensure, in the next 6 years, the transfer of public services to the online mode using remote services, and the document flow between government structures should be carried out exclusively in electronic form. These measures to digitize the work of the authorities will, in the opinion of the President of the Russian Federation, increase the transparency of their work and fight corruption. It should be noted right away that at present, none of the regulatory legal acts, as well as in the strategic planning documents of the Russian Federation, enshrine such concepts as

© The Author(s), under exclusive license to Springer Nature Switzerland AG 2021
R. A. Aliev et al. (Eds.): WCIS 2020, AISC 1323, pp. 580–587, 2021.
https://doi.org/10.1007/978-3-030-68004-6_76

"digitalization" and "digital state", however, UNTKAD experts (United Nations Conference on Trade and development) the following interpretation of the term "digitalization" was proposed - a socio-economic transformation initiated by the massive introduction and assimilation of digital technologies, that is, technologies for creating, processing, exchanging and transmitting information; an attempt to determine the meaning of the concept of "digital state" was undertaken within the framework of the implementation of the direction "Digital state regulation" of the national project "Digital Economy of the Russian Federation", which will be discussed in more detail later in the text of this article. № 632-r of 06.05.2008 approved the "Concept of formation of e-government in the Russian Federation until 2010" (hereinafter-the Concept). According to the Concept, the term "e-government" refers to a new form of organizing the activities of public authorities, which provides a qualitatively new level of efficiency and convenience for organizations and citizens to receive public services and information about the results of the activities of public authorities due to the wide use of information and communication technologies.

The Federal Law of the Russian federation dated by 06.04.2011 № 63-FL "On e-signature" plays also a significant role by regulating the use of electronic signature in the process of state-legal transactions, providing with state and municipal services, fulfilling state and municipal functions.

2 Materials and Method

If we refer to the foreign experience of creating and implementing a digital state, we should immediately specify that the digital state is rather understood as an electronic government, which is a complex of electronic communication devices, computers and the Internet for providing public services to citizens and other persons in a country or region. This term includes digital interactions between a citizen and their government (C2G), between governments and other government agencies (G2G), between government and citizens (G2C), between government and employees (G2E), and between government and business/Commerce (G2B). It is interesting to note that the concept of "digital economy" was first mentioned in 1995 in the book "Being Digital" by Nicholas Negroponte, in which he examines the history of several digital media technologies and makes a forecast of the development of other digital technologies.

In the United States, the U.S. Digital Service is responsible for implementing the policy of providing public services to citizens in electronic form. This service deals with technology projects, relying on a user-oriented design structure to prioritize user needs, as well as modern software development methods that allow for iterative development and rapid response to changes and feedback. In addition, the Digital service is responsible for modernizing procurement procedures, taking into account the latest trends in digitalization.

In Spain, e-government covers any type of communication or interaction between citizens, businesses and public organizations through the use of information and communication technologies for e-government processes. The public sector is constantly making efforts to strengthen the commitment to e-government among Autonomous communities and local authorities and to improve coordination mechanisms between different levels of government in providing e-services to citizens. Examples of these efforts are the local "Avanza" plans, which aim to promote and distribute e-government at the local level.

In connection with the above, for the implementation of breakthrough scientific, technological and socio-economic development of the Russian Federation in accordance with paragraph 2 of the Decree of the President of the Russian Federation dated 07.05.2018. № 204 "on national goals and strategic objectives for the development of the Russian Federation for the period up to 2024" was assigned to the Government of the Russian Federation to develop national projects (programs) in the following areas: demography, health, education, housing and urban environment, ecology, safe and high-quality roads, labor productivity and employment support, science, digital economy, culture, small and medium-sized businesses and support for individual, business initiatives, international cooperation and export.

In the strategy of information society development in Russian Federation to 2017–2030, approved by decree of the President of the Russian Federation from 09.05.2017 № 203 "On the Strategy of information society development in Russian Federation to 2017–2030" (hereinafter - Strategy), the concept of "digital economy" is defined as an economic activity in which the key factor of production is data in digital form, processing large volumes and using the results of analysis of which, in comparison with traditional forms of management, can significantly improve the efficiency of various types of production, technologies, equipment, storage, sale, delivery of goods and services.

One of the tools for implementing the above-mentioned Strategy is the national project "Digital economy of the Russian Federation", the program of which was approved by the order of the Government of the Russian Federation dated 28.07.2017 № 1632-r "on approval of the program "Digital economy of the Russian Federation".

Under this program identified three levels of fields of activity on which the digital economy has an impact on the lives of citizens and companies, namely: markets and industries, where the interaction of specific actors (suppliers and consumers of goods, works and services) platforms and technologies, which formed a competence for the development of markets and industries; an environment that creates conditions for the development of platforms and technologies and effective interaction of market participants and economic sectors, and covers regulatory regulation, information infrastructure, personnel and information security. This influence is expressed through the implementation of the relevant areas of the national project "Digital economy of the Russian Federation", and each of the areas has its own fixed indicators.

Thus, for the direction of "Digital state regulation", characterized by the introduction of digital technologies and platform solutions in the areas of public administration and public services, including in the interests of the population and small and medium-sized businesses, including individual entrepreneurs, such indicators are the achievement of the following results by 2024:

1) 70% - digital interaction of citizens and commercial organizations with state (municipal) bodies and budget institutions;
2) 100% - priority state services and services are provided without the need for personal visits to state bodies and other organizations, using the registry model, online, proactively;

3) 70% - master data has been harmonized (master data compliance);
4) 90% - internal and interdepartmental legally significant electronic document flow of state and municipal bodies and budget institutions.

It is important to note that at the moment, at the level of Federal Executive bodies (hereinafter - FEB), work is also underway to "embed" them in the digital economy, which is expressed, inter alia, through the inclusion of such a position as Deputy Minister for digital development in the structure of key ministries. On the basis of open sources in the "Internet" telecommunications network, schemes were prepared through which an attempt was made to determine the place of the Ministry of transport of the Russian Federation in the digital economy, as well as the relationship between the national projects "Safe and high-quality roads", "international cooperation and export", as well as a Comprehensive plan for the modernization and expansion of the main infrastructure for the period up to 2024.

It is also necessary to keep in mind the risks that can most likely arise in the process of digitalization:

- violation of privacy/potential surveillance of citizens;
- reducing the level of data security;
- reducing the number of low-and medium-skilled jobs;
- increasing the level of complexity of business models and interaction schemes;
- sharp increase in competition in all areas of the economy;
- changes in the behavior patterns of producers and consumers;
- the need to review the administrative and tax codes.

In the context of the "fourth industrial revolution", which implies the integration of computing resources into physical processes, there is almost an annual change in technologies and business models that are applicable, including in traditional sectors of the economy. Back in 2010, China released the strategy "Made in China 2025" and "Internet+", which defined a new vector of development of the country's economy. A special role in this strategy is assigned to digital technologies as a catalyst for the transformation of each of the sectors of the economy. In addition, China is implementing system tools of state support aimed at the development of Internet of things technologies: tax regulation, state financing, development of common standards, and implementation of pilot projects. For example, companies in the high-tech industry are taxed at a preferential rate (15% versus the standard 25%), and software developers are exempt from income tax for two years and pay 50% of their tax liabilities in the next three years. The state's readiness for rapid technological changes is becoming one of the key factors of success on the world stage. But at the same time, the technological advantage should be applied primarily for the benefit of citizens, who are the core and heart of any state.

The United Nations (hereinafter - the UN), whose activities are aimed at solving problems and issues in the field of peace and security, economic and social development, human rights, the environment, international law, humanitarian issues and health, identifies in the block of international law such a problem as the fight against international crime, one of the manifestations of which is corruption.

The UN Convention against corruption (hereinafter - Convention), which was adopted by the UN General Assembly in October 2003, is the only international legal instrument to combat corruption.it outlines the main areas of action to reduce vulnerabilities and risks of corruption. However, the Convention provides a definition of the term corruption, but stands out a certain list of corruption cases of: bribery of officials in the public sector, bribery in the private sector, embezzlement, trading in influence, abuse of position, illicit enrichment.

It is interesting to note that the fight against corruption is also included in the block of problems related to creating conditions for business activity. Thus, we can conclude that the fight against corruption is a global problem, which should be solved using the most advanced technologies and best international practices.

In this regard, the development of digital methods for detecting corruption, preventing it and analyzing its consequences is becoming an important and relevant direction in the policy of any state within the global process of building a digital state. In 2016, the International economic forum highlighted the following promising areas for the use of digital technologies that are successfully used in the fight against corruption in the European Union:

1. Big data. This technology is mainly used in the fields of healthcare, trade, and taxation, where predictive analysis and visualization are used to obtain information, allowing you to determine trends, patterns, and relationships in data sets. An example is the Australian Taxation Office, which uses Big Data technology to search for necessary information in submitted reports in order to disclose evidence of the use of offshore zones, as well as, for example, to calculate online stores that do not meet their obligations to pay tax fees.
2. Data collection/Data mining. In public procurement, data mining is used as an audit tool aimed at tracking the actions of authorities when submitting applications and subsequently to detect collusion and provide false information. This technology also uses data visualization to detect corrupt intentions in various transactions. Experts From the corruption Research Center Budapest have used this technology to identify the following risk zones in public procurement in the European Union: short bidding periods, lack of competition in public procurement, and multiple bids won by one organization. Experts Of the research center on corruption using this technology have identified the following risk zones in the implementation of public procurement by countries of the European Union: short bidding periods, lack of competition in public procurement, and multiple bids won by one organization.
3. Mobile applications. In developing countries, this technology helps citizens from remote areas to access the necessary information. There is every reason to assume that this technology can be successfully used in the fight against corruption. For example, the world Bank has created its own version of the "I paid a bribe" app. The Integrity app is designed to provide citizens with access to world Bank-funded projects and the ability to immediately report fraud and corruption cases. Through the app, users can directly send information related to projects funded by the bank.

4. Expert systems/analytical tools. Technology self-monitoring, analysis and reporting technology (SMART). Along with technological advances, these tools are becoming more sophisticated for data processing, as they can analyze transactions in real time, simulate purchases and payments, and detect violations. Risk assessment algorithms can pre-warn and stop potentially incorrectly generated payments during the procurement process. In order to strengthen internal processes and prevent fraudulent schemes, auditors can periodically investigate transactions in purchasing and payment models, check for violations, and quickly identify illegal financial flows.
5. Fuzzy models and control mechanisms within a system of systems.

System of Systems is a generic name for a collection of target-oriented systems that combine their resources and capabilities to produce a new, holistic, more complex "metasystem" that, in turn, offers more functionality and performance than just the sum of composite systems, meaning it has emergence. Currently, System of Systems (SoS) "it is a research area for which reference systems, thought processes, quantitative analysis, tools and methods of design are incomplete, determined by NON-factors" [12].

Usually the following main types of SoS are distinguished [13]: directed, in which there is a recognized architect who has the right to issue orders to the component systems and manages their resources; acknowledged, in which the recognized architect is present, but he can only recommend that the component systems self – change according to the architecture developed by him; collaborative, in which all systems agree with each other on each problem, but there is no architect, project Manager, or similar dedicated management body.

Linking systems with joint SoS ensures interaction and synergy of enterprise management systems, computer management, communications and information, intelligence, and so on. A system of systems is a large-scale parallel and distributed system whose components are complex systems themselves. The system of system education includes the integration of systems into a system that ultimately contributes to the evolution of social infrastructure. The most important features of SoS that can be associated with the solution of the tasks are their synergy and the initial uncertainty of real functioning situations. The real solution to the country's challenges in the field of digitalization primarily faces questions of uncertainty or under-determination in the management of the economy, social sphere, legal regulation, etc.

Let's consider the influence of NON-factors [12] at the key stages of obtaining and processing information. To adequately account for these factors, it is necessary to conduct:

- selection of a mathematical model for the representation of a specific NON-factor (type of fuzziness);
- selecting the control tool (settings) for this model;
- measurement of actual uncertainty and its comparison with the model uncertainty;
- parameterization of adequate models for real objects and monitoring processes and the formation of computable indicators of uncertainty.

Let's consider the features of SoS on the example of the so-called geoinformation space formed by the entire combination of information flows and processes within a state or region. Spatially distributed information obtained, including by means of remote sensing of the Earth (RES) "consists of three large groups: semantic, metric and topological. A structured set of these groups of information about a specific territory, presented in a form suitable for automated processing, forms a digital model of the territory (DMT). The basis of such a CMT is the ability of one group of information (as a system) to use parts of another group" [13], that is, the ability to mutually use information in each of these groups (systems). In other words, we are talking about geo-interoperability. The international organization for standardization ISO 19119 gives the following definition: "Interoperability is the ability to connect, execute programs, or transmit data among various functional modules in a way that does not require the user to have knowledge of the characteristics of these modules" [2]. This means that two (or more) systems can work together to complete the task at hand, provided they are mutually interoperable. In the framework of geoinformation systems, "geointeropability is the ability of information systems to: freely exchange all types of spatial information about the Earth and about objects and phenomena, as well as above and below the Earth's surface; joint network use of software designed to manage such information" [13]. Obviously, storing and maintaining such a constantly evolving model requires appropriate information "resources available, at this stage, within the framework of the data warehouse concept (WC). The form of organizing a database that combines DMT and relational databases is currently most common. However, the complexity of such an organization caused by a set of tools for creating and maintaining a data topology creates certain problems" [8]. And, although such a model does not always provide the necessary flexibility and is focused only on specific systems for managing databases of fuzzy data [9, 14], there is currently no satisfactory alternative for it.

For the successful implementation of the above-mentioned models and technologies in practice, the state in the process of digital transformation of public services, management decisions, conducting licensing and Supervisory activities should conduct a range of activities that can be combined into four blocks: the introduction of a system of regular management and strategic management; digital transformation of public administration; new quality of personnel management and modern regulatory policy. We must not forget that the civil servants themselves will have to acquire relevant new knowledge, skills and abilities, otherwise all this work will lose any meaning and will remain simple slogans that will not go into action.

Taking into account the above, the concept of "State-as-a-Platform" (hereinafter-the Concept), published By the center for strategic development in April 2018, is of particular interest. This Concept is designed to ensure the transition from irrelevant SOS approaches to planning and monitoring the implementation of plans to accurate and individual indicators of the standard of living of citizens and the development of economic sectors. Updated indicators in the context of management decision-making will clearly show and record the level of responsibility of the individual in this process, as well as allow you to receive prompt feedback from management objects, work more accurately with key indicators of development. The creation of such a platform in the future should bring benefits to citizens, the state and business (Table 1).

Table 1. The advantage of the citizens to establish the concept of "Government-as-Platform"

State	Citizenry	Business
Reducing the level of expenditures on national expenditures by 0.3% of GDP by 2024	High level of satisfaction with the quality of public and municipal services	Reduction of administrative costs for the directivity factor and business reporting burden
Embedding data in decision-making processes, automatic decision-making algorithms, fundamental process reengineering, real-time situation monitoring	Improving the speed and quality of service delivery and minimizing face-to-face contact with government agencies: 1) a high proportion of digital services provided in a proactive mode; 2) personalization of public services; 3) a high percentage of data submitted by citizens once	User satisfaction with data quality and service: 1) high volume of government data used by businesses; 2) high percentage of businesses using/subscribing to government data

3 Conclusion

In connection with the above, it is worth asking whether the Russian Federation is ready to digitalize the economy, taking into account the lack of elaboration of this issue from a regulatory point of view. It seems reasonable that the implementation of such a large-scale program should begin with a normative settlement of this issue, in particular, with the definition of basic terms for the formation of a unified conceptual apparatus.

In addition, taking into account the fact that the digital literacy index of citizens of the Russian Federation according to the Regional public center for Internet technologies has decreased by 14.7% compared to 2018, a lot of work should be done to organize a series of educational events aimed at preparing society for life in the digital economy.

References

1. Novikov, D.A., Belov, M.V.: Teoriya kompleksnoj deyatel'nosti. Chast' I. Struktury kompleksnoj deyatel'nosti. Neopredelyonnost' i porozhdenie kompleksnoj deyatel'nosti J Problemy upravleniya, Moskva **4**, 36–45 (2018). https://doi.org/10.25728/pu.2018.4.5
2. Bednar, J.: The Robust Federation: Principles of Design. Bibliotech Web Service. Cambridge University Press, New York, pp. 373–374 (2009). https://doi.org/10.1017/S1537592709992027
3. Novikov, D.A.: Analiticheskaya slozhnost' i pogreshnost' resheniya zadach upravleniya organizacionno-tekhnicheskimi sistemami. J Avtomatika i telemekhanika **5**, 107–118. Moskva (2018). https://doi.org/10.1134/S0005117918050077
4. Hramov, V.V.: Teoriya informacionnykh processov i sistem, p. 47. Publishing House FGBOU RGUPS, , Rostov-na-Donu (2011)

Control of Deployment and Development of Cluster and Virtual Structures in Industry

A. S. Khokhlov[1(✉)], D. Y. Mishutin[2], and E. S. Baulin[3]

[1] Center of Digital Technologies LLC, Moscow, Russia
khokhlov.as@mipt.ru
[2] AO Honeywell, Moscow, Russia
dmitry.mishutin@honeywell.com
[3] Moscow Institute of Physics and Technology, Moscow, Russia
baulin.es@mipt.ru

Abstract. This report analyzes the origins of M. Porter's recommendations within the context of systemic view on the network nature of life as a social sphere in order to answer the question how to organize and manage the development of cluster structures, and it also defines 5 of its properties. This allows to develop recommendations for the organization of cluster structures, and their management should include modeling of exchange operations between their actors. Optimization modeling of Nizhnekamsk Petrochemical Cluster is considered as an example of implementation of such a model.

Keywords: Emergence · ICT · Feedback · Self-regulation · Autopoiesis · Ecology · Virtual corporation · Modeling · Optimization · Petrochemical cluster · Solomon associates

1 Introduction

The world enters into the post-industrial era of information and communication technologies (ICT) progress that leads to widespread online contacts, setting to zero social distances, huge flows of information and a situation of continuous changes [1]. According to the definition of clusters [1] and in fact, it is socio-economic network structure that unites independent actors (A) in order to make their joint work innovative, thus increasing its effectiveness to the world level. In 2019, Russia remained in the 43rd place among 141 countries in the GCI-Global Competitiveness Index ranking. The analysis of causes and actions (in particular, follow the recommendations of M. Porter) is given in [1]. In the report, the authors analyze the origins of M. Porter's recommendations within the context of systemic view on the network nature of life as a social sphere in order to answer the questions how to organize and manage the development of cluster structures.

2 Key Properties

The paper [2] provides an overview of social sphere research. Let's briefly highlight a number of its key properties related to the issue:

© The Author(s), under exclusive license to Springer Nature Switzerland AG 2021
R. A. Aliev et al. (Eds.): WCIS 2020, AISC 1323, pp. 588–595, 2021.
https://doi.org/10.1007/978-3-030-68004-6_77

1. Emergence is the process of coming into existence. Life is interpreted as an emergent property, which is understood as the presence of special new properties inherent in the network ensemble of elements, and arising in the course of their interaction. New properties mean that none of the elements separately has this property. More generally, the structure of the living and the way it interacts with the environment are two facets of the same phenomenon of life. According to the Theory of Changes [3], these structures are based on interaction (collaboration) of opposites.

2. Networks. Whatever living systems we consider – organisms, parts of organisms, or communities of organisms - their organisms will be ordered in the image of a network. The most important property of these structures, which is common to all living systems, is their network nature.

3. Feedback. The main and most obvious property of any network is its non-linearity: the network diverges in all directions. Thus, the relationships in the network structure are non-linear, in particular, any impact or message can follow a cyclical path, forming feedback loops. In living networks, the concept of feedback is closely related to their network structure.

Wu Hsing (hereinafter, the romanization of Chinese words are according to the Wade–Giles system.) cycle with history of more than 2.5 thousand years from Chinese natural philosophy is an example of such a feedback loop before it formed the foundation of cybernetics. We give a brief description from the paper [3]. Wu Hsing is created by a cycle of five primary elements of human consciousness connected by two sequences: generation and suppression, as in Fig. 1 or 2.

The sequence of generation of primary elements is the following: mind, emotions, desires, creativity, thinking.

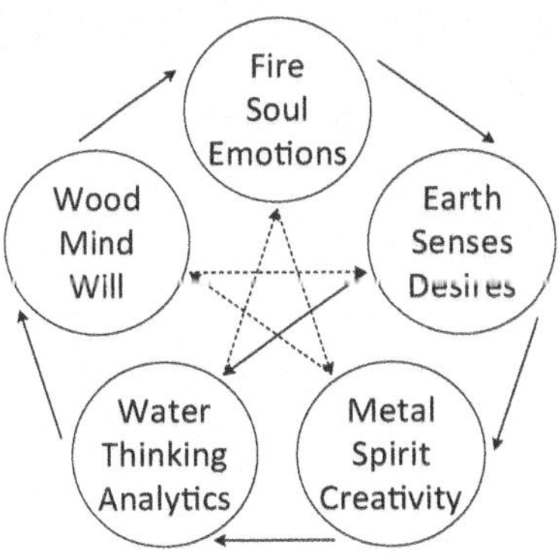

Fig. 1. Wu Hsing cycle

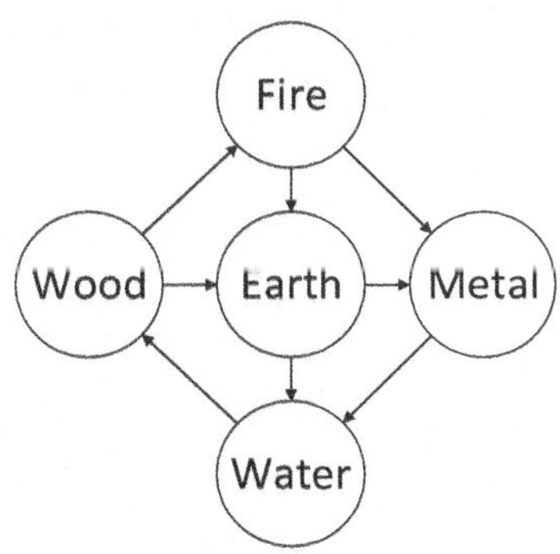

Fig. 2. Representation of Wu Hsing cycle

The sequence of suppression is as follows: mind, desires, thinking, emotions, creativity. Everyone can see with the naked eye the fact that these sequences are the law.

On the one hand, desires make you create, and it is impossible to create without thinking, but thinking cannot be without mind, and things created by mind bring satisfaction – emotions. On the other hand, the mind oppresses desires, desires oppress thinking, and thinking oppresses emotions, then emotions do not allow you to create, which stimulates self-organization.

Each of the five primary elements is assigned a wide range of concepts from various fields [3], and a cycle appears with the emergence of network connections.

4. Self-regulation. Since feedback loops can be formed in communication networks, this makes it possible for them to self-regulate, i.e. to provide the possibility of self-organization (in 1932, Walter Cannon called these self-regulating processes homeostasis). According to Santiago theory of cognition (Humberto Maturana and Francisco Varela, 1970s), there are three different level of organization for complex living systems which should be taken into account when considering them: the first level is a phenomenon of self-organization, i.e. the ability of the system to come to an ordered state determined by its internal rules; the second one is the autopoiesis which means that self-organizing system is capable to restore all its components only on the basis of its own resources (necessary condition of life as such); and in the third level, the autopoiesis is connected to cognition which provides both necessary and sufficient conditions for life existence. As a result, there is a dual role of living systems, where each of the systems is both a part and a whole at the same time, which leads to the existence of two opposite tendencies: to integration, striving to fulfill the role as an integral part within a larger whole, and to self-organization, to self-affirmation and preservation of individuality and autonomy.

5. Ecology. It is interpreted as interaction with the environment. There is no "environment" in any independent or abstract sense. As well as there is no organism without an environment, so there is no environment without an organism. Organisms contribute to the "creation" of the environment. All of us are built into the cyclical processes of nature and, ultimately, are completely dependent on them. Such understanding of processes is also called ecosystemic, and an interactive form of cooperation in the social sphere with the use of ICT is defined as "collaboration".

As a consequence of properties 1–5, life can be represented as a triad in Fig. 3.

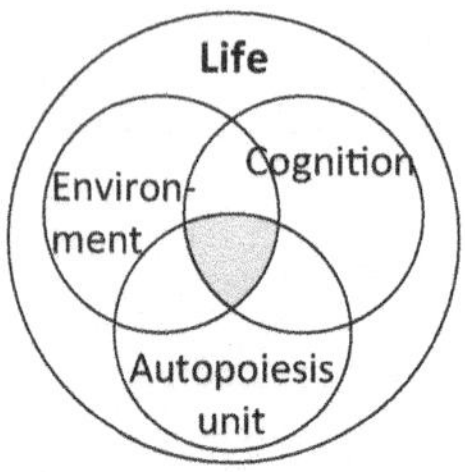

Fig. 3. The triad of life

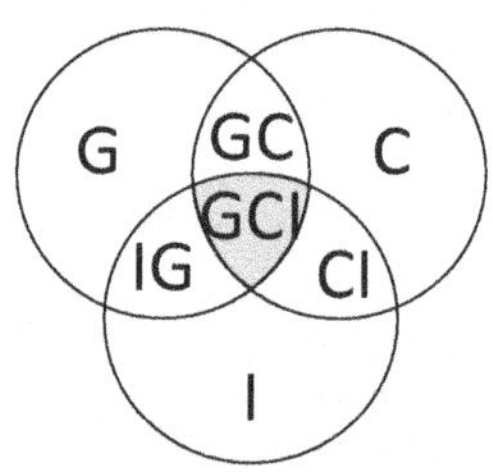

Fig. 4. The triple helix model

The organic, living structure interacts with the environment through the instrument of cognition – this specific product of individual cognition and the evolutionary process. All three areas are closely related according to 1–5, and life can be represented as a synergistic interaction of all three areas [2].

Let's establish an obvious correspondence: Environment G ↔ State/Local authorities; Cognition C ↔ Science/Universities; Autopoiesis unit I ↔ Investors/Companies, and we will get an innovative mechanism in the form of Etzkowitz-Ledersdorff Triple Helix Model with the GCI-index of M. Porter (Fig. 4), so popular among its adherents in the field of clusters.

Now let's turn to Fig. 1 Wu Hsing Cycle and present it in the form of Fig. 2.

The Theory of Changes states (for details see [3]) that the existence of the world and all processes in it is connected with the existence and cyclic interaction of the five primary elements. If we assume that "Earth" is the collaboration insitutes, and "Fire" and "Water" are respectively symbols of conditions for "Competition" and "Cooperation", and the symbols "Wood" and "Metal" are respectively symbols of conditions for innovation, for "Production" and "Demand", then we get the version in Fig. 5. of M. Porter's diamond model. At the same time, the GCI structure contains 12 sections [1], which corresponds to 12 earthly branches of the Theory of Changes [3], and the "Golden rules" of M. Porter's cluster policy (names: State must; State may; State must not [1]) fit into the 9-cell magic square of Lo-Shih from the Theory of Changes as well.

Fig. 5. Porter's diamond

In a number of works on clusters, a reference to Wu Hsing cycle (Fig. 6) is made explicitly, for example, in G.D. Boush's one, or implicitly in the form of the concept of "5I" by M.P. Voynarenko as a necessary condition for creating clusters.

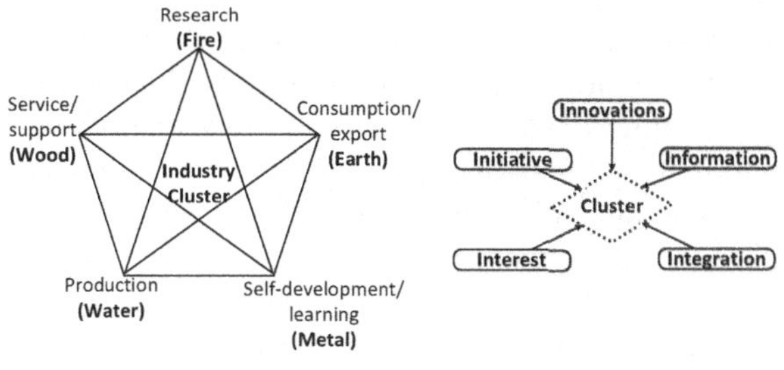

Fig. 6. Interpreting a cluster through the Wu-Hsing cycle

We point out that the application of the Theory of Change in practice is Feng Shui, different styles of internal alchemy in the form of Taijiquan, and in traditional Chinese medicine, for example, it is the technique of acupuncture [3]. These practices are based on the collaboration of opposites and their cyclical nature in the form of a sequence of type generation and overcoming (Fig. 1, 2). But it is the same basis that is in the properties 1–5 discussed above.

Perhaps, M. Porter and his followers understand the origins of their "diamonds and golden rules", but the authors did not find a clear indication to this.

Cluster-type structures are modified in the process of functioning due to properties 1–5 and under the influence of continuous changes: they are expanded or qualitatively transformed, passing properties 1–5 by inheritance. An example and comparative characteristics of such metamorphoses are given in Table 1 [1].

Table 1. Clusters, cluster organizations and networks: comparison of basic characteristics.

1. Cluster	2. Cluster organization	3. Intercompany network (the virtual corporation)
Joint placement Automatic participation	Availability of a platform for collaboration. Participation on the basis of membership. The composition of the parts of the triple helix is determined by the territorial and industry specifics of the cluster. Focus on collaboration is related to the broad task of increasing the competitiveness of the cluster	Availability of a platform for collaboration Participation on the basis of membership. The membership is determined by a specific project and not related to the specifics of the region. Focus on collaboration is related to narrow issues

3 Organization of Industrial Cluster Structures

It follows directly from property 4 (self-regulation) that creating a cluster-type structure is only possible if there is an autopoiesis unit (see Fig. 3). From the correspondence of Fig. 3 and Fig. 4, all statements in "Golden rules" with name "State must not" follow, otherwise there is a clear contradiction with the concept of autopoiesis, where auto is self. From property 5 (ecology), it follows that the necessary condition for the emergence of an autopoiesis unit in the process of cognition is its interaction with the environment, which corresponds to the statements with name "State must", and with name "State may" only depending on the circumstances. The criterion marking that the appearance of the structure as in Table 1 actually took place is the emergence of properties 1–5 in it.

Let's assume that the management of a Petrochemical Company has a plan for reconstruction with the creation of structures like those shown in Table 1, following the global trend in the industry. At this stage, we suggest to refrain from creating something new, and start with researching own complex using the CPA™ methodology of Solomon Associates, as it has been done in Russia for more than 20 years together with the Russian partners Honeywell JSC and Center of Digital Technologies, LLC [4]. Solomon conducts benchmarking analysis of indicators of the entire production and sales chain in the oil and gas industry: from exploration and production of raw materials to logistics, processing and sale of raw materials and products.

Through benchmarking of the Petrochemical Company, its management will be able to understand its actual position according to a wide range of indicators (financial, cost-based, technological) in relation to other enterprises in the world and, in particular, in its region. As a result, management, along with improvement of operational activities, will be able to move to the development of innovative and investment programs for reconstruction in the selected areas with a focus on the Table 1 in network options of the Petrochemical Company development (structures Fig. 7).

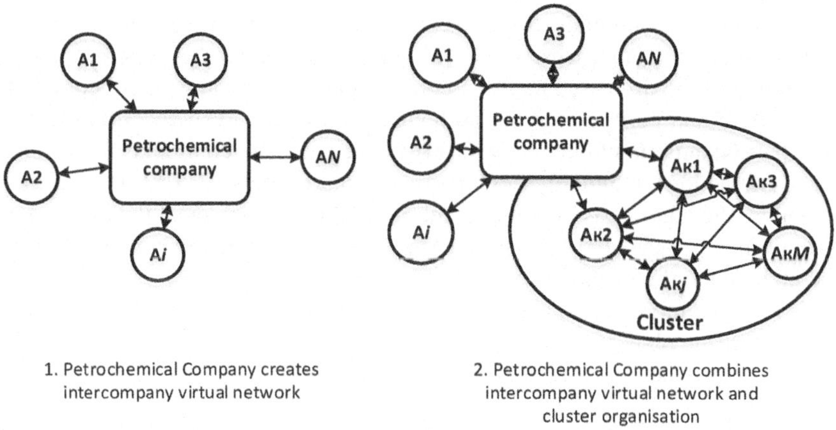

Fig. 7. Network development options for a Petrochemical company as a virtual corporation

According to Fig. 5 the implementation of the structures (Fig. 7) in the form of a virtual corporation assumes creation of an independent institute of collaboration by the Petrochemical Company management, for example, in the form of an outsourcing team with a coordinator (hereinafter - the Coordinator).

First of all, when creating the specified structure, the Coordinator, solves the following three main tasks:

1. Initiating the formation of network options. Initially, the Coordinator examines the companies applying for a partnership for sufficiency and competitiveness of their resources, and determines whether they have the abilities and desire to participate in virtual partnership. At the end of this period, the virtual corporation Coordinator and the partners engaged prepare a memorandum of understanding of the mission status, vision and understanding of the most important rules and legal norms, procedures for forming common costs, common risk and profit.
2. Forming the activities of a virtual corporation. The Coordinator becomes responsible for maintaining and implementing the collaboration immediately after understanding the structure of the virtual corporation. The Coordinator, as a neutral party, facilitates the resolving of conflicts that have arisen among partners, monitors the internal and external environment, and deals with the feedback signals.

3. Supporting the activities of the virtual corporation. The Coordinator examines resources and competencies, identifying the weaknesses and strengths of external partners of the virtual corporation, and develops a supply chain that integrates external consumers/suppliers and internal partners. The effective functioning of the created virtual corporation involves the use of high-performance ICT and modeling systems that are successfully operated in the selected application area.

4 Management Models

The network order, according to sociologist Walter Powell [1], is based on mutually beneficial models of communication and exchange, which are dependent on the presence of several types of mutual trust (unity of intentions, partners' competence, unity of understanding of the discussed ideas and obligations, i.e. 5th property – ecosystem).

The mutual and parity nature of exchanges is the core idea: according to Powell, network participants should be ready to invest and give, as well as take and receive.

Such interactions assume that one party voluntarily depends on resources that are controlled by the other party, and that combining of the resources of the parties under a network project is beneficial for both parties. Therefore, mutually beneficial exchange will be included in the management model.

Let's consider modeling of resources (flows) exchange operations between actors at the example of the objects of NPC (part of the INNOKAM cluster) in the Republic of Tatarstan. The production structure of NPC can be divided into three blocks [5]:

1. OAO TAIF-NK, AO Taneco – continuous production industry – oil refinery;
2. PAO Nizhnekamskneftekhim as far as a number of chemical and petrochemical plants – mainly continuous production industries – the Petrochemical Company and 10 small and medium-sized manufacturers of industrial products.
3. PAO Nizhnekamskshina (hereinafter – NKSH)—mass discrete parts industry, provides production of passenger and truck tires, purchase of raw materials and delivery of tires. Includes OAO Nizhnekamskshina, OOO Tatneft-Neftekhimsnab, OOO Trading house Kama and the Management company of OOO Tatneft-Neftekhim.

At OAO TAIF-NK, AO Taneco, PAO Nizhnekamskneftekhim planning problems were solved at different times using the RPMS (Refinery and Petrochemical Modeling System) system developed by Honeywell Corporation; the Management company of PAO Nizhnekamskshina uses the SOFTYRE – an automated optimization complex for planning a tire association.

In [5] modeling the structure is conducted in three stages. At the first stage all models of the three blocks are updated according to the current technical and economic indicators, at the second stage, a block of continuous productions specified in pp. 1, 2 is created in the form of many factory model (the F_PRESS system – an optimization one), and at the third stage, the solution from the second stage is inputted into the SOFTYRE system, and the production of NKSH is modelled.

5 Conclusion

Thus, viewing the industrial clusters as network social structures, it is possible to draw the following conclusions:

1. It is considered that an ensemble of actors takes place if it has the emergent properties, and the actors are in network interaction, which allows them to provide self-regulation of the ensemble through feedback, and develop an adequate response to environmental influences. Thus, the ensemble will show ecologic properties, which can lead to the correction of the emergent properties of the ensemble and in the cycle. In our opinion, it is the appearance of such cycle that can be considered as the necessary and sufficient conditions for the existence of the ensemble.
2. The paper describes that when creating clusters formed like an ensemble, cluster organizations or virtual corporations, three innovative M. Porter's mechanisms were used: GCI-the Triple Helix, Golden rules and Diamond model. In fact, they start a cycle represented by a sequence of generation and suppression with a collaboration of opposites, which creates the basis of the cycle specified above in p. 1.
3. At the preliminary stage of creating cluster structures, as part of the reconstruction of the industrial complex, it is worth starting with its research, for example, using the ABC Payroll Services (M. Porter) methods, but the key measure is the application of the CPA™ Solomon Associates methodology.
4. The management of the industrial complex, aiming to supplement it with a network structure, should create an independent institution of collaboration, in the form of a team with a coordinator. The coordinator initiates the creation of network options in three stages, conducts the creation of their activities until the stable cycle specified in p.1 is reached, and then supports their functioning as an outsourcing manager.
5. Effective functioning of the created network option (for example, as on Fig. 7 in version 1 or 2) involves the use of high-performance ICT and modeling tools that are successfully used in the selected application area [5, 6].
6. According to property 5, the ecosystem perspective assumes an assessment of the impact of industrial technologies on the environment, both current and future.

References

1. Smorodinskaya, N.V.: Global economy: from hierarchies to network structure, RAS EI, Moscow, p. 344 (2015). https://inecon.org/docs/Smorodinskaya_book_2015.pdf
2. Capra, F., Luisi, P.L.: A systematic view of life. Holistic view. Translated from English, URSS: LENARD, Moscow, p. 504 (2020)
3. Dudorov, O.E.: Theory of changes. The mystery of I Ching, Flinta, Moscow, p. 232 (2006)
4. Ellerington, T., Power, P., Shishorin, Y.R., Aksenova, T.S.: Improving operational efficiency and benchmarking. In: Automation in Industry, vol. 4, pp. 34–41 (2015)
5. Khokhlov, A.S., Mishutin, D.Y., Borodin, P.E.: Optimization tools for modeling petrochemical clusters. In: Modern Science, vol. 3/2 (2019). ISSN 2223–2966
6. Khokhlov, A.S., Baulin, E.S., Konnov, A.I., Mishutin, D.Y.: Complex of integrated planning of vertically integrated oil companies' activities. In: Automation in Industry, vol. 12, pp. 15–26 (2018)

Modeling of Dehydrogenation and Oxidative Dehydrogenation of Isopropyl Alcohol Over Ti-V-O Catalyst

N. I. Dushdurova(✉) and S. A. Mammadkhanova

Azerbaijan State Oil and Industry University, Baku AZ-1010, Azerbaijan
nurlana.dushdurova@mail.ru, sevinc.mammadxanova@asoiu.edu.az

Abstract. The activities of the titanium-vanadium oxide catalysts in the dehydrogenation and oxidative dehydrogenation of isopropyl alcohol to acetone have been studied. It is found that the main products of the dehydrogenation reaction of isopropyl alcohol are acetone and propylene, while the oxidation of isopropyl alcohol in the reaction products also leads to the formation of a significant amount of carbon dioxide. It was shown that the yield and distribution of the reaction products strongly depend on the reaction temperature and the composition of the feedstock. It was found that the activity of titanium-vanadium oxide catalysts in the conversion of isopropyl alcohol to acetone in the presence of oxygen is significantly higher than in the presence of nitrogen, which is in good agreement with the performed thermodynamic calculations. Thus, the highest acetone yield on the studied catalysts reaches 37.4% with a selectivity of 89.3%. Studies have shown that the activity of binary titanium-vanadium oxide catalysts is also depends on their composition. It was found that in the reaction of isopropyl alcohol dehydrogenation to acetone with increasing titanium content in the binary titanium-vanadium oxide catalyst, the acetone yield passes through a maximum on the sample Ti-V = 5-5. When the reaction is carried out in the presence of oxygen, a picture of the dependence of the acetone yield on the atomic ratio of titanium to vanadium is observed with two maxima for the samples V = 1-9 and Ti-V = 7-3. In order to account for imprecision of experimental data, we proposed to use fuzzy rules for modeling of the considered dependence.

Keywords: Isopropanol · Acetone · Catalyst

1 Introduction

One of the methods for producing acetone is the conversion of isopropyl alcohol [1, 2]. Acetone can be obtained from isopropyl alcohol by both dehydrogenation and oxidative dehydrogenation reactions of the initial secondary alcohol [3, 4]. In order to determine the advantages of each of these reactions, we previously carried out comparative thermodynamic calculations of the reactions of producing acetone by the reactions of dehydrogenation and oxidative dehydrogenation of isopropyl alcohol [5]. It was found that the value of the change in the Gibbs energy is negative for both studied reactions

© The Author(s), under exclusive license to Springer Nature Switzerland AG 2021
R. A. Aliev et al. (Eds.): WCIS 2020, AISC 1323, pp. 596–603, 2021.
https://doi.org/10.1007/978-3-030-68004-6_78

at temperatures above 27 °C. It was concluded that in order to obtain acetone, it is preferable to carry out both reactions at temperatures below 327 °C. It is known that vanadium containing catalysts are highly active in the reactions of partial conversion of alcohols. In this regard, in this work we have studied the reactions of dehydrogenation and oxidative dehydrogenation of isopropyl alcohol to acetone over Ti-V-O catalysts. In order to account for imprecision of experimental data, we proposed to use fuzzy rules for modeling of the considered dependence.

2 Experimental Part

Titanium-vanadium oxide catalysts were prepared by mixing aqueous suspensions of titanium oxide and ammonium vanadate. The obtained mixture was evaporated at 95–100 °C, then the formed precipitate was dried at 100–120 °C and then decomposed until the complete release of nitrogen oxides at a temperature of about 250 °C. The resulting solid mass was calcined at a temperature of 600 °C for 10 h. Thus, 9 catalysts with a different atomic ratio of elements from 9:1 to 1:9 were synthesized.

The catalytic activity of the synthesized samples was studied on a flow-through unit with a tubular reactor in the temperature range 100–500 °C. For this, a catalyst with a grain size of 1–2 mm was loaded into a tubular reactor. The dehydrogenation reaction of isopropyl alcohol was carried out by feeding the reaction mixture of the following composition isopropyl alcohol:nitrogen:steam = 10:50:40 into the reactor. The space velocity of the feedstock was 1200 h^{-1}. The oxidative dehydrogenation of isopropyl alcohol was carried out by feeding the reaction mixture of the following composition isopropyl alcohol: air: steam = 10:50:40 into the reactor. The space velocity of the feedstock in this case was also 1200 h^{-1}. Analysis of the feedstock and reaction products was carried out by chromatography method.

3 Results and Discussion

The study of the activity of the synthesized titanium-vanadium oxide catalysts showed that the products of the dehydrogenation of isopropyl alcohol are acetone and propylene, while the formation of a significant amount of carbon dioxide in the reaction products is also observed in the oxidation reaction of isopropyl alcohol. In the studied reactions at temperatures above 350 °C, the formation of decomposition products of isopropyl alcohol is also observed. The results of studying the dehydrogenation reaction of isopropyl alcohol over catalyst Ti:V = 1:9 are shown in Fig. 1. As can be seen, the conversion of isopropyl alcohol on this catalyst begins from a temperature of 150 °C. At this temperature, 3.4% acetone and 4.5% propylene are formed. The acetone yield with increasing temperature increases to 23.8% at 300 °C and then decrease and at 400 °C reaches 11.9%. As can be seen from Fig. 1, the rate of propylene formation also passes through a maximum at 300 °C. The highest propylene yield for this sample reaches 40.5%. Figure 1 also shows that at reaction temperatures above 300 °C, the formation of small amounts of carbon dioxide is observed in the reaction products and its yield does not exceed 7.1%. The conversion of isopropyl alcohol in the dehydrogenation reaction, as seen from the figure, reaches 70%.

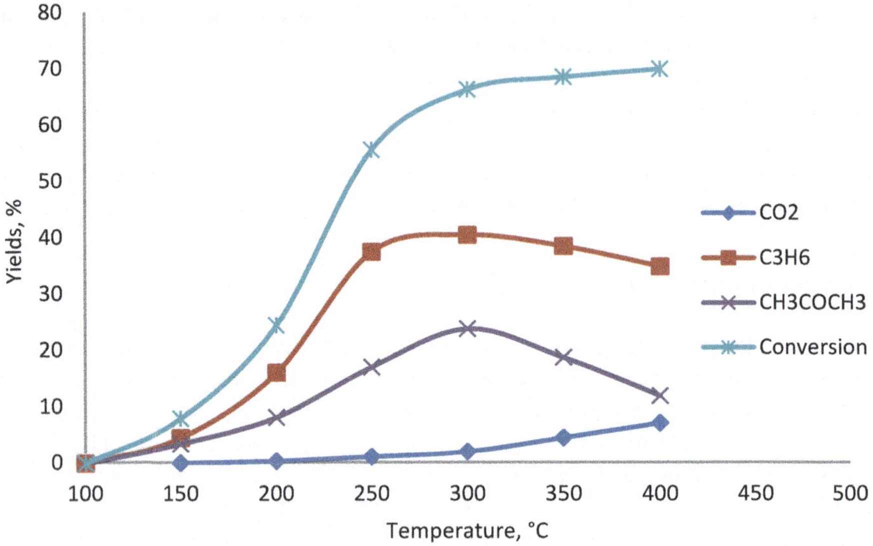

Fig. 1. Influence of temperature on the activity of catalyst Ti:V = 1:9 in the reaction of dehydrogenation of isopropanol.

Figure 2 shows the results of a study of the oxidation reaction of isopropyl alcohol on a Ti:V = 1:9 catalyst.

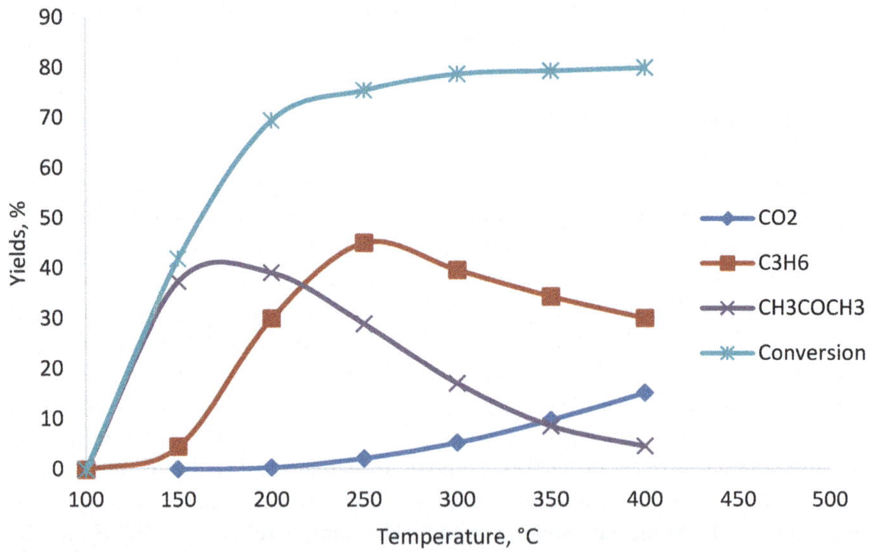

Fig. 2. Influence of temperature on the activity of catalyst Ti:V = 1:9 in the reaction of oxidative dehydrogenation of isopropyl alcohol.

As you can see, the oxidation reaction of isopropyl alcohol, as well as the dehydrogenation reaction, starts from a temperature of 150 °C. At this temperature, 37.4% acetone and 4.5% propylene are formed. The selectivity for acetone at this temperature is equal 87.2%. A further increase in temperature leads to an increase in the acetone yield and at 200 °C reaches 39.1%. The propylene yield also passes through a maximum with an increase in the reaction temperature. The highest propylene yield on the studied catalyst is 45% and is observed at a temperature of 250 °C. In contrast to the dehydrogenation reaction, in the oxidation of isopropyl alcohol in the reaction products, the formation of carbon dioxide is observed in a rather significant amount. The highest carbon dioxide yield reaches 15.1% at 400 °C. From the obtained data, it can be seen that the conversion of isopropyl alcohol in the presence of oxygen is higher than in the presence of nitrogen and reaches up to 80%.

For both of studied reaction we calculated the temperature dependences of the yields of the reaction products using regression equations. Linear, quadratic, cubic, logarithmic and exponential regression equations were calculated. The best description of the experimental data was obtained using cubic regression equations. The values of the yields of the reaction products and the conversion of isopropanol at different temperatures were calculated using the following cubic regression equation.

$$\alpha = aT^3 + bT^2 + cT + d$$

The values of *a, b, c* and *d* for each of the isopropanol dehydrogenation reaction products are given in Table 1 and for oxidative dehydrogenation reaction products are given in Table 2.

Table 1. The values of *a, b, c* and *d* for each of the isopropanol dehydrogenation reaction products.

Reactionproducts	*a*	*b*	*c*	*D*
CO_2	1.3333E−7	−8.9676E-5	0.00512779	3.677877
C_3H_6	−5.2444E−6	0.00753805	−3.3486499	471.17607
CH_3COCH_3	−4.2444E−6	0.0061633	−2.831071	418.79397
Conversion	−7.3111E−6	0.0107849	−4.887203	700.52748

Table 2. The values of *a, b, c* and *d* for each of the isopropanol oxidative dehydrogenation reaction products.

Reactionproducts	*a*	*b*	*c*	*D*
CO_2	1.1111E−7	6.6619048E−5	−0.11309	27.24781033
C_3H_6	−2.08888E−6	0.0021403	−0.3743937968	−53.2338537
CH_3COCH_3	1.23333E−5	−0.02059576	11.159332	−1935.355253
Conversion	7.37777E−6	−0.0132252	7.881685	−1482.910456

We have previously carried out thermodynamic calculations of the dehydrogenation and oxidation reactions of isopropyl alcohol. Table 3 shows the calculated values of the thermodynamically possible yields of acetone in the reactions of oxidation and dehydrogenation of isopropyl alcohol. It can be seen that for the oxidation reaction of isopropyl alcohol, already beginning from 127 °C, the theoretically possible yield of acetone is equal 100%, while in the dehydrogenation of isopropyl alcohol, the yield of acetone increases from 1.5% at 27 °C to 100% at 427 °C.

Table 3. Thermodynamically possible yields of acetone in the dehydrogenation and oxidative dehydrogenation reactions of isopropyl alcohol.

Reactions	Conversions in percent of initial isopropyl alcohol to acetone at various temperatures (in °C)								
	127	177	227	277	327	377	427	477	527
Oxidative dehydrogenation of isopropyl alcohol	100	100	100	100	100	100	100	100	100
Dehydrogenation of isopropyl alcohol	25	56	82	94	98	99	99	100	100

Comparing the results of thermodynamic calculations with the activity of synthesized titanium-vanadium oxide catalysts, we can say that the yield of acetone is higher in the presence of oxygen than in its absence, and indeed the rate of the oxidative dehydrogenation of isopropyl alcohol is higher than the rate of the dehydrogenation reaction. However, the presence of oxygen in the initial reaction mixture, as can be seen from the data obtained, negatively affects the selectivity of the process with respect to acetone, increasing the rate of deep oxidation of isopropyl alcohol to carbon dioxide.

Studies have shown that the activity of binary titanium-vanadium oxide catalysts is also influenced by their composition. Table 4 shows the activities of titanium-vanadium oxide catalysts in the reaction of isopropyl alcohol dehydrogenation.

Table 4. Dependence of the activity of titanium-vanadium oxide catalysts in the dehydrogenation of isopropyl alcohol on the atomic ratio of titanium to vanadium in the catalyst, T = 200 °C.

Reaction products	Yields, %								
Atomic ratio of Ti/V	1:9	2:8	3-7	4:6	5:5	6:4	7:3	8:2	9:1
Carbon dioxide	0	0	0	0	0	0	0	0	0
Propylene	16	25.4	26.4	45.4	51	39	27.1	13.5	8.3
Acetone	8.1	5.1	7.4	13.6	17	8.5	3.4	3.4	1.7
Acetone selectivity	33.6	16.7	21.9	23.1	25	17.9	11.1	20.1	17
Conversion	24.1	30.5	33.8	59	68	47.5	30.5	16.9	10

From Table 4 it is seen that on all catalysts at 200 °C the formation of carbon dioxide is not observed. With an increase in the titanium content in the catalyst, the yields of acetone, propylene and the conversion of isopropyl alcohol pass through a maximum. The highest yields of propylene and acetone are observed for the sample Ti-V = 5-5 and are 51 and 17%, respectively. The conversion of isopropyl alcohol on this catalyst at 200 °C is 68%. A similar dependence is observed at higher reaction temperatures.

Table 5. Dependence of the activity of titanium-vanadium oxide catalysts in the oxidative dehydrogenation of isopropyl alcohol on the atomic ratio of titanium to vanadium in the catalyst. T = 150 °C.

Reaction products	Yields, %								
Atomic ratio of Ti/V	1:9	2:8	3-7	4:6	5:5	6:4	7:3	8:2	9:1
Carbon dioxide	0	0	0	0	0	0	0	0	0
Propylene	4.5	9.6	5.9	6.6	6	7.1	12	0	0
Acetone	37.4	3.4	8.4	5.1	5.1	4.1	26.5	0	0
Acetone selectivity	89.3	26.2	58.7	35.9	26.1	36.6	68.8	0	0
Conversion	41.9	13	14.3	14.2	19.5	11.2	38.5	0	0

Table 5 shows that in the oxidation reaction of isopropyl alcohol, two samples Ti-V = 1-9 and Ti-V = 7-3 have the highest activity in the formation of acetone. For these samples, the yields of acetone are 37.4 and 26.5%, respectively. These samples also have the highest selectivity in the acetone formation reaction. The results show that the catalysts Ti-V = 8-2 and Ti-V = 9-1 are inactive at 150 °C in the oxidation reaction of isopropyl alcohol. Table 3 also shows that on all catalysts at 150 °C during the oxidation of isopropyl alcohol, the formation of carbon dioxide is not observed.

The data on the oxidation reaction (the graphs in Figs. 1, 2) are characterized by imprecision. In view of this, we intend to model the dependencies of the yields on the temperature values by using fuzzy set theory [6]. The fuzzy If-Then rules are to be constructed by using Fuzzy C-means clustering algorithm applied to imprecise data. For simplicity, we will model dependence of the acetone yield (the output variable) on the reaction temperature (the input variable). The obtained fuzzy rules are shown in Fig. 3.

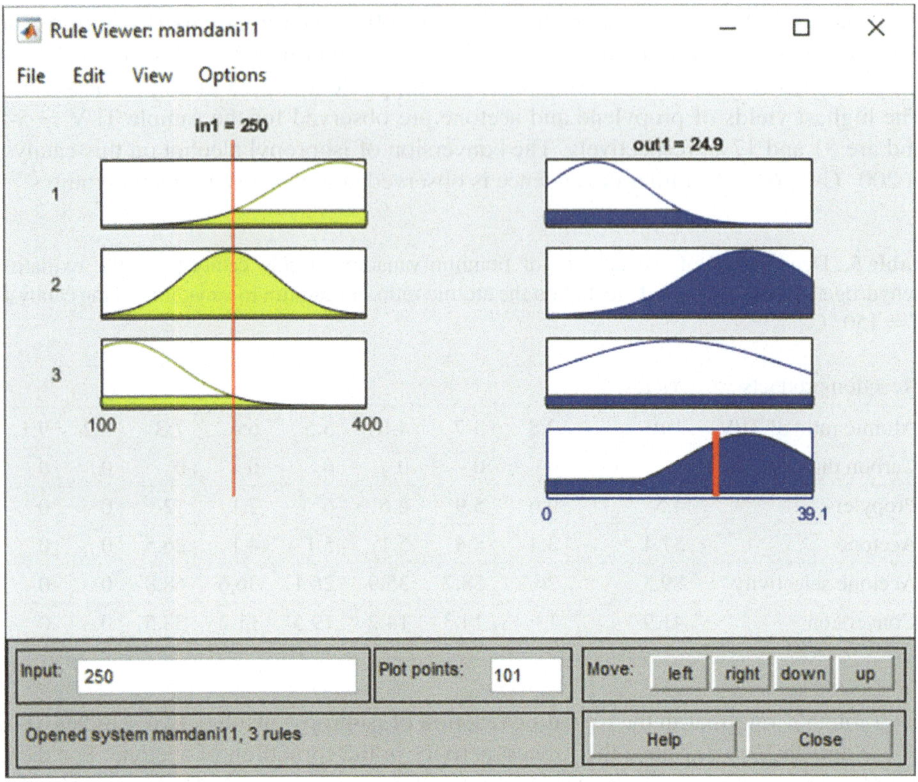

Fig. 3. The membership functions of the fuzzy rules describing a fragment of relationship between acetone yields (colored in yellow) and the temperature (colored in blue).

4 Conclusion

The products of the dehydrogenation of isopropyl alcohol on titanium-vanadium oxide catalysts are acetone and propylene, while in the oxidation of isopropyl alcohol, in addition to these products, a significant amount of carbon dioxide is also formed.

The activity of titanium-vanadium oxide catalysts in the conversion of isopropyl alcohol in the presence of oxygen is much higher than in the presence of nitrogen, which is in good agreement with the performed thermodynamic calculations.

The activity of titanium-vanadium oxide catalysts in the isopropyl alcohol conversion reaction depends on the atomic ratio of titanium to vanadium in the composition of the binary catalyst. Catalyst of the composition Ti-V = 1-9 shows highest activity in the formation of acetone. Over this catalyst the acetone yield reaches 37.4% with a selectivity of 89.3%. Due to imprecision of the experimental data, we proposed to describe the considered process by using fuzzy rules.

References

1. Pato, A.H., Balouch, A., Talpur, F.N., et al.: Synthesis and catalytic practicality of titania@ITO-grown nanoflakes: an excellent candidate for isopropanol conversion to acetone. Appl. Nanosci. **10**, 739–749 (2020). https://doi.org/10.1007/s13204-019-01200-4
2. Wang, J.A., Chen, L.F., López, T., Gómez, R., Novaro, O.: Dehydrogenation and dehydration of isopropanol catalyzed with Sol-Gel MgO-Al_2O_3 oxide. In: López, T.M., Avnir, D., Aegerter, M. (eds.) Emerging Fields in Sol-Gel Science and Technology. Springer, Boston (2003). https://doi.org/10.1007/978-1-4615-0449-8_8
3. Jasińska, E., Krzyżyńska, B., Kozłowski, M.: Activated carbon modified with different chemical agents as a catalyst in the dehydration and dehydrogenation of isopropanol. Catal. Lett. **125**, 145–153 (2008). https://doi.org/10.1007/s10562-008-9536-z
4. Zhang, H., Dai, L., Feng, Y., et al.: A resource utilization method for volatile organic compounds emission from the semiconductor industry: selective catalytic oxidation of isopropanol to acetone Over Au/α-Fe_2O_3 nanosheets. Appl. Catal. B **275**(15), 119011 (2020). https://doi.org/10.1016/j.apcatb.2020.119011
5. Dushdurova, N.I., Mammadkhanova, S.A.: Comparison of the thermodynamic possibility of obtaining low molecular weight ketones by the reactions of dehydrogenation and oxidative dehydrogenation of C_3-C_4 alcohols. Azerbaijan J. Chem. News N3, 17 (2019)
6. Zadeh, L.A.: Fuzzy sets. Inf. Control **8**, 338–353 (1965). https://doi.org/10.1016/S0019-9958(65)90241-X

Regression Model of Composites Based on Epoxy-Diane Functionalized Oligomer

T. M. Naibova[1(✉)], Ahmet Tutar[2], F. A. Amirov[1], G. S. Hasanov[1], and A. A. Mammadova[1]

[1] Azerbaijan State University of Oil and Industry, Baku, Azerbaijan
{tamilla.naibova,fariz.emirov,g.hasanov, aytan.mammadova}@asoiu.edu.az
[2] Sakarya University, Serdivan, Sakarya, Turkey
atutar@sakarya.edu.az

Abstract. The use of existing oligomers as binders in the manufacture of composites for various purposes has become widespread on an industrial scale. However, since these oligomers have certain disadvantages (fragility, imperviousness to heat and shock), the demand for composites made on their basis is not fully justified. To eliminate these disadvantages, the actual problems are their modification by chemical means to improve the functionality of oligomers. In the functional analysis, the main parameters of the new composition of the co-oligomer used as a modifier of an organic compound of the amide group (carbamide) and obtained by co-polycondensation were studied. For the purpose of comparison, the main physical-chemical, physical-mechanical and spectral analyses were carried out in parallel, synthesized from an unmodified epoxy-Diane oligomer under the same conditions. It became known that the main indicators of the functioning epoxy-Diane oligomer have increased, and its shortcomings have been eliminated. Amide and epoxy groups participated in the operation, and chemical modification of co-polycondensation was confirmed. As a new functional binder co-oligomer, and as a filler - from the shell powder (waste) of hazelnuts, nuts, almonds and chestnuts, the composition of the coating was developed, and the main indicators were studied. Mathematical modeling was used in the analysis. The resulting composite coating is recommended for use in equipment used in the oil and gas industry, for corrosion protection and in the furniture industry.

Keywords: Synthesis · Epoxy resin · Carbamide · Oligomers · Modification · Composite · Filler · Coating corrosion

1 Introduction

Compositions of different purposes, made on the basis of epoxy-diane oligomers, become fragile and flimsy. To eliminate such shortcomings, it is possible to eliminate them by upgrading their functionality by method of chemical modification. When examining the results of various research studies, it becomes clear that functional modification of epoxy-Diane oligomer with chemical modification differs sharply from physical modification

© The Author(s), under exclusive license to Springer Nature Switzerland AG 2021
R. A. Aliev et al. (Eds.): WCIS 2020, AISC 1323, pp. 604–612, 2021.
https://doi.org/10.1007/978-3-030-68004-6_79

and preference is given to chemical modification [1, 2]. For this purpose, carbamide produced on an industrial scale was used as a modifier in the research work.

This modifier increases the functionality of the oligomer, strengthens the interfacial bonds in the preparation of composites. Functional groups are also involved in the solidification process of the oligomer and its composite. Special attention was paid to this factor in the process of its functionalization, since the ratio of primary components in the synthesis of the epoxy-diane oligomer directly affects the polycondensation process. The process of functionalization in different proportions of the modifier is carried out, and the main physical, chemical and spectral analyses of the resulting new co-oligomer are investigated. It became known that the main indicators of the functioning epoxy-diane oligomer increased by about 1.0–1.5 times, thereby eliminating the shortcomings. A functionalized epoxy-diane oligomer was used as a binder for the composite, and a powder of the bark (waste) of hazelnuts, nuts, almonds and chestnuts was used as a filler. Adding an oligomer to the filler increases the viscosity, strength, and hardness of the mixture and prevents it from compressing during solidification. Adding a filler to the matrix affects the rate of solidification of the mixture. Depending on the amount of filler, the solidification period is reduced, but when the amount of filler is 5–10 mol. Fractions, the performance indicators of the composite composition produced at its interval correspond to the goal. The composition of waste used as fillers is shown in Table 1.

Below are given non-functionalized and functionalized epoxy-diane oligomers (a, b), fillers (c) and finished composites (d)) (see Fig. 1). The solidification process was carried out by chemical and thermal means. In the process of chemical solidification, polyethylene polyamine was used. The main indicators of the prepared target composition and application areas are defined.

Fig. 1. Non-functionalized and functionalized epoxy-diane oligomers (a, b), fillers (c) and finished composites (d).

Table 1. The composition of the fillers

№	Composition, %	Fillers			
		Hazelnut shell powder	Walnut shell powder	Almond shell powder	Chestnut shell powder
1	Lignin	52.46	48.23	45.82	34.82
2	Cellulose	36.24	38.58	38.86	40.03
3	Oils and resins	3.75	4.45	4.49	8.42
4	Water-soluble compounds	6.68	8.02	10.22	15.96
5	Ash substances	0.87	0.72	0.61	0.60

2 Experimental Part

Taking into account the cost efficiency and environmental friendliness of the components required for the production of an epoxy-diane oligomer-diphenylolpropane and epichlorohydrin, the functionalization of the binder used to produce the target composite was carried out by the method of co-polycondensation [3–5].

The dried, functionalized epoxy-diane oligomer has a light yellow color and is highly soluble in acetone, toluene, alcohol, dioxane and tetrahydrofuran. Does not dissolve in water. The dependence of the yield of the co-oligomer obtained by modification of the epoxy-diane oligomer with carbamide on the amount of the modifier and the reaction time is investigated both experimentally and mathematically (Table 2). The output of co-oligomers obtained when the amount of modifier to 0.25 mol, is 85–90 °C and the duration of the reaction for 2.5 h at 75%, but the relatively reduced output co-oligomers dissolved in an amount of 0.35 mol of modifier. The reason for this is the high activity of the modifier and, as a result, the transformation of the co-polycondensation product into an insoluble solid mass. Conducting a co-oligomerization reaction for more than 2.5 h does not meet the goal, leading to an increase in the amount of insoluble fraction.

Table 2. Dependence of co-oligomer output on various factors

№	Quantity of modifier, mol, (Q)	Temperature, °C, (T)	Reaction time, hour (t)	Reaction yield, % (Y)	Predicted reaction yield, % ($Y_{predicted}$)
1.	0.15	85	1	42	45.384
2.	0.15	85	1.5	52	51.7675
3.	0.15	85	2	56	58.151
...	...	...	...	...	...
6.	0.15	90	1.5	54	51.7675
7.	0.15	90	2	58	58.151
8.	0.15	90	2.5	63	64.5345
...	...	...	...	...	...
23.	0.35	90	2	70	68.651
24.	0.35	90	2.5	72	75.0345

Multiple regression analysis method was used to determine the dependence of co-oligomer output output on the quantity of modifier (Q), temperature (T) and time (t).

The analysis was done using R software and the result of multiple regression is given in Table 3. We find that the adjusted R-squared value is 0.9006 which means that the independent variables explain 90% of the variation in the co-oligomer output.

Table 3. Multiple regression output

Predictors	Estimates	Std. error	t value	Pr (>|t|)
Intercept	−14.6333	20.3841	−0.718	0.4811
Q	52.5000	7.0749	7.421	3.65e−07
T	0.4500	0.2311	1.947	0.0657
t	12.7667	1.0334	12.355	8.11e−11
Adjusted R2	0.9006			

F-statistic: 70.5 on 3 and 20 DF, p-value: 8.23e−11.

The resulting model is given by:

$$Y = -14.6333 + 52.5000 * \mathrm{Q} + 0.4500 * \mathrm{T} + 12.7667 * \mathrm{t}$$

The model calculated values are added to Table 2 as the Predicted Reaction yield in %.

The correlation coefficient between the observed and predicted model was found to be 0.9558 which shows strong correlation.

The number of epoxy groups in functionalized epoxy-Diane co-oligomers of epoxy acid is carried out on the basis of titration of the remaining acid in a reaction involving hydrochloric acid of these groups.

The applied co-oligomer's main physical-chemical and physical-mechanical properties have been identified using classical methods [6, 7]. The results of physical-chemical and physical-mechanical indicators are shown in Table 4.

Table 4. The main indicators of nonfunctionalized and functionalized epoxy-diane oligomers

Oligomers	Indicators						
	Density, kg/m^3	Molecular mass	The degree of hardening, %	Softening temperature, °C	Quantity of epoxy groups, %	Adhesive strength, MPa	Brinell hardness, MPa
Non-functionalized epoxy-Diane oligomer	12–50	380–390	95–96	70–75	20–22	2.4–2.6	110–115
Functionalized epoxy-Diane oligomer	1285	540–620	98–99	85–90	14–16	3.6–4.0	155–160

3 Results and Discussions

IR and NMR spectra of functionalized co-oligomers in accordance with FTIR Alpha Brucker spectrometers 400–4000 cm^{-1} and drawn in the range 0–11 m.d. [8–10]. For the purpose of comparison were also given IR spectrum - modifier, epoxy functionalized oligomer and epoxy oligomer. It was found that in the IR spectrum, there is a 3399.15 cm^{-1} valence oscillation in the band of intense absorption of amide with carbamide, 2929.50 cm^{-1} weak $-NH_2^+$ group = NH_2, 1634.19 cm^{-1} relatively weak - valence vibration in the band of absorption – C = C-C group, 1421.48 cm^{-1} deformation vibration in the band of intense absorption relatively weak bands characteristic of the NH_2 group, finally, 1019.50; 883.20; 619.64 and 476.62 cm^{-1}, are images related to CO, NH, CH and solvent.

In the spectrum of a nonfunctionalized epoxy-Diane oligomer, 3338.28 cm^{-1} OH group with a valence fluctuation in the intensive absorption band, of 1694.73 cm^{-1} 606.93 cm^{-1} benzene ring in diphenylolpropane in the absorption band, of 1592.89; 1556.91 and 1510.31 cm^{-1} NH group with isolated C – C vibration and deformation vibration in weak, medium and intense absorption bands, 1361.44 cm^{-1} -CH2 group, 1231,61; 1151,73; 1097,90; 1014,36 cm^{-1} in intense and medium bands –C-O-C – - HC = CH_2; CO and – CH_2 –O-CH_2- strain vibration groups, 996, 96; 959.75; 890.01; 812.95; 691.09; 639.71; 617.68 cm^{-1} CCl in the middle and weak band; and the groups characteristic of CH were reflected.

By comparing the bands observed in the IR spectrum of the functionalized epoxy-Diane oligomer (Fig. 2), it is known that in contrast to the spectrum of the non-functionalized epoxy-diane oligomer, the addition to the functional oligomer is 3340,91; 2964,04; 2930,41; 2872,09 and 1294,93 cm^{-1} in accordance with intense bands such as -OH; -NH; CH_2; CH and groups that seem to confirm the process of chemical modification involving oligomers of epoxy-Diane and carbamide groups.

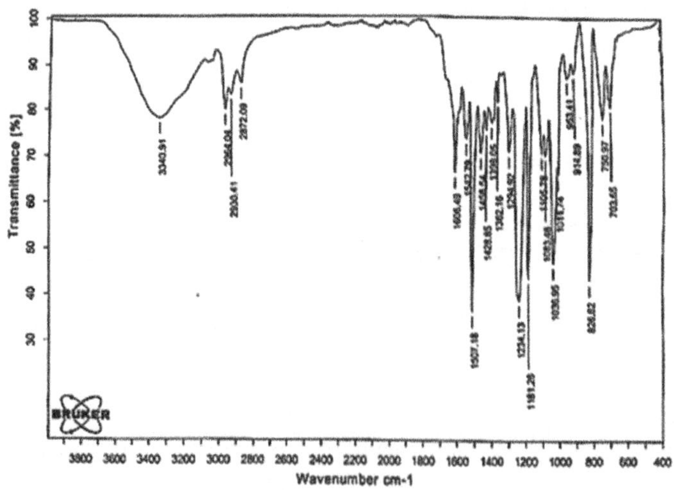

Fig. 2. The spectrum İR of the functionalized epoxy-diane oligomer

From the NMR spectrum of the functionalized epoxy-diane oligomer (Fig. 3) it was found that H1 is 7.4–8.5 m.d. in the NMR spectrum. Peaks NH and OH group, 6.7–7.2 m.d. to the protons of the aromatic ring to the peaks, 3.7–4.3 m.d. it is suitable for the CH_2 group, which combines two aromatic rings of peaks. 2.4–3.3 m.d. and the peaks are suitable for groups CH_2 and CH, which are in the epoxy group, 1.6 m.d. peaks refers to the solvent (acetone – D).

Scanned with CamScanner

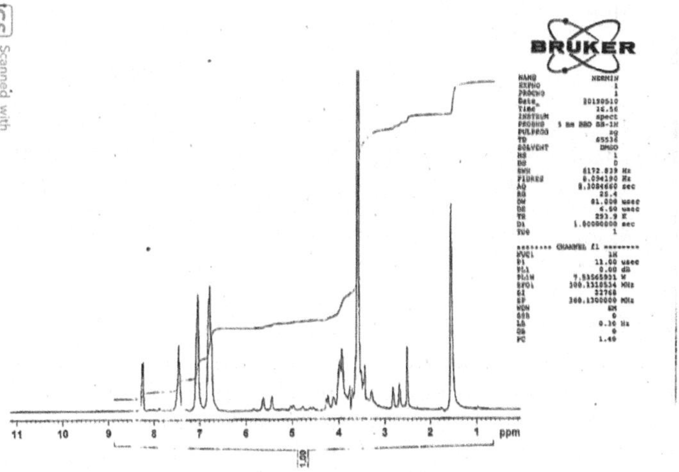

Fig. 3. NMR spectrum of the functionalized epoxy-diane co-oligomer

In the IR and NMR spectra, where the chemical modification of the functionalized epoxy-Diane oligomer process occurs, the confirmed and probable course of the co-oligomerization reaction can be shown as follows:

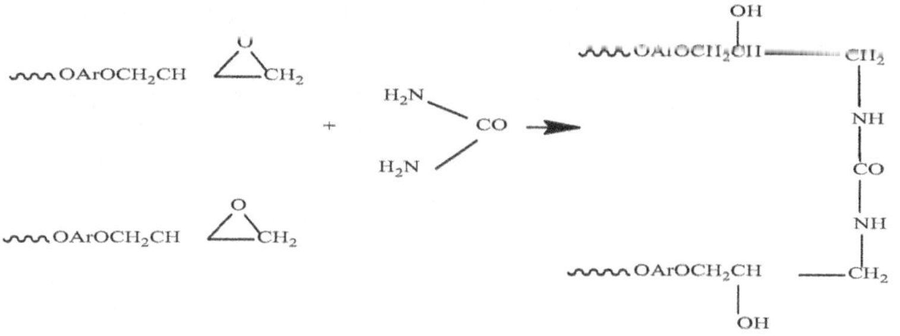

The distribution of the molecular weight of the co-oligomer (MCP) was studied by an exclusive method of chromatography using liquid chromatography produced by Kovo (Czech Republic). The analyses used two column filled with A Separon – SGX adsorbent, 3.3 × 150 mm in size and 7 μ in particle size, with a pore diameter of 100 A^0. The results of the study of the functional chromatography process by exclusive chromatography are shown in Table 5.

The results of the analysis show that 80% of the product of the functionalization process carried out by the co-oligomerization method consists of high-molecular fractions ($V_R = 8$–12) with an average molecular weight ($M_w = 4010$) and high-molecular fractions with an average molecular weight ($M_n = 2570$) and low-molecular masses, ($V_R = 8$–12) with the corresponding parameters $M_w = 610$ and $M_n = 580$ ($V_R = 12$–14), that is, in the process of 80% of the epoxy-Diane oligomer functions.

Table 5. Molecular weight distribution of the functionalized epoxy-diane oligomer

№	Sample	Fraction, %	MKP		
			Mw	*Mn*	*Mw/Mn*
1	Functionalized epoxy-diane oligomer	High-molecular (80)	4010	2570	1.56
		Low-molecular (20)	610	580	1.05
		Total	3340	1530	2.18

Calculated by the MCP chromatogram method [11]. Thus, taking into account the hydroxyl, amine and amide groups that are part of the functional oligomer of epoxy-diane, in the manufacture of composites from it, they were used as a binder, a filler from hazelnut, walnut, almond and chestnut shells (waste), a solvent from acetone and a hardener from polyethylene polyamine. When preparing the composite, first dissolve the co-oligomer in acetone, then add the appropriate amount of filler, mix for 0.5–1.5 h (until a homogeneous mass is obtained) and add a hardener (2.5–3.0 mass fraction). The finished composition is laid out on 2–3 layers of metal samples. The hardening process is completed within 2–3 days at room temperature, 140–150 °C for 5 h.

Elemental analysis of fillers used in the manufacture of composites (C, H, N) was studied using the standard test method ASTM D 5373-02, and sulfur was studied using the standard test method ASTM D 4239-05. The composition and density of the filler element are shown in Table 6. To make hazelnut, walnut, almond and chestnut shells (waste) suitable for use as a filler, they are first washed in water (until the washing water becomes transparent), then dried completely, grinded in a low–speed Shini SJ-1411 mill and passed through a sieve with a hole size of 0.09 mm.

Table 6. Element composition and density of fillers

№	Fillers	C	H	N	S	Density, kg/m^3
1	Hazelnut shell	56.34	5.36	0.52	0.02	1161
2	Walnut shell	51.25	6.52	0.34	–	1218
3	Almond shell	52.32	5.54	0.47	–	1230
4	Chestnut shell	54.28	5.24	0.48	0.01	1188

The main indicators of composites made on the basis of the functionalized epoxy-Diane oligomer and the non-functionalized epoxy-diane oligomer were studied for comparison (Table 7).

As can be seen from the table, the main indicators of a composite based on a functionalized epoxy-diane co-oligomer are higher than those of a non-functionalized composite based on an epoxy-diane, and no changes were observed in seawater for 12 months. To protect equipment operated in seawater from corrosion and use in the furniture industry, it is recommended to use a coating composition based on a functionalized co-oligomer.

Table 7. Main indicators of composites

№	Indicators	A nonfunctionalized epoxy-diane oligomer	A functionalized epoxy-diane oligomer
1	The amount of dry residue, % of the mass	52	54
2	Conditional viscosity in the VZ-4 device, sec	48	56
3	Bending strength, in the ShQ device, mm	3	5
4	Adhesive strength, MPa	3,2	4,2
5	Impact resistance, on the U-1 device, conventional unit	0.85	0.95
6	Stability in seawater, for 12 months	Color change	No change

4 Conclusion

Functionalized by the method of co-polycondensation epoxy-Diane oligomer that is modified with carbamide. The factors influencing the output of the functional co-oligomer, physical-chemical, physical-mechanical and spectral analyses (IR, NMR and MCP) are investigated. It turned out that the functionalized epoxy-diane oligomer by co-polycondensation is a process of chemical modification involving epoxy and amide groups. An epoxy-Diane oligomer was used as a binder, a composite coating was developed from hazelnut, walnut, almond and chestnut shell powder (waste) as a filler, and polyethylene polyamine as a hardener. Some features of dispersed fillers are investigated. The main indicators of the composite are defined. Taking into account the positive indicators of the functionalized epoxy-diane oligomer, as well as the compositions based on it, and the stability of seawater, it is recommended to use them for corrosion protection of equipment and installations operated at sea. It can also be used in the furniture industry as a coating composition.

References

1. Paken, F.M.: Epoxy compounds and epoxy resins. Chemistry, Moscow (1962) https://doi.org/10.1002/pol.1962.1205616408
2. Smirnov, V.A., Korolev, E.V.: Nanomodified epoxy composites **15**, 150–157 (2012)
3. Shiblawi, K.A., Pershin, V.F., Yartsev, V.P.: Modification of epoxy resins: modern condition and prospects, pp. 068–078 (2018). https://doi.org/10.17277/amt.2018.02
4. Naibova, T.M., Amirov, F.A., Rahimova, N.A.: The modification of epoxy oligomers with co-polycondensation method. In: Proceeding of Inter Development Goals, pp. 201–203 (2017)
5. Naibova, T.M., Amirov, F.A., Ragimova, N.A.: Modification of the epoxy oligomer with benzoguanamine. In: International Scientific and Practical Conference "The Role of Economic Science in the Development of Society: Theoretical and Practical Aspects", Belgorod, pp. 160–164 (2018)
6. Saba, N., Jawaid, M.: 3-epoxy resin based hybrid polymer composites. In: Hybrid Composite Materials, Properties and Characterization, pp. 57–82 (2017) https://doi.org/10.1016/B978-0-08-100787-7.00003-2
7. Patnaik, T.K., Nayak, S.S.: Investigation of physico-mechanical and thermo-mechanical analysis of alumina filled needle-punch nonwoven jute epoxy composites. Polym. Compos. **39**, 1553–1556 (2018). https://doi.org/10.1002/pc.24099
8. Yamasaki, H., Morita, S.: Two-step curing reaction of epoxy resin studied by thermal analysis and infrared spectroscopy. Appl. Spectrosc. **66**(8), 926–933 (2012)
9. Tarutina, L.I., Pozdnyakova, F.O.: Spectral analysis of polymers. Chemistry, L (1986)
10. Bektashi, N.R., Mustafaev, A.M., Huseynov, I.A., Mustafaev, S.G.: Chromatographic study of the synthesis patterns of halogen-containing unsaturated polyesters. J. Appl. Chem. **84**(7), 1211–1216 (2011)
11. Naibova, T.M.: Process development of modified phenol-formaldehyde oligomers, and compositions thereof. Herald Azerbaijan Eng. Acad. **4**(4), 84–93 (2012)

Author Index

A
Abdullayev, A., 402
Abdullayev, Khanmurad, 113
Abdullayeva, M. Y., 196
Abdullayevich, Safarov Rustam, 170
Abdurasulov, Fakhritdin, 545
Abizada, Sanan, 452
Abulfas, Habibov Ibrahim, 147
Adilov, Farukh, 545
Agaguseynova, M. M., 442
Agahuseynova, M. M., 427
Ahmadov, Shamil A., 303
Ahmedov, A. H., 504
Akbaraliev, B. B., 103
Akhmetov, D. A., 207
Alakbarov, Y. Z., 127
Aliev, Araz R., 96
Aliyev, Aydin, 55
Aliyev, Elchin, 113
Aliyev, Rafig R., 329
Aliyeva, G. A., 487
Aliyeva, Kamala, 180
Aliyeva, N. F., 520
Alizadeh, Akif V., 72
Amirov, F. A., 487, 604
Asli, Ehsan Faramarzi, 394

B
Babanli, Mustafa, 254
Babayev, A. I., 472
Babayev, Ata, 409
Babek, Iravanli Konul, 147
Baghirli, A. N., 342
Barat, Ahmedov, 402
Barno, Solieva, 280
Baulin, E. S., 588
Baxtiyorovich, Botirov Fayzullajon, 26
Bekmurodov, Ulugbek Bakhrom, 132
Bilgehan, Bülent, 466
Botirov, T. V., 460
Bukhanov, Sergey A., 418

D
Damirova, J. R., 127
Davidov, Fedor, 545
Dilnoz, Muxamedieva, 280
Dmitriy, Chereshkin, 529
Dushdurova, N. I., 596

E
Efendiyev, G. M., 207
Elmuratov, M. K., 309
Ergashbaevich, Iskandarov Zohid, 435
Eshimbetov, A. G., 262

F
Farzalizada, Z. I., 127
Fayziyevich, Bekmuratov Tulkun, 26
Fayzullaev, B. A., 309
Fedorchuk, Volodymyr, 18
Fomenko, A. I., 564

G
Gardashova, Latafat A., 33
Gasanova, L. M., 377
Gaybulov, Q. M., 40
Gazvini, K. A., 512
Giyosov, U. E., 155

© The Editor(s) (if applicable) and The Author(s), under exclusive license to Springer Nature Switzerland AG 2021

R. A. Aliev et al. (Eds.): WCIS 2020, AISC 1323, pp. 613–615, 2021.
https://doi.org/10.1007/978-3-030-68004-6

Gojayev, Tural, 254
Gregory, Royzenson, 529
Gudratova, F. D., 442
Guliyev, Huseyngulu, 55
Guliyeva, A. N., 512
Guliyeva, Nigar, 348
Gulyamov, Sh. M., 246

H
Hajizade, S. G., 385
Hakimovah, Siddikov Isamiddin, 435
Hasanov, A. A., 141, 537
Hasanov, B. Allahverdi, 164
Hasanov, G. S., 520, 604
Hasanov, R. G., 520
Hatamkhanova, Gulnara, 348
Huseynov, Nazim Tofiq oglu, 272
Huseynov, A. B., 427
Huseynova, Farida, 358

I
Ibragimovich, Jumanov Isroil, 170
Igamberdiev, H. Z., 322, 460
Ikramov, Alisher, 81
Imanov, Elbrus, 394
Imanova, Gulara, 348
Imanova, Gunay, 237
Isakova, Sitora, 545
Iskandarov, E. Kh., 385
Iskandarova, F. N., 40
Ismailov, M. A., 504
Ismayilov, Elviz A., 96
Ismayilova, F. B., 385
Ismayilova, H. G., 127, 385
Isroilovich, Djumanov Olim, 170
Ivaniuk, Vitalii, 18

J
Jabbarov, T. G., 316
Jabbarova, Aynur I., 287
Jabbarova, K. I., 287

K
Kadirbergenovich, Khudaybergenov Kabul, 47
Kaipbergenov, B. T., 309
Karazhanova, M. K., 207
Karimov, M. M., 571
Khamdamova, S. M., 571
Khamrakulov, U. Sh., 479
Khasanov, Doston, 186
Khokhlov, A. S., 588
Kholmanov, U. U., 334
Khramov, Vladimir V., 553
Khujamatov, Khalim, 186
Khujayorov, Ilyos, 215
Korotaev, Andrey, 545
Kuliyev, R. H., 207
Kuşaf, Günay, 496

L
Logmanli, O. Azim, 164

M
Makhotenko, M. A., 580
Mallaev, A. R., 64
Mamedkhanova, S. A., 342
Mamirov, U. F., 322
Mamitova, N. V., 580
Mammadkhanova, S. A., 596
Mammadova, A. A., 604
Matyokubov, N. R., 201
Mehdiyev, Nihad, 224
Mirzayan, Kamilov, 280
Mishutin, D. Y., 588
Mukhitdinov, D. P., 229
Muminov, Bahodir Boltayevich, 132
Musaev, Muhammadjon, 215
Mustafayev, Islam, 348

N
Nabiev, A. A., 262
Nabijonov, Kamoliddin, 81
Naibova, T. M., 604
Narzullayev, Sh.N., 559
Nazarov, Kh. N., 201
Nematov, E. H., 504
Nishanov, A. H., 103
Nodirbekovich, Yusupbekov Azizbek, 389
Nourani, Vahid, 1
Nuraliev, F. M., 155

O
Ochilov, Mannon, 215
Okada, Yoshihiro, 155
Ovsiannikov, Mikhail, 418
Özdeşer, Hüseyin, 496

P
Paknezhad, Nardin Jabbarian, 1
Piriverdiyev, I. A., 207
Podkopaev, Sergey A., 418
Primova, H. A., 40
Pulatova, L. T., 294

Q
Qosimov, F. O., 504

R
Rahimova, F. R., 487
Rajabov, A. T., 334
Rakhimov, T. O., 201
Rakhimova, L. S., 262

Rakhimovich, Marakhimov Avazjon, 47
Rasulev, Bakhtiyor, 81
Reypnazarov, Ernazar, 186
Rustamova, G. F., 487
Ruziev, U. A., 121
Rzayev, Ramin, 113

S
Sadikoglu, Fahreddin, 1, 466
Safarov, S. Rza, 164
Saidov, A. A., 294
Salmanova, N. I., 427
Sattarov, O. U., 229
Sayfullaev, Sherzod, 10
Sevinov, J. U., 64
Shahlarli, M. E., 127
Sharghi, Elnaz, 1
Shodiev, M. K., 121
Shodiyevich, Avazov Yusuf, 364
Siddikov, Ilkhom, 186
Sterten, Jo, 18
Sultanova, A. B., 196

T
Tadjitdinov, G. T., 559
Tajibaev, Sh. Kh., 103
Talybov, G. M., 342
Turabdzhanov, S. M., 262
Tutar, Ahmet, 604

U
Ubaydullayev, U. M., 559
Ugli, Turaev Khurshid Salim, 389
Uljayev, E., 559
Usmanova, N. B., 246

V
Valiyev, V. K., 520
Vasil'ev, Ed. A., 564
Verlan, Andriy, 18
Vladimir, Britkov, 529

X
Xakimova, F. A., 294
Xusanov, S. N., 64

Y
Yusifbayli, Nurali, 55
Yusifov, I. Salahaddin, 164
Yusupbekov, A. N., 121
Yusupbekov, N. R., 229, 246

Z
Zaripov, O. O., 479
Zaynutdinova, Mastura, 10
Zeynalov, E. B., 427
Zhetekova, L. B., 207

The manufacturer's authorised representative in the EU is Springer Nature Customer Service Centre GmbH, Europaplatz 3, 69115 Heidelberg, Germany. If you have any concerns regarding our products, please contact ProductSafety@springernature.com

Printed and bound by CPI Group (UK) Ltd, Croydon, CR0 4YY
30/09/2026
02225251-0018